BOTULINUM AND TETANUS NEUROTOXINS

Neurotransmission and
Biomedical Aspects

BOTULINUM AND TETANUS NEUROTOXINS
Neurotransmission and Biomedical Aspects

Edited by
BIBHUTI R. DASGUPTA
University of Wisconsin
Madison, Wisconsin

PLENUM PRESS • NEW YORK AND LONDON

Library of Congress Cataloging-in-Publication Data

Botulinum and Tetanus neurotoxins : neurotransmission and biomedical
aspects / edited by Bibhuti R. DasGupta.
     p.   cm.
     "Proceedings of an International Conference on Botulinum, Tetanus
Neurotoxins: Neurotransmission and Biomedical Aspects, held May
11-13, 1992, in Madison, Wisconsin"--T.p. verso.
   Includes bibliographical references and index.
   ISBN 0-306-44412-7
   1. Botulinum toxin--Congresses.   2. Tetanus toxin--Congresses.
3. Neurotoxic agents--Congresses.   I. DasGupta, Bibhuti R.
II. International Conference on Botulinum, Tetanus Neurotoxins:
Neurotransmission and Biomedical Aspects (1992 : Madison, Wis.)
   [DNLM: 1. Botulinum Toxins--pharmacology--congresses.   2. Tetanus
Toxin--pharmacology--congresses.   3. Neurotoxins--pharmacology-
-congresses.   4. Neural Transmission--physiology--congresses.   QW
630 B751 1992]
QP632.B66B665   1993
615.9'5299--dc20
DNLM/DLC
for Library of Congress   93-3485
                                                             CIP

QP
632
.B66
B665
1993

Proceedings of an International Conference on Botulinum, Tetanus
Neurotoxins: Neurotransmission and Biomedical Aspects,
held May 11–13, 1992, in Madison, Wisconsin

ISBN 0-306-44412-7

© 1993 Plenum Press, New York
A Division of Plenum Publishing Corporation
233 Spring Street, New York, N.Y. 10013

Printed in the United States of America

FOREWORD

Three days in Madison have thoroughly modified my view on clostridial neurotoxins. While still realizing the numerous activating, modifying and protective inputs, I cannot judge the meaningfulness of the meeting impartially. Neither may the reader expect a complete summary of all presentations. Collected in this volume, they speak for themselves without requiring an arbiter. Instead I shall write down my very personal opinions as a researcher who has studied clostridial neurotoxins for nearly 25 years.

Comparable conferences have been rare during this time. A comprehensive symposium on *C. botulinum* neurotoxins has been organized at Ft. Detrick.[4] International conferences on tetanus have been held regularly under the auspices of the World Health Organization. One or maximally two days of these meetings have been devoted to tetanus toxin and its actions whereas the sponsor and the majority of the participants have been interested mainly in epidemiology, prevention and treatment of tetanus as a disease (see refs. 5,6). Some aspects of clostridial neurotoxins have been addressed in the context of bacterial toxins, in particular in the biennial European workshops.[1-3,7,8]

The Madison meeting differed from the previous ones in three aspects. First, it covered both tetanus and botulinum neurotoxins. The fusion was justified because of their huge similarities in primary structure, in their mode of action and in their cellular targets. Second, the meeting was not limited to toxins but drew some lines on which modern neurobiology might proceed. Since long-term inhibition of neurotransmitter release is the ultimate event, better knowledge of endocytosis, exocytosis and neuronal plasticity is mandatory for understanding how tetanus and botulinum neurotoxins might act. Finally the organizer dared to merge the theoretical aspects of neurotoxin research with the clinical use of botulinum neurotoxins.

1. GENERAL NEUROBIOLOGY

The contributions on general neurobiology overwhelmed the participant with an avalanche of data and flow diagrams of events. However their yield was limited for me, being a toxin researcher. They did not furnish novel hints of how to use neurotoxins as analytical tools nor any novel system for measuring their action. I admired the bright experiments and arguments suggesting that quantal neurotransmitter release was not based on the fusion of vesicles but on a membranal sorting of molecules normally spread through the cytoplasm, i.e. the non-vesicular pool. But before lunch I was fed with some presentations demanding that nothing but vesicles represent the morphological and functional entities of neurotransmitter release. Deeply disturbed in this crucial respect, I silently consumed the rich decoration of the synaptic vesicles with a dozen or more proteins which all might serve as toxin targets. I accepted the distinction between Ca^{2+}-dependent and Ca^{2+}-independent neurotransmitter release, without reasons given how the ion might operate the trigger. Understanding the role of Ca^{2+} should enable us to delineate the mode of toxin action, but one has to wait until the next meeting at least.

Neurobiology came out as a special case of general cell biology. The transition became evident, even in morphological terms, by the contributions concerning the origin of vesicular structures, their role for neurotransmitter storage and concerning the recycling of membranes. Vesicles play a dual role. On the one hand they are involved in the toxin-sensitive handling of neurotransmitters. On the other hand, cellular uptake and even transcellular transport are crucial in the pharmacokinetics of clostridial toxins, and endocytosis directs them into vesicular compartments. Provided progress keeps its present fast pace we may expect a flow scheme of the assembly, maturation, fusion and ultimate fate of vesicles pretty soon. The exocytosis–endocytosis cycle will then be embedded into a more thorough knowledge of the role and fate of vesicles in general. Already now some morphologically different subcellular structures such as translucent vesicles and granular structures have been classified according to their composition, their origin, their distribution in different cells, and the electrophysiological correlates of their exocytosis.. Tetanus and botulinum toxins are known to inhibit the release of every transmitter generated by cells of the neural crest but not the Ca^{2+}-evoked release of, for instance, enzymes from endodermal cells. Biochemical differences between the various vesicle populations might be related to their relative sensitivity to toxins. Sorting of vesicles depends on the complementary structures of their surfaces and their targets. The vesicles have to pass the three-dimensional structure of the cytoskeleton, and will interact with them. So far the docking structures are still conjectural, and so is the role of the cytoskeleton. Special attention has been paid to the submembranal actin structures and the microtubules as possible partners of the toxins and of the vesicles. But we are still far away from establishing a cascade of neurotransmitter release resembling the cascade of blood coagulation, and to reserve proper places for the toxins.

The role of cytoskeletal and cytosolic factors can now be assessed by various means. It has been shown that vesicle traffic can be partially mimicked in yeast and influenced by mutations. Soluble components can be removed from cells by permeabilization, and then replaced by purified cytosolic components. The procedure, widely used in chromaffin cells, may be helpful to pinpoint the target of the neurotoxins. However, it is questionable whether a "run down" permeabilized cell can be equated to an intoxicated but otherwise intact cell.

Another important aspect of general neurobiology, namely plasticity of the nervous system is gaining momentum as botulinum A toxin is used for long-term paralysis in man. Plasticity has to be considered whenever neural structures have been given time to adapt to stimuli or to selective elimination. Plasticity can manifest itself even on structures and functions that are not modified primarily. The resulting adaptation may weaken a persistent stimulus, or balance a disequilibrium, or compensate a deficit by functional changes. Plasticity was demonstrated after selective receptor blockade with pharmacological antagonists and after elimination of some neuronal populations. In the future, elective neurosurgery on the ganglia of *Aplysia*, with or without intracellular application of clostridial neurotoxins, may allow the assessment of plasticity in a system consisting of only a few neurons. Botulinum toxins are now used in treatments over several years' time. The resulting long-term changes of the muscular and neural parts of the endplates, the compensatory nerve sprouting and the role of trophic factors can now be studied in man. Plasticity also became evident at a higher level in patients treated with botulinum toxin. It was reported that normalization of the muscle tonus exceeded the injected muscle but involved, for instance, the antagonist muscle too. A complete set of muscles was reset by paralyzing the spastic one, which indicates a high degree of adaptation. Another example was given for the rat brain where local application of tetanus toxin led to long-term epileptiform changes. Here again, remote, non-injected areas were involved. Finally impulses from and to the brain were reported to modulate the symptoms of local tetanus which originate from the spinal cord as the ultimate dispatch station. The interplay between spinal and supraspinal events in early local tetanus and their changes due to neural plasticity in long-term disease needs investigation in appropriate models. A reductionistic approach to neural plasticity will be one item on the agenda of future neurobiology, and clostridial neurotoxins may be useful in this respect.

Presumably most of the participants may have left the meeting with the feeling that neurobiology will equip the toxin researcher with ideas, models and tools for further research. However, at present the target of the toxins is still below our horizon. Other participants have expected that neurotoxins will be required to solve the riddles of neurotransmitter release. Both opinions have a ring of truth, and are not mutually exclusive.

2. BIOCHEMISTRY AND MOLECULAR BIOLOGY OF THE CLOSTRIDIAL NEUROTOXINS

In the good old times toxins outnumbered the scientists interested. Whereas only a few toxins of novel action were discovered during the last years, the scientists multiplied exponentially like well-fed bacteria. Now they compete for food, i.e. for solvable problems in special toxinology or general neurobiology. Crowding around the honeycomb full of natural toxins may be mitigated in future. The meeting has reflected the dawn of two novel approaches to toxins. In the last six years biochemistry of the genes has established itself as the most progressive branch of molecular biology. It already allows the site-directed mutagenesis of the toxin genes too, and even their total synthesis is within reach. By the end of this century more and more toxins, their derivatives and hybrids may be available. Their number and type will depend not only on the technical feasibility but also on the approval by the authorizing agencies. Moreover, self-poisoning of appropriate cells may be triggered by injection of the toxin mRNA instead of toxin, provided translation does work reproducibly. Data for both approaches have been presented during the meeting and may soon be regarded as classical.

Another portion of the meeting addressed the relation between Zn^{2+} content, a common sequence in the primary structure, and the mode of action of clostridial neurotoxins. Their light chains are sufficient to inhibit Ca^{2+}-evoked noradrenaline release from permeabilized adrenomedullary cells, and are mandatory for the inhibition of neurotransmission upon injection of the toxins into *Aplysia* neurons. Some groups have reported that the light chain of tetanus toxin has a Zn^{2+}-binding domain. Its sequence is well preserved in the light chains of all botulinum toxins and largely corresponds to the sequences of notorious Zn^{2+}-proteases. The action of light chain on permeabilized cells was reproduced with a protease, and inhibited by Zn^{2+}-chelators. Evidence was given by another group that the action of tetanus toxin light chain can be prevented by the Zn^{2+}-protease inhibitor phosphoramidon and that the chain possesses proteolytic activity, however the data were not confirmed by others. Cleavage of denatured actin by reduced botulinum E toxin was also introduced as an argument for a proteolytic action. The protease hypothesis is novel and not unreasonable but waits for confirmation by the detection of a functionally relevant substrate. Until then the researcher may, depending to his optimism, select among four alternatives: a) Zn^{2+} serves just as a stabilizer of the light chain. b) Zn^{2+} is required for binding of the light chain to its target. c) Zn^{2+} marks the active proteolytic site which is, however, irrelevant for inhibition of neurotransmitter release. Optimists like the writer prefer the straightforward hypothesis that: d) light chain is and acts as a Zn^{2+} protease.

The models for the assessment of the pharmacology of clostridial neurotoxins have changed in recent years. Studies on living animals were hardly presented during the meeting, or they had merely pilot functions. Only one system preserved the traditions of pharmacology. It was the isolated phrenic nerve-hemidiaphragm of the rodent which is still useful to distinguish between binding, internalization and action, and for electrophysiological analysis. *Aplysia* neurons were preferred whenever intracellular injection of toxins or its constituents had to be combined with recording of single synaptic events. As mentioned, they can produce their autogenous toxin when inoculated with mRNA. Synaptoneurosomes were favored for quantitative measurements of release from preparations with intact membranes. Chromaffin cells were in high esteem. They can be isolated in high yield and homogeneity, cultured and replaced by cell lines which are sensitive to nerve growth factor. They can be sensitized to the toxins with gangliosides, and their responses can be analysed by electrophysiological techniques. Above all, they can be permeabilized by various means. Electroporation, mechanical cracking and digitonin treatment were presented, with results that were not always congruent.

Selection of methods is an indicator of the expected progress. The present consensus favors methods of biochemical neurobiology and molecular biology. Tetanus toxin research in man and mammals appeared to be outdated. The same was true for botulinum toxin research, with the exception of clinical research coherent with its practical use. I found electrophysiology at the analytical level to be underrepresented, despite a splendid promotional presentation of frog oocyte work. I felt that much more effort should be devoted to the response of neuronal networks, for instance in *Aplysia* and in the spinal cord, and to the ubiquitous phenomenon of neural plasticity.

X-Ray crystallography apparently has to wait until well-sized crystals have grown up. Then the unique Zn^{2+} atom may be localized. Classical protein chemistry requires a second wind. At present it has left the field to molecular biology. However, protein complementation between the respective heavy and light chains, hybridization between chains of different origin, and the consequences of site-directed mutagenesis for the three-dimensional protein structure deserve future attention and should be evaluated according to the biochemical and biophysical state of the art.

3. CLINICAL ASPECTS OF BOTULINUM NEUROTOXINS

Laurels were not initially granted to the organizer when he included the clinical use of botulinum toxins into a meeting devoted mainly to basic science. Disappointments were still lingering from previous "mixed" symposia on tetanus toxin. However, he was right for various reasons. Although scientists foster their undercooled image, they nevertheless feel happy if something purposeful emerges from their seemingly purposeless zest for knowledge. Research on botulinum toxin had been the precondition for its application as an antispastic drug until it even achieved the official status of an orphan, i.e. altruistic drug. Retrospectively the relief of burden of handicapped patients justified the death of numerous mice in the course of basic research. Clinical use also acquitted the investigators from the blame for having served not science but the military-industrial complex. The participants welcomed this outcome of their work. Clinical usefulness also opened the way for new research, no longer in animals or biochemical systems but in man. Deliberate and long-term use of botulinum toxin is intended to cause long-term paralysis. As mentioned, plasticity of the nervous system at all levels tends to counteract or at least to compensate the motor deficits inflicted by the physician. The underlying events can and need to be investigated using non-invasive neurological techniques.

Botulinum toxins are large molecules. To them, the pharmacokinetic rules elaborated for drugs of low molecular weight no longer apply. Spread of paralysis around the injection sites reveals puzzling pathways probably because the toxins, unlike the local anesthetics enter the lymphatics for further transport. Elimination also follows particular rules because the toxins are large, hydrophilic and thus cannot pass into the urine. The local and systemic pharmaco-kinetics, once known, of large molecules will furnish the clinician with rules for more reliable injections.

The long-term actions of the toxins must be quantified, which is far from easy. Double-blind studies are necessary, and at least one has been presented. Complex phenomena like relief from motor spasticity and even life quality of handicapped children have to be judged in this context. More has to be learned about the wise use of clinimetrics, i.e. the hardening of soft data, for instance on mobility and well being, for quantitative evaluation.

Above all, basic research in man is unavoidable because at least two sets of data needed for therapeutics with toxin cannot be obtained in animals. One of them might be called minimal destruction. The amount of toxin must be high enough to reach just the desired level of action. The questions are how minimal destruction evolves, can be measured, and disappears, and which functional and morphological scars might remain after single or repetitive application. The other phenomenon is minimal immunization. Contrary to the previously held view, the long-term outcome of repeated local injections of toxin challenges the immune system of many patients which respond with antibody production, thus decreasing the efficacy of treatment. Immunization is inherent to the injection schedule, and so it can be well investigated. In order to achieve controlled destruction and to avoid immunization as far as possible, only highly purified, undenatured toxins should be used. Botulinum toxins for therapeutic purpose should be as free as possible from immunologically cross-reacting agents and from inadvertently toxoided material. Both contaminants will contribute to immunogenicity but not to paralysis.

May I finish this personal and therefore incomplete summary with my initial sentence slightly extended. Three days in Madison have thoroughly modified our views on clostridial neurotoxins, thanks to Bibhuti R. DasGupta, the modest and efficient organizer of this splendid meeting.

Ernst Habermann
Rudolf-Buchheim-Institut für Pharmakologie
Justus-Liebig-Universität
D-6300 Gießen, Federal Republic of Germany

REFERENCES

1. Falmagne P et al, eds. Bacterial Protein Toxins. Second European Workshop. Zbl Bakteriol Mikrobiol Hyg I Abt Suppl 15, 1986.
2. Fehrenbach FJ et al, eds. Bacterial Protein Toxins. First European Workshop. FEMS Symp. No. 24. London: Academic Press, 1984.
3. Fehrenbach FJ et al, eds. Bacterial Protein Toxins. Third European Workshop. Zbl Bakteriol Mikrobiol Hyg I Abt Suppl 17, 1988.
4. Lewis GE, Ed. Biomedical aspects of botulism. New York: Academic Press, 1981.
5. Nisticó G et al, eds. Seventh International Conference on Tetanus. Roma: Gangemi, 1985.
6. Nisticó G et al, eds. Eighth International Conference on Tetanus. Rome: Pythagora Press, 1989.
7. Rappuoli R et al, eds. Bacterial Protein Toxins. Fourth European Workshop. Zbl Bakteriol Internat J Med Microbiol Suppl 19, 1990.
8. Witholt B et al, eds. Bacterial Protein Toxins. Fifth European Workshop. Zbl Bakteriol Internat J Med Microbiol Suppl 23, 1992.

Contents

AN OVERVIEW

NEUROTROPHIC FACTORS AND DENERVATION

NEUROTRANSMITTER SECRETION; POISONING MODELS

ENDOCYTOSIS, ENDOSOMAL PROCESSING

MECHANISM OF ACTION

INTRODUCTION

Bibhuti R. DasGupta

Department of Food Microbiology and Toxicology, Food Research Institute
University of Wisconsin-Madison, Madison, WI 53706

Botulinum and tetanus neurotoxins, products of the bacteria *Clostridium botulinum* and *C. tetani*, are extremely specific for target selection (presynaptic membrane), site of action (intracellular), action (blockade of neurotransmitter release), and they are lethal at extremely low concentrations. Notorious as the most poisonous poison found in nature, botulinum neurotoxin, a potential tool for harmful use (scare during January 1991 Gulf War), is paradoxically improving quality of life (clinical use) and may advance our knowledge in neuroscience. The tetanus neurotoxin shares three of these attributes: extreme lethality, worldwide natural scourge (565,000 children dying from neonatal tetanus every year; source: WHO, 1211 Geneva 27, Switzerland, update May 1991), and pharmacological tool. The first joint appearance of the two prima donnas of the toxin world on stage for 3 days in Madison (May 11-13, 1992) attracted 132 attendees from 13 nations (Argentina, Canada, France, Germany, Italy, Japan, People's Republic of China, Slovenia, South Africa, Spain, Sweden, United Kingdom, and United States).

This conference was conceived by me in mid-1989, after attending the Fourth European Workshop on Bacterial Protein Toxins (Urbino, Italy). The structures, structure-function relationships and mechanisms of action of botulinum and tetanus neurotoxins were becoming known at an accelerating pace. It occurred to me that the investigators in the specialized fields such as anaerobe-bacteriology, virology, biochemistry, pharmacology, neuroscience, medicine, and clinical neurology should get together for cross-fertilization of ideas and critical evaluation of our understanding of the two neurotoxins for their genetics, structures, mechanisms of action, the diseases they cause, and as potential tools to explore uncharted areas of neurobiology and medicine. Thus the title of the conference: *Botulinum, Tetanus Neurotoxins: Neurotransmission and Biomedical Aspects.*

Between Dec. 1989–Feb. 1990 I polled 35 key investigators in the field for their opinion about a 3-day conference in May 1992 via a questionnaire; each responded very positively. Then an advisory group was formed with individuals from diverse research and geographical areas: Stephen Arnon, Patrice Boquet (Pasteur Institute, Paris), Mitchell F. Brin, Ernst Habermann, Charles Hatheway, Christy Ludlow, Morihiro Matsuda, John Middlebrook, Alan Scott, and Stephen Thesleff. The scientific community in general and the specialists were alerted about the conference by announcements in broad-spectrum and specialized international journals: *Science, FASEB Journal, Toxicon, Journal of Neurochemistry,* and *Neuroscience NewsLetters.*

Botulinum and Tetanus Neurotoxins, Edited by
B.R. DasGupta, Plenum Press, New York, 1993

The conference was planned to provide an opportunity for scientists and physicians from specialized disciplines to share their knowledge and insights. Cross-fertilization of ideas was to be accomplished by mixing the basic science and clinical presentations. During the conference clinical and basic science lectures were alternated, to keep the M.D.s and Ph.D.s together. Lectures, including some sharply focused in specialized areas, were planned to integrate the interests of basic research and clinical medicine. Posters were for purely experimental results. The conference—after opening with two overviews: "Current concepts on the mechanism of action of clostridial neurotoxins" and "Disorders with excessive muscle contraction: candidates for treatment with intramuscular botulinum toxin," had the following sessions: neurotransmission; receptors and trophic effects; endocytosis/endosomal processing; model systems; biochemistry of neurotoxins and mechanism of action; neurotoxin, a tool; clinical; bacteriology and food microbiology; antigenic structure/antibody; detection of botulinum neurotoxin; role of the Food and Drug Administration (USA) and contributions from the U.S. Army; and perspective on the conference—an interim summary.

Our knowledge of mechanisms of action of the two neurotoxins is very limited other than that they: i) bind to specific receptors/acceptors on presynaptic membranes; ii) get internalized through the endocytotic/lysosomal vesicle pathway; iii) penetrate through the endosomal membrane into the cytosol; and then iv) the light chain alone, acting as an enzyme (*vide infra*), inhibits secretion of neurotransmitter and thus blocks intercellular communication. Speakers were invited to cover each of these steps. Only the third step could not be addressed on a chemical basis. Dr. Ferenc J. Kezdy (Upjohn Co., Kalamazoo, Michigan), who was to have discussed "How does a water soluble protein interact with lipid membrane and pass through the membrane?" could not participate for reasons beyond his control.

The chemical denervation produced by the neurotoxins has had two inherent features that remained to be explored and exploited: i) induction of controlled paralysis in very narrowly focused and selected anatomical areas and ii) study synapse formation, maintenance and trophic interactions without physically damaging the morphology of the synaptic connections. Exploitation of the former, in the therapeutic application of botulinum neurotoxin, initiated by Alan Scott, M.D., has been lauded as "The most imaginative, technically brilliant, and courageous of such applied research..." (see Lamanna C., in *Microbial Toxins in Foods and Feeds, Cellular and Molecular Modes of Action*. Pohland A.E., et al., eds. Plenum Press, New York, 1990, p. 25). Tetanus neurotoxin is a potential tool for studying epilepsy [see Mellanby J., et al. *J. Physiol. (Paris)* 79:207, 1984]. Several contributions covered this area. To keep the second feature of chemical denervation, i.e. trophic factors and plasticity, in the forefront for further studies, six contributions were included.

Dr. Carl Lamanna, who had first produced crystalline botulinum toxin type A (the forerunner of the preparation used as Oculinum®) provided, *in absentia*, his observations on the reversible cardiac effects of botulinum toxin type A. A few speakers presented observations from their specialties that appear remote from the botulinum and tetanus neurotoxin fields; these fringe areas may have "light houses." Although our attention was on the neurotoxins and their biological effects, the bacteria that produce them were not forgotten, they were also considered in three papers. Wu Jigao (People's Republic of China) brought to the general attention the problems of botulism in China. His poster noted: i) that reports from 15 provinces and autonomous regions of China (out of a total of 30) during 1948-1989, documented 2861 cases of foodborne botulism in 745 outbreaks causing 421 deaths (for further details and diagnostic procedure used see Gao et al., A Review of Botulism in China, *Biomedical and Environmental Sciences* 3:326-336, 1990) and ii) the first report of infant botulism (4 months old, male) caused by type B. The area of immunology covered antigenic structures of the neurotoxins, their detection based on serological techniques, newly developed antibodies (based on molecular biology, from human volunteers and chicken egg), prophylaxis-therapy of botulism, and the relationship of vitamin A to antibody formation in response to tetanus toxoid. In this context the report that appeared in April 1992, "A single human monoclonal antibody that confers total protection from tetanus" [Arunachalam et al., *Hybridoma* 11(2):154, 1992], is notable.

An exciting new news at the conference was that both neurotoxins are zinc proteases. The intracellular poisoning activities of the 150 kDa neurotoxins and/or their 50 kDa light chains have long been presumed, based on analogy with other toxin proteins, such as diphtheria toxin, pseudomonas exotoxin, cholera toxin, and ricin, to be due to an enzymatic activity—the substrate of which was not known. Dr. Heiner Niemann reported that the tetanus neurotoxin is a Zn protein and therefore it is likely to be a metalloprotease. Prof. Ernst Habermann reported actual measurement of Zn content in the tetanus neurotoxin and provided evidence to postulate that its light chain acts as a metalloprotease. These two reports, at the end of the second day, brought out two unscheduled 5-min presentations that complemented the emerging story. Prof. Cesare Montecucco: i) reported that botulinum neurotoxin type A, B, and E each contains one atom of Zn per molecule of the neurotoxin protein; ii) provided the experimental evidence that pinpointed the segment of the light chain that binds the Zn; and iii) predicted that botulinum neurotoxins may act as metalloprotease. Finally DasGupta reported the actual in vitro proteolytic properties of the light chains of the types A, B, and E botulinum neurotoxins. Thus the predictions from three laboratories (Niemann, Habermann, and Montecucco) and actual experimental demonstration (from DasGupta's laboratory) converged, and the discovery of the proteolytic property suddenly opened a new chapter on future work on the two neurotoxins. Subsequent to the conference it became known that tetanus neurotoxin, like botulinum, acts as an endoprotease.

The proteolytic activity, confined to the neurotoxin's light chain, perhaps explains the fragmentation of the neurotoxins in solution under certain conditions (a major headache for those who purify them) due to inter- and/or intra-molecular self-digestion. The "spontaneous breakdown" of the isolated pure preparations of the light chain of botulinum neurotoxin (reported in *Biochemie* 71:1193, 1989) is perhaps the earliest documented symptom of self-digestion and proteolysis.

Fifteen years ago the structural similarity between the botulinum and tetanus neurotoxins was pointed out for the first time on the basis of quite meager structural data (DasGupta, B. R. and Sugiyama, H., in *Perspectives in Toxinology*, Bernheimer A. W., ed. John Wiley & Sons, NY, 1977; p. 87). It is now a beautiful revelation that the two neurotoxins resemble one another even to the detail of having one zinc atom per light chain, which is bound by an homologous amino acid sequence.

The status of our knowledge of the two neurotoxins and the rapid developments of the "good" uses of the two most poisonous poisons is very gratifying to me as a person, as a scientist, and as the organizer of the conference. My sentiments expressed in concluding the conference were joyous: We the scientists develop abstract ideas, make discoveries and invent things. Our discoveries and inventions are put to uses that are inhuman and humanitarian. In 1945, on July 16 at the early morning hours scientists exploded the first atomic bomb. Watching from a safe distance, the project leader, the brilliant physicist J. Robert Oppenheimer, thought of Bhagabat Gita—the old Hindu scripture, "I am become death, the destroyer of world." Against this background this assembly of scientists and physicians is so different. We can say proudly that we have turned an agent of death to betterment of life. Paraphrasing Sir Winston Churchill—this is one of our finest hours.

The speakers and poster presenters were asked to submit written versions of their contributions within 1 month after the conference so that ideas developed following the conference could be incorporated for the proceedings book. Participants were encouraged to speculate, giving bases of speculation, critique the published data, provide needed reinterpretation, and point out what might have been overlooked, as well as new paths. The results were 76 contributions and human delay.

The conference could not have been organized nor these proceedings assembled as a book without the generous financial contributions from Allergan, Inc., Athena Neurosciences, List

Biologicals, Pharmacia (U.S.A.), and the Graduate School of the University of Wisconsin-Madison. Carole Ayres' help in typing nearly 200 letters and Prof. Dean O. Cliver's editorial suggestions on many occasions are deeply appreciated. Barbara Cochrane happily accepted the challenging and tedious work of processing all the manuscripts for the proceedings book and also on my behalf nudged and reminded the late contributors as to their "due." The editorial guidance of Mary Safford, Plenum Press, is also gratefully acknowledged.

CURRENT CONCEPTS ON THE MECHANISM
OF ACTION OF CLOSTRIDIAL NEUROTOXINS

Lance L. Simpson

Division of Environmental Medicine and Toxicology
Departments of Medicine and Pharmacology
Jefferson Medical College
Philadelphia, PA 19107

INTRODUCTION

The purpose of this chapter is to provide a brief review of the literature on botulinum neurotoxin and tetanus toxin. This review will emphasize studies that pertain to cellular and subcellular actions of the toxins on mammalian preparations, and it will deal exclusively with issues that relate to blockade of exocytosis.

Botulinum neurotoxin and tetanus toxin act on nerve endings and other secretory cells to inhibit spontaneous and evoked mediator release.[1-3] The two toxins are thought to act in a somewhat similar manner, with the exception of intracellular trafficking. During the course of natural poisoning, botulinum neurotoxin binds preferentially to motor nerve endings. It is internalized by the process of receptor-mediated endocytosis, and it escapes endosomes by an acidification process. The toxin then acts locally at motor nerve endings to block acetylcholine release, and this produces flaccid paralysis. The scheme of events with tetanus toxin is somewhat different. This toxin also binds to motor nerves, and it too is internalized by receptor-mediated endocytosis, but it does not act locally. Tetanus toxin is carried by retrograde axonal transport to the central nervous system, where it exits the primary neuron, crosses the synaptic space, and preferentially binds to inhibitory nerve endings. At this point the mechanism of action of tetanus toxin is thought to be similar to that of botulinum neurotoxin. Tetanus toxin is endocytosed and released into the cytosol by an acidification mechanism. By virtue of blocking the release of inhibitory transmitters, it produces disinhibition of excitatory activity. This is believed to be the underlying basis for toxin-induced spastic paralysis.

In addition to its well-known ability to evoke spastic paralysis, tetanus toxin can also produce flaccid paralysis.[2] When large amounts of toxin are administered to laboratory animals, or when high concentrations are added to isolated neuromuscular preparations, some fraction of the toxin escapes retrograde axonal transport and acts locally at nerve endings to block exocytosis. In this case, the general features of botulinum neurotoxin poisoning and tetanus toxin poisoning are essentially the same.

Botulinum and Tetanus Neurotoxins, Edited by
B.R. DasGupta, Plenum Press, New York, 1993

5

ORIGIN AND STRUCTURE

Botulinum neurotoxin is produced mainly by *Clostridium botulinum*, but it is also made by *Clostridium baratii* and *Clostridium butyricum*. *Clostridium botulinum* synthesizes seven serotypes of toxin designated A, B, C, D, E, F and G; *Clostridium baratii* synthesizes a single serotype that is similar to type F, and *Clostridium butyricum* synthesizes a single serotype similar to type E. At the moment, *Clostridium tetani* is the only organism known to make tetanus toxin.

Regardless of origin, the synthesis and processing of botulinum neurotoxin and tetanus toxin are identical. Each neurotoxin is synthesized as a single chain polypeptide with a molecular weight of approximately 150,000. In the immediate post-translational stage the neurotoxins have diminished biological activity. When exposed to proteases, the single chain toxins are nicked to yield dichain molecules in which a light chain (Mr ~50,000) is linked by a disulfide bond to a heavy chain (Mr ~100,000). This represents the biologically active form of the toxin.

The relationship between nicking and activation has been studied in some detail.[4] The results indicate that nicking is essential but not sufficient for full activation to occur. Several other possible sites for proteolytic cleavage have been considered, including the aminoterminus of the light chain and the aminoterminus of the heavy chain, but these appear not to be related to activation. On the other hand, Giménez, and DasGupta[5] have obtained evidence that suggests that activation includes proteolytic cleavage at the carboxyterminus of the heavy chain.

Most of the clostridial neurotoxins have been sequenced, beginning with tetanus toxin[6,7] and botulinum neurotoxin type A.[8,9] The data indicate that there is significant sequence homology among the various botulinum neurotoxins, and furthermore the botulinum neurotoxins show sequence homology with tetanus toxin. A comparison of botulinum neurotoxin type A and tetanus toxin helps to illustrate the point. An overall comparison shows that there is 30-40% identity between the toxins, and when conservative changes are taken into account the identity rises to 40-50%. The greatest amount of similarity exists in the aminoterminal regions of the heavy chains, which may be the domains that promote internalization (see below).

The work on the primary structures has confirmed a belief that arose from neuropharmacologic research. The various botulinum neurotoxins and tetanus toxin are all descendants of the same ancestral parent. The primary structure work also confirms certain findings in immunology. Monoclonal antibodies have been found that recognize epitopes in several clostridial neurotoxins.[10-12] The commonality of epitopes is an indicator of commonality of structure.

There is much work that remains to be done on the structure of the toxins, and then to relate structure to function, but at least an elementary scheme has already evolved.[13] It appears that the heavy chain, and more precisely the carboxyterminus of the heavy chain, plays an important role in tissue targeting the toxin. The aminoterminus of the heavy chain appears to be essential for the internalization process. The light chain acts inside mammalian cells to block exocytosis, and therefore this portion of the molecule contains the domain that poisons transmission. There is a prevailing belief that the light chain is an enzyme (viz., protease), but the substrate remains unknown.

BIOLOGICAL ACTIVITY

Definitions

As an antecedent to discussing the mechanism of action of the toxins, it will be useful to introduce the concepts of universal antagonists and differential antagonists. A universal antagonist will be defined as a drug or procedure that delays the actions of all serotypes of botulinum neurotoxin and tetanus toxin. A differential antagonist will be defined as a drug or procedure that substantially inhibits the actions of one or several clostridial neurotoxins, but it is not capable of antagonizing all toxins.

The value of universal antagonists is that they help to clarify events that are common to all toxins, such as binding, internalization and escape from endosomes. By contrast, differential antagonists help to reveal unique characteristics of individual toxins.

Universal Antagonists

1. Binding. The receptors for botulinum neurotoxin and tetanus toxin have not been isolated and characterized (but see the promising work reported by Schiavo et al.[14]). However, work from a variety of laboratories on the binding of iodinated toxins to membrane preparations has revealed one interesting point. The various clostridial neurotoxins do not share the same receptor. Each serotype of botulinum neurotoxin appears to have its own receptor, as does tetanus toxin.

In spite of the apparent differences in receptors, there is one characteristic they may have in common. Each receptor may contain a sialic acid residue, or be closely associated with a membrane determinant that has sialic acid residues. This belief stems from earlier work showing that certain classes of gangliosides can cause loss of biological activity of tetanus toxin[15-17] and botulinum neurotoxin.[18,19] This work showed that clostridial neurotoxins had affinity for complex gangliosides, and sialic acid residues were important to the toxin-ganglioside interaction. Free sialic acid was relatively ineffective in producing loss of toxicity.

A more recent and somewhat different line of research tends to support the proposed involvement of sialic acid residues. A large number of lectins with affinity for different sugars were tested as potential antagonists of botulinum neurotoxin and tetanus toxin.[20] Two types of assays were done. In the first, individual lectins were tested for their ability to antagonize the binding of iodinated toxin to rat brain membrane preparations. In the second, these same lectins were tested for their abilities to antagonize the neuromuscular blocking properties of toxin.

Of the various lectins that were tested, two showed promising levels of activity. *Triticum vulgaris* lectin and *Limax flavus* lectin both antagonized the binding of toxins to membranes and the activity of toxins on isolated tissues. The lectin from *Triticum vulgaris* has affinity for N-acetyl-β-glucosamine and N-acetyl-α-sialic acid, but the lectin from *Limax flavus* has affinity only for N-acetyl-α-sialic acid. The possible involvement of N-acetyl-β-glucosamine as a component of the receptor was ruled out by testing a lectin with affinity for this sugar (*Datura stramonium*).

There were three major points that emerged from the lectin work.[20] First, these compounds were universal antagonists of clostridial neurotoxins. They blocked the activity of all serotypes of botulinum neurotoxin and tetanus toxin in both assays. Second, the lectins were very effective in diminishing the binding of toxins to brain membrane preparations. When tested at adequately high concentrations (e.g., 3×10^{-5} to 10^{-4} M), they virtually abolished binding. And third, the lectins were very active in diminishing the neuromuscular blocking properties of the toxins. For example, at a concentration of 3×10^{-3} M, the lectin from *Triticum vulgaris* diminished the apparent potency of botulinum neurotoxin type B by nearly two orders of magnitude.

Findings such as these tend to implicate sialic acid residues in the binding of clostridial toxins to nerve membranes. However, they do not clarify the nature of the involvement. A conventional interpretation of the data might be that the receptor is a sialoglycoprotein. Hence, preincubation of membranes with lectins that have affinity for sialic acid could lead to occlusion of the receptor and blockade of toxin binding. A more intriguing interpretation has been advanced by Montecucco.[21] He has proposed that binding may represent a sequence of two events. There could be an initial binding step that involves a charged lipid, such as a ganglioside. This relatively low affinity binding would promote association between toxin and the plasma membrane. This would facilitate the subsequent high affinity binding of toxin to its authentic receptor. This is a provocative idea that deserves serious consideration. In the meantime, it should be noted that lectins with affinity for sialic acid could also act as antagonists in the sequential binding model.

2. Internalization

A. Plasma Membrane. Productive internalization of clostridial neurotoxins involves a sequence of events that can be summarized as follows.[13] Toxin molecules bind to membranes of vulnerable cells, after which they are internalized by the process of receptor-mediated endocytosis. This allows toxin to cross the plasma membrane while being retained within an endosome. Passage through the endosome membrane is thought to be initiated by the endosome itself. The membrane possesses a proton pump that progressively lowers pH within the organelle. The toxin molecule possesses a "pH sensor", and when endosomal pH falls to a critical level the sensor triggers a conformational change that exposes a hydrophobic domain that inserts into the endosome membrane. Insertion is the key event in toxin passage across the membrane, although it is not yet known how passage occurs.

The initial work to implicate a membrane penetration step was done on the neuromuscular junction. A series of studies with polyclonal antibody showed that surface-bound toxin disappeared from accessibility to extracellular antibody and that this occurred before onset of paralysis.[1,22] These results suggested that toxin had to be internalized. Subsequent work demonstrated that several drugs known to block receptor-mediated endocytosis were toxin antagonists. The first agent to be tested was chloroquine and its analogs,[23] and this was followed by work with ammonium chloride and methylamine hydrochloride.[24] The results indicated that all three drugs antagonized botulinum neurotoxin, whereas ammonium chloride and methylamine hydrochloride but not chloroquine antagonized tetanus toxin. This work, in combination with the polyclonal antibody studies, was taken as evidence that clostridial toxins were internalized by receptor-mediated endocytosis.

The pharmacological work on internalization was nicely complemented by the morphological work of Black and Dolly,[25,26] who iodinated botulinum neurotoxin and used it to do electronmicroscopic autoradiography at the neuromuscular junction. They were able to show that botulinum neurotoxin binds to the plasma membrane, and membrane-bound toxin could subsequently be localized in endosomes. Movement of toxin across plasma membranes and into endosomes was energy-dependent and accelerated by nerve stimulation. Ammonium chloride, methylamine hydrochloride and chloroquine altered uptake and/or distribution of toxin. In sum, the morphological data were very supportive of the earlier pharmacological data.

B. Endosome Membrane. The current belief is that clostridial neurotoxins possess a pH sensor that induces conformational changes that lead to exposure of a hydrophobic domain. This in turn causes the toxins to insert into the endosomal membrane and eventually penetrate it. Several lines of research support this model.

A representative study is the one conducted by Hoch et al.[27] on pH-induced channels in planar lipid bilayers. These workers added botulinum neurotoxin to one side of an artificial membrane, then altered pH on the cis or trans side of the membrane. They observed relatively few channels with an iso-pH of 7.0 or 4.0, but they saw rapid appearance of channels when pH on the cis-side was lowered to approximately 5.5 or below and pH on the trans-side was maintained at 7.0. Structure-function analyses revealed that the heavy chain was responsible for channel formation.

Subsequent work by Donovan and Middlebrook,[28] Blaustein et al.,[29] and Shone et al.[30] confirmed that botulinum neurotoxin formed pH-dependent channels in lipid bilayers. Donovan and Middlebrook[28] demonstrated that toxin-induced channels inserted permanently into the membrane and fluctuated between open and closed states. Blaustein et al.[29] and Shone et al.[30]

extended the structure-function analyses by localizing channel activity to the aminoterminal half of the heavy chain.

Biochemical techniques have also been used to demonstrate that pH can cause botulinum neurotoxin and its respective chains to partition into a lipid environment. Montecucco et al.[31,32] used photoreactive reagents to monitor toxin insertion into the core of a lipid bilayer. They found that acid pH caused both the heavy and light chains to insert and be labelled. This prompted them to suggest that it was not necessary to view the heavy chain as an agent for promoting passage of a hydrophilic light chain into the cytosol. Instead, the two chains could act cooperatively to achieve translocation.

Kamata et al.,[33] who used a fluorescent reporter group, obtained data that support the notion of cooperativity between the two chains. Their work showed that low pH induced striking conformational changes in the intact toxin leading to exposure of hydrophobic domains. The same result was obtained when the heavy and light chains were examined independently.

There is a closely related body of research that pertains to tetanus toxin insertion into membranes. Indeed, investigation of tetanus toxin began before the corresponding studies with botulinum neurotoxin. Boquet and Duflot[34] prepared asolectin vesicles loaded with potassium, and they showed that at low pH tetanus toxin inserted into the membrane and created a pore that caused release of cation. Borochov-Neori et al.,[35] Gambale and Montal,[36] Menestrina et al.[37] and Rauch et al.[38] have used electrophysiologic techniques to show that tetanus toxin forms channels in membranes.

There is evidence that pH-induced changes in tetanus toxin lead to exposure of a hydrophobic domain. This was demonstrated by measuring [^{3}H]-Triton X-100 association with toxin,[39] photoactivatable phospholipid association with toxin,[40] and fluorescent reporter group association with toxin.[33] Each study has demonstrated that the hydrophobic domain is in the aminoterminal half of the heavy chain and the light chain. In a related study, Cabiaux and her colleagues have shown that low pH causes tetanus toxin to induce fusion of lipid vesicles.[41]

To summarize, there is a consensus among workers that clostridial neurotoxins must enter target cells to produce blockade of exocytosis. Crossing of the plasma membrane is probably due to receptor-mediated endocytosis. Crossing of the endosome membrane is due to a mechanism that is triggered by a fall in pH.

The putative role of an acid-triggered mechanism has led to the discovery of another universal antagonist, a compound known as bafilomycin. This compound is microbial in origin, and it possesses the property of inhibiting membrane ATPase. Interestingly, bafilomycin acts preferentially on vacuolar ATPase rather than plasma membrane ATPase. Therefore, bafilomycin can be used to inhibit the ATPase-dependent proton pump in endosomal membranes without inhibiting ATPase-dependent ion pumps in plasma membranes.

Recent studies (Simpson, in preparation) have shown that bafilomycin is an antagonist of all serotypes of botulinum neurotoxin and tetanus toxin. The drug did not inhibit toxin binding nor did it alter intracellular expression of toxicity. Bafilomycin appeared to inhibit the process of productive internalization. This is in keeping with the action of a drug that inhibits acidification of endosomes.

Differential Antagonists

There are two known groups of differential antagonists, one of which was fully predictable and the other of which emerged quite unexpectedly. Neutralizing antibodies are an example of the former and drugs that promote acetylcholine release, such as the aminopyridines, are an example of the latter.

1. Antibodies. Although antibodies have been identified that cross-react with more than one clostridial neurotoxin (see above), no antibody has been found that neutralizes all neurotoxins. Both the linear and conformational characteristics of the active domains within each serotype are sufficiently unique to preclude the generation of antibodies that bind to active sites in multiple serotypes. Because of this, antibodies have enjoyed little utility as research tools for probing the common structural properties of clostridial neurotoxins.

In spite of the limited value of antibodies in defining common structural elements, they may be helpful in clarifying the sequence of events in poisoning. A study with monoclonal antibodies against botulinum neurotoxin type E helps to illustrate the point.[42] A family of monoclonal antibodies was isolated, and from this family it was possible to identify three that possessed special utility. Each of the antibodies diminished the potency of native toxin, and each interacted with a different domain in the toxin (i.e., the light chain, the aminoterminus of the heavy chain or the carboxyterminus of the heavy chain). Additionally, each of the antibodies recognized a conformational rather than a linear epitope, and it was this property that allowed them to be used as research tools.

Neuromuscular preparations were incubated with toxin under conditions that allowed binding but prevented internalization. These tissues were then exposed to one or more of the monoclonal antibodies. The results showed that monoclonal antibodies were almost as effective in diminishing the potency of bound toxin as they were in diminishing the potency of free toxin. An outcome like this reveals two interesting things about toxin binding to neuronal membranes. First, binding does not cause any of the three domains in the toxin to disappear completely from the cell surface. At least that portion of each domain that possessed the relevant epitope was still exposed. Second, binding may not be associated with major changes in the conformation of the toxin molecule. Each of the antibodies was directed against a conformational epitope, and each had substantial activity against free toxin and bound toxin. At a minimum, this means that the respective epitopes did not undergo major changes during binding.

A larger family of antibodies that recognize all epitopes in the toxin molecule would allow one to determine whether any part of the toxin undergoes conformational changes during binding. In the meantime, the data indicate that binding may be very different from internalization. As explained above, pH-induced internalization is associated with marked changes in conformation.

2. Aminopyridines. Drugs such as 4-aminopyridine and 3,4-diaminopyridine block voltage-dependent potassium channels in nerve endings. By virtue of delaying the process of repolarization, these drugs promote influx of calcium and secondarily increase efflux of acetylcholine.

The ability of these drugs to enhance stimulus-evoked release of acetylcholine led to their evaluation as possible antagonists of botulinum neurotoxin. The initial results with serotype A appeared rather favorable. The drugs slowed the rate of onset of toxin-induced paralysis of isolated neuromuscular preparations, and they even reversed - at least temporarily - the development of mild paralysis. However, the results with other serotypes were less promising. Aminopyridines provided little or no protection against most serotypes of botulinum neurotoxin and tetanus toxin.

When the actions of the aminopyridines were studied at the electrophysiological level, the differences between and among serotypes became more obvious. The disparity between responses to type A, on the one hand, and responses to type B and tetanus toxin were especially clear.[43-45] Neuromuscular preparations poisoned with serotype A had a very low rate of spontaneous miniature endplate potentials, and the probability of evoking an endplate potential was also low. However, the addition of aminopyridines substantially increased the likelihood of evoking an endplate response. By contrast, neuromuscular preparations poisoned with botulinum neurotoxin type B or tetanus toxin showed a different result. Following addition of aminopyridines, only rarely was there an evoked endplate potential.

Closer examination of poisoned tissues appeared to reveal a reason for the disparate responses. In the case of serotype A, the combination of nerve stimulation and aminopyridines resulted in synchronous release of quanta, and thus the increased likelihood of observing an endplate response. In the case of serotype B and tetanus toxin, the release of quanta was desynchronized and therefore quanta could not summate to evoke a response.

One especially interesting finding arose from dual poisoning experiments.[46] Tissues exposed initially to serotype A displayed synchronous release, but when they were subsequently exposed to serotype B or tetanus toxin the release became desynchronized. When toxins were added in the reverse order, release was always desynchronized. These data with aminopyridines, as well as other drugs, toxins and physical procedures, led to the hypothesis that there could be two sites of clostridial neurotoxin action.[46] Botulinum neurotoxin type A was postulated to act at a site somewhat remote from the membrane, whereas serotype B and tetanus toxin acted closer to the membrane, and hence closer to the site of exocytosis.

INTRACELLULAR ACTIONS

There is compelling evidence to show that clostridial neurotoxins act in the cell interior to block exocytosis. Over the period of approximately a decade there has been a progression of research to demonstrate this point. The work began with pharmacologic experiments involving the use of antibody escape and the use of drugs known to antagonize endocytosis.[22,23,24] This was followed by studies on electronmicroscopic localization of toxin at various stages during poisoning.[25,26] Next, it was shown that direct intracellular injection of toxin blocked exocytosis.[47] This prompted experiments in which toxin, or polypeptides derived from toxin, were introduced into cells by permeabilization techniques.[48-51] Most recently, it has been shown that intracellular injection of mRNA that encodes toxin leads to expression of toxin and blockade of transmitter release.[52]

In addition to demonstrating that clostridial toxins act in the cell interior, this work has highlighted an anomaly for which there is currently no explanation. When tested in mammalian preparations, the light chains of clostridial neurotoxins are both necessary and sufficient to block exocytosis. By contrast, when tested in *Aplysia*, the light chains are necessary but not sufficient.[53] There is a domain in the heavy chain that is needed in addition to the light chain to produce poisoning of transmitter release. There are other structure-function differences between toxin action on mammalian preparations and *Aplysia*, and these differences have yet to be explained.

Messenger Systems

The fact that the toxins act inside nerve terminals has encouraged investigators to search for possible targets. There has been a tendency to divide these possible targets into four broad categories: i) messenger systems, ii) the cytoskeleton, iii) vesicles, and iv) the plasma membrane. Relatively little work has been done with the specific intent of evaluating vesicles or membranes as sites of toxin action. However, there is an emerging literature on messenger systems and cytoskeleton.

Several laboratories have evaluated the possibility that clostridial neurotoxins modify messenger systems that govern transmitter release. This effort has been largely negative, but one possible exception is the work on protein kinase C.

Considine and his colleagues have demonstrated that agonist-induced changes in protein kinase C activity in NG-108 cells are inhibited by tetanus toxin. This has been shown both for an artificial agonist (e.g., phorbol ester[54]) and for a natural agonist (e.g., neurotensin[55]). This work raises the possibility that protein kinase C may be a target for clostridial neurotoxins. This possibility was tested by determining the effects of known inhibitors of protein kinase C on transmitter release at the mammalian neuromuscular junction (Considine, Sherwin and Simpson, in preparation). Two inhibitors of the catalytic domain (H7, staurosporine) and two inhibitors of the regulatory domain (calphostin, sphingosine) were added individually to phrenic nerve-hemidiaphragm preparations at concentrations that virtually abolished protein kinase C activity in neuronal cell cultures (e.g., NG-108). Interestingly, the tissues continued to respond for several hours even under conditions in which the enzyme was substantially or completely inhibited. Furthermore, pretreatment of tissues with inhibitors of protein kinase did not predispose them to the poisoning effects of clostridial neurotoxins.

The results with protein kinase C inhibitors give rise to three conclusions. First, the enzyme is not required to sustain short-term neuromuscular transmission. Second, protein kinase C cannot be the target for clostridial toxins. And third, protein kinase C cannot be the target for any toxin that acts rapidly to block neuromuscular transmission.

Cytoskeleton

A number of hypotheses have been advanced that implicate the cytoskeleton, and especially the actin-based cytoskeleton, in the process of transmitter release. One of the more thoroughly studied of these models is one that pertains to mediator release from adrenal chromaffin cells. According to this model, it is envisioned that storage granules are held in place by a lattice-work of actin filaments. When an agonist acts on the cell to trigger mediator release, one of its effects is to stimulate disaggregation of actin lattices. This releases storage granules that can then move toward the plasma membrane and discharge their contents.

It is unclear whether a similar model applies to cholinergic nerve endings. However, if the cytoskeleton does play a role in exocytosis, it might conceivably be the target for clostridial neurotoxins. Therefore, a series of studies have been undertaken to evaluate the possible role of actin and related molecules in cell poisoning. One line of research has focused on actin and the cholinergic neuromuscular junction[56] (Considine and Simpson, unpublished findings); a second line of research has concentrated on microtubules and central nervous system synaptosomes.[57]

The role of actin in exocytosis and toxin action has been evaluated by using the botulinum binary toxin as a research tool.[58] This toxin is composed of two separate and independent polypeptide chains (M_r ~45,000 and 100,000), both of which are necessary to produce cell poisoning. The heavy chain is a tissue targeting domain that binds to receptors on vulnerable cells. The heavy and light chains of the binary toxin have no affinity for one another in solution, but when the heavy chain associates with membranes of vulnerable cells it creates a docking site for the light chain. The heavy and light chains are then internalized by the process of receptor-mediated endocytosis.[59]

The light chain of the botulinum binary toxin is an enzyme that possesses mono(ADP-ribosyl)transferase activity.[60] The substrate for the enzyme is monomeric actin.[61] By virtue of modifying G-actin, the binary toxin disrupts the cytoskeleton of cells. More precisely, the binary toxin exerts two specific effects. Monomeric actin is in dynamic equilibrium with filamentous or F-actin. When the binary toxin ADP-ribosylates G-actin, it irreversibly removes these molecules from the pool of monomeric actin and this in turn promotes dissociation of filamentous actin to maintain equilibrium. A related effect is that ADP-ribosylated actin acts as a capping protein, thus preventing elongation by unmodified actin.

The consequences of binary toxin action are easy to observe in cultured cells. The toxin causes flattened and varigated cells that are attached to the plating surface to retract and become round.[62-64] In essence, the cells collapse on themselves because they lack a cytoskeleton.

The botulinum binary toxin has been used to study exocytosis in a variety of secreting cells, and the results do not encourage a belief that the actin-based cytoskeleton plays either a simple or a universal role in mediator release. For example, treatment of neuromuscular preparations with the binary toxin has no obvious effect on stimulus-evoked transmitter release. Similarly, pretreatment of cells with the binary toxin neither enhances nor inhibits the actions of clostridial neurotoxins.[56] A survey of binary toxin action on all secreting cells that have been studied to date reveals no consistent action on either spontaneous or evoked mediator release.[58] Enhanced mediator release, inhibited mediator release or no effect has been observed in one cell type or another. These results do not indicate a link between actin and clostridial neurotoxins.

Another line of inquiry has been directed at microtubules. Dolly and his colleagues have examined the ability of microtubule dissociating drugs to alter the actions of clostridial neurotoxins on norepinephrine release from rat brain synaptosomes.[57] They have observed that these drugs have little or no effect on certain toxins (e.g., serotype A), but they exert a small and statistically significant effect on others (e.g., serotype B) Interestingly, this disparity of action is the same as that observed with differential antagonists such as aminopyridines, but in the opposite direction. Aminopyridines antagonize serotype A but have little action on serotype B and tetanus toxin.

It is encouraging to see that two lines of research (aminopyridines and other differential antagonists; microtubule dissociating drugs) are coming to the same apparent conclusion. There may be two classes of clostridial neurotoxins, with serotype A being a prototype of one class and serotype B and tetanus toxin being prototypes of the other. Hopefully, the molecular basis for this division will be discovered shortly.

Enzymatic Actions

Although the specific target for clostridial neurotoxins has not been identified, there is reason to suspect that the toxins are enzymes with protease activity. Work on the primary structures of the toxins has revealed that they contain a histidine motif that is characteristic of zinc-containing proteases.[8] This observation has prompted several laboratories to undertake a vigorous search for a substrate. Some tantalizing preliminary observations were presented during the conference, but no authentic substrates for toxin action were identified.

REFERENCES

1. Simpson LL. The origin, structure, and pharmacological activity of botulinum toxin. Pharmacol Rev 1981; 33:155-188.
2. Habermann E, Dreyer F. Clostridial neurotoxins: Handling and action at the cellular level. Curr Topics Microbiol Immun 1986 ;129:93-179.
3. Simpson LL, ed. Botulinum Neurotoxin and Tetanus Toxin. San Diego: Academic Press, 1989.
4. Dasgupta BR. The structure of botulinum neurotoxin. In: Simpson LL, ed. Botulinum Neurotoxin and Tetanus Toxin. San Diego: Academic Press, 1989:53-67.
5. Gímenez JA, Dasgupta BR. Botulinum neurotoxin type E fragmented with endoproteinase Lys-C reveals the site trypsin nicks and homology with tetanus neurotoxin. Biochimie 1990; 72:213-217.
6. Eisel U, Jarausch W, Goretzki K, Henschen A, Engels J, Weller U, Hudel M, Habermann E, Niemann H. Tetanus toxin: primary structure, expression in E. coli, and homology with botulinum toxins EMBO J 1986; 5:2495-2502.
7. Fairweather NF, Lyness VA. The complete nucleotide sequence of tetanus toxin. Nucl Acids Res 1986; 14:7809-7812.
8. Binz T, Kurazono H, Wille M, Frevert J, Wernars K, Niemann H. The complete sequence of botulinum neurotoxin type A and comparison with other clostridial neurotoxins. J Biol Chem 1990; 265:9153-9158.
9. Thompson DE, Brehm JK, Oultram JD, Swinfield TJ, Shone CC, Atkinson T, Melling J, Minton NP. The complete amino acid sequence of the Clostridium botulinum type A neurotoxin, deduced by nucleotide sequence analysis of the encoding gene. Europ J Biochem 1990; 189:73-81.
10. Hambleton P, Shone CC, Wilton-Smitt P, Melling J. A possible antigen on clostridial toxin detected by monoclonal anti-botulinum neurotoxin antibodies. In: Alouf JE, Fehrenbach FJ, Freer JH, and Jeljaszewicz J, eds. Bacterial Protein Toxins. London: Academic Press, 1984.
11. Tsuzuki K, Yokosawa N, Syuto B, Ohishi I, Fujii N, Kimura K, Oguma, K. Establishment of a monoclonal antibody recognizing an antigenic site common to Clostridium botulinum type B, C1, D, and E toxins and tetanus toxin. Infect Immun 1988; 56:898-902.

12. Halpern JL, Smith LA, Seamon KB, Groover KA, Habig WH. Sequence homology between tetanus and botulinum toxins detected by an antipeptide antibody. Infect Immun 1989; 57:18-22.
13. Simpson LL Molecular pharmacology of botulinum toxin and tetanus toxin. Ann Rev Pharmacol Toxicol 1986; 26:427-453.
14. Schiavo G, Ferrari G, Rossetto O, Montecucco C. Tetanus toxin receptor. Specific cross-linking of tetanus toxin to a protein of NGF-differentiated PC12 cells. FEBS Letts 1991; 290:227-230
15. Van Heyningen WE. The fixation of tetanus toxin by nervous tissue. J Gen Microbiol 1959; 20:291-300.
16. Van Heyningen WE. Tentative identification of the tetanus toxin receptor in nervous tissue. J Gen Microbiol 1959; 20:310-320.
17. Van Heyningen WE. Gangliosides as membrane receptors for tetanus toxin, cholera toxin and serotonin. Nature 1974; 249:415-417.
18. Simpson LL, Rapport MM. The binding of botulinum toxin to membrane lipids: sphingolipids, steroids and fatty acids. J Neurochem 1971; 18:1751-1759.
19. Simpson LL, Rapport MM. Ganglioside inactivation of botulinum toxin. J Neurochem 1971; 18:1341-1343.
20. Bakry N, Kamata Y, Simpson LL. Lectins from Triticum vulgaris and Limax flavus are universal antagonists of botulinum neurotoxin and tetanus toxin. J Pharmacol Exp Ther 1991; 258:830-836.
21. Montecucco C. How do tetanus and botulinum toxins bind to neuronal membranes? Trends Biochem Sci 1986; 11:314-317.
22. Simpson LL. Kinetic studies on the interaction between botulinum toxin type A and the cholinergic neuromuscular junction. J Pharmacol Exp Ther 1980; 212:16-21.
23. Simpson LL. The interaction between aminoquinolines and presynaptically acting neurotoxins. J Pharmacol Exp Ther 1982; 222:43-48.
24. Simpson LL. Ammonium chloride and methylamine hydrochloride antagonize clostridial neurotoxins. J Pharmacol Exp Ther 1983; 225:546-552.
25. Black JD, Dolly JO. Interaction of ^{125}I-labeled botulinum neurotoxins with nerve terminals. I. Ultrastructural autoradiographic localization and quantitation of distinct membrane acceptors for types A and B on motor nerves. J Cell Biol 1986; 103:521-534.
26. Black JD, Dolly JO. Interaction of ^{125}I-labeled botulinum neurotoxins with nerve terminals. II. Autoradiographic evidence for its uptake into motor nerves by acceptor-mediated endocytosis. J Cell Biol 1986; 103:535-544.
27. Hoch DH, Romero-Mira M, Ehrlich BE, Finkelstein A, Dasgupta BR, Simpson, LL. Channels formed by botulinum, tetanus, and diphtheria toxins in planar lipid bilayers: relevance to translocation of proteins across membranes. Proc Natl Acad Sci USA 1985; 82:1692-1696.
28. Donovan JJ, Middlebrook JL. Ion-conducting channels produced by botulinum toxin in planar lipid membranes. Biochem 1986; 25:2872-2876.
29. Blaustein RO, Germann WJ, Finkelstein A, Dasgupta BR. The N-terminal half of the heavy chain of botulinum type A neurotoxin forms channels in planar phospholipid bilayers. FEBS Letters 1987; 226:115-120.
30. Shone CC, Hambleton P, Melling J. A 50-kDa fragment from the NH2-terminus of the heavy subunit of Clostridium botulinum type A neurotoxin forms channels in lipid vesicles. Europ J Biochem 1987; 167:175-180.
31. Montecucco C, Schiavo G, Gao Z, Bauerlein E, Boquet P, Dasgupta BR. Interaction of botulinum and tetanus toxins with the lipid bilayer surface. Biochem J 1988; 251:379-383.
32. Montecucco C, Schiavo G, Dasgupta BR. Effect of pH on the interaction of botulinum neurotoxins A, B and E with liposomes. Biochem J 1989; 259:47-53.
33. Kamata Y, Lautenslager G, Simpson LL. Structural changes in the botulinum neurotoxin molecule that may be associated with the process of internalization and expression of pharmacologic activity. Infect Immun (submitted for publication).
34. Boquet P, Duflot E. Tetanus toxin fragment forms channels in lipid vesicles at low pH. Proc Natl Acad Sci USA 1982; 79:7614-7618.
35. Borochov-Neori H, Yavin E, Montal N. Tetanus toxin forms channels in planar lipid bilayers. Biophys J 1984; 45:83-85.
36. Gambale F, Montal M. Characterization of the channel properties of tetanus toxin in planar lipid bilayers. Biophys J 1988; 53:771-783.
37. Menestrina G, Forti S, Gambale F. Interaction of tetanus toxin with lipid vesicles. Effects of pH, surface charge, and transmembrane potential on the kinetics of channel formation. Biophys J 1989; 55:393-405.
38. Rauch G, Gambale F, Montal M. Tetanus toxin channel in phosphatidylserine planar bilayers: conductance states and pH dependence. Europ Biophys J 1990; 18:79-83.

39. Boquet P, Duflot E, Hauttecoeur B. Low pH induces a hydrophobic domain in the tetanus toxin molecule. Europ J Biochem 1984; 144:339-344.
40. Montecucco C, Schiavo G, Brunner J, Duflot E, Boquet P, Roa M. Tetanus toxin is labeled with photoactivatable phospholipids at low pH. Biochem 1986; 25:919-924.
41. Cabiaux V, Lorge P, Vandenbranden M, Falmagne P, Ruysschaert JM. Tetanus toxin induces fusion and aggregation of lipid vesicles containing phosphatidylinositol at low pH. Biochem Biophys Res Comm 1985; 128:840-849.
42. Simpson LL, Kamata Y, Kozaki S. Use of monoclonal antibodies as probes for the structure and biological activity of botulinum neurotoxin. J Pharmacol Exp Ther 1990; 255:227-232.
43. Dreyer F, Schmitt A. Different effects of botulinum A toxin and tetanus toxin on the transmitter releasing process at the mammalian neuromuscular junction. Neurosci Letts 1981; 26:307-311.
44. Dreyer F, Schmitt A. Transmitter release in tetanus and botulinum A toxin-poisoned mammalian motor endplates and its dependence on nerve stimulation and temperature. Pflugers Archiv - Europ J Physiol 1983; 399:228-234.
45. Sellin LC, Thesleff S, Dasgupta BR. Different effects of types A and B botulinum toxin on transmitter release at the rat neuromuscular junction. Acta Physiol Scand 1983; 119:127-133.
46. Gansel M, Penner R, Dreyer F. Distinct sites of action of clostridial neurotoxins revealed by double-poisoning of mouse motor nerve terminals. Pflugers Archiv - Europ J Physiol 1987; 409:533-539.
47. Penner R, Neher E, Dreyer F. Intracellularly injected tetanus toxin inhibits exocytosis in bovine adrenal chromaffin cells. Nature 1989; 324:76-78.
48. Ahnert-Hilger G, Weller U, Dauzenroth ME, Habermann E. Gratzl M. The tetanus toxin light chain inhibits exocytosis. FEBS Letters 1986; 242:245-248.
49. Bittner MA, Holz RW. Effects of tetanus toxin on catecholamine release from intact and digitonin-permeabilized chromaffin cells. J Neurochem 1988; 51:451-456.
50. Bittner MA, Dasgupta BR, Holz RW. Isolated light chains of botulinum neurotoxins inhibit exocytosis. Studies in digitonin-permeabilized chromaffin cells. J Biol Chem 1989; 264:10354-10360.
51. Stecher B, Weller U, Habermann E, Gratzl M, Ahnert-Hilger G. The light chain but not the heavy chain of botulinum A toxin inhibits exocytosis from permeabilized adrenal chromaffin cells. FEBS Letters 1989; 255:391-394.
52. Mochida S, Poulain B, Eisel U, Binz T, Kurazono H, Niemann H, Tauc L. Exogenous mRNA encoding tetanus or botulinum neurotoxins expressed in Aplysia neurons. Proc Natl Acad Sci USA 1990; 87:7844-7848.
53. Poulain B, Tauc L, Maisey EA, Wadsworth JDF, Mohan PM, Dolly JO. Neurotransmitter release is blocked intracellularly by botulinum neurotoxin, and this requires uptake of both toxin polypeptides by a process mediated by the larger chain. Proc Natl Acad Sci USA 1988; 85:4090-4094.
54. Considine RV, Bielicki JK, Simpson LL, Sherwin JR. Tetanus toxin attenuates the ability of phorbol myristate acetate to mobilize cytosolic protein kinase C in NG-108 cells. Toxicon 1990; 28:13-19.
55. Considine RV, Handler CM, Simpson LL, Sherwin JR. Tetanus toxin inhibits neurotensin-induced mobilization of protein kinase activity in NG-108 cells. Toxicon 1991; 29:1351-1357.
56. Simpson LL. A comparison of the pharmacological properties of *Clostridium botulinum* type C1, and type C2 toxins. J Pharmacol Exp Ther 1982; 223:695-701.
57. Ashton AC, Dolly JO. Microtubule-dissociating drugs and A23187 reveal differences in the inhibition of synaptosomal transmitter release by botulinum neurotoxins types A and B. J Neurochem 1991; 56:827-835.
58. Considine RV, Simpson LL. Cellular and molecular actions of binary toxins possessing ADP-ribosyltransferase activity. Toxicon 1991; 29:913-936.
59. Simpson LL. The binary toxin produced by *Clostridium botulinum* enters cells by receptor-mediated endocytosis to exert its pharmacologic effects. J Pharmacol Exp Ther 1989; 251:1223-1228.
60. Simpson LL. Molecular basis for the pharmacological actions of *Clostridium botulinum* type C2 toxin. J Pharmacol Exp Ther 1984; 230:665-669.
61. Aktories K, Barmann M, Ohishi I, Tsuyama S, Jakobs KH, Habermann E. Botulinum C2 toxin ADP-ribosylates actin. Nature 1986; 322:390-392.
62. Miyake M, Ohishi I. Response of tissue-cultured cynomolgus monkey kidney cells to botulinum C2 toxin. Microbiol Pathogen 1987; 3:279-286.
63. Reuner KH, Presek P, Boschek CB, Aktories K. Botulinum C2 toxin ADP-ribosylates actin and disorganizes the microfilament network in intact cells. Eur J Cell Biol 1987; 43:134-140.
64. Zepeda H, Considine RV, Smith HL, Sherwin JR, Ohishi I, Simpson LL. The actions of the Clostridium botulinum binary toxin on the structure and function of Y-1 adrenal cells. J Pharmacol Exp Ther 1988; 246:1183-1189.

STRUCTURE AND FUNCTIONS OF NGF RECEPTORS

Moses V. Chao, Julie Huber, Margaret Berg, Marta Benedetti, Curt Horvath, and Barbara Hempstead

Department of Cell Biology and Anatomy, Hematology/Oncology Division
Cornell University Medical College
1300 York Avenue
New York, New York 10021

INTRODUCTION

Classic studies on nerve growth factor (NGF) revealed its essential role as a survival and differentiation factor in both the peripheral and central nervous systems (Levi-Montalcini, 1987; Thoenen et al., 1987). The biological actions of NGF include survival of embryonic and adult neurons and induction of neurite outgrowth and neurotransmitter enzymes. The primary targets in the peripheral nervous system are sympathetic and populations of sensory neurons derived from the neural crest. In the central nervous system, NGF is capable of influencing cholinergic neurons in the basal forebrain by the induction of choline acetyltransferase (ChAT) activity (Mobley et al., 1985; Martinez et al., 1985), and by reducing, or even preventing degeneration of basal forebrain neurons. A number of differentiative properties are induced by NGF, including increased tyrosine hydroxylase, acetylcholinesterase, and choline acetyltransferase activities, as well as many neuron specific proteins such as synapsin, tau, microtubule associated proteins, neurofilaments, and sodium channels (Chao, 1990).

An essential prerequisite for the actions of NGF is the presence of functional receptor sites on responsive cells. Two NGF receptors exist that are represented by transmembrane glycoproteins of 75,000 (p75^{NGFR}) and 140,000 daltons (p140^{trk}), the product of the proto-oncogene *trk*, a receptor tyrosine kinase. Both receptor proteins bind NGF, and are expressed in the traditional target cells for NGF. Our laboratory has emphasized the role of the functional NGF receptor complex in mediating the trophic and differentiative effects of NGF. To this end, knowledge of the structure, function, and regulation of receptors for NGF is essential to understanding the mechanism of action of NGF in responsive cells.

CROSSLINKING OF NGF RECEPTORS

Affinity crosslinking reagents have historically been used to characterize NGF receptors. The most reactive and widely used reagents have been carbodimide compounds, such as ethyl-3-dimethylaminopropyl-carbodiimide (EDAC). Using affinity crosslinking of ^{125}I-NGF with

Botulinum and Tetanus Neurotoxins, Edited by
B.R. DasGupta, Plenum Press, New York, 1993

EDAC, the major radioiodinated complex isolated from neurons and tumor cell lines has an apparent molecular weight of 100,000 daltons (Grob et al., 1983; Taniuchi et al., 1986; Green and Greene, 1986). Since the ß-dimer of NGF has a M_w of approximately 25 kilodaltons, the apparent molecular weight for this receptor species is about 75,000 daltons (p75[NGFR]).

The p75[NGFR] is a glycosylated transmembrane protein, rich in cysteine residues and negatively charged (Johnson et al., 1986; Radeke et al., 1987). Of 28 cysteine residues in the predicted amino acid sequence, 24 cysteines are found in four canonical repeats in the NH_2-terminal extracellular region. Similar cysteine-rich repeats have been detected in the extracellular domains of a number of proteins including TNF receptors, the B cell activation molecule, CDw40, and the T cell molecule, OX40 (Smith et al.,1990; Mallet and Barclay, 1991). The existence of this family of diverse cell surface molecules implies that common mechanisms may serve to mediate signal transduction.

A second receptor species of 135,000-158,000 daltons can be identified by crosslinking ^{125}I-NGF using the heterobifunctional and relatively lipophilic reagent hydroxysuccinimidyl-4-azido benzoate, HSAB (Massague et al.,1981; Hosang and Shooter, 1985). This crosslinked species was observed only on cells which were responsive to NGF (PC12 cells and sympathetic neurons). This species was originally thought to represent the p75 receptor crosslinked to ^{125}I-NGF and an second protein of smaller size. However, antibodies against the p75 receptor did not recognize the 160 Kd species, indicating that the 160 Kd species represented an independent protein binding complex (Meakin and Shooter, 1991). This higher molecular weight species is the product of the *trk* proto-oncogene, a member of the receptor tyrosine kinase family. The *trk* receptor tyrosine kinase binds directly to NGF (Kaplan et al., 1991ab; Klein et al., 1991a) and results in the autophosphorylation of tyrosine residues on *trk* and cellular substrates.

TRK: THE SIGNALING RECEPTOR FOR NGF

The *trk* proto-oncogene was originally identified as a rearranged oncogene from a colon carcinoma (Martin-Zanca et al., 1986; 1989), and *trk* oncogenes have been discovered in papillary thyroid carcinomas (Bongarzone et al. 1989). *Trk* encodes a receptor tyrosine kinase that exists as a transmembrane glycoprotein of M_w 140,000 (p140[prototrk]), capable of binding to NGF directly. The extracellular sequence of p140[prototrk] differs substantially from the p75 receptor ligand binding domain. In particular, multiple cysteine repeats are absent in p140[prototrk], and affinity crosslinking agents such as EDAC that crosslink ^{125}I-NGF efficiently to p75[NGFR], do not crosslink ^{125}I-labeled NGF to p140[prototrk]. The dramatic difference in the crosslinking pattern between EDAC and HSAB is likely due to the length of the spacer arm and the different mechanism of the crosslinking reaction. EDAC creates protein-protein crosslinks without a spacer arm, but HSAB can crosslink at a distinct to 8 Angstroms. These experiments indicate that NGF is capable of making different contacts with each receptor on the cell surface.

The product of the proto-oncogene product *trk* contains a consensus tyrosine kinase catalytic activity in its cytoplasmic domain (Martin-Zanca et el., 1986; 1989). The *trk* gene was originally identified and cloned from NIH3T3 cells in gene transfer experiments where cells were transfected with high molecular weight DNA from human tumor cell DNA. The *trk* oncogene was found to be the result of a genomic rearrangement wherein tropomyosin sequences were aberrantly fused onto the transmembrane and cytoplasmic domains of *trk*. The truncated *trk* oncogene product presumably unleashed its phosphorylation activities to create the transformed phenotype.

NGF initiates the tyrosine kinase activity of p140[trk] in PC12 cells and in isolated embryonic dorsal root ganglia (Kaplan et al., 1991b; Klein et al., 1991a). The discovery that a member of the tyrosine kinase family is the signal transducer for NGF simultaneously solves a number of questions, including the signalling mechanism for NGF (Maher, 1988), and the way in which specificity of action is generated for the neurotrophic family of factors, including NGF and its relatives, BDNF (brain-derived neurotrophic factor), and other neurotrophins, NT-3 and NT-4, (Barde et al., 1982; Leibrock et al., 1989; Maisonpierre et al., 1990; Hohn et al., 1990; Hallbook et al., 1991; Rosenmeier et al., 1991), which display similar biological effects upon distinct but overlapping populations of neuronal cells.

Related molecules to *trk*, such as *trkB* and *trkC*, have been identified that bind to other neurotrophic factors (BDNF & NT-3), but not NGF (Soppet et al., 1991; Squinto et al., 1991; Klein et al., 1991b; Cordon-Cardo et al., 1991; Lamballe et al., 1991). These *trk* family members bear considerable resemblance to p140^{prototrk} molecule, not only in the consensus tyrosine kinase domain, but also in the extracellular domain. Cysteine residues are conserved among *trk*, *trkB*, and *trkC*, and there exists more than 50% conservation in the ligand binding domains. The catalytic tyrosine kinase domains contain a consensus ATP binding site and is characterized by dual tyrosine residues at the putative autophosphorylation site, and a short C-terminal tail. Among the family of tyrosine kinases (Hanks et al., 1988), the *trk* family shows the most similarity to the insulin receptor. Consistent with the identity of BDNF and NT-3 as ligands for these receptors is the highly specific expression of *trkB* (Klein et al., 1990) and *trkC* (Lamballe et al., 1991) in the central nervous system. The trk family of tyrosine kinase receptors are distinguished by their extremely specific expression in the developing nervous system, a trait not yet observed with any other receptor tyrosine kinases.

THE P75 NGF RECEPTOR

The p75^{NGFR} does not have an intrinsic tyrosine kinase activity and is not an apparent substrate for tyrosine phosphorylation. The p75 receptor binds ^{125}I-NGF with low affinity (K_d = 10^{-9} M), and with rapid association and dissociation rates (Radeke et al., 1987). For p75^{NGFR}, the extracellular cysteine repeats have been directly implicated in NGF binding (Welcher et al., 1991; Yan and Chao, 1991a), and it is likely that similar domains in related cell surface molecules are used for ligand binding.

The p75 receptor was identified not only by direct binding and crosslinking experiments but also by monoclonal antibodies made against human melanoma cells (Ross et al., 1984) and PC12 cells (Chandler et al., 1984) that inhibited or altered NGF binding. These antibodies were specific either for the rat receptor (IgG-192), or the human receptor (ME20.4), and recognized p75^{NGFR}. The Ig-192 is capable of increasing binding of ^{125}I-NGF to p75^{NGFR}, whereas the ME20.4 is an effective at blocking binding. Both of these antibodies do not recognize the p140^{prototrk} receptor.

The cloning of the p75 gene was accomplished by using total genomic DNA in gene transfer experiments with the herpes virus thymidine kinase gene to generate transfected fibroblasts expressing the p75^{NGFR} (Chao et al., 1986; Radeke et al., 1987). The ME20.4 and IgG-192 antibodies were used to identify positive cells by rosette analysis or the fluorescence-activated cell sorter, and the receptor genes were isolated by either subtractive hybridization (Radeke et al., 1987) or by using Alu repetitive sequences as a tag for the receptor sequence (Chao et al., 1986).

In transfected fibroblast cells or Schwann cells or melanoma cells, the p75 receptor binds ^{125}I-NGF with an equilibrium dissociation constant of 10^{-9} M (Radeke et al., 1987; Hempstead et al., 1989; DiStefano and Johnson, 1988). The p75 receptor does not appear to have any direct signal transduction capability in these cells. However, transfection of p75^{NGFR} cDNA in cells of neural crest origin display both high and low affinity sites, and result in NGF responsiveness, such as increased cellular tyrosine phosphorylation (Berg et al., 1991). These results suggested that p75^{NGFR} participated in NGF signaling, but was not sufficient for signal transduction, unless expressed in the appropriate cellular environment. Genetic disruption of the p75^{NGFR}

gene in mice (Lee et al., 1992), together with other gene transfer experiments in cultured cells (Pleasure et al., 1990; Matsushima and Bogenmann, 1990), and with mutant or chimeric forms of p75^{NGFR} (Hempstead et al., 1990; Yan et al., 1991) indicate that NGF binding and responsiveness are influenced by the expression, activity, and stoichiometry of both NGF receptors.

NGF RECEPTOR BINDING

Equilibrium binding of ^{125}I-NGF to cells reveals two distinct affinity states for the NGF receptor (Sutter et al., 1978; Landreth and Shooter, 1980; Schechter and Bothwell, 1981). In most responsive cells, such as sympathetic neurons or pheochromocytoma PC12 cells, approximately 10-15% of the receptors display high affinity binding with a K_d of 10^{-11} M, with the remainder of the receptors possessing a low affinity K_d of 10^{-9} M. Biological responsiveness to NGF are thought to depend upon interactions with the high affinity form. This difference implies that the molecular events that change NGF receptor affinity are involved in the signal transduction mechanism of NGF.

The finding that NGF binds directly to the *trk* tyrosine kinase to increase tyrosine phosphorylation raises an important question concerning the affinity of NGF binding to its two receptors. Previous affinity crosslinking studies predicted that the higher molecular weight species constituting NGF bound to *trk* (M_r 160 kilodaltons) represented the high affinity binding complex (Hosang and Shooter, 1985; Meakin and Shooter, 1991). However, equilibrium binding studies with membranes isolated from fibroblast cells expressing *trk* alone demonstrate that p140^{prototrk} bind NGF with a K_d of 10^{-9} M, a value similar to the p75 NGF receptor (Kaplan et al., 1991b; Hempstead et al., 1991). A similar equilibrium binding constant of 10^{-9} M to 10^{-10} M was measured in whole cell binding studies of ^{125}I-NGF binding to *trk* at 4°C, but higher affinity binding ($K_d = 10^{-11}$ M) was observed at 37°C (Klein et al, 1991a). Since p140^{prototrk} is significantly internalized in fibroblasts at 37°C, binding constants at higher temperatures are more difficult to measure due to internalized ^{125}I-NGF. Taken together, the binding studies indicate that the overwhelming majority of binding sites for NGF on p140^{prototrk} is a relatively low affinity interaction.

Another method for analyzing the binding behavior is to measure binding by kinetic parameters. Analysis of time courses of ^{125}I-NGF binding to cells expressing either p75^{NGFR} or p140^{prototrk} at 4°C is an independent means of obtaining equilibrium dissociation constants. Interestingly, the on and off-rates of ^{125}I-NGF binding differ dramatically for the two receptors, with p140^{prototrk} exhibiting a much slower off-rate that p75^{NGFR} (Meakin et al., 1992). Likewise, the on-rate for NGF binding to *trk* is far slower than the binding to p75^{NGFR}. Since the K_d is a ratio of the rate constants for dissociation and association, the measurement of these kinetic parameters indicates that p140^{prototrk} interacts with NGF with low affinity, but with comparatively slow on- and off-rates (Hempstead et al., unpublished results). Therefore, both p140^{prototrk} and p75^{NGFR} bind ^{125}I-NGF with a similar low affinity (10^{-9} M) K_d, implying that high affinity binding requires co-expression of both receptors.

Reconstitution studies by membrane fusion and transfection of cloned cDNAs together with equilibrium binding analysis indicated that co-expression of both receptors, p75^{NGFR} and p140^{trk}, is required for high affinity binding (Hempstead et al., 1991) and biological function. If high affinity binding is taken as a criteria for biological action for NGF, then these findings suggest that both receptors must be co-expressed in the same cells for functional responses. Other studies, however, have indicated that NGF-induced *trk* or BDNF/NT-3-induced *trkB* tyrosine kinase activities were sufficient for cell proliferation in fibroblasts (Corden-Cardo et al., 1991; Glass et al., 1991), suggesting that the biological actions mediated by neurotrophins are carried out solely by interactions with an appropriate *trk* tyrosine kinase family member.

The major biological functions of NGF in the developing nervous system are to mediate cell survival and maintenance in post-mitotic neuronal populations (Levi-Montalcini, 1987). That NGF exerts mitogenic properties *in vivo* is a widely held misconception, though its name implied that it behaved to stimulate neuronal cell division. NGF does not stimulate neuronal cell division (Black et al., 1990). Functional responses of *trk* family members in fibroblast cells may not be physiologically significant when the neurotrophins do not mediate proliferation *in vivo*. The *trk* receptors can function to initiate increased cell division in heterologous cells, but are more relevant in mediating signaling activities normally associated in cells that are post-mitotic and undergoing terminal differentiation.

In addition, multiple neurotrophin factors are known to interact with each *trk* tyrosine kinase. Autophosphorylation of p140^{prototrk} in fibroblast cells is induced by not only NGF, but also NT-3 and NT-5 (Berkemeier et al., 1991). The promiscuity of neurotrophin action in fibroblasts expressing *trk* or *trkB* (Soppet et al., 1991; Cordon-Cardo et al., 1991; Squinto et al., 1991) suggest that neurotrophin action must be examined more closely in neuronal cell environments. Hence the cellular context that the *trk* tyrosine kinase is expressed is probably a major parameter in determining biological functions of the neurotrophin factor family.

CONCLUSIONS

The presence of two receptors in NGF responsive neurons poses new questions concerning the requirement for both receptors. The role of the *trk* tyrosine kinase is to act as the signalling transducing receptor for NGF, and it has been shown to be necessary for NGF function. Initiation of NGF action depends upon increased tyrosine phosphorylation by p140^{prototrk}. Lack of expression of this receptor leads to NGF unresponsiveness, indicating that p140^{prototrk} function is essential for NGF signal transduction (Loeb et al., 1991). In contrast, the p75 receptor does not appear to signal independently (Hempstead et al., 1988), but can bind to all neurotrophin family members (Rodriguez et al., 1990; 1992). The p75^{NGFR} should be more appropriately considered as a neurotrophin receptor. Its function is to participate in formation of high affinity binding site for NGF. The low affinity p75 receptor may also participate in retrograde transport, and to modulate the tyrosine kinase activities of *trk*.

The expression of *trk* in the developing nervous system (Martin-Zanca et al., 1990) has allowed for the identification of the ligands for the *trk* subfamily of tyrosine kinases, and defined the mechanism of NGF receptor mediated signal transduction. The p75^{NGFR} is widespread throughout development in a variety of diverse cell types (Bothwell, 1990), suggesting that this receptor must acting in a wider manner as a receptor for the family of neurotrophins. A major unanswered question is how specificity of neurotrophin factors is determined by expression of these receptors. It is anticipated that this question will be ultimately answered by examining the pattern of differential expression of *trk* family members and the p75 receptor during development and maturity, and by determining how specificity is built into each neurotrophin-receptor interaction.

BIBLIOGRAPHY

Barde Y-A, Edgar D, Thoenen H. Purification of a new neurotrophic factor from mammalian brain. EMBO J 1982; 1:549-553.
Berg MM, Sternberg D, Hempstead BL, Chao MV. The low affinity p75 nerve growth factor (NGF) receptor mediates NGF-induced tyrosine phosphorylation. Proc Natl Acad Sci USA 1991; 88:7106-7110.
Berkemeier LR, Winslow JW, Kaplan DR, Nikolics K, Goedell DV, Rosenthal A. Neurotrophin-5: A novel neurotrophic factor that activates trk and trkB. Neuron 1991; 7:857-866.
Black IB, DiCiccio-Bloom E, Dreyfus C. Nerve growth factor and the issue of mitosis in the nervous system. Curr Topics Dev Biol 1990; 24:161-192.

Bongarzone I, Pierotti MA, Monzini N, Mondellini P, Manenti G, Donghi R, Pilotti S, Grieco M, Santoro M, Fusco A, Vecchio G, Della Porta G. High frequency of activation of tyrosine kinase oncogenes in human papillary thyroid carcinoma. Oncogene 1989; 4:1457-1462.

Bothwell MA. Tissue localization of nerve growth factor and nerve growth factor receptors. Curr Topics Micro Immun 1990; 165:55-70.

Chandler CE, Parsons LM, Hosang M, Shooter EM. A monoclonal antibody modulates the interaction of nerve growth factor with PC12 cells. J Biol Chem 1984; 259:6882-6889.

Chao MV, Bothwell MA, Ross AH, Koprowski H, Lanahan A, Buck CR, Sehgal A. Gene transfer and molecular cloning of the human NGF receptor. Science 1986; 232:418-421.

Chao MV. Nerve growth factor. In Sporn MB, Roberts AB, eds. Peptide Growth Factors and Their Receptors. Springer-Verlag, 1990: 135-165.

Cordon-Cardo C, Tapley P, Jing S, Nanduri V, O'Rourke E, Lamballe F, Kovary K, Klein K, Jones KR, Reichardt LF, Barbacid M. The trk tyrosine protein kinase mediates the mitogenic properties of nerve growth factor and neurotrophin-3. Cell 1991; 66:173-183.

DiStefano PS, Johnson EM. Nerve growth factor receptors on cultured rat Schwann cells. J Neurosci 1988; 8:231-241.

Glass DJ, Nye SH, Hantzopoulos P, Macchi MJ, Squinto SP, Goldfarb M, Yancopoulos GD. $trkB$ mediates BDNF/NT-3-dependent survival and proliferation in fibroblasts lacking the low affinity NGF receptor. Cell 1991; 66:405-413.

Green SH, Greene LA. A single M_r=103,000 ^{125}I-ß-nerve growth factor-affinity-labeled species represents both the low and high affinity forms of the nerve growth factor receptor. J Biol Chem 1986; 261:15316-15326.

Grob PM, Berlot CH, Bothwell MA. Affinity labeling and partial purification of nerve growth factor receptors from rat pheochromocytoma and human melanoma cells. Proc Natl Acad Sci USA 1983; 80:6819-6823.

Hallbook F, Ibanez CF, Persson H. Evolutionary studies of the nerve growth factor family reveal a novel member abundantly expressed in Xenopus ovary. Neuron 1991; 6:845-858.

Hanks SK, Quinn AM, Hunter T. The protein kinase family: conserved features and deduced phylogeny of the catalytic domains. Science 1988; 241:42-52.

Hempstead BL, Patil N, Olson K, Chao MV. Molecular analysis of the nerve growth factor receptor. Cold Spring Harbor Symp Quant Biol 1988; 53:477-485.

Hempstead BL, Schleifer LS, Chao MV. Expression of functional nerve growth factor receptors after gene transfer. Science 1989; 243:373-375.

Hempstead BL, Patil N, Thiel B, Chao MV. Deletion of cytoplasmic sequences of the NGF receptor leads to loss of high affinity ligand binding. J Biol Chem 1990; 265:9595-9598.

Hempstead BL, Martin-Zanca D, Kaplan D, Parada LF, Chao MV. High affinity NGF binding requires co-expression of the trk proto-oncogene and the low affinity NGF receptor. Nature 1991; 350:678-683.

Hohn A, Leibrock J, Bailey K, Barde Y-A. Identification and characterization of a novel member of the nerve growth factor/brain-derived neurotrophic factor family. Nature 1990; 334:339-341.

Hosang M, Shooter EM. Molecular characteristics of nerve growth factor receptors on PC12 cells. J Biol Chem 1985; 260:655-662.

Johnson D, Lanahan A, Buck CR, Sehgal A, Morgan C, Mercer E, Bothwell M, Chao M. Expression and structure of the human NGF receptor. Cell 1986; 47:545-554.

Kaplan DR, Martin-Zanca D, Parada LF. Tyrosine phosphorylation and tyrosine kinase activity of the trk proto-oncogene product induced by NGF. Nature 1991a; 350:158-160.

Kaplan DR, Hempstead BL, Martin-Zanca D, Chao MV, Parada LF. The trk proto-oncogene product: a signal transducing receptor for nerve growth factor. Science 1991b; 252:554-558.

Klein R, Parada LF, Coulier F, Barbacid M. $trkB$, a novel tyrosine protein kinase receptor expressed during mouse neural development. EMBO J 1989; 8:3701-3709.

Klein R, Jing S, Nanduri V, O'Rourke E, Barbacid M. The trk proto-oncogene encodes a receptor for nerve growth factor. Cell 1991a; 65:189-197.

Klein R, Nanduri V, Jing S, Lamballe F, Tapley P, Bryant S, Cordon-Cardo C, Jones KR, Reichardt LF, Barbacid M. The $trkB$ tyrosine protein kinase is a receptor for brain-derived neurotrophic factor and neurotrophin-3. Cell 1991b; 66:395-403.

Lamballe F, Klein R, Barbacid M. $trkC$, a new member of the trk family of tyrosine protein kinases, is a receptor for neurotrophin-3. Cell 1991; 66:967-979.

Landreth GE, Shooter EM. Nerve growth factor receptors on PC12 cells: Ligand-induced conversion from low- to high-affinity states. Proc Natl Acad Sci USA 1980; 77:4751-4755.

Lee KF, Li E, Huber LJ, Landis SC, Sharpe AH, Chao MV, Jaenisch R. Targeted mutation of the gene encoding the low affinity NGF receptor p75 leads to deficits in the peripheral sensory nervous system. Cell 1992; 69:737-749.

Leibrock J, Lottspeich F, Hohn A, Hofer M, Hengerer B, Masiakowski P, Thoenen H, Barde Y-A. Molecular cloning and expression of brain-derived neurotrophic factor. Nature 1989; 341:149-152.

Levi-Montalcini R. The nerve growth factor: Thirty-five years later. Science 1987; 237:1154-1164.

Loeb D, Martin-Zanca D, Chao MV, Parada LF, Greene LA. The *trk* proto-oncogene rescues NGF responsiveness in mutant NGF-nonresponsive PC12 cell lines. Cell 1991; 66:961-966.

Maher PA. Nerve growth factor induces protein-tyrosine phosphorylation. Proc Natl Acad Sci USA 1988; 85:6788-6791.

Maisonpierre PC, Belluscio L, Squinto S, Ip NY, Furth ME, Lindsay RM, Yancopoulos GD. Neurotrophin-3: a neurotrophic factor related to NGF and BDNF. Science 1990; 247:1446-1451.

Mallet S, Barclay AN. A new superfamily of cell surface proteins related to the nerve growth factor receptor. Immunol Today 1991; 12:220-223.

Martinez HJ, Dreyfus CF, Jonakait GM, Black IB. Nerve growth factor selectively increases cholinergic markers but not neuropeptides in rat basal forebrain in culture. Brain Res 1987; 412:295-301.

Martin-Zanca D, Hughes SH, Barbacid M. A human oncogene formed by the fusion of truncated tropomyosin and protein tyrosine kinase sequences. Nature 1986; 319:743-77.

Martin-Zanca D, Oskam R, Mitra G, Copeland T, Barbacid M. Molecular and biochemical characterization of the human *trk* proto-oncogene. Mol Cell Biol 1989; 9:24-33.

Martin-Zanca D, Barbacid M, Parada LF. Expression of the *trk* proto-oncogene is restricted to the sensory cranial and spinal ganglia of neural crest origin in mouse development. Genes Dev 1990; 4:683-688.

Massague J, Guillette BJ, Czech MP, Morgan CJ, Bradshaw RA. Identification of a nerve growth factor receptor protein in sympathetic ganglia membranes by affinity labeling. J Biol Chem 1981; 256:9419-9424.

Matsushima H, Bogenmann E. NGF induces neuronal differentiation in neuroblastoma cells transfected with the NGF receptor cDNA. Mol Cell Biol 1990; 10:5015-5020.

Meakin SO, Shooter EM. Molecular investigations on the high affinity nerve growth factor receptor. Neuron 1991; 6:153-163.

Meakin SO, Suter U, Drinkwater CC, Welcher AA, Shooter EM. The rat *trk* proto-oncogene product exhibits properties characteristic of the slow NGF receptor. Proc Natl Acad Sci USA 1992; 89:2374-2378.

Mobley WC, Rutkowski JL, Tennekoon GI, Buchanan K, Johnston MV. Choline acetyltransferase activity in striatum of neonatal rats increased by nerve growth factor. Science 1985; 229:284-287.

Pleasure SJ, Reddy UR, Venkatakrishnan G, Roy AK, Chen J, Ross AH, Trojanowski JQ, Pleasure DE, Lee VM. Introduction of NGF receptors in a medulloblastoma cell line results in expression of high- and low-affinity NGF receptors but not NGF-mediated differentiation. Proc Natl Acad Sci USA 1990; 87:8496-8500.

Radeke MJ, Misko TP, Hsu C, Herzenberg LA, Shooter EM. Gene transfer and molecular cloning of the rat nerve growth factor receptor. Nature 1987; 325:593-597.

Rodriguez-Tebar A, Dechant G, Barde Y-A. Binding of brain-derived neurotrophic factor to the nerve growth factor receptor. Neuron 1990; 4:187-192.

Rodriguez-Tebar A, Dechant G, Gotz R, Barde Y-A. Binding of neurotrophin-3 to its neuronal receptors and interactions with nerve growth factor and brain-derived neurotrophin factor. EMBO J 1992; 11:917-922.

Ross AH, Grob P, Bothwell MA, Elder DE, Ernst CS, Marano N, Ghrist BFD, Slemp CC, Herlyn M, Atkinson B, Koprowski H. Characterization of nerve growth factor receptor in neural crest tumors using monoclonal antibodies. Proc Natl Acad Sci USA 1984; 81:6681-6685.

Schechter AL, Bothwell MA. Nerve growth factor receptors on PC12 cells: Evidence for two receptor classes with differing cytoskeletal association. Cell 1981; 24:867-874.

Smith CA, Davis T, Anderson D, Solam L, Beckmann MP, Jerzy R, Dower SK, Cosman D, Goodwin RG. A receptor for tumor necrosis factor defines an usual family of cellular and viral proteins. Science 1990; 248:1019-1023.

Soppet D, Escandon E, Maragos J, Middlemas DS, Reid SW, Blair J, Burton LE, Stanton BR, Kaplan DR, Hunter T, Nikolics K, and Parada LF. The neurotrophin factors brain-derived neurotrophin factor and neurotrophin-3 are ligands for the *trkB* tyrosine kinase receptor. Cell 1991; 65:895-903.

Squinto SP, Stitt TN, Aldrich TH, Davis S, Bianco SM, Radziejewski C, Glass DJ, Masiakowski P, Furth ME, Valenzuela DM, DiStefano PS, Yancopoulos GD. trkB encodes functional receptor for brain-derived neurotrophic factor and neurotrophin-3 but not nerve growth factor. Cell 1991; 65:885-893.

Sutter A, Riopelle RJ, Harris-Warrick RM, Shooter EM. NGF receptors: characterization of two distinct classes of binding sites on chick embryo sensory ganglia cells. J Biol Chem 1979; 254:5972-5982.

Taniuchi M, Schweitzer JB, Johnson EM. Nerve growth factor receptor molecules in rat brain. Proc Natl Acad Sci USA 1986; 83:1950-1954.

Thoenen H, Bandtlow C, Heumann R. The physiological function of nerve growth factor in the central nervous system: comparison with the periphery. Rev Physiol Biochem Pharmacol 1987; 109:145-178.

Welcher AA, Bitler CM, Radeke MJ, Shooter EM. Nerve growth factor binding domain of the nerve growth factor receptor. Proc Natl Acad Sci USA 1991; 88:159-163.

Yan H, Chao MV. Disruption of cysteine-rich repeats of the p75 nerve growth factor receptor leads to loss of ligand binding. J Biol Chem 1991a; 266:12099-12104.

Yan H, Schlessinger J, Chao MV. Chimeric NGF and EGF receptors define domains responsible for neuronal differentiation. Science 1991b; 252:561-563.

THERAPEUTIC POTENTIAL OF NEUROTROPHIC FACTORS: THE ROLE OF CILIARY NEUROTROPHIC FACTOR IN PROMOTING NEURONAL SURVIVAL AND RECOVERY FOLLOWING NEURONAL INJURY

James L. McManaman, Deborah Russell and Frank Collins

Synergen Inc.
Boulder, Colorado 80301

Neurotrophic factors are naturally occurring proteins that promote the survival and functional activities of nerve cells.[1] Traditionally, neurotrophic factors were found in the target cell to which an innervating nerve cell connects. Such target-derived neurotrophic factors regulate the number of contacts formed between the innervating nerve cells and the target cell population.[2] Since the innervating nerve cells are dependent for their survival on the target-derived neurotrophic factor and since their access to the factor is limited, the levels of available factor influence both the survival of nerve cells and the development of synaptic connections.[2]

In addition to maintaining the survival of the innervating nerve cells, target-derived neurotrophic factors promote the functional activities of the innervating nerve cells, particularly their levels of neurotransmitter and neurotransmitter-synthesizing enzymes.[3,4] Examples of such target-derived neurotrophic factors are nerve growth factor (NGF) and the closely related members of the NGF family (brain derived neurotrophic factor, BDNF; neurotrophin-3, NT-3; neurotrophin-4, NT-4; neurotrophin-5, NT-5).[5-7]

Neurotrophic factors are also found in cells that are not innervated. Examples of non-traditional neurotrophic factors are basic fibroblast growth factor (bFGF),[8,9] and ciliary neurotrophic factor (CNTF).[10] Although the biological roles of such factors have not been conclusively established, these factors are thought to be released upon injury to the nervous system[10,11] and may be involved in limiting the extent of neuronal damage.[9,12]

Target-derived and non-traditional neurotrophic factors can protect responsive nerve cells against a variety of unrelated insults. NGF, for example, is not only necessary for the survival of embryonic sensory and sympathetic neurons,[13] but is capable of rescuing a significant portion of sensory nerve cells from death caused by axotomy,[14,15] and damage caused by administration of the cancer chemotherapeutic agent, taxol.[16] NGF also rescues a significant portion of basal forebrain cholinergic nerve cells from damage caused by axotomy,[17] and damage due to natural aging.[18] The general nature of this neuro-protection has lead to the concept that neurotrophic factors may be useful in treating diseases that involve damage to those nerve cells, even though the mechanism of damage is unknown.

Botulinum and Tetanus Neurotoxins, Edited by
B.R. DasGupta, Plenum Press, New York, 1993

EVIDENCE THAT CNTF IS A NEUROTROPHIC FACTOR FOR THE PERIPHERAL NERVOUS SYSTEM NEURONS

Localization and Effects of Injury

CNTF protein and CNTF messenger RNA (mRNA) are localized to peripheral nerve tissue in adult animals.[19,20] Much lower levels of both CNTF protein and mRNA are found in the spinal cord and brain (central nerve tissue); and there is little or no CNTF protein or mRNA in non-nervous tissues.[19,20] CNTF mRNA is found at high levels in peripheral nerves,[19,20] such as the sciatic nerve, from shortly after birth[20] into adulthood,[19,20] and is localized to Schwann cells.[21] The proposed glial site of CNTF synthesis and storage in peripheral nerves is further supported by the observation than Schwann cells contain CNTF-like immunoreactivity.[22] CNTF is probably not normally secreted from cells, since it has no conventional hydrophobic signal sequence for secretion[19,20] and it is not released into the culture medium by animal cells over-expressing the CNTF gene.[19,20]

Although the exact mechanisms are obscure, there is increasing evidence that specific axonal signals govern the synthesis and release of CNTF by Schwann cells. Crush or lesion injuries to peripheral nerve axons produce dramatic decreases in both protein and mRNA levels of CNTF regions both adjacent and distal to the site of injury.[21,23] Peripheral nerve injury also induces Schwann cell proliferation[24] and increases expression of NGF and low affinity (p75[NGFR]) nerve growth factor receptor mRNA in Schwann cells.[25,26] Because these changes are correlated with the extent of axonal degeneration they are thought to be related, in part, to the loss of specific neuronal influences.[27] Support for this conclusion is provided by studies demonstrating that during peripheral nerve regeneration NGF[25] and CNTF[21] mRNA levels return to normal.

Nerve injury also produces a transient release of CNTF into the extracellular space. The level of CNTF-like bioactivity increases approximately 100-fold at the site of injury within 24 hours after transecting the rat sciatic nerve.[10] This increase is transient, and by one week after injury the CNTF-like activity has declined back to undetectable levels.[10] The qualitative and temporal natures of the extracellular changes in CNTF levels, appear to correlate with the decrease of CNTF within Schwann cells of degenerating nerve fibers, and suggest that the two processes may be related.

Effects of CNTF on Motoneuron Survival

Although highly purified recombinant CNTF has been shown to support the survival, in cell culture, of chick embryonic parasympathetic,[28] sympathetic,[28] sensory,[28] and motor neurons,[29] the strongest evidence that CNTF has neurotrophic activity is provided by its ability to prevent both ontogenetic and injury-induced death of motoneurons. The death of primary motoneurons occurs naturally as a normal part of embryonic development. During a specific period in embryonic development about half of the nerve cells of a particular type that were generated earlier in development die.[2] The timing of this "ontogenetic" neuronal death is dependent both on species and the type of nerve cell. In lateral motor column of chick embryos, ontogenetic death of motoneurons begins on the sixth day of development (E6) and is essentially complete by the tenth day of development (E10). During this period approximately 12,000 of the 24,000 motoneurons originally present on E6 die.[2] It is thought that ontogenetic neuronal death results from a competition among the nerve cells for a limited supply of a neurotrophic factor required for their survival. Thus, a protein is considered neurotrophic if its exogenous application can rescue neurons that would have died during the period of ontogenetic death. The administration of recombinant human CNTF to chick embryos from days 5 to 9 of development is able to rescue a significant number of motoneurons that normally would have died during this period.[30] In embryos exposed to the highest dose of CNTF, only about 23% of the neurons died. The ability of CNTF to reduce ontogenetic death of motor neurons was confirmed by counting the number of motor neurons undergoing visible degeneration. The highest dose of CNTF reduced the numbers of degenerating motoneurons 4-fold, from 16 to 4 per thousand and represent a sparing of at least half of the motoneurons that would have died.[30] CNTF is the only known neurotrophic factor able to rescue primary motoneurons from ontogenetic death. Although motoneurons express high affinity receptors for NGF,[31] and are responsive to bFGF in cell culture,[29,32] neither of these agents enhances motoneuron survival during the period of ontogenetic death.[33,34]

The maintenance and survival of post-embryonic neurons also appears to depend on specific neurotrophic factors. Transection of peripheral axons, for instance, leads to rapid and extensive neuronal death of perinatal sensory and motoneurons.[35,36] The administration of NGF to injured perinatal axons is able to prevent the death of sensory neurons, but does not affect motoneurons. CNTF, in contrast, is able to prevent injury-induced death of perinatal motoneurons. Transection of the facial nerve of new born rats leads to the death of about 80% of the motoneurons whose axons have been injured. The application of CNTF to the ends of injured axons reduces the extent of this death to about 20% of the injured axons.[12] Similarly, CNTF is able to reduce the death of postnatal facial motoneurons that occurs in a mutant mouse model of motoneuron disease.[38]

The evidence presented above demonstrates that 1) CNTF is stored in peripheral nerve tissue; 2) the expression of CNTF mRNA and the intracellular levels of CNTF are influenced by axonal integrity; 3) CNTF is released in a bioactive form at the site of a peripheral nerve injury, and 4) CNTF provides neurotrophic support for all major classes of neurons that extend processes into peripheral nerves. Together, this evidence suggests that CNTF may play a role in maintaining normal peripheral nerves and function as a repair factor for damaged or diseased nerves.

To further evaluate CNTF's role as a repair factor, recombinant human CNTF was administered subcutaneously at 0.1 and 0.25 mg/kg/day to young adult Sprague-Dawley rats following a unilateral crush injury to the right sciatic nerve. This injury produced a loss of sensory and motor function in the lesioned limb. Sensory performance was quantitatively assessed by determining the amperage required for perception of an electrical stimulus applied to a region of the foot denervated by sciatic nerve crush. As nerve regeneration progressed, animals were able to perceive smaller electrical stimuli applied to the soles of their feet. Vehicle-treated animals required approximately 30 days for nearly complete restoration of sensory performance. Treatment with CNTF significantly accelerated the recovery by 3 days. CNTF also accelerated the recovery of injured motor axons. Crush injury produces a loss of motor function in the extensor and flexor muscles of the tarsus and digits. Motor performance was quantitatively assessed after sciatic nerve crush by measuring the position of the digits during walking. Regain of a normal spread of digits indicates recovery from injury. Like the recovery of sensory function, the recovery of motor function requires approximately 30 days. CNTF-treatment significantly reduces the time required for recovery. The average time required to achieve 50% recovery was 21.2 ± 0.3 (mean $\pm$ SD) days in the CNTF-treated group compared to 24.1 ± 0.3 days in the control (saline-treated) group. These results indicate that CNTF is able to accelerate the regeneration of sensory and motor function when administered systemically by subcutaneous injection to adult animals.

SUMMARY

There is increasing evidence that neurotrophic factors are necessary for the survival and development of embryonic neurons, and contribute to the maintenance of normal adult neuronal properties. It has also been proposed that a diminished neurotrophic support contributes to the development of neurodegenerative diseases, such as amyotrophic lateral sclerosis (ALS), Parkinsonism and Alzheimer's disease.[37] Although the involvement of neurotrophic factors in the pathophysiology of these diseases remains speculative, their ability to support the survival of the neuronal sub-types at risk in these diseases suggests that they may be useful therapeutic tools. ALS is a progressive, fatal motor neuron disease primarily affecting older adults. The clinical and pathological characteristics of this disease are weakness and muscular atrophy due to the degeneration and death of motoneurons. At present there is no cure for ALS. However, the discovery of CNTF and its ability to prevent ontogenetic and injury-induced death of motoneurons, and enhance recovery of injured sensory and motor neurons raises the possibility that it may also prevent motoneuron death in patients with motor neuron disease. In support of this possibility Sendtner *et al*[38] have recently demonstrated that CNTF can prevent motoneuron degeneration in a mouse mutant with progressive motor neuropathy. Thus, CNTF appears capable of providing neurotrophic support for diseased as well as developing or injured motoneurons.

REFERENCES

1. Barde YA. Trophic factors and neuronal survival. Neuron 1989; 2:1525-1534.
2. Oppenheim RW. The neurotrophic theory and naturally occurring motoneuron death. Trends Neurosci 1989; 12:252-255.
3. Martinez HJ, Dreyfus CF, Jonakait GM and Black IB. Nerve growth factor promotes cholinergic development in brain striatal cultures. Proc Natl Acad Sci USA 1985; 82:7777-7781.
4. Vantini G, Schiavo N, Di Martino A, Polato P, Triban C, Callegaro L, Toffano G, Leon A. Evidence for a physiological role of nerve growth factor in the central nervous system of neonatal rats. Neuron 1989; 3:267-273.
5. Leibrok J, Lottspeich F, Hohn A, Hofer M, Hengerer B, Masiakowski P, Thoenen H. Molecular cloning and expression of brain-derived neurotrophic factor. Nature 1989; 341:149-152, 1989.
6. Hohn A, Leibrock J, Bailey K, Barde YA. Identification and characterization of a novel member of the nerve growth factor/brain-derived neurotrophic factor family. Nature 1990; 344:339-341.
7. Rosenthal A, Goeddel DV, Nguyen T, Lewis M, Shih A, Laramee GR, Nikolikcs K, Winslow JW. Primary structure and biological activity of a novel human neurotrophic factor. Nature 1990; 4:767-773.
8. Baird A, Esch F, Mormede P, Ueno N, Ling N, Bohlen P, Shao-Yao Y, Wehrenberg WB, Guilemin R. Molecular characterization of fibroblast growth factor: distribution and biological activities in various tissues. In: Greep RO, ed. Recent Progress in Hormone Research, 42:144-200, New York: Academic Press, 1986.
9. Anderson KJ, Dam D, Lee S, Cotman CW. Basic fibroblast growth factor prevents death of lesioned cholinergic neurons *in vivo*. Nature 1988; 332:360-361.
10. Manthorpe M, Ray J, Pettmann B, Varon S. Ciliary neurotrophic factors. In: Rush RA, ed. Nerve Growth Factors. New York: John Wiley & Sons Ltd, 1989:31-53.
11. Finkelstein SP, Apostolides PJ, Caday CG, Prosser J, Phillips MF, Klagsbrun M. Increased basic fibroblast growth factor (bFGF) immunoreactivity at the site of focal brain wounds. Brain Res 1988; 460:253-259.
12. Sendtner M, Kreutzberg GW, Thoenen H. Ciliary neurotrophic factor prevents the degeneration of motor neurons after axotomy. Nature 1990; 345:440-441.
13. Hamburger V, Yip JW. Reduction of experimentally induced neuronal death in spinal ganglia of the chick embryo by nerve growth factor. J Neurosci 1984; 4:767-774.
14. Rich KM, Lusqcqynski JR, Osborne PA, Johnson EM. Nerve growth factor protects adult sensory neurons from cell death and atrophy caused by nerve injury. J Neurocytology 1987; 16:261-268.
15. Otto D, Unsicker K, Grothe C. Pharmacological effects of nerve growth factor and fibroblast growth factor applied to the transectioned sciatic nerve on neuron death in adult rat dorsal root ganglia. Neurosci Lett 1987; 83:156-160.
16. Apfel SC, Lipton RB, Arezzo JC, Kessler JA. Nerve growth factor prevents toxic neuropathy in mice. Ann Neurol 1991; 29:87-90.
17. Williams LR, Varon S, Peterson GM, Wictorin K, Fischer W, Bjorklund A, Gage FH. Continuous infusion of nerve growth factor prevents basal forebrain neuronal death after fimbria fornix transection. Proc Natl Acad Sci USA 1986; 83:9231-9235.

18. Fischer W, Wictorin K, Bjorklund A, Williams LR, Varon S, Gage FH. Amelioration of cholinergic neuron atrophy and spatial memory impairment in aged rats by nerve growth factor. Nature 1987; 329:65-68.

19. Lin L-F, Mismer D, Lile JD, Armes LG, Butler ET, Vannice JL, Collins FC. Purification, cloning, and expression of ciliary neurotrophic factor (CNTF). Science 1989; 246:1023-1025.

20. Stockli KA, Lottspeich F, Sendtner M, Masiakowski P, Carroll P, Gotz R, Lindholm D, Thoenen H. Molecular cloning, expression and regional distribution of rat ciliary neurotrophic factor. Nature 1989; 342:920-923.

21. Sendtner M, Stockli KA, Thoenen H. Synthesis and localization of ciliary neurotrophic factor in the sciatic nerve of the adult rat after lesion and during regeneration. J Cell Biol 1992; 118:139-142.

22. Rende M, Muir D, Ruoslahti E, Hagg T, Varon S, Manthorpe M. Immunolocalization of ciliary neuronotropic factor in adult rat sciatic nerve. Glia 1992; 5:25-32.

23. Rabinovsky ED, Smith GM, Browder DP, Shine HD, McManaman JL. Peripheral nerve injury down-regulates CNTF expression in adult rat sciatic nerves. J Neurosci Res 1992; 31:188-192.

24. Abercrombie M, Johnson ML. Qualitative histology of Wallerian degeneration. J Anat 1946; 80:37-50.

25. Heumann R, Lindholm D, Bandtlow C, Meyer M, Radeke MJ, Misko TP, Shooter E, Thoenen H. Differential regulation of mRNA encoding nerve growth factor and its receptor in rat sciatic nerve during development, degeneration, and regeneration: Role of macrophages. Proc Natl Acad Sci USA 1987; 84:8735-8739.

26. Taniuchi M, Clark HB, Johnson EM. Induction of nerve growth factor receptor in Schwann cells after axotomy. Proc Natl Acad Sci USA 1986; 83:4094-4098.

27. Heumann R, Korsching S, Bandtlow C, Thoenen H. Changes in nerve growth factor synthesis in nonneuronal cells in response to sciatic nerve transection. J Cell Biol 1987; 104:1623-1631.

28. McDonald JR, Ko C, Mismer D, Smith DJ, Collins F. Expression and characterization of recombinant human ciliary neurotrophic factor from *Escherichia coli*. Biochim Biophys Acta 1991; 1090:70-80.

29. Arakawa Y, Sendtner M, Thoenen H. Survival effect of ciliary neurotrophic factor (CNTF) on chick embryonic motoneurons in culture: comparison with other neurotrophic factors and cytokines. J Neurosci 1990; 11:3507-3515.

30. Oppenheim RW, Prevette D, Yin QW, Collins F, McDonald JR. Control of embryonic motoneuron survival *in vivo* by ciliary neurotrophic factor. Science 1991; 251:1616-1618.

31. Marchetti D, Haverkamp LJ, Clark RC, McManaman JL. Ontogeny of high- and low-affinity nerve growth factor receptors in the lumbar spinal cord of the developing chick embryo. Dev Biol 1991; 148:306-313.

32. McManaman JL, Crawford F, Clark R, Richker J, Fuller F. Multiple neurotrophic factors from skeletal muscle: Demonstration of effects of basic fibroblast growth factor and comparisons with the 22-kilodalton choline acetyltransferase development factor. J Neurochem 1989; 53:1763-1771.

33. Oppenheim RW, Haverkamp LJ, Prevette D, McManaman JL, Appel SH. Reduction of naturally occurring motoneuron death *in vivo* by a target-derived neurotrophic factor. Science 1988; 240:919-922.

34. McManaman JL, Oppenheim RW, Prevette D, Marchetti D. Rescue of motoneurons from cell death by a purified skeletal muscle polypeptide: Effects of the ChAT development factor CDF. Neuron 1990; 4:891-898.

35. Kashihara Y, Kuno M, Miyata Y. Cell death of axotomized motoneurons in neonatal rats, and its prevention by peripheral reinnervation. J Physiol 1987; 386:135-148.

36. Miyata Y, Kashihara Y, Homma S, Kuno M. Effects of nerve growth factor on the survival and synaptic function of Ia sensory neurons axotomized in neonatal rats. J Neurosci 1986; 6:2012-2018.

37. Appel SH. A unifying hypothesis for the cause of amyotrophic lateral sclerosis, parkinsonism, and Alzheimer's disease. Ann Neurol 1981; 10:499-505.

38. Sendtner M, Schmalbruch H, Stockli KA, Carroll P, Kreutzberg GW, Thoenen H. Ciliary neurotrophic factor prevents degeneration of motor neurons in mouse mutant progressive motor neuronopathy. Nature 1992; 358:502-504.

TETANUS TOXIN AS A TOOL FOR INVESTIGATING
THE STRUCTURAL BASES OF NEUROTROPISM

Bernard Bizzini, Mohamed Khiri, and Martine Carlotti

Department of Molecular Toxinology
Pasteur Institute
75724 Paris Cedex, France

INTRODUCTION

Tetanus toxin binds with high affinity to fixation sites at the nerve terminal membrane. As a result the toxin molecule is internalized and transported retrogradely within the axon to the central nervous system (CNS). This neurotoxin is a "neurotropic" agent, since it is endowed with the capacity to migrate selectively to the CNS.

Neurotropism is a characteristic that is shared by many biological entities as different as neurotoxins (Fedinec, 1965; Stoeckel et al., 1971; Habermann, 1974; Wiegand et al., 1976); neurotropic viruses (Tsiang, 1979; Wildy, 1967; Kristensson et al., 1971); lectins (Stoeckel et al., 1977; Harper et al., 1980); lymphokines (Besançon and Ankel, 1974); antibodies, i.e. dopamine-beta-hydroxylase (Fillenz et al., 1976; Ziegler et al., 1976); hormones, e.g. nerve growth factor (NGF) (Hendry et al., 1974; Stoeckel et al., 1974).

Tetanus toxin (TT) interacts optimally with di- and tri-sialogangliosides (GD_{1b} and GT_1) (Van Heyningen and Miller, 1961). Gangliosides have been considered as being an integral part of the cell receptor for tetanus toxin. The atoxic fragment responsible for conveying the toxin molecule to the CNS has been identified to the half COOH-terminal fragment of the heavy chain, termed fragment C. The transport characteristics of fragment C have been determined and compared to those of other neurotropic agents. It was found that the affinity of fragment C for the uptake system was comparable to that of wheat germ agglutinin (WGA), whereas its transport velocity was close to that of NGF. TT, cholera toxin (ChT) and WGA were shown to be transported in all peripheral neurons, adrenergic, sensory and motor neurons, whereas NGF was transported only in adrenergic and sensory neurons (Stoeckel et al., 1971) and antibodies to dopamine-β-hydroxylase only in adrenergic neurons (Stoeckel et al., 1977). Furthermore, the transport of WGA and NGF in contrast to that of TT and ChT was unaffected by gangliosides (Stoeckel et al., 1977). Therefrom it was concluded that binding and subsequent retrograde transport of TT, ChT and WGA should depend on properties which are common to membranes of all peripheral neurons. However, gangliosides should be involved in the transport of only TT and ChT (Stoeckel et al., 1971). Moreover, since the migration rate of NGF and TT are identical, the specificity of retrograde transport should be determined by specific uptake sites in the neuronal membrane, whereas the retrograde transport system is non-specific. The transport system was found to be a colchicine-sensitive mechanism (Stoeckel et al., 1975).

A similar selective uptake and retrograde transport for TT, ChT, WGA and NGF has further been demonstrated by the capacity of each of these components to block binding and uptake sites of others (Dumas et al., 1979).

We have been interested in investigating the molecular bases of neuro-tropism using as a tool the tetanus toxin molecule or its C fragment. According to Stoeckel et al. (1971, 1975) the retrograde transport of molecules should depend on the recognition by these of a more or less similar structure in the neuronal membrane. Should this hypothesis be right, competition between different molecular entities for binding to this structure should occur at varying degrees in relation to the percentage of homology that is shared by the recognizing sites carried by these entities. In our study we have measured the capacity of TT and fragment C to interfere in vivo or in vitro with 2 neurotropic viruses, herpes and rabies viruses.

Herpes virus has been chosen, because it is also transported retrogradely within the axon (Wildy, 1967; Severin and White, 1968). Replication of mature virions within the neurons and their presence within the axons have been ascertained (Cook and Stevens, 1973). Moreover, like tetanus toxin, it can cross the synaptic cleft. These features have made it a valuable transneuronal labeling agent (Kuypers and Ugolini, 1990). Similarly to TT migration that of herpes virus is impaired by colchicine. Moreover, a 100% lethal herpes virus infection model has been produced in mice (Guillon, 1976), which should allow one to assess the inhibiting effect of fragment C on herpes virus ascent to the CNS.

The choice of rabies virus for the in vitro competition studies with TT was based on the selective binding of rabies virus to neuronal cells and its subsequent transfer from the periphery to the CNS via retrograde axoplasmic flow (Tsiang et al., 1983). In addition, the involvement of polysialogangliosides (GT_{1b} and GQ_{1b}) in binding infective rabies virus particles to the surface of susceptible CER cells has been established (Superti et al., 1986). It has also been shown that CER cells treated with neuraminidase could recover their susceptibility to rabies virus by coating them with gangliosides (Superti et al., 1986; Conti et al., 1986). A dose-dependent inhibition by gangliosides of the capacity of rabies virus to infect neuronal cells has also been observed (Superti et al., 1986). Further, in relation to the observation by Marconi et al. (1982) that the lethal action of tetanus toxin can be counteracted in mice by the injection of Con A from 48 hours before until 24 hours after toxin administration, a dose-dependent inhibition of infection with rabies virus of CER cells was achieved when CER cells had previously been treated with Con A (Conti and Tsiang, 1985).

The results reported in the present paper show that fragment C is capable of preventing death from occurring in a significant number of mice lethally infected with herpes virus. These also point to the capacity of TT to interact with rabies virus for binding to cortical neurons in culture.

Moreover, in an attempt to come closer to the structure involved in binding of TT, crude synaptic membranes were solubilized and the soluble fraction was purified. The binding characteristics of solubilized membranes (sCSM) were determined. It was also found that rabies virus could compete with TT for attachment to sCSM.

MATERIALS AND METHODS

Tetanus Toxin (TT)

It was purified as reported by Bizzini et al. (1969). It assayed 3000 Lf per mg N and 2×10^8 MLDs per mg N.

Biotinylated Tetanus Toxin (b-TT)

Biotinylation of tetanus toxin was carried out with NHS-LC-biotin (Pierce, U.S.A.) according to the manufacturer's recommendations. Briefly, 38mg of TT (6.36×10^{-7} moles) were reacted with 1.2mg of NHS-LC-biotin (1.8×10^{-6} moles) in 0.5ml of 100mM bicarbonate buffer, pH8.5, for 2h in an ice-bath. Thereafter the reaction mixture was dialyzed overnight against a large volume of 70mM sodium phosphate buffer, pH8.2. b-TT exhibited the same toxicity as the original TT sample. It migrated on SDS-polyacrylamide gel electrophoresis as a single band at the same level as native TT.

Tetanus Toxoid (TToid)

It was prepared from purified TT by formaldehyde treatment as described by Bizzini et al. (1969). It assayed 2900 Lf per mg N.

Biotinylated Tetanus Toxoid (b-TToid)

This was prepared in the same way as described above for TT.

Tetanus Toxin Fragments C and AB

These were purified from papain-digested purified tetanus toxin as described elsewhere (Bizzini et al., 1988).

Toxin Subunits

These were prepared according to the procedure of Matsuda and Yoneda (1975) as modified by Taylor et al., (1983).

Isolated Crude Synaptic Membranes (CSM)

CSM were prepared from spinal cords of adult Wistar rats according to the procedure of Young and Snyder (1973) as specified elsewhere (Bizzini et al., 1981). The yield of CSM was of 6.5mg membrane proteins per spinal cord. They were stored at -80°C until use.

Solubilized Crude Synaptic Membranes (sCSM)

The CSM were harvested by centrifugation for 20 min, at $48,000 \times g$, at 4°C. The sediment was resuspended in 2ml of 50mM Tris, pH7.2, containing 4% octyl-b-d-glucopyranoside, 3% Triton X100 and 1% Tween 80. The suspension was sonicated for 25 sec, at 18W (Vibracell, Bioblock) and the sonicated suspension was allowed to stay successively for 1h at 20°C and 45 min at 0°C. Subsequently, the suspension was centrifuged for 1h, at $100,000 \times g$, at 4°C and the solubilized fraction recovered in the supernatant was dialyzed overnight, at 4°C, against a large volume of 50mM Tris, pH7.2, in a dialysis bag previously boiled in a 0.5M EDTA solution. After dialysis the remaining traces of detergent were removed by stirring the solution with 0.64g Bio-Beads (Bio-Rad, U.S.A.) for 2h at 4°C and further 1h at 25°C after addition of 0.5g Bio-Beads per ml solution. The Bio-Beads were allowed to sediment and the supernatant collected with a pipette. The sCSM solution was stored at -80°C until use. Solubilization yield was equal to 58%.

Determination of Bound b-TT to sCSM by an Enzymatic Assay

For this a 96-well polystyrene plate (NUNC) was coated by pipetting into each well 100 µl of 20 µg/ml sCSM solution in 100mM bicarbonate buffer, pH9.5. The plate was incubated for 1h at 37°C and successively overnight at 4°C. The solution was poured out from the wells and the plate was saturated for 1h at 37°C by adding to each well 150 µl of PBS-containing 2% BSA. Thereafter the plate was thoroughly washed with PBS containing 0.1% Tween 20. A calibration curve was run by pipetting into each of 4 wells 100 µl of increasing dilutions of b-TT ranging from 100 µg to 0.1 µg/ml. The reaction was allowed to take place for 30 min at 37°C and, after washing the plate the bound, bTT was allowed to react with a 0.05mg/ml HRP-labeled streptavidin solution (100 µl per well) for 15 min at 37°C. The plate was washed with PBS-Tween 20 and the bound streptavidin reacted for 2 min at room temperature with 100 µl of OPD substrate solution. The reaction was blocked by adding to each well 50 µl of $2\,M\,H_2SO_4$. The plate was read at 412nm. A curve was drawn by plotting the O.D.'s as a function of b-TT concentration. A linear response was obtained in the concentration range of 1 µg/ml to 100 µg/ml in Tris, acetate buffer and 10 µg/ml to 200 µg/ml in Krebs-Ringer solution.

Trypsin Digestion of sCSM

SCSM 1mg/ml were treated with 100 µg per ml of reaction mixture with Trypsin (Porcine pancreas: 15 000 units/mg, SIGMA, U.S.A.) at 37°C, for 30 min, in Krebs-Ringer buffer, pH7.4. The reaction was stopped by addition of 100 µg soybean trypsin inhibitor per ml of reaction mixture.

Neuraminidase Treatment of sCSM

SCSM 1mg/ml were treated with 0.025U of neuraminidase (Behring, *Vibrio cholerae* 1U/ml) per ml of reaction mixture at 37°C, for 1h, in 50mM Tris, acetate buffer containing 4mM $CaCl_2$.

Heat Treatment of sCSM

A 1mg/ml sCSM solution was heated for 5 min at 100°C and immediately cooled down in an ice-bath.

For enzyme and heat-treated sCSM controls were similarly run which contained no sCSM.

Herpesvirus Infection of Mice

Herpesvirus infection was produced according to Guillon (1976) with *Herpesvirus eidolon* isolated from the African bat, *Eidolon helvum* which is similar to *Herpesvirus hominis*. When the virus is inoculated by scarification a 100% lethal infection is generated that reproduces all aspects of neurotropic herpes infection (Guillon et al., 1975). The progress of infection can be quantified by a score method giving a Daily Symptom Index (DSI).

Subcutaneous injection of the virus produces a zonal infection that spontaneously regresses.

Mice were treated with the C fragment, 500 µg per mouse, by either the intravenous (I.V.) or intramuscular (I.M.) route, either before or after virus inoculation.

Rabies Virus

The challenge virus strain (CVS) was used all through this study.

Neuronal Cells in Culture

These were rat primary cortical neurons. They were culture in F10 culture medium containing 7.5% of fetal calf serum and 7.5% of horse serum.

Binding Assays to Cultured Neurons

They were carried out with cortical neurons in culture. Binding was assessed immunohistochemically. Binding of TT took place for 20 min at 4°C and that of CVS for 60 min at 37°C or 4°C, as specified, followed by additional culturing for 40h at 37°C. After each contact with either TT or CVS, cells were washed with F10 medium to remove unbound ligand.

For assessment of TT binding the cultures were reacted with rabbit anti-TT IgG's for 1h at 37°C. The cells were washed and successively treated with paraformaldehyde and acetone. They were subsequently reacted with an anti-rabbit IgG antiserum labeled with rhodamine for 2h at 37°C. The fluorescence-stained cells were examined under the fluorescence microscope.

For assessment of CVS binding the culture medium was discarded at the end of the 40h-incubation period, the cells were fixed successively with p-formaldehyde and acetone and stained with an anti-rabies virus nucleocapside antiserum labeled with fluresceine, for 60 min, at 37°C.

Labeling intensity of the neurons was scored as 1, 2 and 3 crosses.

Competition Assays with Cultured Neurons

These were carried out by treating the cultures, first with TT or CVS and thereafter with either CVS or TT, under the conditions specified under the heading "Binding Assays". The sequence of cell staining always started with TT and then with CVS.

Binding Assays to sCSM

These were carried out by immuno-enzymatic assay using a plate coated with sCSM. Binding assays were performed using two different buffer solutions: 25mM Tris, acetate buffer, pH6, that measures preferentially binding to gangliosides and Krebs-Ringer buffer, pH7.4, that mimics physiological conditions of pH and salt concentration.

These buffers were added for binding with 0.1% BSA. The negative control for b-TT consisted of its binding to a plate coated with BSA instead of sCSM. Non specific binding was determined by reacting first for 30 min at 37°C 100 µl of a 1mg/ml TT solution in wells coated with sCSM and then 100 µl of b-TT for 90 min, at 37°C. Each experimental point is the mean of 3 determinations.

RESULTS

Interaction of Fragment C with Herpesvirus In Vivo

The results schematically represented in Figure 1 show that fragment C of tetanus toxin can interfere with the course and outcome of the 100% lethal *Herpesvirus eidolon* infection produced by scarification. The best protective effect was exercised when the C fragment was given 6 or 24h by the i.m. route or 24h by the i.v. route after virus inoculation. In all experiments carried out the percentage of protection afforded by fragment C has never exceeded approximately 50%. The course of the infection is characterized by successive steps starting with a local lesion (score 1) that progresses through a zonal lesion (score 2) to generalized paralyses (score

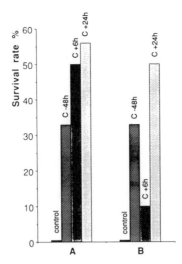

Figure 1. Protective effect of tetanus toxin-derived fragment C on *Herpervirus eidolon* infection produced by scarification. A: fragment C injected i.m.; B: fragment C injected i.v.

3) to end in death (score 4) of mice. Mice which survived did not develop paralyses. As a consequence fragment C should have interacted with the virus ascent before it has reached the CNS (Figure 1).

When the C fragment was injected together with the virus an acceleration of the course of infection was observed and mice died in a shorter time interval than infected and untreated control mice (result not shown). This cooperative effect of C fragment on the spread of herpesvirus is to be related to the cooperative effect of fragment C on the ascent of TT (Dumas et al., 1979).

Figure 2 shows the effect of fragment C injected i.m. on the course of the spontaneously regressing *Herpesvirus eidolon* injection produced by s.c. inoculation of the virus. This infection characterized by development from a local lesion of a zonal lesion regularly regresses within 20 days. Mice given the C fragment 24h postinjection did not develop for most of them a zonal lesion and the infection had fully regressed for the others within 13 days. Again, these results speak for a likely interaction of fragment C with the virus at the stage of binding to fixation sites or of an early step of retrograde transport of the virus.

Interaction of Tetanus Toxin with Rabies Virus for Binding to Cortical Neurons in Culture

Like tetanus toxin, rabies virus exhibits selective binding to polysialogangliosides and retrograde intraaxonal transport. Consequently, it has appeared of interest to us to investigate the ability of the tetanus toxin molecule to interact *in vitro* with rabies virus for attachment to neuronal cells.

Preliminary experiments showed that both TT (500 μg/ml) and CVS (7×10^6 virus particles per ml) strongly damaged the structure of cortical neurons in culture, eventually resulting in cell lysis.

The results summarized in Table 1 show that strong labeling (3+) of cortical neurons in culture was achieved when not damaging concentrations of TT or numbers of virus particles were used. It also results from Table 1 that CVS occupies fixation sites (3+) which are no longer

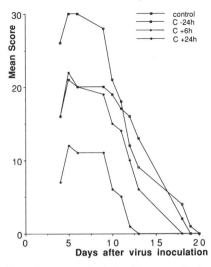

Figure 2. Effect of tetanus toxin-derived fragment C injected i.v. at the dose of 500 µg per mouse, on *Herpesvirus eidolon* infection produced by the s.c. route.

available for the toxin, since only decreased binding of toxin takes place. When neurons were treated 3 successive times with CVS, TT binding was only residual (1±). However, addition to CVS-treated cells of a damaging concentration of TT (500 µg/ml) displaced bound CVS (1+), whereas fairly strong binding of TT occurred (2+) without significant impairment of neurons.

Conversely, treatment of neurons in culture with TT strongly labeled the cells (3+) and prevented strong labeling with CVS (1+). Moreover, when neurons were first treated with an optimal concentration of TT (40 µg/ml) and second with an optimal number of virus particles, subsequent treatment of the cultured neurons with a damaging concentration of TT did not alter the neurons structure.

Table 1. Inhibition of the binding of TT or CVS to primary cortical neurons in culture by CVS or TT, respectively.

Treatments	Neuron labeling	
	TT	CVS
TT (20 up to 100 µg/ml)	+++	-
CVS, once at 37°C or 4°C	-	+++
Thrice at 37°C	-	+++
CVS, once at 4°C followed by TT (20 µg/ml)	+	+++
CVS, thrice at 37°C followed with TT (100 µg/ml)	±	+++
CVS, thrice at 4°C followed with TT (500 µg/ml)	++	+
TT (20 or 40 µg/ml) followed with CVS at 4°C	few cell debris	
TT (40 µg/ml) followed with CVS at 37°C and again with TT (500 µg/ml)	+++	+
CVS + TT (40 µg/ml) 60 min at 4°C)	++	++

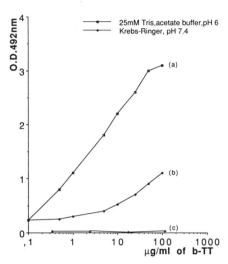

Figure 3. Binding of b-TT to sCSM immobilized on a polystyrene plate.
(a) and (b) = b-TT; (c) = b-TToid.

Simultaneous treatment of neurons in culture with TT and CVS resulted in decreased (2+) but equal labeling of cells with either of these.

Control experiments were carried out using A-B fragment instead of TT for binding experiments. A-B fragment did not label the neurons nor did it influence CVS binding.

Binding of Tetanus Toxin to Solubilized Crude Synaptic Membranes

Crude synaptic membranes were solubilized and the soluble fraction was immobilized onto a polystyrene plate. Binding of TT to the soluble fraction was investigated applying a method similar to ELISA using biotinylated toxin which was revealed with HRP-labeled streptavidin.

Figure 3 shows the binding curves obtained with b-TT either in 25mM Tris, acetate buffer pH6 or Krebs-Ringer buffer pH7.4. Higher binding was observed at pH6 than at pH7.4. In addition, biotinylated toxoid used as a control failed to bind to sCSM.

Similar experiments were conducted with sCSM which had been treated with neuraminidase or trypsin prior to immobilization to the plate.

The binding curves represented in Figure 4 show that neuraminidase treatment completely abolished binding of b-TT to sCSM at pH6 and only slightly so at pH7.4. In contrast, trypsin treatment resulted in complete suppression of b-TT binding to sCSM in Krebs-Ringer, at pH7.4.

As shown in Figure 5 the addition of increasing quantities of mixed gangliosides insolubilized by complex formation with cerebrosides to b-TT almost completely prevented b-TT from binding to sCSM. When b-TT was first bound to sCSM and increasing quantities of complexed gangliosides were subsequently added, b-TT was unbound from sCSM only in Krebs-Ringer, but not in acetate buffer.

The levels of b-TT binding to sCSM which have been either boiled or digested with trypsin prior to their immobilization on a polystyrene plate are represented in Figure 6. This Figure shows that the binding of b-TT to heat or trypsin-treated sCSM was significantly reduced in Krebs-Ringer solution with respect to binding to untreated sCSM, but not in Tris, acetate buffer.

In subsequent experiments it was looked for whether b-TT could compete with CVS for binding to sCSM. For this increasing quantities of b-TT were mixed with a fixed amount of CVS particles (7×10^6) and the mixtures were added to wells of a polystyrene plate coated with sCSM. The results obtained are schematically represented in Figure 7. This Figure shows that in a defined interval of b-TT concentration binding of CVS to sCSM was reduced by about 30%. This reduced binding could be accounted for by the occupation by b-TT of a family of binding sites common to both TT and CVS.

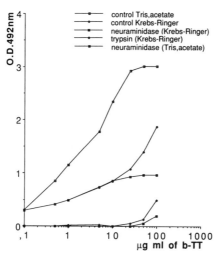

Figure 4. Binding of b-TT to sCSM treated either with trypsin or neuraminidase prior to immobilization on a polystyrene plate.

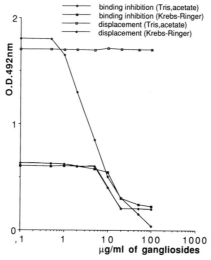

Figure 5. Prevention by gangliosides of the binding of b-TT to and displacement by gangliosides of b-TT from sCSM immobilized on a polystyrene plate.

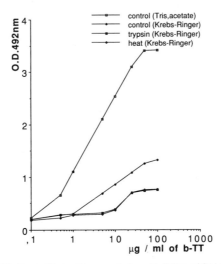

Figure 6. Binding in Krebs-Ringer solution of b-TT to sCSM treated with either heat or trypsin prior to their immobilization on a poly-styrene plate. Heat and trypsin treatments had no effect on binding in Tris, acetate buffer.

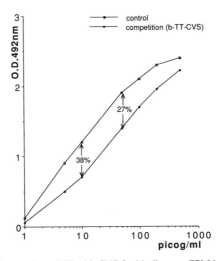

Figure 7. Interaction of TT with CVS for binding to sCSM immobilized on a polystyrene plate.

Attempt to Purify the Binding Component from sCSM

An attempt has been made to purify the binding component from the solubilized crude synaptic membranes. For this sCSM were reacted with b-TT for 1h at 37°C in Krebs-Ringer, pH7.4. Thereafter the complex [b-TT-sCSM] that formed was applied to a column of Agarose-Avidin equilibrated with Krebs-Ringer solution. The sample of the complex was allowed to sink into the column and left in contact with the gel for 30 min, after which time the column was washed with 2 column volumes of Krebs-Ringer solution collecting 1ml fractions. Elution of the adsorbed fraction was carried out by flowing through the column a 1mg/ml immunobiotin solution. One ml fractions were collected and read at 280nm. The fractions absorbing at 280nm were pooled and applied to an Ultrogel AcA22 column (100×0.8cm). One ml fractions were collected and read at 280nm. Two fractions were eluted for which Mr equal to 1000 and 350 KDa, respectively, could be calculated. These fractions were pooled separately and concentrated by dialysis against 30% PEG 20000.

Either of the fractions exhibited binding characteristics for b-TT similar to those of sCSM in Tris, acetate buffer, pH6. In contrast, they failed to bind b-TT in Krebs-Ringer solution, pH7.4.

These 2 fractions were further analyzed by SDS-PAGE (3-12%), in the presence and absence of 2-mercapto-ethanol. In the absence of the reducing agent no component penetrated the gel, whereas in the presence of 2-mercapto-ethanol, in addition to the fraction excluded from the gel, low molecular weight components migrated exhibiting Mr in the range of 150, 100, 65, 45 and 35 KDa.

DISCUSSION

The reported results show that the tetanus toxin-derived fragment C can interfere with either the uptake and/or retrograde transport systems of the neurotropic *Herpesvirus eidolon*, since its injection by either the I.M. or I.V. routes 48h before or 24h after virus inoculation could protect 30 to 55 per cent of mice from death. Since generalized paralyses failed to develop, this indicates that fragment C should have prevented the virus from reaching the CNS. Since protection did never exceed 50%, it appears likely that two different types of receptors are responsible for the virus uptake of which only one is recognized by fragment C or only one type of receptors with differential affinities for herpes virus and fragment C (Figure 1 and 2).

It should also be stressed that fragment C can interfere at the binding level, since a zonal lesion did not develop in a number of mice infected by the s.c. route with the virus (cf. Figure 2).

It was also observed that the tetanus toxin molecule could interact with rabies virus (CVS) for binding to fixation sites in the membrane of cortical neurons in culture. Again, binding inhibition of TT with CVS or of CVS with TT was never complete, although significant. To account for the failure of either agents to inhibit completely the binding at the other, it can be postulated that TT and CVS might either share only a part of fixation sites or exhibit differential affinities for the same fixation sites. In support to the former hypothesis is the observation that simultaneous addition of both TT and CVS to the neurons resulted in strong labeling of the cells with either of them. An additional explanation would be that too low concentrations of TT or too low numbers of virus particles could be used for binding inhibition experiments because of the intrinsic lytic effect of these agents. In favor of such an assumption speaks the fact that CVS-treated cells added with a high concentration of TT (500 µg/ml) were weakly labeled with CVS, but strongly with TT without appreciable damage to the neurons (cf. Table 1). Could higher TT concentrations be used without impairing the cells, it is safe to predict that more CVS would be unbound from the neurons.

The above-reported results further support the hypothesis that neurotropic agents should share common features enabling them to recognize in neuron membrane similar structures.

It is of interest that solubilized crude synaptic membranes (sCSM) were still capable of binding TT. The technique developed for binding studies, that used an enzymatic assay proved to be both reliable and easy to perform. Two different buffers were used for the binding assays: 1) 50mM Tris, acetate buffer, pH6 which preferentially measures the binding to gangliosides and 2) Krebs-Ringer buffer, pH7.4, which is more consistent with the measure of binding to protein fixation sites, while allowing the fixation to gangliosides to be evaluated under physiologic conditions.

The results of the binding studies with b-TT show that more toxin was bound to sCSM in Tris, acetate buffer than in Krebs-Ringer solution, whereas biotinylated tetanus toxoid failed to bind (cf. Figure 3). Pretreatment of sCSM with neuraminidase completely abolished binding of b-TT to the membranes at pH6, whereas it had practically no effect on binding at pH 7.4. In contrast, treatment of sCSM with trypsin resulted in complete inhibition of b-TT binding to the membranes in Krebs-Ringer solution. These results further indicate that in Tris is measured the binding involving gangliosides (neuraminidase-sensitive) and in Krebs-Ringer that involving a proteic structure (trypsin-sensitive) (cf. Figure 4). Pretreatment of b-TT with a mixture of gangliosides resulted, as could be expected, in a dose-dependent inhibition of binding to sCSM in Tris buffer, but not in Krebs-Ringer solution. Again, these results show that in Tris buffer is measured the binding involving gangliosides. The toxin once fixed to sCSM could not be displaced by gangliosides both at pH6 and 7.4, which, in the former case, points to the high-affinity binding to gangliosides in Tris buffer and, in the latter case to the non-implication of gangliosides in binding in Krebs-Ringer buffer (cf. Figure 5). Heat-denaturation and trypsin-digestion of sCSM was effective in inhibiting only the fixation of b-TT to the membranes in Krebs-Ringer solution, but these treatments did not affect the binding in Tris buffer (cf. Figure 6). This result could be expected whether the binding observed in Krebs-Ringer is dependent on a protein containing structure.

It was also possible with TT to inhibit to some extent the binding of CVS to sCSM. This result is to be related to inhibition of rabies virus infection by a soluble CNS membrane fraction (Conti et al., 1988).

Purification of sCSM by a combination of affinity and molecular exclusion chromatography allowed us to separate 2 fractions with Mr in the range of 350 and 1000KDa, respectively, which bound effectively TT. The binding characteristics of the 2 purified fractions showed to be similar to those exhibited by the sCSM in acetate buffer. In contrast, in Krebs-Ringer, binding of TT to the purified fractions failed to occur. In this connection, it should be remembered that the protein matrix of the receptor measured in Krebs-Ringer solution possesses an ordered structure containing a high content of hydrophobic residues. In an aqueous environment the hydrophobic regions are likely to interact with one another resulting in a change of the protein conformation accompanied by a loss of the binding capacity.

Further work is currently being carried out with a view to characterizing chemically these soluble fractions endowed with binding capacity for the tetanus toxin molecule.

Acknowledgments

We are grateful to Mrs. Regine Lambrecht for her secretarial assistance.

REFERENCES

Besançon F, Ankel H. Binding of interferon to gangliosides. Nature 1974; 252:478.
Bizzini B, Turpin A, Raynaud M. Production et purification de la toxine tétanique. Ann Inst Pasteur (Paris) 1969; 116:686.

Bizzini B, Grob P, Akert K. Papain-derived fragment IIc of tetanus toxin: its binding to isolated synaptic membranes and retrograde axonal transport. Brain Res 1981; 210:291.

Bizzini B, Toth P, Fedinec AH. Defining a region on tetanus toxin responsible for neuromuscular blockade. Toxicon 1988; 26:309.

Conti C, Tsiang H. Effect of concanavalin A on the early events of rabies virus infection of CER cells. Intervirol 1985; 24:166.

Conti C, Hauttecoeur B, Morelec MJ, Bizzini B, Orsi N, Tsiang M. Inhibition of rabies virus infection by a soluble membrane fraction from the rat central nervous system. Arch Virol 1988; 98:73.

Cook ML, Stevens JG. Pathogenesis of herpetic neuritis and ganglionitis in mice: evidence for intra-axonal transport of infection. Infect Immun 1973; 34:272.

Dumas M, Schwab ME, Thoenen H. Retrograde axonal transport of specific molecules as a tool for characterizing nerve terminal membrane, J Neurobiol 1979; 10:179.

Fedinec AA. Studies on the mode of the spread of tetanus toxin in experimental animal. In: Raudonat HW, ed. Recent Advances in the Pharmacology of Toxins. Oxford: Pergamon, 1965.

Fillenz M, Gagnon C, Stoeckel K, Thoenen H. Selective uptake and retrograde axonal transport of dopamine beta-hydroxylase antibodies in peripheral adrenergic neurons. Brain Res 1976; 114:293.

Guillon JC. Un modèle expérimental d'infection herpétique: l'infection cutanée *Herpesvirus eidolon* (nov. sp.) chez la souris. Sci Tech Anim Lab 1976; 1, n°2:79.

Habermann E. ^{125}I-labelled neurotoxin from *Clostridium botulinum* A. Preparation binding to synaptosomes and ascent to the spinal cord. Naunyn-Schmiedeberg's Arch Pharmacol 1974; 281:47.

Harper CG, Gonatas JO, Stieber A, Gonatas NK. *In vivo* uptake of wheat germ agglutinin-horse radish peroxydase conjugates into neuronal GERL and lysosomes. Brain Res 1980; 188:465.

Hendry IA, Stoeckel K, Thoenen H, Iversen LL. The retrograde axonal transport of nerve growth factor. Brain Res 1974; 68:103.

Kristensson K, Lycke E, Sjöstrand J. Spread of herpes simplex virus in peripheral nerves. Acta Neuropath 1971; 17:44.

Knypers HGJM, Ugolini G. Viruses as transneuronal tracers. Trends in Neurosci 1990; 13:71.

Marconi P, Pitzurra M, Vecchiarelli A, Pitzurra L, Bistoni F. Resistance induced by concanavalin A and phytohaemagglutinin P against tetanus toxin in mice. Ann. Immunol (Inst Pasteur) 1982; 133D:15.

Matsuda M, Yoneda M. Isolation and purification of two antigenically active "complementary" polypeptide fragments of tetanus neurotoxin. Infect Immun 1975; 12:1147.

Severin MJ, White RJ. The neural transmission of herpes simplex virus in mice. Am J Pathol 1968; 53:1009.

Stoeckel K, Schwab M, Thoenen H. Role of gangliosides in retrograde axonal transport of tetanus and cholera toxin. Experientia 1971; 32:783.

Stoeckel K, Paravicini U, Thoenen H. Specificity of the retrograde axonal transport of nerve growth factor. Brain Res 1974; 76:413.

Stoeckel K, Schwab M, Thoenen H. Comparison between the retrograde axonal transport of growth factor and tetanus toxin in motor, sensory and adrenergic neurons. Brain Res 1975; 99:1.

Stoeckel K, Schwab ME, Thoenen H. Role of gangliosides in the uptake and retrograde axonal transport of cholera and tetanus toxin as compared to nerve growth factor and wheat germ agglutinin. Brain Res 1977; 132:273.

Superti F, Hauttecoeur B, Morelec MJ, Goldoni P, Bizzini B, Tsiang MJ. Involvement of gangliosides in rabies virus infection. J Gen Virol 1986; 67:47.

Taylor CF, Britton P, Van Heyningen S. Similarities in the heavy and light chains of tetanus toxin suggested by their amino acid composition. Biochem J 1983; 209:897.

Tsiang H. Evidence for an intraaxonal transport of fixed and street rabies virus. J Neuropath Exp Neurol 1979; 38:286.

Tsiang H, Koulakoff A, Bizzini B, Berwald-Netter Y. Neurotropism of rabies virus. An *in vitro* study of neurons and glia. J Neuropath Exp Neurol 1983; 42:439.

Van Heyningen WE, Miller PA. The fixation of the tetanus toxin by ganglioside, J Gen Microbiol 1961; 24:107.

Wiegand H, Erdmann G, Wellhöner HH. ^{125}I-labelled botulinum A neurotoxin: pharmacokinetics in cats after intramuscular injection. Naunyn-Schmiedeberg's Arch Pharmacol 1976; 292:161.

Wildy P. The progression of herpes simplex virus to the CNS of the mouse. J Hyg 1967; 65:173.

Young AB, Snyder SH. Strychnine binding associated with glycine receptors of the central nervous system. Proc Natl Acad Sci (Wash) 1973; 70:2832.

Ziegler MG, Thomas JA, Jacobowitz DM. Retrograde axonal transport of antibody to dopamine beta-hydroxylase. Brain Res 1976; 104:390.

PARALYSIS BY BOTULINUM NEUROTOXINS UNCOVERS TROPHIC SECRETIONS AT THE NEUROMUSCULAR JUNCTION

Stephen Thesleff

Department of Pharmacology
University of Lund, Sölvegatan 10
S-22362 Lund, Sweden

It is well established that the motor nerve in addition to spontaneous and nerve impulse evoked phasic release of acetylcholine (ACh) also secretes, but by a different mechanism, a variety of peptides (neuropeptides). For recent publications and reviews on the subject see, De Camilli and Jahn (1990), Verhage et al.(1991) and two conference volumes (Johnson, 1987; Edelman et al., 1987).

The classical transmitter ACh is synthetized in the motor nerve terminal and is subsequently incorporated into small clear synaptic vesicles which exocytos their content at specific release sites in the terminal, so called "active zones". These zones contain arrays of voltage activated calcium channels and the exocytotic process is strictly regulated by the influx of calcium ions. New vesicles are formed in the terminal by endocytotic retrieval of membrane and their subsequent filling with ACh.

A different class of synaptic vesicles are the "large dense core vesicles". These vesicles have about a 10 times larger volume and an electron dense core and are formed at the Golgi apparatus of the motoneuron soma where they are filled by various neuropeptides. Subsequently, they are moved along the axon by proximo-distal axonal transport towards the nerve terminal where ACh is added to their content. There they exocytos their neuropeptide-ACh content but not at "active zones" but at non-specialized regions. The process of exocytosis is not triggered by the phasic influx of calcium but may instead depend on the overall level of intraterminal free calcium or on some other messenger (for detailes see, Johnson, 1987; Verhage et al., 1991). It is believed that the release of neuropeptides controls a number of postjunctional properties like the synthesis and aggregation of the nicotinic cholinergic receptor and the acetylcholinesterase at the muscle endplate (Fischbach et al., 1986; Laufer and Changeux, 1987. For an overview see Ochs, 1988). The morphological and functional specialization of the postsynaptic endplate region seems therefore to be under the control of neuropeptides released by the nerve. This is by no means unique for the neuromuscular junction since neuropeptides within large dense core synaptic vesicles are present at most, if not at all, chemical synapses (Thureson-Klein et al., 1988).

Trophic signals do not flow only in proximo-distal direction but also from the muscle to the nerve. There is ample evidence that the target tissue, muscle, releases (a) factor(s) which is (are) important for the survival, growth and maintenance of the motoneuron. Presumably, they are nerve growth factor (NGF)-like proteins but that remains to be shown.

As documented in several chapters of this monograph botulinum neurotoxins have a very selective mode of action on cholinergic nerves. They block the exocytotic, calcium dependent, release mechanism of synaptic vesicles at active zones without otherwise affecting functional nor morphological features of the neuron and its terminals. In that way impulse transmission across the synapse is blocked with resulting paralysis of the muscle. Trophic effects of activity, which mainly affect extrajunctional muscle properties (Lömo and Gundersen, 1988), are thereby abolished. Since, the effect of the neurotoxin, particularly that of type A, is long-lasting (several weeks) the actions on extrajunctional areas of muscle mimic those of long-term surgical denervation while those of the muscle endplate remain relatively unchanged (Thesleff et al., 1990).

Because the morphology of the synaptic connection is intact in botulinum neurotoxin paralysis it seems possible that trophic secretions across the synapse might still operate and in fact be exaggerated to promote the recovery of transmission. My presentation will deal with such questions both pre- and postsynaptically. Most of the experiments were made in vertebrates, mice, rats or frogs.

At the normal unpoisoned neuromuscular junction ACh is released as quanta giving rise postsynaptically to either spontaneous, small and uniform sized potentials so called miniature endplate potentials (m.e.p.ps) or, following a nerve impulse, to a number of synchronized quanta causing postsynaptically a much larger potential, the endplate potential (e.p.p.). The e.p.p. is normally of sufficient size to initiate an action potential and thereby a muscle twitch (see, Katz, 1969). In addition, as intially shown by Liley (1957), occasional very large spontaneous potentials are recorded at the muscle endplate, so called giant m.e.p.ps (g.m.e.p.ps). Such potentials constitute on the average about 4 % of the total number of m.e.p.ps but their frequency is highly variable between fibres in a muscle and between muscles (Heinonen et al., 1982; Colméus et al., 1982 and Kim et al., 1984). Botulinum neurotoxins block the appearance of m.e.p.ps and e.p.ps but leave the number of g.m.e.p.ps unaffected (Dolly et al., 1987). In fact following a period of 4-5 days of paralysis the frequency of g.m.e.p.ps gradually increases from about 0.01 Hz to 0.5 Hz (Kim et al., 1984). As demonstrated by Liley (1957) and by Molgo and Thesleff (1982) the g.m.e.p.ps are produced by ACh released from the motor nerve. The ACh comes from the same intraterminal transmitter pool as that giving rise to m.e.p.ps and to e.p.ps since blockade of transmitter synthesis by hemicholinium reduces equally all three types of potentials. Furthermore, the ACh responsible for g.m.e.p.ps apparently is of synaptic vesicular origin since blockade of the high affinity transport mechanism for cytoplasmic ACh into synaptic vesicles abolishes all g.m.e.p.ps (Lupa et al., 1986). Figure 1 illustrates one g.m.e.p.p. and several m.e.p.ps in a normal muscle and a number of g.m.e.p.ps recorded in a muscle paralysed for 16 days by botulinum toxin type A. As shown g.m.e.p.ps are not only larger than m.e.p.ps (more than two times the modal amplitude of m.e.p.ps) but frequently also have a slower time-to-peak. Figure 2 are recordings of endplate currents resulting in g.m.e.p.ps and m.e.p.ps respectively. It should be noted that the charge transfer causing a g.m.e.p.p. is several times larger than that causing a m.e.p.p. G.m.e.p.ps are characteristic not only of muscle paralysis caused by botulinum neurotoxins but are also abundant following long-lasting muscle paralysis caused by curare or by tetrodotoxin (Ding et al., 1983; Gundersen, 1990). Thus, the g.m.e.p.p frequency is enhanced by long-term paralysis of neuromuscular transmission and not by toxic actions on the nerve and its terminals.

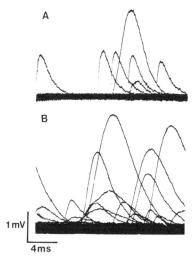

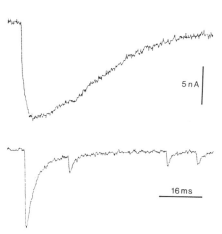

Figure 1. A. Superimposed traces showing spontaneous m.e.p.ps in a normal rat extensor digitorum longus muscle. Note, the appearance of one g.m.e.p.p. B. The frequency of spontaneous g.m.e.p.ps is greatly increased following prolonged blockade of neuromuscular transmission. The record is from a rat extensor digitorum longus muscle paralysed for 16 days by the use of botulinum neurotoxin type A.

Figure 2. Spontaneous miniature (lower tracing) and giant miniature (upper tracing) endplate currents recorded at the same frog neuromuscular junction at a holding membrane potential of -90 mV. The normal Ringer solution contained 10 μM Tacrine and 1 μM tetrodotoxin. By courtesy of J. Molgo.

I have proposed that g.m.e.p.ps result from the exocytotic release of ACh from large dense core synaptic vesicles and not from the release of ACh from small clear synaptic vesicles. The reasons for this proposal are as follows: Unlike m.e.p.ps, g.m.e.p.p. frequency is unaltered by nerve stimulation, except for prolonged high frequency nerve stimulation which enhances their frequency (Heuser, 1974). A similar behaviour is seen for the nervous release of neuropeptides (Hökfelt et al., 1986). Changes in extracellular calcium ion concentrations are not affecting g.m.e.p.p. frequency indicating that their dependence on intraterminal calcium is different from that of m.e.p.ps (Thesleff and Molgo, 1983). That is similar to the release of neuropeptides. Large dense core synaptic vesicles are transported by axoplasmic transport, blockade of this transport by colchicine, metabolic blockers or cooling stops the release of neuropeptides and also abolishes all g.m.e.p.ps (Thesleff et al., 1990). Table I summarizes some of the differences between m.e.p.ps and g.m.e.p.ps. It should be mentioned that while various procedures may increase m.e.p.p. frequency to several hundred Hz, the g.m.e.p.p. frequency never exceeds 1-2 Hz. That is to be expected if g.m.e.p.ps correspond to the release of the content of large dense core vesicles which represent only a few percent of the total number of synaptic vesicles.

Table 1. Summary of the effects of various procedures on the frequency of m.e.p.ps and g.m.e.p.ps in rat muscle.

Agent or procedure	M.e.p.p. freq.	G.m.e.p.p. freq.
Ca^{2+}, 8 mM	Increase 4X[1]	No change
K^+, 20 mM	Increase 20X	Increase 2X
K^+, 0 mM	Increase 15X	No change
Ouabain, 0.2 mM	Increase 50X	No change
Ethanol, 0.5 M	Increase 10X	No change
Mn^{2+}, 10 mM	Increase 8X	No change
Hypertonicity 2X	Increase 80X	Decrease 0.3X
Hypotonicity 0.5X	Decrease 0.3X	Increase 3X
CCCP, 2 μM	Increase 200X	Decrease > 10X

[1] Approximate change times control

In conditions where g.m.e.p.ps are prevalent, i.e. at newly formed endplates (Bennett et al., 1973; Colméus et al., 1982), and at endplates following prolonged neuromuscular blockade the number of large dense core vesicles is increased (Lüllmann-Rauch, 1971; Jirmanova and Thesleff, 1976; Pecot-Dechavassine et al., 1991) while other types of large vesicles, cisternae or aggregates of vesicles are not seen (Pecot-Dechavassine and Molgo, 1982). The variable time-to-peak of g.m.e.p.ps could be explained by exocytotic release at areas outside the active zones, i.e. at variable distances from the postsynaptic nicotinic ACh receptor.

In connection with the possibility that g.m.e.p.ps signal the release of the content of large dense core vesicles it is of interest that a number of drugs selectively enhance the frequency of g.m.e.p.ps without affecting that of m.e.p.ps (Molgo and Thesleff, 1982; Alkadhi, 1989; Thesleff et al., 1990). Figure 3 illustrates effects of 4-aminoquinoline, tetrahydroaminoacridine (Tacrine) and emetine. In addition several tacrine-like compounds have a similar effect (unpublished observations). These findings further emphasize the difference in release mechanism between the vesicles giving rise to g.m.e.p.ps and those responsible for m.e.p.ps. Personally, I find the possibility that certain drugs selectively stimulate the exocytosis of large dense core vesicles containing neuropeptides highly intriguing since it may have therapeutic applications in a variety of neurodegenerative conditions.

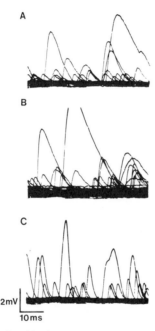

Figure 3. Examples of g.m.e.p.ps induced in the mouse nerve-diaphragm preparation by (A) 4-aminoquinoline (200 μM), (B) tetrahydroaminoacridine (Tacrine, 20 μM) and (C) emetine (20 μM). Note, that unlike 4-aminoquinoline and tacrine, emetine is not an inhibitor of cholinesterase.

It has long been known that paralysis by botulinum neurotoxins causes a marked outgrowth of motor nerve terminals observed as terminal sprouting (Duchen and Strich, 1968; Duchen, 1970; Molgo et al., 1990). Direct electrical stimulation of the paralysed muscles prevents sprouting (Brown et al., 1977; 1981). Thus, it appears that muscle activity in some way controls the nerve terminal outgrowth, possibly, as previously mentioned, by trophic substances. In that connection it is of interest that the former endplate region of surgically denervated muscle fibres has marked endocytotic activity. The endocytotic activity is limited to a segment around the former endplate (Libelius et al., 1979; Tågerud and Libelius, 1984). Figure 4 illustrates such segments, with horseradish peroxidase uptake, in the denervated hemidiaphragm of a mouse. Figure 5 shows the segmental endocytotic uptake of rhodamine-labelled dextran centered around cholin-

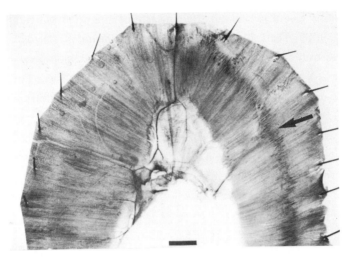

Figure 4. High endocytotic activity in the endplate region of a 12-days denervated mouse hemidiaphragm muscle 2 hrs after an intravenous injection of horseradish peroxidase. Arrow denotes segments with high endocytotic activity in the denervated part of the muscle. Bar = 2 mm.

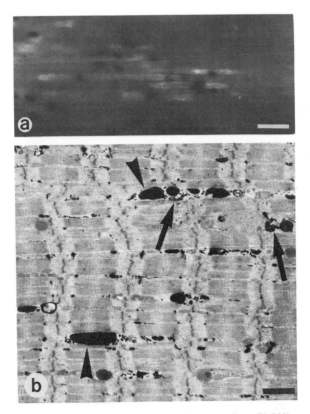

Figure 5. (a) Segmental uptake of rhodamine-labelled dextran (av. mol. wt 70,000) centered around cholinesterase stained endplates in a 12-days denervated mouse hemidiaphragm. Bar = 100 um. (b) Electron micrograph of the endplate region a 14-days denervated mouse muscle fibre showing proliferation of the transverse tubular system (arrows) and large peroxidase positive phagosomes (arrowheads). Bar = 1 μm. By courtesy of R. Libelius.

esterase stained endplates in a 12-days denervated mouse diaphragm and an electron micrograph of a 14-days denervated fibre with proliferation of the transverse tubular system and horseradish peroxidase positive phagosomes. Recent observations suggest that the endocytotic activity occurs periodically in the endplate (Lawoko and Tågerud, 1992). Similar segments with high endocytotic activity are observed in rat myotubes differentiated *in vitro* (Tågerud et al., 1990).

To maintain the cell membrane, endocytosis must be linked to exocytotic activity. It is therefore tempting to suggest that the observed activity at the endplate reflects exo-endocytosis as a sign of secretion from the denervated muscle. Paralysis by botulinum neurotoxin causes similar segmental endocytotic activity in mouse skeletal muscle, although the changes develop more slowly and apparently to a lesser extent than after surgical denervation (Tågerud et al., 1986). Consequently, I propose that an inactive muscle secretes substances from its endplate region which promote the growth of the motoneuron and thereby induce the terminal sprouting characteristic of botulinum neurotoxin poisoning.

The purpose of this presentation has been to focus attention on trophic signals across the neuromuscular synapse. That aspect of synaptic function is far less studied and understood than that of impulse transmission. In fact much of what I have presented on the subject are still hypothetical ideas based mainly on circumstantial evidence. However, I believe that botulinum and tetanus neurotoxins, because of their selective blockade of impulse transmission, are and will be valuable tools in such studies.

REFERENCES

Alkadhi, K.A., 1989, Giant miniature end-plate potentials at the untreated and emetine-treated frog neuromuscular junction, *J. Physiol.*, 412:475-491.

Bennett, M.R., McLachlan, E.M. and Taylor, R.S., 1973, The formation of synapses in reinnervated mammalian striated muscle, *J. Physiol.*, 233:481-500.

Brown, M.C., Goodwin, G.M. and Ironton, R., 1977, Prevention of motor nerve sprouting in botulinum toxin poisoned mouse soleus muscles by direct stimulation of the muscle, *J. Physiol.*, 267:42P.

Brown, M.C., Holland, R.L. and Hopkins, W.G., 1981, Motor nerve sprouting, *Ann. Rev. Neurosci.*, 4:17-42.

Colméus, C., Gomez, S., Molgo, J. and Thesleff, S., 1982, Discrepancies between spontaneous and evoked potentials at normal, regenerating and botulinum toxin poisoned mammalian neuromuscular junctions, *Proc. R. Soc. Lond. B Biol. Sci.*, 215:63-74.

De Camilli, P., and Jahn, R., 1990, Pathways to regulated exocytosis in neurons, *Ann. Rev. Physiol.*, 52:625-645.

Ding, R., Jansen, J.K.S., Laing, N.G. and Tönnesen, H., 1983, The innervation of skeletal muscles in chickens curarized during early development, *J. Neurocytol.*, 12:887-919.

Duchen, L.W., 1970, Changes in motor innervation and cholinesterase localization induced by botulinum toxin in skeletal muscle of the mouse, differences between fast and slow muscles, *J. Neurol. Neurosurg. Psychiat.*, 33:40-54.

Duchen, L.W. and Strich, S.J., 1968, The effects of botulinum toxin on the pattern of innervation of skeletal muscle in the mouse, *Quart. J. Exp. Physiol.*, 53:84-89.

Edelman, G.M., Gall, W.E. and Cowan, W.M., eds., 1987, "Synaptic Function", John Wiley & Sons, New York.

Fischbach, G.D., Harris, D.A., Falls, D.L., Dubinsky, J.M., English, K.L. and Johnson, F.A., 1989, The accumulation of acetylcholine receptors at developing chick nerve-muscle synapses, *in*: "Neuromuscular Junction", L.C. Sellin, R. Libelius, and S. Thesleff, eds., Fernström Foundation Series, Elsevier Science Publ., Amsterdam.

Gundersen, K., 1990, Spontaneous activity at long-term silenced synapses in rat muscle, *J. Physiol.*, 430:399-418.

Heinonen, E., Jansson, S.-E. and Tolppanen, E.-M., 1982, Independent release of supranormal acetylcholine quanta at the rat neuromuscular junction, *Neurosci.*, 7:21-24.

Heuser, J.E., 1974, A possible origin of the "giant" spontaneous potentials that occur after prolonged transmitter release at frog neuromuscular junctions, *J. Physiol.*, 239:106P-108P.

Hökfelt, T., Fuxe, K. and Pernow, P., eds., 1986, Coexistence of neuronal messengers, *Progr. Brain Res.*, 68:1-411.

Jirmanova, I. and Thesleff, S., 1976, Motor end-plates in regenerating rat skeletal muscle exposed to botulinum toxin, *Neurosci.*, 1:345-347.

Johnson, R.G., ed., 1987, Cellular and molecular biology of hormone- and neurotransmitter-containing secretory vesicles, *Ann. N.Y. Acad. Sci.*, 493:1-590.

Katz, B., 1969, The release of neural transmitter substances, *in:*" The Sherrington Lectures X"., Liverpool Univ. Press.

Kim, Y.I., Lömo, T., Lupa, M.T. and Thesleff, S., 1984. Miniature endplate potentials in rat skeletal muscle poisoned with botulinum toxin, *J. Physiol.*, 356:587-599.

Laufer, R. and Changeux, J-P., 1987, Calcitonin gene-related peptide elevates cyclic AMP levels in chick skeletal muscle: possible neurotrophic role for a coexisting neuronal messenger, *EMBO J.*, 6:901-906.

Lawoko, G. and Tågerud, S., 1992, High endocytic activity occurs periodically in the endplate region of denervated mouse skeletal muscle fibres, In press.

Libelius, R., Josefsson, J.-O. and Lundquist, I., 1979, Endocytosis in chronically denervated mouse skeletal muscle. A biochemical and ultrastructural study with horseradish peroxidase, *Neurosci.*, 4:283-292.

Liley, A.W., 1957, Spontaneous release of transmitter substance in multiquantal units, *J. Physiol.*, 136:595-605.

Lömo, T. and Gundersen, K., 1988, Trophic control of skeletal muscle membrane properties, *in:* "Nerve-Muscle Cell Trophic Communication", H.L. Fernandez, ed., CRC Press, Boca Raton.

Lüllmann-Rauch, R., 1971, The regeneration of neuromuscular junctions during spontaneous re-innervation of the rat diaphragm, *Z. Zellforsch.*, 121:593-603.

Lupa, M.T., Tabti, N., Thesleff, S., Vyskocil, F. and Yu, S.-P., 1986, The nature and origin of calcium-insensitive miniature end-plate potentials at rodent neuromuscular junctions, *J.Physiol.*, 381:607-618.

Molgo, J., Comella, J.X., Angaut-Petit, D., Pecot-Dechavassine, M., Tabti, N., Faille, L., Mallart, A. and Thesleff, S., 1990, Presynaptic actions of botulinal neurotoxins at vertebrate neuromuscular junctions, *J. Physiol.* (Paris), 84:152-166.

Molgo, J. and Thesleff, S., 1982, 4-aminoquinoline-induced "giant" miniature endplate potentials at mammalian neuromuscular junctions, *Proc. R. Soc. Lond. B Biol. Sci.*, 214:229-247.

Ochs, S., 1988, An historical introduction to the trophic regulation of skeletal muscle, *in:* "Nerve-muscle Cell Trophic Communication", H.L. Fernandez, ed., CRC press, USA.

Pécot-Dechavassine, M. and Molgo, J., 1982, Attempt to detect morphological correlates for the "giant" miniature end-plate potentials induced by 4-aminoquinoline, *Biol. Cell.*, 46:93-96.

Pécot-Dechavassine, M., Molgo, J. and Thesleff, S., 1991, Ultrastructure of botulinum type-A poisoned frog motor nerve terminals after enhanced quantal transmitter release caused by carbonyl cyanide m-chlorophenylhydrazone, *Neurosci. Lett.*, 130:5-8.

Tågerud, S. and Libelius, R., 1984, Lysosomes in skeletal muscle following denervation. Time course of horseradish peroxidase uptake and increases of lysosomal enzymes, *Cell Tissue Res.*, 236:73-79.

Tågerud, S., Libelius, R. and Shainberg, A., 1990, High endocytotic and lysosomal activities in segments of rat myotubes differentiated in vitro, *Cell Tissue Res.*, 259:225-232.

Tågerud, S., Libelius, R. and Thesleff, S., 1986, Effects of botulinum toxin induced muscle paralysis on endocytosis and lysosomal enzyme activities in mouse skeletal muscle, *Pflügers Arch.*, 407:275-278.

Thesleff, S. and Molgo, J., 1983, Commentary. A new type of transmitter release at the neuromuscular junction, *Neurosci.*, 9:1-8.

Thesleff, S., Molgo, J. and Tågerud, S., 1990, Trophic interrelations at the neuromuscular junction as revealed by the use of botulinal neurotoxins, *J. Physiol.* (Paris), 84:167-173.

Thesleff, S., Sellin, L.C. and Tågerud, S., 1990, Tetrahydroaminoacridine (tacrine) stimulates neurosecretion at mammalian motor endplates, *Br. J. Pharmacol.*, 100:487-490.

Thureson-Klein, Å.K., Klein, R.L., Zhu, P.-C. and Kong, J.-Y., 1988, Differential release of transmitters and neuropeptides co-stored in central and peripheral neurons, *in:* "Cellular and Molecular Basis of Synaptic Transmission", H. Zimmermann, ed., Springer Verlag, Berlin.

Verhage, M., McMahon, H.T., Ghijsen, W.E.J.M., Boomsma, F., Scholten, G., Wiegant, V.M. and Nicholls, D.G., 1991, Differential release of amino acids, neuropeptides, and catecholamines from isolated nerve terminals, *Neuron*, 6:517-524.

MOTOR NERVE TERMINAL MORPHOLOGY FOLLOWING
BOTULINUM A TOXIN INJECTION IN HUMANS

Kathy Alderson

Salt Lake City V.A. Medical Center
and
Department of Neurology
University of Utah
Salt Lake City, UT 84132

INTRODUCTION

Botulinum A toxin injection into muscle is an important form of therapy for focal dystonias, including blepharospasm. The basis for the effect of botulinum toxin is paralysis of muscle by blocking neuromuscular transmission, as discussed elsewhere in this volume. The block of neuromuscular transmission produces functional muscle denervation.   Muscle denervation stimulates changes in terminal motor axons and neuromuscular junctions, which can, in some types of denervation, be a repair process. This paper will discuss the morphology of motor nerve terminals and neuromuscular junctions following two types of denervation, the morphology the neuromuscular junction following botulinum toxin in experimental animals, and the response of the human neuromuscular junction to injection of botulinum toxin.

DENERVATION OF MUSCLE

Muscle denervation can be classified as either anatomic or functional. The two types differ primarily as to cause, but also vary in the response of the terminal motor axon and neuromuscular junction to denervation, and the morphology of the repair process. As will be discussed below, the morphology of terminal motor axons in the human orbicularis oculi following botulinum toxin injection shares features of both. This section will review both anatomic and functional denervation.

Anatomic Denervation

The loss of motor axons to muscle fibers produces anatomic denervation. If all motor axons to a muscle are lost, denervation is **complete**. Complete denervation develops following, for example, cutting or crushing a nerve. All motor axons distal to the site of injury undergo Wallerian degeneration, no motor axons are present in the muscle, and the muscle is completely paralyzed. Loss of some, but not all, motor axons to a muscle produces **partial** denervation; some motor axons remain intact. Partial denervation occurs in progressive human neuromuscular disease, such as anterior horn cell disease[1] (Werdnig-Hoffman disease

Botulinum and Tetanus Neurotoxins, Edited by
B.R. DasGupta, Plenum Press, New York, 1993 .

53

or amyotrophic lateral sclerosis), radiculopathy, many peripheral neuropathies, and compression neuropathies.

Recovery of muscular activity following complete denervation requires regeneration of axons from the site of injury towards the denervated muscle fibers. Regenerating axons follow the remaining Schwann cell tubes towards the end plates of denervated muscle fibers.[2] In contrast, repair from partial denervation is via the development of axonal collaterals (Figure 1).* Axonal processes develop from multiple sites of the remaining intact preterminal axons, including the nodes of Ranvier of preterminal axons (preterminal collaterals), the terminal axon immediately proximal to the end plate (terminal collaterals) and the axonal arborization over the end plate region (ultraterminal collaterals). The collaterals grow towards, and re-innervate denervated end plates on separate, previously denervated muscle fibers.[3-7] Development of axonal collaterals from an intact axon to reinnervate previously denervated end plates occurs in both human and experimental neuromuscular disease,[1] including ALS, polio, peripheral neuropathies, and partial axonotmesis.[5] Axon regeneration may also contribute to repair, in some situations, following partial denervation. However, on the basis of evaluation of terminal motor axons in numerous human and experimental animal nerve and neuromuscular diseases, the author feels the role of regeneration is relatively minor in chronic neuromuscular disease.

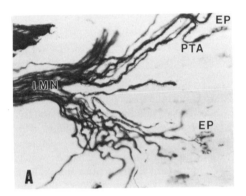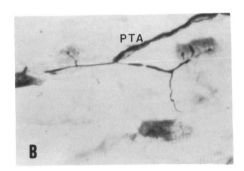

Figure 1. Human terminal motor axons. a) normal terminal motor nerves. The intramuscular nerve (IMN) is a dense collection of myelinated axons. Individual myelinated pre-terminal axons (PTA) leave the intramuscular nerve and proceed, generally unbranched, to an end plate (EP) on a muscle fiber. b) Axonal collaterals in amyotrophic lateral sclerosis. A single myelinated PTA branches into unmyelinated collaterals which extend to end plates on 4 separate muscle fibers (A - X 72.5, B- X 290; both, deltoid, silver-cholinesterase-immunocytochemistry (SCI) stain.[8] Dark, silver-stained axons are surrounded by a lighter immunostained myelin sheath, using an antibody to the myelin protein P_o. Muscle end plates are identified with a reaction product of cholinesterase).

Functional Denervation

Functional denervation develops following a block of neuromuscular transmission with, e.g. botulinum toxin.[9-11] The motor axon remains in anatomic contact with the muscle end plate, but, because neuromuscular transmission is blocked, the muscle is paralysed. In rats and other mammals, two morphological changes develop at the neuromuscular junctions following functional denervation by botulinum toxin. These changes differ between proximal and distal muscles. In proximal muscles, there is conspicuous motor axon sprouting. In more distal muscles, expansion of the end plate region is more pronounced, though both responses can be seen in any muscle.

---

* For purposes of this discussion, axonal collaterals are axonal processes which contact a muscle end plate. Axonal sprouts are processes which do not contact an end plate. Ideally, axonal sprouts end in a growth cone, but, in fact, "collaterals" extending out of the plane of section would be identified here as "sprouts".

NEUROMUSCULAR JUNCTION ALTERATIONS FOLLOWING BOTULINUM TOXIN-INDUCED FUNCTIONAL DENERVATION IN EXPERIMENTAL MAMMALS

Sprouts develop from multiple sites of the preterminal axon following botulinum toxin-induced functional denervation (Figure 2 and [9,10]). Sprouts can arise from the terminal axonal arborization over the end plate region (ultra-terminal sprouts), from the axon immediately proximal to the end plate (terminal sprouts) and from the nodes of Ranvier of pre-terminal axons (pre-terminal sprouts). Axonal sprouting is more abundant in proximal muscles, though is present to a lesser extent in distal muscles. By electron microscopy, the axon sprout follows the inner layer of a basal lamina, either of muscle or of the enveloping Schwann cell[(Alderson, unpublished)]. The author has never encountered sprouts not associated with a basal lamina, which probably reflects the importance of the inner layer of the basil lamina in guidance of nerve terminal outgrowth. The sprouts are otherwise non-directed, growing through the muscle, but not generally terminating at muscle end plates. There has been mention in the literature of these spouts terminating at end plates, but this is, by no means, a well-described phenomenon.[10] In the author's experience, after 28-35 days following botulinum toxin exposure, 5-7 % of botulinum toxin-induced sprouts become myelinated, which the author feels reflects sufficient enlargement of the sprout diameter to stimulate Schwann cell myelin formation (Figure 2b). The sprouts decrease in abundance after several months, though the process of sprout loss is not well understood.

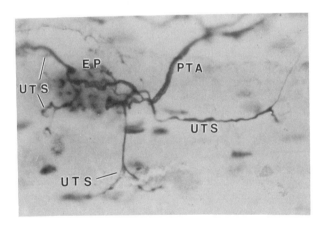

Figure 2. Axonal sprouts in the rat rhomboid following botulinum A toxin injection. Exuberant ultra-terminal axonal sprouts (UTS and arrows) arise from the terminal axonal arborization over the end plate (EP) region 14 days following botulinum toxin injection into the muscle (Silver-cholinesterase, X 410).

Expansion of the End Plate is more apparent in distal muscles following botulinum toxin exposure. As demonstrated in Figure 3, following botulinum toxin exposure in the rat soleus, the terminal axonal arborization over the end plate becomes more extensive, and the cholinesterase-containing end plate region also expands from the normal 30-60 um in length to greater than 120 um. The end plate region is not always continuous, with cholinesterase-containing regions separated by 5 - 15 um. Synaptic vesical antigenicity is present along the extended branches of the axonal arborization, suggesting that the expanded neuromuscular junction is capable of neuromuscular transmission.

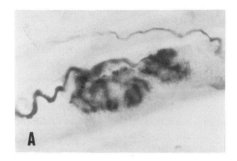 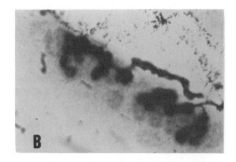

Figure 3. Expansion of the end plate in the rat soleus 14 days following botulinum toxin injection into the muscle. A) The terminal axonal arborization has expanded, and the diameter of the cholinesterase-containing region has increased from the normal 30-60 um to, here, 120 um (X 348). B) Two cholinesterase-containing end plate regions are separated by a non-cholinesterase-containing muscle membrane. An ultra-terminal process from the axonal arborization extends to the second end plate region (X 464). In both, synaptic vesicle antigenicity is the fuzzy area surrounding the axonal processes over the terminal arborization (SCI, using and antibody to the synaptic vesicle protein synaptophysin).

TERMINAL MOTOR AXON CONFIGURATION FOLLOWING BOTULINUM TOXIN INJECTION IN HUMANS

Because of the well-described response of terminal motor axons to botulinum toxin injection in other mammals, it was anticipated that similar abnormalities would develop in humans who receive botulinum toxin injections for treatment of dystonia. However, unique and surprising changes were identified in the orbicularis oculi of patients receiving intramuscular botulinum A toxin injections as treatment of blepharospasm. Orbicularis oculi myectomy is elected as treatment for disabling blepharospasm by some patients in whom medical therapy or botulinum toxin injections are not successful in controlling the spasms, therefore we were able to obtain surgical specimens of this muscle to evaluate the effect of botulinum toxin in humans.

Orbicularis oculi muscle from 3 groups of patients, age 35-81 years, was evaluated: 1) 9 Patients with blepharospasm who had previously received 2 to 19 botulinum A toxin injections 5 weeks to 3 years before undergoing orbicularis oculi myectomy [12] as treatment for blepharospasm. 4 of the 9 patients developed resistance to the botulinum toxin, requiring either increased dose on subsequent injections or not responding to botulinum toxin injection with paralysis. One patient had severe and prolonged paralysis following his second injection, and refused further injections. The other patients had continuing response to injections. 2) 2 Patients with blepharospasm without prior botulinum toxin therapy electing orbicularis oculi myectomy, and 3) 6 "normal" patients without blepharospasm having cosmetic blepharoplasty.

Normal Human Orbicularis Oculi Terminal Motor Axons

The morphology of terminal motor axons and neuromuscular junctions in muscle not exposed to botulinum toxin, both "normals" and patients with blepharospasm, was identical. (Figure 4).[13,14] Single pre-terminal axons exit the intramuscular nerve and are myelinated up to the muscle end plate region. There are no unmyelinated axons. Most preterminal axons are unbranched and innervate a single muscle fiber; less than 10% of preterminal axons branch to innervate more than one muscle fiber. Muscle end plate diameter was between 6-34 um, mean 20 um.[14]

Motor Axons and Neuromuscular Junctions Following Botulinum Toxin Exposure

Sprouts. As expected based on animal data, axonal sprouts develop following botulinum toxin injections into the orbicularis oculi of man (Figure 5).[13] Sprouts were present in all botulinum toxin-treated muscle, though of greater incidence in muscles having more injections (see **Clinical correlation**, below). Sprouts extend from multiple sites along the preterminal axon, including the preterminal nodes of Ranvier, terminal, and ultraterminal regions. The sprouts are almost uniformly unmyelinated. Some sprouts clearly end in a small bulbous dilatation, presumed to be the growth cone.

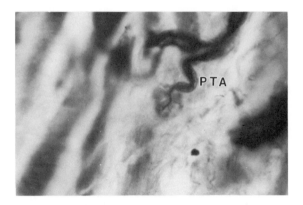

Figure 4. Normal orbicularis oculi terminal motor axons. The preterminal axon (PTA) is myelinated as it leaves the intramuscular nerve, and myelination extends to the muscle end plate region (X 204, SCI using an antibody to P_o).

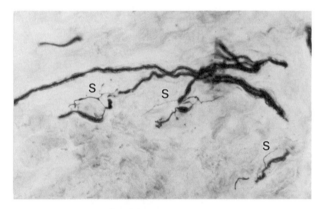

Figure 5. Unmyelinated axonal sprouts (S) extend from nodes of Ranvier of preterminal axons in the orbicularis oculi. Most of these sprouts seem to end blindly. (X 111, SCI using an antibody to P0.

Alteration of End Plate Size and Arrangement.[14] As in animal muscle, some cholinesterase-containing end plate regions expand in the orbicularis oculi following botulinum toxin injections. Segmented end plates form, with cholinesterase-containing regions separated by relatively long areas of non-cholinesterase containing muscle fiber membrane. Each region of the segmented end plate is innervated by axon processes from the terminal or ultra-terminal region of a single axon (Figure 6). It appears that the axonal sprout can induce the formation of an end-plate region on the underlying muscle membrane.

Very small end plates are also present, innervated by thin, unmyelinated axonal processes. End plate diameter ranged from 3 - 65 um. Though the mean end plate diameter in botulinum-treated muscle is about 20 um, similar to untreated muscle, the preponderance of larger and smaller end plates contributes to a significantly larger standard deviation. The number of end plates in an area of muscle increases, and the number of end plate regions identified on an individual muscle fiber increases, from one in normal muscle to five or more in botulinum-treated muscle.

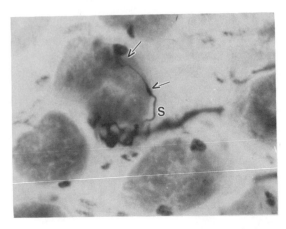

Figure 6. Segmented end plated. An unmyelinated ultraterminal process (s) extends to 2 separate end plate regions (arrows) on the muscle fiber X 370 , SCI using an antibody to P_0).

Axonal Collaterals.[14] Unexpectedly, axonal collaterals are abundant in botulinum toxin-treated muscles (Figure 8), with preterminal, terminal, and ultraterminal collaterals innervating from 2 to as many as 9 muscle fibers (Figure 8). Axonal collaterals are generally thin and unmyelinated. Collaterals extend to end plates of all sizes. Axonal collaterals were present in muscle from all patients receiving botulinum toxin, whether they had few or many injections. Axonal collaterals were present as early as 5 weeks following the second injection and persisted in patients who had their last injection 3 years previously.

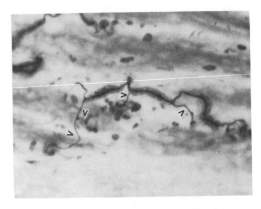

Figure 7. Axonal collaterals. Unmyelinated collaterals originate at nodes of Ranvier of a preterminal axon and from the unmyelinated terminal axon, and extend to end plates on adjacent muscle fibers (X 164, SCI using an antibody to P_0.)

Based on animal data, axonal collaterals are not expected following functional denervation. Why do these axonal collaterals develop? As discussed above, axonal collaterals form readily in partial denervation, associated with loss of motor axons. The reduction in the number of motor axons is, in the author's experience, readily appreciable as decreased density of axons in the intramuscular nerve in experimental partial denervation and in human denervating disease, such as ALS or peripheral neuropathy. It is **unlikely** that there is a significant loss of motor axons following botulinum toxin injection, as the morphology and density of axons in the intramuscular nerve is normal, whereas a reduction would be expected if collateral sprouting developed because of a partial denervation from loss of axons. The following section will discuss a proposed mechanism for the formation of axonal collaterals.

Dual Axonal Innervation of Individual Muscle Fibers.[14] Some muscle fibers in botulinum-toxin-treated muscle contain more than 1 end plate, each innervated by separate pre-terminal

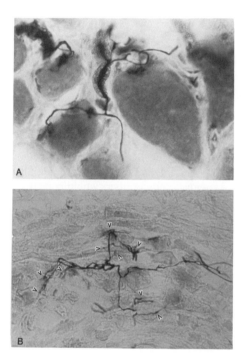

Figure 8. Variability in axonal collaterals. A) A myelinated preterminal axon gives rise to collaterals to end plate regions on adjacent muscle fibers as well as collaterals which leave the plane of section (X 290, SCI using an antibody to P_0). B) A single preterminal axon innervates at least 9 muscle fibers via collaterals (X 58, Silver cholinesterase).

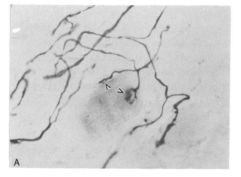 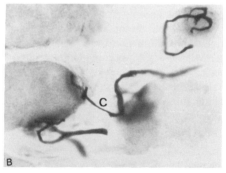

Figure 9. Dual innervation of muscle fibers. A) Two separate preterminal axons extend to separate end plates on a single muscle fiber (X 400, Silver-cholinesterase). B) An ultraterminal collateral extends to an end plate on a muscle fiber also innervated by 1 (or possibly 2) other preterminal axons (X 348, Silver-cholinesterase).

motor axons (Figure 9). Dual innervation of muscle fibers has not been identified previously, neither following functional denervation in experimental animals nor in human neuromuscular disease. Dual innervation of muscle fibers by separate preterminal axons on separate end plate regions may develop because functional denervation makes the muscle "receptive" to responding (in some unknown manner) to an axonal process to form an end plate region. The axonal process can arise from the original innervating axon, producing segmented end plates, or from a separate preterminal axon, producing dual innervation. The ability of functionally denervated muscle to allow formation of a second end plate has been demonstrated experimentally following botulinum toxin administration and subsequent implantation of other motor axons at a distance from the denervated end plate.[15] Such experimental evidence supports the hypothesis that repeated functional denervation in humans allows axonal sprouts to induce formation of new end plates on the functionally denervated muscle fiber, thus forming collateral innervation.

MUSCLE HISTOCHEMISTRY FOLLOWING BOTULINUM TOXIN INJECTION[16]

Muscle fiber atrophy, as well as gross atrophy of the muscle, develops immediately following botulinum toxin administration in both experimental animals and humans.[17-18 & Borodick, this volume] The atrophy reverses readily in experimental animals. After several weeks to months the histological appearance of the muscle returns to "normal". It could be possible that the increased density of cholinesterase-staining end plate regions and small end plates reflect persistence of this atrophy in the human orbicularis oculi. That is, if the muscle contained many small muscle fibers, the end plates on atrophic fibers could be smaller, and end plates on smaller, more compact, muscle fibers would appear denser when more muscle fibers occupy a given area.

To determine if muscle fiber atrophy persists in humans, the histological appearance of the same muscles described above was studied, 5 weeks to 3 years following botulinum toxin administration, and compared to normal orbicularis oculi and muscle from patients with blepharospasm not previously treated with botulinum toxin injection. Muscle fiber size and histological appearance was identical.[16] There are no long term, persistent changes in orbicularis muscle histological appearance following botulinum toxin injections, and botulinum toxin-induced muscle fiber atrophy is reversible as has been demonstrated by others in animals. The absence of alterations in muscle fiber size supports the hypothesis that the increased density of end plates represents the formation of new end plates on functionally denervated muscle fibers.

THE MAINTENANCE OF ALTERED NEUROMUSCULAR INTERACTIONS FOLLOWING BOTULINUM TOXIN INJECTION

The normal neuromuscular junction is not static, but is constantly remodeled: axonal sprouts spontaneously grow from the terminal axonal arborization, and other axonal processes retract. Overall, the area of neuromuscular contact remain relatively constant. In muscles treated with botulinum toxin,in contrast, the potential sites of neuromuscular contact seem to increase: there are more end plates on muscle fibers, and many of the end plates are larger than normal. Why these synapses are not remodeled is not known. Investigation of the persistence of these synapses may have implications for understanding abnormal synapse maintenance in disease.

CLINICAL CORRELATION

There is a striking variation in the degree of terminal motor axon and neuromuscular junction abnormalities among patients, including the relative number of unmyelinated axons, sprouts, and the visual density of end plates in a given region. For the most part, those muscles with fewer injections had less severe abnormalities of the terminal motor axon. The more injections, the more "abnormal" the terminal motor axon and end plates The main exception was the patient who has a severe and prolonged paralysis following his second botulinum A toxin injection, 21 months before myectomy. In this case, it is possible that the longer paralysis allows the development of greater morphological changes. Some patients chose to have a myectomy, in part, because of increasing resistance to botulinum toxin injections, either increasing dose requirement or lack of paralysis following an injection. While speculative, it is possible that increased area of neuromuscular contact could contribute, in part, to resistance following repeated injections.

CONCLUSIONS

Botulinum toxin injection into the human orbicularis oculi produces axonal sprouts, variation in the size of end plates, with expansion of some end plates and the apparent formation of new and smaller end plates, and reorganization of muscular innervation. Repeated and long-standing functional denervation allows axonal sprouts to induce the formation of nascent end plates on the functionally denervated muscle fibers, thus producing axonal collaterals. The number of muscle fibers innervated by a single preterminal axon increases. A single muscle fiber may be innervated by more than one preterminal axon (Table).

Morphological abnormalities of terminal motor axons and neuromuscular junctions increase with the number of injections or the severity of the paralysis, and seem to be persistent. The observations may have implications for understanding both the stability of the neuromuscular junction and may have clinical relevance as well.

Table 1.
MORPHOLOGY OF BOTULINUM TOXIN-INDUCED SPROUTS IN HUMANS

"Non-directed" sprouts following muscle or Schwann cell basal lamina
Expansion of muscle end plate regions
Increased number of end plates on a single muscle fiber
Increased number of end plates in regions of muscle
Axonal collaterals
Increased number of muscle fibers innervated by a single motor axon
More than one axon innervating a single muscle fiber

with no loss of motor axons and normal muscle histochemistry

Acknowledgments

Drs. Richard Anderson and John Holds performed the orbicularis oculi myectomy and blepharoplasty. Dr. Cheryl Harris and Jonathan Nebeker assisted with the muscle histochemistry.

REFERENCES

1. Wohlfart G. Collateral sprouts from residual motor nerves in amyotrophic lateral sclerosis. Neurology 7:124-34 (1957).
2. C. Ide, K. Tohyama, et al. Schwann cell basal lamina and nerve regeneration. Brain Research 288:61-75 (1983).
3. M.V. Edds. Collateral regeneration of residual motor axons in partially denervated muscles. J Exp Zool 113:517-551 (1950).
4. H. Hoffman. Local reinnervation in partially denervated muscle. A histophysiological study. Aust J Exp Biol 28:383-97 (1950).

5. W.G. Hopkins, M.C. Brown, R.J. Keynes. Nerve growth from nodes of Ranvier in inactive animals. Brain Res 222:125-128 (1981).

6. M.C Brown, R.L. Holland, W.G. Hopkins. Motor nerve sprouting. Ann Rev Neurosci 4:17-42 (1981).

7. K. Alderson, W.C Yee, A. Pestronk. Reorganization of intrinsic components in the distal motor axon during outgrowth Journal of Neurocytology, 18:541-552 (1989).

8. K. Alderson, A. Pestronk, W.C. Yee, D.B. Drachman. Silver-Cholinesterase-Immunocytochemistry: A new neuromuscular junction stain. Muscle Nerve 12:9-14 (1989).

9. L.W. Duchen. Changes in motor innervation and cholinesterase localization induced by botulinum toxin in skeletal muscle of the mouse: Differences between fast and slow muscles. J Neurol Neurosurg Psychiatry 33:40-54 (1970).

10 L.W. Duchen, S.J. Stritch. The effects of botulinum toxin on the pattern of innervation of skeletal muscle in the mouse. Q J Exp Physiol 53:84-89 (1968).

11 I. Kao, D.B. Drachman. Motor nerve sprouting and acetylcholine receptors. Science 193:1256 (1978).

12. W.N. Gillum, R.L. Anderson. Blepharospasm surgery: An anatomical approach. Arch Ophthalmol 99:1056-1062 (1981).

13. J.B. Holds, K. Alderson, S.G. Fogg, R.L. Anderson. Terminal nerve and motor end plate changes in human orbicularis muscle following botulinum A injection. Invest Ophthalmol Vis Sci 31:178-181 (1990).

14. K. Alderson, J.B. Holds, R.L. Anderson. Botulinum-induced alteration of nerve-muscle interactions in the human orbicularis oculi following treatment for blepharospasm. Neurology 41:1800-1805 (1991).

15. L.W. Duchen, D.A. Tonge. The effects of implantation of an extra nerve on axonal sprouting usually induced by botulinum toxin in skeletal muscle of the mouse. J Anat 124:205-215 (1977).

16. C.P. Harris, K. Alderson, J. Nebeker, J.B. Holds, R.L. Anderson. Histology of human orbicularis muscle treated with botulinum toxin. Arch Ophthalmol 109:393-395 (1991).

17. R.F. Spencer K.W. McNeer. Botulinum toxin paralysis of adult monkey extraocular muscles. Arch Ophthalmol 105:1703-1711 (1987).

18. S.R. Cohen, J.W. Thompson, F.S. Camilon. Botulinum toxin for relief of bilateral abductor paralysis of the larynx: Histological study in an animal model. Ann Otol Rhinol Laryngol 98:213-216 (1989).

BOTULINUM TOXIN INDUCED MUSCLE DENERVATION:
A MRI STUDY

Dušan Šuput,[1] Jernej Dolinšek,[1] Robert Frangež,[1] and Franci Demšar[2]

[1]School of Medicine, Institute of Pathophysiology, Ljubljana, Slovenia
[2]Institute Josef Stefan, Jamova 39, 61000 Ljubljana, Slovenia

INTRODUCTION

Several serious genetic diseases, nerve crush injuries, and other long-lasting denervations cause muscle atrophy. Previous magnetic resonance (MR) imaging and spectroscopy studies of the spinal muscle atrophy (SMA) and Duchenne muscular dystrophy (DMD) showed patterns of muscle degeneration, which were typical of each type of the disease.[1,2] The aim of our study was to explore animal models of muscle atrophy, which would best correspond to human neurodegenerative diseases. A model of persistent neuromuscular block has been developed by use of the botulinum neurotoxin type A (BoNTxA). Consequent muscular atrophy, followed by regeneration of the affected muscle, was assessed by use of a small MR imaging system, which enables MR imaging and spectroscopy of experimental animals. The importance of presence of the neuromuscular junction, though not secreting Acetylcholine, for the physiological and morphological state of the treated muscle was studied by comparison to other disorders of the neuromuscular junction. Thus, nerve crush injury induced denervation was chosen as a parallel model disease. Investigation of BoNTx induced denervation also seemed to be relevant because of the clinical use of *C. botulinum* toxins[3] - especially with the respect to the duration of denervation and possible side effects.[4]

METHODS

Mechanical and Chemical Denervation of Muscles

All experiments were performed on male albino rats (Wistar) weighing 180-190 g at the beginning of an experiment. Denervation was caused either by pinching sciatic nerve (n=9), or by an intramuscular (i.m.) injection of 5 mouse units (m.u.) of Oculinum (Oculinum Inc, San Francisco)(n=6). Nine animals served as a control. Nerve crush injury was always performed proximally on an easily accessible standard anatomical location. Recovery of the neuromuscular transmission was monitored by measurements of muscular contractions after nerve stimulation, and muscle size was measured by means of MRI.

Nuclear Magnetic Resonance Imaging and Spectroscopy

Measurements were performed on a 2.35 T Bruker tomograph. A spin echo sequence with T_1 weighted images (TR = 600 ms, TE = 34 ms) was used. A field of view was 10-20 cm with data acquisition matrix 256 x 256. Slice thickness was 2 to 5 mm. Images of acceptable quality were obtained within 10 minutes. Muscle atrophy was evaluated by surface measurements of the transversal and sagittal cross sections of rat hind limbs. Untreated leg served as a control. The possible compensatory hypertrophy of untreated legs from the denervated animals was assessed by comparison of cross sections of untreated leg of each animal to the average cross section of the legs of control animals of the same age and weight (see also Fig. 4).

^{31}P spectra were measured on the same magnet. Various organic phosphorous compounds and inorganic phosphates were identified according to their position relative to the position of the phosphocreatine (PCr) peak (position 0). The peak of inorganic phosphate (P_i) represents the sum of 'visible' monobasic and dibasic inorganic phosphates. The chemical shift of this peak (relative to the position of the PCr peak) represents a weighted mean of both types of inorganic phosphates. Intracellular acidosis shifts the P_i peak to the right, and intracellular alkalosis to the left. With an assumption, that the intracellular concentration of Mg^{2+} remains practically constant during the experiment, the chemical shift of P_i enables measurement of intracellular pH.

RESULTS

MR spectra show a shift of intracellular pH towards a more alkaline value during the first two days after application of Oculinum. From the calibrating curve it has been estimated that the shift varied between 0.12 to 0.2 pH units (n=4).

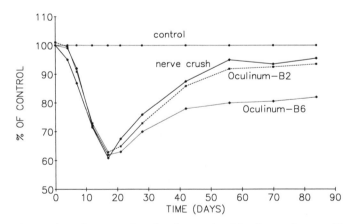

Figure 1. Muscle atrophy and regeneration after denervation. Muscle atrophy caused by nerve crush has a faster onset than the denervation caused by injection of Oculinum. Note the unsuccessful regeneration of injected muscle groups in animal B6.

MR imaging of hind limbs revealed that the nerve crush and Oculinum induced denervations cause a reversible muscle atrophy. The time course of the atrophy and 'regeneration' seems to be slightly faster in the case of nerve crush (Fig. 1) than in the case of BoNTxA induced denervation, but differences were not statistically significant during the first two months of investigation. However, two animals (labeled B1 and B6) showed incomplete recovery of BoNTxA injected musculature (Fig. 1). In the case of B1, the cross section of the injected hind limb was only 80% of the control even 6 months after the application, and in the case of animal B6 the recovery of the injected musculature reached 80% of the control after three months.

Compensatory hypertrophy of untreated hind limbs is difficult to assess due to variations in the normal growth. Nevertheless, it seems that the compensation of loss of force in the denervated muscles by hypertrophy of muscles in the intact leg is more successful in the case of mechanical denervation of muscles (Fig. 2).

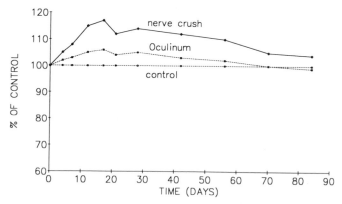

Figure 2. Compensatory hypertrophy of the untreated leg. Note a larger hypertrophy of the untreated leg in mechanically denervated experimental animals.

Results show that each model of muscle denervation causes a rapid muscle atrophy. The onset of the atrophy is delayed for one day in the case of the denervation produced by the use of Oculinum. Similarly, the onset of the recovery of the muscle mass is also slightly delayed. The recovery of muscle size after Oculinum induced paralysis was typically slower when compared to the recovery after the nerve crush injury. None of the models revealed muscle degeneration characterized by fatty infiltrations, typical of neurodegenerative disorders seen in patients.[2,3,4,5] Individual muscles can be identified in the thigh, but not in the calf, where the toxin was injected. Therefore it is not possible to evaluate the time course of atrophy and regeneration of individual muscles. MR images are of homogeneous dark gray color in all muscles. The intensity of the signal does not differ between the treated and untreated leg. This is illustrated in Fig. 3, which shows MR images from the time of injection of Oculinum in the leg (day 0) on. Position of slices is proximal and the same in the first four images. The last two images (day 42 and day 56 after the application of Oculinum) display slices which are more distal compared to the position of slices from the first four images. The intensity of the signal is similar in all images, regardless of the position of the cross section. This reconfirms our observation that the botulinum toxin induced denervation does not induce a fatty degeneration of the affected musculature. The signal intensity of treated and untreated legs is comparable.

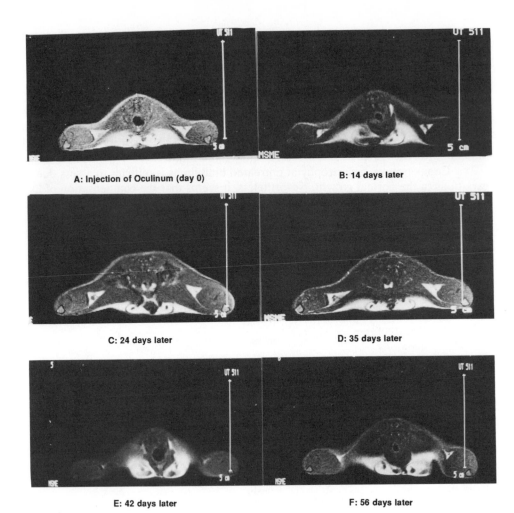

Figure 3. MR imaging of rat calf muscles following the injection of Oculinum. Images were obtained at different time intervals after the nerve crush. Cross section shows atrophy of right calf muscles, with peak after 17 days, which is then followed by a slow regeneration.

DISCUSSION

Neuromuscular diseases represent a relatively broad spectrum of neurological conditions which are difficult to diagnose, especially in the early stage. The primary site of these diseases is most commonly either in the muscle itself (myopathy or muscular dystrophy) or in the nervous tissue (neuropathy). In particular, the juvenile Kugelberg-Welander form of inherited spinal muscular atrophy (SMA) may produce a similar clinical picture as Duchenne muscular dystrophy. Recent study demonstrated the possibility that these disorders could be discriminated by use of MR imaging.[2]

Other authors have also used MR imaging in assessment of disorders of musculo-skeletal system.[5,6,7] The principal finding was that the musculature is being replaced by adipose tissue. The present study on animal models revealed clear differences from the more chronic conditions observed on patients. Brust[8] reported a relative resistance to dystrophy of slow skeletal muscle of the mouse. In our study we also explored the possibility that the neuromuscular junction of slow muscles might be more resistant to the action of BoNTxA, but usually it was not possible to discriminate the individual muscles in the investigated legs undoubtedly. However, it is clear that there was no fatty infiltration in the muscles of the treated legs, which is contrary to our findings in patients with neuromuscular diseases.[9]

There are several reports in the literature showing that the neuromuscular junction is partly functioning even after application of BoNTx, and that nerve terminal sprouting begins after application of BoNTx.[10,11] Our results have revealed only minor differences in the time course of 'mechanical' and 'chemical' denervation. There was no evidence that a partially functioning neuromuscular junction may have contributed to a better preservation of muscular function in the pharmacologically denervated muscles. It seems that BoNTxA suppresses compensatory hypertrophy of muscles in the control leg. This problem needs further examination on a larger number of experimental animals, because it could imply generalized effects of the toxin.

Acknowledgements

This work was supported by the Ministry of Science and Technology of Slovenia (Grant P3-0170-381). Authors appreciate the technical help of A. Sepe and M. Kadunc.

REFERENCES

1. Lamminen AE. Magnetic resonance imaging of primary skeletal muscle diseases: patterns of distribution and severity of involvement. British J Radiol 1990; 63:946-50.
2. Šuput D, Zupan A, Sepe A, and Demsar F. Discrimination between neuropathy and myopathy by use of MRI. Acta Neurol Scand: In press.
3. Scott AB, Rosenbaum A, Collins CC. Pharmacologic weakening of extraocular muscles. Invest Ophtalmol 1973; 12:924-27.
4. Olney RK, Aminoff MJ, Gelb DJ, and Lowenstein DH. Neuromuscular effects distant from the site of botulinum neurotoxin injection. Neurology 1988; 38:1780-3.
5. Murphy WA, Totty WG, Carroll JE. MRI of normal and pathologic skeletal muscle. AJR 1986: 146: 565-74.
6. Schwartz MS, Swash M, Ingram DA, et al. Patterns of selective involvement of thigh muscles in neuromuscular disease. Muscle Nerve 1988: 11: 1240-5.
7. Chance B, Younkin DP, Kelley R, et al. (1986): Magnetic resonance spectroscopy of normal and diseased muscles. Am J Med Gene; 25: 659-79.
8. Brust M. Relative resistance to dystrophy of slow skeletal muscle of the mouse. Am J Physiol 1966: 210: 445-51.
9. Šuput D., Zupan A., Sepe A. in Demsar F. (1990): Assessment of Neuropathies and myopathies by Magnetic Resonance Imaging. 9th SMRM Meeting - Works in Progress, New York, p. 1191
10. Dolly JO, Lande S, and Wray DW. The effects of in vitro application of purified botulinum neurotoxin at mouse motor nerve terminals. J Physiol 1987; 386:475-484.
11. Angaut-Petit D, Molgo J, Comella JX, Faille L, and Tabti N. Terminal sprouting in mouse neuromuscular junctions poisoned with botulinum type A toxin: Morphological and electro-physiological features. Neuroscience 1990; 37:799-808.

CALCIUM-INDEPENDENT NEUROTRANSMITTER RELEASE

Eric A. Schwartz

Department of Pharmacological and Physiological Sciences
The University of Chicago
947 E. 58th Street
Chicago, IL 60637

The brain is not a muscle. And we should not expect communication between neurons to mimic the well studied events that occur during transmission at a neuromuscular junction. Although many neurons use Ca^{+2}-dependent exocytosis to release transmitters, some retinal neurons now appear to use a different mechanism. A motor neuron rapidly mobilizes the release of sufficient transmitter to ensure muscle contraction. The pre-synaptic signal is an action potential, a 100 mV depolarization that lasts several milliseconds. In contrast, photoreceptors, and other neurons in the outer retina, signal the absorption of light with a slow, graded change in voltage. A dark-adapted rod photoreceptor maintains a potential of approximately -35 mV; the absorption of a photon produces a 30 µV hyperpolarization that lasts nearly one second (Schwartz, 1976). The mechanisms of synaptic transmission that operate in motor neurons and in photoreceptors may be optimized to ensure the reliable communication of their very different pre-synaptic signals.

Transmission in the outer retina can persist during conditions commonly assumed to interrupt calcium influx into presynaptic terminals (Schwartz, 1986). For example, continuously superfusing an isolated retina with a saline that lacks Ca^{+2} and contains Co^{+2} blocks transmission at many synapses as soon as the Co^{+2} diffuses into the tissue, usually 3-4 minutes. Yet some synapses continue to operate even 1-2 hours later.

The mechanism of Ca^{+2}-independent transmitter release has been explored in a preparation of isolated horizontal cells, neurons that make reciprocal synapses with photoreceptors and also produce graded responses. A retina can be cultured in a mixture of toxins that allows horizontal cells to survive while other neurons die (Schwartz, 1982). The surviving horizontal cells accumulate and release [^{3}H]GABA. The release of GABA is stimulated by depolarization and is independent of Ca^{+2} in the external medium (see also Yazulla and Kleinschmidt, 1983; Ayoub and Lam, 1984). Two observations indicate that efflux is mediated by a GABA transporter. First, efflux is reduced or occluded if the transporter is forced to spend part of the time mediating influx; second, efflux is blocked by nipecotic acid, a relatively selective transport antagonist.

The electrophysiology of GABA transport has been observed in solitary horizontal cells. Jon Cammack and I have studied horizontal cells with a whole-cell voltage clamp and observed currents produced by GABA influx and efflux (Cammack and Schwartz, unpublished observation). When a horizontal cell is depolarized, influx is decreased and efflux is increased. Thus, GABA transport can have two roles in synaptic transmission. GABA is removed from the extracellular space when a cell is hyperpolarized, and released into the extracellular space when a cell is depolarized. Membrane voltage plays an essential role in regulating the balance between influx and efflux. The net effect can be observed when a horizontal cell is juxtaposed to a GABA-sensitive bipolar cell (Schwartz, 1987). Depolarization of the horizontal cell releases GABA and induces a current in the adjacent bipolar cell.

A slow graded response may be needed to regulate carrier-mediated release. The electrical responses of horizontal cells are well suited to control carrier-mediated release. These cells do not normally produce action potentials. They are depolarized during darkness. A flash of light produces a hyperpolarizing wave that can last several hundred milliseconds. A voltage-dependent transporter controls GABA release. Unlike other neurons, exocytosis appears to make little contribution. Indeed, the pre-synaptic processes of horizontal cells that contact photoreceptors are notably vesicle-poor (see Lasansky, 1980). Other neurons may utilize both Ca^{+2}-dependent exocytosis and Ca^{+2}-independent transport. This situation may occur in photoreceptors. These cells make two morphological types of synapse (Dowling and Boycott, 1966; Lasansky, 1971). One type has a presynaptic aggregation of vesicles and develops exocytotic, membrane pits during depolarization; the other type is vesicle-poor and fails to develop exocytotic pits (see Raviola, 1976). Moreover, one component of the normal synaptic transmission from photoreceptors to post-synaptic cells requires extracellular Ca^{+2}; while another component functions without extracellular Ca^{+2} (Schwartz, 1986). Photoreceptors may use Ca^{+2}-dependent exocytosis at vesicle-rich invaginating junctions and voltage-dependent transporters at vesicle-poor basal junctions. Voltage-dependent transporters may also release transmitters from other neurons that communicate with slow, graded signals.

REFERENCES

Ayoub, G.S. and Lam, D.M.K., 1984, The release of γ-aminobutryic acid from horizontal cells of the goldfish (*Caarassius auratus*). *J. Physiol. (Lond.)* 355: 191.

Dowling, J. E. and Boycott, B. B., 1966, Organization of the primate retina: electron microscopy. *Proc. Roy. Soc. (Lond.)* B170: 205.

Lasaznsky, A., 1971, Synaptic organization of cone cells in the turtle retina. *Phil. Trans. Roy. Soc. (Lond.)*

Lasansky, A., 1980, Lateral contacts and interactions of horizontal cell dendrites in the retina of the tiger larval salamander. *J. Physiol.(Lond.)* 301: 59.

Raviola, E. 1976, Intercellular junctions in the outer plexiform layer of the retina. *Invest. Ophth.*15: 881.

Schwartz, E. A., 1976, Electrical properties of the rod syncytium in the retina of the turtle. *J. Physiol. (Lond.)* 257: 379.

Schwartz, E.A., 1982, Calcium-independent release of GABA from isolated horizontal cells of the toad retina. *J. Physiol. (Lond.)* 323: 211.

Schwartz, E.A., 1986, Synaptic transmission in amphibian retinae during conditions unfavorable for calcium entry into presynaptic terminals. *J. Physiol. (Lond.)* 376: 411.

Schwartz, E.A., 1987, Depolarization without calcium releases GABA from a retinal neuron. *Science* 238: 350.

Yazulla. S. and Kleinschmidt, J. (1983). Carrier-mediated release of GABA from retinal horizontal cells. *Brain Res.* 263: 63.

Ca^{2+} DEPENDENT EVOKED QUANTAL NEUROTRANSMITTER RELEASE DOES NOT NECESSARILY INVOLVE EXOCYTOSIS OF SYNAPTIC VESICLES

Ladislav Tauc and Bernard Poulain

Laboratoire de Neurobiologie Cellulaire et Moléculaire
C.N.R.S.
F-91198 Gif-sur-Yvette Cedex, France

INTRODUCTION

There is now evidence that the target for tetanus (TeTx) and botulinum (BoNT) neurotoxins is located intracellularly (Penner et al., 1986; Poulain et al., 1988) and that they block synaptic transmission without interfering with the influx of calcium at the nerve terminal (Dreyer et al., 1983; Molgo et al., 1989). It seems therefore clear that these toxins act on a component of the nerve terminal implicated in the neurotransmitter release mechanism. Furthermore, when the membrane steps of intoxication are bypassed by different techniques (reviewed in Poulain and Molgo, 1992; Niemann, 1991; Dolly, 1992) the toxins appear nearly equipotent in inhibiting release of acetylcholine, mono-amines or neuropeptides in various preparations. Thus, it may be postulated that their target is common to cells or neurons releasing different transmitters. The above conclusion has to be borne in mind when researching the intracellular target of these toxins. The main complication, however, arises from the fact that the transmitter release mechanism is still obscure despite the rather definitive descriptions found in the textbooks and ascribing this role to the exocytosis of synaptic vesicles.

The view that, in a process analogous to that used by gland cells, synaptic vesicles are involved in the transport of transmitter across the lipid bilayer of the terminal presynaptic membrane that separates the axoplasm from the synaptic cleft, was initially based on the necessity to find a way in which the transmitter can be released in molecular paquets. In the original formulation of the "vesicular hypothesis" presented by Bernard Katz at a meeting in Gif-sur-Yvette in 1955 (Del Castillo and Katz, 1957), it was assumed that the recently discovered synaptic vesicles contained the transmitter substance and that when, in their random movement, they touched the presynaptic membrane at critical contact zones (i.e. active zones), they discharged their content by exocytosis, the probability of their fusion with the plasma membrane increasing with the infux of Ca^{2+}

Botulinum and Tetanus Neurotoxins, Edited by
B.R. DasGupta, Plenum Press, New York, 1993

ions triggered by nerve terminal depolarization. This was not only a satisfactory explanation of both the occurrence of spontaneous miniature potentials and the quantal character of the evoked end-plate potential, but it also explained how the quanta are formed and stored for ready release.

The demonstration that synaptic vesicles do effectively contain neurotransmitter (De Robertis et al., 1962; Whittaker et al., 1964) and ultrastuctural studies showing the reality of the exocytotic process considerably asserted the hypothesis of vesicular exocytosis as the mechanism for release of neurotransmitter. This hypothesis is intellectually very attractive for its ostensible simplicity and for giving to a quantum an easily understandable structural support (for reviews see Ceccarelli and Hurlbut, 1980; Vizi, 1984; Cooper and Meyer, 1984; Thesleff, 1986; Whittaker, 1986; Zimmermann, 1988; Van der Kloot, 1988, 1991; Hurlbut, 1989; Heuser, 1989; Valtorta et al., 1990). Recent elegant works have attributed a role to several presynaptic proteins (small G-proteins, synapsins, synaptoporin, synaptophysin...) in the intracellular trafficking of synaptic vesicles, their docking at the active zone and their fusion with the plasma membrane (De Camilli and Jahn, 1990; Almers, 1990; Almers and Tse, 1990; Valtorta et al., 1990; Kelly, 1991; Matteoli and De Camilli, 1991; Llinas, 1991).

However, despite the large amount of work done, no decisive direct evidence has been brought forward that validates the vesicular hypothesis as the mechanism underlying fast transmitter release at neuro-neuronal synapses or at the skelettal nerve-muscle junction. Moreover, today we know that neurotransmitter is also present in the cytoplasm of the terminal at relatively high concentration, the role of this "free" pool being unaccounted for by the vesicular hypothesis. Observations presented below and in several review articles (Israël et al., 1979; Marchbanks, 1979; Tauc, 1979, 1982; Vizi, 1984; Cooper and Meyer, 1984; Israël and Manaranche, 1985; Tauc and Baux, 1985; Dunant, 1986; Morel and Israël, 1990; Tauc and Poulain, 1991a,b) cannot easily be explained by the vesicular exocytosis hypothesis and justify a reassessement of the model or the elaboration of alternative explanations of the release mechanism.

In this rapid review, we deal essentially with cholinergic synapses, the only ones for which enough experimental information is available regarding the fast release of neurotransmitter (for the release of other neurotransmitters or neuromodulators see Uvnäs, 1985; Nicholls, 1989; De Camilli and Jahn, 1990; Almers, 1990; Maycox et al., 1990).

CRITICAL CONSIDERATIONS ON THE VESICULAR HYPOTHESIS

Morphological correlates of ACh release

The release of a large quantity of transmitter in a quantal form must leave a trace on the membrane. In the hypothesis of the exocytosis of vesicles, the presence of "omega shapes" on electron micrographs of the presynaptic terminal membrane is considered to indicate the opening and membrane fusion of vesicles whereby a quantum of ACh is supposed to be ejected into the synaptic cleft (Couteaux and Pécot-Dechavassine, 1970; for reviews see Ceccarelli et al., 1973; Heuser, 1989). However, it has not been possible to demonstrate a clear correlation between transmitter release and vesicle number (for a review see Ceccarelli and Hurlbut, 1980). Moreover, there are physiological conditions in which transmitter is released in the absence of vesicles. In chick ciliary ganglion, normal synaptic transmission can be observed at a developmental stage where the few morphological synapses present contain few or no synaptic vesicles (Landmesser and Pilar, 1972).

An important observation is the rearrangement of specific intramembrane particles associated with Ca^{2+} dependent evoked ACh release in stimulated preparations of *Torpedo* (Israël et al., 1983; Müller et al., 1987). The same ultrastructural modifications have been observed using reconstituted proteoliposomes free of synaptic vesicles that contain, incorporated into their membrane, only the "mediatophore", a protein found to mediate the release of ACh in a Ca^{2+} dependent manner (Israël et al., 1981, 1988; Birman et al., 1986). These data suggest that a membrane bound macromolecule is responsible for - or at least implicated in - the extrusion of ACh.

Can exocytosis of a vesicle occur fast enough?

Llinas (1977, 1991) used a fast voltage-clamp technique to measure the time elapsed between depolarization induced Ca^{2+} entry into the terminal of the squid giant synapse and the start of the postsynaptic potential. The minimal time was 200 μsec. During this very short time, Ca^{2+} must diffuse, reach strategic sites inside the nerve terminal, and trigger the release mechanism; the transmitter has to cross the synaptic cleft and bind to postsynaptic receptors; and, finally, ions must start flowing across the opened postsynaptic channels. Even if the presynaptic calcium channels implicated in the release of neurotransmitter are very close to the release sites (Llinas et al., 1992), only a part of these 200 μsec can be devoted to activation of the release mechanism. Other synapses can probably work even faster (Hubbard and Schmidt, 1963).

Can triggering and turning on the fusion of synaptic vesicles occur in such a short (200 μsec) period of time? Many theoretical difficulties are involved in the process and no model so far gives a satisfactory explanation, the fastest one takes milliseconds (Llinas and Heuser, 1978; Almers et al. 1991). The formation of a fusion pore (Almers and Tse, 1990; Zimmerberg et al. 1991) that dilates up to the total fusion of the vesicle and plasma membranes would shorten this time. There is no doubt that exocytosis occurs in various secretory cells where incorporation of vesicles into the membrane can be detected by membrane capacitance measurements (Neher and Marty, 1982). But, for instance, the exocytotic process in chromaffin cells (one of the fastest secretory cells), is about 1,000 times slower than that required for release of ACh at motor synapses (Almers, 1990) and no experimental evidence supports an extrapolation of findings from secretory cells to fast releasing neurons.

Correlationship between vesicle content and unitary electrical events

Quantal size and vesicle content The size of the quantum of ACh thatcorresponds to the simultaneous release of several thousand ACh ions: 10,000 at a snake neuromuscular junction (Kuffler and Yoshikami, 1975), 6,000 at the rat neuromuscular junction (Flechter and Forrester, 1975), 8,000-12000 at the frog neuromuscular junction (Miledi et al., 1980, 1983) or 5,000-6,000 at a central synapse in *Aplysia* (Simonneau et al., 1980) - is supposed to reflect the ACh content of individual vesicles (11,000-15,000) at the frog neuromuscular junction, Miledi et al., 1982). However, this correlation does not fit in the case of *Torpedo*. Indeed, vesicles in the electric organ contain 130,000-200,000 ACh molecules (reviewed in Zimmermann, 1988; Dunant, 1986; Van der Kloot, 1991) while the size of a quantum is considered to correspond to only 7,000 to 9,000 molecules (Dunant and Müller, 1986).

A classical quantum is composed of subquanta The discovery of subminiature endplate potentials (S-MEPPs) at the mammalian and amphibian neuromuscular junction

(Kriebel and Gross, 1974; Bevan, 1976; Kriebel, LLados and Matteson, 1976; Wernig and Stirner, 1977; Kriebel, 1978; Vautrin and Membrini, 1981; Kriebel et al., 1990) and in *Torpedo* electric organ (Müller and Dunant, 1985) sheds new light on the supposed relationship between vesicles and quanta. Thanks to improved recording techniques in preparations with high input resistance, a quite large population of "small" miniature potentials has been found with sizes about 7 to 15 times smaller than "normal" MEPPs. The amplitude distribution of the size of these S-MEPPs indicates that they are probably the responses to the release of subunits of ACh quanta (Erxleben and Kriebel, 1988). Release of a full sized ACh quantum would then result from the synchronous release of several subunits. The proportion of S-MEPPs in respect to normal-size MEEPs changes during morphogenesis and can be experimentally modified by several factors like high temperature, tetanic simulation, and also application of botulinum neurotoxin (Kriebel et al. 1976; Dunant et al., 1987; Kriebel et al., 1990).

It becames difficult to explain the presence of S-MEPPs within the framework of the vesicular exocytosis hypothesis, even by proposing that "normal" size MEPPs result from the simultaneous release of several vesicles (Tremblay et al., 1983). Seven to fifteen times more vesicles would be needed to produce a normal size MEPP. This poses additional problems because vesicles contain far more ACh that is needed to produce a S-MEPP. Thus one vesicle apparently cannot correspond to one physiological unit (Kriebel et al., 1980).

Pre-arrangement of quanta and high rate release

In the hypothesis of vesicular exocytosis, the fact that in resting terminals some vesicles appear "docked", ready for release, beneath the active zone (Heuser et al., 1981), is expected to reduce the time required for interaction of the releasable vesicle with the plasma membrane. Their number is compatible with the number of quanta (about 200) that are released by each presynaptic action potential at the vertebrate neuromuscular junction and agrees with the idea that at a given time there is only a small pool of releasable vesicles amongst the 500,000 present in a motor nerve terminal (see reviews quoted in Introduction). Similarly, at the cholinergic synapse between B4 and B3 neurons in the buccal ganglion of *Aplysia*, the evoked response is composed of 150 to 600 quanta (Simonneau et al., 1980; Baux et al., 1986; Poulain et al.,1987; Fossier et al., 1990), probably corresponding to a number of synaptic vesicles not very different from that of the neuromuscular junction.

It is not likely, however, that only pre-arranged vesicles participate in evoked ACh release during intense and prolonged stimulation because i) recent observations shown that synaptic vesicles are randomly redistributed after stimulation (Betz and Bewick, 1992) and ii) under these conditions the number of quanta released exceeds the number of "readily releasable" vesicles adjacent to the presynaptic membrane along the active zone (Katz and Miledi, 1979). In Aplysia, under sustained depolarization, each presynaptic neuron can release up to 2,500,000 quanta within 3 seconds. Thus for exemple B4 can release 250,000 quanta or more only onto B3, one of several postsynaptic followers (Baux et al., 1986; Poulain et al., 1987). Since the number of quanta released exceeds by far the total number of vesicles, this observation argues against the possibility of transmitter release from a pool of releasable vesicles, docked or not at the active zone. Moreover, under the hypothesis of vesicular exocytosis, the calculated amount of vesicle membrane incorporated to the cell membrane during a 3 sec stimulation would be about 10% of the whole neuron surface and hence some hundred fold the active nerve terminal membrane surface (considering that vesicles are 60 nm in diameter and the cell body of B4 neuron is 200 μm in diameter plus the axonal tree).

Pool of ACh affected during intense stimulation

Newly synthesized ACh is released preferentially upon stimulation (Collier, 1969; Potter, 1970) which is contrary to what would be expected if ACh is released from a stable preformed store and excludes any prearrangement of vesicles ready for release at the active zone. It has been demonstrated, using radioactive precursors, that stimulation of the electric organ of *Torpedo* first increases the turnover of ACh in the cytoplasmic compartment whereas vesicular ACh is unchanged for several minutes (Dunant et al., 1974; Dunant, 1986). Moreover, in conditions where re-synthesis of ACh is prevented, stimulation of ACh release leads to a clear decrease of cytoplasmic ACh to near exhaustion while the vesicular ACh is essentially preserved (Israël and Lesbats, 1981). This indicates that, during intense stimulation, there is little or no exchange of ACh between cytoplasmic and vesicular compartments. A correlation between free ACh in the terminal and the amplitude of the evoked postsynaptic synaptic response has been observed in the electric organ of *Torpedo* (Corthay et al., 1982).

ATP is not co-released stoichiometrically with ACh

A stoichiometric release of all soluble contents of vesicles would be expected during exocytosis. ATP is present in cholinergic vesicles (Volknandt and Zimmermann, 1986) and the co-release of ATP and ACh was observed in brain synaptosomes (White et al., 1980) and *Torpedo* electric organ (Meunier et al., 1975). The ATP to ACh ratio was 1:5 in the vesicles and 1:45 in the perfusate (Morel and Meunier, 1981). It is unlikely that this difference could result from the presence of presynaptic ATPase (Unsworth and Johnson, 1990), since application of botulinum neurotoxin blocks the release of ACh but not that of ATP (White et al., 1980, Marsal et al., 1987). These experiments are not consistent with the view that the whole vesicle content is released by exocytosis unless ATP and ACh are confined in different vesicular populations.

Blocking the vesicular uptake of ACh

The finding that potent blocker of vesicular uptake, AH-5183 (vesamicol) (Anderson et al., 1983, Marshall and Parsons, 1987) also blocks the release of newly synthetised ACh, seemed to validate the vesicular theory (Melega and Howards, 1984; Michaelson and Burstein, 1985). However AH-5183 has many other effects including the blockade of the molecular leakage of ACh (Edwards et al., 1985). It is unlikely that this leakage is related to the insertion of the vesicular ACh transporter into the presynaptic membrane (Grinnel et al.,1989) because no leakage of ACh was detected in the vincinity of the quantal release sites, even following a tetanic stimulation expected to produce massive incorporation of the ACh transporter by fusion of vesicle and plasma membranes. In addition, AH-5183 does not block Ca^{2+}-dependent K^+ evoked ACh release from *Torpedo* electric organ synaptosomes (Morot Gaudry-Talarmain et al., 1989). Since facilitation of Ca^{2+} entry by aminopyridines transiently reverses the blockade of transmission induced by AH-5183 (Estrella et al., 1988), AH-5183 probably acts directly on the release mechanism.

Moreover, after blockade of ACh release by AH-5183, addition of black widow spider venom, a toxin known to induce paroxismal quantal release of ACh (Fesce et al., 1986), can induce the release of newly synthetized ACh (Cabeza and Collier, 1988). Since AH-5183 blocks the vesicular uptake of ACh, this above observation strongly suggests that ACh is released from a non-vesicular compartment. Also, in *Aplysia*, presynaptic injection of atropine which like curare is a potent antagonist of vesicular ACh uptake (Anderson et al., 1983) did not alter the release of ACh whereas curare did (Tauc and Baux, 1985, Baux and Tauc, 1987).

Modification of the ACh cytoplasmic compartment

Use of false transmitter. A way to test the origin of released neurotransmitter is to use the false transmitter precursor monoethylcholine, a compound that is taken up by nerve terminals at the neuromuscular junction, acetylated in the cytoplasm and released as the false transmitter, acetylmonoethylcholine (AMECh). The analysis of miniature endplate currents has shown that both transmitters present in the terminal, ACh and AMECh, mixed immediately and are co-released within the same quantum (Large and Rang, 1978). Since a rapid exchange between cytoplasmic and vesicular ACh has not been detected in any biochemical study (Israël et al., 1979 and above reviews) it seems probable that quanta of mixed false and true transmitters came from the cytoplasmic compartment.

Experimental alteration of the cytoplasmic concentration of ACh. The presynaptic ACh content can be decreased either by hydrolysis or by acting on the synthesis of this transmitter. Following the first method, intracellular ACh was destroyed by injecting exogenous AChE into a presynaptic cholinergic neuron in the buccal ganglion of *Aplysia*. As the vesicular pool of ACh appears to be protected against AChE (Whittaker et al., 1963; Whittaker et al., 1972, Marchbanks and Israël, 1972) only the cytoplasmic ACh was expected to be hydrolyzed. After a delay corresponding to the diffusion/transport of the enzyme to the terminal, the size of the postsynaptic response was reduced and eventually abolished (Tauc et al., 1974; Tauc and Baux, 1980, 1982 and 1985; Baux and Tauc, 1983). Accordingly, when ACh in the presynaptic neuron was labeled by injection of radioactive precursor, only the free pool was decreased after introduction of AChE (Tauc, 1977). The reduction of ACh release was not due to an effect on the transmitter release mechanism as the nerve terminal was still able to release a non-hydrolyzable analogue of ACh, carbachol (Baux and Tauc, 1983).

Similar results were obtained when the formation of ACh was prevented by removing the choline necessary for its synthesis (Poulain et al., 1986 a and b) with intracellular injection of choline oxidase. This method, without the requirement of intensive stimulation, destroyed the choline and reduced the cytoplasmic ACh by reversing the reaction of ACh synthesis. Conversely, the enhancement of presynaptic ACh content by intracellular ionophoretic injection of exogenous ACh or choline into the presynaptic neuron, rapidly led (in 4-15 min) to an increase in transmitter release (Tauc and Baux, 1985; Poulain et al., 1986 a and b). Again, a transient modification in the number of quanta released was observed when the osmotic pressure of the extracellular medium was rapidly increased or decreased (Poulain et al., 1986 b) confirming the above results.

In presence of 10 mM of ACh - a concentration close to that found in the cytoplasm (Katz and Miledi, 1977, see below) - the rate of ACh uptake by the vesicles is slow (with a maximal velocity ranging between 1.6 nmol ACh/min/mg of vesicular protein (Bahr and Parsons, 1983) to 31 nmol ACh/h/mg of vesicular protein (Giompres and Luqmani, 1980). Since the ACh vesicular content ranges from 2 to 5 μmol ACh/mg of vesicular protein (for review see Zimmermann, 1982, 1988) this would mean than intact and filled vesicles can accumulate about 1% of their ACh content in 1 h. The vesicular population, however, is heterogenous and comprises VP1, filled and thus considered releasable, and VP2 which are thougt to represent empty vesicles that can change to VP1 upon refilling their ACh content (see review by Zimmermann, 1988). Thus, the modification of cytoplasmic concentration should affect preferentially the most active population (VP2) which appears to accumulate ACh much more efficiently than the VP1 fraction (by a factor 2-6 fold) (Gracz et al., 1988). However, it would be still too slow to account for the large increase of evoked ACh release after injection of ACh into cholinergic neurons in *Aplysia* (see above).

An important observation was that whether the concentration of ACh in the cytoplasm of the presynaptic neuron was increased or decreased, the size of the released quanta remained unchanged (Tauc and Baux, 1982, Baux and Tauc 1983, Poulain et al., 1986 a and b). Since quanta appeared independent of the transmembrane chemical gradient for ACh, this indicated that ACh was not released through a channel as later confirmed by Young and Chow (1987). It can be concluded that, for a given Ca^{2+} influx, the number of quanta released is controlled by the presynaptic cytoplasmic ACh concentration. Such a control mechanism is probably responsible for the parallel evolution of the concentration of free ACh in the terminal and the instant amplitude of the postsynaptic response observed in *Torpedo* electric organ (Corthay et al., 1982; Dunant et al., 1982).

ALTERNATIVE HYPOTHESES TO THE VESICULAR EXOCYTOSIS

A satisfactory transmitter release hypothesis should be in complete agreement with all experimental results and using a minimum of assumptions it should i) explain the formation of quanta, ii) explain the triggering and regulation of release, iii) identify a morphological correlate of release, and iv) provide a mechanism for the translocation of quanta accross the presynaptic membrane.

Since the exocytosis of vesicular content does not account for many of the experimental observations presented above, two possible models are considered:

1) the released ACh originates from vesicles but is delivered to the extracellular space through a membrane structure, without fusion of the vesicle and nerve terminal membranes.

2) the released ACh originates directly from the cytoplasmic compartment. We propose a working hypothesis which attempts to explain experimental data by attributing the roles of quantum formation and of neurotransmitter extrusion from the cytoplasmic compartment to a presynaptic membrane mechanism, termed *vesigate* (Tauc, 1979; 1982).

1) Release of vesicular content through a membrane pore

As noted above, strong arguments against exocytosis of vesicular content are the timing required for the release process (200 μsec) and the impossibility to recycle the terminal membrane fast enough to allow the sustained release of more than 2,500,000 quanta within 3 sec at *Aplysia* synapses (see above). To overcome these problems it was proposed that vesicles release their neurotransmitter content throughout a membrane structure.

Earlier reports from Ceccarelli's group have shown that under experimental conditions where vesicles are not lost (moderate α-Latrotoxin, presence of calcium), synaptophysin (a membrane bound vesicle protein) is not incorporated into the plasma membrane (Ceccarelli et al., 1988). The proposed interpretations were i) that "axolemma and vesicle membranes can fuse with no extensive intermixing of their molecular component" or this could mean ii) that fusion of membranes does not occur during the release process in physiological conditions. The observation that in mast cells, the first step of exocytosis is the formation of a reversible and transient (flickering) pore (see

Almers, 1990) supported the idea that the release of the vesicle content might not necessarily require the fusion of its membrane with that of the nerve terminal.

Both vesicle and the plasma membrane are supposed to contribute to the structural integrity of the pores. A good candidate protein for the formation of the pore at the level of the vesicles could be synaptophysin or the related synaptoporin (see for reviews Almers, 1990; De Camilli and Jahn, 1990; Matteoli and De Camilli; 1991) because these proteins are integral membrane proteins of vesicles with the ability to form channels in synaptic vesicles (Thomas et al., 1988; Knaus et al., 1990). The plasma membrane counterpart could be the "mediatophore" protein identified by Israël's group (Israël et al., 1981, 1988; Birman et al., 1986) as this molecule is able to release ACh in a Ca^{2+} dependent manner.

Because quanta appear to be released at specific fixed sites of the presynaptic membrane (Kriebel et al., 1990), it may be that privileged sites of release are due to the anchoring of "mediatophore" molecules to the presynaptic membrane cytoskeletton of the active zone. In this hypothesis the release of sub-quanta generating S-MEPPS should be the result of the "flickering" of the pore-channel. Contrary to the classical vesicular hypothesis where the vesicle is the morphological support of the quantum, the size of the quantum would be determined by the duration of the opening of the pore-channel. To explain the observation that not only the cytoplasmic concentration of ACh (Baux and Tauc, 1983; Poulain et al., 1986 a and b) but also that of various cholinomimetics (Baux and Tauc, 1983; Tauc and Baux, 1985; Baux et al., 1986; Baux and Tauc, 1987, Fossier et al., 1990) modulate the number of quanta released, it would be necessary to assume that the membrane component of the pore bears ACh binding sites implicated in the modulation of their triggerig sensitivity. As in the classical vesicular hypothesis, since the released ACh comes from a vesicular store, any modification of the availability of vesicles - such as that induced by changing the phosphorylation state of synapsin I (De Camilli et al., 1990; Lin et al., 1990; Llinas, 1991) - would affect the amount of transmitter released.

However, the hypothesis of the release of ACh from vesicles through a channel cannot account for the above described i) diminution of free-ACh but not of the vesicular ACh during intense stimulation (Dunant et al., 1974; Israël and Lesbat, 1981), ii) rapid mixing up of true and false transmitter (Large and Rang, 1978), iii) constancy of the size of a EPPs repetitively appearing at the same spot of the active terminal region (Kriebel et al., 1990) and iv) slow vesicular uptake of ACh.

2) Vesigate hypothesis

i) **Formation of quanta**. As an alternative to the synaptic vesicle hypothesis (the ACh channel hypothesis being discarded), the other mean to package ACh molecules together to be ejected as a quantum was proposed to be a membrane bound structure or *vesigate* (Tauc, 1979 and 1982) bearing a given number of ACh binding sites. This can be envisaged because a large non-vesicular ACh compartment exists in the presynaptic terminal. Indeed, ACh is synthetised in the cytoplasm (for review see Tucek, 1985) and distributed between the vesicular and cytoplasmic compartments, the "free" ACh representing 30% to 50% of the total ACh content (Dunant et al., 1974; Tauc, 1977; Baux et al., 1979). Inside the terminal, the concentration of ACh is estimated to be between 10 to 50 mM (Israël et al., 1979, Morel et al., 1978; Katz and Miledi, 1977).

To comply with all experimental data, the vesigate should be apposed to the inner face or be a part of the presynaptic membrane. It should be loaded with ACh from the cytoplasmic compartment - ACh bound to the vesigate would be in dynamic equilibrium with free cytoplasmic ACh - and must release ACh extracellularly in a non electrogenic manner. The vesigate should be a structure which binds ACh to saturation before it can operate because quantal size is independent of free ACh concentration. This means that a vesigate will be releasable only when all the binding sites are occupied by ACh or a releasable agonist (carbachol for instance); occupation of some binding sites by a presynaptic intracellular antagonist would lead to an incomplete saturation, thereby preventing release. (Vesicular hypothesis also states that only full-sized quanta can be released (Katz 1969).)

It is reasonable to assume that these "presynaptic intracellular ACh receptors", if they exist, must have some pharmacological properties similar to those of the classical postsynaptic ACh receptors. Accordingly, ACh and carbachol are good intrasynaptic agonists since, when introduced presynaptically they potentiate the release of transmitter (Baux and Tauc, 1983, Tauc and Baux, 1985, Poulain et al. 1986 a and b). Other cholinergic drugs like curare, hexamethonium, hemicholinium-3, succinylcholine, acetylmethylcholine, acetylthiocholine or pralidoxime when intracellularly applied in *Aplysia*, decrease or even block synaptic transmission (Tauc and Baux, 1985; Baux et al., 1986; Baux and Tauc, 1987, Fossier et al;, 1990) without altering ACh quantum size, and therefore can be considered intrasynaptic antagonists. Since decrease in ACh release occured even in the absence of stimulation, it appears that the pool of releasable ACh is in equilibrium with cytoplasmic ACh; this is not the case with synaptic vesicles.

In the vesigate hypothesis, the number of releasable ACh molecules per quantum would be determined by the number of ACh binding sites in the vesigate. Assuming that binding of ACh to any site is independent of the bound-free state of the others, the process might be described with conventional Michaelis-Menten enzymatic kinetics of receptor/ligand interactions: at a given concentration of cytoplasmic ACh the probability (**P**) that a vesigate of (**n**) binding sites will be fully loaded by ACh is:

$$P = p^n \qquad \text{with} \qquad p = [ACh]_i / ([ACh]_i + K_m)$$

(**p**) is the probability of binding of one ACh molecule to a binding site; $[ACh]_i$ is the concentration of cytoplasmic ACh., (**n**) the number of binding sites, which must be identical to the number of ACh molecules of a quantum (5,000 to 12,000 - Kuffler and Yoshikami, 1975; Simonneau et al., 1980; Miledi et al., 1983; Dunant and Müller, 1986). K_m is the dissociation constant for a single receptor. If K_m is considered to be similar to the K_{DL} of postsynaptic ACh receptors ($\approx 2.10^{-6}M$) and assuming 5,000 binding sites per vesigate, the cytoplasmic ACh concentration required to occupy 50% sites would be approximately 20 mM. This concentration is comparable to experimentaly determined values (see above).

This delivery mechanism should be more precise than a vesicle in determining the number of releasable ACh molecules in a quantum and the vesigate hypothesis thus predicts that quanta of constant size should be released at a given presynaptic site (i.e. active zone, see below). Accordingly, appearance of MEPPs of identical amplitude and released at the same site were effectively detected by focal microelectrodes (Kriebel et al., 1990). Such an observation cannot be accounted for by the vesicle hypothesis; since vesicles vary sligtly in size they necessarily exhibit variations in their quantal content and would fuse randomly with the presynaptic membrane during transmitter release.

ii) Triggering and regulation of the release. The vesigate hypothesis fully agrees with the role of Ca^{2+} influx and high instantaneous Ca^{2+} concentrations at critical zones as a trigger for transmitter release (see reviews by Augustine et al., 1987; Llinas, 1991). However, contrary to the vesicular hypothesis in which releasable neurotransmitter is prepackaged into a vesicle serving as a stable store, the vesigate hypothesis does not assume the formation of stable quanta in the terminal. Rather, continuous exchange of the transmitter occurs between the vesigate ACh binding sites and the cytoplasmic free ACh pool. As a critical number of binding sites may or may not be occupied by ACh at any moment, the status of a vesigate will continually change from releasable to non-releasable. Statistically, however, the number of releasable vesigates for a given cytoplasmic concentration of transmitter would remain constant. As a consequence the number of quanta released should be regulated by the free ACh concentration as effectively observed (Baux and Tauc, 1983; Poulain et al., 1986 a and b).

iii) **Identification of vesigates.** Because it is unlike that a macromolecule binds simultaneously thousands of ACh molecule, vesigate should necessarily result from the coordination of several subunits creating a kind of presynaptic network. It can be expected that such a complex macromolecular structure would be visible ultrastructurally. It is tempting to speculate that there is a relationship between the vesigate and the active zone. In electron micrographs of central synapses, the prominent feature facing the postsynaptic receptors is indeed the presynaptic densification or active zone which Gray and Akert (Gray, 1976) identified in some terminals and termed "presynaptic grid". The repeated pattern of the "presynaptic grid" would provides a morphological support for the existence of vesigate subunits. These subunits may also explain the sub-miniature postsynaptic potentials observed at neuromuscular junctions and *Torpedo* electric organ that reflect the release of sub-quanta (Kriebel et al., 1990; Müller and Dunant, 1985).

The identification of the vesigate as the active zone implies that one active zone is able to release only one quantum at a time. Indeed, this point was most clearly demonstrated at inhibitory synapses on the Mauthner cell of the goldfish where synaptic contacts consist in only few boutons each having a single active zone. In this preparation, the number of quanta released at a time is never more and usually less than the total number of boutons (Faber and Korn, 1985; Korn and Faber, 1991).

iv) **Translocation of ACh - Role of synaptic vesicles.** The vesigate hypothesis excludes vesicular exocytosis and simple ACh channels, but gives no indication of how the transmitter is translocated accross the presynaptic membrane. The fact that the vesigate is considered to be a structure directly associated with the presynaptic membrane and that it is fully loaded by the neurotransmitter when releasable contribute to the rapidity of the release process. The protein "mediatophore" isolated from the membrane of *Torpedo* synaptosomes (Birman et al., 1986) and from rat brain (Israël et al., 1988) or/and another presynaptic protein studied by Eder-Colli et al. (1989) may function as a part of the vesigate and play a part in the translocation of ACh.

The vesigate hypothesis does not underestimate the role of synaptic vesicles. The control of synaptic transmission requires that the release of transmitter not only starts, but also stops as quickly as possible. Therefore calcium must be removed rapidly from the proximity of the release sites. Synaptic vesicles are strategically located near the active zones in great numbers and can take up Ca^{2+} very efficiently (Michaelson et al., 1980, Israël et al., 1980); their exocytosis will then extrude Ca^{2+} from the terminal. This may explain why under non-physiological conditions that potentiate considerably the entry of Ca^{2+} a correlation was observed between the number of quanta released and the number of exocytic infoldings as seen by electron microscopy (Tarro-Tarelli et al., 1985; Heuser, 1989) and why there is a correlation between vesicle loss and blockade of quantal release in some experimental conditions (reviewed by Ceccarelli and Hurlbut, 1980, Hurlbut, 1989). However, in physiological conditions, sequestred Ca^{2+} can be slowly released from vesicles into the cytoplasm with the vesicles thus recovering its initial condition. This agrees with the observation that in normally stimulated synapses the exocytotic pits are practically never found on electron micrographs (Heuser, 1974) and the membrane bound vesicle protein synaptophysin is not present in the plasma membrane (Ceccarelli et al., 1988).

Modification of the availability of free vesicles by changes in the phosphorylation status of synapsin I affects the release of neurotransmitter (De Camilli et al., 1990; Lin et al., 1990; Llinas, 1991). This could be a consequence of changes in the capacity of vesicles to buffer intraterminal Ca^{2+}. In Aplysia buccal synapses, blocking the vesicular uptake of Ca^{2+} with 2,5-diterbutyl 1,4-benzohydroquinone (tBuBHQ), facilited quantal evoked release (Fossier et al., 1992) thus confirming the important calcium buffering role of synaptic vesicles in neurotransmitter release. This, naturally, does not preclude that vesicles serve as a storage site for transmitter and have some other additional functions.

CONCLUSION

The alternatives to the vesicle exocytosis hypothesis assume that, under normal conditions, vesicles do not translocate the transmitter across the presynaptic membrane at fast synapses. It is plausible that quantal release of neurotransmitter is performed by a more complex process. One possibility is the delivery of vesicle content throughout a pore-channel structure, the duration of the opening of the channel being responsible of the size of the quantum. This hypothesis, however, is not compatible with the slow uptake of ACh by synaptic vesicles limiting the number of fully loaded and thus releasable vesicles during intense stimulation.

Other alternative proposes that transmitter release is performed by a macromolecular membrane structure or *vesigate*, composed of several subunits that possess binding sites for ACh and may be morphologically identified as presynaptic active zones. Contrary to the vesicular hypothesis of neurotransmitter release, for which a quantum is confined to a vesicle, the vesigate hypothesis does not assume that there is formation of stable quanta in the terminal. Rather, continuous exchange of the transmitter occurs between the vesigate ACh binding sites and the ACh cytoplasmic pool, continuously changing the status of a vesigate from releasable to non releasable. In this hypothesis, synaptic vesicles have multiple functions, not only as a storage for ACh but also as a buffer of Ca^{2+} that enters the terminal close to release sites during stimulation.

It is not without interest for scientists on the quest for the intracellular target of the action of clostridial neurotoxins which view they adopt on the transmitter release mechanism. Depending on the hypothesis adopted, the experimental approach will be preferentially directed towards the fate of synaptic vesicles in one case, towards the plasma membrane in the other case. It seems advisable to avoid certainties and look at both sides in order to minimize the chances of finding that, during many years of hard work, one has been barking up the wrong tree.

REFERENCES

Almers, W., 1990, Exocytosis, *Annu. Rev. Physiol.* 52:607.

Almers, W., Breckenridge, L. J., Iwata, A., Lee, A. K., Spruce, A. E. and Tse, F. W., 1991, Millisecond studies of single membrane fusion events, *Ann. N. Y. Acad. Sci.* 635:318.

Almers, W. and Tse, F. W., 1990, Transmitter release from synapses: does a preassembled fusion pore initiate exocytosis, *Neuron* 4:813.

Anderson, D. C., King, S. C. and Parsons, S. M., 1983, Inhibition of ^{3}H-acetylcholine active transport by tretraphenylborate and other anions, *Molecular Pharmacol.* 24:55.

Augustine, G. J., Charlton, M. P. and Smith, S. J., 1987, Calcium action in transmitter release, *Annu. Rev. Neurosci.* 10:633.

Bahr, B. A. and Parsons, S. M., 1986, Acetycholine transport and drug inhibition kinetics in Torpedo synaptic vesicle, *J. Neurochem.* 46:1214.

Baux, G., Poulain, B. and Tauc, L., 1986, Quantal analysis of action of hemicholinium-3 studied at a central cholinergic synapse of Aplysia, *J. Physiol. (Lond.)* 380:209.

Baux, G., Simonneau, M. and Tauc, L., 1979, Transmitter release : ruthenium red used to demonstrate a possible role of sialic acid containing substrates, *J. Physiol. (Lond.)* 291:161.

Baux, G. and Tauc, L., 1983, Carbachol can be released at a cholinergic ganglionic synapse as a false transmitter, *Proc. Natl. Acad. Sci. USA* 80:5126.

Baux, G. and Tauc, L., 1987, Presynaptic actions of curare and atropine on quantal acetylcholine release at a central synapse of Aplysia, *J. Physiol. (Lond.)* 388:665.

Betz, W. J. and Bewick, G. S., 1992, Optical analysis of synaptic vesicle recycling at the frog neuromuscular junction, *Science* 255:200.

Bevan, S., 1976, Subminiature end-plate potentials at contracted neuromuscular junctions, *J. Physiol. (Lond.)* 258:145.

Birman, S., Israël, M., Lesbats, B. and Morel, N., 1986, Solubilization and partial purification of a presynaptic membrane protein ensuring calcium-dependent acetylcholine release from proteoliposomes, *J. Neurochem.* 47:433.

Cabeza, R. and Collier, B., 1988, Acetylcholine mobilization in a sympathetic ganglion in the presence and absence of 2-(4-phenylpiperidino)cyclohexanol (AH5183), *J. Neurochem.* 50:112.

Ceccarelli, B. and Hurlbut, W. P., 1980, Vesicle hypothesis of the release of quanta of acetylcholine, *Physiol. Rev.* 60:396.

Ceccarelli, B., Hurlbut, W. P. and Mauro, A., 1973, Turnover of transmitter and synaptic vesicles at the frog neuromuscular junctions, *J. Cell. Biol.* 57:499.

Ceccarelli, B., Valtorta, F. and Hurlbut, W., 1988, New evidence supporting the vesicle hypothesis for quantal secretion at the neuromuscular junction, *in:* "Cellular and Molecular Basis of Synaptic Transmission" H. Zimmermann, ed., NATO ASI Series 21:51.

Collier, B., 1969, The preferential release of newly synthetized transmitter by a sympathetic ganglion, *J. Physiol. (Lond.)* 205:341.

Cooper, J. R. and Meyer, E. M., 1984, Possible mechanisms involved in the release and modulation of release of neuroactive agents, *Neurochem. Int* 6:419.

Corthay, J., Dunant, Y. and Loctin, F., 1982, Acetylcholine changes underlying transmission of a single nerve impulse in the presence of 4-aminopyridine in Torpedo, *J. Physiol. (Lond.)* 325:461.

Couteaux, R. and Pécot-Dechavassine, M., 1973, Données ultrastructurales et cytochimiques sur le mécanisme de la libération de l'acétylcholine dans la transmission synaptique, *Arch. Ital. Biol.* 3:231.

De Camilli, P., Benfenati, F., Valtorta, F. and Greengard, P., 1990, The synapsins, *Annu. Rev. Cell. Biol.* 6:433.

De Camilli, P. and Jahn, R., 1990, Patway to regulated exocytosis in neurons, *Annu. Rev. Physiol.* 52:625.

Del Castillo, J. and Katz, B., 1957, La base "quantale" de la transmission neuromusculaire, *in:* "Microphysiologie comparée des éléments excitables" Colloq. Int. C.N.R.S. A. Fessard, ed. Paris 67:245.

De Robertis, E. D. P., Pellegrino de Iraldi, A., Rodriguez de Lores Arnais, G. and Salganicoff, L., 1962, Cholinergic and non-cholinergic nerve endings in rat brain I. Isolation and subcellular distribution of acetylcholine and acetylcholinesterase, *J. Neurochem.* 9:23.

Dolly, J. O., 1992, Peptide toxins that alter neurotransmitter release, *in:* "Handbook of Experimental Pharmacology", H. Herkin and F. Hucho, eds., Springer, Berlin.

Dreyer, F., Mallart, A. and Brigant, J. L., 1983, Botulinum A and tetanus toxin do not affect presynaptic membrane currents in mammalian motor nerve endings, *Brain Research* 270:373.

Dunant, Y., 1986, On the mechanism of acetylcholine release, *Progress in Neurobiol.* 26:55.

Dunant, Y., Esquerda, J. E., Loctin, F., Marsal, J. and Müller, D., 1987, Botulinum toxin inhibits quantal acetylcholine release and energy metabolism in the *Torpedo* electric organ, *J. Physiol. (Lond.)* 385:677.

Dunant, Y., Gautron, J., Israël, M., Lesbats, B. and Manaranche, R., 1974, Evolution de la décharge de l'organe électrique de la Torpille et variations simultanées de l'acétylcholine au cours de la stimulation, *J. Neurochem.* 23:635.

Dunant, Y., Jones, G. J. and Loctin, F., 1982, Acetylcholine measured at short time intervals during transmission of nerve impulses in the electric organ of Torpedo, *J. Physiol. (Lond.)* 325:441.

Dunant, Y. and Müller, D., 1986, Quantal release of acetylcholine evoked by focal depolarization at the Torpedo nerve-electroplaque junction, *J. Physiol. (Lond.)* 379:461.

Eder-Colli, L., Briand, P. A., Pellegrinelli, N. and Dunant, Y., 1989, A monoclonal antibody raised against plasma membrane of cholinergic nerve terminals of the Torpedo inhibits choline acetyltransferase activity and acetylcholine release, *J. Neurochem.* 53:1419.

Edwards, C., Dolezal, V., Tucek, S., Zemkova, H. and Vyskocil, F., 1985, Is an acetylcholine transport system responsible for nonquantal release of acetylcholine at the rodent myoneural junction?, *Proc. Natl. Acad. Sci. USA* 82:3514.

Erxleben, C. and Kriebel, M. E., 1988, Subunit composition of the spontaneous miniature end-plate currents at the mouse neuromuscular junction, *J. Physiol. (Lond.)* 400:659.

Estrella, D., Green, K. L., Prior, C., Dempster, J., Halliwell, R. F., Jacobs, R. S., Parsons, S. M., Parsons, R. L. and Marshall, I. G., 1988, A further study of the neuromuscular effects of vesamicol (AH5183) and of its enantiomer specificity, *Br. J. Pharmacol.* 93:759.

Faber, D. S. and Korn, H., 1985, Binary mode of transmitter release at central synapses, *T.I.N.S.* 5:157.

Fesce, R., Segal, J. R., Ceccarelli, B. and Hurlbut, W. P., 1986, Effects of black widow spider venom and Ca^{2+} on quantal secretion at the frog neuromuscular junction, *J. Gen. Physiol.* 88:59.

Flechter, P. and Forrester, T., 1975, The effect of curare on the release of acetylcholine from mammalian nerve terminals and an estimate of the quantum content, *J. Physiol. (Lond.)* 251:131.

Fossier, P., Baux, G., Poulain, B. and Tauc, L., 1990, Receptor-mediated presynaptic facilitation of quantal release of acetylcholine induced by pralidoxime in Aplysia, *Cell. Molecular Neurobiol.* 10:383.

Fossier, P., Baux, G., Trudeau, L. E. and Tauc, L., 1992, Involvement of Ca^{2+} uptake by a reticulum-like store in the control of transmitter release, *Neuroscience*, in press.

Giompres, P. E. and Luqmani, Y. A., 1980, Cholinergic vesicle isolated from *Torpedo marmorata*: demonstration of acetylcholine and choline uptake in an *in vitro* system, *Neuroscience* 5:1041.

Gracz, L. M., Wang, W. C. and Parsons, S. M., 1988, Cholinergic synaptic vesicle heterogeneity: Evidence for regulation of acetylcholine transport, *Biochemistry* 27:5268.

Gray, E. G., 1976, Problems of understanding the substructure of synapses, *in:* ''Progress in Brain Research. Perspectives in brain research'', M.A. Corner and D.F. Swaab, Eds., Elsevier, Amsterdam 45:207.

Grinnell, A. D., Gundersen, C. B., Meriney, S. D. and Young, S. H., 1989, Direct measurement of ACh release from exposed frog nerve terminals: constraints on interpretation of non-quantal release, *J. Physiol. (Lond.)* 419:225.

Heuser, J. E., 1989, Review of electron microscopic evidence favouring vesicle exocytosis as the structural basis for quantal release during synaptic transmission, *Quaterly J. Exp. Physiol.* 74:1051.

Heuser, J. E. and Reese, T. S., 1981, Structural changes after transmitter release at the frog neuromuscular junction, *J. Cell Biol.* 88:564.

Heuser, J. E., Reese, T. S. and Landis, D. M. D., 1974, Functional changes in frog neuromuscular junction studied with freeze fracture, *J. Neurocytol.* 3:109.

Hubbard, J. I. and Schmidt, R. F., 1963, An electrophysiological investigation of mammalian motor nerve terminals, *J. Physiol. (Lond.)* 166:145.

Hurlbut, W. P., 1989, The correlation between vesicle loss and quantal secretion at the neuromuscular junction, *Cell Biol. Int. Reports* 13:1053.

Israël, M., Dunant, Y. and Manaranche, R., 1979, The present status of the vesicular hypothesis, *Progress in Neurobiology* 13:237.

Israël, M., Lesbat, B., Manaranche, R. and Morel, N., 1983, Acetylcholine release from proteoliposomes equipped with synaptosomal membrane constituents, *Biochim. Biophys. Acta* 728:438.

Israël, M. and Lesbats, B., 1981, Continuous determination by a chemiluminescent method of acetylcholine release and compartmentation in Torpedo electric organ synaptosomes, *J. Neurochem.* 37:1475.

Israël, M., Lesbats, B., Morel, N., Manaranche, R. and Le Gal la Salle, G., 1988, Is the acetylcholine releasing protein mediatophore present in rat brain?, *FEBS Letters* 233:421.

Israël, M. and Manaranche, R., 1985, The release of acetylcholine: from a cellular towards a molecular mechanism, *Biol. Cell* 55:1.

Israël, M., Manaranche, R., Marsal, J., Meunier, F. M., Morel, N., Frachon, P. and Lesbats, B., 1980, ATP-dependent calcium uptake by cholinergic synaptic vesicles isolated from Torpedo electric organ, *J. Membrane Biol.* 54:115.

Israël, M., Manaranche, R., Morel, N., Dedieu, J. C., Gulik-Krzywicki, T. and Lesbats, B., 1981, Redistribution of intramembrane particles in conditions of acetylcholine release by cholinergic synaptosomes, *J. Ultrastruct. Res.* 75:162.

Israël, M. and Morel, N., 1990, Mediatophore: a nerve terminal membrane protein supporting the final step of the actetylcholine release process, *in:* ''Progress in Brain Research'', S.M. Aquilonius and P.G. Gillerg, eds., 84:101.

Katz, B., 1969, ''The Release of Neural Substances'' Liverpool University Press, Liverpool.

Katz, B. and Miledi, R., 1977, Transmitter leakage from motor nerve endings, *Proc. R. Soc. Lond. Ser. B* 196:59.

Katz, B. and Miledi, R., 1979, Estimates of quantal content during ''chemical potentiation'' of transmitter release, *Proc. R. Soc. Lond. Ser. B* 205:369.

Kelly, R. B., 1991, Secretory granule and synaptic vesicle formation, *Cell Biol.* 3:654.

Knaus, P., Marquez-Pouey, B., Scherer, H. and Betz, H., 1990, Synaptoporin, a novel putative channel protein of synaptic vesicles, *Neuron* 5:453.

Korn, H. and Faber, D. S., 1991, Quantal analysis and synaptic efficacy in the CNS, *T.I.N.S.* 14:439.

Kriebel, M. E., 1978, Small mode miniature end plate potentials are increased and evoked in fatigued preparations and in high Mg^{2+} saline,, *Brain Research* 148:381.

Kriebel, M. E. and Gross, C. E., 1974, Multimodal distribution of frog miniature endplate potentials in adult, denervated and tadpole leg muscle, *J. Gen. Physiol.* 64:85.

Kriebel, M. E., LLados, F. and Carlson, C. G., 1980, Effect of the Ca^{2+} ionophore X-537A at heat challenge on the distribution of mouse MEPP amplitude histograms, *J. Physiol. (Paris)* 76:435.

Kriebel, M. E., Llados, F. and Matteson, D. R., 1976, Spontaneous subminiature endplate potentials in mouse diaphragm muscle: evidence for synchronous release, *J. Physiol. (Lond.)* 262:553.

Kriebel, M. E., Vautrin, J. and Holsapple, J., 1990, Transmitter release: prepackaging and random mechanism or dynamic and deterministic process, *Brain Res. Rev.* 15:167.

Kuffler, S. W. and Yoshikami, D., 1975, The number of transmitter molecules in a quantum: an estimate from ionophoretic application of acetylcholine at the neuromuscular synapse, *J. Physiol. (Lond.)* 251:465.

Landmesser, L. and Pilar, G., 1972, The onset and development of transmission in the chick ciliary ganglion, *J. Physiol. (Lond.)* 222:691.

Large, W. A. and Rang, H. P., 1978, Factors affecting the rate of incorporation of a false transmitter into mammalian nerve terminals, *J. Physiol. (Lond.)* 285:1.

Lin, J. W., Sugimori, M., Llinas, R. R., McGuinness, T. L. and Greengard, P., 1990, Effect of Synapsin I and Ca^{2+}/Calmodulin dependent protein kinase II on spontaneous neurotransmitter release in the squid giant synapse, *Pro. Natl. Acad. Sci. USA* 87:8257.

Llinas, R. R., 1977, Calcium and transmitter release in squid synapse, *Soc. Neurosci.* 2:139.

Llinas, R. R., 1991, Depolarization release coupling: an overview, *Ann. N. Y. Acad. Sci.* 635:3.

Llinas, R. R. and Heuser, J. E., 1978, Depolarization-release coupling system in neurons, *Res. Program Bull.* 15:557.

Llinas, R., Sugimori, M. and Silver, R. B., 1992, Microdomains of high calcium concentration in a presynaptic terminal, *Science* 256:677.

Marchbanks, R. M., 1979, Role of storage vesicles in synaptic transmission, *Soc. Exp. Biol. Symp. XXXII. "Secretory mechanisms"*, Hopkins and Duncan, eds., :251.

Marchbanks, R. M. and Israël, M., 1972, The heterogeneity of bound acetylcholine and synaptic vesicles, *Biochem. J.* 129:1049.

Marsal, J., Solona, C., Rabasseda, X., Blasi, J. and Casanova, A., 1987, Depolarization-induced release of ATP from cholinergic synaptosomes is not blocked by botulinum toxin type A, *Neurochem. Int.* 10:295.

Marshall, I. G. and Parsons, S. M., 1987, The vesicular acetylcholine transport system, *T.I.N.S.* 10:174.

Matteoli, M. and De Camilli, P., 1991, Molecular mechanisms in neurotransmitter release, *Curr. Opinion Neurobiol.* 1:91.

Maycox, P. R., Hell, J. W. and Jahn, R., 1990, Amino acid neurotransmission: spotlight on synaptic vesicles, *T.I.N.S.* 13:83.

Melega, W. P. and Howard, B. D., 1984, Biochemical evidence that vesicles are the source of the acetylcholine released from stimulated PC12 cells, *Proc. Natl. Acad. Sci. USA* 81:6535.

Meunier, F. M., Israël, M. and Lesbats, B., 1975, Release of ATP from stimulated nerve electroplaque junctions, *Nature* 257:407.

Michaelson, D. M., Avissar, S., Ophir, I., Pinchasi, I., Angel, I., Kloog, Y. and Sokolovsky, M., 1980, On the regulation of acetylcholine release : a study utilizing Torpedo synaptosomes and synaptic vesicles, *J. Physiol. (Lond.)* 76:505.

Michaelson, D. M. and Burstein, M., 1985, Biochemical evidence that acetylcholine release from cholinergic nerve terminals is mostly vesicular, *FEBS Lett.* 188:389.

Miledi, R., Molenaar, P. C. and Polak, R. L., 1980, The effect of lanthanum ions on acetylcholine in frog muscle, *J. Physiol. (Lond.)* 309:199.

Miledi, R., Molenaar, P. C. and Polak, R. L., 1982, Free and bound acetylcholine in frog muscle, *J. Physiol. (Lond.)* 333:189.

Miledi, R., Molenaar, P. C. and Polak, R. L., 1983, Electrophysiological and chemical determination of acetylcholine release at the frog neuromuscular junction, *J. Physiol. (Lond.)* 334:245.

Molgo, J., Siegel, L. S., Tabti, N. and Thesleff, S., 1989, A study of synchronization of quantal transmitter release from mammalian motor endings by the use of botulinal toxins type A and D, *J. Physiol. (Lond.)* 411:195.

Morel, N., Israël, M. and Manaranche, R., 1978, Determination of ACh concentration in Torpedo synaptosomes, *J. Neurochem.* 30:1553.

Morel, N. and Meunier, F. M., 1981, Simultaneous release of acetylcholine and ATP from stimulated cholinergic synaptosomes, *J. Neurochem.* 36:1766.

Morot Gaudry-Talarmain, Y.,Diebler, M. F. and 0'Regan, S,1989, Compared effects of two vesicular acetylcholine uptake blockers AH5183 and cetiedil on cholinergic functions in Torpedo synaptosomes: acetylcholine synthesis choline transport vesicular uptake and evoked acetylcholine release, *J. Neurochem.* 52:822.

Müller, D. and Dunant, Y., 1985, Subminiature electroplaque potentials are present in Torpedo electric organ, *Experientia* 41:824.

Müller, D., Garcia-Segura, L. M., Parducz, A. and Dunant, Y., 1987, Brief occurence of a population of presynaptic intramembrane particles coincides with transmission of a nerve impulse, *Proc. Natl. Acad. Sci. USA* 84:590.

Neher, E. and Marty, A., 1982, Discrete changes of cell membrane capacitance observed under conditions of enhanced secretion in bovine adrenal chromaffin cells, *Proc. Natl. Acad. Sci. USA* 79:6712.

Nicholls, D. G., 1989, Release of glutamate, aspartate, and -aminobutyric acid from isolated nerve terminals, *J. Neurochem.* 52:331.

Niemann, H., 1991, Molecular biology of clostridial toxins, *in:* "Sourcebook of Bacterial Protein Toxin", Alouf J. and Freer J., eds., Academic Press, Lond.

Penner, R., Neher, E. and Dreyer, F., 1986, Intracellularly injected tetanus toxin inhibits exocytosis in bovine adrenal chromaffin cells, *Nature* 324:76.

Potter, L. T., 1970, Synthesis, storage and release of ^{14}C-acetylcholine in isolated rat diaphragm muscles, *J. Physiol. (Lond.)* 206:145.

Poulain, B., Baux, G. and Tauc, L., 1986, Presynaptic transmitter content controls the number of quanta released at a neuro-neuronal cholinergic synapse, *Proc. Natl. Acad. Sci. USA* 83:170.

Poulain, B., Fossier, P., Baux, G. and Tauc, L., 1987, Hemicholinium-3 facilitates the release of acetylcholine by acting on presynaptic nicotinic receptors at a central synapse in Aplysia, *Brain Res.* 435:63.

Poulain, B. and Molgo, J., 1992, Botulinal neurotoxins. Methods used in the study of their mode of action on neuro-transmitter release, *in:* "Methods in Neurosciences", P. M. Conn, ed., Academic Press, San Diego.

Poulain, B., Tauc, L., Maisey, E. A., Wadsworth, J. D. F., Mohan, P. M. and Dolly, J. O., 1988, Neurotransmitter release is blocked intracellularly by botulinum neurotoxin, and this requires uptake of both toxin polypeptides by a process mediated by the larger chain, *Proc. Natl. Acad. Sci. USA* 85:4090.

Simonneau, M., Tauc, L. and Baux, G., 1980, Quantal release of acetylcholine examined by current fluctuation analysis at an identified neuro-neuronal synapse of Aplysia, *Proc. Natl. Acad. Sci. USA* 77:1661.

Tauc, L., 1977, Transmitter release at cholinergic synapses, *in:* "Synapses", G.A. Cottrell and P.N.R. Usherwood, eds., Blackie, Glasgow.

Tauc, L., 1979, Are vesicles necessary for release of acetylcholine at cholinergic synapses ?, *Biochem.Pharmacol.* 27:3493.

Tauc, L., 1982, Nonvesicular release of neurotransmitter, *Physiol. Rev.* 62:857.

Tauc, L. and Baux, G., 1980, Libération du transmetteur synaptique: examen critique de la theorie vésiculaire, *in:* "La transmission neuromusculaire. Les médiateurs et le milieu intérieur", Colloque Claude Bernard, Masson Paris.

Tauc, L. and Baux, G., 1982, Are there intracellular acetylcholine receptors in the cholinergic synaptic nerve terminals ?, *J. Physiol. (Paris)* 78:366.

Tauc, L. and Baux, G., 1985, Mechanism of acetylcholine release at neuro-neuronal synapses, *in:* "Nonvesicular transport", S.S.Rothman and J.J.L. Ho, eds., Willey and Sons, New York.

Tauc, L., Hoffmann, A., Tsuji, S., Hinzen, D. H. and Faille, L., 1974, Transmission abolished on a cholinergic synapse after injection of acetylcholinesterase into the pre-synaptic neurone, *Nature* 250:496.

Tauc, L. and Poulain, B., 1991, The non-vesicular hypothesis of quantal release of neurotransmitters, *in:* "Presynaptic Regulation of Neurotransmitter Release: A Handbook", M. Hanani and J. Feigenbaum, eds., Freund Publishing House Lond.

Tauc, L. and Poulain, B., 1991, Vesigate hypothesis of neurotransmitter release explains the formation of quanta by a non-vesicular mechanism, *Physiol. Res.* 40:279.

Thesleff, S., 1986, Different kinds of acetylcholine release from the motor nerve, *Int. Rev. Neurobiol.* 28:59.

Thomas, L., Hartung, K., Langosch, D., Rehm, H., Bamberg, E., Francke, W. W. and Betz, H., 1988, Identification of synaptophysin as a hexameric channel protein of the synaptic vesicle membrane, *Science* 242:1050.

Torri-Tarelli, F., Grohovaz, F., Fesce, R. and Ceccarelli, B., 1985, Temporal coincidence between synaptic vesicle fusion and quantal secretion of acetylcholine, *J. Cell Biol.* 101:1386.

Tremblay, J. P., Laurie, R. E. and Colonnier, M., 1983, Is the MEPP due to the release of one vesicle or to simultaneous release of several vesicles at one active zone?, *Brain Res. Rev.* 6:299.

Tucek, S., 1985, Regulation of acetylcholine synthesis in the brain, *J. Neurochem.* 44:11.

Unsworth, C. D. and Johson, R. G., 1990, Acetylcholine and ATP are coreleased from the electromotor nerve terminals of Narcine brasiliensis by an exocytotic mechanism, *Proc. Natl. Acad. Sci. USA* 87:553.

Uvnäs, B., 1985, Is exocytosis an initial or final event in the release of biogenic amines?, *In:* "Nonvesicular transport", S.S. Rothman and J.J.L. Ho eds., Willey and Sons, New York.

Valtorta, F., Fesce, R., Grohovaz, F., Haimann, C., Hurlbut, W. P., Iezzi, N., Torri Tarelli, F., Villa, A. and Ceccarelli, B., 1990, Neurotransmitter release and synaptic recycling, *Neuroscience* 35:477.

Van der Kloot, W., 1988, Acetylcholine quanta are released from vesicles by exocytosis (and why some think not), *Neuroscience* 24:1.

Van der Kloot, W., 1991, The regulation of quantal size, *Progress in Neurobiol.* 36:93.

Vautrin, J. and Membrini, J., 1981, Caractéristiques du potentiel unitaire de plaque motrice de la grenouille, *J. Physiol. (Paris)* 77:999.

Vizi, E. S., 1984, Critique. Physiological role of cytoplasmic and non-synaptic release of transmitter, *Neurochem. Int.* 6:435.

Volknandt, W. and Zimmermann, H., 1986, Acetylcholine, ATP, and proteoglycan are common to synaptic vesicles isolated from the electric organs of electric eel and electric catfish as well as from rat diaphragm, *J. Neurochem.* 47:1449.

Wellhöner, H. H., 1992, Tetanus and botulinum Neurotoxins, *in:* "Handbook of Experimental Pharmacology", H. Herkin and F. Hucho, eds., Springer, Berlin.

Wernig, A. and Stirner, H., 1977, Quantum amplitude distributions points to functional unity of the synaptic 'active zone', *Nature* 268:820.

White, T. D., Potter, P. and Wonnacott, S., 1980, Depolarization induced release of ATP from cortical synaptosomes is not associated with acetylcholine release, *J. Neurochem.* 34:1109.

Whittaker, V. P., 1986, The storage and release of acetylcholine, *T.I.P.S.* 7:312.

Whittaker, V. P., Essman, W. B. and Dowe, G. H. C., 1972, The isolation of pure cholinergic synaptic vesicles from the electric organs of elasmobranch fish of the family Torpinidae, *Biochem. J.* 128:833.

Whittaker, V. P., Michaelson, I. A. and Kirkland, R. J. A., 1963, The separation of synaptic vesicles from disrupted nerve ending particles, synaptosomes, *Biochem. Pharmacol.* 12:300.

Young, S. H. and Chow, I., 1987, Quantal release of transmitter is not associated with channel opening on the neuronal membrane, *Science* 238:1712.

Zimmerberg, J., Curran, M. and Cohen, F. S., 1991, A lipid/protein complex hypothesis for exocytotic fusion pore formation, *Ann. N. Y. Acad. Sci* 635:307.

Zimmermann, H., 1982, Biochemistry of the isolated cholinergic vesicles, *in:* "Neurotransmitter vesicles", R.L. Klein, H. Lagercrantz and H. Zimmermann, eds., Academic Press, New York.

Zimmermann, H., 1988, Cholinergic synaptic vesicles, *in:* "The cholinergic synapse", V.P. Whittaker ed., Springer-Verlag, Berlin.

IDENTIFICATION OF PROTEINS REQUIRED
FOR Ca²⁺-TRIGGERED SECRETION

Thomas F.J. Martin, Jane H. Walent, Bruce W. Porter, and Jesse C. Hay

Department of Zoology
University of Wisconsin
Zoology Research Building
Madison, WI 53706

INTRODUCTION

Eukaryotic cells synthesize proteins whose final destinations are the extracellular space, the plasma membrane or intracellular organelles. The constitutive secretory pathway employed to sort proteins consists of a series of inter-organelle transport steps that are vesicle-mediated. The cellular mechanisms employed in this pathway are under active investigation since such studies promise to reveal insights into the poorly understood biochemical events involved in vesicle budding, vesicle targeting and membrane fusion. Individual vesicle transport steps of the constitutive pathway from the endoplasmic reticulum to plasma membrane fusion have been successfully reconstituted in semi-intact cells or in homogenates.[1] These *in vitro* reactions typically exhibit a requirement for ATP and cytosolic proteins[2], and several cytosolic proteins have been purified and characterized, including the 76 kD NSF protein, 35-39 kD SNAP proteins, a 25 kD POP protein, low molecular weight GTP-binding proteins, and components of a non-clathrin coat.[3,4] Genetic approaches in *Saccharomyces cerevisiae* have complemented the *in vitro* reconstitution approach, and a large number of genes required in the yeast secretory pathway have been identified by mutational analysis.[5] Although the biochemical events underlying membrane fusion have eluded detailed description thus far, the identification of several key protein participants represents an important step toward this goal.

All cells possess a constitutive secretory pathway; however, endocrine, neural and exocrine cells also possess a regulated secretory pathway utilized for the export of hormones, neurotransmitters and enzymes.[6] The regulated pathway diverges from the constitutive pathway in the trans Golgi where constitutive vesicles and the immature granules of the regulated pathway are formed.[7] Whereas vesicle-plasma membrane fusion in the constitutive pathway does not require Ca²⁺, the distal steps involving exocytotic fusion of secretory granules in the regulated pathway are Ca²⁺-triggered.[8] It remains to be determined whether secretory granule translocation, fusion or both are Ca²⁺-dependent-processes. Moreover, it is unclear whether

Botulinum and Tetanus Neurotoxins, Edited by
B.R. DasGupta, Plenum Press, New York, 1993

granule targeting and fusion in the regulated pathway involve similar biochemical reactions to those in the constitutive pathway, or whether components are shared between the pathways. Dissection of the molecular requirements of the regulated secretory pathway will provide answers to these questions as well as insights into the biochemical events involved in vesicle/ granule targeting and membrane fusion. This chapter will briefly review our recent progress in identifying several proteins required for regulated secretion in neuroendocrine cells.

Ca^{2+}-ACTIVATED SECRETION IN SEMI-INTACT CELLS

Genetic approaches to the study of the regulated secretory pathway remain limited to protozoans.[9] It has not yet been possible to preserve the Ca^{2+}-regulated fusion of secretory granules with the plasma membrane in homogenates of mammalian secretory cells. It is likely that cytoskeletal or other structural elements required for vectorial fusion are not preserved during homogenization. It has, however, been possible to study regulated secretion in cells permeabilized by either detergents or bacterial lysins.[10-12] In order to eliminate uncertainties associated with the use of membrane-active permeabilizing agents, we developed a mechanical method for permeabilizing cells called cell cracking.[13] A single pass of cells through a narrow clearance ball homogenizer was found to generate a tear in the plasma membrane, creating cells that are freely permeable to solutes including macromolecules.[13-15] Following washing, the cracked cells, termed cell ghosts, are extensively depleted of soluble constituents but retain a normal cytoplasmic ultrastructure with intact secretory granules.[14] We have employed cell cracking with two neuroendocrine cell types: prolactin-secreting GH$_3$ pituitary cells[14] and norepinephrine(NE)-secreting PC12 adrenal cells.[15-18] In both cases, Ca^{2+} stimulated the release of secretory granule contents from cell ghosts. Additional requirements for Ca^{2+}-activated secretion in cell ghosts are described below.

The cumulative evidence indicates that the Ca^{2+}-activated release of secretory granule constituents from GH$_3$ or PC12 cell ghosts faithfully represents the process of regulated secretion observed in intact cells. Release is stimulated by micromolar Ca^{2+} (0.1-3 µM), and it requires millimolar MgATP and elevated temperature.[14-18] The regulated codischarge of several granule constituents has been documented for GH$_3$ cell ghosts (prolactin, carboxypeptidase E, sulfated proteoglycan)[14,16] as well as for PC12 cell ghosts (NE, secretogranin II, chromogranin B).[15,16] In contrast, the lumenal contents of other organelles are not released by cell ghosts during incubations.[14] Multiple passes of cells through the ball homogenizer creates a homogenate which contains intact secretory granules; however, Ca^{2+} fails to stimulate release of granule contents in the homogenate.[14] It is unlikely that a nonspecific mechanism of Ca^{2+}-induced granule rupture can account for the Ca^{2+}-activated release of granule constituents from cell ghosts.

Additional studies document the similarity of Ca^{2+}-activated release in cell ghosts to that of regulated secretion in intact cells. Botulinum neurotoxin light chain of the E type has been found to rapidly and irreversibly inhibit Ca^{2+}-activated NE release from PC12 cell ghosts.[15] The Ca^{2+}-activated release of NE from PC12 cell ghosts under optimal conditions is associated with the disappearance of dense core granules from the cells. Botulinum neurotoxin treatment blocked both Ca^{2+}-activated NE release and the loss of dense core granules. The fate of the granule membrane during Ca^{2+}-activated NE release was determined in electron microscopic immunocytochemical studies of a granule membrane protein. A portion of intragranular chromogranin B in PC12 cells is granule membrane-associated and can be detected on the cell surface following the stimulation of secretion in intact PC12 cells.[19] Immunogold cytochemical studies with specific antibodies demonstrated that chromogranin B is converted to a cell-surface, antibody-accessible form in PC12 cell ghosts under conditions of optimal Ca^{2+}-activated NE release (our unpublished results). The studies reviewed in this section indicate that semi-intact neuroendocrine cells generated by cell cracking represent a valid *in vitro* model for biochemical studies of regulated secretion.

Ca²⁺-ACTIVATED SECRETION IN NEUROENDOCRINE CELLS REQUIRES CYTOSOLIC PROTEINS

In extensively washed GH_3 or PC12 cell ghosts, Ca^{2+} in the presence of MgATP stimulates the release of low levels of prolactin or NE, respectively.[14-16] In contrast, when cytosolic proteins are included in the incubations, Ca^{2+}-activated release is stimulated 2- to 10-fold. For PC12 cell ghosts, optimal Ca^{2+}-activated NE release in the presence of MgATP and cytosol is quantitatively similar to the level of NE release observed in secretagogue-treated intact PC12 cells, corresponding to 60% of the cellular catecholamine pool. The active constituents of cytosol were found to be thermolabile, nondialyzable and protease-sensitive.[14,16,18] The results indicate that full reconstitution of Ca^{2+}-activated secretion in PC12 cell ghosts requires cytosolic proteins.

It was of interest to determine the tissue distribution of cytosolic factors responsible for the reconstitution of regulated secretion. Cytosols were prepared from a number of rat tissues and tested for their ability to support Ca^{2+}-activated NE release from PC12 cell ghosts in the presence of MgATP.[16] Cytosols from brain, anterior pituitary, posterior pituitary, adrenal and pancreas were highly active in the assay. In contrast, cytosols from liver, muscle and lung exhibited little activity. Cytosol-mixing experiments excluded the possibility that cytosols with poor activity contained inhibitors. These studies indicated that cytosols from tissues which exhibit a regulated secretory pathway were much more active than cytosols from tissues which lack such a pathway.[16] For protein purification studies described below, rat brain cytosol was utilized since it exhibited the highest specific activity for reconstitution.

Studies with digitonin-permeabilized PC12[10] and adrenal chromaffin[11] cells or with streptolysin O-permeabilized mast cells[12] indicated that Ca^{2+}-activated secretory responsiveness declined with incubation time. Sarafian et al.[11] demonstrated that the addition of proteins lost from digitonin-permeabilized chromaffin cells could retard the loss of responsiveness. The study with mast cells[12] provided evidence that soluble factors lost following permeabilization of mast cells may be larger than cytoplasmic lactate dehydrogenase (135 kD). Patch clamping studies of adrenal chromaffin cells have also identified a requirement for a soluble macromolecular constituent for the maintenance of agonist-regulated secretion.[20] Based on the kinetics of wash-out, this constituent was estimated to have a molecular mass of less than 680 kD.[20] With cell cracking, soluble macromolecular constituents are lost immediately upon permeabilization and their back-addition is required to reconstitute regulated secretion. Gel filtration studies of brain cytosol indicate that the principal factor in cytosol required for the full reconstitution of regulated secretion in GH_3 and PC12 cell ghosts exhibits an apparent molecular mass of about 300 kD.[14,16-18] This estimate is compatible with those derived from wash-out studies of detergent-permeabilized and patch clamped secretory cells. The presence of this factor in neuroendocrine and exocrine but not in other tissues motivated efforts to purify and characterize it.

PURIFICATION OF A BRAIN CYTOSOLIC PROTEIN (P145) THAT RECONSTITUTES REGULATED SECRETION

To further characterize cytosolic proteins responsible for reconstituting Ca^{2+}-activated secretion, PC12 cell ghosts were utilized as an assay to monitor activity during column chromatographic procedures. PC12 cells accumulate [³H]NE in dense core granules which are retained throughout permeabilization and washing procedures. Incubations of PC12 cell ghosts at 30°C for 15 min in the presence of 1 μM Ca^{2+} and 2 mM MgATP results in a low level (~20% of cellular content) of [³H]NE release which is markedly enhanced by inclusion of cytosolic proteins (~60% of cellular content). Using this assay, a brain cytosolic protein (p145) has been purified to apparent homogeneity in five chromatographic steps:[16] DEAE cellulose, phenyl-Sepharose, Matrex Green, phenyl-TSK and Mono Q. A key step in the purification of p145 was Ca^{2+}-dependent hydrophobic interaction chromatography on phenyl-Sepharose. The cytosolic reconstituting activity was retained by the column in the presence of Ca^{2+} and eluted by EGTA (in the presence of ethylene glycol to minimize Ca^{2+}-independent hydrophobic interactions). As described elsewhere,[16] the activity was also resolved from a potentiating factor which exhibited little activity but was required along with p145 for full reconstitution. During the final stage of purification, reconstituting activity in column fractions from the Mono Q column coincided with a 145 kD protein detected by SDS gel electrophoresis. By activity estimates, reconstituting activity was purified at least 100-fold from brain cytosol.

Specific activity estimates at each purification stage are only approximate since both potentiating and inhibitory factors were evident in the purification. The apparent low yield of reconstituting activity during the p145 purification prompted the development of antibodies to the p145 protein to further evaluate the contribution of this protein to the overall reconstituting activity of cytosol. Rabbit polyclonal antibodies generated with purified p145 were shown to identify a doublet at 145 kD in Western blots of crude brain cytosol following SDS polyacrylamide gel electrophoresis[16]. The p145-specific antibody was used subsequently used in an immunoprecipitation protocol to deplete brain cytosol of p145. p145-depleted cytosol was found to exhibit about 40% of the reconstituting activity of control cytosol. Similar results were obtained when p145-specific antibodies were employed to neutralize activity. The IgG fraction of immune serum was found to inhibit about 60% of the activity of brain cytosol in supporting Ca^{2+}-activated [^{3}H]NE release from cracked PC12 cells; the IgG fraction from nonimmune serum was noninhibitory. Purified p145 reversed the inhibitory effect of immune IgG, indicating that the antibody inhibition was p145-specific.[16] These studies demonstrated that p145 is an essential constituent of brain cytosol required for the full reconstitution of Ca^{2+}-activated secretion in PC12 cell ghosts. The results confirm that the p145 purification resulted in the identification of a factor which accounts for a substantial portion of the reconstituting activity of brain cytosol. The incomplete neutralization of brain cytosol activity with p145 antibody also indicates the presence of additional cytosolic factors with reconstituting activity (see below).

Additional immunoneutralization studies indicated that p145-specific antibody fully blocked the reconstituting activity of adrenal, PC12 cell, pancreatic and pituitary cytosols. These results indicate that p145 is also an essential constituent of cytosols derived from secretory tissues other than brain. Moreover, the relative contribution of p145 to the reconstituting activity of cytosol appeared to be greater than in brain cytosol.

Gel filtration studies with brain, adrenal and PC12 cytosols indicate that the major portion of the recovered reconstituting activity exhibited an apparent molecular mass of ~300 kD.[14,17,18] This high molecular weight activity peak has been found to coincide with p145 immunoreactivity.[17] Moreover, activity in the high molecular weight region of the column is fully inhibited by p145-specific antibodies. Gel filtration analyses of purified brain p145 indicated that this protein eluted as a high molecular weight factor exhibiting a Stokes radius of 50 A. A sedimentation coefficient of 13S determined by sucrose gradient centrifugation allowed the calculation of an apparent molecular mass of 290 kD for the purified protein.[16] Hence, it appears that p145 exists predominantly as a homodimer of 145 kD subunits in the native state. Moreover, p145 appears to account for the majority of reconstituting activity detected in crude cytosols.

P145 IS A NOVEL PROTEIN WITH A TISSUE-SPECIFIC EXPRESSION PATTERN

As indicated, cytosols from tissues which possess a regulated secretory pathway were found to be highly effective in supporting Ca^{2+}-activated [^{3}H]NE release from PC12 cell ghosts, whereas cytosols from tissues lacking this pathway were poorly active. The results from the p145 purification and with the p145-specific antibody suggest that p145 is the primary determinant of cytosolic activity. Hence, we analyzed the p145 content of a variety of rat tissues by Western immunoblotting.[16] The results indicated that p145 was abundant in brain, pituitary, adrenal and pancreatic cytosols. In contrast, a large number of other tissues contained either extremely low (<1% brain) or undetectable levels of p145. These included: muscle, liver, lung, heart, stomach, spleen, intestine, kidney, thymus, diaphragm, placenta, uterus and lymphocytes. Additional studies indicated that adrenal p145 was localized to the medulla rather than cortex. Both pancreatic cell lines of endocrine (RINm5f) and exocrine (AR42J) origin contained readily detectable levels of p145. These studies indicate a tissue expression pattern for p145 restricted to neural, endocrine and exocrine secretory tissues.

Recent effort has been directed toward the further characterization of the p145 protein. Purified brain p145 was proteolytically digested and peptides were resolved by reverse phase HPLC. Automated Edman degradation sequencing provided 90 residues of protein sequence. Searches of protein or translated nucleic acid sequence databases failed to reveal significant amino acid sequence homology between p145 and other characterized proteins. These results indicate that p145 may be a novel protein and efforts are being directed toward acquiring the complete sequence.

RESOLUTION OF REGULATED SECRETION IN PC12 CELLS INTO SEQUENTIAL MGATP-DEPENDENT AND Ca^{2+}-DEPENDENT STAGES

In GH$_3$ cell ghosts, Ca^{2+}-activated, cytosol factor-dependent prolactin secretion is highly dependent upon inclusion of MgATP.[14] In contrast, Ca^{2+}-activated, cytosol-dependent [^{3}H]NE secretion from PC12 cell ghosts is stimulated about 2-fold by MgATP.[15-18] Recent studies have shown that the MgATP dependent events of the regulated pathway can be fully resolved from the Ca^{2+}-dependent events.[18] Preincubation of PC12 cell ghosts without Ca^{2+} in the presence of MgATP markedly enhances the Ca^{2+}-dependent [^{3}H]NE release observed in a subsequent incubation without MgATP.[18] Preincubation with MgATP was found to enhance the rate as well as final extent of Ca^{2+}-activated, MgATP-independent secretion, a process referred to as priming. Priming was stimulated at least 2-fold by inclusion of cytosol, indicating that cytosol contains factors required for optimal priming. MgATP-dependent priming was fully reversed by subjecting primed cell ghosts to incubations which lacked MgATP, a process referred to as depriming. By employing priming or depriming incubations, respectively, Ca^{2+}-activated [^{3}H]NE release could be converted between MgATP-independent and MgATP-dependent states.

Priming and depriming incubations allow little [^{3}H]NE release since they are conducted at low (<1 nM) Ca^{2+}. In contrast, brief (3 min) incubations of MgATP-primed cell ghosts in the presence of Ca^{2+} resulted in [^{3}H]NE release nearly equivalent to that in the longer (15 min) regular incubations which included both Ca^{2+} and MgATP. The brief incubations of primed cell ghosts with Ca^{2+} which resulted in MgATP-independent secretion were referred to as triggering reactions. Ca^{2+}-dependent triggering of [^{3}H]NE secretion was found to be enhanced by the inclusion of cytosol.[18] These studies indicated that cytosol contains triggering factors in addition to priming factors.

Since both priming and triggering reactions were stimulated by cytosolic factors, it was possible to determine whether one type of factor was operative in both steps or distinct cytosolic factors were involved. It was found that cytosolic priming and triggering activities differed in thermolability, likely indicating that distinct factors were involved.[18] Moreover, gel filtration studies confirmed that this was the case. In adrenal medullary cytosol, a low molecular weight (~20 kD) priming factor was identified. Triggering activity eluted in different regions of the column corresponding to high molecular weight (~380 kD) and low molecular weight (~37 kD) triggering factors.[18] The high molecular weight triggering factor likely corresponds to p145 since purified p145 was shown to possess triggering but not priming activity. Conversely, the partially purified ~20 kD priming factor was shown to lack triggering activity. Similar gel filtration studies with brain cytosol indicated a somewhat more complex pattern. Brain cytosol contained factors similar to those of adrenal cytosol but additional triggering factors (~75 kD) and priming factors (~500 kD and 120 kD) were identified.[18]

The results indicate that MgATP-dependent, Ca^{2+}-activated NE secretion in PC12 cell ghosts can be resolved into two distinct stages: priming which is MgATP-dependent and Ca^{2+}-independent, and triggering which is MgATP-independent and Ca^{2+}-dependent. Each stage requires distinct cytosolic protein cofactors, however, there appear to be functionally redundant factors of each type. It is possible to reconstitute full secretory activity in PC12 cell ghosts by employing one factor of each type (e.g., the 20 kD priming factor followed by the p145 triggering factor).[18]

MECHANISMS UNDERLYING REGULATED SECRETION

The priming step in regulated secretion utilizes ATP hydrolysis since nonhydrolyzable nucleotides fail to substitute for ATP. Moreover, priming possesses high specificity for adenine nucleotides (e.g., GTP fails to substitute).[18] Since priming is fully reversible, it is possible that this step involves a phosphorylation of some essential component of the secretory apparatus. Both lipid[21] and protein[22,23] kinases have been suggested to play a role in regulated secretion. Priming in PC12 cell ghosts can be inhibited by a large variety of protein kinase inhibitors;[18] however, the most effective of these exhibit an ATP competitive mode of action. The inhibitor data do not provide compelling evidence that a protein kinase, as opposed to some other type of ATP-dependent enzyme, is involved in priming. Ca^{2+}-dependent protein kinases are apparently not essential for regulated secretion since priming occurs in the absence of Ca^{2+} whereas triggering proceeds in the absence of ATP. The complete characterization of cytosolic priming factors should provide critical new information for understanding the role of ATP in regulated secretion. It is possible that priming factors are also employed in the constitutive secretory pathway, since vesicle trafficking events in this pathway also require ATP and cytosolic proteins.

The biochemical events underlying Ca^{2+}-dependent triggering in the regulated pathway remain undefined. These events are likely to be unique to the regulated secretory pathway since Ca^{2+} is not required for vesicle-plasma membrane fusion in the constitutive pathway.[8] Consistent with this, the principal cytosolic factor required for triggering, p145, was found to be expressed only in tissues which possess a regulated secretory pathway. p145 exhibits Ca^{2+}-dependent changes in hydrophobicity, suggesting that this protein may function in association with membranes during Ca^{2+}-triggered granule exocytosis. Elucidation of the complete primary sequence of p145 should provide insight into the biochemical process which it regulates.

RELATIONSHIP TO BOTULINUM NEUROTOXIN ACTION

We previously reported[15] that brief exposure of PC12 cell ghosts to botulinum neurotoxin effectively inhibits Ca^{2+}-activated [^{3}H]NE release, indicating that the site of toxin inhibition is membrane- or cytoskeleton-associated. We detect virtually no p145 in PC12 cell ghosts, suggesting that the neurotoxin does not act on membrane or cytoskeletal counterparts of the cytosolic factors. In addition, neurotoxin has been found to inhibit Ca^{2+}-triggered secretion in primed PC12 cell ghosts, and to block the Ca^{2+}-dependent disappearance (fusion) of dense core granules. These observations suggest that the toxin acts at a late step in Ca^{2+}-activated secretion, and that elucidation of the mechanism of toxin inhibition will identify an important membrane or cytoskeletal component required for regulated secretion.

REFERENCES

1. Rothman JE, Orci, L. Molecular dissection of the secretory pathway. Nature 1992; 355:409.
2. Goda Y, Pfeffer, SR. Cell-free systems to study vesicular transport along the secretory and endocytic pathways. FASEB J 1989; 3:2488.
3. Wilson DW, Whiteheart SW, Orci L, Rothman JE. Intracellular membrane fusion. Trends Biochem Sci 1991; 16:334.
4. Rothman JE, Orci L. Movement of proteins through the Golgi stack: a molecular dissection of vesicular transport. FASEB J 1990; 4:1460.
5. Hicke L, Schekman R. Molecular machinery required for protein transport from the endoplasmic reticulum to the Golgi complex. BioEssays 1990; 12:253.
6. Burgess TL, Kelly RB. Constitutive and regulated secretion of proteins. Ann Rev Cell Biol 1987; 3:243.
7. Tooze SA, Weiss U, Huttner WB. A requirement for GTP hydrolysis in the formation of secretory vesicles. Nature 1990; 347:207.
8. Miller SG, Moore H-PH. Reconstitution of constitutive secretion using semi-intact cells: regulation by GTP but not calcium. J Cell Biol 1991; 112:39.
9. Turkewitz AP, Madeddu L, Kelly RB. Maturation of dense core granules in wild type and mutant *Tetrahymena thermophila*. EMBO J 1991; 10:1979.
10. Peppers SC, Holz RW. Catecholamine secretion from digitonin-treated PC12 cells. J Biol Chem 1986; 261:14665.
11. Sarafian T, Aunis D, Bader M-F. Loss of proteins from digitonin-permeabilized adrenal chromaffin cells essential for exocytosis. J Biol Chem 1987; 262:16671.
12. Koffer A, Gomperts BD. Soluble proteins as modulators of the exocytotic reaction of permeabilized rat mast cells. J Cell Sci 1989; 94:585.
13. Martin TFJ. Cell cracking: permeabilizing cells to macromolecular probes. Meth Enzymol 1989; 168:225.
14. Martin TFJ, Walent JH. A new method for cell permeabilization reveals a cytosolic protein requirement for Ca^{2+}-activated secretion in GH_3 pituitary cells. J Biol Chem 1989; 264:10299.
15. Lomneth R, Martin TFJ, DasGupta BR. Botulinum neurotoxin light chain inhibits norepinephrine secretion in PC12 cells at an intracellular membranous or cytoskeletal site. J Neurochem 1991; 57:1413.
16. Walent JH, Porter BW, Martin TFJ. A novel 145 kD brain cytosolic protein reconstitutes Ca^{2+}-regulated secretion in permeable neuroendocrine cells. Cell 1992 (in press).
17. Nishizaki T, Walent JH, Kowalchyk JA, Martin TFJ. A key role for p145 in the stimulation of Ca^{2+}-dependent secretion by protein kinase C. J Biol Chem 1992 (in press).
18. Hay JC, Martin TFJ. Resolution of regulated secretion into sequential MgATP-dependent and Ca^{2+}-dependent stages mediated by distinct cytosolic proteins. J Cell Biol 1992 (in press).
19. Pimplikar SW, Huttner WB. Chromogranin B (Secretogranin I), a secretory protein of the regulated pathway, is also present in a tightly membrane-associated form in PC12 cells. J Biol Chem 1992; 267:4110.
20. Augustine GJ, Neher E. Calcium requirements for secretion in bovine chromaffin cells. J Physiol, in press.
21. Eberhard DE, Cooper CL, Low MG, Holz, RW. Evidence that the inositol phospholipids are necessary for exocytosis. Biochem J 1990; 268:15.
22. Wagner PD, Vu N-D. Regulation of norepinephrine secretion in permeabilized PC12 cells by Ca^{2+}-stimulated phosphorylation. J Biol Chem 1990; 265:10352.
23. Howell TW, Kramer IM, Gomperts BD. Protein phosphorylation and the dependence on Ca^{2+} and GTP-γ-S for exocytosis from permeabilized mast cells. Cellular Signalling 1989; 1:157.

INTRACELLULAR CONTROL OF EXOCYTOSIS
IN CHROMAFFIN CELLS

Alan Morgan[1], Isabelle Cenci de Bello[2], Ulrich Weller[3], J. Oliver Dolly[2]
and Robert D. Burgoyne[1]

[1] The Physiological Laboratory, University of Liverpool, P O Box 147,
 Liverpool L69 3BX, UK
[2] Biochemistry Dept, Imperial College, South Kensington, London
 SW7 2AY, UK
[3] Abteilung fur Neuropsychopharmakologie der Freien Universitat,
 1000 Berlin, Germany

INTRODUCTION

Chromaffin cells are endocrine cells of the adrenal medulla which synthesize large quantities of adrenaline and noradrenaline and store these, along with a cocktail of other compounds including bioactive peptides, within membrane-enclosed secretory vesicles known as chromaffin granules. In situations of stress, acetylcholine is released from splanchnic nerve terminals innervating the adrenal medulla and this in turn causes the chromaffin granule membranes to fuse with the plasma membrane of the cell, thereby releasing the granule contents into the blood to effect the 'fight or flight response'. Acetylcholine, then, is the physiological stimulus resulting in the exocytotic secretion of catecholamines from chromaffin cells. Acetylcholine can activate nicotinic and muscarinic receptors; and chromaffin cells possess both types of receptor. Selective activation of either nicotinic or muscarinic receptors results in a similar increase in cytosolic Ca^{2+} concentration (long recognised as the major signal leading to exocytosis in this and many other secretory cell types), but only nicotinic receptor agonists elicit exocytosis, muscarinic agonists being ineffective.[1,2] This paradox can be explained by the source of the increased cytosolic Ca^{2+}. Nicotinic receptor activation depolarises the plasma membrane causing entry of extracellular Ca^{2+} through voltage-dependent Ca^{2+} channels, whereas muscarinic receptor activation causes IP_3 generation and subsequent release of Ca^{2+} from intracellular stores.[3] Thus, exocytosis requires entry of extracellular Ca^{2+}–intracellular Ca^{2+} mobilization is ineffective. That this is the case is evidenced by the observation that nicotinic agonists (and indeed all secretagogues so far tested) fail to elicit a secretory response in the absence of extracellular Ca^{2+}.[3] Although no direct evidence is available to explain the crucial nature of extracellular Ca^{2+} entry in exocytosis, it is likely that the opening of plasma membrane channels would lead to a very high level of Ca^{2+} immediately beneath the plasma membrane—the very region in which the process of exocytotic membrane fusion occurs. Indeed, it has been calculated that the influx of Ca^{2+} through such channels could raise intracellular Ca^{2+} levels within a few nanometres of the plasma membrane to as high as 100μM.[4]

THE CYTOSKELETON IN EXOCYTOSIS

In resting chromaffin cells a small number of chromaffin granules are closely apposed to the plasma membrane. However, the vast majority of granules (approximately 99%) are excluded from this area by the presence of a dense subplasmallemal cytoskeletal network composed predominantly of actin.[5] Clearly, since the process of exocytosis requires granules to fuse with the plasma membrane, some mechanism must exist to allow the granules to traverse this cortical cytoskeletal network. It has been proposed that one such mechanism could be the active transport of vesicles to the plasma membrane by cytoskeletal elements in a manner analogous to the Ca^{2+}-activated sliding filament mechanism of skeletal muscle.[6] However, it has been calculated.[7] that the rate of diffusion of chromaffin granules is sufficient to account for the kinetics of exocytosis and so a role for active transport seems unlikely. Instead, it is now generally accepted that the cortical cytoskeleton plays a passive role by acting as a barrier to exocytosis in resting cells.[3]

If the barrier hypothesis is correct, secretory stimuli should result in the reorganisation of the peripheral cytoskeleton to allow chromaffin granules access to the plasma membrane. Direct evidence of such reorganisations during exocytosis in chromaffin cells has been reported by several groups. Immunofluorescence studies of fodrin—a protein heterodimer believed to mediate the attachment of actin filaments to the plasma membrane—have revealed that the protein has a continuous distribution in the subplasmallemal region of resting cells, but becomes reorganised into subplasmallemal patches upon nicotinic stimulation.[8] We have shown using two different biochemical assays that nicotine-induced secretion is accompanied by a transient reduction in cytoskeletal actin content which occurs over the same time-course as exocytosis.[5,9] This reduction in assembled actin is due to the disassembly of subplasmallemal actin filaments, since the bright ring of staining seen at the periphery of resting cells using rhodamine-phalloidin disappears upon challenge with nicotine only to reappear after secretion has terminated.[5,10]

A full secretory response can be elicited in permeabilized cells (cells whose plasma membrane have been porated to allow the entry of membrane impermeant macromolecules) simply by raising Ca^{2+} to micromolar levels in the presence of MgATP.[11,12] Studies using both digitonin- and streptolysin-O-permeabilized chromaffin cells have demonstrated that Ca^{2+} produces a transient reduction in cytoskeletal actin which occurs over the same concentrations of Ca^{2+} as does exocytosis.[9,13] Clearly, then, Ca^{2+} can directly cause disassembly of the subplasmallemal cytoskeleton, but what are the proteins which Ca^{2+} acts on to bring about this change? One such protein is likely to be the Ca^{2+} and phospholipid-dependent enzyme, protein kinase C (PKC), since the tumour promoter PMA (a direct activator of PKC) can itself promote actin disassembly in both intact and permeabilized chromaffin cells.[9] PKC exerts its cellular effects via the phosphorylation of key protein substrates, and it seems likely that at least some of the proteins mediating disassembly/assembly of the cytoskeleton will be targets for phosphorylation by PKC. In addition, chromaffin cells contain two Ca^{2+}-activated actin filament-severing proteins, gelsolin and scinderin.[14] Stimulation of chromaffin cells with depolarizing secretagogues results in the redistribution of scinderin (but not gelsolin) to leave patches of the cell cortex devoid of the protein.[10] Furthermore, the regions lacking scinderin were the same as those lacking assembled actin and also correlated with exocytotic sites as evidenced by dopamine-β-hydroxylase immunoreactivity.[10] Thus it may be that scinderin is a key protein in the Ca^{2+}-dependent disassembly of cortical actin which accompanies exocytosis. Caldesmon is a chromaffin granule membrane-binding protein which crosslinks

F-actin at low ($<1\mu M$) Ca^{2+} concentrations but not at elevated Ca^{2+} concentrations.[15] It is possible that caldesmon also mediates cortical actin reorganisation since the increase in cytosolic Ca^{2+} which elicits exocytosis may inhibit the tonic crosslinking of actin filaments by this protein in resting conditions, thus reducing the viscosity of the cortical cytoskeleton and facilitating the access of chromaffin granules to the plasma membrane.

More direct evidence for a barrier role of the subplasmallemal cytoskeleton in exocytosis has come from the use of agents which interfere with actin assembly/disassembly. Cytochalasins and DNase I cause disassembly of the F-actin network and have been reported to promote Ca^{2+}-activated exocytosis from both digitonin[16] and streptolysin-O-[13] -permeabilized chromaffin cells; whereas the actin filament stabilizer phalloidin inhibits secretion.[16] Furthermore, Botulinum C2 toxin, which ADP-ribosylates actin and so inhibits actin polymerisation, enhances secretion from PC12 cells.[17] In addition, a direct role for fodrin in exocytosis is supported by the finding that anti-α-fodrin antibodies partially inhibit Ca^{2+}-induced exocytosis in digitonin-permeabilized chromaffin cells.[18]

In summary, there is now considerable evidence to support the notion that the subplasmallemal cytoskeleton plays an important role in the control of exocytosis in chromaffin cells by acting as a barrier to granule movement. Although the removal of this barrier is essential for a full secretory response, this is not in itself sufficient to elicit exocytosis since both PMA treatment[9] and nicotinic stimulation[5] in the absence of external Ca^{2+} result in disassembly of the cortical cytoskeleton but elicit little or no secretion. Therefore, an additional Ca^{2+}-dependent mechanism must also be involved in the late stages of exocytosis (granule docking and/or fusion of granule and plasma membranes) after the disassembly of the subplasmallemal cytoskeleton.

THE EXOCYTOTIC MACHINERY

In the search for the proteins which mediate Ca^{2+}-dependent exocytosis, attention has understandably been focused on proteins known to regulate other Ca^{2+}-dependent cellular events. Many groups have investigated the potential involvement of the Ca^{2+} binding protein calmodulin in the exocytotic process using anti-calmodulin drugs and antibodies [see ref. 3]. However, the results have been contradictory and so it seems safest at present to conclude that calmodulin is not essential for Ca^{2+}-dependent exocytosis. Protein kinase C (PKC) has been suggested to be involved in Ca^{2+}-dependent exocytosis in chromaffin cells. In both electropermeabilized and digitonin-permeabilized cells, activators of PKC increase the affinity of exocytosis for Ca^{2+} and also the extent of exocytosis;[19,20] whereas inhibitors of the enzyme partially reduce Ca^{2+}-dependent exocytosis.[21,22] Although the importance of PKC in the exocytotic process has been controversial, most workers now agree that the enzyme performs a modulatory role rather than being itself essential for exocytosis.[3] As was mentioned earlier, it is likely that one site of action of PKC is the Ca^{2+}-dependent disassembly of the cortical actin network which accompanies exocytosis. Whether PKC also regulates the final stages of exocytosis is at present unknown.

The process of Ca^{2+}-dependent membrane fusion which occurs during secretion from chromaffin cells requires, at least, a Ca^{2+} sensor and a membrane fusogen. Considerable attention has been focused upon Ca^{2+}-dependent membrane binding proteins as putative 'exocytotic fusion proteins' since a single such protein could conceivably mediate both processes. Synexin (Annexin VII) was isolated from the adrenal medulla on the basis of its ability to aggregate chromaffin granules in the presence[23] of Ca^{2+}. It was subsequently found that these aggregated granules could be made to fuse with one another in the presence of arachidonic acid.[24] These observations are consistent with a minimalistic model of exocytosis in chromaffin cells whereby the simultaneous presence of synexin, Ca^{2+} and arachidonic acid would be sufficient to bring about the linkage of chromaffin granules to the plasma membrane (and to other chromaffin granules) and also the subsequent fusion of these membranes. Since synexin is present in chromaffin cells and since Ca^{2+}-activated secretion from these cells correlates with an increase in free arachidonic acid levels[25,26] this hypothesis seemed plausible. However, two criticisms can be levelled at this theory. Firstly, a substantial level of Ca^{2+}-activated exocytosis (around 80%) remains even when the Ca^{2+}-activated increase in cellular free arachidonic acid has been totally abolished,[27] thus demonstrating that arachidonic acid is not essential for Ca^{2+}-dependent exocytosis. Secondly, the *in vitro* effects of synexin require $[Ca^{2+}]$ in the region of 10-100µM or more. Since secretion from permeabilized cells is maximal at around 10µM Ca^{2+}, synexin would not appear to be the most likely candidate for a Ca^{2+}-dependent fusion protein. However, as was mentioned earlier, the $[Ca^{2+}]$ directly beneath the plasma membrane may be as high as 100µM in intact cells after stimulation and so the rather low Ca^{2+} affinity of synexin may not necessarily preclude a role for this protein in exocytosis.

Synexin is a member of a family of proteins known as the annexins.[28] Several members of this family are present in chromaffin cells and share the synexin-like properties of granule aggregation and fusion, but only one, annexin II (calpactin), has a Ca^{2+} affinity similar to that of exocytosis (half-maximal granule aggregation by annexin II occurs at 1.8µM Ca^{2+}, while half maximal secretion from permeabilized chromaffin cells occurs at approximately 1µM Ca^{2+}).[29] Thus annexin II would appear to be a good candidate for a Ca^{2+}-dependent fusion protein. Nevertheless, the ability of annexin II to promote granule aggregation and fusion *in vitro* does not necessarily mean that the protein is involved in exocytosis. Other evidence detailed below does suggest that annexin II is indeed involved in Ca^{2+}-dependent exocytosis in the adrenal chromaffin cell.

CYTOSOLIC PROTEINS IN EXOCYTOSIS

Exocytosis can be directly triggered by micromolar Ca^{2+} in permeabilized chromaffin cells. However, one drawback of certain permeabilizing agents, such as digitonin and streptolysin-O, is that the lesions created in the plasma membrane are so large as to allow the exit (and entry) of cytosolic proteins. This leakage of cytosolic proteins results in a run-down of the ability of the cell to secrete in response to Ca^{2+}. Ironically, this apparent drawback has great potential for identifying those proteins involved in the exocytotic process. It has been shown[30] that reintroduction of the proteins which leak from digitonin-permeabilized chromaffin cells can reconstitute secretion from such run-down cells indicating that cytosolic proteins are essential for exocytosis. The protocol for the run-down-reconstitution approach comprises three stages: stage (1), cells are permeabilized for 10 min with digitonin; stage (2), cells are incubated with proteins (typically for 15 min); stage (3), cells are challenged with micromolar Ca^{2+}. Annexin II (calpactin) was found to leak from permeabilized cells over a time course which paralleled that of secretory run-down,[31,32] implying that annexin II might be the protein whose leakage was responsible for secretory run-down. This possibility was directly tested by using purified annexin II in the run-down-reconstitution system. It was found that exogenous annexin II could indeed retard the secretory run-down phenomenon, thus suggesting an essential role for annexin II in Ca^{2+}-dependent exocytosis in chromaffin cells.[31] This observation was later confirmed by others.[33] Furthermore, it seems likely that endogenous annexin II is required for secretion since a synthetic peptide corresponding to the conserved domain of the annexins inhibited secretion from permeabilized chromaffin cells, with this effect of the peptide being reversed by the addition of exogenous annexin II.[31] Annexin II may exist either as a heterotetramer of two light chains (p10) and two heavy chains (p36) or as monomeric p36. It was found that either form could stimulate Ca^{2+}-dependent exocytosis in the run-down-reconstitution system,[31] indicating that p36 and not p10 is the subunit responsible for the protective effect of annexin II p36 contains a 33kDa core domain with homology to the other annexins and a unique 3kDa N-terminal sequence. Limited proteolysis of p36 gives rise to these fragments, the larger of which (p33) retains the ability to bind Ca^{2+} and phospholipid albeit with lower Ca^{2+} affinity.[29] Incubation of p33 in the run-down-reconstitution system has no effect, even when cells are challenged with 200μM Ca^{2+},[34] indicating that the unique N-terminal sequence of annexin II (which contains the phosphorylation sites for tyrosine and serine/threonine kinases) is essential for the stimulatory effect of the protein on Ca^{2+}-dependent exocytosis. This is consistent with the recent claim[33] that protein kinase C-mediated phosphorylation of the N-terminal domain of annexin II is required for its protective effect in the run-down-reconstitution system.

Although there is a strong case for a role of annexin II in Ca^{2+}-dependent exocytosis in chromaffin cells, it is extremely unlikely that annexin II is the sole 'exocytotic protein' since maximal concentrations of this protein fail to restore secretion from run-down cells after prolonged run-down.[32,33] In other words, annexin II retards the run-down process but cannot reactivate secretion to control levels. Therefore, other essential cytosolic components must leak from permeabilized cells in addition to annexin II. We therefore set out to purify the cytosolic protein(s) able to reactivate exocytosis. Leaked protein preparations do not provide enough material for purification purposes and so whole adrenal medullary cytosol was tested in the run-down-reconstitution system. It was found that cytosol could indeed retard secretory run-down and that this effect was due to a trypsin- and NEM-sensitive protein(s).[35] Fractionation of adrenal medullary cytosol on the anion-exchanger Q-Sepharose revealed the existence of three fractions capable of affecting secretion in the run-down-reconstitution system: two stimulatory and one inhibitory.[35] We termed the stimulatory fractions Exo1 and Exo2. The same three activities were also detected in sheep brain cytosol and so this was used for subsequent purification. Although the identities of Exo1 and Exo2 were at that stage unknown, it was clear that neither was annexin II, since it does not bind to anion exchangers. Exo2 has not yet been purified to homogeneity but is known to run as a single peak of apparent molecular mass 44kDa on gel filtration.[35] Exo1, however, was purified to homogeneity by a combination of hydrophobic interaction chromatography, gel filtration and further anion exchange using Mono Q FPLC. Native Exo1 is a 70 kDa dimer composed of three subunits of 32, 30 and 29 kDa. Unlike annexin II and Exo2, Exo1 could completely reactivate exocytosis in the run-down-reconstitution system back to control levels. The ability of Exo1 to stimulate exocytosis from run-down cells was Ca^{2+}-dependent and was half maximal at 100μg/ml.[35] Although this value may seem rather high, it should be noted that diffusion of antibodies across chromaffin cells takes around 16 min,[36] suggesting that permeability to proteins is restricted. Therefore, the effective concentration of Exo1 at exocytotic sites is likely to be much lower than that added to the cells. Furthermore, we have calculated that, based on the relative abundance of Exo1 in cytosol and from a calculated cytosolic protein concentration of 18mg/ml, the theoretical concentration of Exo1 in intact chromaffin cells is roughly 370μg/ml.

The chromatographic properties of Exo1 are very similar to a family of proteins comprising the brain 14-3-3 proteins and a partially characterised group of protein kinase C inhibitory proteins (KCIPs).[37,38] The 14-3-3 family comprises several widely distributed, homologous gene products whose cellular functions are unknown. Protein sequencing of two randomly selected tryptic peptides of Exo1 revealed strong homology with 14-3-3/KCIPs, thus revealing that Exo1 is a member of this family of proteins.[35] The highly conserved sequences of the 14-3-3 family appear to be unique with the exception of a domain which is very similar to the conserved C-terminus of the annexins, and is most similar to annexin II (calpactin).[38,39] Since annexin II and Exo1 are the only purified proteins which have been shown to stimulate exocytosis from run-down cells, it is intriguing that they should share such a conserved domain. Although Exo2 has not yet been purified to homogeneity, it is tempting to speculate that Exo2 might also possess this putative 'exocytotic sequence'.

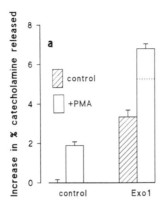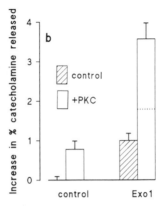

Figure 1. PKC and Exo1 act synergistically in exocytosis. Cultured bovine chromaffin cells were permeabilized for 10 min in Buffer A (139mM potasium glutamate, 5mM EGTA, 2mM $MgCl_2$, 2mM ATP, 20mM PIPES, pH 6.5) containing 20μM digitonin, then incubated for a further 15min with the appropriate purified proteins or dialysis buffer (Buffer A containing 1mM DTT and 0.02% NaN_3) for controls before being challenged by replacement of buffer with Buffer A containing 10μM free Ca^{2+}. (a) 200nM PMA was included in stages 1 and 2 of the above protocol to retain and activate PKC; a supramaximal dose of 1mg/ml Exo1 was used. (b) partially purified rat brain PKC was used at 200μg/ml in stage 2 along with a submaximal dose of 50μg/ml Exo1. The *dotted lines* indicate the theoretical effects of the treatments were their actions additive.

Another of the proteins which leak from digitonin-permeabilized cells during secretory run-down is the putative regulator of exocytosis, PKC.[40] This leakage of PKC can be prevented by prior treatment of the cells with PMA, which causes the enzyme to translocate from cytosol to membranes.[40] Such treatment of chromaffin cells causes a small increase in exocytosis in the run-down-reconstitution system which is additive to the effect of Exo2 but is synergistic with the effect of Exo1 (Fig. 1).[35] However, PMA treatment prior to stimulation will, in addition to preventing PKC leakage, cause enzyme activation and hence phosphorylation of key substrates. Significantly, it has been claimed that PMA pre-treatment of digitonin-permeabilized chromaffin cells also prevents the leakage of annexin II.[33] Clearly, then, PMA treatment is an unsuitable approach to investigate the relationship between PKC and Exo1 in exocytosis, since the observed synergy may result from the retention/activation of proteins other than PKC. We have obviated this problem by using exogenous partially purified PKC along with Exo1 in the run-down-reconstitution system.[41] It was found that Exo1's stimulatory effect was completely dependent on the continuous presence of MgATP, consistent with a requirement for protein phosphorylation in its mechanism of action. Exogenous PKC was found to cause a dose-dependent increase in secretion from run-down cells, further suggesting a role for this protein in exocytosis. Moreover, this effect of PKC was synergistic when combined with exogenous Exo1 (Fig. 1), thus indicating that the previous observation of synergy between PMA treatment and Exo1 incubation[35] was indeed due to PKC retention and not to the prior phosphorylation and retention of key substrates such as annexin II. An obvious explanation for the synergistic relationship between PKC and Exo1 is that Exo1 is phosphorylated by PKC. However, phosphorylation studies revealed Exo1 to be an extremely poor substrate for PKC-mediated phosphorylation[41] and so its seems likely that PKC acts on some other component of the exocytotic process and in so doing facilitates the involvement of Exo1 in exocytosis. Although the sites of action of PKC and Exo1 in exocytosis are unknown it may be that their synergistic interaction is related to the disassembly of the subplasmalemmal cytoskeleton, since PKC is thought to regulate this Ca^{2+}-dependent process (see earlier). Alternatively, PKC and Exo1 may both act to regulate the activity of a single protein (the fusogen?) involved in the late stages of exocytosis.

CLOSTRIDIAL TOXINS AND EXOCYTOSIS

Tetanus and botulinum neurotoxins are the causative agents of the fatal diseases tetanus and botulism which are characterized by spastic and flaccid paralysis, respectively. These clinical symptoms are due to the specific neuronal site of action of the toxins: botulinum neurotoxins preferentially inhibit acetylcholine release from peripheral motor neurons, whereas tetanus toxin inhibits glycine release in the central nervous system [for review see ref. 42]. Despite these differences, both toxins act at an intracellular site(s) to inhibit the exocytotic release of neurotransmitters from nerve terminals. Although the identity of the intracellular targets for clostridial neurotoxins are completely unknown, they are almost certainly also present in a variety of regulated secretory cells since exocytosis is inhibited by toxin treatment in neurohypophyseal cells,[43] PC12 cells[44,45] and chromaffin cells.[46] In order to inhibit exocytosis, the toxins must gain access to the cell interior. For this reason chromaffin cells permeabilized with either digitonin or streptolysin-O have been used by several groups to investigate the effects of the toxins on Ca^{2+}-activated exocytosis.[45,47-49] The results are fairly consistent in that botulinum neurotoxin A (BoNT A) typically results in a 40-60% (but never complete) inhibition of secretion whereas the tetanus toxin (TeTX) inhibition approaches 100%. This is consistent with the notion that BoNT A and TeTX inhibit exocytosis by acting at two distinct sites in the exocytotic process and that BoNT A probably acts on a non-essential regulatory component whereas TeTX acts on an essential component in exocytosis—possibly the fusion protein itself. This hypothesis is supported by the recent finding that the block of exocytosis in intact chromaffin cells due to BoNT A could be partially overcome by increasing nicotine concentrations whereas the block due to tetanus toxin could not be reversed.[50] Although the mechanism of action of clostridial neurotoxin-induced inhibition of exocytosis in chromaffin cells is unknown recent evidence suggests the possible involvement of the subplasmalemmal cytoskeleton, based on the observation that both TeTX and BoNT A inhibit carbachol-induced cortical actin disassembly.[51] It may be significant that various botulinum neurotoxins have been shown to inhibit exocytosis in PC12 cells by acting on an intracellular membranous or cytoskeletal site.[52]

Unfortunately, the search for the key proteins which BoNT A and TeTX act on to inhibit exocytosis has not yet borne fruit. This is mainly because of the paucity of information on the proteins which mediate exocytosis in control cells. However, as was mentioned earlier, the run-down-reconstitution system is allowing the identification and characterization of putative regulators of exocytosis. Clearly, such proteins are potential targets for neurotoxin-mediated inactivation and so this approach has great promise for systematically determining the mechanism of action of the toxins. We have some data on the effects of BoNT A and TeTX in the run-down-reconstitution system which will be summarised below. In our hands, TeTX (either intact toxin or light chain) and BoNT A (intact toxin) could produce approximately 80% and 40% inhibition of Ca^{2+}-activated exocytosis, respectively, which is in line with results from other groups. In order to see whether Exo1 was the target for neurotoxin-mediated inactivation, BoNT A and $TeTX_{LC}$ were included alongside Exo1 in the run-down-reconstitution system at concentrations of the toxins which produced a similar (approximately 40%) inhibition of exocytosis in control cells. It was found that the ability of Exo1 to stimulate exocytosis from run-down cells was completely abolished by $TeTX_{LC}$ but was virtually unaffected by BoNT A. In fact, TeTX could prevent Exo1 action under conditions when it has little effect on residual control secretion (Fig. 2). This finding raised the exciting possibility that Exo1 was the target for TeTX action. However, in vitro incubation of Exo1 with $TeTX_{LC}$ and subsequent separation of the proteins using FPLC revealed that this Exo1 fraction was just as active in stimulating exocytosis as the control Exo1. Therefore, it seems unlikely that Exo1 is the target for TeTX, but rather that the target is a protein(s) with which Exo1 interacts to promote exocytosis. The observation that Exo1 is unaffected by BoNT A adds more weight to the concept that BoNT A and TeTX have different mechanisms of action in inhibiting

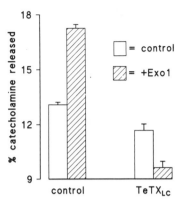

Figure 2. Tetanus toxin abolishes the stimulatory effect of Exo1 in exocytosis.Cultured bovine chromaffin cells were permeabilized for 10min in Buffer A containing 20μM digitonin; then incubated for a further 15min with partially purified Exo1 at 2mg/ml, tetanus toxin light chain at 26μg/ml or dialysis buffer (Buffer A containing 1mM DTT and 0.02% NaN₃) for controls before being challenged by replacement of buffer with Buffer A containing 10μM free Ca^{2+}.

exocytosis. We aim to follow up these investigation by extending them to other proteins active in the run-down-reconstitution system when available in a purified form. Such an approach, coupled with investigation of the interactions between Exo1, Exo2 and annexin II and their cellular target proteins, may eventually lead to an understanding of the mechanism of clostridial neurotoxin-induced inhibition of exocytosis and also to a more detailed understanding of the protein chemistry of exocytosis.

REFERENCES

1. Cheek TR, Burgoyne RD. Effect of activation of muscarinic receptors on intracellular free calcium and secretion in bovine adrenal chromaffin cells. Biochim Biophys Acta 1985; 846:167-174.
2. Kao LS, Schneider AS. Calcium mobilisation and catecholamine secretion in adrenal chromaffin cells. A quin 2 fluorescence study. J Biol Chem 1986; 261:4881-4888.
3. Burgoyne RD. Control of exocytosis in adrenal chromaffin cells. Biochim Biophys Acta1991; 1071:174-202.
4. Smith SJ, Augustine GJ. Calcium ions, active zones and synaptic transmitter release. Trends Neurosci 1988; 11:458-464.
5. Cheek TR, Burgoyne RD. Nicotine evoked disassembly of cortical actin filaments in bovine adrenal chromaffin cells. FEBS Lett 1986; 207:110-113.
6. Trifaro J-M. Contractile proteins in tissues originating in the neural crest. Neuroscience 1978; 3:1-24.
7. Burgoyne RD, Cheek TR, O'Sullivan AJ, Richards CR. Control of the cytoskeleton during secretion. In: Thorn NA, Treiman M, Petersen OH, eds. Molecular Mechanisms In Secretion. Copenhagen: Munksgaard, 1988:612-627.
8. Perrin D, Aunis D. Reorganisation of α-fodrin induced by stimulation in secretory cells. Nature 1985; 314:589-592.
9. Burgoyne RD, Morgan A, O'Sullivan AJ. The control of cytoskeletal actin and exocytosis in intact and permeabilized chromaffin cells: Role of calcium and protein kinase C. Cell Signal 1989; 1:323-334.
10. Vitale ML, Rodriguez Del Castillo A, Tchakarov L, Trifaro J-M. Cortical filamentous actin disassembly and scinderin redistribution during chromaffin cell stimulation precede exocytosis, a phenomenon not exhibited by gelsolin. J Cell Biol 1991; 113:1057-1067.
11. Baker PF, Knight DE. Calcium control of exocytosis and endocytosis in bovine adrenal medullary cells. Phil Trans R Soc Lond [Biol] 1981; 296:83-103.

12. Dunn LA, Holz RW. Catecholamine secretion from digitonin treated adrenal medullary chromaffin cells. J Biol Chem 1983; 258:4989-4993.
13. Sontag JM, Aunis D, Bader M-F. Peripheral actin filaments control calcium-mediated catecholamine release from steptolysin O-permeabilized chromaffin cells. Eur J Cell Biol 1988; 46:316-326.
14. Rodriguez Del Castillo A, Lemaire S, Tchakarov L, et al. Chromaffin cell scinderin, a novel calcium-dependent actin filament-severing protein. EMBO J 1990; 9:43-52.
15. Burgoyne RD, Cheek TR, Norman KM. Identification of a secretory granule-binding protein as caldesmon. Nature 1986; 319:68-70.
16. Lelkes PI, Friedman JE, Rosenheck K, Oplatka A. Destabilization of actin filaments as a requirement for the secretion of catecholamines from permeabilized chromaffin cells. FEBS Lett 1986; 208:357-363.
17. Matter K, Dreyer F, Aktories F. Actin involvement in exocytosis from PC12 cells: Studies on the influence of botulinum toxin C2 on stimulated noradrenaline release. J Neurochem 1989; 52:370-376.
18. Perrin D, Langley OK, Aunis D. Anti α-fodrin inhibits secretion from permeabilized chromaffin cells. Nature 1987; 326:498-501.
19. Knight DE, Baker PF. The phorbol ester TPA increases the affinity of exocytosis for calcium in leaky adrenal medullary cells. FEBS Lett 1983; 160:98-100.
20. Pocotte SL, Frye RA, Senter RA, Terbush DR, Lee SA, Holz RW. Effects of phorbol ester on catecholamine secretion and protein phosphorylation in adrenal medullary cell cultures. Proc Natl Acad Sci USA 1985; 82:930-934.
21. Knight DE, Sugden D, Baker PF. Evidence implicating protein kinase C in exocytosis from electro-permeabilized bovine chromaffin cells. J Memb Biol 1989; 104:21-34.
22. Burgoyne RD, Morgan A, O'Sullivan AJ. A major role for protein kinase C in calcium activated exocytosis in permeabilized adrenal chromaffin cells. FEBS Lett 1988; 238:151-155.
23. Creutz CE, Pazoles CJ, Pollard HB. Identification and purification of an adrenal medullary protein (synexin) that causes calcium-dependent aggregation of isolated chromaffin granules. J Biol Chem 1978; 253:2858-2866.
24. Creutz CE. Cis-unsaturated fatty acids induce the fusion of chromaffin granules aggregated by synexin. J Cell Biol 1981; 91:247-256.
25. Frye RA, Holz RW. The relationship between arachidonic acid release and catecholamine secretion from cultured bovine adrenal chromaffin cells. J Neurochem 1984; 43:146-150.
26. Frye RA, Holz RW. Arachidonic acid release and catecholamine secretion from digitonin-permeabilized chromaffin cells. Effects of micromolar calcium, phorbol ester, and protein alkylating agents. J Neurochem 1985; 44:265-273.
27. Morgan A, Burgoyne RD. Relationship between arachidonic acid release and calcium-dependent exocytosis in digitonin permeabilized bovine adrenal chromaffin cells. Biochem J 1990; 271:571-574.
28. Burgoyne RD, Geisow MJ. The annexin family of calcium-binding proteins. Cell Calcium 1989; 10:1-10.
29. Drust DS, Creutz CE. Aggregation of chromaffin granules by calpactin at micromolar levels of calcium. Nature 1988; 331:88-91.
30. Sarafian T, Aunis D, Bader M-F. Loss of proteins from digitonin-permeabilized adrenal chromaffin cells essential for exocytosis. J Biol Chem 1987; 262:16671-16676.
31. Ali SM, Geisow MJ, Burgoyne RD. A role for calpactin in calcium-dependent exocytosis in adrenal chromaffin cells. Nature 1989; 340:313-315.
32. Burgoyne RD, Morgan A. Evidence for a role of calpactin in calcium-dependent exocytosis. Biochem Soc Trans 1990; 18:1101-1104.
33. Sarafian T, Pradel L-A, Henry J-P, Aunis D, Bader M-F. The participation of Annexin II (calpactin I) in calcium-evoked exocytosis requires protein kinase C. J Cell Biol 1991; 114:1135-1147.
34. Ali SM, Burgoyne RD. The stimulatory effect of calpactin (annexin II) on calcium-dependent exocytosis in chromaffin cells, requirement for both N-terminal and core domains of p36 and ATP. Cell Signal 1990; 2:765-776.
35. Morgan A, Burgoyne RD. Exo1 and Exo2 proteins stimulate calcium-dependent exocytosis in permeabilized adrenal chromaffin cells. Nature 1992; 355:833-835.
36. Schweizer FE, Schafer T, Tapparelli C, et al. Inhibition of exocytosis by intracellularly applied antibodies against a chromaffin granule-binding protein. Nature 1989; 339:709-712.
37. Ichimura T, Isobe T, Okuyama T, et al. Molecular cloning of cDNA coding for brain-specific 14-3-3 protein, a protein kinase-dependent activator of tyrosine and tryptophan hydroxylases. Proc Natl Acad Sci USA 1988; 85:7084-7088.

38. Toker A, Ellis CA, Sellers LA, Aitken A. Protein kinase C inhibitor proteins. Purification from sheep brain and sequence similarity to lipocortin and 14-3-3 protein. Eur J Biochem 1990; 191:421-429.

39. Aitken A, Ellis CA, Harris A, Sellers LA, Toker A. Kinase and neurotransmitters. Nature 1990; 344:594.

40. Terbush DR, Holz RW. Effects of phorbol esters, diglyceride and cholinergic agonists on the subcellular distribution of protein kinase C in intact or digitonin-permeabilized adrenal chromaffin cells. J Biol Chem 1986; 261:17099-17106.

41. Morgan A, Burgoyne RD. Interaction between protein kinase C and Exo1 and its relevance to exocytosis in permeabilized adrenal chromaffin cells. Biochem J 1992; in press.

42. Dolly JO. Functional components at nerve terminals revealed by neurotoxins. In: Vincent A, Wray DW, eds. Neuromuscular Transmission—Basic and Applied Aspects. Manchester: Manchester University Press, 1990:107-131.

43. Dayanithi G, Weller U, Ahnert-Hilger G, Link H, Nordmann J-J, Gratzl M. The light chain of tetanus toxin inhibits calcium-dependent vasopressin release from permeabilized nerve endings. Neuroscience 1992; 46:489-493.

44. McInnes C, Dolly JO. Calcium-dependent noradrenaline release from permeabilized PC12 cells is blocked by botulinum neurotoxin A or its light chain. FEBS Lett 1990; 261:323-326.

45. Ahnert-Hilger G, Bader M-F, Bhakdi S, Gratzl M. Introduction of macromolecules into bovine medullary chromaffin cells and rat pheochromocytoma cells (PC12) by permeabilization with streptolysin O: inhibitory effect of tetanus toxin on catecholamine secretion. J Neurochem 1989; 52:1751-1758.

46. Penner R, Neher E, Dreyer F. Intracellularly injected tetanus toxin inhibits exocytosis in bovine adrenal chromaffin cells. Nature 1986; 324:76-78.

47. Bittner MA, Holz RW. Effects of tetanus toxin on catecholamine release from intact and digitonin-permeabilized chromaffin cells. J Neurochem 1988; 51:451-456.

48. Stecher B, Gratzl M, Ahnert-Hilger G. Reductive chain separation of botulinum A toxin—a prerequisite to its inhibitory action on exocytosis in chromaffin cells. FEBS Lett 1989; 248:23-27.

49. Bittner MA, DasGupta BR, Holz RW. Isolated light chains of botulinum neurotoxins inhibit exocytosis. J Biol Chem 1989; 264:10354-10360.

50. Marxen P, Bartels F, Ahnert-Hilger G, Bigalke H. Distinct targets for tetanus and botulinum A neurotoxins within the signal transducing pathways in chromaffin cells. Naunyn-Schmiedeberg's Arch Pharmacol 1991; 344:387-395.

51. Marxen P, Bigalke H. Tetanus and botulinum A neurotoxin inhibit carbachol-induced F-actin rearrangement in chromaffin cells. Neuroreport 1991; 2:33-36.

52. Lomneth R, Martin TFJ, DasGupta BR. Botulinum light chain inhibits norepinephrine secretion in PC12 cells at an intracellular membranous or cytoskeletal site. J Neurochem 1991; 57:1413-1421.

PC12 CELLS AS A MODEL
FOR NEURONAL SECRETION

Paul D. Wagner, Ngoc-Diep Vu, and You Neng Wu

Laboratory of Biochemistry
National Cancer Institute
National Institutes of Health
Bethesda, Maryland 20892

INTRODUCTION

PC12 cells are widely used as models for both neuroendocrine secretion and neuronal differentiation. In this article we will briefly review the origin of these cells and the various types of experiments in which they are employed. We will then discuss in some detail their use as a model for neuronal secretion. There are several extensive and authoritative reviews on PC12 cells which provide much more detailed information on the origin, differentiation, and biochemical properties of PC12 cells.[1-4]

PC12 cells were cloned by Greene and Tischler[5] from a pheochromocytoma tumor that had arisen in the adrenal medulla of an irradiated rat. In the developing adrenal medulla, neural crest stem cells differentiate into sympathetic neurons, adrenal chromaffin cells, or small intensely fluorescent (SIF) cells. In some respects SIF cells are intermediate between chromaffin cells and sympathetic neurons. PC12 cells when cultured in the absence of nerve growth factor are morphologically similar to rat chromaffin cells in a late stage of fetal development. They are small round (about 10 μm in diameter) or polygonal shaped cells, have few if any neurite-like processes (Figure 1), and contain numerous dense core granules (40-350 nm in diameter). These secretory vesicles which contain catecholamines are similar to those found in both chromaffin cells and sympathetic neurons. PC12 cells synthesize, store, secrete, and take up catecholamines. Unlike rat chromaffin cells, cultured PC12 cells contain more dopamine than norepinephrine. The relatively low level of norepinephrine appears to result from a lack of ascorbic acid rather than from a low level of dopamine β hydroxylase.[6] PC12 cells do not contain phenylethanolamine-N-methyl transferase and, therefore, do not synthesize epinephrine. Stimulation of PC12 cells with nicotine or by K^+-depolarization results in the Ca^{2+}-dependent secretion of both dopamine and norepinephrine. These cells also synthesize and store acetylcholine. Unlike in nerve terminals where acetylcholine is stored in clear synaptic vesicles, acetylcholine in PC12 cells appears to be stored in dense core granules.[7] These granules appear to be distinct from those that store catecholamines. PC12 cells also secrete acetylcholine in a Ca^{2+}-dependent manner.

Botulinum and Tetanus Neurotoxins, Edited by
B.R. DasGupta, Plenum Press, New York, 1993

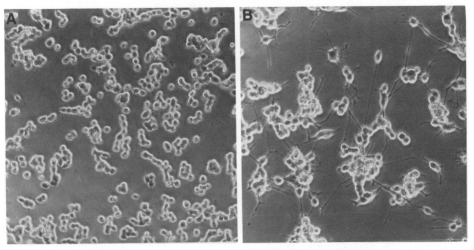

Figure 1. PC12 cells before (A) and after treatment with 75 ng/ml nerve growth factor for 7 days (B).

Treatment of PC12 cells with nerve growth factor (NGF) for 7-10 days results in a cessation of cell division and the extension of processes (Figure 1). Morphologically these processes are very similar to those of cultured sympathetic neurons. After treatment with NGF, PC12 cells show electrical excitability[8] and can form functional cholinergic synapses when co-cultured with a rat muscle cell line.[9] Like the untreated cells, NGF-treated PC12 cells synthesize, store, and take up catecholamines. NGF treatment increases the level of choline acetyltransferase[10,11] resulting in an increased acetylcholine content.[12] In response to K^+-depolarization or stimulation by nicotine NGF-treated PC12 cells secrete both catecholamines and acetylcholine.

In contrast to the effect of NGF, treatment with glucocorticoids or sodium butyrate or coculture with adrenal medullary endothelial cells causes PC12 cells to become more chromaffin-like.[4,13]

PC12 cells are generally considered one of the best if not the best cell line for the study of the effects of NGF and for the study of neuronal differentiation. It is for these purposes that PC12 cells are most commonly used. The usefulness of PC12 cells comes from their ability to grow in culture in the absence of NGF. The addition of NGF to these untreated cells allows one to examine the effect of NGF on naive cells. In contrast, sympathetic and sensory neurons can not survive without NGF which makes it very difficult to study the effects of NGF. There are a large number of papers on the effects of NGF on various aspects of PC12 cell differentiation and on the regulation of metabolic enzymes, cytoskeletal proteins, protein phosphorylation, and gene regulation. PC12 cells are also frequently used to the study of catecholamine biosynthesis and to examine the action of neurotransmitters and hormones on Ca^{2+} influx and second messenger formation. While all of these areas are of interest, they are beyond the focus of this paper, and we refer the reader to the review article mentioned above.[1-4]

EFFECTS OF NGF ON SECRETORY ACTIVITY

PC12 cells are used as a model system to study neuronal secretion. While NGF-treated PC12 cells more closely resemble sympathetic neurons than do the untreated cells, most of the experiments which have examined the secretory activity of PC12 cells have used the undifferentiated cells. Here we compare the secretory activity of NGF-treated cells with that of the untreated cells. We will also present some data on bovine chromaffin cells which are widely used to study catecholamine secretion. In general, while differentiation of PC12 cells by NGF may cause some quantitative differences in secretory activity, there does not appear to be any significant change in the basic secretory mechanism.

Secretion by intact PC12 cells

Secretion of catecholamines and acetylcholine by PC12 cells both before and after differentiation with NGF can be induced by depolarization with high K^+ or stimulation by nicotine. These releases result from an increase in Ca^{2+} influx through voltage operated Ca^{2+} channels. Like sympathetic neurons, NGF-differentiated PC12 cells contain voltage-sensitive Ca^{2+} channels which are insensitive to dihydropyridines. In contrast, undifferentiated PC12 cells, like chromaffin cells, do not contain dihydropyridine-insensitive Ca^{2+} channels.[14] Consequently, catecholamine releases from undifferentiated PC12 cells and from normal chromaffin cells are inhibited by dihydropyridines, while catecholamine releases from NGF-differentiated PC12 cells and from sympathetic neurons are not inhibited by dihydropyridines.

Although NGF causes profound changes in the morphology of PC12 cells, its effects on secretory activity are more subtle, and different laboratories have come to different conclusions as to whether NGF-treated cells release more or less catecholamines than do the control cells. Greene and Rein[15] reported that after NGF-treatment, PC12 cells release a smaller fraction of their catecholamines than do the untreated cells, and Kongsamut and Miller[14] found that the NGF-treated cells released significantly less norepinephrine in response to high K^+ than did the untreated cells. In contrast, Meldolesi et al.[16] found no difference in K^+-stimulated catecholamine release from NGF-treated and control PC12 cells. Sandberg et al.[17] found that NGF-treated cells released more acetylcholine when stimulated with veratridine than did control cells.

Muscarinic agonists also stimulate the release of catecholamines by PC12 cells, but the amount released is much less than that obtained with nicotine. Muscarinic agonists appear to act through a receptor-activated Ca^{2+}-channel that is distinct from voltage operated Ca^{2+} channels.[18,19] The amount of catecholamine released by muscarinic-stimulation of NGF-treated cells is comparable to that of the untreated cells.[16]

In Table 1, we show data obtained in our laboratory for the release of norepinephrine from NGF-treated PC12 cells and from untreated PC12 cells. These data were chosen as they give close to the average values we have obtained for the amounts of norepinephrine released by the various stimulants. However, the extent of release can vary significantly from one experiment to the next. For example, we have obtained values for K^+-stimulated release from control cells that range from less than 10% to more than 20% [^3H] norepinephrine released. Some, but not all, of this variation appears to depend on the passage number of the cells. Other investigators have also observed significant variation in the secretory activity of PC12 cells depending on passage number, and different laboratories using what appear to be very similar conditions have reported quite different levels of catecholamine secretion.

The phorbol ester TPA, 12-O-tetradecanoylphorbol 13-acetate, stimulates norepinephrine release by PC12 cells (Table 1). Phorbol esters activate protein kinase C. As has been observed by Meldolesi et al.,[16] the NGF-treated cells were less responsive to TPA than were the untreated cells. The reason for this change is unclear. Meldolesi et al.[16] and Matthies et al.[20] have found that differentiation of PC12 cells with NGF causes no major changes in protein kinase C levels.

As shown in Table 1, ATP can also stimulate catecholamine release by both control and NGF-treated cells. For the untreated cells this stimulation results primarily from an influx of Ca^{2+} through a receptor operated ion channel rather than through a voltage operated Ca^{2+} channel.[22,23] The nature of the Ca^{2+} channel of the NGF-differentiated cells has not been reported.

Table 1. Comparison of [³H] norepinephrine release from control and NGF-treated cells.

	% [³H] norepinephrine released[a]	
	Control Cells[b]	NGF-treated Cells[c]
Basal	4	4
60 mM K^+	14	20
10 μM nicotine	7	6
250 μM MgATP	18	19
20 nM TPA	12	8

[a]The amount of [³H] norepinephrine released was measured as described in Wagner and Vu.[21]
[b]Cells plated for 2-3 days on polylysine coated dishes with no added NGF
[c]Cells plated for 7-10 days on polylysine coated dishes in the presence of 75 ng/ml NGF.

Secretion by permeabilized PC12 cells

Permeabilized cells are widely used to investigate the mechanism by which an increase in intracellular Ca^{2+} induces secretion and to examine the role of intracellular proteins in this process. The techniques used are designed to permeabilize the plasma membrane but leave the secretory vesicles intact. The release of secretory products by these permeabilized cells appears to result from exocytosis, fusion of the secretory vesicles with the plasma membrane. Permeabilization of the plasma membrane allows one to control Ca^{2+} and nucleotide concentrations, and depending on the method of permeabilization to introduce macromolecules such as proteins into the cell. Permeabilized cells are very useful to determine whether the effect of some agent on secretion results from an alteration in Ca^{2+} influx or of some process that occurs after Ca^{2+} influx.

Two different permeabilization methods have been reported for PC12 cells; treatment with α-toxin[24-26] or with digitonin.[21,27-32] While there are some differences in the secretory activities of α-toxin-permeabilized and digitonin-permeabilized PC12 cells, the overall secretory properties of these two types of permeabilized cells are similar. PC12 cells can also be permeabilized by disruption with a ball homogenizer.[33] The release of catecholamines by these "cracked cells" appears to be similar to that from the digitonin-permeabilized cells.

α-Toxin makes small, 2-3 nm, holes in the plasma membrane. Small molecules such as Ca^{2+} and ATP can diffuse through these holes, but proteins can not.[24] Release of catecholamines by these cells is Ca^{2+}-dependent, but the Ca^{2+}-concentration required for half maximum release, about 20 μM, is rather high.[24,26] Stimulation of intact cells increases intracellular Ca^{2+} from about 0.1 μM to 0.5 μM. Secretion of catecholamines by these α-toxin-permeabilized cells does not require exogenously added ATP.[24,26] An advantage of permeabilization with α-toxin, as opposed to with digitonin, is that membrane associated proteins which may be involved in the secretory response (for example certain G-proteins) are retained in the cell.[25]

Treatment of PC12 cells with a low concentration of digitonin permeabilizes the plasma membrane but leaves the secretory vesicles intact. The holes made by digitonin are larger than those made by α-toxin, and proteins as well as Ca^{2+} and ATP can diffuse into and out of the permeabilized cells.[27,28,32] Catecholamine secretion by digitonin-permeabilized PC12 cells is Ca^{2+}-dependent. Half maximum release occurs in about 1.0 μM Ca^{2+} which is close to the Ca^{2+} concentration in stimulated intact PC12 cells. Immediately after permeabilization, catecholamine secretion by digitonin-permeabilized cells is almost independent of exogenous

ATP.[21,27,28] However, if the permeabilized cells are incubated for 15 to 30 min in the absence of ATP, then exogenously added ATP causes a large increase in Ca^{2+}-dependent secretion.[21,27,28] An advantage of the large holes created by digitonin is that it is possible to diffuse proteins such as enzymes and antibodies into the digitonin-permeabilized cells. This allows one to examine the role of specific proteins. For example, we have diffused protein phosphatases[32] into digitonin-permeabilized PC12 cells to examine the role of protein phosphorylation in Ca^{2+}-dependent norepinephrine secretion.

Although proteins readily diffuse from digitonin-permeabilized PC12 cells,[27] the permeabilized cells retain a high level of secretory activity[21,28,29,31,32] even after 30-60 min incubations (Table 2). This suggests that the secretory machinery of these cells may not require the participation of soluble proteins. In contrast, incubation of digitonin-permeabilized bovine chromaffin cells in the absence of Ca^{2+} results in a progressive loss of Ca^{2+}-dependent secretion.[34] Addition of cytosolic proteins prevents this loss of Ca^{2+}-dependent secretion[34-36] (Table 2). As the precise roles of these cytosolic proteins in chromaffin cells are not known, it is unclear whether this reflects some fundamental difference in the secretory machinery of these two similar types of cells.

Table 2. Comparison of [^{3}H] norepinephrine release from digitonin-permeabilized PC12 and bovine chromaffin cells.

	% [^{3}H] norepinephrine released[a]	
	No Ca^{2+}	10 μM Ca^{2+}
PC12 cells		
no wash	5	27
30 min wash	4	22
Bovine chromaffin cells		
no wash	4	21
30 min wash	4	9
30 min wash containing cytosolic proteins.	4	18

[a]The amounts of [^{3}H] norepinephrine released were measured as described in Wagner and Vu[21] and Wu and Wagner.[37]
[b]PC12 cells were permeabilized by a 6 min incubation in 7.5 mM digitonin. The release and wash solutions did not contain digitonin.
[c]Chromaffin cells were permeabilized by a 6 min incubation in 15 mM digitonin. The release and wash solutions did not contain digitonin.

Cracked cells offer an another method to gain access to the secretory machinery. The cells are homogenized using a ball homogenizer which tears open the plasma membrane but leaves the secretory vesicles intact. Soluble proteins can be separated from what are referred to as cell ghosts by centrifugation. Norepinephrine release by these PC12 cell ghosts is Ca^{2+}-dependent; 0.5 μM Ca^{2+} gives about half maximum stimulation.[33] Secretion from these cell ghosts is enhanced by the addition of MgATP and cytosolic proteins. The dependence of the secretory activity of the PC12 cell ghosts on the addition of cytosolic proteins is more like the results obtained with digitonin-permeabilized bovine chromaffin cells than with digitonin-permeabilized PC12 cells. However, these PC12 cell ghosts even in the absence of exogenous protein retain a significant amount of Ca^{2+}-dependent secretory activity which suggests that the basic secretory response may not require cytosolic proteins. Cracked PC12 cells appear to be a promising system to examine the role of both membrane and cytosolic proteins in the secretory response.

Effect of NGF-treatment on secretion by digitonin-permeabilized PC12 cells

Most of the results in the literature for secretion by digitonin-permeabilized PC12 cells are for cells which have not been treated with NGF. To determine whether these results are applicable to the differentiated cells, we examined the secretory activity of PC12 cells which have been cultured for 7-10 days in the presence of NGF. Figure 2 shows the Ca^{2+}-dependence of [^{3}H] norepinephrine release from digitonin-permeabilized cells. For both control and NGF-treated cells, half maximum release occurred in 1-2 μM Ca^{2+}, and maximum release was obtained in 10 μM Ca^{2+}. A similar result has been reported by Peppers and Holz.[27] As shown in Figure 2, the amount of norepinephrine secreted by the permeabilized NGF-treated cells is usually less than that secreted by the permeabilized control cells. However, as observed with the intact cells, the amount of norepinephrine released by the digitonin-permeabilized cells is somewhat variable. Ca^{2+}-dependent secretion by the undifferentiated cells varies from 15 to 25% [^{3}H] norepinephrine released depending in part on cell passage number, and we have obtained as much as 22% [^{3}H] norepinephrine release from the NGF-differentiated PC12 cells in the presence of 10 μM Ca^{2+}.

The time course of norepinephrine release from control and NGF-treated cells are shown in Figure 3. For both types of cells the rate of norepinephrine release during the first 5 min after the addition of Ca^{2+} is about twice the rate of release during the next 10 min. While the overall amounts of NE released from the permeabilized cells are comparable to those from the ATP- or K^+-stimulated intact cells, the initial rates of norepinephrine release from intact cells are faster than those from the permeabilized cells. For example, rate of norepinephrine release by intact cells during the first 1 min of stimulation by ATP is about 10 times that during the next 5 min of stimulation.

GTPγS-stimulation of norepinephrine release from permeabilized PC12 cells

Norepinephrine secretion in digitonin-permeabilized PC12 cells can also be stimulated by the addition of hydrolysis resistant GTP analogs, GTPγS, guanosine 5-'O-3-thiotriphosphate,

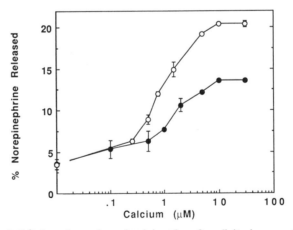

Figure 2. Ca^{2+}-dependence of norepinephrine release from digitonin-permeabilized PC12 cells. (○) cells plated for 2-3 days on polylysine coated dishes with no added NGF. (●) cells plated for 7-10 days on polylysine coated dishes in the presence of 75 ng/ml NGF. Cells were permeabilized for 6 min with 7.5 μM digitonin in the absence of Ca^{2+}, and then [^{3}H] norepinephrine released measured in the presence of the indicated Ca^{2+} concentration as described in Wagner and Vu.[21]

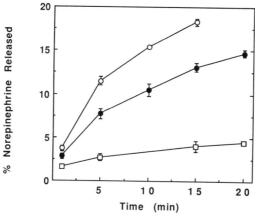

Figure 3. Time course of norepinephrine release from digitonin-permeabilized PC12 cells. (O) cells plated for 2-3 days on polylysine coated dishes with no added NGF. (●, ▢) cells plated for 7 days on polylysine coated dishes in the presence of 75 ng/ml NGF. Cells were permeabilized for 6 min with 7.5 μM digitonin in the absence of Ca^{2+}, and then [^{3}H] norepinephrine released measured in the presence of 10 μM Ca^{2+} (O, ●) or in the absence of Ca^{2+} (▢) as described in Wagner and Vu.[21]

and GMP-PNP, guanylyl imidodiphosphate.[31] These GTP analogs bind to and activate heterotrimeric G-proteins, and they also bind to low molecular weight GTP binding proteins such as the *ras* proteins. While secretion by digitonin-permeabilized PC12 cells in the presence of Ca^{2+} is not affected by GTPγS or GMP-PNP, secretion in the absence of Ca^{2+} is stimulated 2 to 3 fold[31] (Table 3). This stimulation does not appear to result from the release of Ca^{2+} from intracellular stores, activation of either protein kinase C or cAMP-dependent protein kinase, or stimulation of phospholipase A_2.[31,38] GTPγS has also been shown to stimulate Ca^{2+}-independent secretion in a number of other types of permeabilized cells (reviewed by Gomperts[39]), but as with PC12 cells the mechanism by which GTPγS induces release in these cells is not known.

In Table 3, we compare the effects of GTPγS on norepinephrine release from digitonin-permeabilized control PC12 cells, PC12 cells differentiated with NGF, and bovine chromaffin cells. GTPγS caused a large increase in Ca^{2+}-independent secretion but not in Ca^{2+}-dependent secretion from both types of PC12 cells. Ca^{2+}-independent secretion in the digitonin-permeabilized bovine chromaffin cells was also stimulated by GTPγS, but the extent of stimulation was much less than that observed with the PC12 cells. Others have previously reported that GTPγS and GMP-PNP stimulate Ca^{2+}-independent secretion in digitonin-permeabilized chromaffin cells.[40-42] While there is some disagreement as to the amount of stimulation GTPγS causes in digitonin-permeabilized chromaffin cells, the extent is significantly less than that observed with PC12 cells.

The data shown in Table 3 were obtained using digitonin-permeabilized cells. Very different results are obtained when α-toxin permeabilized cells are used. The addition of GTPγS to α-toxin-permeabilized PC12 cells inhibits Ca^{2+}-stimulated release and has no effect on Ca^{2+}-independent secretion.[25] This inhibition of Ca^{2+}-stimulated release by GTPγS is prevented if the cells are first modified with pertussis toxin. Pertussis toxin ADP-ribosylates two classes of G-proteins, G_i and G_o. Ca^{2+}-stimulated release in the absence of GTPγS by α-toxin-permeabilized PC12 cells is not affected by pertussis toxin.[25] These results suggest that catecholamine secretion by PC12 cells might be modulated by a pertussis toxin sensitive G-protein. In contrast to these results with α-toxin-permeabilized cells, GTPγS causes only a very slight increase in

Table 3. GTPγS-stimulation of norepinephrine secretion by PC12 cells and bovine chromaffin cells.

	% [³H] norepinephrine released			
	Basal	200 µM GTPγS	10 µMCa²⁺	10 mM Ca²⁺ and 200 mM GTPγS
PC12 cells[a]	4	15	24	26
NGF-treated PC12 cells[b]	6	12	22	24
Bovine chromaffin cells	4	7	17	17

The cells were permeabilized by a 6 min incubation in the absence of Ca²⁺ with 7.5 µM digitonin for PC12 cells and 15 µM digitonin for bovine chromaffin cells.[21,37] The permeabilization buffers were removed and [³H] norepinephrine release measured in the presence and absence of 10 µM Ca²⁺ and in the presence and absence of 200 µM GTPγS.
[a]Cells plated for 2-3 days on polylysine coated dishes with no added NGF.
[b]Cells plated for 7-10 days on polylysine coated dishes in the presence of 75 ng/ml NGF.

Ca²⁺-stimulated release by digitonin-permeabilized PC12 cells and a large increase in Ca²⁺-independent secretion (Table 3). Pretreatment of PC12 cells with pertussis toxin has little or no effect on either Ca²⁺-stimulated or GTPγS-stimulated norepinephrine release from the digitonin-permeabilized PC12 cells.[31] As digitonin makes larger pores in the plasma membrane than does α-toxin[24], it is possible that the G-protein responsible for the inhibition of Ca²⁺-dependent secretion in the α-toxin-permeabilized cells is lost from the digitonin-permeabilized cells.

The effect of GTP analogs on Ca²⁺-dependent catecholamine secretion by permeabilized bovine chromaffin cells also depends on the method of permeabilization. In electropermeabilized bovine chromaffin cells, GTPγS inhibits Ca²⁺-stimulated catecholamine release[43], but in α-toxin permeabilized bovine chromaffin cells, GTPγS enhances Ca²⁺-stimulated release.[44] Stimulation of Ca²⁺-dependent secretion by GTPγS in α-toxin permeabilized cells appears to result from an activation of protein kinase C.[43] As both α-toxin and electric fields[45] make small holes in the plasma membrane, it is not obvious why cells permeabilized by these two different techniques respond so differently to GTPγS.

In chromaffin cells permeabilized with digitonin, GTPγS has been reported to have little effect on Ca²⁺-stimulated release[40], enhance Ca²⁺-stimulated release[46], and to inhibit Ca²⁺-stimulated release.[42] As shown in Table 3, our results are similar to those of Bittner et al.[40]; GTPγS had little effect on Ca²⁺-stimulated release. At this time, it is difficult to reconcile these differences.

Effects of tetanus and botulinum neurotoxins

PC12 cells and bovine chromaffin cells have been used to study the mechanism of action of tetanus toxin. While incubation of NGF-differentiated PC12 cells with a low concentration of tetanus toxin inhibits evoked acetylcholine secretion,[17] acetylcholine secretion by the undifferentiated cells is not inhibited by tetanus toxin. This difference does not appear to result from a difference in tetanus toxin binding as both undifferentiated and NGF-differentiated PC12 cells contain high affinity receptors for tetanus toxin.[47] However, NGF treatment does increase the level of expression of high affinity tetanus toxin receptors. The addition of tetanus toxin light chain to permeabilized chromaffin cells inhibits Ca²⁺-dependent norepinephrine secretion[48,49] confirming that the light chain is responsible for the intracellular effects of this toxin on exocytosis. Permeabilized cells should offer a good system to investigate the mechanism by which tetanus toxin inhibits exocytosis and to identify its intracellular targets.

PC12 cells and bovine chromaffin cells have also been used to study the mechanism of action of botulinum neurotoxins. Basal and K^+-stimulated norepinephrine releases by PC12 cells are not affected by preincubation with botulinum A toxin.[50] However, the addition of the whole toxin or its isolated light chain to digitonin-permeabilized PC12 cells inhibits Ca^{2+}-dependent norepinephrine secretion.[50] The addition of the isolated light chains of botulinum neurotoxin serotypes A, B and E to cracked PC12 cells also inhibits Ca^{2+}-dependent norepinephrine release.[33] The amount of inhibition by type A light chain is significantly less than that obtained with the other two types of light chains indicating that these neurotoxins may have different intracellular targets. The sites of action of these light chains appear to be located either on intracellular membranes or on the cytoskeleton.[33] Botulinum neurotoxin light chains also inhibit Ca^{2+}-dependent catecholamine secretion by permeabilized bovine chromaffin cells.[51,52]

CONCLUSIONS

PC12 cells appear to be a valid system for examining the molecular processes underlying neuronal secretion. Permeabilized PC12 cells are useful as they allow one to examine directly the roles of Ca^{2+}, nucleotides (ATP and GTP), and proteins (for example G-proteins, kinases, and phosphatases) in the final steps of exocytosis. As shown here, the secretory activities of permeabilized NGF-treated cells are very similar to those of the undifferentiated cells. Thus, although NGF causes a large change in the morphology of PC12 cells and alters many of the biochemical properties of these cells,[1-4] differentiation of PC12 cells with NGF probably does not result in any large changes in the proteins directly involved in the fusion of the secretory vesicles with the plasma membrane.

REFERENCES

1. Greene LA, Tischler AS. PC12 pheochromocytoma cultures in neurobiological research. In: Fedorff S, Hertz L, eds. Advances in cellular neurobiology, vol. 3. New York: Academic Press, 1982:373-414.
2. Greene LA. The importance of both early and delayed responses in the biological actions of nerve growth factor. Trends Neurosci 1984; 7:91-94.
3. Guroff G. PC12 cells as a model of neuronal differentiation. In: Bottenstein JE, Sato G, eds. Cell culture in the neurosciences. New York: Plenum, 1985:245-271.
4. Fujita K, Lazarovici P, Guroff G. Regulation of the differentiation of PC12 pheochromocytoma cells. Envir Health Perspec 1989; 80:127-142.
5. Greene LA, Tischler AS. Establishment of a noradrenergic clonal line of rat adrenal pheochromocytoma cells which respond to nerve growth factor. Proc Natl Acad Sci USA 1976; 73:2424-2428.
6. Greene LA, Rein, G. Short-term regulation of catecholamine synthesis in an NGF-responsive clonal line of rat pheochromocytoma cells. J Neurochem 1978; 30:549-555.
7. Rebois RV, Reynolds EE, Toll L, Howard BD. Storage of dopamine and acetylcholine in granules of PC12, clonal pheochromocytoma cell line. Biochemistry 1980; 19:1240-1248.
8. Dichter MA, Tischler, AS, Greene LA. Nerve growth factor-induced increase in electrical excitability and acetylcholine sensitivity of a rat pheochromocytoma cell line. Nature 1977; 268:501-504.
9. Schubert D, Heinemann S, Kidokoro Y. Cholinergic metabolism and synapse formation by a rat nerve cell line. Proc Natl Acad Sci USA 1977; 74:2579-2583.
10. Lucas CA, Czlonkowska A, Kreutzberg GW. Regulation of acetylcholinesterase by nerve growth factor in the pheochromocytoma PC12 cell line, Neurosci Lett 1980; 18:333-337.
11. Rieger F, Shelanski ML, Greene, LA. The effects of nerve growth factor on acetylcholinesterase and its multiple forms in cultures of rat PC12 pheochromocytoma cells: increased total specific activity and appearance of the 16S molecular form. Develop Biol 1980; 76:238-243.
12. Greene LA, Rein G. Synthesis, storage and release of acetylcholine by a noradrenergic pheochromocytoma cell line. Nature 1977; 268:349-351.

13. Mizrachi Y, Naranjo JR, Levi B-Z, Pollard HB, Lelkes PI. PC12 cells differentiate into chromaffin cell-like phenotype in coculture with adrenal medullary endothelial cells. Proc Natl Acad Sci USA 1990; 87:6161-6165.

14. Kongsamut S, Miller RJ. Nerve growth factor modulates the drug sensitivity of neurotransmitters release from PC12 cells. Proc Natl Acad Sci USA 1986; 83:2243-2247.

15. Greene LA, Rein G. Release of (^{3}H)-norepinephrine from a clonal line of pheochromocytoma cells (PC12) by nicotinic cholinergic stimulation. Brain Res 1977; 138:521-528.

16. Meldolesi J, Gatti G, Ambrosini A, Pozzan T. Second messenger control of catecholamine release from PC12 cells: role of muscarinic receptors and nerve-growth-factor-induced differentiation. Biochem J 1988; 255:761-768.

17. Sandberg K, Berry C, Rogers TB. Studies on the intoxication pathway of tetanus toxin in the rat pheochromocytoma (PC12) cells line: binding, internalization, and inhibition of acetylcholine release. J Biol Chem 1989; 264:5679-5686.

18. Inoue K, Kenimer JG. Muscarinic stimulation of calcium influx and norepinephrine release in PC12 cells. J Biol Chem 1988; 263:8157-8161.

19. Takashima A, Kenimer JG. Muscarinic-stimulated norepinephrine release and phosphoinositide hydrolysis in PC12 cells are independent events. J Biol Chem 1989; 264:10654-10659.

20. Matthies HJG, Palfrey HC, Hirning LD, Miller RJ. Down regulation of protein kinase C in neuronal cells: effects on neurotransmitter release. J Neurosci 1987; 7:1198-1206.

21. Wagner PD, Vu N-D. Thiophosphorylation causes Ca^{2+}-independent norepinephrine secretion from permeabilized PC12 cells. J Biol Chem 1989; 264:19614-19620.

22. Inoue K, Nakazawa K, Fujimori K, Takanaka A. Extracellular adenosine 5'-triphosphate-evoked norepinephrine secretion not relating to voltage gated Ca channels in pheochromocytoma PC12 cells. Neurosci Lett 1989; 106:294-299.

23. Sela D, Ram E, Atlas D. ATP receptor: a putative receptor-operated channel in PC-12 cells. J Biol Chem 1991; 266:17990-17994.

24. Ahnert-Hilger G, Bhakdi S, Gratzl M. Minimal requirements for exocytosis: a study using PC12 cells permeabilized with staphylococcal α-toxin. J Biol Chem 1985; 260:12730-12734.

25. Ahnert-Hilger G, Brautigam M, Gratzl M. Ca^{2+}-stimulated catecholamine release from α-toxin-permeabilized PC12 cells: biochemical evidence for exocytosis and its modulation by protein kinase C and G proteins. Biochemistry 1987; 26:7842-7848.

26. Ahnert-Hilger G, Gratzl M. Further characterization of dopamine release by permeabilized PC12 cells. J Neurochem 1987; 49:764-770.

27. Peppers SC, Holz RW. Catecholamine secretion from digitonin-treated PC12 cells: effects of Ca^{2+}, ATP, and protein kinase C activators. J Biol Chem 1986; 261:14665-14669.

28. Schafer T, Karli UO, Gratwohl EK-M, Schweizer FE, Burger MM. Digitonin-permeabilized cells are exocytosis competent. J Neurochem 1987; 49:1697-1707.

29. Schafer T, Karli UO, Schweizer FE, Burger MM. Docking of chromaffin granules; a necessary step in exocytosis. Biosci Rep 1987; 7:269-279.

30. Matthies HJG, Palfrey CH, Miller RJ. Calmodulin- and protein phosphorylation-independent release of catecholamines from PC12 cells. FEBS Lett 1988; 229:238-242.

31. Carroll AG, Rhoads AR, Wagner PD. Hydrolysis-resistant GTP analogs stimulate catecholamine release from digitonin-permeabilized PC12 cells. J Neurochem 1990; 55:930-936.

32. Wagner PD, Vu N-D. Regulation of norepinephrine secretion in permeabilized PC12 cells by Ca^{2+}-stimulated phosphorylation: effects of protein phosphatases and phosphatase inhibitors. J Biol Chem 1990; 265:10352-10357.

33. Lomneth R, Martin TFJ, DasGupta BR. Botulinum neurotoxin light chain inhibits norepinephrine secretion in PC12 cells at an intracellular membranous or cytoskeletal site. J Neurochem 1991; 57:1413-1421.

34. Sarafian T, Aunis D, Bader M-F. Loss of proteins from digitonin-permeabilized adrenal chromaffin cells essential for exocytosis. J Biol Chem 1987; 262:16671-16676.

35. Wu YN, Wagner PD. Calpactin-depleted cytosolic proteins restore Ca^{2+}-dependent secretion to digitonin-permeabilized bovine chromaffin cells. FEBS Lett 1991; 282:197-199.

36. Morgan A, Burgoyne RD. Exo1 and Exo2 protein stimulate calcium-dependent exocytosis in permeabilized adrenal chromaffin cells. Nature 1992; 355:833-836.

37. Wu YN, Wagner PD. Effects of phosphatase inhibitors and a protein phosphatase on norepinephrine secretion by permeabilized bovine chromaffin cells. Biochim Biophys Acta 1991; 1092:384-390.

38. Yang YC, Vu N-D, Wagner PD. Guanine nucleotide stimulation of norepinephrine secretion from permeabilized PC12 cells: effects of Mg^{2+}, other nucleotide triphosphates and N-ethylmaleimide. Biochim Biophys Acta 1992; 1134:285-291.

39. Gomperts BD. G_E: a GTP-binding protein mediating exocytosis. Annu Rev Physiol 1990; 52:591-606.
40. Bittner MA, Holz RW, Neubig RR. Guanine nucleotide effects on catecholamine secretion from digitonin-permeabilized adrenal chromaffin cells. J Biol Chem 1986; 261:10182-10188.
41 Morgan A, Burgoyne RD. Stimulation of Ca^{2+}-independent catecholamine secretion from digitonin-permeabilized bovine adrenal chromaffin cells by guanine nucleotide analogues. Biochem J 1990; 269:521-526.
42. Athayde CM. Guanosine-5'-[3-O-thio]triphosphate inhibits secretion from bovine chromaffin cells. Biochem Soc Trans 1990; 18:452-453.
43. Knight DE, Baker PF. Guanine nucleotides and Ca-dependent exocytosis: studies on two adrenal cell populations. FEBS Lett 1985; 189:345-349.
44. Bader M-F, Sontag J-M, Thierse D, Aunis D. A reassessment of guanine nucleotide effects on catecholamine secretion from permeabilized chromaffin cells. J Biol Chem 1989; 264:16426-16434.
45. Baker PF, Knight DE. Calcium control of exocytosis and endocytosis in bovine adrenal medullary cells. Phil Trans R Soc Lond B 1981; 296:83-103.
46. Burgoyne RD, Morgan A, O'Sullivan AJ. The control of cytoskeletal actin and exocytosis in intact and permeabilized adrenal chromaffin cells: role of calcium and protein kinase C. Cell Signalling 1989; 1:323-344.
47. Walton KM, Sandberg K, Rogers TB, Schnaar RL. Complex ganglioside expression and tetanus binding by PC12 pheochromocytoma cells. J Biol Chem 1988; 263:2055-2063.
48. Ahnert-Hilger G, Weller U, Dauzenroth M-E, Habermann E, Gratzl M. The tetanus toxin light chain inhibits exocytosis. FEBS Lett 1989; 242:245-248.
49. Bittner MA, Habig WH, Holz RW. Isolated light chain of tetanus toxin inhibits exocytosis: studies in digitonin-permeabilized cells. J Neurochem 1989; 53:966-968.
50. McInnes C, Dolly JO. Ca^{2+}-dependent noradrenaline release from permeabilized PC12 cells is blocked by botulinum neurotoxin A or its light chain. FEBS Lett 1990; 261:323-326.
51. Bittner MA, DasGupta BR, Holz RW. Isolated light chains of botulinum neurotoxins inhibit exocytosis: studies in digitonin-permeabilized chromaffin cells. J Biol Chem 1989; 264:10354-10360.
52. Strecher B, Habermann E, Gratzl M, Ahnert-Hilger G. The light chain but not the heavy chain of botulinum, A toxin inhibits exocytosis from permeabilized adrenal chromaffin cells. FEBS Lett 1989; 255:391-394.

TRANSMITTER RELEASE IN *APLYSIA*: APPLICABILITY OF QUANTAL MODELS AND EVIDENCE FOR POSTSYNAPTIC CONTROL

Daniel Gardner

Department of Physiology and Biophysics, and
Department of Neuroscience
Cornell University Medical College
New York, New York 10021

INTRODUCTION

Neurobiology in Molluscan Ganglia

The use of molluscan ganglia for neurobiological studies combines multiple technical advantages with demonstrated versatility. These benefits derive not only from the accessibility of individual neurons, and consequent ease of study, but also from the network properties of these neuronal arrays. A well-recognized advantage is that many molluscan neurons are distinct individuals. As a consequence, using criteria such as shape, size, position, electrophysiology and connectivity, molluscan neurons in several species have been identified. The use of identified cells permits incrementally assembling a library of specific information by the repeated study of the same cell in successive preparations. It further permits examination of those specific features which distinguish individual neurons, rather than studying only those general properties which are common to classes of similar cells. Similarly, specific network properties which characterize interconnections of identified cells may be used to examine or contrast particular elements, permitting analyses not available in such systems as cultured cell lines, chromaffin cells, or synaptosomes.

Using both cellular and network properties of identified neurons in the buccal ganglia of the marine mollusc *Aplysia californica*, this chapter presents three findings relevant to our picture of how molluscan cells release transmitter, and how the nervous system controls this release. The first finding shows synaptic strength to vary from synapse to synapse, even between identical synapses from preparation to preparation. The second shows these differences in synaptic strength to be specified by the postsynaptic neuron. Finally, the postsynaptic neuron is found to exert its influence by telling the presynaptic neuron how much transmitter to release. These results are presented not only for their inherent significance, but also as aids to examination of models for transmitter release at central synapses that may be relevant to studies of neurotoxins.

The Buccal Ganglia of *Aplysia*

Although the *Aplysia* buccal ganglia are an acknowledged standard preparation for analysis of the mechanisms of action of presynaptic neurotoxins[1,2], this use is only part of the long history of the utilization of this gangion for studies of transmitter release and presynaptic actions, as well as postsynaptic and network properties. Studies of presynaptic mechanisms using these ganglia are facilitated by two similar, paired, cells first discovered by Strumwasser[3]. This finding was extended by Gardner[4] and Gardner and Kandel[5], who identified several cells forming a complex network in each ganglion, including the two multiply-presynaptic neurons B4 and B5, as well as other neurons which receive cholinergic input from them. This network features not only synaptic convergence and divergence but also several different types of synaptic potential[5,6]. For example, divergent presynaptic terminals permit B4 and B5 to produce not only cholinergic excitation and inhibition on distinct identified postsynaptic neurons, but also dual excitation-inhibition on a third type of postsynaptic cell. An early exploitation of synaptic convergence from this cell pair was by Tauc et al[7], who modified transmitter release by injecting cholinesterase into one of the presynaptic cells, using the other as a control.

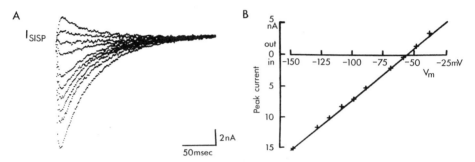

Figure 1. A self-inhibitory synaptic potential (SISP) is mediated by presynaptic autoceptors. A: curare-sensitive currents underlying the SISP. Each trace is the difference between a membrane current recorded in control sea water and a corresponding current recorded in the same cell at the same clamp potential (V_m) in tubocurarine. Currents were digitized at a 1 ms rate and subtracted digitally. The ten traces show the shape of the SISP at each of ten values of V_m, from −37 mV (top) to −147 mV (bottom). Inward currents shown as downward deflections. B: peak synaptic conductance is voltage–independent over this 110 mV range. The peak current of each trace is plotted (+) as a function of V_m. The line is a least-squares fit with slope of 0.18 μS. Reprinted from Gardner[8].

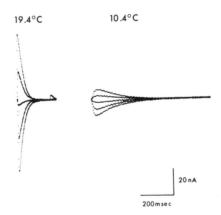

Figure 2. Cooling reduces peak amplitude and slows decay of buccal ganglia inhibitory postsynaptic currents. Left: five postsynaptic currents recorded at 19.4 °C at five different clamp potentials (V_m) with 20 mV spacing. The small displacement at the end of one trace is an additional synaptic current from an unidentified presynaptic neuron. Right: same cell and V_m as left traces, but recorded at 10.4 °C. Each set of five records is recorded on a common baseline.

The synaptic connectivity of this ganglion includes modulatory presynaptic receptors, first noted by Gardner[8], in the form of a self-inhibitory synaptic potential. Presynaptic autoceptors, responding to acetylcholine (ACh) released with each action potential, produced a synaptic conductance change in the same cell that released the transmitter. Figure 1 shows the underlying synaptic currents in the conjoint pre-and-postsynaptic cell, revealed as the difference currents obtained by subtraction of tubocurarine-blocked afterpotential currents from intact controls. White and Gardner[9] analyzed the effect of this self-inhibitory potential on neuronal firing and synchrony. Subsequently, Baux and Tauc[10] and Baux et al[11] identified both nicotinic and muscarinic autoceptors, as well as additional receptors to other neuromodulators and related effects on transmitter release.

Transmitter release in the buccal ganglia displays marked temperature dependence. In a study of conventional inhibitory synapses, Gardner and Stevens[12] presented evidence suggesting that synaptic current decay is determined by a postsynaptic-channel-closing conformational change, additionally showing that this decay process was markedly prolonged by cooling, with a Q_{10} of approximately 5. Figure 2 shows that cold not only slows postsynaptic currents, it also reduces synaptic current peak amplitude. Typically, peak current at 10 °C is only one-quarter of that seen at 22 °C. However, the time integral of synaptic conductance is reduced only slightly over the same temperature range[13]. Depending upon the degree of prolongation of the release process, these data suggest a marked reduction in transmitter release with cooling which may serve to balance partially the increased duration of postsynaptic action.

Analysis of postsynaptic current fluctuations has been used to examine both presynaptic and postsynaptic properties in this preparation. Using noise analysis of exogenous ACh, Gardner and Stevens[12] made predictions about postsynaptic channel function. Simonneau et al[14] first used another form of noise analysis – fluctuations in the amplitude of long-duration postsynaptic currents produced by prolonged presynaptic depolarization – to derive presynaptic properties related to quantal release of the neurotransmitter at the conventional inhibitory synapse.

METHODS

Recording

The analyses described in this chapter depend upon simultaneous recording from the four-cell network of B4, B5, and a pair of common postsynaptic cells, featuring four similar inhibitory synapses. As diagrammed in Figure 3, four neurons are simultaneously penetrated by six electrodes in order to record from the two presynaptic cells, and to voltage-clamp the two postsynaptic cells. Standard methods for recording and voltage clamping are summarized in Gardner[15]. Each action potential in presynaptic cell B4 or B5 produces a postsynaptic current (PSC) in the postsynaptic neuron. The two presynaptic cells are stimulated alternately, at 30-second intervals, and the postsynaptic currents produced by each, in alternation, are recorded simultaneously in the two postsynaptic neurons.

Figure 4 presents examples of four single, unaveraged, PSCs produced by this protocol. In this system, PSCs can be recorded with very low noise under a spatially well-controlled voltage clamp. Properties of the PSCs are described in Gardner and Stevens[12] and Gardner[17]. Briefly, the PSC is of large amplitude, often of 10 to 30 nA peak current. It decays with single time constant, with both synaptic conductance and synaptic decay independent of membrane potential. The two PSCs on the left of Figure 4 are recorded simultaneously in postsynaptic cells B3 and B6 following an action potential in presynaptic cell B4; the PSCs on the right are produced simultaneously in the same pair of postsynaptic cells to an action potential in the other presynaptic cell, B5. Characteristically, these PSCs are of different amplitude, reflecting different synaptic strengths at each of these four synapses. Additionally, patterns of greater or lesser synaptic efficacy are not fixed by cell type, but vary at the same identified synapses from animal to animal, suggesting that the synaptic strengths seen result from a post-developmental plastic process.

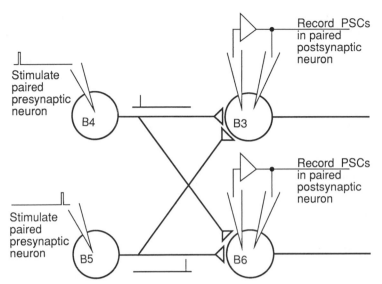

Figure 3. Diagram of experimental protocol for comparing PSCs produced in paired postsynaptic cells by paired presynaptic neurons B4 and B5. Cells B4 and B5 are stimulated alternately by means of individual intracellular electrodes, producing action potentials shown schematically. Two of the follower cells innervated by B4 and B5 (here, B3 and B6) are each voltage-clamped with two electrodes to permit recording and comparison of simultaneously-produced PSCs. Adapted from Gardner[16], with permission of the American Physiological Society.

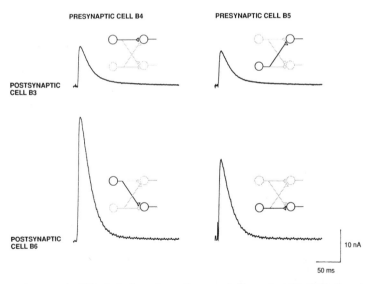

Figure 4. Synaptic strengths differ in the buccal ganglia network. Examples of individual unaveraged PSCs recorded at four synapses in the same preparation. *Left:* PSCs recorded simultaneously in postsynaptic cells B3 and B6 in response to stimulation of presynaptic cell B4. *Right:* PSCs recorded simultaneously in post-synaptic cells B3 and B6 in response to stimulation of presynaptic cell B5. Both postsynaptic cells are record-ed at the same gain; the larger noise observed in the B6 trace reflects a greater sensitivity of one voltage clamp to interference radiated by the PDP-11 computer. Reprinted from Gardner[15], with permission of the American Physiological Society

In a volume stressing mechanisms of action of presynaptic neurotoxins, it is of interest to ask if these different synaptic strengths result from differences in presynaptic transmitter release. A related question asks which element of the synapse – presynaptic or postsynaptic neuron – has greater influence in the specification of synaptic strength.

Quantal Analysis

Relating synaptic strength to presynaptic or postsynaptic mechanisms utilizes a technique first introduced by Del Castillo and Katz[18] at neuromuscular junction and subsequently modifed for the study of central synapses (for recent examples, see Bekkers and Stevens[19], Malinow and Tsien[20], and Redman[21]). These techniques for quantal analysis examine transmitter released by conventional presynaptic action potentials, evoked in or near the neuronal soma and propagating into divergent presynaptic terminals. The methods are thus distinct from the two other techniques for examining fluctuations in postsynaptic current amplitudes used by Gardner and Stevens[12] and by Simonneau et al[14] and noted above.

Quantal analysis requires not only selection of a particular model, but also assurance that experimental data conform to the model. In order to avoid unwarranted arbitrary constraints, the analysis procedure selected is relatively model-independent. Furthermore, like previous examinations of these synapses[14], it is an indirect quantal analysis in which individual low-noise miniature PSCs are not directly recorded. As a consequence, the method cannot be used to derive independent values of such traditional quantal parameters as the quantal conductance γ, the releaseable store n, and the probability of release p. However, these analyses, properly employed, can be used to determine if larger PSCs result from presynaptic factors releasing more transmitter, or from postsynaptic factors affecting the number or sensitivity of receptors.

At a minimum, it is necessary to assume that neurotransmitter release is quantal, with independent release of quanta characterized by a uniform probability, and that the parameters characterizing quantal release display stationarity. (It should be noted that these assumptions are often used to characterize the neuromuscular junction, but their applicability to central synapses is unproven.) These assumptions permit equating the mean peak amplitude M of the postsynaptic conductance with the product of the three quantal factors introduced above:

$$M = n p \gamma \tag{1}$$

Analysis utilizing fluctuations in synaptic current amplitude, rather than its mean value alone, requires examination of sources of variance. The assumption that the short-term variability in the number of quanta released per action potential is substantially more than the variability in the number of molecules packed in each quantum, and also more than the variability in postsynaptic receptor number or in desensitization is consistent with synaptic biophysics. As discussed below, the data examined in this study can be used to provide partial verification of this view. With this assumption, the variance σ^2 of synaptic conductance is:

$$\sigma^2 = n p \gamma^2 (1 + (CV_\gamma)^2 - p) \tag{2}$$

where CV_γ is the coefficient of variation of an individual quantum. The mean M and variance σ^2 of sets of PSCs at each synapse are experimentally-determined data, and can be used to derive the following factor:

$$\frac{M^2}{\sigma^2} = \frac{\gamma^2 n^2 p^2}{\gamma^2 n p (1 + (CV_\gamma)^2 - p)} = \frac{m}{(1 + (CV_\gamma)^2 - p)} \tag{3}$$

where $m = n p$, the number of quanta released per action potential, conventionally called the quantal content. This computation is useful because M^2/σ^2 is an essentially presynaptic factor, dependent upon processes located presynaptically, rather than postsynaptically. It is proportional, but not identical, to quantal content m.

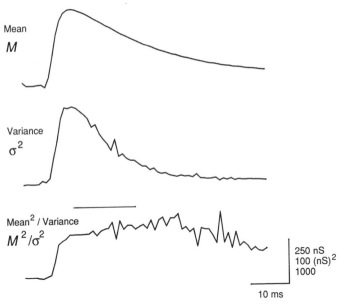

Figure 5. Ensemble averaging of postsynaptic currents yields an index that reflects presynaptic quantal factors. *Top:* the mean M of 31 successively-recorded postsynaptic currents at a single synapse shows the time course of the underlying conductance. *Middle:* the ensemble variance σ^2 is near zero preceding the PSC, with a time course approximating the square of the mean response. *Bottom:* M^2/σ^2 is equal to $m/(1 + (CV_\gamma)^2 - p)$, and thus reflects presynaptic factors. The horizontal bar shows the 16 ms following the PSC peak during which the average value of M^2/σ^2 was computed; for this synapse, M^2/σ^2 averaged 1075. For each trace, only the initial 64 ms is shown. Reprinted from Gardner[22], with permission of the American Physiological Society.

Substituting equations (2) and (3) in (1) shows that M can be expressed as the product of two factors:

$$M = \frac{m}{(1 + (CV_\gamma)^2 - p)} \cdot \gamma (1 + (CV_\gamma)^2 - p) \tag{4}$$

The first factor is M^2/σ^2, while the second, equivalent to σ^2/M, is a partially postsynaptic factor which is proportional to the postsynaptic quantal conductance change.

Typical experimental derivation of these quantally-derived parameters for one synapse is shown in Figure 5. The top trace is the ensemble average of 31 successively recorded PSCs, showing the time course of this mean PSC, whose peak conductance is the value of M characterizing the synapse. The 31 PSCs contributing to this average have approximately the same amplitude, and this amplitude must not drift during collection of the data set. This requirement for stationarity – for a stable mean value – is essential for the statistical analysis; disregarding it can introduce serious error. At all synapses, the number of PSCs recorded per epoch has been small, in order to avoid errors in estimation of quantal parameters due to nonstationarity. Analyses of longer runs produced essentially similar correlations and conclusions. Similarly, data sets showing drift in peak amplitude over time have been discarded.

Each of the 31 PSCs in this typical data set contributing to the mean does not have exactly the same amplitude, presumably because of quantal fluctuations, and the time course of the resulting variance is shown in the middle trace of Figure 5. This trace additionally provides validation of the assumption that PSCs fluctuate because of differences in the number of released quanta. Succinctly, the basis of this quantal analysis is that as the average size of a PSC gets smaller, variations due to individual quanta become increasingly visible. However, such variations might instead be ascribed not to quantal fluctuations, but to a contaminating artifact from membrane noise. Any such variance from a non-quantal source would distort the analysis and produce artificially low values of M^2/σ^2. However, the time course of the variance – approximating the square of the PSC – suggests that the observed variance originates from the synaptic conductance change, consistent with predictions of the quantal model.

The bottom trace of Figure 5 shows the derived presynaptic M^2/σ^2. Because this factor is well-behaved and reasonably constant during the PSC, contributions of individual time points of the ensemble variance to M^2/σ^2 are reduced by averaging for 16 ms following the PSC peak.

To summarize, the quantal analysis thus permits using mean and variance of sets of PSCs collected at a single synapse to be used to express mean amplitude, or synaptic strength, as the product of two factors, one of which is essentially presynaptic, reflecting parameters related to neurotransmitter release.

RESULTS

Synaptic Strength

The network in the buccal ganglia can be used to analyze both synaptic strength and presynaptic quantal parameters. I first show that synapses differ in synaptic strength, as displayed in Figure 4, and that these differences in synaptic strength are specified by the postsynaptic neuron.

Each synapse is characterized by parameters derived from sets of PSCs recorded at that synapse. Because the network circuit includes paired presynaptic and paired postsynaptic neurons, it permits internally-controlled comparisons of these parameters. Synaptic convergence and divergence permit pairing the same data in each of several ways: by common postsynaptic or presynaptic cells, or by cross-pairing.

Figure 6 plots mean synaptic conductance M and mean time constant τ for each of 64 paired synapses. The top pair of graphs present two pairings of M. Within each graph, each point compares mean synaptic conductance between two synapses, with values of M at one synapse, plotted on the horizontal axis, against M at the paired synapse, plotted on the vertical axis. The graph in the top left, comparing synapses sharing a common postsynaptic cell, shows that the paired values in this case are similar to one another, with points near the diagonal. In contrast, no similarity is shown by the same data when plotted by common presynaptic cell, in the top right. Some metrics of synaptic strength include decay time constant τ as well as peak amplitude[13]; the bottom pair of graphs of Figure 6 show that this parameter also displays greater similarity with postsynaptic pairing.

Figure 6 thus shows similar synaptic strengths of different inputs from different presynaptic cells converging on a postsynaptic cell, as well as different synaptic strengths of synapses made by divergent terminals of the same presynaptic neuron on different postsynaptic cells. These findings[15] imply that it is the postsynaptic neuron which determines or specifies M.

Transmitter Release

I next show that these differences in synaptic strength, although postsynaptically specified, reflect differences in presynaptic transmitter release. The quantal methods described above are employed in order to test two related hypotheses. The first states that synaptic terminals converging from different presynaptic neurons onto a common postsynaptic neuron produce similar-sized postsynaptic conductances because they release similar numbers of quanta per action potential. The second states that branches of the same presynaptic neuron synapsing on different postsynaptic neurons produce different-sized postsynaptic conductances because they release different numbers of quanta per action potential.

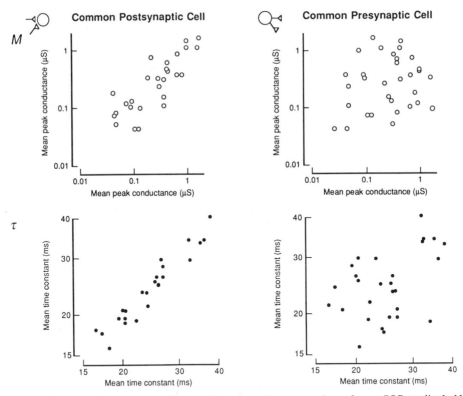

Figure 6. Postsynaptic neurons specify synaptic strength. *Top:* comparison of mean PSC amplitude M at paired synapses. Each data point compares M for paired sets of synapses, plotting one value on the horizontal axis against the paired value on the vertical axis. The same data are presented in the two upper graphs, which differ only in the pairing scheme. *Top left:* values of M paired by common postsynaptic cell are significantly more similar than randomly paired ($P = 0.0001$). *Top right:* pairing the same data by common presynaptic cell reveals no similarity ($P = 0.68$). *Bottom:* similar comparison of PSC decay time constant τ. Each data point compares τ for paired sets of synapses, plotting one value on the horizontal axis against the paired value on the vertical axis. *Bottom left:* values of τ paired by common postsynaptic cell are significantly more similar than randomly paired ($P = 0.0001$). *Bottom right:* pairing the same data by common presynaptic cell also reveals some similarity ($P = 0.002$).

These analyses were carried out for each of forty-six synapses, deriving values of presynaptic M^2/σ^2 and partially postsynaptic σ^2/M for each synapse. These presynaptic and postsynaptic factors can be compared to the synaptic strength of that same synapse, to see which factor is varied to produce stronger or weaker synapses[22].

Figure 7 displays a direct comparison of pre and postsynaptic factors to synaptic strength. In each graph, the horizontal axis plots mean peak conductance M for this wide ranging set of data. On the left, presynaptic M^2/σ^2 is plotted against M, with correlation coefficient $r = 0.85$ and slope near one. These data are consistent with the idea that different synaptic strengths reflect differences in presynaptic factors such as quantal content. On the right, there is no correlation between σ^2/M and mean amplitude, showing that this largely postsynaptic value, proportional to the postsynaptic sensitivity, is not varied systematically so as to produce synapses of varying efficacy.

Utility of the presynaptic metric M^2/σ^2 as an approximate index of quantal content m is enhanced for small values of p. Figure 8 plots variance σ^2 versus mean M for the same synapses shown in Figure 7. If larger M were associated with significantly larger p, this curve would show pronounced downward curvature toward the right. Instead, the approximately linear relation suggests that p for this action-potential-evoked release is not large, perhaps smaller than that seen with prolonged presynaptic depolarization[23].

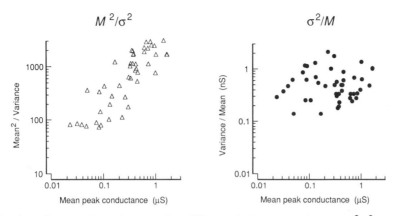

Figure 7. Values of mean peak conductance reflect differences in the presynaptic factor M^2/σ^2. For each of 46 synapses, values of M^2/σ^2, $\triangle$ and of σ^2/M, $\bullet$ are plotted against the corresponding values of M, the mean peak conductance. *Left*: M^2/σ^2 is well-correlated with M ($r = 0.85$, $P = 0.0001$ for the regression line of slope 0.96 ± 0.10 on this log-log plot). *Right*: σ^2/M, which reflects postsynaptic sensitivity as well as some presynaptic factors, is uncorrelated with M ($r = 0.07$, $P = 0.64$). Reprinted from Gardner[22], with permission of the American Physiological Society.

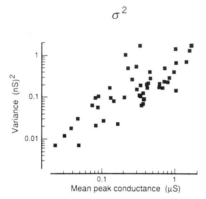

Figure 8. Variance of synaptic conductance increases with increasing mean value. For each of the synapses shown in Figure 7, values of σ^2 are plotted against the corresponding values of M. Variance σ^2 is well-correlated with M ($r = 0.81$, $P = 0.0001$ for the regression line of slope 1.02). Reprinted from Gardner[24].

Figure 6 showed that synaptic strengths are similar at convergent synapses, while Figure 7 correlated presynaptic M^2/σ^2 with synaptic strength for unpaired synapses. Finally, Figure 9 shows that synapses fed by convergent terminals have similar strength because they have similar values of presynaptic M^2/σ^2, not because of any postsynaptic similarity. Pairing is here used to examine quantal parameters, rather than synaptic strength. In the top row of Figure 9, presynaptic M^2/σ^2 is paired in order to detect similarities or differences at different terminals of the same presynaptic cell, and at terminals of different presynaptic neurons synapsing on the same postsynaptic target. The same data are presented in each of three plots which differ only in the pairing scheme. In each of the graphs, similarity is shown by a histogram of absolute log ratios of paired values, such that greater similarity is represented by values closer to the origin. The top left histogram of Figure 9 shows that even though M^2/σ^2 is a presynaptic factor, it shows pairing similarity in the common postsynaptic cell case, which pairs terminals of different presynaptic cells synapsing on the same postsynaptic neuron, when compared to the unpaired case seen in the top right.

125

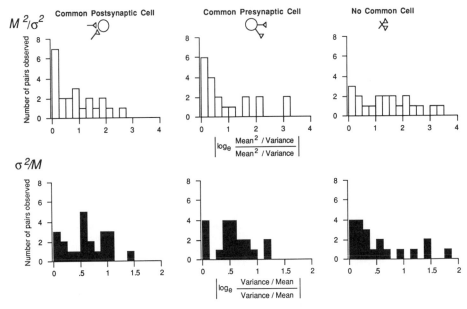

Figure 9. Similar amplitudes at synapses converging on a common postsynaptic cell result in part from similar values of the presynaptic parameter M^2/σ^2 at terminals of different presynaptic neurons. Similarity of synaptic pairs is shown by histograms of the absolute values of the log ratios of a synaptic parameter, with the same data presented in each of the three panels in a row, differing only in the pairing scheme. *Top*: histograms compare paired values of the presynaptic factor M^2/σ^2. *Top left*: M^2/σ^2 paired by a common postsynaptic cell are similar, with average log ratio of 0.87 ± 0.16. *Top center*: pairing the same data by common presynaptic cell again shows greater similarity, average log ratio = 0.94 ± 0.22, than *top right*: cross paired data, average log ratio = 1.41 ± 0.22. *Bottom*: histograms compare paired σ^2/M. *Bottom left*: pairing σ^2/M by common postsynaptic cell, average log ratio = 0.64 ± 0.08, or *bottom center*: pairing by common presynaptic cell, average log ratio = 0.54 ± 0.08, is no more similar than *bottom right*: cross-paired data, average log ratio = 0.56 ± 0.12. Reprinted from Gardner[22], with permission of the American Physiological Society.

In the bottom row, σ^2/M is similarly paired to compare synapses sharing common presynaptic or postsynaptic elements. This parameter, which includes postsynaptic sensitivity, is no different for any pairing, and so does not contribute to similarities or differences in synaptic strength at paired synapses.

DISCUSSION

Applicability of Quantal Models

The data presented in this chapter suggest that the simple binomial quantal model is largely applicable to *Aplysia*. Since M^2/σ^2 depends not only upon quantal content m, but also upon additional factors, it is necessary to ask if these could contribute to observed synapse-to-synapse differences in M^2/σ^2. Although Faber and Korn[25] have argued that slopes near one in potentiation curves analogous to Figure 7 *left* could be accounted for in some cases by decreases in CV_γ, Gardner[22] has pointed out that such decreases could not account for the large range of agreement observed in this preparation between M^2/σ^2 and M. The additional finding by Poulain *et al*[1] that direct injection of botulinum toxin into the presynaptic neuron reduces both M and M^2/σ^2 strengthens this association of M^2/σ^2 with presynaptic factors. Similarly, changes in the probability of release p alone are insufficient to explain the range of M^2/σ^2 and correlated M, unless accompanied by parallel changes in m. In summary, even though this preparation does not fulfill all criteria for a 'proper' quantal analysis, as defined by Korn and Faber[26], the many-fold parallel differences in M and in M^2/σ^2 appear to provide relatively unambiguous evidence for varying quantal content underlying varying synaptic strength.

Use of the appropriate quantal model is essential for interpretation of experimental data and subsequent evaluation of effects of extrinsic agents on neurotransmitter release. The reduction in M^2/σ^2 by botulinum toxin observed by Poulain et al[1] and their subsequent conclusion that the toxin acts by blocking release, without changing quantal size, is consistent with the model presented above. However, it must be stressed that in the absence of low-noise recordings of isolated quantal events, analyses of this type can not yield exact values of the probability of release p, the quantal conductance change γ, or other quantal parameters which can be detected at neuromuscular junction and elsewhere. Similarly, M^2/σ^2 can not be equated to quantal content, nor σ^2/M to γ.

Adoption of a simplifed Poisson model, in which p is assumed to be close enough to zero that $(1 - p)$ is effectively 1, and in which contributions of CV_γ are assumed insignificant, may be inappropriate in some cases. As a general rule, apparent decreases in σ^2/M coupled to increases in M^2/σ^2 and to M should be interpreted with caution. For example, although Baux and Tauc[10] were unable to find a satisfactory explanation for the apparent decline in calculated γ with increasing presynaptic depolarization (also noted by Simonneau et al[14]), this decline may result from increasing p. If σ^2/M is incorrectly equated with γ, increasing p with greater presynaptic depolarization may account for the observed decrease in σ^2/M and consequent underestimation of γ. Data in Simonneau and Tauc[23] would seem to be consistent with this interpretation, although their equation (3), by eliminating p, makes explicit reference only to number of release sites in accounting for the varying relation between macroscopic and microscopic currents.

Postsynaptic Specification of Quantal Release

The data show that synaptic strength M is largely specified by the postsynaptic neuron, whereas differences in M reflect differences in the presynaptic quantal parameter M^2/σ^2. This analysis argues that synapse-to-synapse correlated differences in M and presynaptic transmitter release are partially specified by postsynaptic neurons.

The mechanism for this postsynaptic specification of presynaptic release remains unknown. However, each of three general classes of mechanism could in principle account for the data presented above. These are: 1) a retrograde postsynaptic-to-presynaptic messenger substance, 2) heterosynaptic modulation of individual presynaptic terminals, or 3) postsynaptic modulation of cell adhesion molecules and subsequent presynaptic sprouting. Evidence for each of these exists in part in *Aplysia*, reviewed by Gardner[22,24]. For example, the first mechanism could be mediated by the previously-mentioned presynaptic autoceptors for acetylcholine, while the second could reflect receptors for extrinsic neuromodulators.

Finally, the apparent utilization by this nervous system of mechanisms for extrinsic modulation of neurotransmitter release from individual terminals may be thought of as a benign analog of actions of presynaptic neurotoxins.

ACKNOWLEDGEMENTS

I thank Drs. O.S. Andersen, E.P. Gardner, and C.F. Stevens for many helpful discussions, and B.R. DasGupta for urging me to relate these data to neurotoxin research. Supported by NS-11555 from the National Institutes of Health and N00014-90-J-1460 from the Office of Naval Research.

REFERENCES

1. Poulain B, Tauc L, Maisey EA, Wadsworth JDF, Mohan PM, Dolly JO. Neurotransmitter release is blocked intracellularly by botulinum neurotoxin, and this requires uptake of both toxin polypeptides by a process mediated by the larger chain. Proc Natl Acad Sci USA 1988;85:4090-4094.

2. Poulain B, Mochida S, Weller U, Högy B, Habermann E, Wadsworth JDF, Shone C, Dolly JO, Tauc L. Heterologous combinations of heavy and light chains from botulinum neurotoxin A and tetanus toxin inhibit neurotransmitter release in *Aplysia*. J Biol Chem 1991;266:9580-9585.

3. Strumwasser F. Types of information stored in single neurons. In: Wiersma CAG, ed. Invertebrate Nervous Systems. Chicago: University of Chicago Press, 1967:291-319.

4. Gardner D. Bilateral symmetry and interneuronal organization in the buccal ganglia of *Aplysia*. Science 1971;173:550-553.

5. Gardner D, Kandel ER. Diphasic post-synaptic potential: a chemical synapse capable of mediating conjoint excitation and inhibition. Science 1972;176:675-678.

6. Gardner D, Kandel ER. Physiological and kinetic properties of cholinergic receptors activated by multi-action interneurons in the buccal ganglia of *Aplysia*. J Neurophysiol 1977;40:333-348.

7. Tauc L, Hoffmann A, Tsuji S, Hinzen DH, Faille L. Transmission abolished on a cholinergic synapse after injection of acetylcholinesterase into the presynaptic neurone. Nature 1974;250:496-498.

8. Gardner D. Voltage-clamp analysis of a self-inhibitory synaptic potential in the buccal ganglia of *Aplysia*. J Physiol (London) 1977;264:893-920.

9. White RL, Gardner D. Self-inhibition alters firing patterns of neurons in *Aplysia* buccal ganglia. Brain Res 1981;209:77-83.

10. Baux G, Tauc L. Presynaptic actions of curare and atropine on quantal acetylcholine release at a central synapse of *Aplysia*. J Physiol (London) 1987;388:665-680.

11. Baux G, Fossier P, Tauc L. Histamine and FLRFamide regulate acetylcholine release at an identified synapse in *Aplysia* in opposite ways. J Physiol (London) 1990;429:147-168.

12. Gardner D, Stevens CF. Rate-limiting step of inhibitory post-synaptic current decay in *Aplysia* buccal ganglia. J Physiol (London) 1980;304:145-164.

13. Gardner D. Time integral of synaptic conductance. J Physiol (London) 1980;304:181-191.

14. Simonneau M, Tauc L, Baux G. Quantal release of acetylcholine examined by current fluctuation analysis at an identified neuro-neuronal synapse of *Aplysia*. Proc Natl Acad Sci USA 1980;77:1661-1665.

15. Gardner D. Paired individual and mean postsynaptic currents recorded in four-cell networks of *Aplysia*. J Neurophysiol 1990;63:1226-1240.

16. Gardner D. Sets of synaptic currents paired by common presynaptic or postsynaptic neurons. J Neurophysiol 1989;61:845-853.

17. Gardner D. Membrane-potential effects on an inhibitory post-synaptic conductance in *Aplysia* buccal ganglia. J Physiol (London) 1980;304:165-180.

18. Del Castillo J, Katz B. Quantal components of the end-plate potential. J Physiol (London) 1980;124:560-573.

19. Bekkers JM, Stevens CF. Presynaptic mechanism for long-term potentiation in the hippocampus. Nature 1990;346:724-729.

20. Malinow R, Tsien RW. Presynaptic enhancement shown by whole-cell recordings of long-term potentiation in hippocampal slices. Nature 1990;346:177-180.

21. Redman S. Quantal analysis of synaptic potentials in neurons of the central nervous system. Physiol Rev 1990;70:165-198.

22. Gardner D. Presynaptic transmitter release is specified by postsynaptic neurons of *Aplysia* buccal ganglia. J Neurophysiol 1991;66:2150-2154.

23. Simonneau M, Tauc L. Properties of miniature postsynaptic currents during depolarization-induced release at a cholinergic neuroneuronal synapse. Cell Molec Neurobiol 1987;7:175-189.

24. Gardner D. Static Determinants of Synaptic Strength. In: Gardner D, ed. The Neurobiology of Neural Networks. Cambridge: MIT Press, 1993 in press.

25. Faber DS, Korn H. Applicability of the coefficient of variation method for analyzing synaptic plasticity. Biophys J 1991;60:1288-1294.

26. Korn H, Faber DS. Quantal analysis and synaptic efficacy in the CNS. Trends in Neurosci 1991;14:439-445.

THE MAMMALIAN NEUROMUSCULAR JUNCTION AS A TARGET TISSUE FOR PROTEIN TOXINS THAT BLOCK EXOCYTOSIS

Lance L. Simpson

Division of Environmental Medicine and Toxicology
Departments of Medicine and Pharmacology
Jefferson Medical College
Philadelphia, PA 19107

INTRODUCTION

There are two major classes of protein toxins that act selectively on motor nerve endings to block exocytosis, these being clostridial neurotoxins and phospholipase A2 neurotoxins. There is a considerable amount of information about the sequence of events in clostridial neurotoxin-induced blockade of transmitter release.[1-3] Pharmacological, morphological and related studies indicate that botulinum neurotoxin and tetanus toxin proceed through a series of three steps to produce their poisoning effects.[1,4] There is a membrane binding step, an internalization step, and an intracellular poisoning step. By contrast, much less is known about the sequence of events in phospholipase A2 neurotoxin-induced blockade of acetylcholine release. For example, it has not been established whether these toxins act at the cell surface, within the membrane, or in the cytosol.

A series of well described and widely used techniques have helped investigators to decipher the events in clostridial toxin action (see OVERVIEW, this volume). Many of these techniques have not been applied to the study of phospholipase A2 neurotoxins. Therefore, it seemed appropriate to approach the study of phospholipase A2 neurotoxin action by adopting research strategies that have been used advantageously with botulinum neurotoxin and tetanus toxin.

ORIGIN AND STRUCTURES

Phospholipase A2 neurotoxins can be obtained from several sources, but the ones in this study were all isolated from snake venoms. The toxins and their chain structures were as follows: notexin (one chain), pseudexin (one chain), crotoxin (two chains not covalently linked), β-bungarotoxin (two chains covalently linked), taipoxin (three chains), and textilotoxin (five chains; two copies of one chain, and one copy each of three chains). The complete primary structures for these toxins have been determined, and the data can be found in Rosenberg.[5]

Botulinum and Tetanus Neurotoxins, Edited by
B.R. DasGupta, Plenum Press, New York, 1993

EXPERIMENTAL TECHNIQUES

The techniques for studying clostridial neurotoxin action on mammalian neuromuscular junctions have been described elsewhere. In brief, phrenic nerve-hemidiaphragms were excised from female mice weighing 20 to 30 g. These preparations were added either to incubation tubes (~1 ml) or to tissue baths (~40 ml) containing a physiological solution of the following composition (mM): NaCl, 137; KCl, 2.8; $CaCl_2$, 1.8; $MgCl_2$, 1.1; NaH_2PO_4, 0.33; $NaHCO_3$, 11.9; and glucose, 11.2. The solution was bubbled with a mixture of 95% O_2 and 5% CO_2. Phrenic nerves were stimulated supramaximally with bipolar electrodes (0.2 Hz, 0.1 msec), and muscle twitch was recorded with a strain-gauge transducer. Poisoning of transmission, which was measured as toxin-induced neuromuscular blockade, was defined as a 90% reduction in muscle response to nerve stimulation.

Two paradigms were used to study toxin action on neuromuscular transmission. The first paradigm involved addition of toxin to neuromuscular preparations in tissue baths (35°C; nerve stimulation). Drugs or antibodies were added as described under Results, and paralysis times were monitored.

In the second paradigm, toxin was added to tissues at low temperature (4°C, 60 min) and without nerve stimulation. These conditions allowed toxin to bind, but they retarded energy-dependent mechanisms in the tissue (e.g., receptor-mediated endocytosis) or in the toxin (e.g., phospholipase activity). Tissues were washed three times to remove unbound toxin, then suspended in physiological solution without added toxin. In some cases temperature was raised, nerve stimulation was applied and paralysis times were monitored. In other cases drugs or antibodies were added at various times after incubation but before paralysis (see Results).

Conformational changes induced by low pH were monitored by using a dye, 2-p-toluidinyl-naphthalene sulfonate (TNS), that associates with the hydrophobic domains of proteins. The methods were similar to those of Roberts and Goldstein[6] as previously described.[7] Briefly, neurotoxins (1×10^{-7} M) were added to a saline solution with dimethylglutaric acid buffer. The pH of the solution (see Results) was adjusted with HCl, and TNS was added to give a final concentration of 1.5×10^{-4} M. Proteins were excited at 350 nm, and emission spectra were obtained from 300 nm to 600 nm. Each protein was analyzed at least twice, and experiments were done in duplicate.

The data were analyzed by determining the area beneath each fluorescence scan. The data were then normalized by using the following equation:

$F = [(F_s/F_b)-1]/[Protein]$
F = Specific Fluorescence
F_s = Fluorescence of Sample (area beneath curve)
F_b = Fluorescence of Blank (area beneath curve)
[Protein] = Protein Concentration (M)

The procedures described above were applied both to phospholipase A2 neurotoxins and to clostridial neurotoxins. The text provides representative examples of fluorescence scans (e.g., Figure 3).

RESULTS

The purpose of the studies was to compare the neuromuscular blocking properties of clostridial neurotoxins and phospholipase A2 neurotoxins. The results of the studies indicated that the two groups of toxins differed in at least 6 ways.

1. Clostridial neurotoxins bind with high affinity to the neuromuscular junction, but phospholipase A2 neurotoxins bind with variable affinity

Botulinum neurotoxin and tetanus toxin have been shown to diffuse into the endplate region and bind tightly to the plasma membrane of nerve terminals.[4,8,9] The initial studies, which were done with botulinum neurotoxin, involved tissue preparations incubated at physiological temperatures.[10] Although it was not appreciated at the time, this introduced a complicating factor. At physiological temperature, clostridial neurotoxins can both bind to and be internalized by nerve endings. Subsequent studies were done at 4°C.[4,9] At this temperature the toxin diffuses to the nerve ending and binds, but it cannot be internalized. Experiments done at 4°C have confirmed that botulinum neurotoxin and tetanus toxin bind essentially irreversibly to the neuromuscular junction.

A similar paradigm was used to study phospholipase A2 neurotoxins.[11,12] Tissues maintained at 4°C were incubated for 60 minutes with various concentrations of toxin. After incubation, tissues were washed and transferred to baths at 35°C. Phrenic nerves were stimulated and the onset of paralysis was monitored. The results are shown in Figure 1.

For crotoxin, β-bungarotoxin, taipoxin and textilotoxin, there was evidence of association with the neuromuscular junction. However, for notexin and pseudexin the outcome was different. When incubated with tissues at 4°C and at reasonable concentrations (10^{-9} to 10^{-7} M), these toxins had little ability to poison transmission. The results suggest that notexin and pseudexin bind with low affinity and that this binding is reversed by washing.

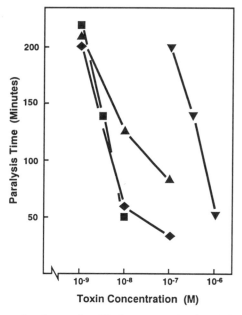

Figure 1. Mouse phrenic nerve-hemidiaphragms were incubated with toxin for 60 min at 4°C, then washed and suspended in baths at 35°C. Nerve stimulation was applied and paralysis times were monitored. The results with notexin and pseudexin are not illustrated because paralysis times were excessively long. For crotoxin (▼), each data point represents an *n* of 9 or more; the standard error of the mean is equal to or less than 22% of the value for each point. For β-bungarotoxin (▲), textilotoxin (♦) and taipoxin (■), each data point represents an *n* of 5 or more, and the standard of error of the means are equal to or less than 9% of the values for the respective means. The results are extracted from Reference 12.

It will be necessary to study a larger number of phospholipase A2 neurotoxins before drawing firm conclusions, but the limited data suggest that there may be a structure-function relationship. The only two toxins that did not bind tightly are the single chain toxins. This could mean that single chain toxins have not evolved efficient tissue-targeting domains, but multiple chain toxins possess efficient binding domains.

2. Clostridial neurotoxins bind to receptors that contain sialic acid residues. Phospholipase A2 neurotoxins bind to receptors that do not have sialic acid in or near the binding domain

Clostridial neurotoxins are known to interact with sialic acid residues, such as those in gangliosides. This was first demonstrated for tetanus toxin by van Heyningen and his colleagues[13-15] and for botulinum neurotoxin by Simpson and his associates.[16,17] More recently, evidence has been provided that clostridial neurotoxins bind to the neuromuscular junction and to brain membrane preparations at or close to a site that has exposed sialic acid residues.[18] This was demonstrated by using lectins with affinity for sialic acid as antagonists of iodinated toxin binding to brain membranes and antagonists of unlabeled toxin binding to phrenic nerve-hemidiaphragm preparations.

A similar experimental approach was used to study the actions of phospholipase A2 neurotoxins. Lectins from *Triticum vulgaris* and *Limax flavus* were individually incubated with neuromuscular preparations at 4°C for 60 minutes. Phospholipase A2 neurotoxins were added to the tissues and incubation was continued for an additional 60 minutes. Tissues were then washed, transferred to baths at 35°C and with nerve stimulation, and paralysis was monitored.

Notexin and pseudexin were not included in the study, because as explained above these single chain toxins did not associate tightly with tissues at 4°C. When tissues were preincubated with lectins and later incubated with crotoxin or β-bungarotoxin, the paralysis times were the same as those for tissues incubated only with toxin. These results indicate that the lectins provided no protection. When tissues were preincubated with lectins and subsequently incubated with taipoxin or textilotoxin, the paralysis times were 40 to 70 percent longer than those of tissues incubated only with toxin.

These results could be interpreted in two ways. Either the membrane has an exposed sialic acid in or near the receptor domain, or the toxin has a sialic acid in the vicinity of its binding domain. The first possibility was tested by treating tissues with neuraminidase, then repeating the protocol described above. The paralysis times in this situation were no different from those obtained with normal tissues. However, when taipoxin and textilotoxin were treated with neuraminidase, the ability of the lectins to afford protection was abolished. This indicates that the lectins were delaying neuromuscular blockade by binding to the toxins and impairing their ability to associate with tissues.

There was an ancillary finding in these studies that was of some interest. The dose-response characteristics of native taipoxin and textilotoxin were the same as those of neuraminidase-treated toxins. This result implies that the toxins may be decorated with sialic acid residues, but these residues are not necessary either for binding or for expression of toxicity.

3. Clostridial neurotoxins that are bound to the cell membrane remain accessible to neutralizing antibodies. After short periods of nerve stimulation, the toxins disappear from accessibility to antibodies. This occurs before onset of paralysis. A similar phenomenon is observed with phospholipase A2 neurotoxins and neutralizing antibodies

One of the early observations supporting the concept of receptor-mediated endocytosis of clostridial neurotoxins involved the use of polyclonal antibodies as research tools.[4] It was found that antibodies mixed with toxin in solution completely neutralized toxin, as expected. It was then found that addition of antibodies to tissues that had bound toxin at 4°C also diminished toxicity. This result demonstrated that bound toxin remained at the cell surface. The final series of experiments proved to be the most revealing. Tissues were incubated with toxin at 4°C to obtain binding, then washed to remove all unbound toxin. Tissues were transferred to baths at 35°C and stimulated for varying amounts of time. Following stimulation, tissues were returned to incubation tubes at 4°C and exposed to neutralizing antibodies. After treatment, tissues were again placed in baths at 35°C, phrenic nerves were stimulated and paralysis times were measured. It was found that tissues with no interval of stimulation at 35°C had maximum protection from antibodies. Tissues that were stimulated for only 5 to 10 minutes had diminished protection, and tissues stimulated for 15 minutes or longer had virtually no protection. These results were interpreted to mean that tissues stimulated at 35°C endocytosed toxin, thus removing it from accessibility to antibody.[4] The endocytosis event occurred well before onset of paralysis.

Identical experiments were performed with crotoxin, β-bungarotoxin, taipoxin and textilotoxin, and the results were qualitatively the same as those obtained with botulinum neurotoxin.[12] Phrenic nerve-hemidiaphragms were incubated individually with the four phospholipase A2 neurotoxins (4°C, 60 minutes). The tissues were then divided into three groups. The first group was exposed immediately to neutralizing antibody (4°C, 60 minutes). The second group was transferred to baths at 35°C and stimulated for 5 minutes, and the third group was similarly transferred and stimulated for 15 minutes. At the end of stimulation, the latter two groups were returned to incubation tubes and treated with antibody (4°C, 60 minutes). At the end of this procedure, tissues were again moved to baths at 35°C, phrenic nerves were stimulated, and the development of paralysis was measured.

The results, which are illustrated in Figure 2, show that phospholipase A2 neurotoxins underwent a time-dependent change that removed them from accessibility to antibodies. When tissues were treated with antibody before stimulation at physiological temperature, the antibodies provided significant protection. When antibodies were added after 5 minutes stimulation at physiological temperature, the magnitude of protection was diminished; and after 15 minutes of stimulation, there was no protection.

4. Clostridial toxins are antagonized by drugs that neutralize endosomal pH or drugs that inhibit vacuolar ATPase. These drugs afford no protection against phospholipase A2 neurotoxins

Studies with polyclonal antibodies provided the initial clue that clostridial neurotoxins are internalized to exert their poisoning effects. Additional evidence was provided by drugs that neutralize endosomal pH. Drugs such as ammonium chloride, methylamine hydrochloride and chloroquine antagonized the neuromuscular blocking effects of botulinum neurotoxin and tetanus toxin.[19,20]

More recent work has exploited the drug bafilomycin, an agent that inhibits vacuolar ATPase. The proton pump in endosome membranes is ATPase-dependent. Therefore, treatment of tissues with bafilomycin inhibits acidification of the intraluminal space. It has been found that bafilomycin is a universal antagonist of clostridial neurotoxins. Pretreatment of tissues with this drug delays the onset of poisoning due to all serotypes of botulinum neurotoxin and tetanus toxin.[21]

Experiments were done to determine whether drugs that neutralize endosomal pH or drugs that inhibit vacuolar ATPase would antagonize the effects of phospholipase A2 neurotoxins. The drugs that were evaluated were ammonium chloride (5 to 15 mM), methylamine hydrochloride (5 to 15 mM), and bafilomycin (0.5 to 2.0 μM). Drugs were added to tissues at 35°C for 30 minutes, after which toxins were added to the tissues. The results were highly consistent and definitive. None of the drugs antagonized the neuromuscular blocking effects of notexin, pseudexin, crotoxin, β-bungarotoxin, taipoxin or textilotoxin.

5. Clostridial neurotoxins undergo marked changes in conformation that are triggered by low pH. Some, but not all, phospholipase A2 neurotoxins undergo similar changes

There is abundant evidence that clostridial neurotoxins respond to low pH. The nature of the response is that of exposing hydrophobic domains that can insert into membranes. The

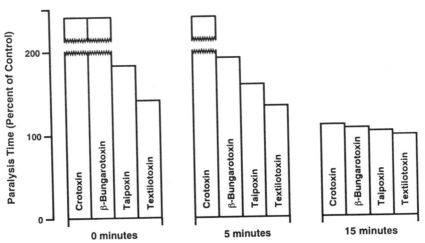

Figure 2. Tissues were incubated with toxin for 60 min at 4°C, then washed free of unbound toxin. Tissues were exposed to neutralizing amounts of antibody either immediately or after 5 min or 15 min of nerve stimulation at 35°C. Tissues were washed again to remove antibody, then transferred to baths to monitor onset of paralysis. Each data point represents the average value obtained in five or more experiments, and the standard errors of the mean are equal to or less than 11% of the respective means. The concentrations of toxin were: crotoxin, 3×10^{-7} M; β-bungarotoxin, 1×10^{-8} M; taipoxin, 3×10^{-9} M; and textilotoxin, 3×10^{-9} M. The data are expressed as percent of control, in which control tissues were not exposed to antibody. Breaks in the curves indicate that experimental tissues required more than twice the time of control tissues to become paralyzed. The results are extracted from Reference 12.

assumption is that pH-induced changes are relevant to the process of translocation. When intraluminal pH falls to levels of 5.5 or lower, botulinum neurotoxin and tetanus toxin are induced to insert into and eventually penetrate endosomal membranes.

Many early studies used indirect techniques to show pH-induced changes. For example, Boquet and Duflot[22] loaded asolectin vesicles with potassium, and they showed that at low pH tetanus toxin caused the vesicles to release the cation. They concluded that low pH induced the toxin to insert into membranes, creating a channel or some other perturbation that allowed calcium to leak from vesicles. Later workers have used electrophysiological approaches to show that tetanus toxin[23-26] and botulinum neurotoxin[27-30] can be induced by low pH to insert into membranes.

A number of investigators have used biochemical techniques to show pH-induced changes in toxins. Boquet et al.[31] demonstrated that low pH increased the amount of Triton X-100 that became associated with tetanus toxin. Montecucco and his colleagues[32,33] used photoactivatable phospholipids to show that tetanus toxin and botulinum neurotoxin can be induced to insert into the lipid core of membranes. More recently, Kamata et al.[7] used a fluorescent reporter group to show that low pH dramatically increases the amount of exposed hydrophobicity in clostridial neurotoxins.

The use of a fluorescent reporter group has now been applied to the study of phospholipase A2 neurotoxins.[12] Toxins were exposed to TNS (1.5×10^{-4} M) at physiological pH and at acidic pH. Simultaneous experiments were done with botulinum neurotoxin type A, botulinum neurotoxin type B, and tetanus toxin. Representative data are presented in Figure 3.

There were four interesting points that emerged from the data. First, the amount of hydrophobicity displayed by the phospholipase A2 neurotoxins at pH 7.0 varied somewhat, from 3.9×10^5 for notexin to 4.4×10^6 for textilotoxin. The actual rank order was notexin < β-bungarotoxin < taipoxin < crotoxin < textilotoxin. Second, when measured at pH 7.0, the amount of exposed hydrophobicity of phospholipase A2 neurotoxins (average specific fluorescence = 1.5×10^6) did not differ significantly from that of clostridial neurotoxins (1.1×10^6). Third, some phospholipase A2 neurotoxins did respond to changes in pH; the hydrophobic character of the proteins increased as pH decreased. However, this effect was not universal. β-Bungarotoxin, taipoxin and textilotoxin responded to low pH, whereas notexin and crotoxin did not. Fourth, the average response of phospholipase A2 neurotoxins to changes in pH (~2.87) was substantially less than that of clostridial neurotoxins (~14.15).

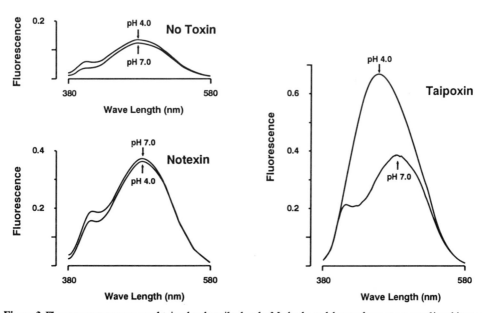

Figure 3. Fluorescence scans were obtained as described under Methods, and the results are expressed in arbitrary units. The scans were obtained at pH 7.0 and pH 4.0. The figure illustrates the results with three different solutions: no toxin, notexin and taipoxin.

6. Strontium produces time-dependent inhibition of phospholipase A2 neurotoxins, but it has no effect on clostridial neurotoxins

Phospholipase A2 neurotoxins, like phospholipase enzymes in general, are calcium-dependent enzymes that are inhibited by strontium. By contrast, clostridial neurotoxins do not have phospholipase activity and are not inhibited by strontium.

There are a variety of ways in which strontium can be shown to antagonize phospholipase A2 neurotoxins. The most straightforward of these is to suspend tissues in baths in which strontium has replaced calcium, then add phospholipase A2 neurotoxins and monitor the development of paralysis. This approach has been used frequently, but it has one serious drawback. If strontium is present throughout an experiment, one cannot determine the precise step in phospholipase A2 neurotoxin action that is antagonized by the cation. Therefore, a series of experiments were done that were intended to clarify the timecourse and mechanism of the interaction between phospholipase A2 neurotoxins and strontium.[12]

In the first series of studies, phrenic nerve-hemidiaphragm preparations were washed and suspended in calcium-free medium that contained strontium (7.0 mM). Phospholipase A2 neurotoxins (crotoxin, β-bungarotoxin, taipoxin and textilotoxin) were added, and tissues were incubated for 60 minutes at 4°C. After incubation, tissues were again washed and then suspended in medium that contained calcium (1.8 mM) and no toxin. Temperature was raised to 35°C, nerves were stimulated, and paralysis times were measured. The results from these tissues were compared with those of tissues that had been incubated with toxin in medium that contained calcium, and it was found that there was no difference. Suspension of tissues in strontium-containing medium during the binding step had virtually no effect on the eventual development of paralysis.

In the next series of experiments the order of additions was reversed. Tissues were incubated (4°C, 60 minutes) in medium that contained calcium (1.8 mM) and phospholipase A2 neurotoxins. After incubation, tissues were washed and transferred to baths with strontium (7.0 mM). As before, temperature was raised, phrenic nerves were stimulated and paralysis times were monitored. The results, which are summarized in Table 1, indicate that strontium had a pronounced effect.

Table 1. Paralysis times of tissues exposed to phospholipase A2 neurotoxins and strontium.[a]

	Paralysis times (min)[b]	
Toxin	Control[c]	Strontium[d]
Crotoxin (3×10^{-7} M)	157 ± 16	>200
β-Bungarotoxin (3×10^{-8} M)	133 ± 12	194 ± 21
Taipoxin (3×10^{-9} M)	121 ± 13	185 ± 20
Textilotoxin (3×10^{-9} M)	141 ± 9	>200

[a]Data extracted from Reference 12.
[b]The values are the mean ± SEM; the number of observations for each data point equals 3 or more.
[c]Tissues were immersed in calcium-containing medium throughout the experiment.
[d]Tissues were transferred to strontium-containing medium 15 min before addition of toxin.

An effort was made to further resolve the timecourse of strontium action. Tissues were incubated with toxin in calcium-containing medium (4°C, 60 minutes), then washed free of unbound toxin. Tissues were transferred to baths at 35°C and stimulated for 15 minutes. As discussed earlier, 15 minutes of stimulation was adequate for bound toxin to disappear from accessibility to neutralizing antibody. At this point, tissues were washed in calcium-free medium and then suspended in strontium-containing medium (7.0 mM). The results of the experiment, which are given in Table 2, reveal an interesting outcome. Even after tissues had been stimulated for 15 minutes at 35°C, phospholipase A2 neurotoxins were still antagonized by strontium.

Experiments are still in progress to determine the length of time that can elapse before addition of strontium will fail to provide protection. However, the data for one phospholipase A2 neurotoxin are already available. To the extent that findings for β-bungarotoxin (3×10^{-8} M) are representative of all phospholipase A2 neurotoxins, the results indicate that strontium was fully active if added after a delay of 15 to 30 minutes. If the cation was added after a delay of 30 minutes or more its antagonistic effect was diminished, and if it was added after a delay of 60 to 70 minutes its antagonistic effects were lost. Strontium did not restore transmission to tissues that were already paralyzed.

Table 2. Paralysis times of tissues exposed to phospholipase A2 neurotoxins and delayed addition of strontium.[a]

Toxin	Paralysis times (min)[b]	
	Control[c]	Strontium[d]
Crotoxin (3×10^{-7} M)	161 ± 29	>200
β-Bungarotoxin (3×10^{-8} M)	131 ± 15	>200
Taipoxin (1×10^{-9} M)	129 ± 6	191 ± 14
Textilotoxin (3×10^{-9} M)	125 ± 19	194 ± 22

[a]Data extracted from Reference 12.
[b]The values are the mean ± SEM; the n for each data point is 3 or more.
[c]Tissues were immersed in calcium-containing medium throughout the experiment.
[d]Tissues were transferred to strontium-containing medium after a delay; see text for details.

DISCUSSION

A paradigm has been developed that allows one to separate the various steps in toxin-induced paralysis of neuromuscular transmission. This paradigm can distinguish the initial, diffusion-mediated binding step from any subsequent energy-dependent step. The paradigm can also distinguish between toxin that acts at the cell surface and toxin that inserts into or penetrates the plasma membrane. This paradigm has been used successfully to deduce the sequence of events in clostridial neurotoxin-induced blockade of neuromuscular transmission.[4] It has now been applied to the study of phospholipase A2 neurotoxins.[11,12]

1. Binding step

There is mounting evidence to show that receptors for phospholipase A2 neurotoxins are voltage-dependent potassium channels.[34,35] The toxins block these channels, and through a mechanism similar to that for aminopyridines they produce a transient increase in acetylcholine release. There is some subsequent action of the toxins that accounts for blockade of transmitter release and paralysis of neuromuscular transmission.

The data in this study indicate that toxin binding to potassium channels occurs at the cell surface. This binding occurs at low temperature, and it leaves the toxins accessible to the neutralizing effects of antibodies. Interestingly, there may be a structure-function relationship to the binding step. One chain toxins bind poorly, whereas multiple chain toxins bind more efficiently. As noted earlier, it is tempting to speculate that multiple chain toxins have evolved more efficient tissue-targeting domains.

There was one particularly striking difference between clostridial toxin binding and phospholipase A2 toxin binding. All clostridial neurotoxins are antagonized by lectins with affinity for sialic acid.[18] The underlying mechanism is that clostridial toxin receptors, or the microenvironment of these receptors, have exposed sialic acid residues. When lectins bind to these residues, they impair clostridial toxin binding. Lectins with affinity for sialic acid antagonize some (taipoxin, textilotoxin) but not all phospholipase A2 neurotoxins. The underlying mechanism is that the lectins bind to glycosylated phospholipase A2 neurotoxins and thereby hinder toxin association with neural receptors. When phospholipase A2 neurotoxins are treated enzymatically to remove sialic acid, the lectins no longer behave as antagonists.

2. The question of internalization

After binding has occurred, clostridial neurotoxins are internalized by nerve-endings. It would be of interest to know whether phospholipase A2 neurotoxins are internalized by the same mechanism as that used by clostridial toxins.

Entry of clostridial neurotoxins into vulnerable cells involves a sequence of two events.[4] The initial event is receptor-mediated endocytosis. This allows toxin to cross the plasma membrane, but it leaves toxin inside endosome membranes. The second event is an acid-dependent mechanism that triggers translocation into the cytosol. Clostridial neurotoxins are thought to have a domain that detects a fall in pH. When the proton pump in endosome membranes lowers intraluminal pH, the detector moiety induces a conformational change in the structure of the toxin, the most important feature of which is exposure of a hydrophobic domain. The hydrophobic portions of the toxin insert into endosome membranes, and this is probably the event that initiates translocation.

Several properties of the two events in clostridial toxin internalization have been determined. Receptor-mediated endocytosis is characterized by toxin disappearing from the cell surface, and this can be monitored by using neutralizing antibodies as research tools. Acid-dependent penetration of the endosome membrane hinges on functional expression of the proton pump. This can be antagonized by drugs that neutralize endosomal pH or by drugs that inhibit the ATPase-dependent proton pump. Finally, the concept of pH-induced translocation implies that the toxin should undergo a conformational change. There are a variety of techniques for detecting such changes, including the use of TNS as a reporter group.

The phospholipase A2 neurotoxins have been investigated by using the methods that have evolved for the study of clostridial neurotoxins. The principal results were:

a) Phospholipase A2 neurotoxins do disappear from accessibility to extracellular antibody, but
b) Phospholipase A2 neurotoxins are not antagonized by drugs that neutralize endosomal pH, such as ammonium chloride or methylamine hydrochloride,
c) They are not antagonized by drugs that inhibit the ATPase-dependent proton pump, such as bafilomycin, and
d) They do not, as a group, show the pH-induced conformational changes that would be expected of toxins that use an acid mechanism to achieve translocation.

It is interesting that the phospholipase A2 neurotoxins disappear from accessibility to antibodies, but they do not show any of the other properties that would be expected of substances that are internalized by the same mechanism as that used by clostridial neurotoxins. This could mean that phospholipase A2 neurotoxins are endocytosed, but they use a non-acid mechanism to escape endosomes. There is, however, another and perhaps more likely explanation.

3. Insertion into the membrane

There was one particularly interesting difference that emerged during the study of phospholipase A2 neurotoxin antagonists. When neutralizing antibodies were added immediately after the toxin binding step, they were maximally active. If neutralizing antibodies were added after 10 to 15 minutes of stimulation at physiological temperature, they were no longer active. When strontium was added immediately after the toxin binding step, it displayed maximum activity as an antagonist. If strontium was added after 10 to 15 minutes of stimulation at physiological temperature, it was still active. Strontium ceased to be an antagonist only after the paralytic effects of toxin began to be expressed.

Neutralizing antibodies and strontium have in common that neither diffuses across the membrane to any significant extent. There is no known mechanism by which antibodies could reach the cell interior at concentrations adequate to block the toxin. Strontium can reach the cell interior by utilizing calcium channels, but it is unlikely that this would produce intracellular concentrations of the cation sufficient to inhibit phospholipase A2 activity (millimolar range). To act as antagonists, neutralizing antibody and strontium must both be working on the outside of the membrane.

There is a way to reconcile the different temporal actions of neutralizing antibodies and strontium. This explanation calls for a model of phospholipase A2 neurotoxin action that involves at least three discrete steps. There is an initial binding step to a receptor on the cell surface. As indicated earlier, this receptor may be a voltage-dependent potassium channel. When the binding step is complete, the toxin remains accessible to the antagonistic effects of antibodies and strontium. The binding step is followed by a molecular rearrangement step, during which some or all of the toxin molecule inserts into the membrane. This molecular rearrangment causes epitopes to disappear from accessibility to antibody, either because the epitopes undergo conformational changes or because the epitopes are shielded by the membrane. However, strontium can still reach and antagonize the catalytic site. During the final step, the toxin acts in the membrane to exert an effect that produces blockade of exocytosis. This step could be cleavage of certain membrane phospholipids. As this action proceeds, the toxin loses its susceptibility to strontium.

There is much that remains to be done to confirm this proposed three-step model for phospholipase A2 neurotoxin action. However, if the model should prove to be correct there would be one interesting side note. Investigators have often borrowed from other fields of toxin research to develop the model for clostridial neurotoxin action. Research on phospholipase A2 neurotoxins would be the first time investigators have borrowed from the clostridial neurotoxin field to help solve the mechanism of action of another class of toxins.

Acknowledgement

The work described in this chapter was a product of collaborative efforts with Drs. Ivan Kaiser, Gregory Lautenslager, John Middlebrook, and Shephali Trivedi.

REFERENCES

1. Simpson LL. The origin, structure, and pharmacological activity of botulinum toxin. Pharmacol Rev 1981; 33:155-188.
2. Habermann E, Dreyer F. Clostridial neurotoxins: Handling and action at the cellular level. Curr Topics Microbiol Immunol 1986; 129:93-179.
3. Simpson LL. Botulinum Neurotoxin and Tetanus Toxin. San Diego: Academic Press, 1989:1-422.

4. Simpson LL. Kinetic studies on the interaction between botulinum toxin type A and the cholinergic neuromuscular junction. J Pharmacol Exp Ther 1980; 212:16-21.

5. Rosenberg P. Phospholipases. In: Shier WT, Mebs D, eds. Handbook of Toxinology. New York: Marcel Dekker, 1990:67-277.

6. Roberts DD, Goldstein IJ. Hydrophobic binding properties of the lectin from lima beans (*Phaseolus lunatus*). J Biol Chem 1982; 257:11274-11277.

7. Kamata Y, Lautenslager G, Simpson LL. Structural changes in the botulinum neurotoxin molecule that may be associated with the process of internalization and expression of pharmacologic activity. Infect Immun (submitted for publication).

8. Simpson LL. Pharmacological experiments on the binding and internalization of the 50,000 dalton carboxyterminus of tetanus toxin at the cholinergic neuromuscular junction. J Pharmacol Exp Ther 1985; 234:100-105.

9. Schmitt A, Dreyer F, John C. At least three sequential steps are involved in the tetanus toxin-induced block of neuromuscular transmission. Naun-Schmiedebergs Archiv Pharmacol 1981; 317:326-330.

10. Burgen ASV, Dickens F, Zatman LJ. The action of botulinum toxin on the neuromuscular junction. J Physiol 1949; 109:10-24.

11. Trivedi S, Kaiser II, Tanaka M, Simpson LL. Pharmacologic experiments on the interaction between crotoxin and the mammalian neuromuscular junction. J Pharmacol Exp Ther 1989; 251:490-496.

12. Simpson LL, Lautenslager GT, Kaiser II, Middlebrook JL. Identification of the site at which phospholipase A2 neurotoxins localize to produce their neuromuscular blocking effects. Toxicon 1992; (in press).

13. Van Heyningen WE. The fixation of tetanus toxin by nervous tissue. J Gen Microbiol 1959a; 20:291-300.

14. Van Heyningen WE. Tentative identification of the tetanus toxin receptor in nervous tissue. J Gen Microbiol 1959b; 20:310-320.

15. Van Heyningen WE. Gangliosides as membrane receptors for tetanus toxin, cholera toxin and serotonin. Nature 1974; 249:415-417.

16. Simpson LL, Rapport MM. The binding of botulinum toxin to membrane lipids: sphingolipids, steroids and fatty acids. J Neurochem 1971a; 18:1751-1759.

17. Simpson LL, Rapport MM. Ganglioside inactivation of botulinum toxin. J Neurochem 1971b; 18:1341-1343.

18. Bakry N, Kamata Y, Simpson LL. Lectins from *Triticum vulgaris* and *Limax flavus* are universal antagonists of botulinum neurotoxin and tetanus toxin. J Pharmacol Exp Ther 1991; 258:830-836.

19. Simpson LL. The interaction between aminoquinolines and presynaptically acting neurotoxins. J Pharmacol Exp Ther 1982; 222:43-48.

20. Simpson LL. Ammonium chloride and methylamine hydrochloride antagonize clostridial neurotoxins. J Pharmacol Exp Ther 1983; 225:546-552.

21. Simpson LL. Inhibition of vacuolar ATPase antagonizes the effects of clostridial neurotoxins but not phospholipase A2 neurotoxins. J Pharmacol Exp Ther 1992; (submitted for publication).

22. Boquet P, Duflot E. Tetanus toxin fragment forms channels in lipid vesicles at low pH. Proc Natl Acad Sci USA 1982; 79:7614-7618.

23. Borochov-Neori H, Yavin E, Montal N. Tetanus toxin forms channels in planar lipid bilayers. Biophys J 1984; 45:83-85.

24. Gambale F, Montal M. Characterization of the channel properties of tetanus toxin in planar lipid bilayers. Biophys J 1988; 53:771-783.

25. Menestrina G, Forti S, Gambale F. Interaction of tetanus toxin with lipid vesicles. Effects of pH, surface charge, and transmembrane potential on the kinetics of channel formation. Biophys J 1989; 55:393-405.

26. Rauch G, Gambale F, Montal M. Tetanus toxin channel in phosphatidylserine planar bilayers: conductance states and pH dependence. Europ Biophys J 1990; 18:79-83.

27. Hoch DH, Romero-Mira M, Ehrlich BE, Finkelstein A, Dasgupta BR and Simpson, LL. Channels formed by botulinum, tetanus, and diphtheria toxins in planar lipid bilayers: relevance to translocation of proteins across membranes. Proc Natl Acad Sci USA 1985;82:1692-1696.

28. Donovan JJ, Middlebrook JL. Ion-conducting channels produced by botulinum toxin in planar lipid membranes. Biochem 1986; 25:2872-2876.

29. Blaustein RO, Germann WJ, Finkelstein A, Dasgupta BR. The N-terminal half of the heavy chain of botulinum type A neurotoxin forms channels in planar phospholipid bilayers. FEBS Letters 1987; 226:115-120.

30. Shone CC, Hambleton P, Melling J. A 50-kDa fragment from the NH2-terminus of the heavy subunit of Clostridium botulinum type A neurotoxin forms channels in lipid vesicles. Eur J Biochem 1987; 167:175-180.

31. Boquet P, Duflot E, Hauttecoeur B. Low pH induces a hydrophobic domain in the tetanus toxin molecule. Eur J Biochem 1984; 144:339-344.
32. Montecucco C, Schiavo G, Brunner J, Duflot E, Boquet P, Roa M. Tetanus toxin is labeled with photoactivatable phospholipids at low pH. Biochem 1986; 25:919-924.
33. Montecucco C, Schiavo G, Gao Z, Bauerlein E, Boquet P, Dasgupta BR. Interaction of botulinum and tetanus toxins with the lipid bilayer surface. Biochem J 1988; 251:379-383.
34. Dreyer F. Peptide toxins and potassium channels. Rev Physiol Biochem Pharmacol 1990; 115:93-136.
35. Strong PN. Potassium channel toxins. Pharmacol Therapeut 1990; 46:137-162.

ACTION OF BOTULINUM TOXIN ON ARTIFICIAL
MODELS FOR ACETYLCHOLINE TRANSPORT

C. Solsona[1], E. López-Alonso[2], J. Blasi[1], J. Canaves[2], J. Canals[1],
M. Arribas[1], L. Ruiz-Avila[1], J.M. González-Ros[2] and J. Marsal[1]

[1]Laboratori de Neurobiologia Cel.lular i Molecular. Facultat de
Medicina, Hospital de Bellvitge, Universitat de Barcelona, Barcelona
Spain
[2] Departamento de Neuroquímica, Facultad de Medicina, Universidad
de Alicante, Alicante, Spain

SUMMARY

Two artificial models for acetylcholine secretion have been used in order to
study if they are sensitive to the action of Botulinum toxin type A. We have used the
electric organ of Torpedo as the source to obtain subcellular components of the
cholinergic nerve terminals. Giant proteoliposomes which have incorporated
presynaptic plasma membrane proteins are able to translocate acetylcholine in a
calcium dependent manner, in the presence of A23187 calcium ionophore. This
transport is sensitive to BoNT/A. In addition the fluidity of such artificial membrane
is also sensitive to BoNT/A, suggesting that BoNT/A may inhibit ACh release by
making synaptic membranes more rigid. The second model utilized is Xenopus
oocytes injected with cholinergic synaptic vesicles and presynaptic plasma membrane.
These injected cells are able to release ACh under potassium depolarization or by
A23187 ionophore. BoNT/A is capable to inhibit such depolarization-induced release.

INTRODUCTION

Botulinum toxin type A (BoNT/A) is able to block neurotransmitter release at
the neuromuscular junction and at central nervous system synapses (Dolly, 1991). It
has been claimed a great specificity of Botulinum toxin type A to interact with
peripheral cholinergic nerve-terminals. For these reason we have chosen a tissue
exclusively innervated by cholinergic neurons, the electric organ of Torpedo
marmorata. This organ is derived from the skeletal muscle. The miocytes have lost
their cytoplasmic filaments but they are innervated by thousands of hundreds of

Botulinum and Tetanus Neurotoxins, Edited by
B.R. DasGupta, Plenum Press, New York, 1993

cholinergic nerve terminals, which in morphological terms are very similar to motor nerve terminals. The electromotor nerve terminals are covered by a flat and thin process of Schwann cell and they are enriched with large amounts of electronlucent synaptic vesicles. However a difference with respect to neuromuscular junction is that there is no apparent structural active zones, in addition the membrane of the electrocyte facing the nerve terminals has not the characteristic infolding structures. We have previously demonstrated that Botulinum toxin type A interact with nerve terminals of the electric organ of Torpedo (Dunant et al., 1987) blocking the evoked release of ACh and depressing the spontaneous activity of nerve terminals, affecting the frequency and amplitude of miniature end plate currents generated by quantal release of ACh.

We have utilized a subcellular fraction enriched in nerve terminals to know how BoNT/A may interact with nerve terminals (Marsal et al., 1989). BoNT/A does not affect the resting presynaptic membrane potential, the uptake of calcium into the nerve terminals, and the subcellular distribution of ACh. Nevertheless it is able to block the increase of phosphorylation of some presynaptic proteins (Guitart et al., 1987). In addition it is able to block the potassium evoked release of ACh but does not modify the release of ATP (Marsal et al., 1987) which is co-stored with ACh in synaptic vesicles isolated from the electric organ of Torpedo.

It has been proposed that BoNT/A may act like other bacterial toxins acting on three step sequence. The first is the binding step, the second is the internalization of the receptor-toxin complex and the third is the translocation of a toxin fragment from the endocytic vesicle to the cytoplasm where it can exert directly or indirectly its action.

In the present communication we have tried to simplify the experimental model using only parts of the nerve terminals and trying to reconstitute all or part of ACh release machinery.

METHODS

Botulinum toxin was purified, by means of FPLC system, from cultures of Clostridium botulinum type A (NTC 2916) according to DasGupta and Sathyamoorthy (1984) and Woody and DasGupta (1988). The purified neurotoxin had a mouse lethality of 0.66 ng/kg per LD_{50}

Light and heavy chains of BoNT/A were obtained by incubating the neurotoxin with dithiothreitol and urea and using ionic exchange chromatography (performed in QAE Sephadex columns), according to Sathyamoorthy and DasGupta (1985).

Presynaptic plasma membranes were obtained following the method of Morel et al., (1985) which is based on the osmotic shock of electric organ fragments, homogenization and differential centrifugation on sucrose gradients. Liophylized membranes were stored at -80 °C for several weeks.

Giant liposomes were obtained following the method developed by Riquelme et al., (1990): Asolectin lipid was suspended (100 mg/mL) in 1% of CHAPS and lipid vesicles were formed by exhaustive dialysis. Lipid vesicles were mixed with 400 μg of presynaptic membrane fraction, centrifuged and the resultant pellet resuspended in 5% ethylene-glycol solution. Small volumes were placed on a glass microscope slide and submitted to a partial dehydration at 4°C for 3 hours in a desiccator. Drops were rehydrated overnight in a humid chamber by adding a solution containing Tris/HCl 10 mM, pH 7.4, ACh 20 mM, eserine 2 mg/L. The formation of giant liposomes was followed under a light microscope.

ACh release was monitored as follows: aliquots of giant liposomes were diluted 1/100 in 50 mM NaCl and Tris 10 mM, pH 7.4. 10 μL of this suspension was placed in a tube whit chemiluminescent reaction (Israël and Lesbats, 1981) which contain acetylcholinesterase, choline oxidase, peroxidase and luminol in Tris/HCl 50 mM, pH 8.6, NaCl, 50 mM. This chain reaction transforms ACh to choline and acetate, the choline to betaine and H_2O_2, and then to O_2 which reacts with luminol. The light emitted was continuously monitored with an LKB luminometer connected to a pen recorder. The release of ACh was elicited by the addition of A23187 calcium ionophore at different calcium concentrations. At the end of each determination an external dose of ACh was added to be compared with the peak of light detected. Triton X-100 (3 %) was added to evaluate the total ACh content of the liposomal preparation.

Cholinergic synaptic vesicles were isolated from the electric organ of Torpedo according to the method of Israël et al.,(1980). Frozen electric organ fragments were homogenized in liquid nitrogen, synaptic vesicles were obtained by isopicnic centrifugation in a discontinuous sucrose gradients. A flotation gradient step (Diebler and Lazereg, 1985) was introduced in order to increase homogeneity of the preparation.

Synaptic vesicles were loaded with ACh by incubating aliquots of the fraction with ACh 100 mM, DFP 1 mM, KCl 400 mM, HEPES 10 mM, pH 7.2. After overnight incubation they were washed in a Bio-Gel A-5m column (Bio-Rad Laboratoires).

Injection of synaptic vesicles to Xenopus oocytes was made by a method developed by Miledi and col. (submitted paper). Oocytes were excised from Xenopus females under MS 222 anaesthesia. Oocytes were incubated in a frog Ringer solution: NaCl, 115 mM; KCl, 2 mM; CaCl$_2$, 1.8 mM; and HEPES, 5 mM, pH 7.0. and injected with a Dummont pipette, 50 nL of a suspension of synaptic vesicles or presynaptic plasma membrane or both. Oocytes were treated with collagenase 0.5 mg/mL, for 45 min at 18 ºC. Finally oocytes were stored at 18 ºC, for different time periods, in Barth solution: NaCl, 88 mM; KCl, 1 mM; CaCl$_2$ mM, 0.41 mM; Ca(NO$_3$)$_2$, 0.33 mM; MgSO$_4$, 0.82 mM; NaCO$_3$, 2.4 mM and HEPES, 10 mM, pH 7.4.

Liophylized membranes were resuspended in NaCl, 50 mM; CaCl$_2$, 1 mM; Tris/HCl, 10 mM, pH 7. to a final concentration of 50 μM of phospholipid. The suspension was sonicated at 4ºC. BoNT/A was added at a final concentration of 2 nM, for 10 min. 1,6-Diphenyl-1,3,5-hexatrien (DPH) was added to have a final ratio DPH/phospholipid 1:500 and incubated for 60 min at room temperature and in a dark chamber. Membrane fluidity was measured as a relation between fluorescence intensities of vertical and horizontal positions since this technique is based on the excitation of probe molecules with polarized light and the measurement of the degree of polarization of the emitted fluorescence. Light polarization was monitored in temperatures ranging from 20 to 40 ºC.

RESULTS AND DISCUSSION

Effects of Botulinum toxin type A on membrane fluidity

Low polarization values reflects high fluidity, we observed that wild presynaptic membranes had high values of polarization which represents a rigidity of the membrane. Fig 1 shows a negative correlation of polarization with temperature. In the presence of BoNT/A (2 nM), there was a decrease in membrane fluidity, which was maintained in all scanned temperatures. It seems necessary the presence of the two chains of the toxin to obtain such effect since the heavy and the light chain alone were not able to reproduce this effect when added at 2 nM, final concentration. Since the steady-state fluorescence polarization of diphenylhexatriene in bilayers, particulary in heterogeneous bilayers in natural membranes, is the sum of varied contributions representing the interior of the bilayer (Hachisuka et al., 1990) we have also studied the effect of BoNT/A in the fluidity of proteoliposomes containing presynaptic plasma membrane. In this case this artificial membrane preparation had a higher fluidity than natural membranes. The addition of BoNT/A decreased the fluorescence polarization indicating again, that the membrane become more rigid. This result fit very well with previous morphological observations on the effect of BoNT/A on the intramembrane structures (Marsal et al., 1989; Egea et al., 1989).

Among different possible explanations we assume that one of the actions of the toxin may be to alter the arrangement of structural components of the membrane. This fact may contribute to block the molecular steps that led the fusion of synaptic vesicles to the plasma membrane. On the other hand this result suggest that, in this model of ACh release, BoNT/A must not follow necessarily a three step way to block ACh transport.

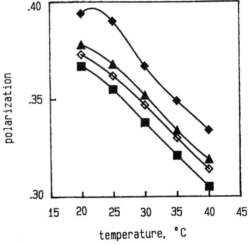

Figure 1. Fluidity of presynaptic plasma membranes. Fluidity of presynaptic plasma membranes (PSPM) was measured at different temperatures. Low values of polarization represent a high fluidity of the membrane. Empty diamonds: non treated membranes. Filled diamonds: 2nM BoNT/A treatment. Triangles: light chain of BoNT/A (2 nM). Squares: heavy chain of BoNT/A (2 nM).

Acetylcholine transport from giant liposomes

Giant liposomes have been obtained submitting lipid vesicles and natural presynaptic membranes to a cycle of dehydration and rehydration. They look very similar to those obtained using postsynaptic membranes of Torpedo electric organ.

The first goal was to investigate if ACh may be entrapped by this giant liposomes, since ACh was present during rehydration. A liposomal suspension was placed in a cuvette containing the chemiluminescent mixture for ACh detection, the addition of Triton X-100 at a final concentration of 3% releases ACh in a variable amount, depending on the batch of liposomes utilized.

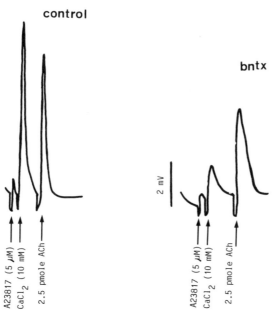

Figure 2. ACh translocation in giant liposomes. Giant liposomes which had incorporated presynaptic plasma membranes and containing ACh, have been stimulated by the calcium ionophore A23187. The addition of calcium is sufficient to trigger ACh translocation. An external doses of ACh has been added at the end of each determination. BoNT/A (2 nM) was able to inhibit the calcium dependent translocation.

These giant proteoliposomes were also able to translocate ACh in calcium dependent manner. The addition of the calcium ionophore A23187, does not increase the presence of ACh in the extraliposomal medium. However, if calcium was added after A23187 ionophore there was an increase of the ACh released. Nevertheless, if calcium is added alone to the liposomal suspension there was not any ACh translocation. This fact suggest that calcium must access to the internal space of the liposome to trigger ACh transport. The amount of ACh released can vary from one to other liposomal preparations, representing from 10 to 30% of the total ACh encapsulated by the liposomes. Magnesium was not able to maintain such translocation. Increasing amounts of external calcium (Fig. 2) increases the levels of ACh translocated, a saturation was reached at 12 mM external calcium concentration.

Another control that we had performed is to make giant liposomes in the absence of presynaptic membrane. These structures were able to retain ACh inside them but they could not translocate ACh in a calcium dependent fashion.

We do not know the physiological significance of this ACh translocating system, because the conditions were far from those present in the presynaptic membrane of the nerve terminal. Moreover we are not capable to know the quantal nature of such transport. However, we feel that it probably represents the non-quantal release machinery of nerve terminals, since it is working in the absence of synaptic vesicles.

In summary, the giant proteoliposomes obtained by dehydration and rehydration have a similar performance to those obtained with other procedures (Israël et al., 1984)

Fig. 3 shows that BoNT/A (1 nM) is affecting this artificial system for ACh translocation. The curve is shift on the right indicating a change in the availability of calcium for the ACh transport loci incorporated to the giant liposome. In addition the maximal release induced by calcium is also lowered. Taking into account that we are utilizing a calcium ionophore, such effect must not be related to an action on the ionic channels, on the contrary, according to the Ca^{2+} receptor model (Silinsky, 1985) it must be reflecting an actual decrease of the affinity of calcium for the transport ACh reconstituted system.

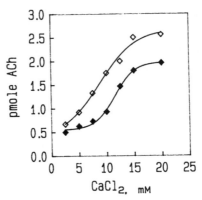

Figure 3. Effect of BoNT/A on calcium dependent ACh translocation. The figure represents the result obtained in one experiment using a batch of giant liposomes. Empty diamond represent non treated giant liposomes. Filled diamonds BoNT/A (2nM) treated liposomes.

These result may represent that, at least for non quantal release (Dolezal et al., 1983), the action of BoNT/A may act as the second mode of action proposed by Simpson (1986), in which the toxin may act directly on the membrane releasing a toxin fragment to the nerve terminal cytoplasm. Moreover, Lomneth et al., (1991) using PC12 ghosts have demonstrated that cytosolic proteins are not essential for the action of the light chain of different types of BoNT/A.

Release of ACh from transformed frog oocytes

It has been recently shown that <u>Xenopus</u> oocytes can acquire the competence to release neurotransmitters. They can release ACh (Cavalli et al.,1991) after the injection of mRNAs from the cholinergic electromotoneurons of the <u>Torpedo</u> encephalon. When chromaffin granules are injected to <u>Xenopus</u> oocytes (Scheuner et al., 1992) they may release noradrenaline and concomitantly there is a fusion of granules to the plasma membrane of the oocyte.

We have injected ACh filled synaptic vesicles into the oocytes and we have tried to induce their fusion to the oocyte plasma membrane by incubating with A23187 calcium ionophore and 1 mM $CaCl_2$ in the medium and we had not detected any release of acetylcholine. However, when synaptic vesicles are co-injected with presynaptic plasma membranes there was a detectable release induced by the calcium ionophore. Then, to induce the release of ACh, the synaptic vesicle population must recognize some specialized zone of the presynaptic plasma membrane like the suggested docking proteins. In addition, the detection of a synaptic vesicle marker, as synaptophysin, can be observed at the oocyte surface (Fig. 4). We believe that this reconstituted system may reproduce the quantal ACh secretion. In fact, it has been shown that proteins of synaptic vesicles recognize presynaptic plasma membrane proteins "in vitro", in a calcium fashion manner (Henkel and Zimmermann, 1992).

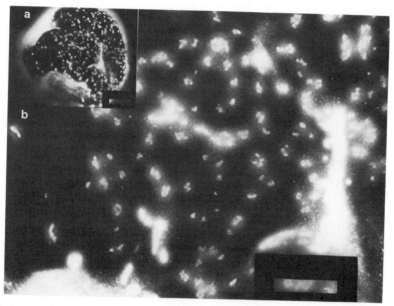

Figure 4. Synaptophysin detection as a vesicle marker in the surface of Xenopus oocytes. Oocytes injected with cholinergic synaptic vesicles and presynaptic plasma membranes exhibit synaptophysin immunoreactivity in their surface. Synaptophysin was detected using immunofluorescence conventional methods. Bars = 200 μm.

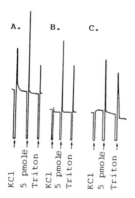

A. Medium with Calcium

B. Medium without Calcium

C. BoNTx inhibited

Figure 5. Effect of BoNT/A on the ACh translocation from Xenopus oocytes injected with filled synaptic vesicles and presynaptic plasma membranes.
(A) Injected synaptosomes were depolarized in a medium containing high potassium concentration and calcium. (B) When the depolarizing medium was calcium-free. (C) The preincubation of injected oocytes with BoNT/A (2nM) in a calcium containing medium inhibits ACh release.

The reconstituted secretory system is also sensitive to potassium depolarization, which is less effective than A23187 ionophore. The loss of content induced by the different stimulus were: A23187 (5 μM) 51.4 ± 3.4, KCl stimulation 44.7 ± 3.3 %, KCl stimulation in a free calcium medium, 11.9 ± 2.3 %.

In a preliminary set of experiments, when injected oocytes were pretreated with BoNT/A (1 nM), for 5 min before potassium. stimulation, the amount of released ACh was only a 13.5 ± 3.51 % of the content of ACh per oocyte (Fig. 5). This value is very close to that obtained when depolarization was done in the absence of extracellular calcium. Our results suggest that BoNT/A may act on presynaptic plasma membrane to inhibit ACh release from these artificial models of synaptic terminals.

Acknowledgments

This work is supported by grants from DGICYT of the Spanish Government to J. Marsal, C. Solsona and to J.M. González-Ros.

REFERENCES

Cavalli, A., Eder-Colli, L., Dunant, Y., Loctin, F., and Morel, N., 1991, Release of acetylcholine by Xenopus oocytes injected with mRNAs from Cholinergic neurons, *EMBO J.*10:1671.

DasGupta, B.R., and Sathyamoorthy, V., 1984, Purification and aminoacid composition of Type A botulinum neurotoxin, *Toxicon.*22:415.

Diebler, M.F., and Lazereg, S., 1985, Mg-ATPase and cholinergic synaptic vesicles, *J. Neurochem.* 44:1633.

Dolezal, V., Vyskocil, F., and Tucek, S., 1983, Decrease of the spontaneous non-quantal release of acetylcholine from the phrenic nerve in botulinum-poisoned rat diaphragm, *Pflügers Arch.* 397:319.

Dolly, J.O., 1991, Peptide toxins that alter neurotransmitter release, *in:* "Handbook of Experimental Pharmacology," H. Herken and F. Hucho, eds., Springer-Verlag, Berlin.

Dunant, Y., Esquerda, J.E., Loctin, F., Marsal, J., and Muller, D., 1987, Botulinum toxin inhibits quantal acetylcholine release and energy metabolism in the Torpedo electric organ, *J. Physiol.* 385:677.

Egea, G., Marsal, J., Solsona, C., Rabasseda, X., and Blasi, J., 1989, Increase in reactive cholesterol in the presynaptic membrane of depolarized Torpedo synaptosomes: blockade by botulinum toxin type A, *Neurosci.* 31:521.

Guitart, X., Egea, G., Solsona, C., and Marsal, J., 1987, Botulinum neurotoxin inhibits depolarization-stimulated protein phosphorylation in pure cholinergic synaptosomes, *FEBS Lett.* 219:219.

Hachisuka, H., Nomura, H., Mori, S., Nakano, S., Okubo, K., Kusuhara, M., Karashima, M., Tanikawa, E., Higuchi, M., and Sasai, Y., 1990, Alterations in membrane fluidity during keratinocyte differentiation measured by fluorescence polarization, *Cell and Tiss Res.* 260:207.

Henkel, A., and Zimmermann, H., 1992, "In vitro" binding of isolated synaptic vesicles to presynaptic plasma membranes: activation by Ca^{2+} and protein kinase C, *Neurochem. Int.* 20:55.

Israël, M., Manaranche, R., Marsal, J., Meunier, F., Morel, N., Frachon, P., and Lesbats, B., 1980, ATP dependent calcium uptake by cholinergic synaptic vesicles isolated from Torpedo electric organ, *J. Membr. Biol.* 54:115.

Israël, M., and Lesbats, B., 1981, Continuous determination by a chemiluminescent method of acetylcholine release and compartmentation in Torpedo electric organ synaptosomes, *J. Neurochem.* 37:1475.

Israël, M., Lesbats, B., Morel, N., Manaranche, R., Gulik-Krzywiki, T., and Dedieu, J.V., 1984, Reconstitution of a functional synaptosomal membrane possessing the protein constituents involved in acetylcholine translocation, *Proc. Natl. Acad. Sci. USA* 81:277.

Lomneth, R., Martin, T.F.J., and Dasgupta, B.R., 1991, Botulinum neurotoxin light chain

Inhibits Norepinephrine secretion in PC12 Cells at intracellular membranous or cytoskeletal site, *J. Neurochem.* 57:1413.

Marsal, J., Solsona, C., Rabasseda, X., Blasi, J., and Casanova, A., 1987, Depolarization-induced release of ATP from cholinergic synaptosomes is not blocked by botulinum toxin type A, *Neurochem. Int.* 10:295.

Marsal, J., Egea, G., Solsona, C., Rabasseda, X., and Blasi, J., 1989, Botulinum toxin type A block the morphological changes induced by chemical stimulation on the presynaptic membrane of Torpedo synaptosomes, *Proc. Natl. Acad. Sci. USA* 86:372.

Morel N., Marsal J., Manaranche R., Lazereg S., Mazie J.C., and Israël M. (1985). Large-scale purification of presynaptic plasma membranes from Torpedo electric organ. *J.Cell Biol.*, 101, 1757-1762.

Riquelme, G., López, E., García-Segura, L.M., Ferragut, J.A., and González-Ros, J.M., 1990, Giant liposomes: A model system in which to obtain Patch-clamp recordings of ionic channels, *Biochemistry* 29:11215.

Sathyamoorthy, V., and DasGupta, B.R., 1985, Separation, purification, and partial characterization and comparison of the heavy and light chains of Botulinum neurotoxin types A,B and E, *J. Biol. Chem.* 260:10461.

Silinsky, E.M., 1985, The biophysical pharmacology of calcium-dependent acetylcholine secretion, *Pharmacol Rev.* 37:81.

Simpson, L.L., 1986 Molecular pharmacology of botulinum and tetanus toxin, *Ann. Rev. Pharmacol. Toxicol.* 26:546.

Scheuner, D., Logsdon, C.D., and Holz, R.W., 1992, Bovine chromaffin granule membranes undergo Ca^{2+}-regulated exocytosis in frog oocytes, *J. Cell. Biol.* 116:359.

Woody, M.A., and DasGupta, B.R., 1988, Fast protein liquid chromatography of botulinum neurotoxin types A, B and E, *J. Chromatography.* 430:279.

EXO-ENDOCYTOTIC RECYCLING OF SYNAPTIC VESICLES IN DEVELOPING NEURONS

Michela Matteoli[1] and Pietro De Camilli[2]

[1]CNR Center of Cytopharmacology and Department of Medical Pharmacology, University of Milano
[2]Howard Hughes Medical Institute and Department of Cell Biology, Yale University Medical School

INTRODUCTION

Regulated secretion of neurotransmitters from neurons involves a cocktail of neurotransmitter molecules and at least two classes of secretory organelles: synaptic vesicles (SVs) and large dense core vesicles (LDCVs). SVs are small vesicles highly homogeneous in size (about 50 nm) which are clustered under the presynaptic plasmalemmma and contain non-peptide neurotransmitters.Their exocytosis is responsible for the fast, point-to-point signalling typical of synaptic transmission. LDCVs are larger organelles with an electron dense core which contain neuroactive peptides and may also contain amines. Their exocytosis is involved in a slower, modulatory intercellular signalling.

SVs and LDCVs have many different properties. SV exocytosis takes place only at active zones while LDCV exocytosis does not appear to be restricted to specialized regions of the nerve terminal surface. SV and LDCV exocytosis are differentially regulated: while a single action potential is an effective stimulus for SV exocytosis, a train of closely spaced action potentials is the most effective stimulus for LDCV exocytosis. LDCVs represent the equivalent organelle in neurons of secretory granules of the well established "regulated secretory pathway" of endocrine cells. Like endocrine secretory granules, they are assembled in the region of the Golgi complex. SVs, instead, are secretory organelles with several unique properties. They are continously reformed in nerve terminals by a process of exo-endocytotic recycling which appears to involve endosomal intermediates [1-3].

Many of the major protein components of synaptic vesicles have been identified and characterized over the last few years (fig.1). The properties and putative functions of these proteins have been discussed in several recent reviews [3-6].

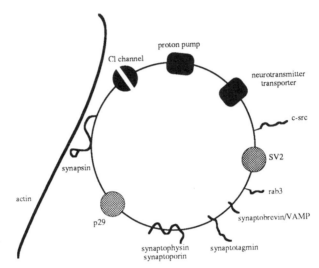

Fig.1   Cartoon illustrating some of the best characterized SV proteins. The putative topology in the membrane of some of the proteins is shown.

Most of the information accumulated until recently on the biochemical and functional properties of SVs was obtained from the study of SVs of mature neurons, i.e. fusion-ready SVs clustered at presynaptic sites. Much less was known about SV proteins at stages of neuronal development preceding synaptogenesis. Are SV proteins already assembled in 'bona fide' SVs even before a synapse is formed? Do SVs of immature neurons undergo exo-endocytotic recycling? Is their exocytosis regulated? What are the stages which lead to the formation of a presynaptic cluster? We have began to address these questions using primary cultures of embryonic hippocampal neurons as the model system. Some of this work is summarized below.

SYNAPTIC VESICLES ARE PRESENT IN NEURONAL PROCESSES BEFORE SYNAPTOGENESIS

Primary cultures of hippocampal neurons offer unique advantages for the study of neuronal development and synaptogenesis [7,8]. Through a series of well characterized developmental stages in vitro, they acquire a specific axonal and dendritic polarity [9,10].

Eventually, they form physiologically active synaptic contacts, which have all the features of a nerve terminal in vivo, including the characteristic presynaptic accumulation of SVs [7,8]. SV proteins are expressed in these hippocampal neurons not only before the formation of synaptic contacts, but even before neurons establish a well differentiated axon. Later in development, when the axon forms and elongates, SVs become concentrated in this process. Eventually, when the axon makes a contact with an appropriate target, SV proteins rapidly redistribute to sites of synaptic contacts [11].

Immunocytochemical studies have demonstrated that, throughout these immature stages, at least a large pool of SV proteins are already colocalized in vesicular structure, many of which have the size of SVs [11]. These vesicles in turn, are scattered throughout distal neuronal processes (primarily axons) rather than being accumulated at hot spots along neurites, as in the case of synapses.

Proteins which are found on these immature SVs include those peripheral membrane proteins which are thought to play an important role in docking and/or fusion of SVs with the plasma membrane. One such protein is synapsin [11], a phosphoprotein which may act as a regulated link between SVs and the actin based cytoskeletal matrix of nerve terminals [4,12-15]. Another such protein is rab3A, a small GTP-binding protein of the Sec4/Ypt1/rab subfamily of the Ras superfamily [16-18]. These findings are consistent with the possibility that SVs present in growing neurons are already functional and already able to undergo exocytosis.

SYNAPTIC VESICLES UNDERGO EXO-ENDOCYTOTIC RECYCLING IN DEVELOPING NEURONS EVEN BEFORE THE FORMATION OF SYNAPSES

The classical assay to monitor SV exocytosis is post-synaptic electrical recording. This assay cannot be used to study secretion from isolated developing neurons. To overcome this limitation, we have developed an exocytic assay based on the immunocytochemical detection of lumenal epitopes of SV proteins which become exposed at the cell surface as a result of exocytosis [19]. More specifically, the assay relies on antibodies directed against the lumenal N-terminus of synaptotagmin I (Syt I), a major SV protein (see fig. 1). Syt I is a member of the synaptotagmin family [20-23], which is represented by at least three isoforms, differentially expressed in the nervous system [24,25]. Syt I is an intrinsic membrane protein, with a single transmembrane region, a short N-terminal intralumenal domain and a cytoplasmic C terminal domain. The cytoplasmic domain of Syt I, which is the portion of the molecule more highly conserved among the different isoforms, contains two homologous repeats which are in turn homologous to the regulatory domain C2 of protein kinase C [21].

To detect exo-endocytosis, antibodies directed against the N-terminus of Syt I (Syt$_{lum}$-Abs) are used at a concentration sufficient to allow optimal labeling of SVs which become exposed to the cell surface, but insufficient to label non-specifically the endocytic pathway by fluid-phase endocytosis. In both highly immature cultures, as well as in mature cultures rich in synaptic contacts, virtually no Syt$_{lum}$-Abs bind to the neuronal surface if the cultures are kept at 0°C, in agreement with the known lack of compositional overlap between SVs and the presynaptic plasmalemma.

Addition of Syt$_{lum}$-Abs (for 1 hr at 37 °C) to cultures rich in synapses results in their rapid and specific internalization at synaptic sites [19]. This is expected considering the occurrence of spontaneous electrical and synaptic activity previously demonstrated in these neurons (see 11 for references). When Syt$_{lum}$-Abs were applied (1 hr 37 °C) to developing isolated neurons (stages 2 and 3 of Dotti et al. [9]), Syt$_{lum}$-Abs were found to be internalized by all neuronal processes and more prominently by axons (fig.2). This finding demonstrated a very active exo-endocytosis of SVs even in these immature neurons.

In double-immunofluorescence experiments a striking colocalization of internalized Syt$_{lum}$-Abs and of SV2 (fig.1), another major protein of SVs, was observed. Such a colocalization indicates first that internalized Syt$_{lum}$-Abs are retained within SVs and, second, that a large fraction of SVs undergo exo-endocytosis during 1 hr [19]. Since the cultures were not subjected to any stimulus, the high rate of Syt$_{lum}$-Abs endocytosis should reflect an high level of constitutive exocytosis in these neurons.

To determine whether Syt I and SVs undergo recycling in developing neuronal processes, cells previously exposed to Syt$_{lum}$-Abs were exposed to a second incubation with protein A-gold particles and then examined by electron microscopy. The rationale of these experiments is that SV recycling should result in the reexposure at the cell surface of Syt$_{lum}$-Abs and therefore in the binding, and internalization within SVs, of protein A-gold particles. The labeling of SVs produced by this procedure demonstrated that SVs undergo exo-endocytic recycling in developing neurons as they do at synapses [19,26,27].

In some experiments, neurons were exposed to Syt$_{lum}$-Abs for 1 hr and subsequently allowed to grow in control media for as long as 6 days. Syt$_{lum}$-Abs were still present in axons after 6 days (fig.2). In contrast, another endocytic marker, wheat germ agglutinin, was rapidly cleared from the processes and transported to the cell body throughout this time. Labeled SVs had the same distribution as the overall population of SVs, demonstrating that presence of Syt$_{lum}$-Abs bound to Syt I does not affect its sorting during the SV exo-endocytic cycle. If during this time the processes had formed synaptic contacts, both SVs and Syt$_{lum}$-Abs became clustered at presynaptic sites.

The occurrence of exo-endocytotic recycling of SVs before synaptogenesis suggests that secretion of neurotransmitters via SVs may precede synaptogenesis. These findings complement previous studies carried out by electrophysiological techniques on motor neurons maintained in primary culture [28-31]. In these studies release of acetylcholine from developing axons had been demonstarated. The quantal nature (i.e. mediated by SV exocytosis) of this release had been inferred but not demonstrated [29].

Neurotransmitters which are released from developing neurons may play some trophic/modulatory role on surrounding glial and neuronal cells. The morphology of a neuron results from both intrinsic properties and environmental inputs. Non-peptide neurotransmitters secreted via SVs may cooperate with growth factors, neuropeptides, extracellular matrix molecules, cell surface molecules and other factors in the regulation of neuronal development and, therefore, in the development and stabilization of neuronal circuits (Mattson, 1988).

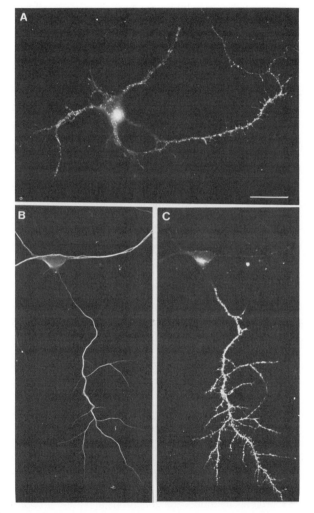

Fig.2   Uptake of Syt$_{lum}$-Abs by living hippocampal neurons grown in isolation. A: 3 days old neurons were incubated in the presence of Syt$_{lum}$-Abs for 1 hr at 37°C, washed, fixed, detergent permeabilized and reacted with rhodamine-conjugated goat anti rabbit IgGs. Syt$_{lum}$-Abs produce a bright staining with a finely punctate appearance in neuronal processes. B and C: 3 days old neurons were incubated for 1 hr at 37°C in the presence of Syt$_{lum}$-Abs. 2 days later they were fixed, detergent permeabilized, reacted with rhodamine-conjugated goat anti rabbit IgGs (C) and counterstained for beta-tubulin (B). Syt$_{lum}$-Abs are still present after 2 days and they are preferentially localized in the axon (Calibration bars: A, 16 μm; B,C, 20 μm).

REGULATION OF SYNAPTIC VESICLES EXO-ENDOCYTOSIS BEFORE SYNAPTOGENESIS

As mentioned above, SV exocytosis from the processes of isolated neurons, as revealed by binding and internalization of Syt_{lum}-Abs, takes place at very high rate in the absence of any stimulus. Preliminary experiments have suggested that this exocytosis still occurrs in Ca^{++}-free media. A high rate of Ca^{++} independent quantal neurotransmitter release from isolated axons of motor neurons in cultures has been demonstrated previously [31]. These findings were obtained by recording from a reporter cell attached to a recording micropipette and micromanipulated in proximity of the axon.

At mature synapses, depolarization evoked release of neurotransmitters is crucially dependent on extracellular Ca^{++}, while basal release is only minimally affected by lack of extracellular Ca^{++} [33,34]. One can hypothesize that SV exocytosis may occur constitutively from developing neurons and that regulatory mechanisms tightly controlled by elevations in cytosolic Ca^{++} may develop only after the formation of a presynaptic specialization. Alternatively, in developing neurons, as in fully differentiated neurons, constitutive and regulated exocytosis may coexist, although maturation of presynaptic nerve endings may coincide with some change in the fine regulation of synaptic vesicle exocytosis.

It is interesting to note that in neurons isolated from the buccal ganglion of Helisoma neurotransmitter release can be detected even prior to the contact with the appropriate muscle target. However, action potential-evoked release of neurotransmitters can be detected only following hours of muscle contact [35]. This finding indicates that the contact with the appropriate target enhances the responsiveness of the secretory machinery of Helisoma's neurons to internal Ca^{++} levels, thereby making the presynaptic cell able to couple action potentials with neurotransmitter release. It has been speculated that appropriate target contact may induce the synthesis of a specific presynaptic protein which may act as a "calcium-detector" of the secretory machinery, playing a critical role for the excitation-secretion coupling in Helisoma neurons [35].

POLARIZED PATHWAYS OF EXO-ENDOCYTOSIS

Expression of SV proteins by hippocampal neurons in culture precedes differentiation of axons and dendrites [11]. Prior to the development of an axon, SV recycling appears to partially overlap with the recycling of transferrin receptor. This was demonstrated both by immunocytochemical staining for transferrin receptor as well as by loading cells with fluorescent iron-saturated transferrin ([36] and our unpublished observations). Transferrin receptor is a well established marker of the receptor-mediated recycling pathway [37]. At this developmental stage, SVs and transferrin receptor may recycle through the same endosomes. When an axon differentiates, the recycling of SV proteins becomes progressively restricted to this process, while transferrin receptor recycling remains confined to dendrites ([36] and our unpublished observations). However, a reduced degree of Syt_{lum}-Abs internalization can be observed also in the dendritic compartment of immature neurons. Only at later stages, SV recycling becomes exclusively restricted to axons [19]. Dendritic uptake of Syt_{lum}-Abs in immature neurons can be observed also when microtubules are previously depolymerized by nocodazole treatment. This indicates that dendritic labeling is not dependent upon uptake of antibodies in the axon followed by retrograde transport along microtubules (our unpublished observations).

Taken together these results indicate that the establishment of structurally and functionally distinct axonal and dendritic compartments occurrs gradually during neuronal development.

RECYCLING SYNAPTIC-LIKE MICROVESICLES ARE PRESENT IN ENDOCRINE CELLS

The ability of SVs to undergo exocytosis irrespectively of their clustering at presynaptic sites is not surprising considering the results recently emerged from studies on endocrine cells. Immunocytochemical and subcellular fractionation studies have shown that endocrine cells contain a population of pleiomorphic microvesicles (referred to as synaptic-like microvesicles, SLMVs), with the same size range of SVs of neurons and with a similar protein composition [38-40]. SLMVs, like bona fide neuronal SVs, undergo exo-endocytotic recycling [36,38,41,42]. More recently, it has been shown that SLMVs of pancreatic β-cells are able to take up and store the neurotransmitter GABA, suggesting that SVs and SLMVs may also have similar function [40]. It is still unclear whether and how exo-endocytosis of SLMVs is regulated.

CONCLUSIONS AND IMPLICATIONS FOR THE MECHANISM OF ACTION OF BOTULINUM AND TETANUS TOXIN

The secretory apparatus which is found at mature nerve terminals and which involves SVs appears to represent the special adaptation for synaptic transmission of an organelle which may have more general functions. SVs are present and undergo exocytosis in immature neurons. Synaptic-like microvesicles, with properties similar to SVs of neurons, are found also in endocrine cells. Botulinum and tetanus toxins appear to have a very specific action on neuronal secretion. Their main well established action is on the Ca^{++} dependent exocytosis of SVs at mature nerve terminals [43,44]. Conflicting results have been reported on their effects on endocrine granules secretion. Nerve terminals completely paralized by the action of botulinum toxin are still able to extend new growth cones [45], indicating that constitutive exocytosis of vesicles which are involved in axonal growth is not blocked by the toxin. In order to further elucidate the site(s) of action of these toxins, it will be important to further elucidate the specificity of their effects. For example, can botulinum and tetanus toxin block the active rate of SV exocytosis observed in developing neurons? Do they have a differential effect on axonal and dendritic exocytosis? Do these toxins exhibit a differential effect on secretion from SVs and LDCVs? Do they block exocytosis of SLMVs from endocrine cells? The definition of the processes which are selectively affected by the toxins may help to identify their substrate proteins.

REFERENCES

1. De Camilli P, Jahn R. Pathways to regulated exocytosis in neurons. Ann Rev Physiol 1990; 52, 625-645.
2. Matteoli M, Thomas Reetz A, De Camilli P. Small synaptic vesicle and large dense core vesicles: secretory organelles involved in two modes of neuronal signalling. In: Volume transmission in the brain: novel mechanisms for neural transmission. New York: Raven Press, 1991: 181-193.
3. Südhof TC , Jahn R. Proteins of synaptic vesicles involved in exocytosis and membrane recycling. Neuron 1991; 6, 665-677.
4. De Camilli P, Benfenati F, Valtorta F, Greengard P. The synapsins. Ann Rev Cell Biol 1990; 6: 433-460.
5. Trimble WS, Linial M, Scheller RH. Cellular and molecular biology of the presynaptic nerve terminal. Ann Rev Neurosci 1991; 14: 93-122.
6. Matteoli M, De Camilli P. Molecular mechanisms in neurotransmitter release. Curr Opinion in Neurobiol 1991; 1: 91-97.
7. Bartlett WP, Banker GA. An electron microscopic study of the development of axon and dendrites by hippocampal neurons in culture. I. Cells which develop without intracellular contacts. J Neurosci 1984a; 4, 1944-1953.
8. Bartlett WP, Banker GA. An electron microscopic study of the development of axon and dendrites by hippocampal neurons in culture. II Synaptic relationships. J Neurosci 1984b; 4, 1954-1965.
9. Dotti CG, Sullivan CA, Banker GA. The establishment of polarity by hippocampal neurons in culture. J Neurosci 1988; 8: 1454-1468.
10. Goslin K, Banker G. Experimental observations on the development of polarity by hippocampal neurons in culture. J Cell Biol 1989: 108: 1507-1516.
11. Fletcher TL, Cameron PL, De Camilli P, Banker, G. The distribution of synapsin I and synaptophysin in hippocampal neurons developing in culture. J Neurosci 1991; 11: 1617-1626

12. De Camilli P, Cameron R, Greengard P. Synapsin I (protein I), a nerve terminal specific phosphoprotein. I. Its general distribution in synapses of the central and peripheral nervous system demonstrated by immunofluorescence in frozen and plastic sections. J Cell Biol 1983; 96, 1337-1354.

13. Hirokawa N, Sobue K, Kanda K, Harada A, Yorifuji H. The cytoskeletal architecture of the presynaptic terminal and molecular structure of synapsin I. J Cell Biol 1989; 108: 111-126.

14. Benfenati F, Valtorta F, Greengard P. Computer modeling of synapsin I binding to synaptic vesicles and F-actin: implications for regulation of neurotransmitter release. Proc Natl Acad Sci 1991; 88: 575-579.

15. Benfenati F, Valtorta F, Chieregatti E, Greengard P. Interation of free and synaptic vesicle-bound synapsin I with F-actin. Neuron 1992; 8: 377-386.

16. Fischer v. Mollard G, Mignery G, Baumert M, Perin MS, Hanson TJ, Burger PM, Jahn R, Südhof TC. Rab3 is a small GTP-binding protein exclusively localized to synaptic vesicles. Proc Natl Acad Sci USA 1990; 87: 1988-1992.

17. Fischer v. Mollard G, Südhof TC, Jahn R. A small G protein dissociates from vesicles during exocytosis. Nature 1991; 349: 79-82.

18. Matteoli M, Takei K, Cameron R, Hurlbut P, Johnston PA, Südhof TC, Jahn R, De Camilli P. Association of rab3A with synaptic vesicles at late stages of the secretory pathway. J.Cell Biol 1991; 115: 625-633.

19. Matteoli M, Takei K, Perin MS, Sudhof TC, De Camilli P. Exo-endocytotic recycling of synaptic vesicles in developing processes of cultured hippocampal neurons. J Cell Biol 117: 849-861.

20. Matthew, W.D., Tsavaler, L. and Reichardt, L.F. (1981) Identification of a synaptic vesicle-specific membrane protein with a wide distribution in neuronal and neurosecretory tissue. J. Cell Biol. 91, 257-269.

21. Perin MS, Fried VA, Mignery GA, Jahn R, Südhof TC. Phospholipid binding by a synaptic vesicle protein homologous to the regulatory region of protein kinase C. Nature 1990; 345: 260-263.

22. Perin MS, Brose N, Jahn R, Südhof T. Domain structure of synaptotagmin (p65). J Biol Chem 1991; 266: 623-629.

23. Perin MS, Johnston P., Ozcelik T, Jahn R, Francke U, Sudhöf TC. Structural and functional conservation of synaptophysin (p65) in Drosophila and humans. J Biol Chem 1991; 266: 615-622.

24. Geppert M,Archer BT, Südhof TC. Synaptotagmin II: a novel differentially distributed form of synaptotagmin I. J Biol Chem 1991; 266,:13548-13552.

25. Wendland B, Miller KG, Schilling J, Scheller RH. Differential expression of the p65 gene family. Neuron 1991; 6: 993-1007.

26. Ceccarelli B, Hurlbut WP, Mauro A. Turnover of transmitter and synaptic vesicles at the frog neuromuscular junction. J Cell Biol. 1973; 57: 499-524.

27. Heuser JE, Reese TS. Evidence for recycling of synaptic vesicles membranes during transmitter release at the frog neuromuscular junction. J Cell Biol 1973; 57: 315-344.

28. Hume RI, Role LW, Fishbach GD. Acetylcholine release from growth cones detected with patches of acetylcholine rich membranes. Nature 1983; 305: 632-634.

29. Young SH, Poo MM. Spontaneous release of transmitter from growth cones of embryonic neurones. Nature 1983; 305, 634-637.

30. Sun Y, Poo MM. Evoked release of acetylcholine from the growing embryonic neuron. Proc Natl Acad Sci. 1987; 84: 2540-2544.

31. Evers J, Laser M, Sun YA, Xie ZP, Poo MM. Studies of nerve-muscle interactions in Xenopus cell cultures: Analysis of early synaptic currents. J Neurosci 1989; 9: 1523-1539.

32. Mattson MP. Neurotransmitters in the regulation of neuronal cytoarchitecture. Br Res Rev 1988; 13: 179-212.

33. Fatt, P. and Katz, B. (1952) Spontaneous subtreshold activity at motor nerve endings. J. Physiol. (London) 117, 109-128.

34. Hubbard, J.I., Jones, S.F. and Landau, E.M. (1968) On the mechanism by which calcium and magnesium affect the spontaneous release of transmitter from mammalian motor nerve terminals. J. Phyisiol. 194, 355-380.

35. Zoran MJ, Doyle RT, Haydon PG. Target contact regulates the calcium responsiveness of the secretory machinery during synaptogenesis. Neuron 1991; 6: 145-151.

36. Cameron PL, Südhof TC, Jahn R, De Camilli P. Colocalization of synaptophysin with transferrin receptors: implications for synaptic vesicle biogenesis. J Cell Biol 1991; 115: 151-164.

37. Fuller SD, Simons K. Transferrin receptor polarity and recycling accuracy in "Tight" and "Leaky" strains of Madine-Darby canine kidney cells. J Cell Biol 1986; 103: 1767-1779.

38. Navone F, Jahn R, Di Gioia G, Stukenbrok H, Greengard P, De Camilli P. Protein p38: an integral membrane protein specific for small vesicles of neurons and neuroendocrine cells. J Cell Biol 1986; 103: 2511-2527.

39. Wiedenmann B, Rehm H, Knierim M, Becker CM . Fractionation of synaptophysin-containing vesicles from rat brain and cultured PC12 pheochromocytoma cells. FEBS Lett. 1988; 240: 71-77.

40. Reetz AT, Solimena M, Matteoli M, Folli F,Takei K, De Camilli P. GABA and pancreatic beta cells: colocalization of glutamic acid decarboxylase (GAD) and GABA with synaptic-like microvesicles suggests their role in GABA storage and secretion. EMBO J. 1991; 10: 1275-1284.

41. Johnston PA, Cameron PL, Stukenbrok H, Jahn R, De Camilli P, Südhof TC. Synaptophysin is targeted to similar microvesicles in CHO and PC12 cells. EMBO J 1989; 8: 2863-2872.

42. Linstedt AD, Kelly RB. Synaptophysin is sorted from endocytotic markers in neuroendocrine PC12 cells but not transfected fibroblasts. Neuron 1991; 7: 309-317.

43. Kim YI, Lomo T, Lupa MT, Thesleff S. Miniature end-plate potentials in rat skeletal muscle poisoned with botulinum toxin. J Physiol 1984; 356: 587-599.

44. Molgo J, Comella JX, Angaut-Petit D, Pecot-Dechavassine M, Tabti N, Faille L, Mallart A, Thesleff S. Presynaptic actions of botulinal neurotoxins at vertebrate neuromuscular junctions. J Physiol Paris 1990; 84: 152-166.

45. Alderson K, Holds JB, Anderson RL. Botulinum-induced alteration of nerve-muscle interactions in the human orbicularis oculi following treatment for blepharospasm. Neurology 1991; 41:1800-1805.

ENDOSOME PROCESSING: STRUCTURAL, FUNCTIONAL AND KINETIC INTERRELATIONS

Lutz Thilo

Dept. of Medical Biochemistry
University of Cape Town Medical School
Observatory, Cape Town
7925 South Africa

INTRODUCTION

Clostridial neurotoxins exert their effect from within the cytoplasm of nerve cells. Rather than gaining direct access through the cell-surface membrane, the toxin exploits the process of endocytosis to gain entry into cells. As this would still leave the toxin on the exoplasmic side of the endocytic membrane, a translocation step through the membrane into the cytoplasm must follow at some stage along the endocytic pathway. An awareness of the steps which are involved in endosome processing can be helpful to assess the mechanism by which the toxin can reach its site of action in the cell.

This article summarises major events during the process of endocytosis, with emphasis on dynamic aspects, while keeping in mind how individual steps might be of relevance for the entry of toxins into nerve cells. Endocytosis is believed to occur in all eukaryotic cells, but relatively little is known about this process in nerve cells.

Endocytosis involves the constitutive formation of endocytic vesicles which pinch off into the cytoplasm at the cell surface (reviewed in ref. 1). Topologically similar, but functionally distinct, is the process for retrieval of secretory membrane after insertion into the cell surface during secretion (reviewed in ref. 2). Endocytosis serves the uptake of extracellular material for cell feeding, for control of cell function, for immunological functions, for transfer across the cell (transcytosis), and for maintaining cell-surface integrity.

BINDING TO CELL-SURFACE RECEPTORS

Under physiological conditions most substrates for endocytic uptake are present at concentrations much below the micromolar range. To allow efficient uptake, these substrates are enriched at the cell surface by binding specifically and with high affinity to receptors (reviewed in ref. 3). The efficiency of receptor-mediated uptake depends on the binding affinity, on the number of cell-surface receptors, and on their rate of internalization.

Botulinum and Tetanus Neurotoxins, Edited by
B.R. DasGupta, Plenum Press, New York, 1993

For some typical receptor types, ligand binding occurs with an affinity such that at nanomolar concentrations of ligand 50% of the total number ($\sim 10^5$) of cell-surface receptors will be occupied.

The majority of cell-surface proteins, including receptor molecules, are glycoproteins. It is not clear what role glycosylation plays in receptor-ligand interaction. In general, glycomoieties on the cell surface can provide sites for binding of lectins,[4] various toxins[5] and viruses (reviewed in ref. 6). The endocytic uptake of toxins via binding to glycomoieties on the surface has been used to select for glycosylation-defective mutants, surviving in the presence of toxin.[7]

COATED AND NON-COATED ENDOCYTIC VESICLES

The formation of endocytic vesicles at the cell surface can occur at specialised sites, called coated pits due to the morphologically observed coat underlying the plasma membrane on the cytoplasmic side (reviewed in ref. 8). Coated vesicle-mediated endocytosis does not seem to be the only mode of endocytic uptake. Recently the formation of non-coated vesicles as well as of coated vesicles has been observed in the same cells.[9] The particular functions of non-coated endocytic uptake are not known. However, when cholera toxin is taken up by fibroblasts, this occurs by means of non-coated vesicles, and these follow the same intracellular pathway as ligand entering via coated pits.[10] In particular, tetanus toxin, in cultured liver cells, binds preferentially to non-coated membrane invaginations prior to internalization.[11]

Coated pits act as sorting sites on the cell surface, eg. for the enrichment of receptors to be internalized. The coat consists of a major protein component, clathrin, which assembles via a tetrameric complex of assembly proteins, also called adaptors because they mediate binding to the cytoplasmic tails of cell-surface receptors (reviewed in ref. 12). Although the same conserved assembly proteins operate in many different cell types, two distinct clathrin-binding proteins (AP180 and auxilin) have so far only been observed in neuronal cells.[13]

Cell-surface receptors move by lateral diffusion ($D < 10^{-9}$ cm^2/s, reviewed in ref. 14) to allow frequent encounters with coated pits. Here they get trapped due to specific amino-acid sequences of their cytoplasmic tails which act as signals for binding with the assembly proteins in the coat (summarised in ref. 15). Some receptor types will only be trapped in coated pits once they have bound their ligand (eg. EGF-R, cf. ref. 16). In the case of the receptor for low-density lipoprotein in fibroblasts, 60-80% of about 10^5 cell-surface receptors are trapped in coated pits which make up about 2% of the plasma membrane area.[17,18] This implies an enrichment of 120-fold against the depleted regions (cf. Fig. 1).

RATES OF INTERNALIZATION

Due to the sorting events on the cell surface, individual membrane constituents can be internalized at different rates. For various receptor types, the average residence time on the surface is about 3 to 7 min, before taking part in a round of internalization and recycling (reviewed in ref. 19). This can be compared to the internalization rate of more general membrane constituents.

Most cell-surface constituents participate in membrane internalization, as has been observed for iodinated (^{125}I) plasma-membrane proteins[20] and for glycoconjugates labelled with radioactive galactose (cf. ref. 19). Kinetics of internalization of these covalent membrane markers, as well as observations for internalized membrane as a morphological entity, suggest that the average membrane constituent spends between about 20 and 120 min on the cell surface, depending on cell type (cf. refs. 1,18,19). This is, therefore, from 3 to 40-fold longer than for typical cell-surface receptors (cf. above).

HIGH AFFINITY UPTAKE

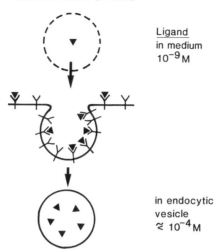

Ligand
in medium
10^{-9} M

in endocytic
vesicle
$\cong 10^{-4}$ M

Figure 1. Concentrated uptake of ligand by receptor-mediated endocytosis. A 10^5-fold increase of ligand concentration in endocytic vesicles as compared to extracellular concentrations (say 10^{-9} M) can be calculated: 1000 coated pits per cell surface giving rise to 1000 coated vesicles of diameter 100 nm (cf. ref. 18); these contain 70% (17,18) of the total 10^5 receptors per cell surface, binding ligand with an affinity according to $K_M \sim 10^{-9}$M (summarised in ref. 19).

Little information is available for the rate at which the lipids of the plasma membrane are internalized. When a fluorescent analogue of sphingomyelin is inserted into the plasma membrane of fibroblasts, it redistributes to intracellular membranes with a half-life of 15 min with 50% remaining on the cell-surface at steady state.[21] This suggests that this lipid marker spends an average of about 50 min on the surface prior to internalization. This rate of internalization corresponds to that of protein constituents on the surface of fibroblasts (cf. above).

INTRACELLULAR RECYCLING TIMES

Internalized membrane constituents make up an intracellular pool from which recycling occurs. Because endocytosis is an ongoing process, with the rate of internalization equalling the recycling rate, a steady state exists with the intracellular pool size remaining constant relative to the plasma membrane pool. The intracellular recycling time for each type of membrane constituent relates to its residence time on the cell surface (cf. above and Fig.2) in the same ratio as their respective pool sizes.

For some common receptor types the intracellular receptor pool is more than twice as large as the cell-surface pool.[19] Accordingly, receptors spend about 7 to 20 min on their intracellular route as compared to about 3 to 7 min on the cell surface (cf. above; the receptor for LDL in fibroblasts is an exception, with not more than 15% of the total in an intracellular pool,[22] implying an intracellular recycling time of less than 1.8 min).

In comparison, internalized membrane glycoconjugates make up an intracellular pool only about 15% that on the cell surface and, therefore, are recycled after only 2-5 minutes (cf. ref. 19; for data based on the internalization of lectin by fibroblasts, significantly different values have been observed, viz. 70% and 24 min, resp. cf. ref. 4).

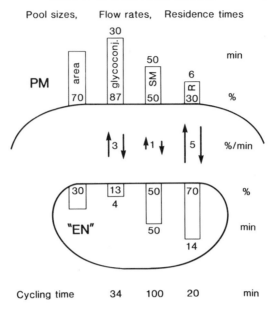

Figure 2. Pool sizes (m, % of total), flow rates (k, % per min.), and residence times (τ = m/k, min.) as compared for the plasma membrane (PM) and for intracellular (endosomal) membranes ("EN"), with respect to cell-surface receptors (R), a membrane lipid (sphingomyelin, SM), membrane glycoconjugates (glycoconj.) and morphometrically observed membrane area.

Internalized sphingomyelin in fibroblasts spends about 50 min in a pool 50% that of the cell surface.[21]

Endocytic contents, destined for delivery into secondary lysosomes for degradation, spend about 15 min or 5 min in a prelysosomal compartment in fibroblasts[18] or macrophages,[19] respectively.

ENDOCYTOSIS VS. RETRIEVAL OF SECRETORY MEMBRANE

Stimulated secretion involves fusion of secretory-vesicle membrane with plasma membrane, leading to a substantial increase in cell-surface area. Retrieval (re-internalization) of this secretory membrane occurs by a process which is topologically similar to endocytosis. Nevertheless, endocytosis and membrane retrieval seem to be independent processes.

In adrenal chromaffin cells, secretory membrane constituents disappear from the cell surface within less than about 50 min[23,24] with the major part of this process being completed within 5 to 10 min.[25,26] It is much faster than membrane internalization due to endocytosis and occurs as a result of the stimulated secretory event. Furthermore, secretory membrane does not mix with plasma membrane after insertion at the cell surface and is retrieved by a selective process which does not involve plasma membrane constituents.[23,27,28,29] In mast cells, selective retrieval of granule membrane occurs about 5-times faster than endocytosis-related membrane internalization.[29]

The binding and uptake of radiolabelled botulinum neurotoxin has been compared for stimulated and non-stimulated motor neuron terminals.[30] Stimulation did not affect the number of toxin molecules binding at the cell surface, but resulted in a doubling of intracellular toxin.[30] This suggests increased toxin uptake due to retrieval of secretory membrane. In so far as retrieval should be selective in these cells (cf. above), it can be concluded that toxin binds to membrane constituents which are present on both plasma and secretory membrane.

MORPHOLOGICAL AND FUNCTIONAL ASPECTS OF ENDOCYTIC ORGANELLES

Morphologically distinct organelles have been identified along the endocytic pathway, as based on their time-structured labelling with endocytic markers and on characteristic staining with antibodies to distinct membrane antigens (summarized in ref. 18), Fig. 3.

Early endosomes (≤5 min) appear as tubulo-vesicular structures. The tubules comprise up to 90% of the membrane area and, accordingly, are noticed as harbouring receptors which are destined for recycling to the cell surface. The vesicular parts may serve to transfer endocytic contents, including receptor-dissociated ligands, to later stages of the endocytic pathway.

Late endosomes (5 to ≤15 min) appear as spherical, multivesicular bodies, diameter about 0.5 μm. The intraluminal vesicles (50 nm) seem to arise by invagination and scission from the vesicle surface and may be involved in capturing membrane constituents destined for degradation in lysosomes (cf. ref. 31).

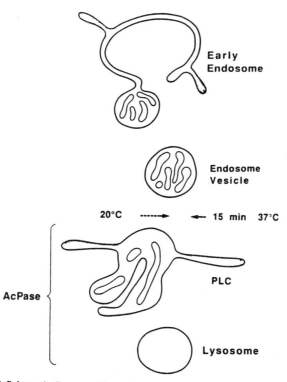

Figure 3. Schematic diagram of the main compartments of the endocytic pathway. The stage reached after 15 min at 37°C is the site where low temperature (20°C) blocks further progress along the pathway. Compartments which stain for the lysosomal marker, acidphosphatase, are indicated (AcPase). From Griffiths G, Back R, Marsh M, J. Cell Biol. 1989; 109:2703. With permission.

Prior to delivery to secondary lysosomes, endocytic markers appear in a prelysosomal compartment which is also of a tubulo-vesicular nature and noticeable for the large areas of internal membrane.[32] It is distinguished from lysosomes by the presence of abundant mannose 6-phosphate receptors which are absent from lysosomal membranes. These receptors serve to sort lysosome-directed enzymes from the biosynthetic, secretory pathway. The prelysosomal compartment, accordingly, is considered as the site of communication between the endocytic and exocytic pathways.[32]

Complementary to the morphological observations at the level of the electron microscope, light-microscopic observations on living cells have shown that the endocytic structures consist of a continuous tubulo-vesicular reticulum.[31] Membrane-bound marker which is destined for recycling moves rapidly (0.5 to 1 μm/s) and bidirectionally along a network of tubules. Lysosomally destined marker accumulates in multivesicular bodies which are observed as varicosities moving along the tubular network at a lower speed (0.05 μm/s) towards the perinuclear region. It is not known whether the discrete morphological entities which are seen at the electron-microscope level are remnants of the continuous network which has broken down during specimen preparation. Recent observations on thick-section electron microscope preparations have indicated numerous tubular endosomal structures.[33] These structures, however, have been interpreted as representing early endosomes (and recycling organelles) only, not being in continuity with late endosomes.

Lysosomes, as the final compartment for endocytic delivery, have also been observed as a tubulo-reticular network.[34] However, this morphological appearance depends on endocytic load and cell type, being most obvious in macrophages.[34]

BIOCHEMICAL CHARACTERISTICS OF ENDOSOMES

Endocytic vesicles which are internalized from the cell surface rapidly fuse and become part of a compartment of early endosomes.[18] It can, therefore, be expected that early endosomes are very similar to the plasma membrane in terms of their biochemical and biophysical properties. Because of their biophysical similarity to both plasma membrane and Golgi membrane, only partial purification and biochemical characterisation of endosomes has been achieved (reviewed in ref. 1).

Endosomes have a similar but not identical composition as the plasma membrane in terms of polypeptides, glycoproteins and lipids and, therefore, do not seem to be entirely cell-surface derived.[1] Endosomes comprise only about 15-30% the area of the cell surface, yet they accommodate 2–3-times the cell-surface receptor pool (cf. above). This would suggest that receptors are enriched about 6–20-fold on the endosome membrane as compared to the cell surface. Part of this enrichment may be the direct result of concentrated endosomal delivery of receptors via coated vesicles (cf. Fig. 1 and 2).

Intraendosomal acidification (reviewed in ref. 35) is a major element of endosomal function, important for receptor-ligand dissociation, regulation of endosomal enzyme activity and, opportunistically, for viral and toxin entry into the cytoplasm. pH-sensitive fluorescent endocytic markers indicate a drop to pH 6 at early stages and to pH 5.5 at later stages of the endocytic pathway (eg. refs. 36,37). The endosomal proton pump is an ATP-ase which operates without translocation of other, charge-compensating ions. Acidification, therefore, involves the generation of an intraendosomal positive membrane potential. Early endosomes also contain the Na+,K+-ATPase derived from the cell surface. This enzyme also generates an intraendosomal positive membrane potential, thus counteracting the acidification process. It has been proposed that further acidification to pH 5 occurs only at later stages of the endocytic pathway when the Na+,K+-ATPase has been recycled from early endosomes.[37]

Low-pH endocytic compartments are the most obvious targets for action of exogenous amines (reviewed in ref. 38). By raising intraendosomal pH, amines can interfere with receptor-ligand dissociation which results in trapping of receptors. Accumulation of the protonated, membrane impermeant amines leads to osmotic distortion of endosome morphology. This can be expected to indirectly affect endocytic membrane flow. It is not clear to what extent, if at all, the presence of amines inhibits fusion between organelles along the endocytic pathway: First, the extent of fusion, during endosomal processing and transfer of endocytic contents towards lysosomes, is uncertain (cf. below) and, second, endosome fusion in vitro is not affected by amines (cf. Philip Stahl elsewhere in this book). The action of clostridial neurotoxins can be antagonised by ammonium chloride and methylamine and it has been proposed that the amines act at the internalisation step.[39] However, the results as reported are also compatible with delayed endosomal processing after internalization of toxin.

Some proteolytic enzymes are not restricted to lysosomes but are encountered also at early stages of the endocytic pathway (summarized in ref. 40). Examples of endosomal protease activity suggest that it is cell-type specific and regulated by endosomal pH. An insulin-specific protease is present already at the cell surface of hepatocytes, with endosomes containing also a carboxypeptidase-B like activity. Endosomes of macrophages contain a cysteine protease (cathepsin B) which operates also at neutral pH, and substantial amounts of an aspartyl protease (cathepsin D) which requires a lower pH. The latter enzyme is associated with the endosomal membrane. Regulated acidification and protease activity seem to be highly important for controlled processing in endosomes, as compared to the more indiscriminate degradation in lysosomes. The step-wise processing of various toxins upon endocytic uptake, and as monitored via their toxic effects on cells, has provided a useful tool to study enzymatic activity in endosomes.[1,40]

ENDOCYTIC RECYCLING ROUTES

The major sites for membrane recycling are the tubular parts of early endosomes. The tubulo-reticular structures have a 4-fold higher surface to volume ratio than the vesicular parts of endosomes,[18] and preferential recycling from tubules may provide a crude mechanism of sorting membrane from contents in the vesicular parts. Even if an individual sorting step of this kind may not seem to be highly efficient, it can provide very effective sorting upon iteration.[41]

When different receptor types are internalized and recycled, they travel together in the same vesicle.[42] Not all simultaneously internalized receptors follow the same recycling route, depending on receptor function. Whereas most receptors recycle from early endosomes, the mannose 6-phosphate receptor parts from these, to be recycled through the prelysosomal compartment where it operates to sort lysosomally destined enzymes from the biosynthetic pathway.[43] Also noteworthy is the observation that a large "silent" pool of asialoglycoprotein receptors exists in the trans-Golgi network and receptors from this pool do not participate in the endocytic traffic of this receptor type.[43] A kinetic model has been presented which can account for ligand-receptor sorting for a variety of receptor behaviours where these are based on the effect of ligand on the rate at which receptors diffuse into tubules and on the efficiency at which they become trapped there.[44]

Endocytosed membrane glycoconjugates in general are recycled more rapidly than cell surface receptors (cf. above) and this suggests that recycling also occurs from a very early stage of the endocytic pathway. A small fraction of internalized glycoproteins, also of cell-surface receptors, are recycled through secondary lysosomes[45] and through the Golgi apparatus.[46,47] This can account for the observation that 5% of internalized ricin becomes localised in the Golgi complex.[48]

Recycling of secretory membrane after retrieval at the cell surface seems to involve a major route to the Golgi complex (discussed in ref. 49), quite distinct from the quantitatively minor role played by this pathway for the recycling of plasma membrane, Fig. 4 (see also above). The separate pathway for membrane retrieval may provide for retrograde transport of TeTx. Unlike

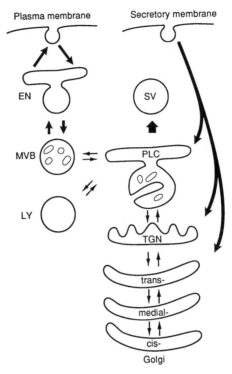

Figure 4. Schematic diagram of the pathway for retrieval of secretory membrane as a process distinct from endocytosis (cf. Fig. 3). TGN: trans-Golgi network, PLC: prelysosomal compartment, SV: secretory vesicle, MVB: multi-vesicular body, EN: endosome, PM: plasma membrane, LY: lysosome. Note the central role of the PLC as the point of communication between the secretory and endocytic pathways as proposed previously.[32]

endocytic vesicles, retrieval vesicles will not have an acidic interior and, therefore, will allow the toxin to escape from low pH-dependent transfer into the cytoplasm.

ENDOSOME PROCESSING: VESICULAR TRAFFIC VS. MATURATION

Transfer of endocytic contents along the membrane compartments of the endocytic pathway requires inter- and intra-compartmental interactions. Therefore, the question arises concerning the structural development between early and late endosomes. Two models are presently being considered to describe these dynamic aspects of endosome processing, Fig. 5. Both are only partially suited to explain all observations (reviewed in refs. 50,51).

According to the vesicular-traffic model, the morphologically distinct organelles represent permanent membrane compartments along the endocytic pathway. These communicate by the exchange of vesicles which shuttle between them: vesicles for the transfer of endocytic substrate, and vesicles for membrane recycling and delivery of enzymes into the endocytic compartments. This mode of operation would require abundant fusion/fission events, accompanied by a sophisticated mechanism for sorting of contents and membrane constituents. The model has been postulated in analogy to the vesicular membrane traffic which operates in the

DYNAMICS

Vesicular Traffic   Maturation

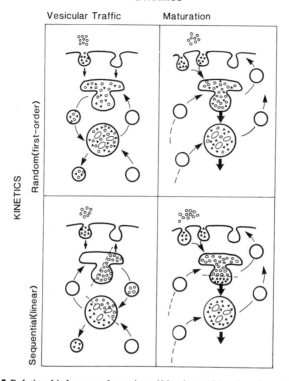

KINETICS
Random(first–order)
Sequential(linear)

Figure 5. Relationship between dynamic and kinetic considerations for endosome processing. Sequentially administered endocytic marker is indicated at the upper left of each panel (▲, first marker; o, second marker). Although the markers as indicated suggest contents markers, they can also be seen as membrane markers. In this case the schematic diagrams would be read as representing intact organelles with markers on their membrane, rather than as cross-sections through organelles indicating contents marker. Dashed lines through organelles suggest that structured progression occurs within an individual organelle. For further explanation, see text.

compartments of the Golgi complex for the biosynthetic, secretory pathway.[52] It can explain the absence of intermediates between distinct compartments and is supported by the detection of a potential carrier vesicle type.[50]

The maturation model suggests a gradual modification of early endosomes to finally become secondary lysosomes (summarized in refs. 51,53). Newly internalized endocytic vesicles would fuse to form an early endosome. To the extent that early endosomes are not entirely cell-surface derived (cf. above), recycling vesicles from later stages of the pathway would also contribute to endosome formation. Subsequent maturation would involve a continuous addition of constituents characteristic of the later stages and a recycling of constituents typical of earlier ones. As is the case for the vesicular-traffic model, the maturation model depends on numerous fusion/fission events between membrane vesicles and the maturing organelle, and this interaction would require specific sorting mechanisms. The maturation model is compatible with the finding that endosomes older than 5 min rapidly lose their fusogenicity, that proteolytic enzymes are delivered to early endosomes, and that no direct evidence for the fusion between endosomes and lysosomes exists (summarized in ref. 53, but cf. ref. 54). A gradual decrease in the abundance of cell surface-derived receptors on endosomes has been interpreted as evidence for the maturation process.[51]

The recent observation of endosomes operating as a continuous tubulo-vesicular network[31] (cf. above) would obviate both of the above models, leaving only the questions concerning the sites and mechanisms of entry and exit to and from the network structure.

ENDOSOME PROCESSING: SEQUENTIAL VS. RANDOMIZATION

The question concerning the kinetic aspects of endosome processing has also been phrased as whether or not sequentially formed endosomes fuse and intermingle their contents,[55,56] Fig. 5.

For sequential processing, the endosome can be seen as a time-structured system of organelles through which membrane and contents move in linear fashion: transfer and recycling operates on a "first come, first served" basis. Sequentially endocytosed substrates will not become intermingled.

For random processing, the endosome can be seen as a mixed bag from which transfer and recycling occurs on a random basis: sequentially endocytosed substrates will intermingle and the relative rate of transfer and recycling will be proportional to the relative abundance of the respective substrate in the randomised compartment (first-order kinetics).

A priori, these considerations do not depend on whether endosomes operate via maturation or by the vesicular traffic model. Also, the kinetics of processing may vary between different segments of the endocytic pathway. For example, common delivery to lysosomes as a final stage can be expected to result in intermingling. Different kinetics may be observed, depending on whether the inspected time intervals are long or short in comparison with the transfer times for various steps along the endocytic pathway.

On a time scale which is small compared to the time of formation of individual maturing endosomes, intermingling due to fusion can be expected when newly formed endocytic vesicles fuse to assemble a de novo endosome for subsequent maturation. Once maturation results in uncoupling from further addition of incoming vesicles, observations on this larger time scale would indicate sequential processing, regardless of whether maturing endosomes within the same age group intermingle their contents by fusion.

The most obvious way to consider the vesicular traffic model for its kinetic implications would be that it results in randomisation within each individual compartment, where incoming vesicles deliver their contents. This in turn would result in randomised transfer to the next compartment. The possibility of sequential kinetics would be less obvious for the vesicular traffic model. It would require that endocytic contents and membrane move sequentially within each organelle, entering at a particular site and leaving at another.

A continuous endosomal network as recently observed[31] (cf. above) could accommodate both sequential as well as randomised processing kinetics.

Present evidence indicates that intermingling due to endosome fusion occurs for very early stages of the endocytic pathway.[55,56] A final choice between the various dynamic, structural, and kinetic combinations can not yet be made.

Fusion between early endocytic vesicles has been observed in several cell-free assays. These in-vitro systems allow access to components which are involved during endosome fusion (see Philip Stahl, elsewhere in this book). Since endosomes also fuse with inverted plasma-membrane vesicles,[57] such in-vitro systems may also provide access to the mechanism by which surface-bound neurotoxin can enter endosomes and from these into the cytoplasm.

CONCLUDING REMARKS

Present knowledge of the endocytic process is based on results obtained for non-neuronal cells. However, a comparison of endocytosis in cells specialized for different functions has revealed no fundamental differences. Only the rates and the sizes of endocytic membrane compartments differ between fibroblasts (connective tissue), macrophages (professional phagocytes), and secretory cells. Concerning the mode of entry of neurotoxins into nerve cells, endocytosis may provide one possibility. Another, more particular to secretory cells, would be the process of retrieval of secretory membrane after its insertion into the cell surface during secretion. Although topologically identical, membrane retrieval occurs independently from endocytosis and may provide a specific mode of uptake of neurotoxin into nerve cells.

REFERENCES

References as given do not include all published articles on a particular topic. In most cases only the most recent publications have been chosen in order to allow convenient access to the literature.

1. Courtoy PJ. Dissection of endosomes. In Steer C, Hanover J, eds. Intracellular trafficking of proteins. Cambridge University Press, 1991:103.
2. Burgess TL, Kelly RB. Constitutive and regulated secretion of proteins. Ann Rev Cell Biol 1987; 3:243.
3. Goldstein JL, Brown MS, Anderson RGW, Russell DW, Schneider, WJ. Receptor-mediated endocytosis: Concepts emerging from the LDL receptor system. Ann Rev Cell Biol 1985; 1:1.
4. Raub TJ, Koroly MJ, Roberts RM. Rapid endocytosis and recycling of wheat germ agglutinin binding sites on CHO cells: Evidence for two compartments in a nondegradative pathway. J Cell Physiol 1990; 144:52.
5. Sandvig K, Olsnes S, Brown JE, Petersen OW, van Deurs B. Endocytosis from coated pits of Shiga toxin: a glycolipid-binding protein from Shigella dysenteriae 1. J Cell Biol 1989; 108:1331.
6. Marsh M. The entry of enveloped viruses into cells by endocytosis. Biochem. J. 1984; 218:1.
7. Green SA, Kelly RB. Endocytic membrane traffic to the Golgi apparatus in a regulated secretory cell line. J. Biol. Chem. 1990; 265:21269.
8. Brodsky FM, Hill BL, Acton SL, Näthke I, Wong DH, Ponnambalam S, Parham P. Clathrin light chains: arrays of protein motifs that regulate coated-vesicle dynamics. Trends Biochem Sci 1991; 16:208.
9. Hansen SH, Sandvig K, van Deurs B. The pre-endosomal compartment comprises distinct coated and noncoated endocytic vesicle populations. J Cell Biol 1991; 113:731.
10. Tran D, Carpentier J-L, Sawano F, Gordon P, Orci L. Ligands internalized through coated and noncoated invaginations follow a common intracellular pathway. Proc Natl Acad Sci USA 1987; 84:7957.
11. Montesano R, Roth J, Robert A, Orci L. Non-coated membrane invaginations are involved in binding and internalization of cholera and tetanus toxins. Nature 1982; 296:651.
12. Keen JH. Clathrin and associated assembly and disassembly proteins. Ann Rev Biochem 1990; 59:415.
13. Ahle S, Ungewickell E. Auxilin, a newly identified clathrin-associated protein in coated vesicles from bovine brain. J Cell Biol 1990; 111:19.
14. McCloskey M, Poo M-M. Protein diffusion in cell membranes: Some biological implications. Int Rev Cytol 1984; 87:19.
15. Hopkins CR. Selective membrane protein trafficking: vectorial flow and filter. Trends Biochem Sci 1992; 17:27.
16. Dunn WA, Hubbard AL. Receptor-mediated endocytosis of epidermal growth factor by hepatocytes in the perfused rat liver: ligand and receptor dynamics. J Cell Biol 1984; 98:2148.
17. Anderson RGW, Brown MS, Goldstein JL. Role of the coated endocytic vesicle in the uptake of receptor bound low density lipoprotein in human fibroblasts. Cell 1977; 10:351.

18. Griffiths G, Back R, Marsh M. A quantitative analysis of the endocytic pathway in Baby Hamster Kidney cells. J Cell Biol 1989; 109:2703.

19. Thilo L. Quantification of endocytosis-derived membrane traffic. Biochim Biophys Acta 1985; 822:243.

20. Mellman IS, Steinman RM, Unkeless JC, Cohn ZA. Selective iodination and polypeptide composition of pinocytic vesicles. J Cell Biol 1980; 86:712.

21. Koval M, Pagano RE. Lipid recycling between the plasma membrane and intracellular compartments: Transport and metabolism of fluorescent sphingomyelin analogues in cultured fibroblasts. J Cell Biol 1989; 108:2169.

22. Bretscher MS, Lutter R. A new method for detecting endocytosed proteins. EMBO J 1988; 7:4087.

23. Phillips JH, Burridge K, Wilson SP, Kirshner N. Visualization of the exocytosis/endocytosis secretory cycle in cultured adrenal chromaffin cells. J Cell Biol 1983; 97:1906.

24. Patzak A, Böck G, Fischer-Colbrie R, Schauenstein K, Schmidt W, Lingg G, Winkler H. Exocytic exposure and retrieval of membrane antigens of chromaffin granules: Quantitative evaluation of immunofluorescence on the surface of chromaffin cells. J Cell Biol 1984; 98:1817.

25. van Grafenstein H, Roberts CS, Baker PF. Kinetic analysis of the triggered exocytosis/endocytosis secretory cycle in cultured bovine adrenal medullary cells. J Cell Biol 1986; 103:2343.

26. Bonifacino JS, Perez P, Klausner RD, Sandoval IV. Study of the transit of an integral membrane protein from secretory granules through the plasma membrane of secreting rat basophilic leukemia cells using a specific monoclonal antibody. J Cell Biol 1986; 102:516.

27. Patzak A, Winkler H. Exocytotic exposure and recycling of membrane antigens of chromaffin granules: Ultrastructural evaluation after immunolabeling. J Cell Biol 1986; 102:510.

28. Torri-Tarelli F, Villa A, Valtorta F, De Camilli P, Greengard P, Ceccarelli B. Redistribution of synaptophysin and synapsin I during α-lactrotoxin-induced release of neurotransmitter at the neuromuscular junction. J Cell Biol 1990; 110:449.

29. Thilo L. Selective internalization of granule membrane after secretion in mast cells. Proc Natl Acad Sci USA 1985; 82:1711.

30. Black JD, Dolly JO. Interaction of ^{125}I-labeled Botulinum neurotoxins with nerve terminals. II. Autoradiographic evidence for its uptake into motor nerves by acceptor-mediated endocytosis. J Cell Biol 1986; 103:535.

31. Hopkins CR, Gibson A, Shipman M, Miller K. Movement of internalized ligand-receptor complexes along a continuous endosomal reticulum. Nature 1990; 346:335.

32. Griffiths G, Hoflack B, Simons K, Mellman I, Kornfeld S. The mannose 6-phosphate receptor and the biogenesis of lysosomes. Cell 1988; 52:329.

33. Tooze J, Hollinshead M. Tubular early endosomal networks in AtT20 and other cells. J Cell Biol 1991; 115:635.

34. Knapp PE, Swanson JA. Plasticity of the tubular lysosomal compartment in macrophages. J Cell Sci 1990; 95:433.

35. Mellman I, Fuchs R, Helenius A. Acidification of the endocytic and exocytic pathways. Ann Rev Biochem 1986; 55:663.

36. Yamashiro DJ, Maxfield FR. Acidification of morphologically distinct endosomes in mutant and wild-type Chinese Hamster Ovary cells. J Cell Biol 1987; 105:2723.

37. Cain CC, Sipe DM, Murphy RM. Regulation of endocytic pH by Na^+,K^+-ATPase in living cells. Proc Natl Acad Sci USA 1989; 86:544.

38. Dean RT, Jessup W, Roberts CR. Effects of exogenous amines on mammalian cells, with particular reference to membrane flow. Biochem J 1984; 217:27.

39. Simpson LL. Ammonium chloride and methylamine hydrochloride antagonise clostridial neurotoxins. J Pharmacol Exp Ther 1983; 225:546.

40. Blum JS, Fiani ML, Stahl PD. Proteolytic cleavage of ricin A chain in endosomal vesicles. J Biol Chem 1991; 266:22091.

41. Dunn KW, McGraw TE, Maxfield FR. Iterative fractionation of recycling receptors from lysosomally destined ligands in an early sorting endosome. J Cell Biol 1989; 109:3303.

42. Ward DM, Ajioka R, Kaplan J. Cohort movement of different ligands and receptors in the intracellular endocytic pathway of alveolar macrophages. J Biol Chem 1989; 264:8164.

43. Stoorvogel W, Geuze HJ, Griffith JM, Schwartz AL, Strous GJ. Relations between the intracellular pathways of the receptors for transferrin, asialoglycoprotein, and mannose 6-phosphate in human hepatoma cells. J Cell Biol 1989; 108:2137.

44. Linderman JJ, Lauffenburger DA. Analysis of intracellular receptor/ligand sorting in endosomes. J Theor Biol 1988; 132:203.

45. Haylett T, Thilo L. Limited and selective transfer of plasma membrane glycoproteins to membrane of secondary lysosomes. J Cell Biol 1986; 103:1249.

46. Woods JW, Doriaux M, Farquhar MG. Transferrin receptors recycle to cis and middle as well as trans Golgi cisternae in Ig-secreting myeloma cells. J Cell Biol 1986; 103:277.

47. Jin M, Sahagian GG, Snider MD. Transport of surface mannose 6-phosphate receptor to the Golgi complex in cultured human cells. J Biol Chem 1989; 263:7675.

48. van Deurs B, Sandvig K, Petersen OW, Olsnes S, Simons K, Griffiths G. Estimation of the amount of internalized ricin that reaches the trans-Golgi network. J Cell Biol 1988; 106:253.

49. Farquhar MG. Progress in unraveling pathways of Golgi traffic. Ann Rev Cell Biol 1985; 1:447.

50. Griffiths G, Gruenberg J. The arguments for pre-existing early and late endosomes. Trends Cell Biol 1991; 1:5.

51. Stoorvogel W, Strous GJ, Geuze HJ, Oorschot V, Schwartz AL. Late endosomes derive from early endosomes by maturation. 1991; Cell 65:417.

52. Rothman JE, Orci L. Molecular dissection of the secretory pathway. Nature 1992; 355:409.

53. Roederer M, Barry JR, Wilson RB, Murphy RF. Endosomes can undergo an ATP-dependent density increase in the absence of dense lysosomes. Eur J Cell Biol 1990; 51:229.

54. Mullock BM, Branch WJ, van Schaik M, Gilbert LK, Luzio JP. Reconstitution of an endosome-lysosome interaction in a cell-free system. J Cell Biol 1989; 108:2093.

55. Ward DM, Hackenyos DP, Kaplan J. Fusion of sequentially internalized vesicles in alveolar macro-phages. J Cell Biol 1990; 110:1013.

56. Salzman NH, Maxfield FR. Fusion accessibility of endocytic compartments along the recycling and lysosomal endocytic pathways in intact cells. J Cell Biol 1989; 109:2097.

57. Mayorga LS, Diaz R, Stahl PD. Plasma membrane-derived vesicles containing receptor-ligand complexes are fusogenic with early endosomes in a cell-free system. J Biol Chem 1988; 263:17213.

IDENTIFICATION OF PROTEINS INVOLVED IN ENDOSOME FUSION: IMPLICATIONS FOR TOXIN ACTIVITY

James M. Lenhard, Maria I. Colombo, Michael Koval,
Guangpu Li, Luis S. Mayorga and Philip D. Stahl

Department of Cell Biology and Physiology
Washington University School of Medicine
St. Louis, MO 63110

INTRODUCTION

Endocytosis of proteins is a fundamental aspect of protein transport required for cell survival. Binding of a number of ligands including hormones, growth factors, nutrients and antigens to cell-surface receptors initiates internalization of these macromolecules. The receptor-ligand complexes that form at the cell surface sequester within clathrin-coated pits and the pits pinch off the plasma membrane to form coated vesicles[1]. After uncoating, the vesicles fuse with other endocytic vesicles, the compartment acidifies and the ligands dissociate from their receptors[2]. The lumenal contents of the endocytic vesicles are commonly delivered to lysosomes or the Golgi apparatus while the majority of the membrane components are recycled to the plasma membrane[3].

Several bacterial and plant toxins follow a similar pathway as other ligands internalized by receptor-mediated endocytosis[3,4]. However, after internalization these toxins may be translocated across the endosomal membrane and into the cytosol were they act as biological response modifiers[4]. Numerous lines of evidence show that internalization into an acidic endosomal compartment of some toxins is a prerequisite for the pathogenic effects of these proteins. Diphtheria toxin, which is normally released into the cytoplasm where it ADP-ribosylates elongation factor Tu, has little effect on mutant cell lines defective in endosomal acidification[5]. Neuroparalysis by botulinum is inhibited by lysosomotropic amines indicating that low pH is required for neurotoxin activity[6]. Studies of ricin A chain, a plant toxin that kills cells by inactivating ribosome function, indicate pH may regulate endosomal proteolysis and toxicity of this protein[7]. Ricin A chain is proteolysed in endosomes by cathepsin D at acidic pH and cathepsin B at acidic to neutral pH[7]. Acidic pH and/or proteolysis is proposed to control conformational changes in these toxins. These conformational changes may activate translocation of the toxins across the endosome membrane which ultimately results in a biochemical lesion.

Since toxin internalization often precedes a pathological effect, studies of receptor-mediated endocytosis can be used to understand, in part, the mechanism leading to intoxication. Although the pathway that proteins follow during endocytosis is fairly well established, less is known about the formation, delivery and fusion of transport vesicles with other organelles. Identification of key components of the transport machinery that regulate early stages in endocytosis may help to understand the pathway leading to the pathogenic effects of toxins. Thus, the remainder of this chapter will review the potential regulatory role of several proteins that have been proposed to regulate the early steps during endocytosis.

IN VITRO ASSAYS FOR STUDYING ENDOCYTOSIS

As the intracellular environment in which endosomes are exposed is difficult to manipulate in intact cells, a number of assays have been developed to study endosome fusion in cell-free systems. Fusion can be defined as the merging of two or more vesicles to form a single membrane-bound compartment. In general, fusion of two sets of endosomes containing two probes with high affinity for each other results in a single continuous compartment that allows for formation of a complex between the probes. One of the probes is typically an enzyme that can be immunoprecipitated only after formation of the complex. Formation of the immune complex is measured as an enzymatic reaction which represents membrane fusion. Based on this principle, several *in vitro* assays have been developed that reconstitute fusion between early endosomes.

The method used by Stahl and co-workers[8,9] to demonstrate endosome fusion requires internalization of two ligands that mutually recognize each other and bind to the macrophage mannose-receptor. These ligands are dinitrophenol-derivitized β-glucuronidase (DNP-β-gluc) and mannose-derivitized monoclonal anti-DNP IgG (Man-IgG). Under the proper conditions, incubation of endosomes from two sets of cells containing either ligand results in vesicle fusion and formation of DNP-β-gluc Man-IgG complexes. The complex is immunoprecipitated and fusion is measured by assaying for precipitated β-glucuronidase activity. In a similar fashion Gruenberg and co-workers[10] used biotinylated-horse radish peroxidase (bHRP), localized in early endosomes, as probe for fusion with endosomes containing avidin. Fusion was assessed by immunoprecipitating the bHRP-avidin complex with anti-avidin antibodies and measuring the activity of bHRP in the complex. Woodman and Warren[11] designed an assay using radioactive transferrin and anti-transferrin antibodies as probes to measure fusion between endosomes. Braell[12] has also developed an assay that is based on high affinity interaction of avidin-linked β-galactosidase and biotinylated IgG. Each of these assays are very sensitive and share common features. For example, these fusion assays are dependent upon temperature, ions, ATP and require proteins present in the cytosol and on membranes of endosomes.

PROTEINS INVOLVED IN ENDOSOME FUSION

GTP-binding Proteins (G proteins, ARFs and Rabs)

Based on subunit composition there are two major groups of GTP-binding proteins. The larger group consists of greater than 50 monomeric small molecular weight GTP-binding proteins (smgs) ranging from 20-30 kDa[13]. The second group consists of the heterotrimeric GTP-binding proteins (G proteins) that are composed of G-α (40-50 kDa), G-β (35 kDa) and G-γ (10 kDa) subunits[14]. Both smgs and G proteins use nucleotide exchange and hydrolysis as molecular switches to control physiological events[15]. In the GTP-bound form these proteins are active and in the GDP-bound form they are inactive. The GTP-GDP exchange reaction is coupled to interaction of the GTP-binding proteins with other proteins such as receptor-ligand complexes, nucleotide exchange factors, GTPase-activating proteins and downstream effectors. Controlled interaction between these proteins regulates cellular activities such as protein synthesis, cell motility, translocation of newly synthesized proteins across the endoplasmic reticulum and signal transduction[14,15].

Studies of intact cells and cell-free systems indicate that GTP-binding proteins are involved in vesicular transport[16]. It has been proposed that the GTP-binding proteins may regulate transport by one of two mechanisms[17]. First, these proteins may transduce sensory signals originating from outside the cell. Subsequently, an intracellular signal would be generated that regulates vesicular transport[17]. This model is analogous to the transmembrane signaling pathway, which includes interaction of G proteins with upstream receptors and downstream effectors, that has been shown to regulate the transduction of hormonal signals and sensory stimuli[14]. Alternatively, these proteins may mediate unidirectional transport of vesicles between appropriate intracellular organelles[17]. In this model the GTP-binding proteins are proposed to be a part of the constitutive transport machinery. This model is similar to the role proposed for the GTP-binding protein elongation factor Tu in controlling protein synthesis[15]. Recent data shows that G proteins[18] and smgs belonging to the rab[10] and ARF (ADP-ribosylation factor) subfamilies[19] are involved in endocytosis indicating that both models may be correct.

Evidence that GTP-binding proteins are involved in endocytosis comes from studies using GTPγS, a non-hydrolyzable analog of GTP[20,21]. Depending on the cytosol concentration, GTPγS has different effects on *in vitro* endosome fusion. GTPγS stimulates fusion at cytosol concentrations that alone are unable to support fusion[22]. At cytosol concentrations that fully support fusion, GTPγS inhibits fusion activity[22,23]. As GTPγS stimulates or inhibits endosome fusion depending on the cytosol concentration, it was proposed that two or more GTP-binding proteins might be active in fusion[22]: one or more that stimulates fusion (i.e. positive regulation) and one or more that inhibits fusion (i.e. negative regulation).

GTPγS activates both G proteins and smgs indicating a potential role for both types of proteins in endocytosis. Several pieces of data indicate that G proteins are involved in endosome fusion. The first line of evidence comes from the observation that aluminum fluoride, an activator of G proteins but not smgs, inhibits endosome-endosome fusion in cell-free systems[20]. Further support comes from experiments using mastoparan[18], a toxic tetradecapeptide from wasp venom that increases nucleotide exchange on G proteins. Mastoparan reverses the effects of GTPγS on *in vitro* endosome fusion[18]. The most conclusive evidence that G proteins regulate endocytosis comes from the observation that purified G$\beta\gamma$ subunits inhibit endosome fusion[18]. Incubating endosomes with GTPγS before addition of G$\beta\gamma$ antagonizes the inhibitory effect. This is consistent with the reduced affinity of G$\beta\gamma$ for Gα subunits bound to GTPγS[14]. These results demonstrate that Gα, in part, may confer sensitivity of endosome fusion to GTPγS.

The second type of GTP-binding protein that is thought to be involved in endocytosis is ARF[19]. ARF was originally identified as a cofactor for cholera-toxin catalyzed ADP-ribosylation of Gα_s[24]. It was recently shown that a synthetic peptide from the amino terminal 16 amino acids of human ARF1 inhibits ribosylation of Gα_s by cholera toxin and ARF[25]. The peptide also reverses the effects of GTPγS on *in vitro* endosome fusion[19]. This result shows that ARF, like Gα, may impart sensitivity of endosome fusion to GTPγS. Further evidence that ARF is involved in endocytosis comes from experiments using purified recombinant myristoylated or non-myristoylated ARF[19]. Myristoylation of the amino terminal glycine of a number of viral and cellular proteins is required for biological activity[26]. Myristoylated ARF but not non-myristoylated ARF inhibits endosome fusion in the presence but not in the absence of GTPγS[19]. A possible interpretation for this observation is that myristoylated ARF acts as a negative regulatory element that inhibits endosome fusion when activated by binding to GTPγS. The peptide from the amino terminus of ARF antagonizes the inhibitory effect of myristoylated ARF on endosome fusion indicating that the peptide may block interaction of ARF with an effector regulating endosome fusion[19]. At high concentrations the peptide also inhibits endosome fusion indicating limited interaction of ARF with its effector is required for optimum fusion activity[19]. *In vivo*, interaction of ARF with its effector is proposed to be controlled by the rate of guanine nucleotide exchange and hydrolysis[19].

Like mastoparan, the amino terminal ARF peptide forms a cationic amphiphilic α-helix in the presence of phospholipid vesicles[25]. An intriguing hypothesis is that the amino terminus of ARF mimics mastoparan and binds to Gα subunits. Thus, Gα and ARF may interact with each other to regulate endosome fusion. Consistent with this hypothesis are the observations that Gβγ prevents association of ARF with Golgi membranes[27] and that ribosylation of Gα$_s$ by cholera toxin requires ARF[24].

In addition to requiring Gα and ARF, cholera toxin action requires internalization and processing in the lumen of endosomes[28,29]. It is interesting that this toxin selectively affects GTP-binding proteins involved during protein transport. In this context, it is worthy to note that the various other proteins (see below) that are involved during endosome fusion may be potential targets for the other toxins that require internalization before exerting their biological effects.

The rab proteins are the third type of GTP-binding protein implicated in regulation of endocytosis. So far, more than 20 different rabs have been identified from different sources[16]. Among them, rab4[30] and rab5[31] are associated with early endosomes, suggesting their possible roles during early events in the endocytic pathway. It has been shown that endosome fusion *in vitro* is blocked by anti-rab5 antibody and this inhibition can be reversed by adding cytosol containing overexpressed rab5[10]. In addition, a mutant rab5 deficient in GTP-binding is a dominant inhibitor of *in vitro* endosome fusion[10]. These results indicate that rab5 is required for endosome fusion. Rab4 is thought to regulate an early endocytic event which may be distinct from that controlled by rab5[30]. Rab4 is enriched in a population of vesicles that recycle the transferrin receptor to the plasma membrane[30]. Overexpression of rab4 in CHO cells decreases the intracellular pool of endocytic markers including HRP and transferrin receptor and results in accumulation of transferrin receptor in a population of endosomes that are morphologically distinct from early endosomes[32]. These results indicate that rab4 may regulate the production or function of endosomes involved in receptor recycling[30,32].

Both rab4 and rab5 are synthesized as cytosolic precursors which are later converted to membrane-associated mature forms by post-translational processing[33]. The post-translational processing, which includes isoprenylation (geranylgeranylation), proteolysis and possibly other modifications[33,34], may play an important role in the biological function and localization of the rab proteins. Isoprenylation alone is necessary but not sufficient for the membrane-association of rab4 or rab5[33]. The highly variable C-terminal domain of the rab proteins contains the structural elements needed for the association of the rabs with their specific membrane targets[35]. Thus, post-translational modifications and protein sequences are critical determinants for the subcellular localization of the rab proteins.

The above observations indicate that multiple subfamilies of GTP-binding proteins are involved during endocytosis. A balance between nucleotide exchange and hydrolysis by G proteins, ARFs, and rabs is thought to control the rate of individual processes regulating endocytosis. These proteins may transduce signals that allow for subsequent assembly and disassembly of the transport machinery or be integral components of the transport machinery. Only further experimentation will lead to a full understanding of the relationship between each of the subfamilies of GTP-binding proteins implicated in protein transport.

Protein Phosphatases and Kinases

The phosphorylation state of proteins is known to regulate processes such as glycogen metabolism, calcium transport, muscle contraction, protein synthesis and cell division[36]. The reversible phosphorylation of serine, threonine and tyrosine residues results in structural changes in the proteins that regulate these cellular processes. A balance between the activities of protein kinases and protein phosphatases determines the phosphorylation state and activity of the proteins required for these events.

Analysis of yeast mutants has allowed for the identification of the cell division cycle gene cdc2[37], a serine/threonine kinase, that is now known to control multiple events during mitosis including endocytosis[38]. During cell division membrane transport along the endocytic pathway is blocked[39]. Tuomikoski et al.[38] found that cdc2 kinase inhibits cell-free fusion between endosomes in interphase cytosol. Inhibition of endosome fusion is nearly complete when the cdc2 activity is in the concentration range estimated for that found in mitotic cells[38]. This indicates that during mitosis cdc2 kinase probably acts as a negative regulator of endocytosis by phosphorylating serine/threonine residues of target protein(s).

An interesting observation is that rab4 is post-translationally phosphorylated by cdc2 kinase[40]. Rab4 is phosphorylated only during mitosis and the phosphorylated rab4 is released from the membrane into the cytosol[40,41]. Dephosphorylation and reassociation of soluble rab4 with membranes occurs upon exit of cells from mitosis[41]. These data, together with the observation that cdc2 kinase inhibits endosome fusion *in vitro*, indicate that phosphorylation of rab4 may play a role in the arrest of endocytosis during mitosis.

Woodman et al.[42] found that okadaic acid and microcystin, inhibitors of serine/threonine protein phosphatases (PP), also block *in vitro* endosome fusion. Protein kinase inhibitors and purified PP2A reversed inhibition of fusion by okadaic acid[42]. This suggests that the degree of protein phosphorylation and the regulation of vesicle fusion is controlled by opposing phosphatase and kinase activities. Surprisingly, inhibition of fusion by okadaic acid is not reversed when cdc2 kinase is depleted from cytosol[42]. This indicates a second distinct kinase from cdc2 may regulate endosome fusion in interphase extracts. Like cdc2 kinase, this protein is likely to be a serine/threonine kinase that phosphorylates an unknown target protein(s) involved in endocytosis.

Tyrosine phosphorylation has also been implicated in regulating receptor-mediated endocytosis[43,44]. The phosphorylation state of some cell surface receptors is thought to coordinate their internalization by the action of tyrosine kinases[43]. For example in non-lymphoid cells, which do not express the protein tyrosine kinase p56*lck*, the glycoprotein CD4 is endocytosed through clathrin-coated pits[45]. However CD4 is excluded from endocytic vesicles in lymphoid cells by interaction with p56*lck*[45]. The effect of p56^{lck} on endocytosis did not include other receptor-mediated or fluid-phase endocytic processes[45]. Thus, tyrosine phosphorylation of some individual cell-surface receptors may mediate their internalization.

Taken together the above observations indicate that cyclic phosphorylation-dephosphorylation of proteins may regulate endocytosis by at least two distinct mechanisms. First, endocytosis as a whole may be regulated by the activity of distinct serine/threonine kinases during interphase and mitosis. Second, endocytosis of some receptors may be regulated by tyrosine kinases that specifically recognize the individual cell-surface proteins.

Identification of a N-ethylmaleimide Sensitive Factor (NSF) Required for Endosome Fusion

The alkylating agent N-ethylmaleimide (NEM) inhibits membrane fusion in cell-free assays that measure transport through endocytic[8,21] and secretory pathways[46,47]. A well characterized protein affected by NEM is the so-called NEM-sensitive fusion protein (NSF) which has been shown to be involved in a wide range of intracellular membrane fusion events [45,47] including endosome fusion[48]. NSF is found in cytosol and on membranes as a homotetramer of 76 kDa subunits[45]. By sequence analysis, NSF is homologous to the yeast gene product sec18[49], known to be required for secretion[50] and endocytosis[51] in Saccharomyces cerevisiae. NSF is an ATPase, in fact, the protein is highly unstable in the absence of ATP[52]. An important aspect of NSF function is the association of this protein with vesicle membranes[53]. This requires an integral membrane receptor, presumably a protein, which has not been identified[53]. Also, binding of NSF to membranes requires interaction of the tetramer with a variety of cofactors including a family of proteins[54,55] termed α-, β-, and γ-SNAPs (soluble NSF attachment proteins). α- and β-SNAP appear to be isoforms of the same protein, while γ-SNAP is more likely a distinct gene product[55]. Any one of the SNAP proteins, in the presence of other undefined, cytosolic factors, is sufficient to mediate NSF binding to the membrane receptor[54,55]. Further, α/β-SNAP and γ-SNAP appear to bind to distinct sites on the NSF/membrane complex, suggesting that both types of SNAP are required for optimal binding of NSF to the membrane[56].

The precise role for an NSF/SNAP/membrane complex is not known at present, but it is believed that this complex is involved in catalyzing bilayer fusion[52,54]. The point at which NSF and SNAP become associated with the membrane is not apparent and is dependent upon the assay conditions used. Some of the evidence supports the notion that NSF binds to Golgi vesicles that have been completely formed[52]. However, recent results suggest that SNAP and NSF may actually be incorporated into the donor membrane at an earlier step and may actually play a role in Golgi vesicle formation[57]. In the case of *in vitro* endosome fusion, where the donor vesicles have already been formed, NSF is able to restore the fusion activity of preparations treated with NEM[48]. This suggests that NSF can associate with more mature vesicles or, alternatively, there may be fundamental differences between the endosome fusion reaction and other fusion reactions *in vitro*.

Fatty acyl-CoA is thought to act as a cofactor to NSF[58]. Fatty acylation has been shown to play a role in vesicle formation[59] and fusion[60] using an *in vitro* Golgi transport assay. This requirement for acyl-CoA turnover is consistent with the involvement of acylated proteins in membrane transport, such as GTP-binding proteins. However, acylation-deacylation cycles of membrane lipids could also be involved[61]. To date, a direct role for acyl-CoA and SNAP in endosome fusion has not been demonstrated. Clearly, further work will be required to elucidate the role of NSF and peripheral co-factors in endosome fusion events.

Characterization of a Trypsin-Sensitive Factor (TSF) Required for Endosome Fusion

Endosome fusion requires both cytosolic and membrane associated factors that are inactivated by treatment with the protease trypsin[9,62]. Similar to NEM-treatment, preincubation of endosomes with trypsin completely abolishes fusion, but unlike NEM-treatment, fusion activity is not restored by the subsequent addition of untreated cytosol[62]. Therefore, fusion between endosomes requires trypsin-sensitive factors that reside on the cytoplasmic face of membranes and that are not present in cytosol in their active form. Fusion activity of trypsinized vesicles (TrV) is restored by the addition of untreated endosomes suggesting that some factors required for fusion may be transferred from one set of vesicles to the other[62]. However, the restorative effect of the untreated endosomes is not observed in the absence of cytosol[62]. These results indicate that untreated vesicles can provide the membrane proteins needed to reconstitute endosomal fusion but not all of the cytosolic factors required for this process.

To assess whether peripheral membrane proteins are responsible for restoration of fusion using TrV, a high salt extract (KE) is prepared from a crude endosomal fraction[62]. Addition of KE restored the fusion activity of TrV. Therefore, the membrane-associated protein(s) that are removed or inactivated by trypsin treatment can also be restored by addition of an extract containing peripheral membrane proteins. The requirements for fusion of TrV in the presence of KE are similar to the normal fusion assay except that this fusion does not require cytosol[62]. It is known that some factors can cycle between cytosolic and membrane-associated form. The results indicate that most of the factors required for fusion are present in KE. Thus, KE may contain both the membrane-associated and recycling cytosolic components required for fusion of TrV.

The restorative activity of KE is abolished by trypsinization and heat inactivation but is only slightly sensitive to NEM[62]. These results suggest that KE contains protein(s) that are responsible for its restorative activity. Therefore, the fusion promoting activity present in the KE is referred to as TSF (trypsin-sensitive factor) due to its sensitivity to proteolysis and its ability to restore fusion of trypsinized endosomes[62].

Preliminary fractionation studies indicate that the restorative activity of KE consists of one or more high molecular weight proteins (200 kDa) that can be resolved from a large protein peak by gel filtration chromatography[62]. The identity of this high molecular weight component (TSF) is not known, although several potential candidates can be ruled out. Because of its large size, TSF is unlikely to correspond to the low molecular weight protein SNAP[54] involved in the Golgi vesicular transport and to the GTP-binding proteins involved in endosome fusion[18,19]. Furthermore, TSF activity is not restored by Fr1, a high molecular weight cytosolic protein that is required for transport through the Golgi stacks[55]. As TSF is not sensitive to NEM and is entirely membrane associated, it also appears to be distinct from NSF[48]. Consistent with this observation, pure NSF does not restore fusion of TrV[62]. Taken together, these observations indicate that TSF is a novel protein(s) that is distinct from other previously described factors required for fusion. This protein(s) appears to play an essential role in fusion among early endosomes[9].

PERSPECTIVES

Numerous proteins have been identified that are required for early events during endosome fusion. Several of these proteins may behave as molecular switches that turn endocytosis on or off. Guanine nucleotide exchange and hydrolysis and/or the opposing activities of phosphatases and kinases probably act as regulatory components of endocytosis. On the other hand, proteins such as NSF or TSF may behave as constitutive factors that make up part of the fusion machinery. At present the relationship between these multiple factors remains unclear.

Some evidence suggests that the proteins required for endosome fusion act at distinct stages during the fusion process. For example, binding of cytosolic proteins to the membrane of endosomes and vesicle aggregation are distinct events preceding endosome fusion[22]. Furthermore, preincubation experiments show that endosomes can form intermediates that are resistant to inhibition by ARF[19] and $G\beta\gamma$ subunits[18] supporting a role for these proteins during a step leading to endosome fusion. Wessling-Resnick and Braell[21] provide direct evidence that a GTPγS-sensitive step occurs before a NEM-sensitive step, indicating these reagents affect two distinct stages before the fusion process. Taken together, these observations may reflect the stepwise assembly of a multisubunit complex or distinct stages in the maturation of endosomes that are required for subsequent docking and membrane fusion events. A future direction of research includes elucidating the relationship between the many factors required for endosome fusion.

ACKNOWLEDGMENTS

Dr. Lenhard is a recipient of a research fellowship from the Cancer Research Institute. Dr. Koval is an Amgen fellow of the Life Sciences Research Foundation. Dr. Li is a fellow supported by the Parker B. Francis foundation. Dr. Mayorga is a recipient of a Rockerfeller Foundation Biotechnology Career Fellowship. Our research is also supported in part by grants AI20015 and GM42259-19 from the National Institute of Health.

REFERENCES

1. B. M. Pearse and M. S. Robinson, Clathrin, adaptors, and sorting, *Annu. Rev. Cell Biol.* 6:151 (1990).

2. S. Kornfeld and I. Mellman, The biogenesis of lysosomes, *Annu. Rev. Cell Biol.* 5:483 (1989).

3. B. Van Deurs, O. Peterson, S. Olsnes and K. Sandvig, The ways of endocytosis, *Int. Rev. Cyt.* 117:131 (1989).

4. J. E. Alouf and J. H. Freer. "Sourcebook of Bacterial Protein Toxins," Academic Press, San Diego (1991).

5. S. Merion, P. Schlesinger, R. Brooks, J. Moehring, T. Moehring, and W. Sly, Defective acidification of endosomes in chinese hamster ovary cell mutants "cross resistant" to toxins and viruses, *Proc. Natl. Acad. Sci. USA* 80:5315 (1983).

6. L. I. Simpson, The binary toxin produced by clostridium botulinum enters cells by receptor-mediated endocytosis to exert its pharmacologic effects. *J. Pharmacol. Exp. Ther.* 251:1223 (1989).

7. J. S. Blum, M. L. Fiani, and P. D. Stahl, Proteolytic cleavage of ricin A chain in endosomal vesicles, *J. Biol. Chem.* 266:22091 (1991).

8. R. Diaz, L. Mayorga, and P. Stahl, In vitro fusion of endosomes following receptor-mediated endocytosis, *J. Biol. Chem.* 263: 6093 (1988).

9. M. Colombo, J. Lenhard, L. Mayorga and P. Stahl, Reconstitution of endosome fusion: Identification of factors necessary for fusion competency, *Meth. Enz.* 219:32 (1992).

10. J. P. Gorvel, P. Chavrier, M. Zerial, and J. Gruenberg, rab5 controls early endosome fusion in vitro, *Cell* 64: 915 (1991).

11. P. G. Woodman and G. Warren, Fusion of endocytic vesicles in a cell-free system, *Methods Cell Biol.* 31:197 (1989).

12. W. A. Braell, Fusion between endocytic vesicles in a cell-free system, *Proc. Natl. Acad. Sci.* 84:1137 (1987).

13. G. Drivas, S. Palmieri, P. D'Eustachio and M. Rush, Evolutionary grouping of the ras-protein family, *Biochem. and Biophys. Res. Comm.* 176:1130 (1991).

14. A. Gilman, G proteins: Transducers of receptor-generated signals, *Ann. Rev. Biochem.* 56:615 (1987).

15. H. Bourne, D. Sanders, and F. McCormick, The GTPase superfamily: a conserved switch for diverse cell function, *Nature* 348:125 (1990).

16. S. Pfeffer, GTP-binding proteins in intracellular transport, *Trends in cell Biol.* 2:41 (1992).

17. H. Bourne, Do GTPases direct membrane traffic in secretion?, *Cell* 53:669 (1988).

18. M.I. Colombo, L.S. Mayorga, P.J. Casey, and P.D. Stahl, Evidence of a role for heterotrimeric GTP-binding proteins in endosome fusion, *Science* 255:1695 (1992).

19. J.M. Lenhard, R.A. Kahn, and P.D. Stahl, Evidence for ADP-ribosylation factor (ARF) as a regulator of in vitro endosome-endosome fusion, *J. Biol. Chem.* 267:13047 (1992).

20. L.S. Mayorga, R. Diaz, and P.D. Stahl, Regulatory role for GTP-binding proteins in endocytosis, *Science* 244:1475 (1989).

21. M. Wessling-Resnick and W. Braell, Characterization of endocytic vesicle fusion in vitro, *J. Biol. Chem.* 265:16751 (1990).

22. L.S. Mayorga, R. Diaz, M.I. Colombo, and P.D. Stahl, GTPγS stimulation of endosome fusion suggests a role for a GTP-binding protein in the priming of vesicles before fusion, *Cell Regul.* 1:113 (1989).

23. J.M. Lenhard, L.S. Mayorga, and P.D. Stahl, Characterization of endosome-endosome fusion in a cell-free system using Dictyostelium discoideum, *J. Biol. Chem.* 267:1896 (1992).

24. R. Kahn and A. Gilman, Purification of a protein cofactor required for ADP-

ribosylation of the stimulatory regulatory component of adenylate cyclase by cholera toxin, *J. Biol. Chem.* 259:6228 (1984).

25. R. Kahn, P. Randazzo, T. Serafini, O. Weiss, C. Rulka, J. Clark, M. Amherdt, P. Roller, L. Orci, and J. Rothman, The amino terminus of ADP-ribosylation factor (ARF) is a critical determinant of ARF activities and is a potent and specific inhibitor of protein transport, *J. Biol. Chem.* 267:13039 (1992).

26. D. Towler, J. Gordon, S. Adams, and L. Glaser, Amino terminal processing of proteins by N-myristoylation, *Annu. Rev. Biochem.* 57:69 (1988).

27. J. Donaldson, R. Kahn, J. Lippincott-Schwartz and R. Klausner, Binding of ARF and β-COP to Golgi membranes: possible regulation by a trimeric G protein, *Science* 254:1197 (1991).

28. M. Janicot, F. Fouque and B. Desbuquois, Activation of rat liver adenylate cyclase by cholera toxin requires toxin internalization and processsing in endosomes, *J. Biol. Chem* 266:12858 (1991).

29. W. Lencer, C. Delp, M. Neutra and J. Madara, Mechanism of cholera toxin action on a polarized human epithelial cell line: role of vesicular traffic, *J. Cell Biol.* 117:1197 (1992).

30. P. Van Der Sluijs, M. Hull, A. Zahraoui, A. Tavitian, B. Goud, and I. Mellman, The small GTP-binding protein rab4 is associated with early endosomes, *Proc. Natl. Acad. Sci. USA* 88:6313 (1991).

31. P. Chavrier, R. Parton, H. Hauri, K. Simons, and M. Zerial, Localization of low molecular weight GTP binding proteins to exocytic and endocytic compartments, *Cell* 62:317 (1990).

32. P. van der Sluijs, M. Hull, P. Webster, P. Malle, B. Goud and I. Mellman, The small GTP-binding protein rab4 controls an early sorting event of the endocytic pathway, *Cell* In Press (1992).

33. G. Li and P. D. Stahl, Post-translational processing of the two early endosome-associated rab GTP-binding proteins (rab4 and rab 5), submitted (1992).

34. B. T. Kinsella and W. Maltese, rab GTP-binding proteins with three different carboxyl-terminal cysteine motifs are modified in vivo by 20-carbon isoprenoids, *J. Biol. Chem.* 267:3940 (1992).

35. P. Chavrier, J. P. Gorvel, E. Stelzer, K. Simons, J. Gruenberg, and M. Zerial, Hypervariable C-terminal domain of rab proteins acts as a targeting signal, *Nature* 353:769 (1991).

36. M. Bollen and W. Stalmans, The structure, role, and regulation of type 1 protein phosphatases, *Crit. Rev. Biochem. Mol. Biol.* 27:227 (1992).

37. P. Nurse, Universal control mechanism regulating onset of M-phase, *Nature* 344:503 (1990).

38. T. Tuomikoski, M. Felix, M. Doree, and J. Gruenberg, Inhibition of endocytic vesicle fusion in vitro by the cell-cycle control protein kinase cdc2. *Nature* 342:942 (1989).

39. R. Berlin, J. Oliver, and R. Walter, Surface functions during mitosis. I. Phagocytosis, pinocytosis, and mobility of surface-bound Con-A, *Cell* 15:327 (1978).

40. E. Bailley, N. Touchet, A. Zahraoui, B. Goud, and M. Bornens, p34cdc2 protein kinase phosphorylates two small GTP-binding proteins of the rab family, *Nature* 350:715 (1991).

41. P. van der Sluijs, M. Hull, P. Male, B. Goud, and I. Mellman, Reversible phosphorylation-dephosphorylation determines the localization of rab4 during the cell cycle, *EMBO J.* In Press (1992).

42. P. Woodman, D. Mundy, P. Cohen, and G. Warren, Cell-free fusion of endocytic vesicles is regulated by phosphorylation, *J. Cell Biol.* 116:331 (1992).

43. A. Ullrich and J. Schlessinger, Signal transduction by receptors with tyrosine kinase activity, *Cell* 61:203 (1990).

44. N. T. Kistakis, D. Thomas, and M. G. Roth, Characterization of the tyrosine recognition signal for internalization of transmembrane surface glycoproteins, *J. Cell Biol.* 111:1393 (1990).

45. A. Pelchen-Matthews, I. Boulet, D. Littman, R. Fagard, and M. Marsh, The protein tyrosine kinase p56^{lck} inhibits CD4 endocytosis by preventing entry of CD4 into coated pits, *J. Cell Biol.* 117:279 (1992).

46. M. Block, B. Glick, C. Wilcox, F. Wieland and J. Rothman, Purification of an N-ethylmaleimide-sensitive protein catalyzing vesicular transport, *Proc. Natl. Acad. Sci.* 85:7852 (1988).

47. C. Beckers, M. Block, B. Glick, J. Rothman and W. Balch, Vesicular transport between the endoplasmic reticulum and the Golgi stack requires the NEM-sensitive fusion protein, *Nature* 339:397 (1989).

48. R. Diaz, L. Mayorga, P. Weidman, J. Rothman and P. Stahl, Vesicle fusion following receptor-mediated endocytosis requires a protein active in Golgi transport, *Nature* 339:398 (1989).

49. D. Wilson, C. Wilcox, G. Flynn, E. Chen, W. Kuang, W. Henzel, M. Block, A. Ullrich and J. Rothman, A fusion protein required for vesicles-mediated transport in both mammalian cells and yeast, *Nature* 339:355 (1989).

50. P. Novick, S. Ferro, and R. Schekman, Order of events in the yeast secretory pathway, *Cell* 25:461 (1981).

51. H. Reizman, Endocytosis in yeast: Several of the yeast secretory mutants are defective in endocytosis, *Cell* 40:1001 (1985).

52. V. Malhotra, L. Orci, B. Glick, M. Block and J. Rothman, Role of an N-ethylmaleimide-sensitive transport component in promoting fusion of transport vesicles with cisternae of the Golgi, *Cell* 54:221 (1988).

53. P. Weidman, P. Melancon, M. Block and J. Rothman, Binding of an N-ethylmaleimide-sensitive fusion protein to Golgi membranes requires both a soluble protein(s) and an integral membrane receptor, *J. Cell Biol.* 108:1589 (1989).

54. D. Clary, I. Griff, and J. Rothman, SNAPs, a family of NSF attachment proteins involved in intracellular membrane fusion in animal and yeast, *Cell* 61:709 (1990).

55. D. Clary and J. Rothman, Purification of three related peripheral membrane proteins needed for vesicular transport, *J. Biol. Chem.* 265:10109 (1990).

56. D. Wilson, S. Whiteheart, M. Wiedmann, M. Brunner and J. Rothman, A multisubunit particle implicated in membrane fusion, *J. Cell Biol.* 117:531 (1992).

57. B. Wattenberg, T. Raub, R. Hiebsch and P. Weidman, The activity of Golgi transport vesicles depends on the presence of the N-ethylmaleimide sensitive factor (NSF) and a soluble NSF attachment protein (SNAP) during vesicle formation, *J. Cell Biol.* In press (1992).

58. B. Glick and J. Rothman, Possible role for fatty acyl-coenzyme A in intracellular protein transport, *Nature* 326:309 (1987).

59. N. Pfanner, L. Orci, B. Glick, M. Amherdt, S. Arden, V. Malhotra and J. Rothman, Fatty Acyl-Coenzyme A is required for budding of transport vesicles from the Golgi Cisternae, *Cell* 59:95 (1989).

60. N. Pfanner, B. Glick, S. Arden and J. Rothman, Fatty acylation promotes fusion of transport vesicles with Golgi cisternae, *J. Cell Biol.* 110:955 (1990).

61. M. Koval, Plasma membrane lipid transport in cultured cells: studies using lipid analogs and model systems, *in*: "Advances in Cell and Molecular Biology of Membranes," B. Storrie and R. Murphy, ed., JAI Press, Greenwich (1992).

62. M.I. Colombo, S. Gonzalo, P. Weidman, and P.D. Stahl, Characterization of trypsin-sensitive factor(s) required for endosome-endosome fusion, *J. Biol. Chem.* 266:23438 (1991).

FACTORS UNDERLYING THE CHARACTERISTIC INHIBITION OF THE NEURONAL RELEASE OF TRANSMITTERS BY TETANUS AND VARIOUS BOTULINUM TOXINS

Anthony C. Ashton[1], Anton M. de Paiva[1], Bernard Poulain[2],
Ladislav Tauc[2] and J. Oliver Dolly[1]

[1]Department of Biochemistry,
Imperial College of Science, Technology and Medicine,
London, S.W.7, 2AY, U.K.
[2]Laboratoire de Neurobiologie Cellulaire et Moléculaire,
CNRS, 91198 Gif-sur-Yvette Cedex, France

INTRODUCTION

Although the gross structures of all serotypes of botulinum neurotoxin (BoNT) and tetanus toxin (TeTX) are similar - proteins composed of a disulphide-linked heavy chain (H; M_r ~100 kDa) and light chain (L; M_r~50 kDa) - dissimilarities exist in their amino acid sequences [1, 2, 3]. Each of the BoNT types produce flaccid neuromuscular paralysis due to a preferential inhibition of acetylcholine (ACh) release from peripheral nerves whilst TeTX gives rise to a spastic paralysis resulting from a blockade of inhibitory transmitter release at central synapses[4]. However, it is intriguing that characteristics of the neuroparalysis caused by type A BoNT differ from those of all the other serotypes and TeTX, the latter exhibiting many common features at motor nerve terminals[5]. For example, asynchronous release can be elicited by intense neural stimulation of muscle endplates treated with TeTX[6, 7] or BoNT/B[8], /D[9], /E[10] (but see [11]) whilst tissue incubated with BoNT/A yield detectable levels of synchronous release. Furthermore, double-poisoning experiments revealed that BoNT/A is unable to alter the pattern of inhibition already produced by type B or TeTX. Moreover, elevation of the intracellular Ca^{2+} concentration (using 4-aminopyridine [4-AP][8, 12, 13, 14] or black widow spider venom[15, 16, 17]) overcomes (at least temporarily) the blockade of the quantal release of ACh caused by limited exposure to BoNT/A (see later)

Botulinum and Tetanus Neurotoxins, Edited by
B.R. DasGupta, Plenum Press, New York, 1993

whereas much less extensive reversal of poisoning by any of these other toxins [8, 17] and F [18] is seen. Such notable differences have also been reported [19] for central nerve terminals in terms of a more pronounced reversal of the action of BoNT/A relative to B using the Ca^{2+} ionophore, A23187; additionally, dissassembly of microtubules antagonised the ability of type B (but not A) to reduce catecholamine release. Data is presented in this chapter showing that the microtubule involvement also applies to other transmitters and is also the case for TeTX and BoNT/E or /F; the importance of the cytoskeletal elements in secretion and involvement with the action of the latter toxins is discussed.

Another facet addressed herein is the spectrum of potency of BoNT types in blocking the synaptosomal release of different neurotransmitters. Related to this is the question of whether each toxin uses distinct ecto-acceptors to target motor nerve endings and gain entry to their intracellular site of action. This is investigated successfully with BoNT/E, /F and TeTX using a reduced, alkylated derivative of BoNT/A (Red-A-BoNT/A) that antagonizes the parent toxin's binding, thereby allowing "productive" toxin ecto-acceptor interaction to be monitored rather than total binding. As both the inter- and intra-disulphides are reduced and alkylated in Red-A-BoNT/A, it was utilised to establish that one or both these bonds are essential for uptake into nerve endings. Finally, H of BoNT/A, in which the thiol that links to L is alkylated, was employed in experiments on *Aplysia* cholinergic neurons (for details see accompanying chapter; Poulain *et al.*) to show that the latter group or the inter-chain disulphide is essential for H-mediated neuronal uptake of L. These and other factors contributing to the subtle differences in the actions of BoNT types will be dealt with.

DISTINCT "PRODUCTIVE" ECTO-ACCEPTORS FOR BoNT SEROTYPES AND TeTX AT MAMMALIAN MOTOR NERVE ENDINGS

Although ^{125}I-labelled BoNT/A and /B have been found to bind to separate sites in peripheral [20] and central [21, 22] nerve preparations, their relevance to the intoxication process has not been proven conclusively. This situation prevails because of the lack of pharmacological antagonists for the binding of each toxin. Thus, it is unknown if different BoNT types share "functional" acceptors at motor nerve terminals, an extremely important issue for the increasing clinical applications of BoNT[23]. This question can now be resolved because dithiothreitol reduction of BoNT/A followed by alkylation using iodoacetamide produces a derivative, Red-A-BoNT/A, with greatly diminished toxicity in mice (Table 1), that is ineffective in reducing ACh release from *Aplysia* neurons (Fig 1A) or at the murine neuromuscular junction[24] when applied extra-cellularly. Yet, it antagonizes the neuroparalytic effects of the parent toxin on both these tissues (Fig 1C, 2) by competing with BoNT/A for ecto-acceptor binding that mediates this action. As detailed in the legends to Figs 1 and 2, the conditions used allowed toxin interaction with the plasma membrane to be monitored. All of the SH groups had been modified in the derivative, as ascertained colorimetrically with the

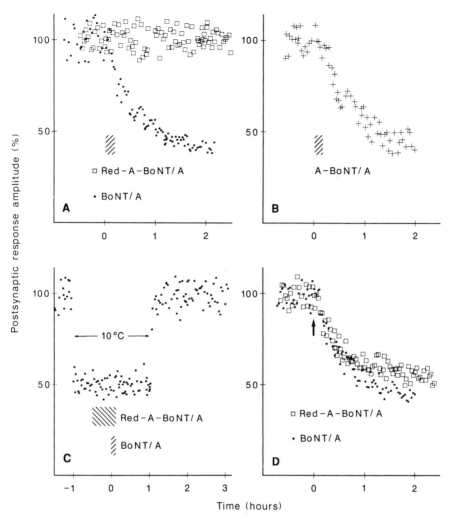

Fig 1. Red-A-BoNT/A binds to ecto-acceptors but is not internalized into cholinergic *Aplysia* unlike alkylated BoNT/A. Postsynaptic responses were elicited by a presynaptic action potential at 22°C, except when specified. After measuring control postsynaptic responses (100%), toxin samples were added to the bath for a period indicated by the hatched area or injected (arrow) into a presynaptic cholinergic neuron of the buccal ganglion. (A) Red-A-BoNT/A, even at a high extracellular concentration (100 nM), did not affect transmission, compared with 10 nM native BoNT/A or (B) 10 nM alkylated BoNT/A (A-BoNT/A). (C) Following measurement of control responses, the temperature was lowered to 10°C in order to arrest intoxication at the binding step. The preparation was then pre-incubated for 30 min with Red-A-BoNT/A (100 nM) followed by a 10 min application of BoNT/A (10 nM) prior to extensive washing of unbound toxin for 1 h. Upon return to 22°C, no depression was observed with BoNT/A alone. This indicated that the disappearance of extracellular toxicity of Red-A-BoNT/A is not due to the absence of binding to BoNT/A acceptors. (D) When injected (each at a final concentration of 10 nM), Red-A-BoNT/A appeared only fractionally less toxic than BoNT/A. These collective findings show that the lack of extracellular toxicity of Red-A-BoNT/A resulted from disruption of its internalization.

Table 1. Reduction and alkylation of BoNT /A and its H.

	THIOL CONTENT (mol/mol protein)		TOXICITY (% of BoNT/A)
BoNT/A	5.6 ± 0.6 (5.0)		100
Red-A-BoNT/A	0.6 ± 0.2 (0.0)		0.03-0.10
A-BoNT/A[a]	4.5 ± 0.5 (4.0)		100
Ren-H	4.3 ± 0.4 (4.0)		-
Ren-H[a]	5.9 ± 0.3 (6.0)		-
Ren-A-H	0.2 ±0.1 (0.0)		-
Ren-A-H[a]	2.7 ± 0.6 (2.0)		-

Protein thiol contents were quantified by a modified Ellman/s reaction described elsewhere [25] and are averages of at least three readings ± SD. The number of thiols taken from previously published sequence data [3] are given in parentheses. Samples were prepared as described in Fig 1 and Fig 3. *a* indicates samples further reduced with 100 mM DTT (60 min at 37°C) before thiol determination. The thiol values for Ren-A-H were measured with (*a*) and without cleavage of the intra-chain disulphide. Toxicities were measured in mice.

Ellman's reaction (Table 1). Notably, non-reduced alkylated BoNT/A (A-BoNT/A) was fully active in all the assay systems. Hence, neither the toxin's free thiols nor disulphides are essential for "effective" interaction with ecto-acceptors; the reasons for its loss of toxicity will be given below.

When equivalent experiments were carried out with BoNT/F or TeTX, no significant antagonism of their blockade of neuromuscular transmission was produced by Red-A-BoNT/A (Fig 2). A minimal prolongation of the paralysis time for BoNT/E resulted from inclusion of a 12.5 -fold excess of the derivatised BoNT/A (Fig 2); this may indicate a limited degree of overlap between acceptors for BoNT/A and /E, as observed electrophysiologically in *Aplysia* neurons[26]. Overall, it can be deduced that motor nerve terminals possess distinct ecto-acceptors for most of these toxins. Thus, it will be intriguing to establish the physiological functions of such numerous components residing on cholinergic presynaptic nerve membranes.

DISULPHIDES OR THEIR PARTICIPATING THIOLS ARE REQUIRED FOR BoNT INTERNALIZATION BUT NON-ESSENTIAL IN THEIR INTRA-NEURONAL ACTION

Red-A-BoNT/A which lacks both the disulphides (the inter-chain and one within H) displays minimal toxicity (0.03-0.10% of the native toxin; Table 1) and fails to alter transmitter release but is capable of binding to neuronal acceptors (see above). Therefore, this derivative must be unable to undergo efficient neuronal uptake or be devoid of intracellular activity. The latter was excluded because when injected inside *Aplysia* neurons in the buccal

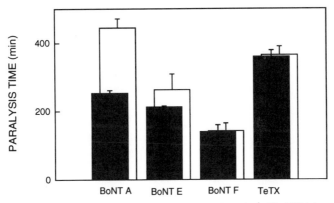

Fig 2. Red-A-BoNT/A significantly antagonizes the neuroparalytic action of BoNT/A but not that of type E, F or TeTX. Phrenic nerve-hemidiaphragms were dissected from mice and bathed in a modified Krebs-Ringer buffer (low Ca^{2+} / high Mg^{2+}) containing BoNT/A and / or its derivatives at 4^OC - a condition that minimizes internalization but does not prevent binding of BoNT at motor nerve-endplates. The diaphragms were washed in physiological Krebs Ringer and the temperature raised to 24^OC before supramaximal stimulation of the phrenic nerve. Red-A-BoNT/A was prepared by incubating BoNT/A with 50 mM DTT in 50 mM Tris/150 mM NaCl, pH 8.0 for 60 min at 37^OC followed by alkylation with a five-fold excess of iodoacetamide for 30 min at 4^OC in the dark. Addition of 0.2 nM BoNT/A to diaphragms for 15 min following a 45 min pre-incubation with Red-A-BoNT/A (2.5 nM; open bar) delayed the action of the native toxin compared to that of treatment with 0.2 nM BoNT/A alone (filled bar). The efficacies of 0.2 nM BoNT /E and /F or 5 nM TeTX (filled bars) were unaffected by pretreatment of the nerve tissues with 2.5 nM Red-A-BoNT/A (open bars). The values shown represent the average (± S.D) of at least three experiments.

ganglion, it inhibited ACh release with only slightly reduced potency relative to unmodified BoNT/A (Fig 1D). Thus, the presence of one or both of the toxin's disulphides must be a pre-requisite for the internalization step. To decipher the relative importance of two disulphides, unsuccessful attempts were made to exclusively reduce (and alkylate) the inter-chain disulphide of BoNT/A using thioredoxin/thioredoxin reductase[24], a system that works well for TeTX[25]. Consequently, *Aplysia* neurons were used because the H can internalize L (reviewed in Poulain *et al.* ; accompanying chapter), unlike the situation in mammalian motor nerves where uptake leading to intoxication is only obtained with intact BoNT (i.e. L disulphide-linked to H). Furthermore, to aid this study renatured, alkylated H of BoNT/A (Ren-A-H) was prepared with its intra-chain disulphide reformed but all free thiols alkylated (including the cysteine previously forming the inter-chain S-S bond in BoNT). These properties were established from the different thiol contents obtained for Ren-H and Ren-A-H (Table 1). As expected from results of the aforementioned studies on Red-A-BoNT, Ren-

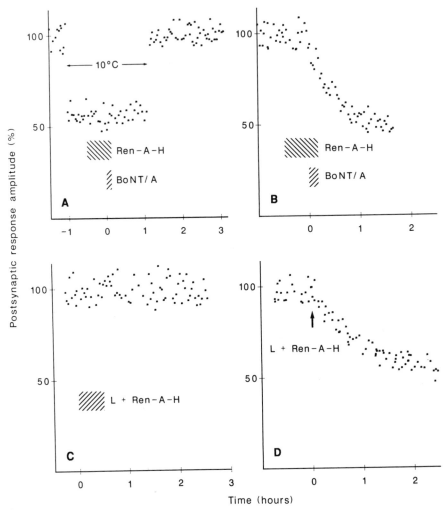

Fig 3. Ren-A-H of BoNT/A cannot mediate the internalization of its L. Conditions were as described in Fig 1. H was isolated from the di-chain native toxin using a chromatographic procedure detailed previously[27] and renatured by dialysis at 4°C over 24 h into 50 mM Tris/150 mM NaCl, pH 8.0 before alkylation with 150 mM iodoacetamide for 30 min in the dark. Ren-A-H was dialysed into artificial sea water before application to *Aplysia* neurons. Pre-incubation of the tissue with 100 nM Ren-A-H for 30 min before the addition of 10 nM BoNT/A - with the internalization process arrested at lowered temperature - prevented inhibition of transmission by the toxin (A). In contrast, when the latter experiment was performed at 22°C, neuroparalysis was seen (B), probably because BoNT/A acceptors recycle at this temperature. Consistent with the inability of Red-A-BoNT/A to internalize (cf. Fig. 1), bath application of 100 nM L plus Ren-A-H of BoNT/A (C) proved ineffective whereas intra-cellular injection (D, arrow) of this mixture at 10 nM (intra-cellular final concentration) depressed transmission.

A-H retained its ability to bind to ecto-acceptor on *Aplysia* cholinergic neurons because it prevented BoNT/A from blocking ACh release provided the temperature was reduced to 10°C during the binding step (Fig 3A,B). Possibly, acceptor recycling occurs at room temperature (Poulain *et al.*, accompanying chapter), thereby precluding any antagonism being detected; additionally, Ren-A-H may have a reduced affinity for the acceptors. Most importantly, bath application of Ren-A-H and L gave no decrease in neuronal transmission, even though this mixture inhibited ACh release when applied intra-neuronally (Fig 3C, D). These experiments indicate that Ren-A-H, unlike Ren-H (see Poulain *et al.*; accompanying chapter), is unable to mediate L internalization revealing that alkylation of the SH group in H participating in the toxin's inter-chain S-S link precludes L uptake. In contrast, the intracellular activity of L can be afforded even when this thiol is alkylated.

BoNT TYPES AND TeTX DISPLAY A SPECTRUM OF PROPERTIES IN BLOCKING THE RELEASE OF TRANSMITTERS FROM BRAIN SYNAPTOSOMES

Inhibitory Potencies of the Toxins on the Release of Different Neurotransmitters

It is well established that much higher concentrations of BoNT/A are required to block ACh release from brain synaptosomes relative to those effective at peripheral motor nerves; in fact, central cholinergic nerve terminals show only marginally greater susceptiblity than other transmitter types [28]. Although inefficient uptake has been postulated for this low potency of BoNT/A [29], type B is considerably more effective in blocking synaptosomal ACh release (Fig 4B) whilst being near-equipotent with A and F towards noradrenaline (NA) efflux (Fig 4A). Even more noteworthy is the remarkable ability of BoNT/E to abolish K^+-evoked, Ca^{2+}-dependent release of NA relative to types A, B, F or TeTX (Fig 4A, C). Such a striking range of potencies could be attributable of one or more steps of the triphasic intoxication process (i.e. binding, uptake and intracellular action). In addition to the demonstrated existence of distinct ecto-acceptors for several of these toxins in both peripheral and central nerves (see above), BoNT/A has been found to differ from TeTX in the temperature sensitivity of their uptake into *Aplysia* neurons[30]. Moreover, in the latter study a less strict temperature dependence was noted for the intracellular blockade of transmitter release by BoNT/A relative to TeTX. These characteristic features extend the number of dissimilarities noted earlier for the toxins (see Introduction) and are elaborated below.

Reversal of toxin action by elevated Ca^{2+} concentrations

An increase in extracellular Ca^{2+} concentration (up to 20 mM) does not interfere with either the action of BoNT/A on resting and evoked ACh release or its inhibition of K^+-stimulated NA secretion from rat cerebrocortical synaptosomes[19,28]. Alleviation of the intoxication was obtained by the Ca^{2+} ionophore, A23187, under optimal conditions the absolute level of release recovered being very similar to that from non-toxin treated samples

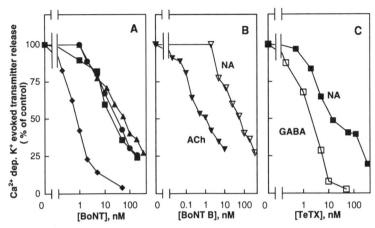

Fig 4. Comparison of the action of BoNT types on NA release. (A) Synaptosomes were incubated with [^{3}H]NA and various BoNT/A (■), /B (▲), /E (◆) or /F (●) for 2h at 37°C. The toxin and excess radiolabelled transmitter were removed by washing in Ca^{2+} free buffers. Evoked release was measured over a 5 min period in buffer containing 1.2mM Ca^{2+} and 25mM K$^+$. An equivalent sample was stimulated in the absence of Ca^{2+} and this amount of release subtracted from that in Ca^{2+} containing medium to yield a measure of Ca^{2+} dependent release. Values for secretion are expressed as a % of that obtained in the absence of toxin. (B) BoNT/B inhibition of NA and ACh release, the latter was detected after prior incubation of synaptosomes with [^{3}H]choline (assay details given elsewhere[28]). (C) Difference in potency of TeTX in inhibiting release of NA and GABA. Amino acid release was measured by pretreating the synaptosomes with radiolabelled GABA in the presence of 1mM amino-oxyacetic acid and release was measured in Na$^+$ free buffers (Na$^+$ replaced by choline). This condition was employed to reduce the amount of non-exocytotic GABA secretion that occurs through the reversal of the Na$^+$ dependent GABA uptaker upon depolarization of the nerve terminals [31]. Results for all panels represent the mean of many experiments and for ease of comparison of the results error bars are not shown.

(Fig 5). However, BoNT/B poisoned terminals gave only a partial reversal upon addition of A23187 with 1.2 mM or higher external Ca^{2+} concentration. This difference between types A and B on noradrenergic central nerve endings is analogous to that obtained with cholinergic terminals in the periphery when the Ca^{2+} concentration was increased by other means[8] (see Introduction).

In view of the aforementioned similarities between TeTX and BoNT/B, /E or /F (see Introduction), it was pertinent to ascertain if they also behaved differently from BoNT/A. Having established dose response curves for the individual toxins (Fig 4), supra-maximal doses of each were used to intoxicate synaptosomes. Fig 5 indicates that only type A intoxication can be fully reversed with A23187 even though the amount of toxin employed (750 nM) was ~4-fold higher than required to produce maximal inhibition. This was an attempt to mimic studies on the late stages of botulinisation of the neuromuscular junction in which 4-AP was found to be less effective as an antagonist on tissues exposed for a long

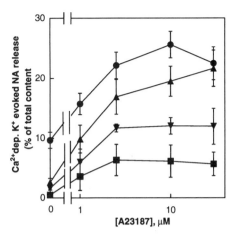

Fig 5. Influence of A23187 on the inhibitory action of BoNT types and TeTX on evoked NA release from synaptosomes. All conditions are as outlined in Fig 4 except that the stimulation buffers contained the appropriate concentration of ionophore. ▲, 750 nM BoNT/A; ■, 400 nM BoNT/E; ▼, 360 nM TeTX, and ●, non-toxin treated controls. Results are the means (+ SD) from 3 independent experiments.

time to BoNT/A[32]. In contrast, only partial reversal (35-40% of the control) of TeTX-induced blockade of NA release was seen (Fig 5), a level similar to that found with BoNT/B or /F. Type E gave the most striking result in that even less recovery could be obtained, reaching a plateau at 20-25% of the control value (Fig 5). Such a result could have arisen because /E is by far the most potent inhibitor of NA release (see Fig 4); 8-80 times less BoNT/E (6-60 nM) than used, herein, produces almost the same blockade of K^+-evoked NA secretion, yet a similar minimal reversal was seen with A23187. Collectively, our results are in accord with the data for ACh release at the motor endplates; namely, in both cases the effects of BoNT/B and /F or TeTX can only be overcome partially by increasing the intracellular Ca^{2+} concentration, whereas type A blockade is readily reversed. Although our data with /E toxin suggests that its intoxication is the least reversible and is, therefore, more like BoNT/B, /F and TeTX, there remains the possibility that one was observing a very advanced state of intoxication like the late stage of botulinisation with /A at the neuromuscular junction[32]. Clearly, use of additional criteria will be necessary to determine the complete spectrum of actions of /E (e.g. see [11]).

The data found with synaptosomes differs from that from experiments on neurosecretosomes[33] and hormone secretory (chromaffin) cells[34, 35] where BoNT-induced inhibition of secretion could not be counteracted by increasing Ca^{2+} concentration (nicotine or veratridine produced a Ca^{2+}-independent reversal[35]). Thus, the differences between these models were examined, particularly because all these hormone secreting preparations release their contents from less specialized regions of the cell than the well defined active zones of synapse-forming nerve terminals.

Inhibition by *Clostridial* Neurotoxins of Synaptosomal NA release from SSVs and LDCVs at Active and Non-Active Zones

Many nerves possess SSVs (45-55nm) and LDCVs (75nm or more) that contain amino acid neurotransmitters/ACh and neuropeptides, respectively[36,37,38]; other neurotransmitter substances such as catecholamines can be found in both types of organelle[38]. SSVs preferentially release their content at synaptic active zones whilst the LDCVs exocytose at ectopic non-specialized regions outside these areas[38,39,40]. Active zones contain a high density of Ca^{2+} channels[41] and upon stimulation they allow the Ca^{2+} concentration to reach $100\mu M$ in the vicinity of this region (only 100 nm from plasma membrane)[42]. The latter can be blocked by use of high Mg^{2+} concentration whilst the release of LDCV's can be triggered simultaneously by ionophores that raise bulk cytosolic Ca^{2+} concentration; subtraction of non-active zone release from the total (measured under normal K^+ stimulation), gives a measure of active zone secretion. Fig.5 indicates the combined amount of NA release from active and non-active zones due to exocytosis of SSVs and LDCVs. High concentrations of BoNT/E, /A and TeTX completely blocked the small amount of non-active zone release apparent in the absence of ionophore (Fig.6A). This is consistent with the abilities of BoNT/A and TeTX to block the release of met-enkephalin-like material from rat striatal particles[43], a peptide that is secreted from LDCVs outside the active zones. A recent study indicates that when release of synaptosomal glutamate from SSVs is blocked by TeTX, there is a corresponding inhibition of peptide release from LDCVs[44]. These results also correlate with the toxins' inhibition of exocytotic release of vasopressin and oxytocin from neurohypothesis and their action on the non-synaptic secretion from endocrine cells (see earlier).

Increasing A23187 concentration (and therefore the level of intracellular Ca^{2+}) augments this type of secretion and partially reverses the action of all these toxins. In fact, the ionophore may be inducing some extra non-active zone secretion that is insensitive to the toxins. The lack of Ca^{2+} induced reversibility of chromaffin cell intoxication by these *Clostridial* toxins may be due to the fact that these cells are unable to induce this extra release (see above). Notably the toxins' inhibition of exocytosis fom outside the active zones may help explain the puzzling phenomenon that BoNT/A can block spontaneous, non-quantal release of ACh[45,46]. The latter can be detected post-synaptically as a slight depolarization[47], and appears to represent neurotransmitter leakage. Careful electrophysiological recording has been unable to detect this type of release from the active zones of nerve terminals[48]. However, exocytosis of ACh containing SSVs can occur outside the active zone[49,50], releasing transmitter into the intracellular space; some of this may diffuse into the synaptic cleft, activate postsynaptic ACh receptors and induce membrane depolarization.

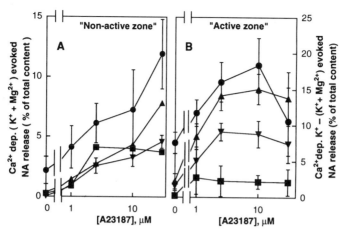

Fig 6. Active and non-active zone release of NA: blockade by BoNT types or TeTX and varying degrees of reversal with A23187. All conditions are given in Figs 4 and 5, except where highlighted. (▲), 750 nM BoNT/A ; (■), 400 nM BoNT/E; (▼), 360 nM TeTX, and (●), control. (A) Non-active zone release was determined in the presence of 20 mM Mg^{2+} which prevents Ca^{2+} entering through the Ca^{2+} channels at active zones. The actual amount of ionophore employed was 1.4 - fold higher than plotted so that the exact intracellular Ca^{2+} concentration induced by the ionophore would be equivalent in the presence or absence of 20 mM Mg^{2+} which reduces the ionophore efficacy by this factor[36]. The points are the mean of 3 independent experiments and the error bars shown are the S.D for those 2 curves where all points are significantly different. Certain points on other curves do not differ statistically. (B) Values for active zone release were calculated by subtracting the values of Ca^{2+} dep.K^+ + high Mg^{2+} evoked release ("non-active zone' release, A) from the values of Ca^{2+} dep. K^+ evoked release (total release, in Fig.5). Results are means (+ SD) for 3 independent experiments.

Although the ACh would be released as discrete packages, the amount reaching the receptors would vary and, thus, release could appear to be non-quantal in nature. Such an interpretation also leads to the conclusion that BoNT types are specific probes for exocytotic secretion only.

Fig 5 and Fig 6A represent increases in NA release triggered by equivalent rises in intracellular Ca^{2+} concentration; subtraction of these yields values for active zone release (Fig 6B). Increasing the intracellular Ca^{2+} level with as little as 1 µM A23187 can nearly completely overcome the action of BoNT/A on the active zone release of NA whilst much higher Ca^{2+} concentration only partially reverse the action of TeTX on this type of release. BoNT/E blocks active zone NA secretion almost completely and A23187 could only antagonise this to a minimal extent. Again, these findings on active zone release are comparable to those reported at the neuromuscular junction (see above). Thus, it would appear that the inhibition of active zone release produced by the various toxins differ in the levels of reversiblilty, suggesting that these observations are not due to non-specific effects of the Ca^{2+} ionophore; otherwise, one would expect that the inhibitory actions of all the neurotoxins would display the same characteristics.

ROLE OF CYTOSKELETON IN SECRETION AND NEUROTRANSMITTER RELEASE: IMPLICATIONS IN THE ACTION OF *CLOSTRIDIAL* NEUROTOXINS?

Actin Microfilaments

Removal of microfilaments by pretreatment of synaptosomes with cytochalasin B failed to interfere with the action of both BoNT/A and /B[19]. Likewise, the toxins did not simply lower the sensitivity of these microfilaments to regulation by cAMP because increasing the level of this second messenger failed to antagonise the intoxication[14,19]. In chromaffin cells actin breakdown apparently preceeds exocytosis by 15-20 s[51] but this slow time course of secretion is partly artifactual due to the assay employed; electrophysiological methods have shown that granules fuse with a latency of about 50 ms after stimulation[52] compared with ~1ms for synaptic vesicles. Indeed, exocytosis can be visualized in adrenal cells[53] and the initial steps of secretion are quicker than 16ms. In any case, this time is about 3 orders of magnitude faster than the measurable breakdown of actin microfilaments. On the other hand, these observed changes in cytoskeleton occur in the same time scale as the membrane retrieval step of 1-60 s[53]; thus, an alternative postulate is that endocytosis requires F-actin redistribution. This idea does not explain why cytochalasin (leads to breakdown of actin) may promote catecholamine release in permeabilised chromaffin cells but it might account for the report that BoNT/A and TeTX prevent F-actin redistribution in chromaffin cells[54]. The exocytotic-endocytotic cycle is closely linked and toxin-induced blockade of secretion means there is a reduced requirement for endocytosis and, therefore, cytoskeleton rearrangement would be lessened. Indeed, it has been shown that when type D toxin blocks exocytosis there is a consequential inhibition of endocytosis[55]. Clearly, the idea that disassembly of cortical actin filaments allows secretory vesicles access to exocytotic sites on the plasma membrane is far from proven and, anyhow, actin microfilaments are not the targets for BoNT/A or /B.

Microtubules

These may deliver vesicles directly to their release sites since synaptic vesicle-microtubule association has been visualized near to the presynaptic membranes (only 40 nm away)[56,57]. Axonal and synaptosomal microtubules can be disrupted by colchicine although high doses are required since the neuronal cytoskeleton is particularly insensitive to this drug[58]; it acts by preventing polymerization of tubulin leading to the disappearance of this cytoskeletal element. Incubation of synaptosomes with taxol (an agent that stabilizes microtubules and prevents their disassembly) increases the number of microtubule coils present[59] and prevents the normal Ca^{2+}-dependent breakdown of this cytoskeleton component induced by A23187 or veratridine. The outcome of studies suggested that, at least in synaptosomes, turnover of microtubules is not required for secretion because the latter secretagogues still induced a normal amount of ACh release when the tubulin-based cytoskeleton was stabilised. In view of these contradictory findings, experiments were performed to determine whether microtubules could be the target for any of the *Clostridial* toxins.

Pretreatment of synaptosomes with colchicine prior to addition of BoNT/A did not interfere with the toxin's blockade of NA release[19] and, therefore, intact microtubules are not required for its action. However, there was some antagonism of BoNT/B intoxication with the same treatment. Variation in incubation conditions including longer pretreatment with

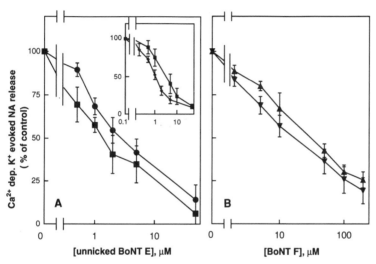

Fig 7. Colchicine attenuates the inhibitory action of unnicked BoNT/E and /F on synaptosomal release of NA. Conditions are as in Fig 4 except that synaptosomes were treated for 40 mins with (●, ▲) or without 1mM colchicine (■ , ▼)prior to additon of toxin. (A) Insert illustrates the action of unnicked (■) or nicked BoNT/E (◆) on NA secretion. Type E was fully nicked by incubation with trypsin (final conc. of 10 ug/ml) for 30 mins at 37 °C before trypsin was inactivated by the addition of a 5-fold excess of soy bean trypsin inhibitor. Results are the mean (+ SD) of 3 (A) and 5 (B) experiments.

colchicine or its inclusion throughout the experiment (wash and release buffers) failed to improve the amount of antagonism. To ensure that this effect was due to colchicine disrupting the cytoskeleton and not to secondary effects, other agents which perturb microtubules by different mechanism were also tested[19]. Nocodazole and griseofulvin antagonised BoNT/B to the same extent; yet all combinations of these three drugs failed to improve the level of protection. Similarly, cytochalasin D plus colchicine treatment of synaptosomes was no more effective than colchicine alone. Stabilization of microtubules by preincubation of synaptosomes with taxol prevented colchicine from antagonising BoNT/B[19]. None of these latter conditions interfered with the action of type A. As BoNT has a triphasic mechanism of action[20,60,61], experiments were carried out to decipher at what stage the microtubule dissociating drugs were acting. Colchicine did not interfere with the measurable binding of this toxin or the internalization step [19]. Thus, there is a requirement for intact microtubules for BoNT/B, but not /A, to exert its full toxicity at an intracellular site.

As BoNT/E, /F and TeTX appeared to be more like type B than /A, the effect of microtubule disrupting agents on their action was investigated. Colchicine pretreatment did partially interfere with the action of unnicked E (Fig 7A) and F (Fig 7B) on synaptosomal NA release. Although, there was a difference in efficacy between unnicked and nicked E toxin (Fig.7A insert), colchicine still only partially antagonised the action of the latter (data

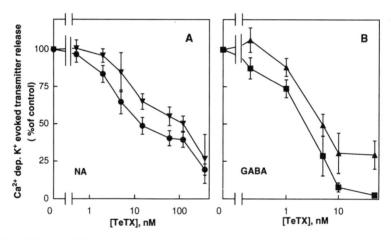

Fig 8. The inhibition of NA or GABA release by TeTX and its partial antagonism by colchicine. Synaptosomes were treated for 40 mins in the absence (●,■) or presence of 1mM colchicine (▼,▲) prior to the addition of toxin. Incubations and the release assay were carried out as oulined in Fig 4. The results are the mean (+ SD) of 5 independent experiments.

not shown). Importantly, the fact that most BoNT types have distinct ecto-acceptors (see earlier) reaffirms that colchicine is not simply acting at the level of toxin binding. Considering that TeTX and BoNT/B seem to act on the same target in peripheral cholinergic nerve terminals[8], it is not surprising that colchicine antagonizes the action of TeTX upon NA release in synaptosomes (Fig 8A). Furthermore, as the internalization processes for BoNT/B and TeTX are quite distinct[20, 61, 62, 63, 64](see earlier), they are unlikely to be affected by microtubule dissociating drugs; instead, these act inside the nerve terminal. Consistent with

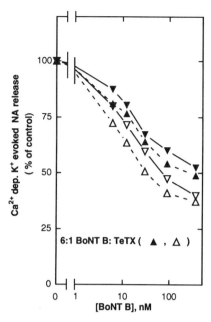

Fig 9 Effect of colchicine on NA release from synaptosomes double poisoned with TeTX and BoNT/B. Experimental conditions were as detailed for Fig. 7 except that the synaptosomes were treated with BoNT/B alone (▼,▽) or the latter plus TeTX (▲,△), in the absence or presence of 1 mM colchicine (closed and open symbols, respectively). Mean values from 2 separate experiments are shown.

both toxins acting at a common locus which is perturbed by these agents, no additivity was observed in the level of antagonism with colchicine (Fig 9).

Accordingly, microtubule dissociating agents should interfere with the action of these toxins on any neurotransmitter if these findings bare any relevance to their molecular mechanism. Colchicine counteracts the action of TeTX upon the release of GABA from synaptosomes.(Fig 8B); it also attentuates the effect of BoNT/B upon ACh release from synaptosomes (data not shown). These results suggest that TeTX and BoNT/B do have a common target in many different terminal types that are sensitive to microtubule dissociating agents.

These findings were duplicated with the use of other secretagogues; 4-AP more closely mimics the physiological mechanism of terminal depolarization than does elevated K^+ concentration because it produces spontaneous action potentials involving the repetitive activation of Na^+ channels[65] leading to an increase in Ca^{2+} dependent secretion from synaptosomes. However, the antagonism by colchicine of TeTX or BoNT/E intoxication was still apparent when release was stimulated in this way (Fig10). Alone, this concentration of K+ channel blocker did not reduce the efficacy of the toxins.

Possible Implication of Microtubules and Associated Components in the Action of BoNT/B, /E, /F and TeTX

The observed partial antagonism of these toxins' activity when microtubules are disassembled suggests that they do not play a *direct* role in their action. However, the drugs

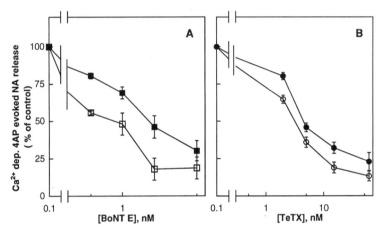

Fig 10. Influence of colchicine on the reduction of 4-AP evoked NA release from synaptosomes produced by BoNT/E or TeTX . All conditions were as outlined in Fig 7 except that release was stimulated with 0.6mM 4-AP. The points represent the mean (+SD) of 3 (A) and 6 (B) experiments in which synaptosomes were incubated in the presence (■,●) or absence (□,O) of 1 mM colchicine.

used do not act on all the microtubules inside nerves. Indeed, colchicine is unable to breakdown all the microtubules in synaptosomes[58]. Many isotypes of tubulin exist in brain with different affinities for colchicine[66]. Some of these variations are due to multiple β-tubulin subunits[67], whilst differences between labile and stable tubules have been attributed to the presence of tyrosinated α-tubulin which increases susceptibility to disrupting drugs[68]. These subtypes of microtubules may play different roles inside the nerve, and actually reside in discrete areas of the cytosol[69]. Some may act in vesicle transport whilst others serve a role in the architecture of the nerve[70]. Colchicine-sensitive and -resistant microtubules transport different cellular components[71]. It can be speculated that these *Clostridial* toxins act on all types of microtubules and so they can act to some extent upon secretion even when the labile fraction has been removed. The simplest explanation for the toxins' action would be that they prevent synaptic vesicle disattachment from the microtubules and these organelles are then unable to exocytose their neurotransmitter content. It is envisaged that the toxins modify (by an enzymic mechanism) a synaptic vesicle protein important in the attachment/disattachment to this cytoskeleton. If such a protein played a dual role in attaching vesicle to additional subcellular structures inside the nerve terminal (similar to certain microtubule-associated proteins which link microtubules and organelles to various other cellular components[72, 73, 74]) then these *Clostridial* toxins could still act at the other loci when microtubules are removed.

Studies carried out on the involvement of microtubules in secretion from non-neuronal cells suggest that there is a very highly ordered distribution of various membraneous organelles and vesicle transport towards the plama membrane[75, 76]. Disruption of the microtubular cytoskeleton interferes with the directionality of the secretory vesicles[76] and may even disturb the final maturation step of these organelles[77]. Therefore, *Clostridial* toxins may act on the synaptic vesicles at a site within the nerve terminal which is perturbed by the microtubule disrupting agents. The forthcoming enzymic mechanism of these toxins' action should be reconciled with the microtubule involvement.

CONCLUSION

It has been established that the free thiols play no role in the intoxication by BoNT/A whilst neither the inter- or intra-chain disulphides contribute to toxin binding or its intracellular action. However, the inter-chain is essential for toxin internalization. The physiologically relevant acceptors for BoNT/A, /E, /F and TeTX are distinct at the neuromuscular junction and, also, in the CNS. BoNT/A differs from the other toxins in that a Ca^{2+} ionophore is able to reverse its action to a much greater extent; also, intact microtubules are not involved in the action of type A whereas they seem to be required for full intoxication with BoNT/B, /E, /F and TeTX. Notably, the difference in Ca^{2+} reversal of poisoning in synaptosomes is very similar to findings at the motor nerve endings.

ACKNOWLEDGEMENTS

This research is supported by USAMRIID (Contract No. DAMD 17-91-Z-1035 to J.O.D.), Wellcome Trust, MRC and SERC studentship (to A. de P.). We thank Prof.J.Melling and Dr. C. Shone for provision of BoNT, and Dr.N.Fairweather and Dr.U.Weller for supply of TeTX.

REFERENCES

1. U. Eisel, W. Jarausch, K. Goretzki, A. Henschen, J. Engels, U. Weller, M. Hudel, E. Habermann, and H. Niemann, Tetanus toxin: primary structure, expression in *E.coli*, and homology with botulinum toxins, *EMBO J.* 5: 2495-2502 (1986).

2. H. Niemann, Molecular biology of Clostridial toxins, in : "Sourcebook of Bacterial Protein Toxins," J. Alouf and J. Freer ed., Academic Press, London (1991).

3. D. E. Thompson, J. K. Brehm, J. D. Oultram, T.-L. Swinfield, C. C. Shone, T. Atkinson, J. Melling, and N. P. Minton, The complete amino acid sequence of the *Clostridium botulinum* type A neurotoxin deduced by nucleotide sequence analysis of the encoding gene, *Eur.J.Biochem.* 189: 73-81 (1990).

4. L. L. Simpson (ed.). "Botulinum Neurotoxin and Tetanus Toxin", Academic Press, New York (1989).

5. J. O. Dolly, Peptide toxins that alter neurotransmitter release, in : "Handbook of Experimental Pharmacology, Vol. 102. Selective Neurotoxicity," H. Herken and F. Hucho eds., Springer-Verlag, Berlin (1992).

6. F. Dreyer and A. Schmitt, Different effects of botulinum A toxin and tetanus toxin on the transmitter releasing process at the mammalian neuromuscular junction, *Neuroscience Lett.* 26: 307-311 (1981).

7. F. Dreyer and A. Schmitt, Transmitter release in tetanus and botulinum A toxin-poisoned mammalian motor endplates and its dependence on nerve stimulation and temperature, *Pflügers Archiv Eur.J.Physiol.* 399: 228-234 (1983).

8. M. Gansel, R. Penner, and F. Dreyer, Distinct sites of action of Clostridial neurotoxins revealed by double-poisoning of mouse motor nerve terminals, *Pflugers Arch.* 409: 533-539 (1987).

9. J. Molgó, L. S. Siegel, N. Tabti, and S. Thesleff, A study of synchronization of quantal transmitter release from mammalian motor endings by the use of botulinal toxins type A and D, *J.Physiol.* 411: 195-205 (1989).

10. L. Wieszt and F. Dreyer, Mode of action of botulinum type E on the transmitter release process at the mouse neuromuscular junction, *Naunyn-Schmeideberg's Arch Pharmacol.* 344(suppl.): R74 (1991).

11. J. Molgó, B. R. DasGupta, and S. Thesleff, Characterization of the actions of botulinum neurotoxin type E at the rat neuromuscular junction, *Acta Physiol Scand.* 137: 497-501 (1989).

12. L. C. Sellin, S. Thesleff, and B. R. DasGupta, Different effects of types A and B botulinum toxin on transmitter release at the rat neuromuscular junction, *Acta Physiol.Scand.* 119: 127-133 (1983).

13. L. S. Siegel, A. D. Johnson-Winegar, and L. C. Sellin, Effect of 3,4-diaminopyridine on the survival of mice injected with botulinum neurotoxin type A, B, E, or F, *Toxicol.appl.Pharmac.* 84: 255 (1986).

14. L. L. Simpson, Use of pharmacologic antagonists to deduce commonalities of biologic activity among Clostridial neurotoxins, *J.Pharmacol.Exp.Ther.* 245: 867-872 (1988).

15. I. Kao, D. B. Drachman, and D. L. Price, Botulinum toxin: mechanism of presynaptic blockade, *Science.* 193: 1256-1258 (1976).

16. S. G. Cull-Candy, H. Lundh, and S. Thesleff, Effects of botulinum toxin on neuromuscular transmission in the rat, *J.Physiol.* 260: 177-203 (1976).

17. F. Dreyer, F. Rosenberg, C. Becker, H. Bigalke, and R. Penner, Differential effects of various secretagogues on quantal transmitter release from mouse motor nerve terminals treated with botulinum A and tetanus toxin, *Naunyn-Schmiedebergs Arch.Pharmacol.* 335: 1-7 (1987).

18. J. A. Kauffman, J. F. Way, L. S. Siegel, and L. C. Sellin, Comparison of the action of types A and F botulinum toxin at the rat neuromuscular junction, *Toxicol.Appl.Pharmaocol.* 79: 211-217 (1985).

19. A. C. Ashton and J. O. Dolly, Microtubule-dissociating drugs and A23187 reveal differences in the inhibition of synaptosomal transmitter release by botulinum neurotoxins type A and B., *J.Neurochem.* 56: 827-835 (1991).

20. J. D. Black and J. O. Dolly, Interaction of [125]-I labelled botulinum neurotoxins with nerve terminals. I. Ultrastructural autoradiographic localization and quantitation of distinct membrane acceptors for types A and B on motor nerves, *J.Cell Biol.* 103: 521-534 (1986).

21. R. S. Williams, C. K. Tse, J. O. Dolly, P. Hambleton, and J. Melling, Radioiodination of botulinum neurotoxin type A with retention of biological activity and its binding to brain synaptosomes, *Eur.J.Biochem.* 131: 437-445 (1983).

22. D. M. Evans, R. S. Williams, C. C. Shone, P. Hambleton, J. Melling, and J. O. Dolly, Botulinum neurotoxin type B. Its purification, radioiodination and interaction with rat brain synaptosomal membranes, *Eur.J.Biochem.* 154: 409-416 (1986).

23. J. S. Elston, Botulinum toxin A in clinical medicine, *J.Physiol.(Paris).* 84: 285-289 (1990).

24. J. O. Dolly, A. de Paiva, B. Poulain, P. Foran, A. Ashton, and L. Tauc, Insights into the neuronal binding, uptake and intracellular activity of botulinum neurotoxins, *in* : "Bacterial Protein Toxins, Zbl. BaKt. Suppl. 23," Witholt et al. ed., Gustav Fischer, Stuttgart, Jena, New York (1992).

25. G. Schiavo, E. Papini, G. Genna, and C. Montecucco, An intact inter-chain disulphide bond is required for the neurotoxicity of tetanus toxin, *Infect.Immun.* 58: 4136-4141 (1990).

26. B. Poulain, J. D. F. Wadsworth, C. C. Shone, S. Mochida, S. Lande, J. Melling, J. O. Dolly, and L. Tauc, Mutiple domains of Botulinum neurotoxin contribute to its inhibition of transmitter release in *Aplysia* Neurons, *J. Biol. Chem.* 264: 21928-21933 (1989).

27. E. A. Maisey, J. D. F. Wadsworth, B. Poulain, C. C. Shone, J. Melling, P. Gibbs, L. Tauc, and J. O. Dolly, Involvement of the constituent chains of botulinum neurotoxins A and B in the blockade of neurotransmitter release, *Eur.J.Biochem.* 177: 683-691 (1988).

28. A. C. Ashton and J. O. Dolly, Characterization of the inhibitory action of botulinum neurotoxin type A on the release of several transmitters from rat cerebrocortical synaptosomes, *J.Neurochem.* 50: 1808-1816 (1988).

29. J. D. Black and J. O. Dolly, Selective location of acceptors for botulinum neurotoxin A in the central and peripheral nervous systems, *Neuroscience.* 23: 767-779 (1987).

30. B. Poulain, A. de Paiva, J. O. Dolly, U. Weller, and L. Tauc, Differences in the temperature dependencies of uptake of botulinum and tetanus toxins in *Aplysia* neurons, *Neuroscience Lett.* 139: 289-292 (1992).

31. M. S. Santos, P. P. Goncalves, and A. P. Carvalho, Release of gamma-[^{3}H]aminobutyric acid from synaptosomes: effect of external cations and of ouabain, *Brain Res.* 547: 135-141 (1991).

32. J. Molgó, M. Lemeignan, and S. Thesleff, Aminoglycosides and 3,4-diaminopyridine on neuromuscular block caused by botulinum type A toxin, *Muscle and Nerve.* 10: 464-470 (1987).

33. J. L. Halpern, W. H. Habig, H. Trenchard, and J. T. Russel, Effect of tetanus toxin on oxytocin and vasopressin release from nerve endings of the neurohypophysis, *J.Neurochem.* 55: 2072-2078 (1990).

34. D. E. Knight, D. A. Tonge, and P. F. Baker, Inhibition of exocytosis in bovine adrenal medullary cells by botulinum toxin type D, *Nature.* 317: 719-721 (1985).

35. P. Marxen, F. Bartels, G. Ahnert-Hilger, and H. Bigalke, Distinct targets for tetanus toxin and botulinum A neurotoxins within the signal transducing pathway in chromaffin cells, *Naunyn-Schmeideberg's Arch.Pharmacol.* 344: 387-395 (1991).

36. M. Verhage, H. McMahon, W. E. J. M. Ghijsen, F. Boomsma, G. Scholter, V. M. Wiegant, and D. G. Nicholls, Differential release of amino acids, neuropeptides, and catecholamines from isolated nerve terminals, *Neuron.* 6: 517-524 (1991).

37. M. Matteoli, C. Haimann, F. Torri-Tarelli, J. M. Polak, B. Ceccarelli, and P. De Camilli, Differential effect of α-latrotoxin on exocytosis from small synaptic vesicles and from large dense-core vesicles containing calcitonin gene-related peptide at the frog neuromuscular junction, *Proc.Natl.Acad.Sci.USA.* 85: 7366-7370 (1988).

38. Å. K. Thureson-Klein, R. L. Klein, P.-C. Zhu, and J.-Y. Kong, Differential release of transmitters and neuropeptides co-stored in central and peripheral neurons, *in* : "Cellular and Molecular Basis of Synaptic Transmission," H. Zimmermann ed., Springer-Verlag, Berlin, Heidelberg (1988).

39. D. V. Pow and D. W. Golding, "Neurosecretion" by aminergic synaptic terminals *in vivo* - a study of secretory granule exocytosis in the corpus cardiacum of the flying locust, *Neuroscience.* 22: 1145-1149 (1987).

40. J. Cuadras, Non-synaptic release from dense-cored vesicles occurs at all terminal types in crayfish neuropile, *Brain Res.* 477: 332-335 (1989).

41. R. Robitaille, E. M. Adler, and M. P. Charlton, Strategic location of calcium channels at transmitter release sites of frog neuromuscular synapses, *Neuron.* 5: 773-779 (1990).

42. H. T. McMahon and D. G. Nicholls, The bioenergetics of neurotransmitter release, *Biochim.Biophys.Acta.* 1059: 243-364 (1991).

43. P. K. Janicki and E. Habermann, Tetanus and botulinum toxins inhibit, and black widow spider venom stimulates the release of methionine-enkephalin-like material in vitro, *J.Neurochem.* 41: 395-402 (1983).

44. H. T. McMahon, P. Foran, J. O. Dolly, M. Verhage, V. M. Wiegant, and D. G. Nicholls, Tetanus and botulinum toxins type A and B inhibit glutamate, GABA, aspartate and met-enkephalin release from synaptosomes: clues to the locus of action, *J.Biol.Chem.* In Press (1992).

45. F. Vyskocil, E. Nikolsky, and C. Edwards, An analysis of the mechanisms underlying the non-quantal release of acetylcholine at the neuromuscular junction, *Neuroscience.* 9: 429-435 (1983).

46. C. B. Gundersen and D. J. Jenden, Spontaneous output of acetylcholine from rat diaphragm preparations declines after treatment with botulinum toxin, *J.Pharmacol.Exp.Ther.* 224: 265-268 (1983).

47. F. Vyskocil, Inhibition of non-quantal leakage by 2(4-phenylpiperidine)cyclohexanol in the mouse diaphragm, *Neuroscience Lett.* 59: 277-280 (1985).

48. S. D. Merriney, S. H. Young, and A. D. Grinell, Constraints on the interpretation of nonquantal acetylcholine release from frog neuromuscular junctions, *Proc.Natl.Acad.Sci.USA.* 86: 2098-2102 (1989).

49. B. Ceccarelli, R. Fesce, F. Grohavaz, and C. Haimann, The effect of potassium on exocytosis of transmitter at the frog neuromuscular junction, *J.Physiol.(Lond.).* 401: 163-183 (1988).

50. E. S. Vizi, K. Gyires, G. T. Somogyi, and G. Ungváry, Evidence that transmitter can be released from regions of the nerve cell other than presynaptic axon terminal: axonal release of acetylcholine without modulation, *Neuroscience.* 10: 967-972 (1983).

51. M. L. Vitale, A. Rodriguez del Castillo, L. Tchakarov, and J.-M. Trifaro, Cortical filamentous actin disassembly and scinderin redistribution during chromaffin cell stimulation precede exocytosis, a phenomenon not exhibited by gelsolin, *J.Cell Biol.* 113: 1057-1067 (1991).

52. R. H. Chow, L. von Rüden, and E. Neher, Delay in vesicle fusion revealed by electrochemical monitoring of single secretory events in adrenal chromaffin cells, *Nature.* 356: 60-63 (1992).

53. S. Terakawa, J.-H. Fan, K. Kumakuru, and M. Ohara-Imaizumi, Quantitative analysis of exocytosis directly visualized in living chromaffin cells, *Neuroscience Lett.* 123: 82-86 (1991).

54. P. Marxen and H. Bigalke, Tetanus toxin and botulinum A toxins inhibit stimulated F-actin rearrangements in chromaffin cells, *Neuroreport.* 2: 33-36 (1991).

55. H. von Grafenstein, R. Borges, and D. E. Knight, The effect of botulinum toxin type D on the triggered and constitutive exocytosis/endocytosis cycles in cultures of bovine adrenal medullary cells, *FEBS Lett.* 298: 118-122 (1992).

56. E. G. Gray, Presynaptic microtubules and their association with synaptic vesicles, *Proc.Roy.Soc.Lond.[Biol.].* 190: 369-372 (1975).

57. N. Hirokawa, K. Sobue, K. Kanda, A. Maruda, and H. Yorifugi, The cytoskeletal architecture of the presynaptic terminal and molecular structure of synapsin I, *J.Cell Biol.* 108: 111-126 (1989).

58. P. R. Gordon-Weeks, R. D. Burgoyne, and E. G. Gray, Presynaptic microtubules: organization and assembly/ disassembly, *Neurosciece.* 7: 739-749 (1982).

59. R. D. Burgoyne and R. Cumming, Taxol stabilizes synaptosomal microtubules without inhibiting acetylcholine release, *Brain Res.* 280: 190-193 (1983).

60. L. L. Simpson, The origin, structure, and pharmacological activity of botulinum toxin, *Pharmacol.Rev.* 33: 155-188 (1981).

61. J. D. Black and J. O. Dolly, Interaction of ^{125}I-labeled botulinum neurotoxins with nerve terminals. II. Autoradiographic evidence for its uptake into motor nerves by acceptor-mediated endocytosis, *J.Cell Biol.* 103: 535-544 (1986).

62. R. G. Parton, C. D. Ockleford, and D. R. Critchley, Tetanus toxin binding to mouse spinal cord cells: an evaluation of the role of gangliosides in toxin internalization, *Brain Res.* 475: 118-127 (1988).

63. R. G. Parton, C. D. Ockleford, and D. R. Critchley, A study of the mechanism of internalization of tetanus toxin by primary mouse spinal cord cultures, *J.Neurochem.* 49: 1057-1068 (1987).

64. K. A. Manning, J. T. Erichsen, and C. Evinger, Retrograde transneuronal transport properties of fragment C of tetanus toxin, *Neuroscience.* 34: 251-263 (1990).

65. G. R. Tibbs, A. P. Barrie, F. J. E. Van Mieghem, H. T. McMahon, and D. G. Nicholls, Repetitive action potentials in isolated nerve terminals in the presence of 4-aminopyridine: effects on cytosolic free Ca^{2+} and glutamate release, *J.Neurochem.* 53: 1693-1699 (1989).

66. A. Banerjee and R. F. Luduena, Kinetics of association and dissociation of colchicine-tubulin complex from brain and renal tubulin. Evidence for the existence of multiple isotypes of tubulin in brain with differential affinity to colchicine, *FEBS Lett.* 219: 103-107 (1987).

67. A. Banerjee and R. F. Luduena, Distinct colchicine binding kinetics of bovine brain tubulin lacking the type III isotype of beta-tubulin, *J.Biol.Chem.* 266: 1689-1691 (1991).

68. P. W. Baas and M. M. Black, Individual microtubules in the axon consist of domains that differ in both composition and stability, *J.Cell Biol.* 111: 495-509 (1991).

69. P. W. Baas, T. Slaughter, A. Brown, and M. M. Black, Microtubule dynamics in axons and dendrites, *J.Neurosci.Res.* 30: 134-153 (1991).

70. R. H. Miller, R. J. Lasek, and M. J. Katz, Preferred microtubules for vesicle transport in lobster axons, *Science.* 235: 220-222 (1987).

71. H. Koike, M. Matsumoto, and Y. Umitso, Selective axonal transport in a single cholinergic axon of *Aplysia-* role for colchicine-resistant microtubules, *Neuroscience.* 32: 539-555 (1989).

72. F. F. Severin, N. A. Shanina, S. A. Kuznetsov, and V. I. Gelfand, MAP2-mediated binding of chromaffin granules to microtubules, *FEBS Lett.* 282: 65-68 (1991).

73. Z. Luo, B. Shafit-Zagardo, and J. Erlichman, Identification of the MAP2- and P75-binding domain in the regulatory subunit (RII beta) of type II cAMP-dependent protein kinase, *J.Biol.Chem.* 265: 21804-21810 (1990).

74. R. Donato, I. Giambanco, and M. C. Aisa, Molecular interaction of S-100 proteins with microtubule proteins *in vitro*, *J.Neurochem.* 53: 566-571 (1989).

75. O. Shimada, H. Ishikawa, and K. Wakabayashi, Role of microtubules in hormone secretion function of the rat anterior pituitary, *Protoplasma.* Suppl.2: 145-157 (1988).

76. K. Parczyk, W. Haase, and C. Kondor-Koch, Microtubules are involved in the secretion of proteins at the apical cell surface of the polarized epithelial cell Madrin-Darby canine kidney, *J.Biol.Chem.* 264: 16837-16846 (1989).

77. G. Herman, G. Busson, M. J. Gorbunoff, P. Mauduit, S. N. Timasheff, and B. Rossignol, Colchicine analogues that bind reversibility to tubulin define microtubular requirements for newly synthesized protein secretion in rat lacrimal gland, *Proc.Natl.Acad.Sci.USA.* 86: 4515-4519 (1989).

PARTIAL CHARACTERIZATION OF BOVINE SYNAPTOSOMAL PROTEINS ADHERED TO BY BOTULINUM AND TETANUS NEUROTOXINS

Cara-Lynne Schengrund,[1] Bibhuti R. DasGupta,[2] and Nancy J. Ringler[1]

[1]Department of Biological Chemistry
The M.S. Hershey Medical Center
Hershey, PA 17033
[2]Department of Food Microbiology and Toxicology
University of Wisconsin
Madison, WI 53706

INTRODUCTION

Binding of botulinum (BTx) and tetanus (TTx) neurotoxins to synaptic termini has been hypothesized to require a protein component(s)[1-3] in addition to gangliosides of the G1b series[4,5] (nomenclature described by Svennerholm[6]). Results that we obtained in studies of inhibition by gangliosides of the binding of BTxA and TTx to ganglioside GT1b-coated plastic wells indicate that G1b gangliosides do not function as high affinity ligands for the neurotoxins. We observed that the concentration of ganglioside GT1b needed to block binding of either BTxA or TTx to GT1b-coated plastic wells was in the μM range.[7] In contrast, when the same assay system was used to determine the concentration of ganglioside GM1 needed to block binding of cholera toxin to GM1-coated plastic wells, nM values were obtained.[8]

The aforementioned observations led us to investigate the ability of BTxA and TTx to adhere to blots of synaptosomal proteins that had been separated by sodiumdodecylsulfate polyacrylamide gel electrophoresis (SDS-PAGE); a technique used to identify the epidermal growth factor receptor,[9] and for monitoring the binding of BTxA to proteins present in the presynaptic membrane of the Torpedo electric organ.[10] Intact synaptosomes were used because the specific molecules with which these neurotoxins interact during the sequence of events that result in inhibition of neurotransmitter release is unknown. Since synaptosomes contain both the presynaptic and synaptic vesicle components, we hypothesized that their use would permit identification of any proteins with which the neurotoxins interacted.

Early reports[11,12] in which the adherence of ^{125}I-labeled TTx to blots of synaptosomal proteins was assessed, suggested that the only component

Botulinum and Tetanus Neurotoxins, Edited by
B.R. DasGupta, Plenum Press, New York, 1993

adhered to by the neurotoxin was the lipid that migrated at the front of the gel. Our results were similar[13]. However, if the TTx was preincubated with GT1b, prior to its addition to a blot of bovine synaptosomal proteins, then adherence to two protein doublets was reproducibly observed. W e also found that ^{125}I-labeled BTxA adhered to the same proteins, and its adherence was enhanced by preincubation with GT1b. The protein doublet with an apparent molecular weight of ~80 kDa was always the most intense, and the second doublet having an apparent molecular of ~116 kDa was generally the next most intense band present upon examination of autoradiographs of the blots that had been overlaid with labeled neurotoxin. Analogous experiments done using ^{14}C-labeled fragments of the BTxA molecule indicated that the N-terminal portion of the heavy chain mediated adherence to the ~116 kDa proteins while the C-terminal portion of the heavy chain and the light chain adhered preferentially to the ~80 kDa proteins[13].

Specificity of the adherence of the neurotoxins to the synaptosomal proteins was indicated in several ways: A) ^{125}I-labeled cholera toxin, preincubated with GM1, did not adhere to the same proteins and did not block the adherence of labeled BTxA, B) comparable results were not obtained when labeled BTxA, preincubated with GT1b was used to probe blots of membrane proteins from liver, spleen, or kidney, and C) excess cold BTxA or TTx, preincubated with GT1b, could block the adherence of labeled BTxA preincubated with GT1b[13].

PARTIAL CHARACTERIZATION OF THE ~80 kDA PROTEINS

The pronounced binding by both neurotoxins to the ~80 kDa protein doublet provided us with a rationale for trying to characterize it. As a first step, experiments were carried out to ascertain whether the proteins were cytosolic or membrane-associated synaptosomal components. Overlay of blots of SDS-PAGE separated proteins with ^{125}I-BTxA (Fig. 1), was used to monitor fractions containing proteins adhered to by the neurotoxins. The results indicated that the ~80 kDa protein doublet was solubilized with 3M KCl, suggesting that the proteins are peripherally-associated membrane components. Ammonium sulfate fractionation of the 3M KCl-solubilized proteins indicated that the ~80 kDa proteins were precipitated at a concentration between 20 and 30%. Densitometric analysis (Hewlett Packard Scan Jet Plus with the Collage Analysis System from Fotodyne, Inc.) of a less exposed autoradiograph than the one shown in Fig.1, indicated that over seven-fold more labeled BTxA was associated with the ~80 kDa protein doublet in the 30% ammonium sulfate pelleted fraction than in the initial synaptosomal preparation.

The two proteins that appeared to comprise the ~80 kDa doublet were isolated by excision and electroelution of the appropriate area of the gel. Overlay of the isolated material with ^{125}I-wheat germ agglutinin indicated that they were glycosylated (Fig. 2). Treatment of the isolated proteins with *Vibrio cholera* sialidase resulted in a slight reduction in molecular weight (Fig. 2) and a decrease in binding by ^{125}I-WGA to blots of the sialidase-treated proteins of ~65% as determined by densitometric scanning. This indicated that the sialidase did catalyze the loss of some of the sialic acid residues and supported the conclusion that the proteins are glycosylated. Interestingly, adherence of ^{125}I-BTxA to blots of the desialylated proteins (Fig. 2) was found by densitometric scanning to be decreased by only 10-15%.

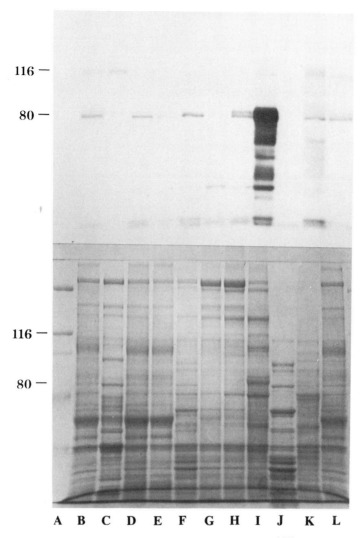

Figure 1: Solubilization of synaptosomal proteins adhered to by ^{125}I-BTxA. The left portion of the figure shows the Coomassie Blue stained proteins recovered in each fraction. The right portion of the figure shows an autoradiograph of a blot of a replicate gel that was exposed to labeled BTxA. The autoradiograph was overexposed in order to show that protein adhered to by the BTxA was present in fractions other than that precipitated by 30% ammonium sulfate. While multiple proteins in the 30% fraction are adhered to by the neurotoxin, densitometric analysis of a less exposed autoradiograph indicated that about five-fold more ^{125}I-BTxA adhered to the ~80 kDa doublet than to any other component. Protein samples in each lane are as follows: A) high molecular weight standards, B and L) bovine brain synaptosomes, C) water-soluble, D) water-insoluble, E) 3M KCl-insoluble, F) 3M KCl-soluble, and G) - K) the 10%-, 20%-, 30%- and 50%-insoluble and 50%-soluble ammonium sulfate fractions, respectively, isolated from the 3M KCl-soluble material. ~125μg of protein was in each sample applied to lanes B-L. Samples were incubated for 15 min at 37°C in sample buffer containing SDS and β-mercaptoethanol prior to SDS-PAGE. The blot was blocked in the cold, overnight, by soaking in 10 mM tris-chloride (pH 7.2) containing 1% BSA. ^{125}I-BTxA that had been preincubated with GT1b was used to identify possible ligands.

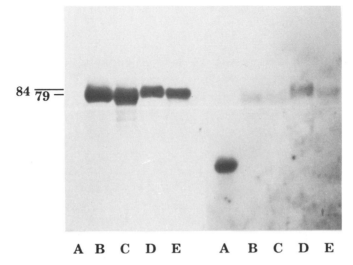

84 $\overline{79}$ =

A B C D E   A B C D E

Figure 2: Autoradiographs of blots of the isolated ~79 and ~84 kDa proteins exposed to either ^{125}I-BTxA (left panel), or ^{125}I-WGA (right panel). Samples in each lane are: A) high molecular weight standards, B) isolated ~79 kDa protein, C) sialidase-treated ~79 kDa protein, D) isolated ~84 kDa protein, and E) sialidase-treated isolated ~84 kDa protein. The blot overlaid with ^{125}I-BTxA was treated as described in the legend to Figure 1. The blot overlaid with labeled-WGA was blocked overnight by soaking in phosphate buffered saline, pH 7.4, containing 2.5% BSA and 2% polyvinylpyrrolidone. The same solution was used for the overlay with ^{125}I-WGA.

While the evidence indicates that BTxA and TTx adhere to synaptosomal protein, it is not yet clear whether they recognize the glycosylated or peptide portions of the proteins. We also need to identify the portion of the synaptosome with which the proteins are associated. Because of the specificity of binding by the neurotoxins, we believe that the proteins adhered to should be studied further to ascertain whether they may function as receptor/acceptor molecules or at a stage after the adherence of the neurotoxins to their cell surface receptors.

REFERENCES

1. K. Stoeckel, M. Schwab, and H. Thoenen, Role of gangliosides in the uptake and retro-grade axonal transport of cholera and tetanus toxin as compared to nerve growth factor and wheat germ agglutinin, Brain Res. 132:273 (1977).
2. E.J. Pierce, M.D. Davison, R.G. Parton, W.H. Habig, and D.R. Critchley, Character-ization of tetanus toxin binding to rat brain membranes, Biochem. J. 236:845 (1986).
3. E. Yavin and A. Nathan, Tetanus toxin receptors on nerve cells contain a trypsin-sensitive component, Eur. J. Biochem. 154:403 (1986).
4. W.E. vanHeyningen, Gangliosides as membrane receptors for tetanus toxin, cholera toxin and serotonin, Nature, Lond. 249:415 (1974).
5. J. Holmgren, H. Elwing, P. Fredman, and L. Svennerholm, Polystyrene adsorbed gang-liosides for investigation of the structure of the tetanus-toxin receptor, Eur. J. Biochem. 106:371 (1980).
6. L. Svennerholm, Ganglioside designation, Adv. Exp. Med. Biol. 125:11 (1980).
7. C-L. Schengrund, B.R. DasGupta, and N.J. Ringler, Binding of botulinum and tetanus neurotoxins to ganglioside GT1b and derivatives thereof, J. Neurochem. 57:1024 (1991).

8. C-L. Schengrund and N.J. Ringler, Binding of *Vibrio cholera* toxin and the heat-labile enterotoxin of *Escherichia coli* to GM1, derivatives of GM1, and nonlipidoligosaccharide polyvalent ligands, *J. Biol. Chem.* 264:13233 (1989).

9. P.H. Lin, R. Selinfreund, and W. Wharton, Western blot detection of epidermal growth factor receptor from plasmalemma of culture cells using ^{125}I-labeled epidermal growth factor, *Anal. Biochem.* 167:128 (1987).

10. C. Solsona, G. Egea, J. Blasi, C. Casanova, and J. Marsal, The action of botulinum toxin on cholinergic nerve terminals isolated from the electric organ of *Torpedo marmorata*. Detection of a putative toxin receptor, *J. Physiol. (Paris)* 84:174 (1990).

11. D.R. Critchley, W.H. Habig, and P.H. Fishman, Reevaluation of the role of gangliosides as receptors for tetanus toxin, *J. Neurochem.* 47:213 (1986).

12. A. Nathan and E. Yavin, Periodate-modified gangliosides enhance surface binding of tetanus toxin to PC12 pheochromocytoma cells, *J. Neurochem.* 53:88 (1989).

13. C-L. Schengrund, N.J. Ringler, and B.R. DasGupta, Adherence of botulinum and tetanus neurotoxins to synaptosomal proteins, *Brain Res. Bull.* (in press).

THE NEUROSPECIFIC BINDING OF TETANUS TOXIN IS MEDIATED BY A 20 kDa PROTEIN AND BY ACIDIC LIPIDS

Giampietro Schiavo,[1] Ornella Rossetto,[1] Giovanna Ferrari[2] and
Cesare Montecucco[1]

[1]Centro C.N.R. Biomembrane and Dipartimento di Scienze Biomediche,
Università di Padova, Via Trieste 75, I-35121 Padova, Italy
[2]Laboratori di Ricerche FIDIA S.p.a., Via Ponte della Fabbrica, Abano Terme,
Italy

INTRODUCTION

The clostridial protein neurotoxins responsible for tetanus and botulism are the most potent bacterial toxins. This extreme potency is at least in part attributed to their absolute neurospecificity. Tetanus toxin (TeTx) acts at the CNS by impairing neuroexocytosis at the inhibitory spinal cord interneurons thus causing a spastic paralysis. The botulinum neurotoxins (BoNT) act at the PNS by blocking neurotransmitter release at the neuromuscular junctions and cause a flaccid paralysis.

This similar mechanism of action is paralleled by a closely similar structure. The clostridial neurotoxins are released as a single polypeptide chain by cell lysis and are readily cleaved at an exposed loop by different proteases[14,39] to generate two chains held together by a single interchain disulfide bridge and by non covalent forces.[31] The heavy chain (H, 100 kDa) is involved in cell binding and in the cytosol penetration of the light chain (L, 50 kDa) which is responsible for the intracellular action of the toxin.[1,18,19,29]

The cellular mechanism of action of the clostridial neurotoxins is proposed to consist of three steps: 1) binding, 2) membrane translocation and 3) intracellular target modification.[35] None of these passages is understood at the molecular level. Very recently TeTx has been demonstrated to be a zinc endopeptidase which cleaves specifically a protein involved in the control of neuroexocytosis.[34] Of the second step it is only known that it must occur because the toxin binds on the cell surface and acts inside the cell. Recent evidence based on the inhibition of clostridial neurotoxins intoxication of the hemidiaphram preparation by bafilomycin A1 (see Simpson, OVERVIEW, this volume) provides an experimental evidence that these toxins enter into the neuronal cytosol via a low pH intracellular compartment, as first proposed by Boquet and Duflot.[3]

The binding step has been investigated in a large number of studies (references quoted in 22, 24, 26) without reaching a definitive conclusion about the cell structure(s) that can account for its two fundamental aspects: the absolute neurospecificity and the high affinity necessary to bind the toxin at the minute concentrations clinically active. A large number of investigations were focused on the lipid components of the membrane, particularly on those lipids enriched in the nerve tissue. Tetanus toxin interacts with acidic lipids with a preference versus polysialogangliosides.[32] Moreover, treatment of cells with neuraminidase reduces, without abolishing, toxin binding.[9] On the other hand, as discussed in detail previously,[20,21,24,25,36] this TeTx lipid interaction cannot account by itself for the neurospecificity and high affinity of toxin binding to neuronal cells. Other studies have shown that TeTx binding to cells is protease sensitive thus implying the existence of a toxin-protein interaction.[4,17,28,42]

Recently we demonstrated by chemical cross-linking the existence on NGF-differentiated PC12 cells of a 20 kDa protein specifically involved in the binding of tetanus toxin.[33] We present here evidence that both acidic lipids and a proteinic component are both involved in the binding of tetanus toxin to neuronal cells.

EXPERIMENTAL PROCEDURES

Tetanus toxin and fragments

Dichain tetanus toxin was purified from culture filtrates of *Clostridium tetani* as detailed before.[32] Tetanus toxin was fragmented by digestion with papain and fragments B and C were purified by HPSELC as described by Weller et al.[39] The mouse lethal doses (MLD) was 2 ng/Kg, as determined with a intraperitoneal route of administration.[38] [^{14}C]-TeTx was prepared by reductive methylation with [^{14}C]-formaldheyde (specific activity 55 Ci/mol) and sodium cyanoborohydride.[13] This method was chosen because it does not alter the protein charge. Residual toxicity, tested as above, was 5.5 ng/Kg mouse body weight. Iodination of TeTx and fragments H_C and L-H_N were obtained by using the Bolton-Hunter reagent. [^{125}I]-TeTx had a specific activities of 4 mCi/mg and retained a residual toxicity higher than 85%.

Lipid monolayers

Lipid monolayers were spread from 1 mM stock solutions in chloroform over 5 ml of phosphate-citrate buffer. Surface pressure was measured with the Wilhelmy method[5] and different lipid amounts were spread to obtain different initial pressures. TeTx was added in the subphase through a lateral subhole and the subphase was magnetically stirred. When needed, pH of the subphase was lowered by addition of the appropriate amount of 8.5% phosphoric acid. The radiolabelled toxin present at the interface was detected by following the surface radioactivity with a gas-flow counter.[5]

Cell cultures and tetanus toxin cell binding

The PC 12 subclone previously described[30,37] was cultured on rat collagen in Dulbecco's modified Eagle's medium (DMEM) supplemented with 44 mM sodium bicarbonate, 7.5% horse serum and 7.5% fetal calf serum, pH 7.4.[8] Experiments were performed on cells seeded on 24 mm dishes (20,000 cells per dish) and differentiated with 50 ng/ml of NGF extracted from mouse submammillary glands.[23]

Cells, treated with NGF for 10 days, were cooled on ice, washed with cold Hanks balanced solution (HBS) containing BSA 0.25% and preincubated at 0°C with increasing amounts of cold TeTx, H_C and L-H_N ranging from 10^{-11} to 10^{-6} M in the same buffer. After 30 minutes 40.000 cpm/sample of [^{125}I]-TeTx were added (30 fMoles) and incubated for 3 hours at 0°C. Unbound [^{125}I]-TeTx was eliminated by a double washing with cold HBS containing 0.25% BSA and cells were dissolved with 0.4 ml of 0.1 N KOH. Radioactivity associated to the samples were counted in a Packard Multi-Prias 1 and the results, after subtraction of unspecific binding, were expressed as percent binding in the absence of unlabeled TeTx or fragments.[28]

Tetanus toxin cell cross-linking

Cells, treated or not treated with NGF, were cooled on ice, washed with cold HBS pH 7.5 and incubated with [^{125}I]-TeTx (0.5-1 nM) for 2 hours at 0°C. After two washings with cold HBS, cells were treated with DSS or BSOCOES or DSP dissolved in HBS (200 µg/ml) at 0°C for 20 min. The medium was removed and replaced with HBS containing 20 mM glycine-NaOH, pH 7.5 for 5 min. Cells were washed two times with HBS and dissolved in 1 ml of HBS containing 0.4% octyl glucoside, 0.05% SDS, 1 mM iodoacetamide, 1 mM benzamidine, 0.1 mM PMSF, 0.1 mM o-phenantroline, and 10 µM TPCK. Supernatants were made 6.5% in TCA. Pellets derived from TCA precipitation were recovered by centrifugation and dissolved in 8% SDS, 10 mM Tris-acetate, 0.1 mM EDTA pH 6.8 containing 5% 2-mercaptoethanol. Samples were boiled for 2 min and analyzed by SDS gel electrophoresis on a discontinuous system (6%-12% polyacrylamide gradient) according to Laemmli.[15] After Coomassie Blue staining, gels were dried and autoradiographed. In some experiments, before incubation with [^{125}I]-TeTx, cells were treated with pronase (1-10 µg/ml) or with neuraminidase (25-50 mU/ml) in HBS for 30 minutes at 37°C or with a mixture of trypsin (0.5 mg/ml) and chymotrypsin (0.5 mg/ml) in HBS for 6 minutes at 37°C. After two washings with 1 ml of cold HBS, cells were incubated with HBS containing protease inhibitors for 5 minutes at 0°C and then treated with BSOCOES as described above. Cells were counted on lysates by using 4',6-diamidine-2'-phenylindole dihydrochloride (DAPI).[33]

RESULTS and DISCUSSION

Interaction of tetanus toxin with lipid monolayers

Several studies indicate that tetanus toxin interacts strongly with polysialogangliosides, particularly with those of the G1b series. These studies were performed with gangliosides in micellar form or adsorbed to paper, plastic or t.l.c. plates (references in 32). These systems are characterized by a high density of surface negative charges and by other unphysiological aspects. The model system that most closely meets the conditions experienced by the toxin on the surface of the membrane is represented by lipid monolayers spread over a buffered water phase.[5] In this system the lipid composition and density can be varied at will. A lipid-protein interaction can be monitored as a surface pressure increase by a Wilhelmy plate or by measuring the surface radioactivity of a radioactively labelled protein with a gas flow counter.[5]

Fig. 1 shows the results of an experiment performed with a lipid monolayer of a composition mimicking that of the outer leaflet of the plasma membrane of a neuronal cell. Also the initial surface pressure value is in the range estimated for living cells. The figure shows that at neutral pH tetanus toxin adsorbs onto the surface of the lipid monolayer and inserts protein segments into the monolayer, as deduced by the increase in lateral surface pressure. That a interdigitation of protein segments among the lipid hydrocarbon chains does occur is demonstrated by the increase of surface radioactivity shown in panel B of fig. 1. A maximal effect was already found at TeTx concentration of 88 nM: such a low concentration of the toxin speaks in favour of a further relevance of the present findings to the in vivo situation.

Having documented the reliability of the method we assayed various polysialogangliosides in search of a differential effect of a particular class of gangliosides such as for ganglioside G_{M1} for cholera toxin. Fig. 2 shows the results of experiments where the ganglioside component of the lipid monolayer was varied. It appears that all gangliosides are able to promote the lipid interaction of TeTx irrespectively of their oligosaccharide structure.

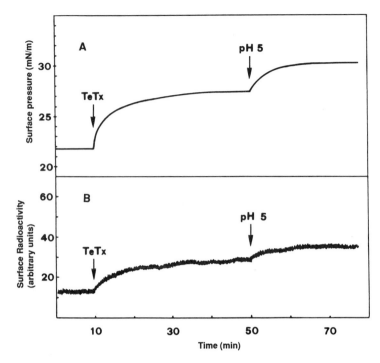

Figure 1. TeTx binding and effect on the surface pressure of phospholipid monolayers at neutral and acid pH. A) Tetanus neurotoxin (final concentration: 88 nM), doped with [^{14}C]-labelled toxin, was injected in the subphase of a monolayer of Ole$_2$GroPCho:Ole$_2$GroPEtn:cholesterol: asolectin:G$_{D1b}$ (35:10:35:20:5, mol:mol), a mixture that is intended to mimic the composition of the outer layer of a non neuronal cell plasma membrane; B) adsorption of tetanus toxin to the same lipid monolayer of A) monitored by a gas-flow radioactivity counter.

It appears that G$_{M1}$, which is also present on non nerve cells, promotes the adhesion of TeTx nearly as well as gangliosides that were considered to be potential neuroreceptors of TeTx such as GD1b or GT1b. Moreover, the equipotency of a bovine brain ganglioside extract with the purified gangliosides is against the hypothesis that a minor ganglioside component is the true receptor of TeTx.

Binding of tetanus toxin and its fragments to NGF-differentiated PC12 cells

NGF is able to induce a remarkable differentiation of PC12 cells in culture. Within few days one observes the outgrowth of long neurite like projections and this has made NGF treated-PC12 cells a model of neuronal differentiation.[8] This cell line is also characterized by a tendency to generate variants, i.e. subclones, that differ morphologically and functionally from the parent cells. Recently a subclone of PC12 has been characterized that, upon NGF treatment, develops high affinity binding sites for TeTx that are able to mediate the entrance of the toxin in the cytosol with consequent inhibition of catecholamine release.[30,37] Fig. 3 shows the remarkable change in appearance of this subclone of PC12 cells after seven days of growth in the presence of NGF.

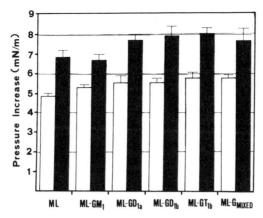

Figure 2. Effect of TeTx on lipid monolayers containing different gangliosides. Tetanus toxin was added to lipid monolayers with a plasma membrane-like mixture: $Ole_2GroPCho:Ole_2GroPEtn:cholesterol:asolectin$ (35:10:35:20, mol:mol) (ML) or of ML plus 4.8% (mol:mol) of the indicated ganglioside. The figure compares the increases in surface pressure caused by the toxin at pH 7.4 (empty bars) and at pH 5.0 (filled bars) on the different monolayers.

The NGF-differentiated cells express high affinity binding sites for TeTx as shown in fig. 4. Also the TeTx 50 kDa carboxy-terminal fragment (H_C) binds strongly to these cells, while the remaining toxin portion L-H_N does not. The strong binding of H_C is expected on the basis of its ability to mediate the uptake and retroaxonal transport of various biological markers.[2,7] However it should be noticed that binding of TeTx is stronger than that of H_C and hence additional interactions must be displayed by the whole toxin with respect to its H_C fragment.

Cross-linking of tetanus toxin and its fragments to differentiated PC12 cells

Covalent linking of a protein ligand to its cellular receptor with bifunctional reagents has been widely used to identify receptors for hormones and cytokines.[6,10-12,16,27]

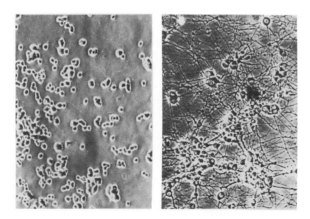

Figure 3. NGF-induced neuronal differentiation of a subclone of PC12. Cells were grown on collagen coated plastic in normal DMEM (left panel) or DMEM supplemented for 7 days with 50 ng/ml of NGF (right panel).

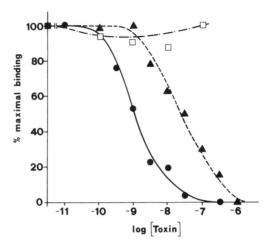

Figure 4. Effect on $[^{125}I]$-TeTx binding to NGF-differentiated PC12 cells of cold TeTx (circles), of fragment H_C (triangles) and of L-H_N (squares).

Fig. 5 reports the structural formulae of the three reagents used here: BSOCOES, DSS, DSP. They differ from the length of the spacer arm between the two reactive groups, for their solubility and for the fact that DSP is cleaved by reducing agents.

PC12 cells were incubated with $[^{125}I]$-TeTx in the cold to allow binding without toxin internalization. Unbound toxin was washed away and cells were incubated in the cold with crosslinker. After washings, proteins were recovered, dissolved in a SDS buffer and electrophoresed. Fig. 6 shows that DSS and BSOCOES added to $[^{125}I]$-TeTx bound to cells induce the formation of additional bands with respect to the H and L chains. The 150 kDa and larger bands are products of cross-linking of the toxin chains among themselves.

The formation of bands above 150 kDa are in agreement with the possibility that part of the toxin is present in form of dimer. That all bands above the H chain are due to cross-linking is demonstrated by lane 4 of fig. 6 which shows that only the two toxin chains are present in a sample incubated in the presence of DSP and reduced before SDS-PAGE.

The presence of cross-linkers induce an additional 120 kDa band, which cannot be ascribed to TeTx alone, but must derive from the cross-linking of the toxin with a cellular component. The formation of the 120 kDa band is prevented by the prior incubation with cold TeTx (lane 5) thus indicating that it is a specific product of the presence of the toxin. Immunoblots with anti L and anti H chain affinity purified antibodies, not shown here, demonstrate that the 120 kDa band contains the H chain of the toxin. Similar experiments performed with undifferentiated PC12 cells (lanes 6 of fig. 6) do not lead to the formation of the 120 kDa band. These results are in agreement with a correlation between the high affinity binding sites, documented in fig. 4 and the 20 kDa protein expressed by this subclone of PC12 cells upon differentiation with NGF.

Effect of neuraminidase and proteases on the cross-linking of tetanus toxin to cells

TeTx binding to nerve cells is sensitive to the action of proteases and of neuraminidase.[17,28,41,42] We determined the effect of treatment of NGF-differentiated PC12 cells with neuraminidase, pronase or a mixture of trypsin and chymotrypsin on the amount of cross-linked product (Fig. 7).

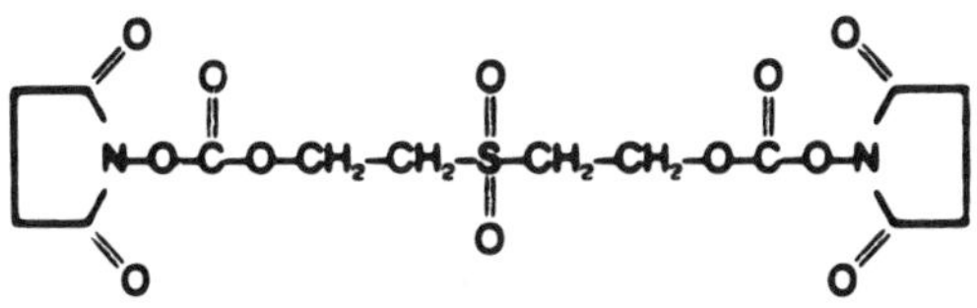

BSOCOES

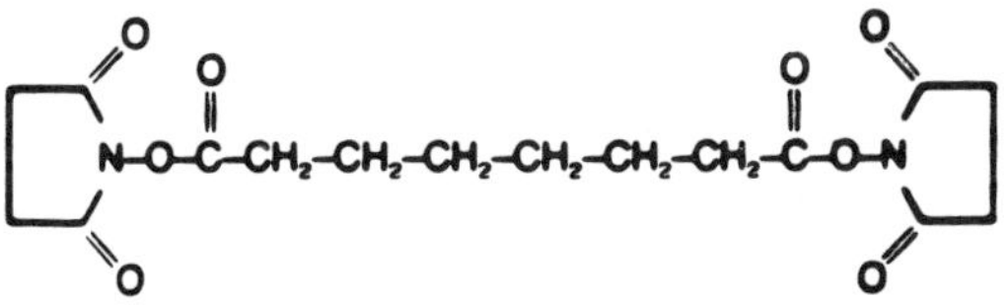

DSS

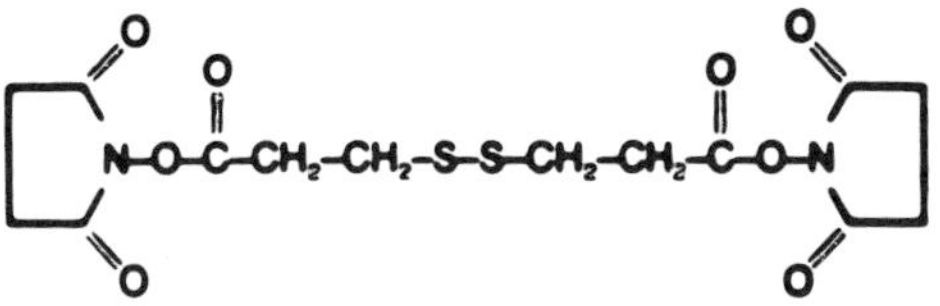

DSP

Figure 5. Structural formulae of the cross-linking agents used in this set of experiments: BSOCOES, *Bis*-2-(succinimidooxycarbonyloxy)ethyl sulfone; DSS, disuccinimidyl suberate; DSP, dithio*bis*(succinimidylpropionate).

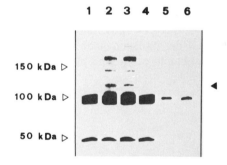

Figure 6. Cross-linking of [^{125}I]-TeTx to PC12 cells. Cells were cultured in the presence (lanes 1 to 5) or in the absence (lane 6) of NGF. Cells were incubated with [^{125}I]-TeTx; the sample of lane 5 contained an excess of cold toxin. After washing, cross-linkers were added: none in the sample of lane 1, DSS (lane 3), BSOCOES (lanes 2, 5, 6), DSP (lane 4). Proteins were electrophoresed under reducing conditions and autoradiographed. The two major radioactive bands represent the heavy (100 kDa) and the light chain (50 kDa) of TeTx. The black triangle indicates a radioactive protein band (120 kDa [^{125}I]TeTx-conjugate) which is formed uniquely in the presence of non-cleavable cross-linkers when the [^{125}I]-TeTx is incubated with the NGF-differentiated PC12 subclone.

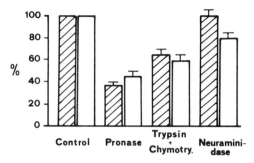

Figure 7. Effect of proteases and neuraminidase treatment on the amount of 120 kDa crosslinking product and on the total binding of TeTx. Hatched bars represent the percentage of 120 kDa [^{125}I]TeTx-conjugate and empty bars the percentage of [^{125}I]-TeTx bound to PC12 cells before and after pretreatment of the cells with pronase, a mixture of trypsin and chymotrypsin and neuraminidase.

Pronase was found to be more effective because it reduced the amount of 120 kDa band to 40% of the value found with untreated cells while the trypsin/chymotrypsin mixture reduced it to around 65%. Neuraminidase did not have any effect on the amount of cross-linked 120 kDa product. These results indicate that a protein is involved in the binding of TeTx to the NGF-differentiated PC12 subclone. Moreover they indicate that if the 20 kDa protein is a sialoglycoprotein, the sialyl moiety does not provide a major contribution to binding the toxin.

Kinetic of expression of tetanus toxin binding protein upon differentiation of PC12 cells with NGF

The morphological differentiation of PC12 cells induced by NGF follows kinetics, which are variable from subclone to subclone. Fig. 8 shows the progressive expression on the cell surface of the protein cross-linked to TeTx with the time of NGF differentiation. After three days of culture in the presence of NGF the formation of the 120 kDa cross-linked product begins to be detected. Its amount increases rapidly to reach a plateau value at eight days.

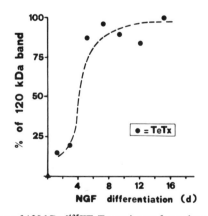

Figure 8. Amount of 120 kDa [^{125}I]TeTx-conjugate formed as a function of the time of exposure of PC12 cells to NGF.

The parallelism between the amounts of neurite outgrowth and of 120 kDa band suggests the possibility of an uneven distribution of the 20 kDa protein involved in TeTx binding on the cell surface with a larger presence on neurites and a lower distribution on the cell body.

CONCLUSIONS

The experiments described here provide evidence for the involvement of a protein component in the neurospecific binding of TeTx to nerve cells. This protein has an approximate molecular weight of 20 kDa and interacts with the H chain of the toxin. We also show that gangliosides, interdispersed among other lipids at a concentration mimicking that of the nerve plasma membrane, promoted the binding of tetanus toxin. Taken together these results are in favour of a double receptor binding of tetanus toxin to cells.[24,26] This model proposes that the productive cell binding, i.e. a binding followed by cell intoxication, is mediated by an interaction of the toxin with a protein component of the cell, responsible for the neurospecific binding, and an interaction with acidic lipids. This double interaction is at the origin of the high affinity binding of the toxin necessary to fix the minute amounts of tetanus toxin sufficient to intoxicate mammals. Such a double mode of binding accounts for all the available physiological and pharmacological data, as discussed *in extenso* elsewhere.[25,40]

Acknowledgements

The present work was supported by a grant from the CNR Target Project on Biotechnology and Bioinstrumentation and it is in partial fulfillment of the doctorate degree of the University of Padova in "Molecular and Cellular Biology and Pathology" of O.R.

REFERENCES

1. Ahnert-Hilger G, Weller U, Dauzenroth M-E, Habermann E, Gratzl M. The tetanus toxin light chain inhibits exocytosys. FEBS Lett 1989; 242:245.
2. Beaude P, Delacour A, Bizzini B, Domuado D, Remy M-H. Retrograde axonal transport of an exogenous enzyme covalently linked to B-IIb fragment of tetanus toxin. Biochem J 1990; 271:87.
3. Boquet P, Duflot E. Tetanus toxin fragment forms channel in lipid vesicles at low pH. Proc Natl Acad Sci USA 1982; 79:7614.
4. Critchley DR, Habig WH, Fishman PH. Revaluation on the role of gangliosides as receptors for tetanus toxin. J Neurochem 1986; 47:213.
5. Demel RA. Monolayers. Description of use and interaction. Methods Enzymol 1974; 32:539.
6. Dukovich M, Wano Y, Bich Thuy L, Katz P, Cullen BR, Kehrl JH, Greene WC. A second human interleukin-2 binding protein that may be a component of high-affinity interleukin-2 receptors. Nature 1987; 327:518.
7. Fishman PS, Savitt JM. Transynaptic transfer of retrogradely transported tetanus protein-peroxidase conjugates. Exp Neurol 1989; 106:197.
8. Greene LA, Aletta JM, Rukenstein A, Green SH. PC12 pheochromocytoma cells: culture, nerve growth factor treatment, and experimental exploitation. Methods Enzymol 1987; 147:207.
9. Habermann E, Dreyer F. Clostridial neurotoxins: handling and action at the cellular and molecular level. Curr Top Microbiol Immunol 1986; 129:93.
10. Howard AD, de La Baume S, Gioannini TL, Hiller JM, Simon EJ. Covalent labeling of opioid receptors with radioiodinated human β-endorphin. J Biol Chem 1985; 260:10833.
11. Hempstead BL, Martin-Zanca D, Kaplan DR, Parada LF, Chao MV. High affinity NGF binding requires coexpression of the trk proto-oncogene and low-affinity NGF receptor. Nature 1991; 350:678.
12. Ioannides CG, Itoh K, Fox FE, Pahwa R, Good RA, Platsoucas CD. Identification of a second T-cell antigen receptor in human and mouse by an anti-peptide gamma-chain-specific monoclonal antibody. Proc Natl Acad Sci USA 1987; 84:4244.
13. Jentoff N, Deaborn DG. Labeling of proteins by reductive methylation using sodium cyanoborohydride. J Biol Chem 1979; 254:4359.

14. Krieglstein KG, Henschen AH, Weller U, Habermann E. Limited proteolysis of tetanus toxin. Eur J Biochem 1991; 202:41.
15. Laemmli UK. Cleavage of structural proteins during the assembly of the head of bacteriophage T4. Nature 1970; 227:680.
16. Laburthe M, Breant B, Rouyer-Fessard C. Molecular identification of receptors for vasoactive intestinal peptide in rat intestinal epithelium by covalent cross-linking. Eur J Biochem 1984; 139:181.
17. Lazarovici P, Yavin E. Affinity purified tetanus neurotoxin interaction with synaptic membranes: properties of a protease-sensitive receptor component. Biochemistry 1986; 25:7047.
18. Lomneth R, Martin TFJ, DasGupta B. Botulinum neurotoxin light chain inhibits norepinephrine secretion in PC12 cells at an intracellular membranous or cytoskeletal site. J Neurochem 1991, 57:1413.
19. McInnes C, Dolly JO. Ca^{2+}-dependent noradrenaline release from permeabilised PC12 cells is blocked by botulinum neurotoxin A or its light chain. FEBS Lett 1990; 261:323.
20. Mellanby J. Comparative activities of tetanus and botulinum toxins. Neuroscience 1984; 11:29.
21. Mellanby J, Green. How do tetanus toxin acts? Neuroscience 1981; 6:281.
22. Middlebrook JL. Cell surface receptors for protein toxins. In Simpson LL, ed, Botulinum Neurotoxin and Tetanus Toxin. London: Academic Press, 1989.
23. Mobley WC, Schenker A, Shooter EM. Characterization and isolation of proteolytically modified nerve growth factor. Biochemistry 1976; 15:5533.
24. Montecucco C. How do tetanus and botulinum neurotoxins bind to neuronal membranes. Trends Biochem Sci 1986; 11:314.
25. Montecucco C. Theoretical considerations on the cellular mechanism of action of clostridial neurotoxins. In Nistico G, Bytchenko B, Bizzini B, Triau R, eds, Eighth International Conference on Tetanus. Rome: Pythagora Press, 1989.
26. Niemann H. Molecular biology of clostridial neurotoxins. In Alouf JE, Freer JH, eds, A Sourcebook of Bacterial Protein Toxins. London: Academic Press, 1991.
27. Park LS, Friend D, Gillis S, Urdal DL. Characterization of cell surface receptor for a multi-lineage colony-stimulating factor (CSF-2) J Biol Chem 1986; 261:205.
28. Pierce EJ, Davison MD, Parton RG, Habig WH, Critchley DR. Characterization of tetanus toxin binding to rat brain membranes. Biochem J 1986; 236:845.
29. Poulain B, Mochida M, Wadsworth JDF, Weller U, Habermann E, Dolly JO, Tauc L. Inhibition of neurotransmitter release by botulinum neurotoxins and tetanus toxin at Aplysia synapses: role of the constituent chains. J Physiol (Paris) 1990; 84:247.
30. Sandberg K, Berry CJ, Rogers TR. Studies on the intoxication pathway of tetanus toxin in the rat pheochromocytoma (PC12) cell line. J Biol Chem 1989; 264:5679.
31. Schiavo G, Papini E, Genna G, Montecucco C. An intact disulfide bridge is required for the neurotoxicity of tetanus toxin. Infect Immun 1990; 58:4136.
32. Schiavo G, Demel R, Montecucco C. On the role of polysialoglycosphyngolipids as tetanus toxin receptors. Eur J Biochem 1991; 199:705.
33. Schiavo G, Ferrari G, Rossetto O, Montecucco C. Tetanus toxin receptor. Specific cross-linking of tetanus toxin to a protein of NGF-differentiated PC12 cells. FEBS Lett 1991; 290:227.
34. Schiavo G, Poulain B, Rossetto O, Benfenati F, Tauc L, Montecucco C. Tetanus toxin is a zinc protein and its inhibition of neurotransmitter release and protease activity depend on zinc. EMBO J 1992; in press.
35. Simpson LL. Botulinum Neurotoxin and Tetanus Toxin. New York: Academic Press, 1989.
36. van Heyningen WE, Mellanby J. A note on specific fixation, specific deactivation and non-specific inactivation of bacterial toxins by gangliosides. Naunyn-Schmiedeberg's Arch Pharmacol 1973; 276:297.
37. Walton KM, Sandberg K, Rogers TB, Schnaar RL. Complex ganglioside expression and tetanus toxin binding by PC12 pheochromocytoma cells. J Biol Chem 1988; 263:2055.
38. Weller U, Mauler F, Habermann E. Tetanus toxin: biochemical and pharmacological comparison between its protoxin and some isotoxins obtained by limited proteolysis. Naunyn-Schmiedeberg's Arch Pharmacol 1988; 338:99.
39. Weller U, Dauzenroth ME, Meyer Zu Heringdorf U, Habermann E. Chains and fragments of tetanus toxin. Eur J Biochem 1989; 182:649.
40. Wellhoner HH. Tetanus and botulinum neurotoxins. In Herken H, Hucho F, eds. Handbook of Experimental Pharmacology, Vol 102. Berlin: Springer-Verlag, 1992,
41. Yavin E. Arch Biochem Biophys 1984; 230:129.
42. Yavin E, Nathan A. Tetanus toxin receptors on nerve cells contain a trypsin-sensitive components. Eur J Biochem 1986; 154:403.

MOLECULAR BASIS OF LOW pH-DEPENDENT MEMBRANE TRANSLOCATION OF BOTULINUM AND TETANUS NEUROTOXINS

Jennifer Doyle and Bal Ram Singh

Department of Chemistry
University of Massachusetts-Dartmouth
North Dartmouth, MA 02747

INTRODUCTION

Botulinum and tetanus neurotoxins, structurally related proteins (150 kDa), attack the nervous system very rapidly, once inside the body; the toxins are quite soluble at pH 7 and are readily transported throughout the body. However, these proteins are known to form membrane-spanning channels at one point in the pathway of their toxic activity.[1] A question is raised as to how a water soluble protein can integrate itself in a non-polar membrane bilayer. The protein must undergo some structural change, allowing it to assume enough hydrophobic or amphiphilic character to interact adequately with the membrane. Furthermore, this channel-forming activity is associated with the pH 4 environment of toxin-containing endosomes. The condition of pH 4 seems to be the incident in conferring this hydrophobic character upon the toxins.

Our objective is to describe in more detail the nature of the structural change and to determine if the change is indeed induced by the drop in pH. It has been previously shown from Fourier transform infrared studies[2] that there is no drastic change in secondary structure (as amount of α-helix, β-sheet, random coil) from pH 7 to pH 4. Therefore, we have turned our attention to hydrophobic moment and hydrophobicity calculations as predictors of surface seeking (amphiphilic) and transmembrane domains. These characteristics are indeed dependant upon pH. The goal is to identify the actual amino acid sequences within the toxins which show amphiphilic and hydrophobic character and to determine how and to what extent these characteristics differ from pH 7 to pH 4. This could lend insight into the molecular mode of translocation of botulinum and tetanus neurotoxin into the target cells.

HYDROPHOBICITY AND HYDROPHOBIC MOMENT

When the amino acid residues of α-helix are oriented such that the hydrophobic residues are on one side of the helix and hydrophilic residues are on the opposite side, a large hydrophobic moment results. Such helices will orient themselves parallel to an aqueous-apolar phase

Botulinum and Tetanus Neurotoxins, Edited by
B.R. DasGupta, Plenum Press, New York, 1993

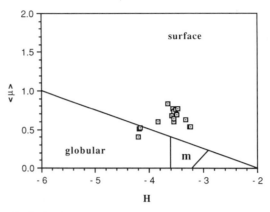

Figure 1. Hydrophobicity (H) versus hydrophobic moment plot of melittin. Because melittin does not have any Asp or Glu in its sequence, the plot remains same for pH 7 and pH 4. m represents the transmembrane region.

boundary or surface with the hydrophilic side facing the aqueous medium and the hydrophobic half facing the non-polar medium. Thus, helices which show large hydrophobic moments (μ) are predicted to be amphiphilic and seek membrane surfaces. Helices which have large average hydrophobicities (H), but small hydrophobic moments are predicted to be transmembrane peptides.

In a plot of $<\mu>$ vs. H, sequences of a given character (globular, surface-seeking, transmembrane) will tend to cluster in distinct regions of the graph.[3] One can visualize the general locations of these regions by generating such plots for α-helical peptides of known structure and hydrophobicity. Having established these criteria, lines of reference are drawn; points falling above the line are designated as surface-seeking, points in the lower left region are predicted to be globular, and points falling within the lower right triangle will be transmembrane (see Figure 1). Various proteins may then be analyzed for the quantity and magnitude of predicted amphiphilic and hydrophobic domains.[4]

METHODS

In order to calculate the average hydrophobic moment of short peptides within a protein, we have used a hydrophobicity scale based on a calculation of free energies of exchange of amino acids from a nonpolar solvent to water. The scale considers both the anionic and protonated (at low pH) forms of aspartic acid and glutamic acid.[5] Hydrophobic moment calculations were carried out according to Singh and Be,[4] assuming α-helical folding, and using a moving window of 11 amino acids. Thus, the hydrophobicities of the 5 amino acids to each side of a given amino acid are incorporated into the calculation. This technique has provided accurate predictions of membrane interacting sites of several proteins of known structure.[3,6]

The reference peptides used in this study were taken from melittin (surface-seeking), influenza HA$_2$ haemagglutinin (membrane), and diphtheria toxin (surface-seeking). Hydrophobic moment vs. hydrophobicity plots were constructed for these and, approximating Eisenberg's method, lines of reference were determined. The N-terminal halves of botulinum and tetanus heavy chains were then analyzed at pH 4 and at pH 7, because these segments are known to form membrane channels at low pH. Amino acid sequences were derived from Binz et al.[7] and from Eisel et al.[8]

RESULTS

Hydrophobic moment vs. hydrophobicity plots of the reference peptides using Abraham and Leo's scale[5] was consistent with Eisenberg's data from similar plots.[3] Using our data as a reference frame, lines of demarcation were drawn on the plots to define the amphiphilic (surface-seeking), hydrophilic (globular), and hydrophobic (transmembrane) regions. Figure 1 shows such data for surface-seeking helix of melittin.

The hydrophobicity vs. residue number plots for botulinum are similar in shape, but there is a definite shift up (more hydrophobic) in the plot from pH 7 to pH 4 (Figures 2 and 3). Results for tetanus show similar trends. This effect on hydrophobicity at low pH is mirrored in the $<\mu>$ vs. H plot. Both toxins showed the same trend of a shift to the right of all points on the $<\mu>$ vs. H plots, along with a general downshift of all points. More specifically, a distinct increase in the number of predicted transmembrane peptides is seen for both toxins at pH 4. However,

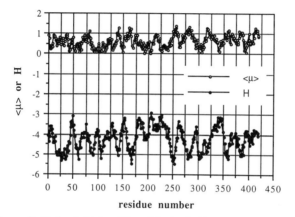

Figure 2. Hydrophobicity (H) and hydrophobic moment (μ) versus residue number (for actual residue number, one should add 457 to number shown on graph) of the N-terminal half of the type A botulinum neurotoxin. Hydrophobicity scale used for the anionic Asp and Glu (pH 7).

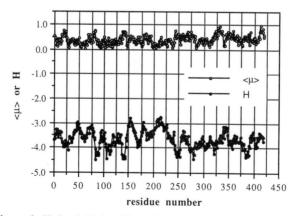

Figure 3. Hydrophobicity (H) and hydrophobic moment (μ) versus residue number (for actual residue number, one should add 457 to number shown on graph) of the N-terminal half of the type A botulinum neurotoxin. Hydrophobicity scale used for the protonated Asp and Glu (pH 4).

it appears that the number of potential surface-seeking regions remains relatively constant. Table 1 shows the specific residues that are predicted to be transmembrane for both toxins and at both pHs.

Table 1. Predicted transmembrane peptides of botulinum and tetanus at pH 7 and pH 4. The numbers refer to the residue numbers of the respective heavy chains of botulinum and tetanus neurotoxins. Amino acid sequences were taken from Binz et al.[7] and Eisel et al.[8]

Botulinum		Tetanus	
pH 7	pH 4	pH 7	pH 4
	451-470		
	485-523		465-480
	542-554		483-498
635-647	565-577	680-692	603-616
651-663	610-663		673-695
670-694	667-683		817-832
786-799	786-799		867-881

DISCUSSION

There is, indeed, a pH-induced effect on the calculated hydrophobicity and hydrophobic moment of the constitutive peptides of botulinum and tetanus proteins. Calculations at low pH predict an increase in average hydrophobicity, and a general decrease in amphiphilicity. These observations are consistent with the knowledge that these toxins form membrane-spanning channels at pH 4. Here, we have identified the specific peptides which are predicted to play a major role in channel forming activity.

This observed low pH-induced change in hydrophobic character is seen to be very similar in tetanus and botulinum neurotoxins. This is consistent with the common mode of action proposed for these neurotoxins.[1] Our results also indicate that a drastic conformational change (at least at the secondary level) is not necessary to explain the role of pH in the translocation of the neurotoxin into target cells. Though secondary structure may not be changing, it is quite possible that the altered amphiphilic/transmembrane characteristics of peptide segments would allow the toxins to assume this interesting and essential function.

Acknowledgements

This work was in part supported by a grant from UMass Dartmouth Foundation.

REFERENCES

1. Simpson LL. Molecular pharmacology of botulinum toxin and tetanus toxin. Annu Rev Pharmacol Toxicol 1986; 26:427-453.
2. Singh BR, Fuller MP, Schiavo G. Molecular structure of tetanus neurotoxin as revealed by Fourier transform infrared and circular dichroic spectroscopy. Biophys Chem 1990; 36:155-166.
3. Eisenberg D, Wilcox W, Eshita S. Hydrophobic moments as tool for the analysis of protein sequences and structures. In: L'Italian JJ, ed. Protein Structure and Function. New York: Plenum Press, 1987:425-436.

4. Singh BR, Be X. Use of sequence hydrophobic moment to analyze membrane interacting domains of botulinum, tetanus and other toxins. In: Angeletti RH, ed. Techniques. Protein Chem. Orlando, FL: Academic Press, 1992:373-383.

5. Abraham D, Leo A. Extension of the fragment method to calculate amino acid zwitterion and side chain partition coefficients. Proteins 1987; 2:130-152.

6. Eisenberg D, Weiss RM, Terwilinger TC. The helical hydrophobic moment: a measure of amphiphilicity of a helix. Nature 1982; 299:371-374.

7. Binz T, Kurazano H, Wille M, Frevert J, Wernars K, Neimann H. The complete sequence of botulinum neurotoxin type A and comparison with other clostridial neurotoxins. J Biol Chem 1990; 265:9153-9158.

8. Eisel U, Jarausch W, Goretzki K, Henschen A, Engels J, Weller U, et al. Tetanus toxin: primary structure, expression in E. coli, and homology with botulinum toxins. EMBO J 1986; 5:2495-2502.

MEMBRANE CAPACITANCE MEASUREMENT: RESTORATION OF CALCIUM-DEPENDENT EXOCYTOSIS BLOCKED BY BOTULINUM A NEUROTOXIN IN BOVINE CHROMAFFIN CELLS

T. Binscheck and H. Bigalke

Institute of Toxicology, Medical School
of Hanover, 3000 Hanover 61, Germany

INTRODUCTION

Botulinum A neurotoxin (BoNtx A) is a potent inhibitor of calcium-dependent exocytosis in bovine chromaffin cells[1]. The intracellular target or mechanism of action are not yet known. Exocytosis is characterized by the fusion of vesicles with the plasma membrane followed by the release of their contents into the extracellular space. The signal transducing pathway connecting receptor stimulation and vesicle fusion have not been elucidated in full detail. There is evidence, however, that the intracellular Ca^{++} concentration plays a key role[2]. Moreover, exocytosis can be stimulated in staphylococcal α-toxin permeabilised chromaffin cells by elevating the intracellular calcium concentration[1,3]. Though the intracellular increase of the Ca^{++} concentration occurred at a late stage within the signal transducing pathway, BoNtx A-treated chromaffin cells failed to secret noradrenaline. The block of exocytosis was resistant even to an enforced stimulation with $100\mu M$ Ca^{++}. In contrast to these findings, however, exocytosis in BoNtx A-treated cerebrocortical synaptosomes was restored by intracellular Ca^{++} in a concentration dependent manner, if there was Ca^{++} diffusion into the cells through Ca^{++}-ionophore-induced channels[4]. Similar results were obtained in poisoned motor end-plates. In this preparation the increase in the intracellular Ca^{++} concentration was achieved by potassium channel blockers[5,6]. To gain more information about the role of Ca^{++} in the action of BoNtx A we have taken a different approach. Chromaffin cells were electroporated in BoNtx A- containing culture medium so that the toxin was able to use the short-lived[7] membrane openings for diffusion into the cytosol[8,9]. Exocytosis was determined by measuring changes in membrane capacitance during perfusion of the cytosol with Ca^{++}[10,11]. The resulting inhibition of hormone release could be overcome by elevated Ca^{++} concentrations. However, to be effective the Ca^{++} concentration had to be increased at an early stage of intoxication.

METHODS

Electroporation

Chromaffin cells were purified by density gradient centrifugation (Percoll, Sigma), and they were electroporated with a BioRad Gene Pulser in the presence or absence of toxin. A suspension of 4×10^6 cells in 800µl Dulbecco's Modified Eagle Medium containing dithiothreitol (DTT, 2mM) was dispensed into a plastic cuvette which had two metal plates fixed to the inner surface on opposite sides. A preloaded capacitor (960 µF) was discharged with the cuvette serving as a resistor, yielding a transient electrical field of 625volts/cm in the cell suspension. The time constant of the capacitive current flowing through the cell suspension was approximately 12.5 ms. Then the cell suspension was diluted with an appropriate amount of growth medium and seeded on plastic dishes (2×10^6 cells/dish).

Membrane Capacitance Measurements

Cells were transferred into a solution consisting of 140mM NaCl, 5mM KCl, 2.5mM $CaCl_2$, 1mM $MgCl_2$, 20mM HEPES, and 5mM glucose (pH 7.35; 23.5°C). Patch pipettes were made from borosilicate glass in a two-stage pulling process yielding pipette resistances of approximately 3.3 MΩ. A constant concentration of unbound calcium ions was maintained in the pipette solution by using a buffer containing 140mM KCl, 12mM NaCl, 10mM HEPES, 1.8mM $MgCl_2$, 0.5mM ATP (disodium salt) and 2mM EGTA (pH 7.2). $CaCl_2$ was added in an appropriate amount to obtain definite Ca^{++} concentrations. The pipette solution was filtered (Millex, 0.2µm) before use. Experiments were performed on the stage of an inverted phase contrast microscope. The pipette was backfilled with pipette solution and connected to a patch clamp amplifier. After establishing the whole cell configuration cells were voltage clamped at a potential of -70 mV, while an 800 Hz sinusoidal wave of 10 mV amplitude was superimposed. Pipette current was fed into a two-phase lock-in amplifier for on line measurement of membrane capacitance. After proper setting of the phase angle, changes of C_m and R_a were monitored on a digital oscilloscope and stored for further analysis.

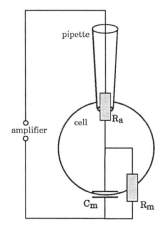

Figure 1. The whole cell configuration and its equivalent circuit. The basic electric elements of the whole cell configuration consist of the access resistance R_a, the membrane resistance R_m and the membrane capacitance C_m. The latter is proportional to the membrane area which increases by the fusion of exocytotic vesicles with the plasma membrane.

RESULTS AND DISCUSSION

The monitoring of exocytosis could begin as soon as the chromaffin cells were firmly attached to the culture plastic dishes, usually 2hr after electroporation. Whereas 100nM Ca^{++} failed to stimulate exocytosis (not shown), $1\mu M$ Ca^{++} was sufficient to reach saturation of membrane capacitance increase (Fig. 2a). Ca^{++} concentrations up to$100\mu M$ did not further increase the membrane capacitance (Fig. 2b). This concentration-dependency could be observed independent of the length of incubation period that followed electroporation. Exocytosis, however, evoked by $1\mu M$ Ca^{++} was totally suppressed after 2hr when electroporation was performed in the presence of 6.6nM BoNtx A (Fig. 2c). The block could be lifted by increasing the intracellular Ca^{++} concentration to $100\mu M$ (Fig. 2d). 24 hr after electroporation, however, even this high Ca^{++} concentration failed to increase the membrane capacitance, reflecting that the block of noradrenaline release by BoNtx A was well established.

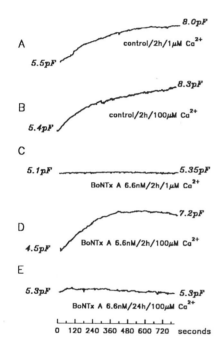

Figure 2. Digitised traces of membrane capacitance reflecting the actual membrane surface area. Chromaffin cells were electroporated in the presence of 6.6nM BoNtx A (C,D,E) or in its absence (A,B). DTT (2mM) was added to cleave the di-chain BoNtx A into its heavy and light chains, the latter being responsible for inhibition of exocytosis[12]. Cells were stimulated 2 hr (A-D) or 24 hr (E) after electroporation with $1\mu M$ Ca^{++} (A,C) or $100\mu M$ Ca^{++} (B,D,E), respt. The number at the beginning of each trace gives the capacitance of resting cells, and the number at the end gives the final capacitance when saturation was reached.

A common feature of cerebrocortical synaptosomes and isolated nerve-muscle preparations is a survival time of only a few hours. Therefore, stimulation of exocytosis had to be initiated shortly after the application of toxin. In contrast,when chromaffin cells were manipulated with low ionic strength buffer[13] a much longer period of time could be allowed to pass after toxin treatment, before stimulation of exocytosis was performed. During this prolonged period the toxin may have been able to damag its target to such a degree that even $100\mu M$ Ca^{++} was no longer capable of restoring exocytosis. We conclude that noradrenaline release, when blocked by BoNtx A, can be restored by elevated intracellular Ca^{++} only in the early stages of intoxication. This shows that BoNtx A interferes with a calcium-dependent process within the pathway of exocytosis in a time-dependent manner. Time-dependency is a common feature of enzymatic processes and underlines the hypothetical proteolytic activity of BoNtx A as reported by DasGupta on this conference.

REFERENCES

1 Penner, R., Neher, E., and Dreyer, F. (1986). Intracellularly injected tetanus toxin inhibits exocytosis in bovine adrenal chromaffin cells. *Nature* **324**, 76-78.

2 Knight, D. E., Grafenstein, v. H., and Athayde, C. M. (1989). Calcium-dependent and calcium-independent exocytosis. *TINS* **12**, 451-458.

3 Marxen, P., Bartels, F., Ahnert-Hilger, G., and Bigalke, H. (1991). Distinct targets for tetanus and botulinum A neurotoxins within the signal transducing pathway in chromaffin cells. *Naunyn-Schmiedeberg's Arch. Pharmacol.* **344**, 387-395.

4 Ashton, A. C., and Dolly, J. O. (1991). Microtubule-dissociating drugs and A23187 reveal differences in the inhibition of synaptosomal transmitter release by botulinum neurotoxins types A and B. *J. Neurochem.* **56**, 827-835.

5 Sellin ,L. C., Thesleff, S., and DasGupta, B.R. (1983). Different effects of types A and B botulinum toxin on transmitter release at the rat neuromuscular junction. *Acta Physiol. Scand.* **119**, 127-133.

6 Gansel, M., Penner, R., and Dreyer, F. (1987). Distinct sites of action of clostridial neurotoxins revealed by double-poisoning of mouse motor nerve terminals. *Pflügers Arch.* **409**, 533-539.

7 Bigalke, H., Bartels, F., Binscheck, T., Erdal, E., Erdmann, G., Weller, U., and Wever, A. (1992), Activation and inactivation of tetanus toxin in chromaffin cells. *This issue*.

8 Lambert, H., Pankow, R., Gauthier, J., and Hancock, R. (1990). Electroporation-mediated uptake of proteins into mammalian cells. *Biochem. Cell Biol.* **68**, 729-734.

9 Bartels, F., and Bigalke, H. (1992). Restoration of exocytosis occurs after inactivation of intracellular tetanus toxin. *Infect. and Immun.* **60**, 302-307.

9 Neher, E. (1988). The use of the patch clamp technique to study second messenger-mediated cellular events. *Neuroscience* **26**, 727-734.

10 Penner, R., and Neher, E. (1989). The patch-clamp technique in the study of secretion. *TINS* **12**, 159-163.

11 Lindau, M. (1991). Time-resolved capacitance measurements: monitoring exocytosis in single cells. *Quart. Rev. of Biophys.* **24**, 75-101.

12 Stecher,B., Gratzl, M., and Ahnert-Higer,G. (1989). Reductive chain separation of botulinum A toxin - a prerequisite to its inhibitory action on exocytosis in chromaffin cells. *FEBS Lett.* **248**, 23-27.

13 Marxen, P., and Bigalke,H. (1989). Tetanus toxin: inhibitory action in chromaffin cells is initiated by specified types of gangliosides and promoted low ionic strength solution. *Neurosci. Lett.* **107**, 261-266.

ACTIVATION AND INACTIVATION OF TETANUS TOXIN

IN CHROMAFFIN CELLS

H. Bigalke[1], F. Bartels[1], T. Binscheck[1], E. Erdal[1], G. Erdmann[1], U. Weller[2], and A. Wever[3]

[1]Institute of Toxicology, Medical School of Hannover, 3000 Hannover 61, Germany
[2]Institute of Microbiology, Johannes Gutenberg University, 6500 Mainz, Germany
[3]Institute of Nutrition Res., Nuclear Res. Center Karlsruhe, 7514 Eggenstein, Germany

INTRODUCTION

Tetanus toxin is produced by Clostridium tetani as a single chain, almost non-toxic, protein with a molecular weight of approximately 150.000 representing 1315 amino acids. Bacterial proteases cleave the molecule between positions A 457 and S 458 (extracellular activation), yielding a heavy chain (MW 100.000) and a light chain (MW 50.000) tetanus toxin (HC-TeTx, LC-TeTx). Both chains remain connected to each other by a disulphur bond between positions C 439 and C 467 (Dichain-TeTx)[1]. The cleavage or nicking dramatically increases the biological activity[2]. HC-TeTx is involved in binding DC-TeTx to gangliosides lodged in the plasma membrane, which is a prerequisite for incorporation into the cytosol[3]. Whether DC-TeTx diffuses into the cells through pores formed by H-CTetx[4] or is taken up by receptor-mediated endocytosis is not yet clear. Within the cells, the disulphur bond between the two chains is reduced to separate LC-TeTx from DC-TeTx (intracellular activation). Toxicity is carried exclusively by the reduced form of LC-Tetx[5]. The intracellular target of the toxin is present in several types of neuronal and endocrinal cells, such as cholinergic, noradrenergic, glycinergic, GABAergic, and other neurons[6] and also chromaffin cells[7]. Therefore they all respond to the toxin. However, since chromaffin cells lack gangliosides in their plasma membrane[8] they cannot take up the toxin unless special manipulations are performed. Nevertheless, chromaffin cells have firmly been established as useful tools for studying the interference of TeTx with exocytosis, because they represent a homogenous population and offer an excellent access to structures controling hormonal release. There are several ways to introduce TeTx into their cytosol. It can diffuse through plasma membrane pores induced by cytolysins[5], and it can be injected[7], or taken up by exogenous gangliosides incorporated into the plasma membrane[9]. Cytolysine-permeabilised chromaffin cells, in contrast to ganglioside-enriched or micropipette-porated cells, respond only to LC-TeTx or to chemically reduced DC-Tetx[5], indicating that, during their short survival time, these cells are not able to cleave the sulphur bonds.

Botulinum and Tetanus Neurotoxins, Edited by
B.R. DasGupta, Plenum Press, New York, 1993

Electroporation provides definite advantages over permeabilisation by means of cytolysins, because those cells that survive the actual procedure, continue to live for days[10]. Using this technique we were able to show that chromaffin cells have the enzymatic tools to activate DC-TeTx and that the toxin's effect on exocytosis is not irreversible. However, neutralisation or degradation of the toxin is a necessity for the recovery of exocytosis.

RESULTS AND DISCUSSION

Exocytosis from Permeabilised and Intact Chromaffin Cells

Permeabilisation had a damaging effect on chromaffin cells. Cytolysin formed permanent pores in the plasma membrane causing a continuous decrease of exocytotic activity (Fig. 1). Electroporation, on the other hand, produced transient pores with a diameter of up to 1μm in those areas of the plasma membrane that faced the electrodes of the poration cuvette during current flow (Fig. 2). Proteins, like tetanus toxin or antibodies, diffuse through these openings and remain trapped inside the cells when the pores are closed by membrane fusion. The closure allows a complete recovery from the damage as judged from the restoration of exocytosis which increased to the level found in control cells (Fig. 1).

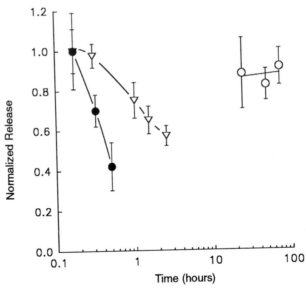

Figure 1. Exocytosis in permeabilised chomaffin cells. Chromaffin cells were loaded with ^{3}H-noradrenaline for two hours. After washing they were permeabilised with 20μM digitonin (closed circles) for 10 min or with 200U/ml of stahpylococcal α-toxin (open triangles) for 30 min. Then they were further incubated in plain medium as indicated (abscissa). The release of ^{3}H-noradrenaline was stimulated with 10μM Ca^{++} [11]. The release was compared with the release from a twin sample stimulated immediately after permeabilisation. Other sets of cells were electroporated[10] or left intact. After different periods of maintenance (abscissa) in monolayer culture ^{3}H-noradrenaline release was stimulated with carbachol. The release from electroporated cells was compared to the release from the non-permeabilised control cells (open circle). The normalized release of 1.0 means that the release from permeabilised cells equals that from intact cells.

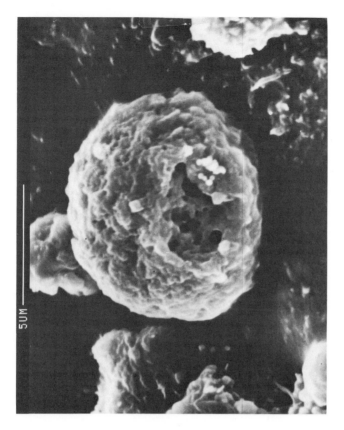

Figure 2. Scanning electron micrograph showing an electroporated chromaffin cell. Electroporation was performed by the capacitor discharge method[10] with a Bio-Rad Gene Pulser. The cell suspension (800μl) was transferred into a Gene-Pulser cuvette in which the poration was carried out at 20 °C. Permeabilisation was achieved by using a capacitance of 960μF and an electric field of 625 V/cm which decreased to 1/e in 10ms. The percentage of viable cells was in the range of 50%. The fixation solution (Karnovsky) was injected into the poration cuvette immediately after the collapse of the electric field.

Exocytosis from TeTx-Treated Cells

When cells were electroporated in the presence of TeTx they failed to restore hormone release. Exocytosis was blocked in a concentration-dependent fashion (Fig. 3) indicating that the toxin incorporated into the cytosol had been able to reach its target. Since chromaffin cells lack binding sites[8], the toxin could not have gained access to the cytosol by any other route than by diffusion through the electrogenerated pores. Furthermore, DC-TeTx must have been split into its constituent chains, because only the reduced LC-TeTx had been shown to be active[5]. Exocytosis was blocked only when the toxin was present during the discharge of the capacitor generating the pores. If dispensed into the poration cuvette only a few milliseconds after the collapse of the electric field the toxin had no effect, indicating that the pores had rapidly shrunk below the size necessary for the passage of the toxin (Fig. 4). The holotoxin was as effective as the purified LC-TeTx in inhibiting exocytosis (Fig.5). Thus cleavage was not the limiting step in the action of TeTx.

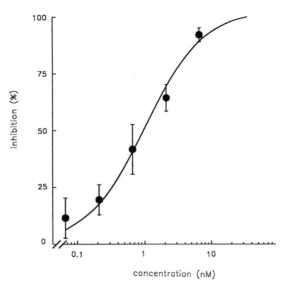

Figure 3. Concentration-dependent inhibition of exocytosis in TeTx-treated, electroporated chromaffin cells. Chromaffin cells were electroporated in the presence of the indicated concentrations of TeTx (abscissa). Exocytosis was induced by carbachol. The inhibition of exocytosis by the toxin (ordinate) is expressed as percentage of the release from electroporated control cells.

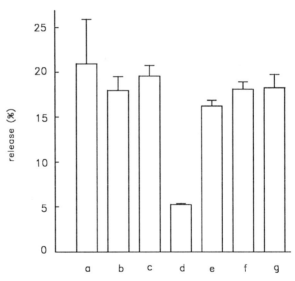

Figure 4. Duration of open-state of electrogenerated pores. Chromaffin cells were electroporated (c-g) in the presence of 6.6nM TeTx (d) or in its absence (c,e,f,g). The toxin was dispensed into the samples (e,f,g) 10ms (e), 1s (f), and 10s (g) after poration. A twin set of samples was left intact (a,b). All samples were seeded on culture dishes and TeTx was added to the medium of one set of intact cells (b). Exocytosis was evoked by carbachol and is expressed as percentage of the total content of ^{3}H-noradrenaline[11].

244

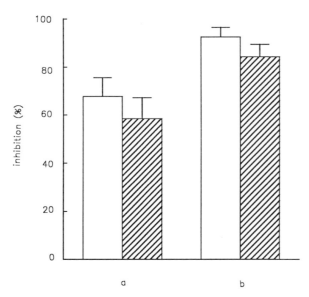

Figure 5. Inhibition of exocytosis by DC- and LC-TeTx. Chromaffin cells were electroporated in the presence of 6.6nM (a) and 66nM (b) DC-TeTx (open colums) and LC-TeTx (hatched colums). Then the cells were cultured, and exocytosis was determined after 48h. The inhibition of exocytosis (ordinate) is expressed as a percentage of exocytosis in electroporated control cells.

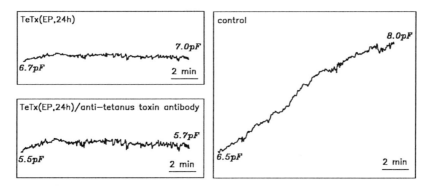

Figure 6. Capacitance measurement of TeTx-treated cells during injection of specific antibodies. Chromaffin cells were electroporated (EP) in the presence of 6.6nM TeTx. After 24h in culture Ca^{++}-evoked exocytosis was ascertained by measuring changes in cell capacitance. Single cells were voltage clamped at -70mV while an 800Hz sinusoidal voltage with a 10mV amplitude was superimposed. The resulting pipette current was measured by a patch clamp amplifier and fed into a two-phase lock-in amplifier allowing high resolution measurement of the electric membrane capacitance[13]. The fusion of vesicles with the plasma membrane increased the surface of the cell and resulted in an increase in membrane capacitance. Exocytosis was triggered by calcium diffusing out of the recording pipette into the cell. The final Ca^{++} concentration was 1µM. The cells were stimulated during the recording time of 60min. Only the first 15min are shown. The three traces show recordings of membrane capacitances from cells electroporated in the absence of TeTx (trace on the right) and in the presence of 6.6nM TeTx (traces on the left) during stimulation with 1µM Ca^{++}. The lower trace on the left represents a recording during infusion of the cell with 1U/ml of a specific anti-tetanus toxin antibody.

Neutralisation of Intracellular Tetanus Toxin

Specific anti-tetanus toxin antibodies prevent the entry of TeTx into neuronal cells by forming antibody-toxin complexes that can no longer bind to gangliosides. However, when specific antibodies were introduced into toxin-treated chromaffin cells by diffusion through chemically induced pores[12] or by injection through a micropipette (Fig. 6) the block of exocytosis could not be lifted. Similar to cytolysin-treated cells, pipette-perforated cells also allowed only short term observation of cellular functions. In both cases exocytosis was evoked by elevating the intracellular Ca^{++} concentration after the antibodies had been applied. In cytolysin-perforated cells exocytosis was stimulated by elevating the extracellular Ca^{++} concentration, and exocytosis could be quantified by measuring the release of ^3H-noradrenaline. In pipette-perforated cells, however, where a definite intracellular Ca^{++} concentration was achieved by direct intracellular injection, exocytosis was determined by measuring the increase in cell capacitance caused by the fusion of vesicles with the plasma membrane. In all cases, in which antibodies were unable to restore exocytosis, the reason for the failure may have been that the damage inflicted by the toxin on the target was irreversible. Therefore, for the effect to persist, the presence of toxin would not be required. Alternatively, antibodies may have bound to a region of the toxin molecule that was irrelevant to the intrinsic activity though important to the affinity.

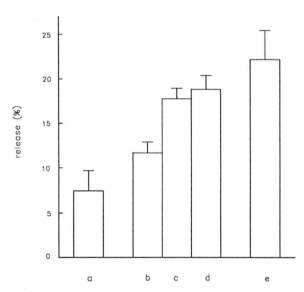

Figure 7. Concentration-dependent restoration of exocytosis by intracellular anti-tetanus toxin antibodies. Ganglioside-enriched chromaffin cells[8] were exposed to 66nM TeTx (a-d). After 24h they were washed twice with saline and dissociated with trypsin[10]. The cell suspensions were electroporated in the absence (a,e) or presence of 0.5 (b), 5 (c), and 50 (c) U/ml of anti-tetanus toxin antibodies and further maintained in culture. After 72h exocytosis was determined. The release of catecholamines from chromaffin cells is given as a percentage of total ^3H-noradrenaline content.

Since humans and animals infected with Clostridium tetani do recover, though very slowly, from tetanus, there should be a mechanism by which the neuronal target of the toxin can be repaired. This putative mechanism may also be active in chromaffin cells. To test this assumption chromaffin cells were electroporated in the presence of antibodies after exocytosis had been blocked by tetanus toxin incorporated via exogenous gangliosides. The hormone release was observed over a period of several days and it appeared that the release increased when the cells were treated with the antibodies. This indicated that the antibodies had succeeded in reaching the same compartment as the toxin. They were able to neutralise the toxin, though the two proteins had passed through the plasma membrane by different routes. The restoration of catecholamine release depended on the concentration of the specific antibodies (Fig. 7). However, a complete restoration could not be achieved over a one-day period, even with a tenfold increase in the concentration of the antibodies. It took 3 to 4 days before a full reconstitution of the hormone release was observed (Fig. 8). Toxin-treated chromaffin cells electroporated in the absence of antibodies recovered more slowly. Exocytosis was only slightly increased three weeks later, and even after five weeks the hormone release was still partially blocked (Fig. 8). The continued presence of LC-TeTx is a prerequisite for maintaining the block of exocytosis. Since the hormone release did not recover after neutralisation of tetanus toxin with injected antibodies and a complete recovery could only be achieved after several days, it can be concluded that the toxin irreversibly altered a constituent of the exocytotic machinery that had to be reconstituted before exocytosis was restored.

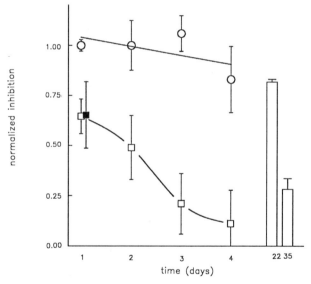

Figure 8. Time-dependent restoration of exocytosis by intracellular anti-tetanus toxin antibodies. Ganglioside-enriched chromaffin cells[8] were exposed to 66nM TeTx. After 24h they were washed twice and dissociated with trypsin[10]. The cell suspensions were exposed to the electric field in the absence (circles and columns) or presence of either 45U/ml (open squares) or 450U/ml (closed square) of anti-tetanus toxin antibodies. They were further maintained in culture and exocytosis was determined after different periods of time (abscissa). Inhibition of exocytosis, expressed as a percentage of the release from toxin-untreated cells, was in the range of 70%. The inhibition measured 24h after electroporation was normalised to 1.0, and the other values were expressed as a fraction of it.

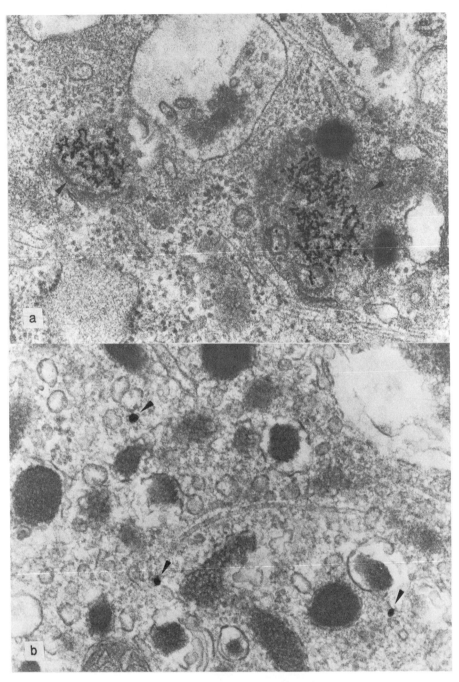

Figure 9a,b. Transmission electron micrographs showing intracellular distribution of gold-labelled TeTx. Chromaffin cells were electroporated in the presence of TeTx bound to 10nm gold particles. 24h later the cells were fixed in glutardialdehyde and processed for electron microscopy. Due to silver enhancement, gold particles in (b) are greatly enlarged over particles in (a).

Metabolism of Tetanus Toxin in Chromaffin Cells

At least two metabolic reactions of DC-TeTx have to occur to explain our findings. First, DC-TeTx has to be reduced to gain intrinsic activity, and second, LC-TeTx must be inactivated to allow recovery of exocytosis. To investigate the metabolism and the distribution of TeTx, chromaffin cells were electroporated in the presence of either ^{125}I-labelled or gold-labelled TeTx. After different periods of time the radioactivity was estimated and analysed by SDS-gel electrophoresis. As the outer cell membrane lacks binding sites for the toxin the radioactivity should be located within the cell. From the very beginning LC-TeTx could be detected, though only a small portion. The reductive cleavage of DC-TeTx may be due to thioredoxin-reductase, because this enzyme is present in chromaffin cells (Habermann, personal communication) and also in neuronal cells[14]. In vitro, at least, the enzyme is capable of reducing the toxin thus yielding disconnected LC-TeTx and HC-TeTx. The decrease in ^{125}I-TeTx followed first order kinetics. The biological half life was approximately 3 days. On the basis of the cell number, which was constant during the experimental period of seven days, and the specific radioactivity of the toxin the number of toxin molecules present in a single cell could be estimated. Taken into account that only a small portion of the total radioactivity represented LC-TeTx, it appeared that less than 100 molecules of LC-TeTx are sufficient to maintain the block of exocytosis. Recovery occurred when the toxin was split into smaller fragments. By far most of the internalised gold-labelled DC-TeTx was distributed in autophagosomes (Fig. 9a). Any toxin located in these structures can be considered inactive. Since only a small amount of the label is freely distributed within the cytosol (Fig. 9b) and may represent the toxin that has access to its putative target, we conclude that even fewer molecules than calculated from the internalised radioactivity were necessary to maintain the block of exocytosis. The target may be a cytosolic structure or molecule that cannot be identified by electron microscopy and that is inactivated by traces of cytosolic LC-TeTx, possibly due to the proteolytic activity of LC-TeTx[15]. After the physiological degradation of LC-TeTx or its neutralisation by antibodies this target may have to be resynthetised before exocytosis can be restored.

ACKNOWLEDGMENT

We are very much indebted to Prof. E. Reale for providing access to the transmission electron microscope of the Dept. of Cell Biology and Electron Microscopy, Medical School of Hannover. We also wish to thank Ms. G. Antrag, Ms. U. Fuhrmann, and Ms. I. Klose for their excellent technical assistance. This work was supported by the Deutsche Forschungsgemeinschaft.

REFERENCES

1. Eisel, U., Jarausch, W., Goretzki, K., Henschen, A., Engels, J., Weller, U., Hudel, M., Habermann, E., and Niemann, H. (1986). Tetanus toxin: Primary structure, expression in E. coli, and homology with botulinum toxins. EMBO J. **5**, 2495-2502
2. Bergey, G., Habig, W., Bennett, J., and Lin, C. (1989). Proteolytic cleavage of tetanus toxin increases activity. J. Neurochem. **53**, 155-161

3. Morris, N.P., Consiglio, E., Kohn, L., Habig, W., and Hardegree, E. (1980). Interaction of fragment-B and fragment-C of tetanus toxin with neural and thyroid membranes and with gangliosides. J. Biol. Chem. **225,** 6071-6076

4. Högy, B., Dauzenroth, M.E., Hudel., Weller, U., and Habermann, E.(1992). Increase of permeability of synaptosomes and liposomes by the heavy chain of tetanus toxin. Toxicon **30,** 63-76

5. Ahnert-Hilger, G., Weller, U., Dautzenroth, M.E., Habermann, E., and Gratz, M. (1989). The tetanus toxin light chain inhibits exocytosis. FEBS Lett. **242,** 245-248

6. Bigalke, H., Heller, I., Bizzini, B., and Habermann, E. (1981). Tetanus toxin and botulinum A toxin inhibit release and uptake of various transmitters as studied with particulate preparations   from rat brain and spinal cord. Naunyn-Schmiedeberg´s Archiv of Pharmacology **316,** 244-251

7. Penner, R., Neher, E., and Dreyer, F., (1986). Intracellular injected tetanus toxin inhibits exocytosis in bovine adrenal chromaffin cells. Nature **324,** 76-78

8. Marxen, P., Fuhrmann, U., and Bigalke, H. (1989). Gangliosides mediate inhibitory effects of tetanus and botulinum A neurotoxins on exocytosis in chromaffin cells. Toxicon **27,** 849-859

9. Marxen, P., and Bigalke, H. (1989). Tetanus toxin: Inhibitory action in chromaffin cells is initiated by specified types of gangliosides and promoted in low ionic strength solution. Neurosci. Lett. **107,** 33-36

10. Bartels, F., and Bigalke, H. (1992) Restoration of exocytosis occurs after inactivation of intracellular tetanus toxin. Infection and Immunity, **60,** 302-307

11. Marxen, P., Bartels, F., Ahnert-Hilger, G., and Bigalke, H. (1991). Distinct targets for tetanus and botulinum A neurotoxins within the signal transducing pathway in chromaffin cells. Naunyn-Schmiedeberg´s Archiv of Pharmacology **344,** 387-395

12. Marxen, P., Ahnert-Hilger, G., Wellhöner, H.H., and Bigalke, H. (1990). Tetanus antitoxin binds intracellular tetanus toxin in permeabilised chromaffin cells without restoring Ca^{++}-induced exocytosis. Toxicon **28,** 1077-1082

13. Lindau, M., and Neher, E. (1988). Patch-clamp techniques for time-resolved capacitance measurements in single cells. Pflüger´s Arch. **411,** 137-146

14. Kistner, A., and Habermann, E. (1992). Reductive cleavage of tetanus toxin and botulinum A neurotoxin by the thioredoxin system from brain. Naunyn-Schmiedeberg´s Archiv of Pharmacology **345,** 227-234

15. Habermann, E., and Sanders, D. (1992). Evidence for a link between proteolysis and the inhibition of ^{3}H-noradrenaline release by the light chain of tetanus toxin. (this issue)

TETANUS TOXIN AND PROTEIN KINASE C

Robert V. Considine[1], Lance L. Simpson[2], and Joseph R. Sherwin[1]

[1]Department of Physiology
[2]Departments of Medicine and Pharmacology
Jefferson Medical College
Philadelphia, PA 19107

INTRODUCTION

The pathogenesis of tetanus toxin poisoning is characterized by an inhibition of neurotransmitter release in the central nervous system; however, the exact mechanism through which the toxin inhibits exocytosis is not yet understood. A suitable neuronal model in which to study the biochemical effects of tetanus toxin on secretory processes is the differentiated neuroblastoma x glioma hybrid cell line NG-108.[1] We have recently reported that tetanus toxin pretreatment of differentiated NG-108 cells attenuated the ability of phorbol myristate acetate (PMA) and neurotensin to mobilize cytosolic protein kinase C (PKC) to the plasma membrane.[2,3] One implication of these observations is that the ability of tetanus toxin to alter PKC metabolism is related to its ability to inhibit exocytosis. In this study we have used a permeabilized cell/synthetic peptide assay to directly measure membrane PKC activity following tetanus toxin pretreatment. Further, we have used this assay to determine the efficacy of four putative inhibitors of membrane PKC activity; staurosporine and H-7 inhibit the catalytic site of the kinase, calphostin and sphingosine block the regulatory site. Finally, experiments were performed on neuromuscular preparations to test the hypotheses that PKC is required for short-term neuromuscular transmission and that the kinase could be the direct target of tetanus toxin.

MATERIALS AND METHODS

Cell and Neuromuscular Preparations

The NG-108 and N1E-115 cell lines were maintained in DMEM with 10% fetal bovine serum, 100 μM hypoxanthine, 16 μM thymidine, and 1 μM aminopterin. NG-108 cells were differentiated by reducing the serum content of the medium to 5% and adding 1 mM dibutyryl cyclic AMP. Cells were incubated in differentiation medium 6-10 days prior to use in experiments.

Tissues were excised and suspended in physiological buffer[4] at 35 C and bubbled with 95%O_2-5%CO_2. Phrenic nerves were stimulated continuously (0.2 Hz square waves of 0.1-0.3 msec duration) and muscle twitch recorded. Paralysis was measured as a 90% reduction in muscle twitch response to nerve stimulation

Synthetic Peptide/Permeabilized Cell Assay

The KRTLRR peptide sequence is based on the PKC phosphorylation site of the EGF receptor and has been demonstrated to be a specific PKC substrate.[5] For the

Botulinum and Tetanus Neurotoxins, Edited by
B.R. DasGupta, Plenum Press, New York, 1993

251

determination of membrane PKC activity cells were pretreated with tetanus toxin for 2 h followed by addition of a concentrated solution of PMA or neurotensin directly into the toxin containing pretreatment medium. Thereafter, the medium was removed and replaced with 200 µl of salt solution containing 137 mM NaCl, 5.4 mM KCl, 0.3 mM NaH_2PO_4, 0.4 mM KH_2PO_4, 5.6 mM dextrose, 20 mM HEPES (pH 7.2), 80 µM digitonin, 10 mM $MgCl_2$, 25 mM ß-glycerophosphate, 100 µM ^{32}P-ATP (1000 cpm/pmol), 5 mM EGTA, 2.5 mM $CaCl_2$, and 600 µM KRTLRR peptide. Following a 10 min incubation at 30 C the reaction was terminated by addition of 25% trichloroacetic acid (40 µl) and the phosphorylated peptide was blotted onto P81 phosphocellulose paper. The P81 paper squares were washed in 75 mM H_3PO_4 (3x) and 75 mM NaH_2PO_4, pH 7.5 (1x). Background phosphorylation, defined as phosphorylation of substrates other than KRTLRR that also bind to the filter paper, was subtracted in each experiment.

RESULTS

Membrane PKC Activity

Basal membrane PKC activity in NG-108 cells was 7.44±1.76 pmol/mg/10 min. A 15 min exposure to 0.1 µM PMA increased the membrane PKC activity to 13.80±2.52 pmol/mg/10 min, a statistically significant increase in activity of 6.36±1.23 pmol/mg/10 min. Pretreatment of NG-108 cells for 2 h with 10^{-10}M tetanus toxin did not alter basal membrane PKC activity; however, toxin pretreatment significantly attenuated the PMA-mediated increase in membrane PKC activity (4.37±1.10 pmol/mg/10 min; p<0.05, Student's paired t-test). A 5 min exposure to 10 µM neurotensin increased membrane PKC activity from 6.40±1.40 to 9.65±2.37 pmol/mg/10 min, a significant 1.4 fold increase. Pretreatment with 10^{-10}M tetanus toxin for 2 h completely abolished the neurotensin-induced increase in PKC activity.

As shown in Table 1, a 1 h exposure to the inhibitors of the regulatory domain of PKC (calphostin or sphingosine) did not alter the basal membrane PKC activity in either NG-108 or N1E-115 neuroblastoma cells. However, at these same concentrations, calphostin and sphingosine did significantly attenuate the ability of PMA to increase

Table 1. Inhibition of membrane PKC activity in neuroblastoma cells

| | NG-108 cells Membrane PKC Activity (pmol/mg/10 min) | | N1E-115 cells Membrane PKC Activity (pmol/mg/10 min) | |
Group	- PMA	+ PMA	- PMA	+ PMA
Control	11.9 ±2.1	22.6* ±4.1	11.1 ±2.2	21.5* ±1.9
Calphostin 2 µM; 60 min	12.3 ±2.4	14.1 ±0.6	8.3 ±2.0	15.1 ±1.7
Control	12.0 ±1.8	21.2* ±3.19	14.6 ±2.5	27.0* ±2.6
Sphingosine 1 µM; 60 min	9.0 ±1.7	14.2 ±2.9	15.4 ±2.1	19.5# ±4.2
Staurosporine 0.2 µM; 60 min	N.D	N.D.	N.D.	N.D.
H-7 1 mM; 60 min	N.D.	N.D.	N.D.	N.D.

Values represent the mean±SEM of the following number of experiments for the NG-108 and N1E-115 cells respectively: calphostin 4 and 5; sphingosine 5 and 6. N.D. indicates the activity was not detectable

* p<0.025; Student's paired t-test comparison to cells not exposed to PMA
# p<0.05; Student's paired t-test comparison to cells exposed to sphingosine

membrane PKC activity. Staurosporine and H-7, inhibitors of the catalytic domain of the kinase, abolished both basal and PMA-stimulated membrane PKC activity in both cell lines.

PKC and the Neuromuscular Junction

Phrenic nerve-hemidiaphragm preparations were exposed to PMA at concentrations that stimulated enzyme activity in neuronal cells in culture, (0.1 and 1.0 μM) and the twitch response monitored for 300 min. PMA did not augment normal transmission, nor did it restore transmission to preparations bathed in low calcium (0.4-0.6 mM) medium.

The effects of PKC inhibitors on neuromuscular transmission are illustrated in Table 2. With all four drugs there was a marked dissociation between the concentrations that inhibited enzyme activity in cultured cells and the concentrations that blocked tissue responses in phrenic nerve-hemidiaphragm preparations. The IC100 for H-7 induced blockade of basal enzyme activity was approximately 1 mM, but this concentration of drug had no effect on short-term neuromuscular transmission. Increasing the H-7 concentration two-fold had only a negligible effect. Similarly, the IC100 for staurosporine in neuronal cells was 0.2 μM, but it was necessary to add a 10-fold molar excess to obtain a notable effect on neuromuscular transmission. The IC80 to IC90 dose of both calphostin and sphingosine had only a negligible effect on short-term neuromuscular transmission.

Table 2. Effect of PKC inhibitors on neuromuscular preparations

Inhibitor	Response (% of control)			
	30 min	60 min	90 min	120 min
H-7				
1mM	100	100	100	100
2 mM	100	100	85	70
Staurosporine				
0.4 μM	100	100	100	100
1.0 μM	100	100	95	90
4.0 μM	100	75	40	35
Calphostin				
1 mM	100	100	100	100
4 mM	100	90	75	60
16 mM	80	50	40	35
Sphingosine				
10 μM	100	100	100	100
50 μM	100	100	85	80
200 μM	80	50	30	20

Neuromuscular preparations (n=5 per group) were treated with PKC inhibitors and transmission monitored for 120 min.

The interaction between drugs that modify PKC activity and several clostridial toxins was also assessed. Tissues were pretreated with PMA (0.1 μM), H-7 (1 mM), staurosporine (0.2 μM), calphostin (2 μM) or sphingosine (1 μM) for 100 min. The tissues were then exposed to botulinum neurotoxin type A ($1 \times 10^{-11} M$), botulinum neurotoxin type B ($3 \times 10^{-10} M$), or tetanus toxin ($3 \times 10^{-9} M$) and the paralysis times determined. The paralysis times of the control tissues were: neurotoxin type A, 136±17 min; neurotoxin type B, 118±12 min; and tetanus toxin, 129±14 min. Pretreatment of the neuromuscular preparations with PMA or PKC inhibitors did not affect the toxin-induced paralysis of neuromuscular transmission.

DISCUSSION

It has been found that tetanus toxin blocks acetylcholine release from NG-108 cells[1] and that the toxin also blocks agonist-stimulated levels of PKC activity in these same cells.[2,3] These findings could mean that tetanus toxin exerts its effects on exocytosis via PKC. Although several authors have found that stimulation of PKC activity is associated with an increase in frequency of miniature endplate potentials and an increase in quantum content,[6,7,8] there are no published reports showing that these effects lead to increases in duration or strength of muscle twitch. Further, no one has studied the effect of inhibitors of PKC activity on neuromuscular transmission. Therefore, a determination of the role of PKC activity in transmitter release would contribute to the understanding of normal neuromuscular physiology as well as resolve whether the kinase is a direct target for tetanus toxin.

In the present study, we were unable to demonstrate an effect of PMA on short-term neuromuscular transmission. This observation suggests that the magnitude of effect that phorbol esters have on transmitter release is not adequate to effect a macroscopic response such as muscle twitch. Further, at concentrations of PKC inhibitors considerably higher than predicted to block enzyme activity, there was never complete blockade of transmission. Taken together, these results argue that PKC is not essential to maintain short-term neuromuscular transmission. Pretreatment of neuromuscular preparations with PMA or PKC inhibitors did not alter the amount of time necessary for botulinum neurotoxin type A, type B or tetanus toxin to paralyze transmission. This observation argues against the possibility that the clostridial toxins exert their effects on exocytosis through PKC.

In summary, although PKC metabolism is altered in NG-108 cells following exposure to tetanus toxin, PKC is not necessary for short-term neuromuscular transmission and therefore cannot be the direct target through which tetanus toxin blocks exocytosis.

REFERENCES

1. Wellhoner HH, Neville DM. Tetanus toxin binds with high affinity to neuroblastoma x glioma hybrid cells NG 108-15 and impairs their stimulated acetylcholine release. J Biol Chem 1987; 262:17374-17378.
2. Considine RV, Bielicki JK, Simpson LL, Sherwin JR. Tetanus toxin attenuates the ability of phorbol myristate acetate to mobilize cytosolic protein kinase C in NG-108 cells. Toxicon 1990; 28:13-19.
3. Considine RV, Handler CM, Simpson LL, Sherwin JR. Tetanus toxin inhibits neurotensin-induced protein kinase C activity in NG-108 cells. Toxicon 1991; 29:1351-1357.
4. Simpson LL. Kinetic studies on the interaction between botulinum toxin type A and the cholinergic neuromuscular junction. J Pharm Expt Therap 1980; 212:16-21.
5. Heasley LE, Johnson GL. Regulation of protein kinase C by nerve growth factor, epidermal growth factor, and phorbol esters in PC12 pheochromocytoma cells. J Biol Chem 1989; 264:8646-8652.
6. Van Der Kloot W. Down-regulation of quantal size at frog neuromuscular junctions: Possible roles for elevated intracellular calcium and for protein kinase C. J Neurobiol 1991; 22:204-214.
7. Shapira R, Silberberg SD, Ginsburg S, Rahamimoff R. Activation of protein kinase C augments evoked transmitter release. Nature 1987; 325:58-60.
8. Murphy RLW, Smith ME. Effects of diacylglycerol and phorbol ester on acetylcholine release and action at the neuromuscular junction in mice. Brit J Pharmacol 1987; 90:327-334.

TETANUS TOXIN BIOTINYLATION AND LOCALIZATION OF BINDING SITES IN CATECHOLAMINERGIC CULTURES AND GRANULES

Philip Lazarovici,[1] Alexander Fedinec[2] and Bernard Bizzini[3]

[1]Department of Pharmacology and Experimental Therapeutics,
 School of Pharmacy, Faculty of Medicine, The Hebrew University
 of Jerusalem, Jerusalem 91120, Israel
[2]Department of Anatomy and Neurobiology, University of Tennessee,
 Memphis, Tennessee 38163, USA
[3]Department of Molecular Toxinology, Institut Pasteur, Paris, France

INTRODUCTION

Tetanus toxin (TT), the 150-kD protein secreted by *Clostridium tetani* bacteria, is a very powerful neurotoxin which produces spastic paralysis and death in humans in the range of ng/kg body weight.[1] The pharmacokinetics of TT entry to the central nervous system target[2] include the following main steps: (1) selective binding of TT on the surface of the peripheral motor neuron terminals;[3] (2) TT internalization into a vesicular compartment which delivers TT by retrograde axonal transport to the spinal cord perikarya and dendrites;[4] (3) TT release from the motor neuron in the spinal cord;[5] (4) trans-synaptic transfer of TT and uptake into the presynaptic terminals of the inhibitory interneurons.[5] (5) Thereafter, the constitutive release of gamma amino butyric acid and/or glycine is blocked, resulting with a syndrome of motor neuron dysinhibition;[6] (6) the results are muscular spastic contractions, spasms, and convulsive seizures leading to musculature rigidity, respiratory failure, and death.[7] It is believed that the binding of TT to neuronal G1b polysialogangliosides is involved in these pharmacokinetic steps and is the initial step triggering axonal transport and inhibition of neurotransmitter release.[8-9] For many years this TT-specific recognition has been used as a diagnostic and basic research tool for marking central and peripheral neurons,[10] as well as neurosecretory cells such as thyroid C, anterior pituitary, pancreatic islet cells, and derived tumors (insulinoma, pheochromocytoma, neuroblastoma).[11-13] All these cells express on their surface G1b polysialogangliosides of different sugar compositions, therefore binding TT.

Although these studies have provided insight into the complex intoxication process and provided a selective probe for neuronal cells, the identification of a cell line of a neural origin which is sensitive to TT would provide a cellular model to dissect the subcellular molecular steps of TT action. Adrenal medullary chromaffin cells, originating from the neural crest, are believed to be modified, postganglionic, sympathetic neurons. Because they synthesize and release catecholamines, these cells have served as a model system over the past two decades in studying the mechanisms(s) of neurotransmitter release.[14] Chromaffin cells, however, do not bind TT,[15] and the Ca^{2+}-dependent, stimulus secretion coupling of catecholamines is not affected by externally applied TT.[16] The pioneering experiments of Penner et al. demonstrating inhibition of catecholamine exocytosis in bovine adrenal chromaffin cells upon intracellular injection of TT,[17] have firmly established the now widely accepted concept that the chromaffin cells, although unable to bind and/or internalize TT, possess an intracellular target for TT action.[15,18] The important discovery that the differentiated pheochromocytoma (the tumor counterpart of chromaffin cells) PC12 cells were very sensitive to TT[19] upon cell-surface appearance of G1b polysialogangliosides receptors[20] provided for the first time a suitable cell line model to examine the molecular pathways by which TT blocks catecholamine release. We have exploited this system to localize and characterize TT binding sites on pheochromocytoma plasma membranes after differentiation with nerve growth factor, and we found for the first time their presence on the chromaffin granules, which represent TT intracellular target. For this purpose, we have prepared and characterized a biotinylated, functional derivative of TT suitable for modern molecular and cytology studies.

Biotinylation of Tetanus Toxin

Homogeneous preparations of TT (Fig. 1A, TT shadowed peak) were biotinylated with biotinyl-N-hydroxy-succinimidyl ester using a described method.[20] After dialysis and separation of unreacted reagent, the derivatized TT fractions were isolated by chromatofocusing (Fig. 1B, Biot-TT$_a$ shadowed fraction) and finally purified by ganglioside-affinity chromatography (Fig. 1C). When TT was biotinylated at a biotin/toxin molar ratio of 4:1, the final purified derivative (Biot-TT$_b$, Fig. 1C) was ~100-fold less toxic than the native toxin, with ganglioside binding affinity of 60% compared to the original toxin.[20]

The presence of ^{125}I-label on the derivatized toxin enabled us to assess the purity of the preparation after biotinylation. As shown in Fig. 2A (lane 2), the SDS-PAGE pattern of the biotinylated derivative (under nonreducing conditions) shows homogeneity and similar electrophoretic mobility as the untreated toxin (lane 1). The biotinylated toxin exhibits a high affinity of binding to avidine. After 15-min incubation in the presence of avidine (Mr=68,000), the molecular weight of the complex is shifted to a value of about Mr=210,000 (lane 3) as expected. A small fraction of biotinylated toxin radioactivity eluted at the front (asterisk, lane 3) may indicate some fragmentation presumably produced during the biotin coupling reaction. To verify the covalent coupling of biotin to the TT derivative, it was submitted for analysis on avidine-sepharose chromatography (Fig. 2B). As expected, native toxin did not bind to the column matrix (Fig. 2B, TT). Biotinylated (Biot-TT$_b$) toxin, however, bound strongly to the matrix poorly eluted by Tris-acetate buffer (TAB, first peak, Fig. 2B). When stringent conditions were applied (i.e., 6M guanidine and 70% formic acid, arrows, Fig. 2B), two additional peaks (shadowed, ~5%) were eluted. However, >90% of the protein, toxicity, and binding activity remained associated to the column matrix, indicative of a covalent reaction between the biotin group and TT which were not broken by the very high affinity of avidine binding.[21]

Immunological properties of biotinylated toxin were estimated by both immunoaffinity column chromatography and immunoprecipitation techniques, using a monoclonal mouse antitoxin antibody.[22] The majority of the native toxin or biotinylated toxin could not be retained by the antibody affinity gel (Fig. 2C). After elution with formic acid, there was a greater proportion (30%) of native toxin eluted off the gel. In contrast, only 10% of the biotinylated toxin was eluted, suggesting a relative loss in the immunoaffinity properties of the TT. This is more clearly evident by comparing the immunoprecipitation properties of native and biotinylated TT (Fig. 2C insert).

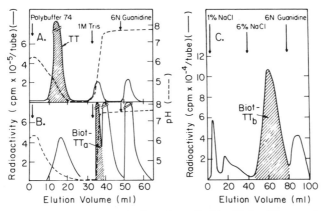

Figure 1. Chromatofocusing (A, B) and ganglioside-affinity (C) separation of biotinyl derivatives of tetanus toxin. (A) Native ^{125}I-TT was applied to a 3-ml PBE-94 gel column, equilibrated with 0.025 histidine HCl buffer, pH 6.2. The gel was eluted with polybuffer 74 (0.0075 mmole/pH unit/ml) pH 4.0 (first arrow), at a flow rate of 90 ml/hr and 1.0 ml fractions were collected. The gel was subsequently washed by a stepwise elution with 1.0 M Tris-HCl, pH 7.5 (second arrow), followed by 6 M Guanidine HCl, pH 8.0 (third arrow). The pH and radioactivity were monitored. (B) ^{125}I-TT fraction isolated from chromatofocusing as in panel A (shadowed peak) was derivatized with biotin (biotin/toxin, molar ratio 4:1) and rechromatographed on the same conditions. Biot-TT$_a$ fraction (shadowed peak) was further separate. (C) 4x10^{-8} cpm sample of Biot-TT$_a$ fraction was separated by affinity chromatography on Spherosil-DEAE dextran beads derivatized with polysialogangliosides. A 5-ml gel was equilibrated with 0.25 M glycine buffer and stepwise elution was performed with 1% NaCl, pH 5.5, 6% NaCl, pH 9.0, and 6 N Guanidine-HCl in glycine buffer. 3-ml fractions were collected and counted; Biot-TT$_b$ iodinated or unlabeled fractions were used in our biological studies.

The strong decreased toxicity and impaired immunogenicity of the biotinylated derivative was previously reported after chemical modifications of TT lysyl residues.[23] They are not proportional to the loss of binding (40%), supporting the established concept that TT contains different, separate site mediating toxicity, antigenicity, and receptor binding.[23] Since this derivative retains almost 60% of the binding affinity, it is a suitable probe for histochemical purposes.

Tetanus Toxin Binding to Pheochromocytoma PC12 Culture Increases after Differentiation with Nerve Growth Factor

Undifferentiated PC12 cell cultures bind poorly the tetanus toxin (Fig. 3, Time 0). In order to measure the time course of the developmental appearance of the binding sites, PC12 cell monolayers were treated with nerve growth factor (NGF) for periods of time up to 2 weeks. Every second day, selected cultures were incubated with an iodinated TT known to have a high affinity for polysialogangliosides.[24] Binding of this TT probe to PC12 cells, induced to differentiate with NGF, shows a 10-fold increase in binding sites from a value of 0.07 nmol/ng protein on untreated cells to 0.8 nmol/mg protein after 14 days of treatment (Fig. 3). The syalosyl nature of these TT binding sites is emphasized by the partial sensitivity of these sites

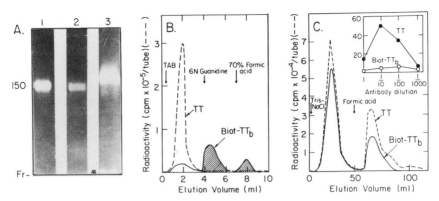

Figure 2. Characterizations of biotinyl derivative of ^{125}I-tetanus toxin. (A) Fluorographs of biotinylated (lane 2), nonderivatized (lane 1), and biotinylated ^{125}I-TT after incubation with avidine (lane 3) on SDS-polyacrylamide gel electrophoresis under nonreductive conditions. (B) Avidine-Sepharose analysis of ^{125}I-TT. Biotinyl derivatives of TT after purification on the ganglioside affinity column (Fig. 1C) were subjected to separation on avidine-Sepharose column (1 ml) preequilibrated with 0.02 M Tris-acetate buffer (TAB), containing 0.1% gelatin pH 6.8. Radioactively labeled (full line) or unlabeled (broken line) (0.6 mg) biotinylated toxins (0.6 mg) dissolved in equilibrium buffer were applied for separation. The gel was eluted with TAB (first arrow) followed by 6 N Guanidine-HCl, pH 1.5 (second arrow), and 70% formic acid (third arrow). Fractions (0.25 ml) at a flow rate of 5 ml/hr were collected, neutralized with 1 N NaOH, and absorbance or radioactivity determined. TT=native toxin; Biot-TT$_b$ as described in Fig. 1C. (C) Monoclonal antibody affinity separation and immunoprecipitation of biotinyl ^{125}I-TT. Equal amounts of radioactive (^{125}I) biotinylated and nonderivatized TT obtained as in Fig. 1C were dissolved in equilibration buffer and applied on 5-ml Sepharose-coupled monoclonal antibody columns. The gel was equilibrated with 0.15 M NaCl in 0.02 M Tris-HCl buffer and subsequently eluted with 0.15 M NaCl in 0.5 M formic acid, pH 2.0. Fractions of 5 ml were collected at a flow rate of 50 ml/hr. Insert depicts bulk immunoprecipitation (percentage of total radioactivity) of each toxin using the antibody at the mentioned dilutions.

to pretreatment of the cells with neuraminidase, before TT binding (Fig. 3). For staining, 14 days, NGF-treated PC12 cultures were incubated with the biotinylated (Biot-TT$_b$, Fig. 1) derivative and Texas-red conjugated streptavidin (Fig. 3, left insert). The differentiated cultures showed an intense fluorescence distributed on the cell body, neurites, and growth cones. The intensive staining on the cell body might represent binding and internalized toxin (Fig. 3, left insert). This staining was intensively reduced by a pretreatment of the cultures with neuraminidase (data not shown) emphasizing the similarity between the biotinylated and iodinated TT derivatives. Most probably, the neuraminidase sensitivity indicates these tetanus toxin binding sites detected on PC12 cells during NGF-induced differentiation are polysialoglycolipids. Previous chemical studies on the ganglioside composition of PC12 cells before and after NGF induced differentiation indicated a several fold-increase in the cellular amount of GD1b/GT1b gangliosides, already after 24 hr of NGF treatment (Fig. 3, right insert).[25] The delayed developmental appearance of these TT binding sites on the pheochromocytoma cell surface in contrast to early (24-48 hr) induction of neurite outgrowths and G1b gangliosides synthesis suggests that some signal(s), in addition to NGF, is necessary.

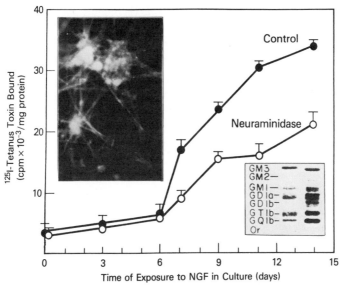

Figure 3. Kinetics of tetanus toxin binding sites appearance on nerve growth factor treated PC12 cells. ^{125}I-TT was added to PC12 cell monolayers that had been treated with NGF for different periods of time. The binding assay was carried out at 37°C for 60 min. Filled circles, binding to control cultures; open circles, binding to monolayers pretreated at 37°C for 10 min with 10 mU of neuraminidase. Left insert: a typical 12 days, NGF-treated PC12 culture stained with biotinylated TT and Texas red-conjugated streptavidin. Note intense staining of neurite outgrowths and cell bodies (x 175); Right insert: ganglioside profile from undifferentiated (left lane) and 1 day NGF-differentiated (right lane) PC12 cells. Cells were labeled with D-[14C]glucosamine, the gangliosides were extracted and separated by thin-layer chromatography in one direction. GM_3, GM_2, GM_1-monosialo-gangliosides; GD1a, GD1b-disialo-gangliosides; GT1b-trisialo-gangliosides; GQ1b-tetrasialo-gangliosides.

Subcellular Localization of Tetanus Toxin Binding Sites on NGF-Differentiated PC12 cells indicates a Major Distribution on the Neurite Outgrowths

PC12 cell monolayers were treated with NGF for 2 weeks. Such cultures, as well as untreated monolayers were incubated with the ganglioside affinity purified, biotinylated tetanus toxin (Biot-TT_b, Fig. 1C) the cells were then treated with streptavidin-gold, fixed, processed and analyzed by electron microscopy.[20] The electron microscopic analysis of the binding of TT to these cultures indicates an asymmetrical distribution of toxin binding sites (Table 1). Relatively few gold particles appear on the surface of the cell body. However, gold particles were seen within intracellular vesicles, suggesting that the biotinylated TT can be internalized.[20] The gold particles were scattered along the plasma membrane of the neurites[20] and were frequently found in large numbers (Table 1) at the neurite outgrowth varicosities, perhaps indicating larger numbers of binding sites at points of attachment to the dish. The growth cones of the cells were also labeled but to a lesser extent that were the neurites (Table 1).

Table 1. Asymmetric distribution of tetanus toxin binding sites on nerve growth factor-differentiated PC12 cells.

Structure	No of gold particles per micron surface area	
	+ biotinylated toxin	- biotinylated toxin
Cell body	0.01	0
Neurite	1.10	0
Growth cone or terminal	1.04	0

The living cultures were incubated with or without biotinylated TT, and then with streptavidin-gold as described.[20] The cells were washed and processed for electron microscopy without postfixation. The gold particles associated with the various cellular structures were counted on the micrographs. The total surface measured for each structure was greater than 0.5 mm. In a particular experiment, 10 gold particles were counted on the surface of 20 cell bodies, 124 gold particles on 38 neurites, and 22 gold particles on 12 growth cones and terminals. Several hundred cells were evaluated.

Priming PC12 Cells in Suspension with Nerve Growth Factor Does Not Affect the Subsequent Development of Tetanus Toxin Binding Sites in Monolayer Cultures

PC12 cells can be grown in suspension, and they respond biochemically to NGF in some of the same ways the cells do when they are in monolayer cultures. But, since in suspension they cannot attach to matrix, the treatment with NGF does not result in the formation of neurites. However, the cells substitute cell-matrix contact by cell-cell interactions producing large clusters. Figure 4 (A-C) presents typical cell sorter analyses of the aggregation process by priming PC12 cells in suspension with NGF. An electron microscopy examination of such an aggregate (Fig. 4D) indicates a close contact between the cells in the aggregate without nuclei and plasma membrane disappearance and/or a change to a heterokaryon phenotype. The effect of treating the cells with NGF in suspension on the subsequent development of TT binding was evaluated by measuring the binding of the radioactive TT as an index (Fig. 5, numbers). Treatment with NGF in suspension for up to 12 days did not lead to the appearance of TT binding sites on the cell surface (Fig. 5, numbers).

When such primed cells were plated under standard culture conditions, they rapidly separated from the clusters to individual cells and they grew neurites more rapidly and robustly than did naive cells plated under similar conditions (Fig. 5).[20] Nevertheless, the level of TT binding sites and the kinetics of their appearance were similar in both cultures (Fig. 5). Thus, treatment of PC12 cells in suspension with NGF did not lead to the formation of TT binding sites on the cell body and did not accelerate the appearance of these sites on the neurite outgrowths when they were plated. The long time course of appearance of these sites in monolayers or even upon transfer of primed cells to monolayers culture indicates that some cellular cell-matrix attachment signal besides NGF is also required for expression of these polysialoganglioside binding sites.

Localization of Tetanus Toxin Binding Sites on Chromaffin Granules

To identify the intracellular sites of action of TT on chromaffin and pheochromocytoma cells, subcellular fractions were prepared by differential preparative ultracentrifugation using discontinuous sucrose[26] or Ficoll[27] gradients. The obtained fractions were tested for ^{125}I-TT binding. The major binding fraction was identified as the catecholamine secretory organelle (chromaffin granule.)[15,18] As previously detailed,[15,18] TT bound in a neuraminidase-sensitive fashion to chromaffin cells, intact granules, and isolated granules membranes, as assayed

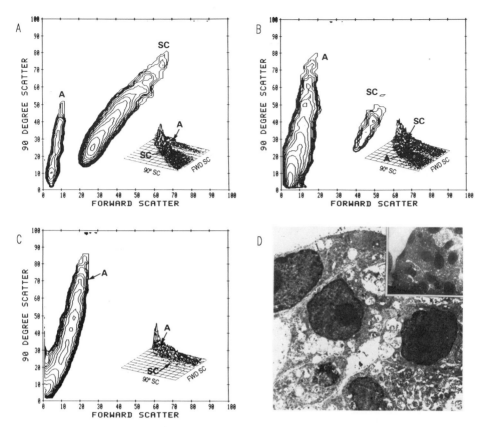

Figure 4. Aggregation of PC12 cells treated in suspension with nerve growth factor (A, B, C) and electron microscopy visualization of an aggregate (D). PC12 cells untreated (A) or treated in suspension for 5 days (B) and 10 days (C) with nerve growth factor (50 ng/ml) were analyzed by cytofluorography scattering on a cell sorter for size and statistical distribution (inserts). A = aggregated cells; SC = single cells. Note the gradual effect of NGF, shifting the cell population from single cells (A) to only aggregated cells (C). (D) Transmission electron micrographs of a 10-day, NGF-induced, PC12 aggregate (insert x 115). Note the distinction of individual cells within the aggregate based on the presence of individual nuclei and cell membrane (~x 6950).

biochemically and visualized by electron microscopy. Recently, we have extended these characterizations to chromaffin granules separated from PC12 cells differentiated for 7-14 days with nerve growth factor (Fig. 6). In line with our previous reports,[15,18] PC12 chromaffin granules specifically bound ^{125}I-TT, ganglioside affinity purified derivative. This binding was saturable at concentrations of about 10^{-9}M free toxin, with an estimated binding capacity of about 0.6 nmol/mg granule membrane protein. Hydrolysis of the chromaffin granules sialic acid residues by neuraminidase caused a significant loss of TT binding sites at all toxin concentrations (Fig. 6) and at pHs higher than 6 (Fig. 6, right insert) as previously noticed with chromaffin cell granules,[15] brain synaptosomes,[23] and primary brain neuronal cultures.[24] Displacement of bound TT from PC12 chromaffin granules by unlabeled TT or the relevant ganglioside receptor GD1b was observed only with intact granules (IC_{50} about 0.5×10^{-8} M) but not neuraminidase-treated granules (Fig. 6). These values are ten times higher than the Kd values for the binding of the toxin to the granules and are consistent with the concept of some undisplaceable interaction between the bound toxin and its sites, described as toxin sequestration and/or internalization.[24] To visualize the location of TT binding sites, we used the biotinylated toxin

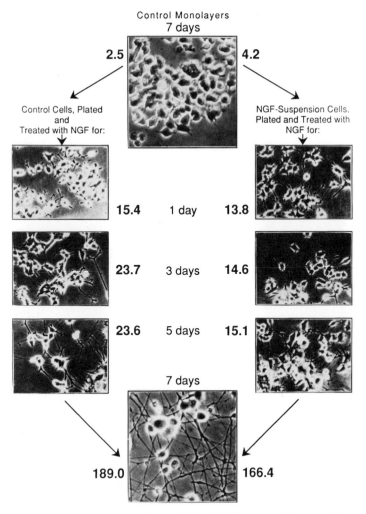

Figure 5. The morphology and lack of effect of NGF treatment in suspension on the subsequent development of tetanus binding sites on PC12 cells. PC12 cells in suspension cultures were transferred to monolayers treated for the mentioned time with NGF and binding assays with ^{125}I-TT were performed. The binding to PC12 cells in suspension was performed under identical conditions using 10^6 cells/assay. The numbers shown are mean values of triplicates (fmol/mg protein) with standard deviations of 2-6 fmol/mg protein (10^6 cells/assay). Control or NG-treated suspension cultures of PC12 cells (up to 14 days treatment) show binding values of 12-15 fmol/mg protein. Before binding, the cultures were photographed and typical fields are presented (x 125).

derivative (Biot-TT$_b$, Fig. 1). After incubation of intact, freshly prepared chromaffin granules with this derivative, streptavidin-gold (20 nm) was added and the preparation processed for transmission electron microscopy. A typical electron micrograph (Fig. 6, left insert) revealed specific sites on the cytoplasmic face of PC12 cell's vesicle or vesicular membrane (broken light vesicle and electron dense vesicle, Fig. 6, left insert). This visualization further supports the concept[15] that the sugar-sialic acid conformation recognized by the "G1b"-tetanus toxin is strictly localized on the external, cytoplasmic surface of chromaffin granules.

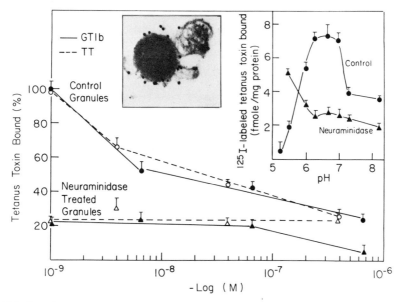

Figure 6. Binding, visualization, and displacement of affinity-purified tetanus toxin to control and neuraminidase-treated PC12 chromaffin granules. Binding as a function of pH (right insert). Left insert: localization of affinity-purified TT; binding sites, visualized with streptavidin-gold on isolated chromaffin granules (electron micrograph, x 61900 magnification).

CONCLUSIONS

PC12 cells induced to differentiate with NGF offer an excellent cellular model with which to study the synthesis and mechanisms of cell surface exposure of TT recognized G1b polysialogangliosides as well as their role and/or involvement in chromaffin granule exocytosis.

The biotinylated TT derivative developed, when used with streptavidin detection, provides a new method to identify and visualize G1b polysialogangliosides in the nervous system and to study their role in neuron adhesion and neurotransmitter release processes.

REFERENCES

1. Gill, DM. Bacterial toxins: a table of lethal amounts. Microbiol Rev 1982; 46:86.
2. Lazarovici P, Yavin E, Bizzini B, Fedinec AA. Retrograde transport in sciatic nerves of ganglioside-affinity-purified tetanus toxins. In: Dolly JO, ed. Neurotoxins in Neurochemistry. Chichester: Ellis Horwood Ltd., 1988:100.
3. Schwab ME, Thoenen H. Selective binding, uptake, and retrograde transport of tetanus toxin by nerve terminals in the rat iris. An electron microscope study using colloidal gold as a tracer. J Cell Biol 1978; 77:1.
4. Stoeckel K, Schwab M, Thoenen H. Comparison between the retrograde axonal transport of nerve growth factor and tetanus toxin in motor, sensory and adrenergic neurons. Brain Res 1975; 99:1.
5. Schwab, ME, Suda K, Thoenen H. Selective retrograde transsynaptic transfer of a protein, tetanus toxin, subsequent to its retrograde axonal transport. J Cell Biol 1979; 82:798.
6. Curtis DR, De Groat WC. Tetanus toxin and spinal inhibition. Brain Res 1968; 10:208.
7. Kryzhanovsky GN. Pathophysiology. In: Veronesi R, ed. Tetanus Important New Concepts. Excerpta Medica. Amsterdam: Casparie, 1981:109.
8. Mellanby J, Greene J. How does tetanus toxin act? Neuroscience 1981; 6:281.

9. Wellhoner HH. Tetanus neurotoxin. Rev Physiol Biochem Pharmacol 1982; 93:1.
10. Mirsky R. The use of antibodies to define and study major cell types in the central and peripheral nervous system. In: Brookes J, ed. Neuroimmunology. New York: Plenum Press, 1982:141.
11. Pearse AGE. The cytochemistry and ultrastructure of polypeptide hormone-producing cells of the APUD series and the embryologic, physiologic and pathologic implications of the concept. J Histochem Cytochem 1969; 17:303.
12. Heymanns J, Neumann K, Haremann K. Tetanus toxin as a marker for small-cell lung cancer cell lines. J Cancer Res Clin Oncol. 1989; 115:537.
13. Berliner P, Unsicker K. Tetanus toxin labeling as a novel, rapid and highly specific tool in human neuroblastoma differential diagnosis. Cancer 1985; 56:419.
14. Rosenheck K, Lelkes P. Stimulus-secretion coupling in chromaffin cells. Vol. I, II. Boca Raton, FL: CRC Press, 1987.
15. Lazarovici P, Fujita K, Contreras ML, Diorio JP, Lelkes PI. Affinity purified tetanus toxin binds to isolated chromaffin granules and inhibits catecholamine release in digitonin-permeabilized chromaffin cells. FEBS Lett 1989; 253:121.
16. Bittner M, Holz R. Effects of tetanus toxin on catecholamine release from intact and digitonin-permeabilized chromaffin cells. J Neurochem 1988; 51:451.
17. Penner R, Neher E, Dreyer F. Intracellularly injected tetanus toxin inhibits exocytosis in bovine adrenal chromaffin cells. Nature 1986; 324:76.
18. Lazarovici P. Characterization and visualization of tetanus toxin acceptors on adrenal chromaffin granules. J Physiol (Paris) 1990; 84:197.
19. Sandberg K, Berry CJ, Rogers T. Studies on the intoxication pathway of tetanus toxin in the rat pheochromocytoma (PC12) cell line. Binding, internalization, and inhibition of acetylcholine release. J Biol Chem 1989; 264:5679.
20. Fujita K, Guroff G, Yavin E, Goping G, Orenberg R, Lazarovici P. Preparation of affinity-purified, biotinylated tetanus toxin, and characterization and localization of cell surface binding sites on nerve growth factor-treated PC12 cells. Neurochem Res. 1990; 15:373.
21. Bayer EA, Wilchek M. The use of avidin-biotin complex as a tool in molecular biology. In: Glick D, ed. Methods of Biochemical Analysis. New York: John Wiley & Sons, 1980; 26:1.
22. Kenimer JG, Habig WH, Hardegree MC. Monoclonal antibodies as probes of tetanus toxin structure and function. Infect Immun 1983; 42:942.
23. Bizzini B. Tetanus toxin structure as a basis for elucidating its immunological and neuropharmacological activities. In: Cuatrecasas P, ed. Receptors and Recognition. London: Chapman and Hall, 1977; 1:175.
24. Lazarovici P, Yavin E. Affinity-purified tetanus neurotoxin interaction with synaptic membranes: properties of a protease-sensitive receptor component. Biochemistry 1986; 25:7047.
25. Yavin E, Lazarovici P, Nathan A. Molecular interactions of ganglioside receptors with tetanus toxin on solid supports, aqueous solutions and natural membranes. In: Wirtz KWA, ed. Membrane Receptors, Dynamics, and Energetics. New York: Plenum Publishing Co., 1987:135.
26. Pollard HB, Zinder O, Hoffman PG, Nikodijevic O. Regulation of the transmembrane potential of isolated chromaffin granules by ATP, ATP analogs, and external pH. J Biol Chem 1976; 251:4544.
27. Wagner JA. Structure of catecholamine secretory vesicles from PC12 cells. J Neurochem 1985; 45:1244.

TETANUS TOXIN INHIBITS A MEMBRANE
GUANYLATE CYCLASE TRANSDUCTION SYSTEM

Alexander A. Fedinec,[1] Teresa Duda,[2] Bernard Bizzini,[3]
Nigel G. F. Cooper[4] and Rameshwar K. Sharma[2]

[1]Department of Anatomy and Neurobiology, The University of Tennessee,
Memphis, TN 38163, U.S.A.
[2]Unit of Regulatory and Molecular Biology, Pennsylvania College
of Optometry, Philadelphia, PA 19141, U.S.A.
[3]Department of Molecular Toxicology, Institute Pasteur, 75724 Paris, France
[4]Department of Anatomical Sciences and Neurobiology, School of Medicine,
University of Louisville, KY 40292, U.S.A.

INTRODUCTION

Original studies suggested that tetanus toxin (TT) in vivo exerts its paralytic effect on cholinergic nerve terminals in rabbit iris by interfering with cyclic GMP pathway (Fedinec, et al. 1976 and King, et al. 1978). One possible target site of TT for this effect is guanylate cyclase, the enzyme that catalyzes the formation of cyclic GMP. There are two forms of guanylate cyclases—soluble and membrane bound. Thus, if TT exerts its paralytic activity through the interference with guanylate cyclase activity, then the toxin could cause such interference by acting on soluble or membrane bound guanylate cyclase. In this study we have considered the latter possibility. Specifically, using the combined tools of immunology and genetically-tailored mutants, we have studied the effect of TT directly on the basal activity of plasma membrane guanylate cyclase, GCα (Duda et al., 1991), and on the ANF-dependent cyclase activity of the genetically-constructed ANF receptor guanylate cyclases, GCα-DmutGln^{338}Leu^{364} and GCα-SmutLeu^{364}.

Initially, two monomeric proteins, the first a 120-140 kD (Kuno, et al. 1986, Takayanagi, et al. 1987 and Meloche et al. 1988) and the other, 180 kD (Paul et al. 1987) were described. They were both, atrial natriuretic factor (ANF) receptors and guanylate cyclases. Atrial natriuretic factor (ANF) is one of the family of structurally related natriuretic peptides that regulate the hemodynamics of the physiological processes of diuresis, water balance, blood pressure and steroidogenesis (reviewed by Rosenzweig and Seidman, 1991). ANF is also widely distributed in the central and peripheral nervous system not specifically involved in cardiovascular homeostasis suggesting a role in neuromodulation (Rosenzweig and Seidman, 1991). Other known members of this family are brain natriuretic peptide (BNP) (Sudoh et al. 1988) and cardiac natriuretic peptide (CNP) (Flynn, et al. 1989) which, like ANF, stimulate membrane

Botulinum and Tetanus Neurotoxins, Edited by
B.R. DasGupta, Plenum Press, New York, 1993

guanylate cyclase activity (Hashiguchi, et al. 1988). Molecular cloning studies have identified two structurally related natriuretic peptide receptor guanylate cyclases, GC-A (Chinkers, et al. 1989) and GC-B (Chang et al., 1989). The natural ligand for GC-A appears to be ANF, and for GC-B is CNP (Chinkers et al., 1989). The originally purified 180 kD membrane guanylate cyclase (Paul et al., 1987) is immunologically and functionally identical to GC-A (Marala et al., 1992, Marala and Sharma, 1992). There is also an another form of plasma membrane guanylate cyclase, GCα, which differs from GC-A in only two amino acids residues and is not a receptor of the known natriuretic peptides (Duda, et al. 1991). These two amino acids are: at the position 338 His instead of Gln and at the position 364 Pro instead of Leu. Leu[364] is critical in ANF signalling (Duda, et al. 1991).

Using rabbit polyclonal antibody to the 180 kD ANF receptor guanylate cyclase (Paul et al., 1987; for antibody specificity see Sharma et al., 1989) and immunoperoxidase microscopy (Cooper et al., 1985), we showed immunolabelling of neuronal and glial elements in specific regions of the rat spinal cord, medulla, cerebellum and brain. The intensity of immunoperoxidase labeling varies, i.e. neurons in the lumbal cord, hippocampus, cerebellar cortex and the retina show a high degree of labelling (Fedinec, et al., 1988). Light and electron microscopy shows immunolabelling of subcellular structures such as membranes of the rat outer segments and chicken cone outer and inner segments and synaptosomes in the inner and outer plexiform layers (Cooper et al., 1989, Cooper et al., 1990). Pretreatment of nerve tissue sections (brain and spinal cord) with tetanus toxin blocks the antibody binding sites on the ANF receptor membrane guanylate cyclase, interfering with its localization (Fedinec et al., 1988).

Now we demonstrate, that tetanus toxin inhibits basal and ANF-dependent guanylate cyclase activities in crude and solubilized adrenocortical carcinoma membranes. We further show that TT inhibits the cyclase activities of GCα and ANF-dependent guanylate cyclases, indicating that one potential mechanism by which TT exhibits its paralytic activity is through the interference with guanylate cyclase transduction system.

METHODS

Immunocytochemical Localization of Atrial Natriuretic Factor (ANF) Dependent 180 kD Membrane Guanylate Cyclase in Rat Cerebellum

Rats (240-260 g) were anesthetized with Ketamine-Xylazine and transcordally perfused with ice-cold TBS (10 mM Tris-HCl, 0.9% NaCl, pH 7.4), followed by ice-cold 4% paraformaldehyde and 0.1% glutaraldehyde in the same buffer. The cerebellum was post-fixed overnight in 4% paraformaldehyde at 4°C.

Vibratome sections (20-40 μm) were washed in TBS and TBS-2% bovine serum albumin (BSA), then cut and processed for peroxidase histochemistry (Cooper et al., 1985). Rabbit anti-guanylate cyclase IgG (diluted 1:100 to 5000) served as primary antiserum. Peroxidase-conjugated Fab fragments of antirabbit IgG (diluted 1:500-1:750) served as the second antibody. Following reaction with diaminobenzidine and hydrogen peroxide, sections were prepared for examination with the microscope to determine the presence or absence of a dark brown reaction product. Control slides were incubated with nonspecific horse or preimmune rabbit IgG or without the primary antisera. Some material was postfixed in 2% osmium tetroxide and processed for electron microscopy.

To localize tetanus toxin binding sites, sections of cerebellum were incubated with 1-7 μg of tetanus toxin in TBS/2% BSA/ml for 2 hr at room temperature or 12 hr at 4°C. Sections were washed in TBS/2% BSA, then was added rabbit IgG antitoxin in equivalent Lf/ml. To determine whether toxin interferes with localization of guanylate cyclase, the same toxin solution was applied first. After washing with buffer, rabbit antiguanylate cyclase antibody was applied. Peroxidase-labeled goat antibody was the secondary antibody.

Preparation of Rat Adrenocortical Carcinoma Plasma Membranes

Rat adrenocortical carcinoma plasma membranes were prepared as described in Paul et al., 1987.

GCα and Its Mutants

GCα was cloned from rat adrenal library and its double (GCα-DmutGln^{338}Leu^{364}) and single (GCα-SmutLeu^{364}) mutants were constructed as in Duda et al. (1991). GCα-DmutGln^{338} Leu^{364} is structurally and functionally identical to the cloned wild-type ANF receptor guanylate cyclase (Chinkers et al., 1989); GCα-SmutLeu^{364} is functionally identical but structurally different in a single amino acid residue from GCα-DmutGln^{338}Leu^{364} (i.e. His^{338} instead of Gln [Duda et al., 1991]). GCα and its mutants used in this study are graphically represented in Fig. 1. GCα and its mutants cDNAs were individually ligated into the mammalian expression vector pSVL, in which the coding region was under the transcriptional control of the simian virus 40 late promoter and expressed in COS-2A cells using the calcium phosphate technique.

Guanylate Cyclase Activity Assay

Guanylate cyclase activity was assayed as described previously (Nambi et al., 1982). Briefly, samples (10-20 μg of adrenocarcinoma plasma membrane protein, crude or solubilized with 3-[(3-cholaimidopropyl)dimethylammonio]-1 propane sulfonate [CHAPS]) were preincubated at 37°C in a 50 mM Tris-HCl buffer pH 7.6 containing 10 mM theophylline (cyclic nucleotides phosphodiesterase inhibitor) 15 mM creatine phosphate, 20 U/ml creatine phosphokinase (250 U/mg protein) in a total volume of 100 μl in the presence or absence of tetanus toxin (potency 3125 Lf/mg protein N and 1.2x108 LDs/mg protein N) followed by 10 min incubation on ice with or without ANF (10^{-6} M). The reactions were started with the addition of substrate (final concentrations 1 mM GTP and 4 mM MnCl$_2$) continued for 10 min at 37°C and terminated by the addition of 0.9 ml of ice-cold 50 mM sodium acetate buffer, pH 6.2, followed by heating for 3 min in a boiling water bath. The cyclic GMP formed was measured by radioimmunoassay and expressed as pmol cGMP formed per mg protein per min. In control experiments toxin was heated to 100°C for 10 min or neutralized with 10% excess of rabbit IgG to TT (rabbit IgG antitoxin had 2000 Lf/ml (Bizzini et. al., 1968) prior to addition to the reaction mixture.

In experiments with transfected COS-2A cells, 60 hrs after transfection, cells (10^4 cells/ 1.5 cm dish) were washed twice with Hank's Balanced Salt Solution (HBSS), incubated in 1 ml

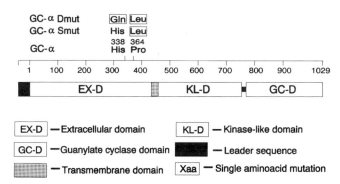

Figure 1. Graphical representation of GC and its mutants.

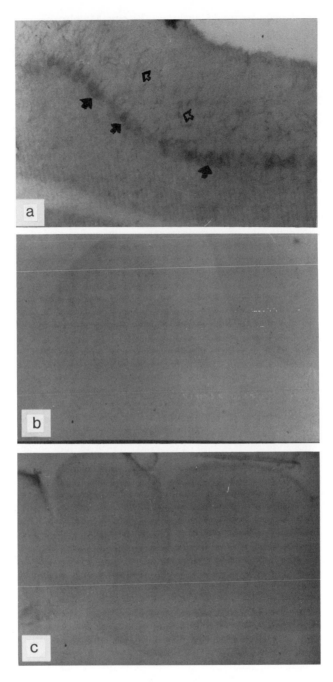

Figure 2. Guanylate cyclase immunocytochemistry. A) Rat cerebellum (X 100). Purkinje cells (closed arrows) and their dendrites (open arrows) show distinct localization of guanylate cyclase; B) Rat cerebellum (X 47). Control section with no primary antibody to guanylate cyclase incubated with goat anti rabbit IgG peroxidase. Absence of reaction product indicates failure of binding of the second antibody; C) Rat cerebellum—effect of TT (X 47). Incubation of sections with TT prior to incubation with antibody to guanylate cyclase prevents its binding to guanylate cyclase in the cerebellum.

of HBSS for 20 min at 37°C with toxin followed by 10 min with ANF. Cells were washed with HBSS, scraped into 2 ml of ice-cold buffer (10 mM Mg_2^+, 50 mM Tris-HCl pH 7.6) and membranes prepared as in Duda et al., 1991. 1-2 μg of membrane protein was used for the cyclase assay.

RESULTS AND DISCUSSION

A specific antibody-labeling pattern in rat cerebellum using rabbit polyclonal antibody to the 180 kD ANF receptor guanylate cyclase is shown in Fig. 2A (compare with control sections which were incubated without antibody to the 180 kD ANF receptor guanylate cyclase—Fig. 2B). Pretreatment of tissue sections with TT blocks the antibody binding sites (Fig. 2C).

The molecular mechanism of tetanus toxin action is unknown. The target of this clostridial toxin is assumed to be on the cytosolic side of the plasma membrane. In the present study the probes of GCα mutants and the ANF receptor guanylate cyclase antibody were used to determine if TT directly interferes with the ANF-dependent cyclase transduction system. ANF-dependent guanylate cyclase in rat adrenocortical carcinoma membranes is well characterized (Paul et al., 1987) and these membranes were chosen as the first model to study the effect of TT. When the crude rat adrenocortical carcinoma membranes were incubated in the presence of increasing concentrations of TT (Fig. 3), the significant inhibition of guanylate cyclase activity was observed and the inhibition was dose- and time-dependent. However, when membranes were incubated with heat inactivated toxin or toxin neutralized with rabbit G globulin to TT, guanylate cyclase activity was not significantly inhibited (Fig. 3).

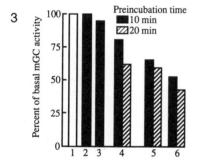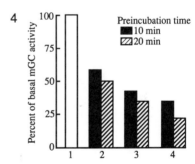

Figure 3. Effect of tetanus toxin (TT) on guanylate cyclase activity in particulate fractions of rat adrenocortical carcinoma membranes. 1: Baseline guanylate cyclase activity (basal mGC activity is expressed as 100% of cGMP formed [8.9 pmol/mg protein/min]); 2: Baseline guanylate cyclase activity in the presence of heat inactivated TT; 3: Baseline guanylate cyclase activity in the presence of TT neutralized with rabbit IgG to TT; 4-6: Guanylate cyclase activity in the presence of: 1.0 μg/ml of TT (4); 3.0 μg/ml of TT (5); 5.0 μg/ml of TT (6). Membranes were incubated with indicated amounts of TT for 10 min (solid bars) and 20 min (stripped bars). Experiments were conducted as described in Methods and performed in triplicate. Depicted results are from one representative experiment.

Figure 4. Effect of TT on guanylate cyclase activity in solubilized-particulate fractions of rat adrenocortical carcinoma membranes. 1: Baseline guanylate cyclase activity (basal mGC activity is expressed as 100% of cGMP formed [30.0 pmol/mg protein/min]); 2-4: Guanylate cyclase activity in the presence of 1.0 μg/ml of TT (2); 3.0 μg/ml of TT (3); 5.0 μg/ml of TT (4). Membranes were incubated with indicated concentrations of TT for 10 min (solid bars) and 20 min (stripped bars). Experiments were conducted as described in Methods and performed in triplicate. Results are from one representative experiment.

These results show, that inhibition of guanylate cyclase activity was due to the direct effect of TT on ANF-dependent cyclase. A similar effect of TT on guanylate cyclase activity was observed when solubilized membranes were tested. (Fig. 4).

Guanylate cyclase activity in adrenocortical carcinoma membranes is stimulated by ANF (Paul et al., 1987). As it is shown in Fig. 5, ANF-dependent cyclase activity in membranes incubated with TT is completely inhibited, but toxin neutralized with specific antibody has no effect on hormonally stimulated guanylate cyclase activity. This indicates that TT inhibits both, basal and ANF-dependent guanylate cyclase activities.

To determine whether TT selectively inhibits the basal and hormonally-dependent cyclase activity of ANF receptor guanylate cyclase, GCα was genetically tailored to create two mutants GCα-Dmut and GCα-Smut (Duda et al., 1991). GCα-Dmut is structurally and functionally identical to GC-A. GCα-Smut differs only in one amino acid from GC-A, there being His^{338} instead of Gln; functionally it is identical to GC-A.

COS-2A cells were transfected with GCα-pSVL expression vector and were used to study the effect of TT on the activity of the over-expressed protein. 60 hr after transfection, cells were incubated with increasing concentrations of tetanus toxin in the presence or absence of ANF. Toxin inhibited the basal cyclase activity (Fig. 6A) and the inhibition was dose-dependent (data not shown). As shown previously, ANF had no effect on the guanylate cyclase activity (Fig. 6A). These results demonstrate that TT selectively inhibits the baseline activity of GCα.

The control experiments were conducted with COS-2A cells transfected with genetically tailored ANF-responsive membrane guanylate cyclases—GCα-DmutGln^{338}Leu^{364} and GCα-SmutLeu^{364}. Cells, individually expressing these proteins, were incubated with or without TT and with or without ANF. Results are summarized in Fig. 6B, 6C. ANF-stimulated guanylate

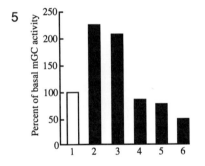 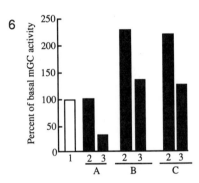

Figure 5. Stimulation of non-solubilized membrane guanylate cyclase by ANF and its inhibition by TT in particulate fractions of rat adrenocortical carcinoma membranes. 1: Baseline guanylate cyclase activity (expressed as 100% of cGMP formed [8.9 pmol/mg protein/min]); 2: ANF-stimulated guanylate cyclase activity; 3: Guanylate cyclase activity in the presence of TT neutralized with rabbit IgG to TT and ANF; 4-6: guanylate cyclase activity in the presence of ANF (10^{-6} M) and 1.0 μg/ml of TT (4); 3.0 μg/ml of TT (5) and 5.0 μg/ml of TT (6). Experiments were conducted as in Methods and done in triplicate. Values are from one representative experiment.

Figure 6. Effect of ANF and TT on membrane guanylate cyclase activity in COS-2A cells transfected with GCα (A), GCα-Dmut (B) and GCα-Smut (C). 1: baseline guanylate cyclase activity (expressed as 100% of cGMP formed [0.4 pmol/min/10^4 cells transfected with GCα, 0.55 pmol/min/10^4 cells transfected with GCα-Dmut, 0.5 pmol/min/10^4 cells transfected with GCα-Smut]); 2: ANF (10^{-6})-stimulated guanylate cyclase activity; 3: TT (1 μg/ml) inhibition of ANF stimulated guanylate cyclase activity. Experiments were conducted as described in Methods. Values are from one representative experiment done in triplicate.

cyclase activity was in both cases (GCα-Dmut and GCα-Smut) almost completely abolished upon the treatment with TT and the inhibition was dose-dependent (data not shown).

It is thus established that TT directly inhibits the basal and ANF-dependent guanylate cyclase activities. According to the present concept guanylate cyclase protein is a multi-module functional component (Goraczniak et al., 1992). The inhibitory-site of TT could therefore be at any one of the modules, including those which have not been identified so far.

Tetanus is a polysystemic disease affecting not only the nervous system, but many other organs and functions as well. Pathophysiological effects of tetanus that received most of the attention concerns central and peripheral neurotransmission. However, important changes occur in the autonomic system: in respiratory function, in cardiovascular system such as blood pressure changes and cardiac arythmias, in hormonal balance such as inhibition of the neurosecretory hypothalamo-hypophyseal system, in natriuretic capacity of the kidneys, in energy metabolism, electrolyte balance and fluid balance (Kryzhanovsky, 1981; Aleksevich, 1989). These effects are important in pharmacologic considerations of patient treatment (Dastur, 1989). Atrial natriuretic factor receptor guanylate cyclase may be functionally related to many physiological functions altered in tetanus disease. Thus tetanus toxin can be used as a probe for further elucidating this hormonally responsive guanylate cyclase signaling system and its role in tetanus disease.

Acknowledgements

This work was supported by the grants from National Institutes of Health (NS 23744, EY 08522) and National Science Foundation (DCB-83-0050).

REFERENCES

Aleksevich YI. Peculiarities of oxidative metabolism in tetanus. In: Nistico G, Bizzini B, Bytchenko B, Triau R, eds. Eighth International Conference on Tetanus. Rome: Pythagora Press, 1989:34.

Bizzini B, Turpin A, Raynaud M. Production et purification de la toxine tetanique. Ann Inst Pasteur 1968; 101:1212.

Bizzini B. Axoplasmic transport and transsynaptic movement of tetanus toxin. In: Simpson LL, ed. Botulinum neurotoxin and tetanus toxin. New York: Academic Press, Inc., 1989:203.

Bizzini B, Fedinec AA. Structural and functional characterization of tetanus toxin. In: Nistico G, Bizzini B, Bytchenko B, Triau R, eds. Eighth International Conference on Tetanus. Rome: Pythagora Press, 1989:37.

Chang MS, Lowe DG, Lewis M, Hellmis R, Chen E, Goeddel DV. Differential activation by atrial and brain natriuretic peptides of two different guanylate cyclases. Nature 1989; 341:68.

Chinkers M, Garbers DL, Chang MS, Lowe DG, Goeddel DV, Schultz S. A membrane form of guanylate cyclase is an atrial natriuretic peptide receptor. Nature 1989; 338:78.

Cooper NGF, McLaughlin BJ, Tallant EA, Cheung WY. Calmodulin-dependent protein phosphatase: Immunocytochemical localization in Chick retina. J Cell Biol 1985; 101:1212.

Cooper NGF, Fedinec AA, Sharma RK. Immunocytochemistry of the 180-kDa membrane guanylate cyclase in the retina. ARVO Supplement to Investigative Ophthalmology and Visual Science 1989; 30:295.

Cooper NGF, Fedinec AA, Sharma RK. ANF-receptor coupled guanylate cyclase in CNS synapses. Soc Neurosci 1990, 16:537.

Dastur FD. Tetanus-pathophysiology as a guide to treatment. In: Nistico G, Bizzini B, Bytchenko B, Triau R, eds. Eighth International Conference on Tetanus, Rome: Pythagora Press, 1989:630.

Dryer F. Peripheral actions of tetanus toxin. In: Simpson LL, ed. Botulinum neurotoxin and tetanus toxin, London: Academic Press, 1989:179.

Duda T, Goraczniak RM, Sharma RK. Site-directed mutational analysis of a membrane guanylate cyclase cDNA reveals the atrial natriuretic factor signaling site. Proc Natl Acad Sci USA 1991; 88:7882.

Fedinec AA, King LE Jr, Latham WC. Glycine, theophilline and antitoxin effects on rabbit sphincter pupillae muscle paralyzed by tetanus toxin. In: Ohaska A, Hayashi K, Sawai Y, eds. Animal, Plant and Microbial Toxins. New York: Plenum Publ. Corp., 1976; 2:351.

Fedinec AA, Cooper NGF, Sharma RK. Immunocytochemical localization of the 180-kDa guanylate cyclase in the rat brain. Soc Neurosci 1988; 14:(1)363.

Fedinec AA, Duda T, Cooper NGF, Sharma RK, Bizzini B. Inhibition of ANF-dependent 180-kDa membrane guanylate cyclase by tetanus toxin: a potential mechanism of action. Toxicon 1989; 27:44.

Flynn TGB, Anoop B, Tremblay L, Sarda I. Isolation and characterization of ISO-rANP, a new natriuretic peptide from rat atria. Biochem Biophys Res Commun 1989; 161:830.

Goraczniak RM, Duda T, Sharma RK. A structural motif that defines the ATP-regulatory module of guanylate cyclase in atrial natriuretic factor signalling. Biochem. J 1992; 282:533.

Habermann E. Clostridial neurotoxins and the central nervous system: Functional studies on isolated preparations. In: Simpson LL, ed. Botulinum neurotoxin and tetanus toxin. London: Academic Press, 1989, 255.

Hashiguchi T, Higuchi K, Ohask M, Minamino N, Kangawa K, Matsuo H, Nawata H. Porcine brain natriuretic peptide, another modulator of bovine adrenocortical steroidogenesis. FEBS Lett 1988; 236:455.

King LE Jr, Fedinec AA, Latham WC. Effects of cyclic nucleotides on tetanus toxin paralyzed rabbit sphincter pupillae muscle. Toxicon 1978; 16:625.

Kryzhanovsky GN. Pathophysiology. In: Veronesi R, ed. Tetanus. Important new concepts. Casparie-Amsterdam: Excerpta Medica, 1981:109.

Kuno T, Anderson W, Kamisaka Y, Waldman SA, Chang LY, Saheki S, Leitman DC, Nakane M, Murad F. Copurification of an atrial natriuretic factor receptor and particulate guanylate cyclase. J Biol Chem 1986; 261:5817.

Marala RB, Sharma RK. Ubiquitous and bifunctional 180 kDa atrial natriuretic factor-dependent guanylate cyclase. Mol Cell Biochem 1991; 100:25.

Marala R, Duda T, Goraczniak RM, Sharma RK. Genetically tailored atrial natriuretic factor-dependent guanylate cyclase. Immunological and functional identity with 180 kDa membrane guanylate cyclase and ATP signaling site. FEBS Lett 1992; 296(3):254.

Meloche S, McNocoll N, Liu B, Ong H, DeLean AD. Atrial natriuretic factor R1 receptor from bovine adrenal zona glomerulosa: purification, characterization, and modulation by amiloride. Biochemistry 1988; 27:8151.

Nambi P, Aiyar NV, Roberts AN, Sharma RK. Relationship of calcium and membrane guanylate cyclase in adrenocorticotropin-induced steroidogenesis. Endocrinology 1982; 111:196.

Paul AK, Marala RB, Jaiswal RK, Sharma RK. Co-existence of guanylate cyclase and atrial natriuretic factor receptor in a 180 kD protein. Science 1987; 235:1224.

Rosenzweig A, Seidman CE. Atrial natriuretic factor related peptide hormones. Annu Rev Biochem 1991; 60:229.

Sharma RK, Marala RB, Duda T. Purification and characterization of the 180-kDa membrane guanylate cyclase containing atrial natriuretic factor receptor from adrenal gland and its regulation by protein kinase C. Steroids 1989; 53:437.

Sudoh T, Kangawa K, Minamino N, Matsuo H. A new natriuretic peptide in porcine brain. Nature (London) 1988; 332:78.

Takayanagi KS, Snajdan RM, Imada T, Timura M, Pandey L, Misono KS, Inagami T. Purification and characterization of two types of atrial natriuretic factor receptors from bovine adrenal cortex: Guanylate cyclase-linked and cyclase free receptors. Biochem Biophys Res Commun 1987; 144:244.

TETANUS NEUROTOXIN: (1) IMMUNOLOGICAL ROLES OF FRAGMENTS OF THE TOXIN IN PROTECTION, AND (2) ATTEMPTS TO IDENTIFY TARGET SITE(S) OF ITS TOXIC ACTION

Morihiro Matsuda,
Toshio Okabe, and Nakaba Sugimoto

Department of Tuberculosis Research (Bacterial Toxinology)
Research Institute for Microbiol Diseases,
Osaka University
3-1, Yamadaoka, Suita, Osaka, 565, Japan

INTRODUCTION

We have been working on the structure-function relationship of tetanus neurotoxin, not only from a fundamental (biological) point of view, as a basis for understanding the mechanism of action of tetanus neurotoxin, but also from a practical (medical) point of view, as a basis for improvement of prophylaxis and treatment of tetanus.

In the first section of this paper, we report the immunological activities of fragments of tetanus toxin in various combinations, including fragment [B] (H_N) we recently isolated, and the protective effects of human anti-tetanus monoclonal antibodies directed against the three functional domains of the tetanus toxin molecule. In the second section of this paper, we describe the results of our attempts to identify the target site(s) of tetanus toxin in inhibiting exocytotic secretion using digitonin-permeabilized bovine adrenal chromaffin cells and in vitro systems for analysis of fusion process in exocytosis.

MATERIALS AND METHODS

Preparation of Tetanus Toxin and Fragments of Tetanus Toxin

Tetanus toxin was purified from culture filtrates of *Clostridium tetani* strain Harvard A47 (Biken substrain) by the method of Matsuda and Yoneda[1]. Fragments [A](L) and [B•C] (H_N•H_C)[1], and [C](H_C)[2] and Fragment [A-B] (L-H_N)[3] were prepared as described previously. Fragment [B] (H_N) was isolated and purified from purified Fragment [A-B] by ion-exchange chromatography as described by Matsuda et al.[4].

Tests of Immunogenic Activity of Toxin and Fragments of Tetanus Toxin

Mice (strain ddY or ICR) were immunized with each formalin-treated[2] fragment(s) or the whole toxin-toxoid by a single subcutaneous injection and were challenged with 20 LD_{50} toxin 4 weeks later as described by Matsuda et al.[5]. The protective activity was evaluated using score system described by Murata et al.[6] by observing symptoms for 7 days.

Human Anti-Tetanus Monoclonal Antibodies(MAbs)

Stable hybrid cell lines G2, G4 and G6 producing anti-tetanus human MAbs of the IgG type with high toxin-neutralizing activity were generated as described previously[7] by fusion of hyperimmunized human peripheral lymphocytes with a mouse/human heteromyeloma SHM D-33 or RF-S1. MAbs were produced in serum-free RDF medium and purified by affinity chromatography on a Protein A-Cellulofine (Seikagaku Kogyo Co., LTD., Tokyo, Japan) column and by ion-exchange chromatography on a Mono S column (Pharmacia LKB Biochemistry) as described previously[7].

Toxin-Neutralization and Protection of Mice from Tetanus by Monoclonal Antibodies

In vitro toxin-neutralization and in vivo protection tests in OF1 mice by the purified MAbs were performed by recording symptoms of mice in scores over 3 weeks as described previously[8] using a human polyclonal IgG preparation, Tetanobulin (Green Cross Co., Osaka, Japan) of known international units (IU) as a standard.

Bovine Adrenal Chromaffin Cell Culture

Chromaffin cells were prepared by the method of Kilpatrick et al.[9]. The cells were permeabilized at 37°C with 20 μM digitonin in KG solution. [3H] norepinephrine ([3H]NE) release was assayed by the method of Dunn and Holz[10].

Preparation of Calmodulin and Other Proteins

Calmodulin was prepared and purified by the method of Kakiuchi et al.[11]. Caldesmon and actin were purified from chicken gizzard by the methods of Bretscher[12] and Strzelecka-Golaszewska et al.[13], respectively. Calpactin I complex (annexin II) was purified from bovine lung by the method of Glenny et al.[14].

Assay of Chromaffin Granule Aggregation

Chromaffin granules were prepared from bovine adrenal medullary tissue by differential centrifugation by the method of Pollard et al.[15]. Granule aggregation was monitored as increase in turbidity (A_{540}) in 8 min at room temperature in a buffer containing 30 mM KCl, 240 mM sucrose, 28 mM HEPES (pH 7.4), 2.5 mM EGTA in the presence or absence of 2.5 mM $CaCl_2$.

Assay of Vesicle Fusion

Chromaffin granules and plasma membrane were prepared by the method of Schäfer et al.[16] and lyzed in 10 mM Tris-HCl (pH 7.0). Fusion of plasma membranes with chromaffin granule membranes was monitored by following the relief of self-quenching of the fluorescent probe octadecyl rhodamine B chloride (R_{18}) by the method of Hoekstra et al.[17].

R_{18} was incubated with chromaffin granule membranes at room temperature for 1 h in the dark. Non-incorporated R_{18} was removed by gel permeation chromatography on a Sephadex G-50 column. R_{18} fluorescence was monitored in a 3-ml quartz cuvette at 560-nm excitation wavelength and 590-nm emission wavelength at room temperature in a buffer containing 139 mM K glutamate, 20 mM PIPES (pH 7.2), 5 mM glucose, 2 mM EGTA in the presence or absence of 2 mM $CaCl_2$. The fluorescence signal increase, due to dequenching upon fusion of the labeled chromaffin granule membranes with unlabeled plasma membrane vesicles, is expressed as a percentage of maximal dequenching obtained with 0.1% Triton X-100.

Materials

[3H]NE (30 Ci/m mol) was purchased from Amersham, MgATP and catalytic subunit of cAMP-dependent protein kinase from Sigma, digitonin from Wako Pure Chemicals (Osaka, Japan), R_{18} from Molecular Probes, Inc. (Eugene, Ore.), phospholipase A_2 from Boehringer Mannheim. Other chemicals, unless otherwise indicated, were purchased from Nacalai Tesque Inc., Kyoto, Japan).

RESULTS AND DISCUSSION

Immunological Roles of Fragments of the Toxin in Protection

Figure 1 shows the localization of the fragments and the three functional domains of the toxin in the tetanus toxin molecule according to our [A-B•C] nomenclature[18] and [L-Hn•Hc] nomen clature recently proposed[19]. Localization, in the toxin molecule, of the epitopes against which human MAbs[7,8] we prepared are directed is also illustrated in Fig. 1.

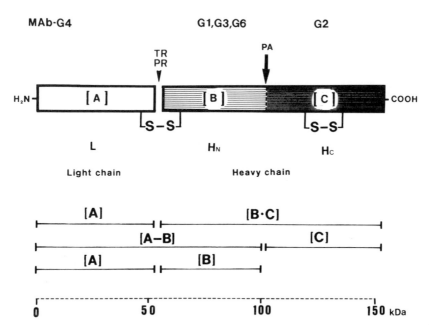

Figure 1. Subunit structure of tetanus neurotoxin prepared from culture filtrates and the localization of the epitopes for anti-tetanus human monoclonal antibodies (MAbs) and the isolated fragments in the tetanus toxin molecule. PR, endogeneous protease; TR, trypsin; PA, papain and MAb, monoclonal antibody.

Table 1. Relative immunopotencies of tetanus toxin-toxoid and toxin fragments.

Antigen	Potency
Toxoid[1](Whole toxin)	1.0
[A] (L)	0.050
[B] (H_N)	0.51
[C] (H_C)	0.026
[A-B]	0.052
[B•C]	0.72
[A-B]+[C]	0.026
[A]+[B•C]	0.25
[A] + [B] + [C]	0.20
[A] + [B]	0.059
[A] + [C]	0.048
[B] +[C]	0.088

[1]Toxin-toxoid was used as a reference for potency test.

Immunopotency of fragments of tetanus toxin. We previously isolated two complementary fragments [A](L) and [B•C](H_N•H_C)[1], and a subfragment of the latter, fragment [C](H_C)[2] and then fragment [A-B](L-H_N)[3]. Recently we succeeded in isolating the N-terminal half of the heavy chain, fragment [B](H_N), from fragment [A-B] (L-H_N)[4]. Thus now all the fragments corresponding to the three functional domains and the complexes of adjacent fragments of tetanus toxin are available at our hands. So, in order to identify the portions of the toxin molecule essential to elicit sufficient protective antibodies against tetanus, we examined the immunopotencies of whole toxin-toxoid, fragments and the mixtures of the fragments by immunizing mice and challenging with tetanus toxin. The results are summarized in Table 1. The heavy chain, fragment [B•C] (H_N•H_C), of the toxin has the highest immunogenic activity comparable with that of the whole toxin-toxoid on an equal weight basis (Table 1). Fragment [B] has a pretty high potency. The other individual fragments corresponding to the single domains or their mixtures showed much lower immunopotency (Table 1). Since currently used tetanus toxoid composed of formalin-treated whole toxin is known to elicit frequently local and rarely generalized adverse effects, these results indicate that the heavy chain (fragment [B•C]), a smaller molecules than whole toxin, can be useful for active immunization in place of the currently used whole toxin-toxoid, and will provide an effective tetanus vaccine with lesser adverse reactions than the current vaccine.

Protective effects of anti-tetanus human monoclonal antibodies(MAbs) directed against the three functional domains of the tetanus toxin molecule. Since the discovery of antitoxin therapy, horse polyclonal antitoxin serum preparations have long been used and then human polyclonal antitoxin antibody in the form of an immunoglobulin preparation is currently usually used for prevention and therapy of human tetanus. However, the supply of the human antitoxin preparations is limited and use of donor blood plasma as a source involves the risk of possible infection with viruses such as HIV. Therefore, hybrid cell lines that secrete human MAb against tetanus in vitro should be a new and better sources of antibody for clinical use. Recently we established[7] five hybrid cell lines for the first time that suffice all the requirements which must be satisfied at least to be used clinically: i.e. the hybridomas secrete (1) human MAbs (2) with high neutralizing activity and affinity, (3) and produce MAbs stably, (4) in serum-free medium, (5) sufficiently high amounts for large scale production in continuous culture.

The individual MAbs we prepared were found to neutralize tetanus toxin completely and rescue mice at sufficient doses. However, below the survival doses, in contrast to the neutralization with polyclonal antitoxin, with the individual MAb delayed intoxication, suppressed development of the symptoms and slowed rate of their development and greatly delayed death of mice much later than 4~7 days (in some cases ca. 3 weeks) after the injection, occurred. Thus we found it necessary to observe the symptoms of mice at least 3 weeks after the injection of the mixture of toxin and MAb, for evaluating the protective effects of anti-tetanus MAbs. We determined international units (IU) equivalence of MAbs using minimum survival dose (MSD) against the test toxin 20 MLD by comparing with human polyclonal IgG preparation of known IU.

Table 2. Neutralizing activities of human anti-tetanus monoclonal antibodies.

Antibody	Domain	Affinity[1] (Ka) ($\times 10^{10}$ M-1)	Minimum survival dose[2] (MSD) (µg IgG)	IU[3] per 70 µg IgG
G2	[C] (H$_C$)	11	0.89	1
G4	[A] (L)	14	28	0.03
G6	[B] (H$_N$)	10	0.28	3
Mixture[4]				
[G2, G4]	[C], [A]		0.28	3
[G2, G6]	[C], [B]		0.028	30
[G4, G6]	[A], [B]		0.89	10
[G2, G4, G6]	[C], [A], [B]			30

[1] Ka was estimated by ELISA and Scatchard plot analysis.
[2] Minimum survival dose(MSD) against 20 MLD of toxin.
[3] Tetanobulin (human polyclonal IgG antitoxin of known IU) used as a standard.
[4] Equal amounts of each MAb.

Table 2 summarizes the results on the neutralizing activities of representative human anti-tetanus MAbs, MAb-G2, MAb-G4 and MAb-G6, against the three domains of the toxin in comparison with human polyclonal IgG antitoxin. Although the binding affinities of these MAbs to tetanus toxin are equally very high (Table 2), there were great differences in their neutralizing activities on a basis of the amount of IgG. The fact that the single individual MAbs alone directed against each of the three domains of the toxin could neutralize toxin completely supports, on the other hand, the existence of the multiple steps in the intoxication process by tetanus toxin. The mixtures of the MAbs showed 3 to 10-fold increase of the neutralizing activity on a basis of the amount of IgG (Table 2). A mixture of the particular two kinds of MAbs, MAb-G2 and G6 that are directed against domains [C](H$_C$) and [B](H$_N$) of the heavy chain respectively showed the highest neutralizing activity similar to that of the mixture of three kinds of MAbs directed against the three functional domains of the tetanus toxin molecule (Table 2). The result indicates that the heavy chain of the toxin molecule plays an important role in immunogenic activity and agrees well with the result obtained on the immunopotency of the fragments of the toxin described above.

The highest neutralizing activity obtained with an appropriate combination of the MAbs (43 IU/100μg IgG) is considered to be comparable with that of the human polyclonal antitoxin preparation currently used clinically on the basis of toxin-specific IgG content and reach the theoretically maximal neutralizing activity[7]. These MAbs were very effective both for treatment and prophylaxis against tetanus in mice. Thus these human MAbs should be useful clinically for treatment and prevention of tetanus in place of currently used polyclonal antitoxin preparations.

In summary, studies in mice on the immunogenic activity of the fragments of tetanus toxin showed that the heavy chain, fragment [B•C] (H$_N$•H$_C$) was almost as efficient as whole toxin toxoid in eliciting a protective immune response on an equal weight basis, whereas other fragments or mixtures of other fragments had relatively poor immunogenic activity.

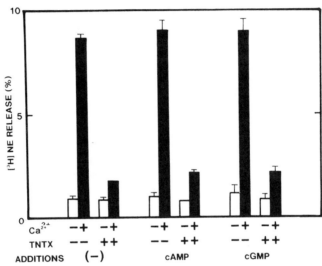

Figure 2.   Effects of exogeneously added cyclic AMP and cyclic GMP on the inhibition of norepinephrine(NE) release by tetanus toxin (TNTX) in digitonin-permeabilized chromaffin cells.

Studies on the neutralizing and protective effects of anti-tetanus human MAbs with high toxin-binding affinity showed that the single individual MAbs directed against each of the three domains of the tetanus toxin molecule neutralized toxin and rescued mice at sufficient doses and that an appropriate combination of two kinds of MAbs each directed against the two domains, [B] (H$_N$) and [C] (H$_C$), of the heavy chain had the highest neutralizing activity comparable with that of the mixture composed of three kinds MAbs each directed against the three domains of the toxin molecule. These results indicate that the heavy chain plays immunologically important roles in protection and support the existence of multiple steps, corresponding to the function of the three domains of the toxin, in the process of intoxication by tetanus toxin.

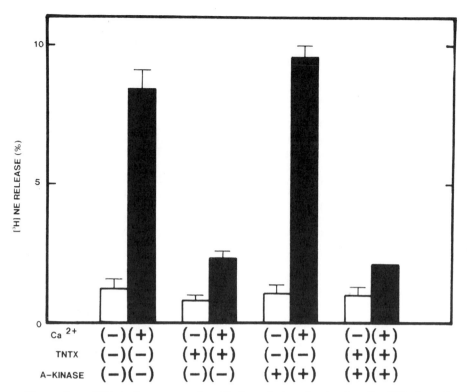

Ca $^{2+}$	(−)(+)	(−)(+)	(−)(+)	(−)(+)
TNTX	(−)(−)	(+)(+)	(−)(−)	(+)(+)
A−KINASE	(−)(−)	(−)(−)	(+)(+)	(+)(+)

Figure 3.   Effect of exogeneously added catalytic subunit of cyclic AMP-dependent protein Kinase(A-kinase) on the inhibition of norepinephrine(NE) release by tetanus toxin(TNTX) in digitonin-permeabilized chromaffin cells.

Studies on the Mechanism of Action of Tetanus Neurotoxin Using In Vitro Systems for Exocytosis.

The underlying mechanism of tetanus intoxication is inhibition of release of neurotransmitters from nerve terminals. However, molecular mechanism of inhibition of neurotransmitter release by tetanus toxins is yet unknown.   The difficulties in studying the mechanism of action of tetanus toxin are that suitable systems for biochemical analysis of exocytosis were not available on one hand, and on the other hand, that exact molecular mechanism leading to exocytotic secretion which would be a basis for studying the toxic action is unknown.   Recently digitonin-permeabilized bovine adrenal chromaffin cells have been shown to be useful for studying the biochemical mechanism of exocytotic secretion[10] and for biochemical analysis of inhibition of catecholamine release by tetanus toxin[20]. In vitro model systems for studying the mechanism of exocytosis have also been reported[17,21,22].   Therefore we tried to identify the target site(s) of tetanus toxin using these in vitro systems for exocytosis.   As for the mechanism of exocytotic secretion, entry of Ca^{2+} into cells is known to trigger the secretory process.   However, little is known on the subsequent sequence of events leading to the final step of exocytosis.   Some informational molecules, including cyclic nucleotides, cyclic nucleotide-dependent protein kinases, various Ca^{2+} binding proteins, have been suggested to be involved in intracellular signal transduction related to exocytotic process.   We focussed our efforts to examine the molecules, which have been suggested to play important roles in exocytotic secretion or have been reported to be influenced by tetanus toxin but their roles in intoxication are still controversial, whether they are target sites of tetanus toxin, using digitonin-permeabilized adrenal chromaffin cells. We also tested the effects of tetanus toxin on in vitro model systems for exocytosis. Thus we studied using digitonin-permeabilized chromaffin cells on the effects of tetanus toxin on (1) cGMP, cAMP and cAMP-dependent protein kinase(A-kinase), (2) protein kinase C (3) calmodulin, and on (4) calpactin I-dependent aggregation of chromaffin granules and (5) phospholipase A2-dependent fusion of chromaffin granule and plasma membranes.

Effects of cGMP, cAMP, and A-kinase on the inhibition of norepinephrine release from digitonin-permeabilized chromaffin cells by tetanus toxin. Injection of cGMP was reported to transiently reverse the paralytic effects of tetanus toxin on sphincter pupillae muscle in rabbits by King et al.[23] Recently it is suggested that the mechanism of action of tetanus toxin involves alterations of cGMP metabolism in neural cells[24,25] and adrenal cortical carcinoma[26].

As shown in Figure 2., exogeneously added cGMP and cAMP did not change the norepinephrine release and addition of cGMP or cAMP did not

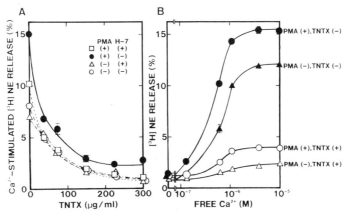

Figure 4.   Effects of phorbol myristate acetate(PMA) and H-7, an inhibitor of protein kinase C on the inhibition by tetanus toxin(TNTX) of Ca²⁺-dependent norepinephrine(NE) release from digitonin-permeabilized chromaffin cells.

recover the inhibition of norepinephrine release by tetanus toxin. We also examined the effect of addition of cAMP-dependent protein kinase (A-kinase) on the release of norepinephrine and on its inhibition by tetanus toxin.

Figure 3 shows that exogeneously added catalytic subunit of A-kinase (10 units/ml) did neither influence the norepinephrine release from digitonin-permeabilized cells nor recover its inhibition by tetanus toxin. Therefore, it is unlikely that in digitonin-permeabilized chromaffin cells, action of tetanus toxin is directly related to the alterations of cAMP and cGMP metabolism, including adenylate and guanylate cyclases, contrarily to the reports on the decrease in intracellular cGMP as decrease in acetylcholine release[24], and alteration of cGMP-specific phosphodiesterase[25] in tetanus toxin-treated PC12 cells, and inhibition of ANF-dependent guanylate cyclase by tetanus toxin in adrenal cortical carcinoma cells[26].

Effect of tetanus toxin on protein kinase C activity. Protein kinase C has been suggested to be involved in the action of tetanus toxin but its involvement has been controversial[20,27~30]. Therefore we examined if tetanus toxin inhibits exocytotic secretion by affecting protein kinase C in digitonin-permeabilized chromaffin cells. Addition of phorbol myristate acetate (PMA), an activator of protein kinase C, increased Ca^{2+}-dependent release of catecholamine in a dose-dependent manner and reached a plateau at 300 ng/ml of PMA. The amount of catecholamine increased by 50 to 90% of the control (Fig. 4A). H-7, an inhibitor of protein kinase C, inhibited the PMA-enhanced release of catecholamine but had no effect on the release of catecholamine from the PMA-untreated cells (Fig. 4A). The above result was confirmed using PKC(19-36), a more specific inhibitor of protein kinase C: PMA-enhanced release was inhibited by PKC(19-36), almost completely at 30 μM.

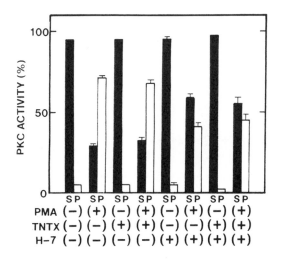

Figure 5.   Effects of tetanus toxin(TNTX) and H-7 on the translocation of protein kinase C(PKC) activity by phorbol myristate acetate(PMA) in digitonin-peremabilized chromaffin cells. S, cytosol fraction and P, particulate fraction.

These results suggest that protein kinase C is involved in the secretory process in the digitonin-permeabilized cells but is not necessarily essential for the secretory process and acts as a modulator of the exocytotic secretion. Tetanus toxin inhibited Ca^{2+}-induced norepinephrine secretion from digitonin-permeabilized chromaffin cells, irrespective of the presence or absence of PMA (Fig. 4A) in agreement with the report by Bittner and Holz[20]. The effect of tetanus toxin was different from those of the inhibitors of protein kinase C which inhibited only the part of secretion which had been enhanced by PMA.   Tetanus toxin inhibited norepinephrine release at Ca^{2+} concentrations of micromolar levels in the presence and absence of PMA (Fig 4B).

We further studied the effect of tetanus toxin on the intracellular distribution of protein kinase C. The intracellular distribution of protein kinase C changed by treating the cells with digitonin for 10 min: 50 to 60% of the total protein kinase C activity was leaked into the extracellular medium. Addition of PMA or tetanus toxin did not change the amount of leakage of protein kinase C from the permeabilized cells. Tetanus toxin did not influence significantly the changes in intracellular distribution of protein kinase C activity in the presence or absence of PMA in the digitonin-permeabilized cells (Fig. 5).

Thus we concluded that tetanus toxin inhibits catecholamine secretion from digitonin-permeabilized chromaffin cells without affecting protein kinase C activity. The present results show that tetanus toxin does not inhibit or activate protein kinase C and that the target site(s) of tetanus toxin is not directly related to the site(s) in which protein kinase C is involved in exocytotic secretion.

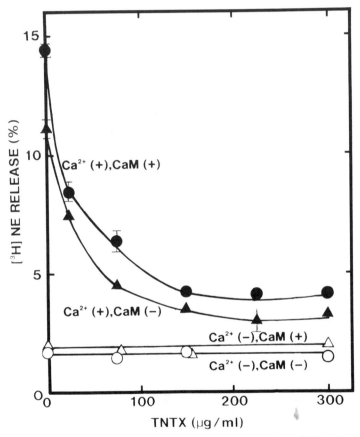

Figure 6. Effects of tetanus toxin(TNTX) on norepinephrine(NE) release from digitonin-permeabilized cells in the presence or absence of calmodulin and Ca^{2+}.

Effect of tetanus toxin on calmodulin. Indirect evidence obtained by experiments using calmodulin angatonists and anti-calmodulin antibody has suggested that calmodulin(CaM) is involved in the secretory process. So we firstly examined the effect of calmodulin on norepinephrine release from chromaffin cells by introducing calmodulin directly into digitonin-permeabilized cells. We found that the addition of calmodulin to the permeabilized cells caused concentration-dependent increase in norepinephrine release with a maximal effect at 30 μM calmodulin: Ca^{2+}-dependent secretion was enhaced 50~60% (Fig. 6).

Calmodulin (20 μM) at its physiological concentration in cells enhanced Ca^{2+}-dependent secretion of norepinephrine submaximally and calmodulin enhanced only at the free Ca^{2+} concentrations of more than 10^{-6} M. In contrast to the effect of calmodulin, caldesmon actin and bovine serum albumin did not enhance the secretion. W-7, an inhibitor of calmodulin inhibited the calmodulin-enhanced secretion of norepinephrine from the permeabilized cells.

These results indicate that calmodulin plays an important role in exocytotic secretion. Tetanus toxin inhibited calmodulin-dependent secretion of catecholamine (Fig. 6). We further examined the possibility that tetanus toxin binds to calmodulin and apparently inhibits calmodulin-enhanced secretion. Binding studies by gel permeation chromatography and by using a zero-length crosslinker, 1-ethyl-3-(3-dimethylaminopropyl) carbodiimide revealed that calmodulin and tetanus toxin did not form any complex. The inhibition of catecholamine release by tetanus toxin was not recovered by exogeneously supplied excess calmodulin. Therefore calmodulin is not directly related to the target site(s) of tetanus toxin.

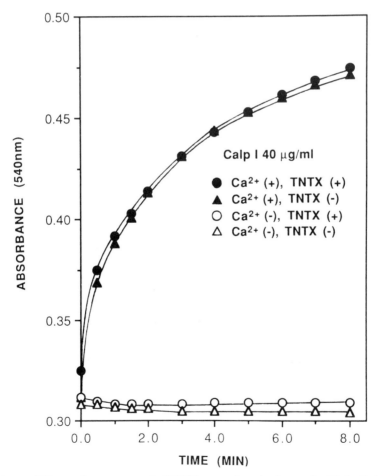

Figure 7. Effects of tetanus toxin(TNTX) on calpactin I complex(Calp I)-dependent aggregation of chromaffin granules.

Effect of tetanus toxin on aggregation of chromaffin granules in vitro.
Calpactin I (annexin II) has been suggested to play an important role in exocytotic secretion from chromaffin cells[31,32] and was reported to aggregate chromaffin granules in Ca^{2+}-dependent manner and this aggregation is considered as a model for a step preceding to the fusion process for exocytosis[21,22]. We examined the effect of tetanus toxin on the aggregation of chromaffin granules by calpactin I.

Figure 7 shows that aggregation of chromaffin granules occurred in the presence of calpactin I in concentration-dependent and calcium dependent manners. Tetanus toxin did not influence the calpactin I-dependent aggregation of chromaffin granules (Fig. 7). Thus we concluded that the target site(s) of tetanus toxin is different from calpactin I and from the site(s) where calpactin I acts in the aggregation process.

Effect of tetanus toxin on the fusion of chromaffin granule membrane and plasma membrane in vitro. Recently using R_{18}, fusion of chromaffin granule membranes with plasma membranes in the presence of Ca^{2+}-dependent phospholipase $A_2(PLA_2)$ was reported[17] and the important role of PLA_2 in the fusion of membranes near the final step of the exocytotic secretion was suggested. We examined the effect of tetanus toxin on this model system for fusion. Figure 8 shows that tetanus toxin had no effect on the PLA_2-dependent fusion of the membranes. Thus it is likely that tetanus toxin did not interact with PLA_2 nor with the process to which PLA_2 is involved in fusion of granule membrane and plasma membrane vesicles in vitro.

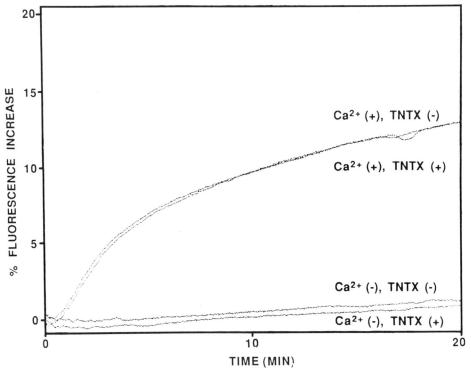

Figure 8.   Effect of tetanus toxin(TNTX) on phospholipase A_2-dependent fusion of chromaffin granule membrane and plasma membrane vesicles.

In summary, we examined the effects of tetanus toxin on the various molecules, which have been suggested to be involved in the process of exocytosis, and on the putative candidate target molecules of tetanus toxin, which have been so far reported and some of the results of which have been controversial, using in vitro systems of digitonin-permeabilized chromaffin cells, aggregation of chromaffin granules and fusion of membrane vesicles. In spite of a number of reports on the subject, the molecular mechanism of tetanus toxin action is still elusive. During the experiments, we realized that very little is known on the sequence of events of exocytosis beginning with Ca^{2+} influx and leading to the final step. However, through our experiments described above, from the other point of view using tetanus neurotoxin as a specific biological inhibitor of exocytotic secretion, we could show directly that calmodulin is involved in the exocytotic secretion and plays an important role in exocytosis.

REFERENCES

1. M. Matsuda and M. Yoneda, Isolation and purification of two antigenically active, "complementary" polypeptide fragments of tetanus neurotoxin, *Infect. Immun.* 12:1147-1153. (1975).
2. M. Matsuda and M. Yoneda, Antigenic substructure of tetanus neurotoxin, *Biochem. Biophys. Res. Commun.* 77:268-274. (1977).
3. K. Ozutsumi, D.-L. Lei, N. Sugimoto,and M. Matsuda, Isolation and purification by high performance liquid chromatography of a tetanustoxin Fragment (Fragment [A-B]) derived from mildly papain-treated toxin, *Toxicon* 27:1055-1057. (1985).
4. M. Matsuda, D.-L. Lei, N. Sugimoto, K. Ozutsumi, and T. Okabe, Isolation, purification, and characterization of Fragment B, the NH_2-terminal half of the heavy chain of tetanus toxin, *Infect. Immun.* 57:3588-3593. (1989).
5. M. Matsuda, G. Makinaga, and T. Hirai, Studies on the antibody composition and neutralizing activity of tetanus antitoxin sera from various species of animals in relation to the antigenic substructure of the tetanus toxin molecule, *Biken J.* 26:133-143. (1983).
6. R. Murata, E. Wada, A. Yamamoto, and K. Kubota, Studies on the standardization of tetanus toxoid: differences in the relative potency by animal species, *Jpn. J. Med. Sci. Biol.* 14:121-129. (1961).
7. M. Kamei, S. Hashizume, N. Sugimoto, K. Ozutsumi, and M. Matsuda, Establishment of stable mouse/human-human hybrid cell lines producing large amounts of anti-tetanus human monoclonal antibodies with high neutralizing activity, *Eur. J. Epidemiol.* 6:386-397. (1990).
8. M. Matsuda, M. Kamei, N. Sugimoto, Y. Ma, and S. Hashizume, Characteristics of toxin-neutralization by anti-tetanus humanmonoclonal antibodies directed against the three functional domains [A], [B] and [C] of the tetanus toxin molecule and a reliable method for evaluating the protective effects of monoclonal antibodies, *Eur. J. Epidemiol.* 8:1-8. (1992).
9. D.L. Kilpatrik, F.H. Ledbetter, K.A. Carson, A.G. Kirshner, R. Slepstis, and N. Kirshner, Stability of bovine adrenal medulla cells in culture, *J. Neurochem.* 35:679-692. (1980).
10. L.A. Dunn, and R.W. Holz, Catecholamine secretion from digitonin-treated adrenal medullary chromaffin cells, *J. Biol. Chem.* 258:4989-4993. (1983).
11. S. Kakiuchi, K. Sobue, R. Yamazaki, J. Kambayashi, M. Sakon, and G. Kosaki, Lack of tissue specificity of calmodulin: A rapid and high-yield purification method, *FEBS Lett.* 126:203-207. (1981).
12. A. Bretscher, Smooth muscle caldesmon: Rapid purification and F-actin cross-linking properties, *J. Biol. Chem.* 259:12873-12880. (1984).
13. H. Strzelecka-Golaszewska, E. Próchniewicz, E. Nowak, and S. Zmorzynski, Chicken-gizzard actin: Polymerization and stability, *Eur. J. Biochem.* 104:41-52. (1980).
14. J.R. Glenny Jr., B. Tack, and M.A. Powell, Calpactins: Two distinct Ca^{++}-regulated phospholipid- and actin-binding proteins isolated from lung and placenta, *J. Cell Biol.* 104:503-511. (1987).
15. H.B. Pollard, O. Zinder, P.G. Hoffman, and O. Nikodejevic, Regulation of the

transmembrane potential of isolated chromaffin granules by ATP, ATP analogs, and external pH, *J. Biol. Chem.* 251:4544-4550. (1976).

16. T. Schäfer, U.O. Karli, E.K.-M. Gratwohl, F.E. Schweizer, and M.M. Burger, Digitonin-permeabilized cells are exocytosis competent, *J. Neurochem.* 49:1697-1707. (1987).

17. D. Hoekstra, T. Boer, K. Klappe, and J. Wilschut, Fluorescence method for measuring the kinetics of fusion between biological membranes, *Biochem.* 23;5675-5681. (1984).

18. B. Bizzini and A.A. Fedinec, Structural and functional characterization of tetanus toxin. Nomenclature, *in*: "The Eighth International Conference on Tetanus" G. Nisticò, B. Bizzni, B. Bytchenko, R. Triau, eds., Pythagora Press, Rome-Milan. p.37-42. (1989).

19. J. Aguilera, G. Ahnert-Hilger, H. Bigalke, B.R. DasGupta, O. Dolly, E. Habermann, J. Halpern, S. van Heyningen, J. Middlebrook, S. Mochida, C. Motecucco, H. Niemann, K. Oguma, M. Popoff, B. Poulain, L. Simpson, C.C. Shone, D.E. Thompson, U. Weller, H.H. Wellhöner, S.M. Whelan, Clostridial neurotoxins-Proposal of a common nomenclature, *FEMS Microbiol. Lett.* 90:99-100. (1992).

20. M.A. Bittner and R. W. Holz, Effects of tetanus toxin on catecholamine release from intact and digitonin-permeabilized chromaffin cells, *J. Neurochem.* 51:451-456 (1988).

21. U.O. Karli, T. Schäfer, and M.M. Burger, Fusion of neurotransmitter vesicles with target membrane is calcium independent in a cell-free system, *Proc. Natl. Acad. Sci. USA* 87:5912-5915. (1990).

22. D.S. Drust, and C.E. Creutz, Aggregation of chromaffin granules by calpactin at micromolar levels of calcium, *Nature* 331:88-91. (1988).

23. L.E. King Jr., A.A. Fedinec, and W.C. Latham, Effects of cyclic nucleotides on tetanus toxin paralyzed rabbit sphincter pupillae muscle, *Toxicon.* 16:625-631. (1978).

24. K. Sandberg, C.J. Berry, E. Eugster, and T.B. Rogers, Studies on the molecular mechanism of action of tetanus toxin: A role for cGMP in the release of acetylcholine from PC12 cells, *in*: "The Eighth International Conference on Tetanus" G. Nisticò, B. Bizzini, B. Bytchenko, and R. Triau, eds., Pythagora Press, Rome-Milan. p.114-126. (1989).

25. K. Sandberg, C.J., Berry, and T.B. Rogers, Studies on the intoxication pathway of tetanus toxin in the rat pheochromocytoma (PC12) cell line. Binding, internalization, and inhibition of acetylcholine release, *J. Biol. Chem.* 264:5679-5686. (1989).

26. A.A. Fedinec, T. Duda, B. Bizzini, N.G.F. Cooper, and P.K. Sharma, Tetanus toxin inhibits the ANF receptor guanylate cyclase singnaling systemat its guanylate cyclase domain, Abstracts of 9th International Congress on Tetanus and Anti-Infectious Defense (Oct. 8-11, Granada, Spain), p.112. (1991).

27. J. Aguilera and E. Yavin, *In vivo* translocation and down-regulation of protein kinase C following intraventricular administration of tetanus toxin, *J. Neurochem.* 54:339-342. (1990).

28. J.L. Ho and M.S. Klempner, Diminished activity of protein kinase C in tetanus toxin-treated macrophages and in the spinal cord of mice manifesting generalized tetanus intoxication, *J. Infect. Dis.* 157:925-933. (1988).

29. R.V. Considine, J.K. Bielicki, L.L. Simpson, J.R. Sherwin, Tetanus toxin attenuates the ability of phorbol myristate acetate to mobilize cytosolic protein kinase C in NG-108 cells. *Toxicon* 28:13-19. (1990).

30. R.V. Considine, C.M. Handler, L.L. Simpson, and J.R. Sherwin, Tetanus toxin inhibits neurotensin-induced mobilization of cytosolic protein kinase C activity, *Toxicon* 11:1351-1357. (1991).

31. S.M. Ali, M.J. Geisow, and R.D. Burgoyne, A role for calpactin in calcium-dependent exocytosis in adrenal chromaffin cells, *Nature* 340:313. (1989).

32. T. Nakata, K. Sobue, and N. Hirokawa, Conformational change and localization of calpactin I complex involved in exocytosis as revealed by quick-freeze, deep-etch electron microscopy and immunocytochemistry, *J. Cell Biol.* 110:13-25. (1990).

EVIDENCES FOR A LINK BETWEEN PROTEOLYSIS AND THE INHIBITION OF [³H]-NORADRENALINE RELEASE BY THE LIGHT CHAIN OF TETANUS TOXIN

Ernst Habermann and Dagmar Sanders

Rudolf-Buchheim-Institut für Pharmakologie
Justus-Liebig-Universität
D-6300 Gießen, Federal Republic of Germany

The mode of action of tetanus toxin (TeTx) light chain was studied on permeabilized bovine adrenomedullary cells in culture. Pretreatment with digitonin allowed free access of proteins together with other agents into the interior of the cell. Ca^{2+}-evoked release of [³H]-noradrenaline, basal outflow in the absence of Ca^{2+}, and total content of radioactivity was measured.[1] Light chain inhibited the release only. We then compared numerous enzymes to light chain with respect to the three parameters mentioned. Phospholipase A_2 from *Crotalus terrificus*, *C. botulinum* C2 toxin together with NAD, and neuraminidase from *V. cholerae* were without effect. Some proteinases, for instance trypsin, chymotrypsin and papain, destroyed the cells as evident from the increased basal outflow and decrease of total content of radioactivity. Others such as thermolysin, collagenase from *C. histolyticum*, endoproteinase Lys-C or leucine amino peptidase were inactive in the range investigated. However, endoproteinase glu-C depressed the Ca^{2+}-evoked release while it increased the basal outflow only slightly over a wide range of concentrations. This proteinase was about 10 times less potent than the light chain of TeTx. Thus light chain at least resembled a proteinase in its action on adrenomedullary cells, although it did hydrolyse none among the proteins and peptides tested. The putative substrate should be limited to neurons and adrenomedullary cells since exocytosis was not influenced in permeabilized pancreatic acinar cells (personal communication from I. Schulz, Institute of Physiology, Homburg/Saar). Searches for specific binding of ¹²⁵I-labelled light chain to cultured cells or subcellular fractions from adrenal medulla were fruitless.

The protease hypothesis was corroborated by inhibition experiments. Light chain was preincubated with a series of protease inhibitors for 1 hour at 37°C and then applied to the permeabilized cells. It was resistant to diisopropylfluorophosphate and phenylmethane-sulfonylfluoride and hence not a serine protease. Light chain also did not interact with leupeptin, pepstatin or bestatin. Neither did it carry a thiol in its active site because it was not inhibited by trans-ethoxysuccinic acid (also called E 64). Nevertheless overall modification of SH groups in light chain by iodocetamide, N-ethylmaleimide or dithionitropyridine decreased or abolished its pharmacological activity. Inhibition by dithionitropyridine which is known to convert two SH groups into a disulfide link was reverted by dialysis against 1 mmol/l DTT. It was concluded that SH-groups are not directly involved, the more as light chain was equiactive in the presence or absence of DTT. This experiment also excluded a disulfide shift between light chain and permeabilized cells. However, inhibition of light chain (0.2 µmol/l) was complete with EDTA (2.5 mmol/l) or the chelator o-phenanthroline (1 mmol/l). Taken altogether the inhibitor spectrum (Table 1) points to a relationship between light chain and metalloproteases carrying Zn^{2+} in their active site. The ion (0.6-0.9 mol/mol protein) was found in thoroughly dialysed two-chain tetanus toxin by atomic absorption spectroscopy (Dr. Scheuermann, Hoechst AG, Frankfurt). It remains attached to the protein even when dialysed against phenanthroline (1 mmol/l).

Table 1. Effects of proteinase inhibitors and SH-reagents on the activity of TeTx light chain.

Inhibitor	Concentration µmol/l	% Inhibition
EDTA[a]	2500	100
EDTA[a]	250	50
1,10-o-Phenanthroline[a]	1000	100
Phosphoramidon[a]	700	n.i.
Thiorphan[a]	20	n.i.
Diisopropylfluorophosphate[a]	1000	n.i.
Phenylmethanesulfonylfluoride[a]	100	n.i.
Leupeptin[a]	1	n.i.
Bestatin[a]	130	n.i.
Pepstatin[a]	1	n.i.
trans Epoxysuccinic acid[a]	28	n.i.
Iodoacetamide[b]	4000	85
N-Ethylmaleimide[b]	2000	100
Dithionitropyridine[b,c]	2000	100

n.i. = not inhibited

[a]Light chain (200 and 20 nmol/l, respectively) was incubated in complete potassium glutamate buffer for 1 hour at 37°C with the inhibitor concentrations given in the table. The protease inhibitors did not influence basal outflow, Ca^{2+}-evoked release or total radioactivity in the absence of light chain.

[b]Incubation of light chain (20 µmol/l) with the SH-reagents mentioned for 2 hours at 25°C was followed by dialysis against K^+-glutamate (139 mmol/l) because permeabilized cells were also sensitive to N-ethyl-maleimide and dithionitropyridine. The modified preparation was then diluted in complete glutamate buffer before use.

[c]A parallel sample was inactivated with dithionitropyridine and completely reactivated by subsequent dialysis against buffer with DTT (1 mmol/l).

All light chains from clostridial neurotoxins contain highly homologous sequences[6] which might be suspected as binding sites for Zn^{2+}. They can be superposed to the active sites of at least eleven proteases[3] and to an aminopeptidase.[4] All of them start with preferentially hydrophobic amino acid residues, followed (numbering for TeTx) by histidine (233) which is known to bind Zn^{2+} in thermolysin,[2] glutamic acid (234) which is involved in hydrolysis of peptide bonds and a very hydrophobic residue, namely leucine, isoleucine or phenyl-alanine. The next, variable amino acid precedes the second histidine (237) of the active site. The five light chains so far sequenced from neurotoxins share a third histidine (240) residue and the triplet leucine-tyrosine-glycine which might contribute to their particular substrate specificity. An additional glutamic acid residue from a more C-terminal position should be brought to the active site by appropriate folding as predicted from the known primary structure of thermolysin.[2] Particular attention should be directed in this context to the two strictly homologous[6] glutamyl residues in the positions 271 and 360 of the tetanus light chain (Table 2).

Table 2. Sequence homologies in the active sites of Zn-proteinases and a Zn-exopeptidase compared to the related sequences of the light chains of clostridial neurotoxins.

Light chains of clostridial neurotoxins[6]															
BoNT E	(209)	T	L	M	H	E	L	I	H	S	L	H	G	L	Y	G
BoNT A	(220)	T	L	A	H	E	L	I	H	A	G	H	R	L	Y	G
BoNT C	(226)	I	L	M	H	E	L	N	H	A	M	H	N	L	Y	G
BoNT D	(226)	A	L	M	H	E	L	T	H	S	L	H	Q	L	Y	G
TeTx	(230)	L	L	M	H	E	L	I	H	V	L	H	G	L	Y	G
Zn-proteinases[3]																
Proteinase Ht-d		T	M	A	H	E	L	G	H	N	L	G	M			
Human type IV collagenase		V	A	A	H	E	F	G	H	A	M	G	L			
Human collagenase		V	A	A	H	E	L	G	H	S	L	G	L			
Human stromelysin		V	A	A	H	E	I	G	H	S	L	G	L			
Stromelysin-2		V	A	A	H	E	L	G	H	S	L	G	L			
Pump-1		A	A	T	H	E	L	G	H	S	L	G	M			
Transin		V	A	A	H	E	L	G	H	S	L	G	L			
Transin-2		V	A	A	H	E	L	G	H	S	L	G	L			
Serratia zinc protease		T	F	T	H	E	I	G	H	A	L	G	L			
Rabbit enkephalinase		V	I	G	H	E	I	T	H	G	F	D	D			
Thermolysin BT		V	V	A	H	E	L	T	H	A	V	T	D			
Thermolysin BS[4]		V	V	G	H	E	L	T	H							
Zn-exopeptidase[4]																
Amino peptidase N		V	I	A	H	E	L	A	H	Q	W	F	G			

Single-chain TeTx and two-chain TeTx were about 100 times less potent than light chain. Full potency was achieved by reduction of two-chain but not of single-chain toxin. The toxins behave like proenzymes which are activated to the mature enzyme by combination of limited proteolysis and reduction.[1,5]

The three evidences given are independent. Together they confirm and extend the starting hypothesis that the light chain of tetanus toxin, and probably of all clostridial neurotoxins inhibits [^{3}H]-noradrenaline release from adrenomedullary cells by degradation of (a) specific, still unknown protein(s) involved in exocytosis. It remains to be assessed if the same process is necessary and, if so, sufficient to inhibit the neuronal transmitter release too.

REFERENCES

1. Ahnert-Hilger G, Weller U, Dauzenroth ME, Habermann E, Gratzl M. The tetanus toxin light chain inhibits exocytosis. FEBS Lett 1989; 242:245-248.
2. Kester WR, Matthews BW. Comparison of the structures of carboxypeptidase A and thermolysin. J Biol Chem 1977; 252:7704-7710.
3. Shannon JD, Baramova EN, Bjarnason JB, Fox JW. Amino acid sequence of a crotalus atrox venom metalloproteinase which cleaves type IV collagen and gelatin. J Biol Chem 1989; 264:11575-11583.
4. Watt VM, Yip CC. Amino acid sequence deduced from a rat kidney cDNA suggests it encodes the Zn-peptidase aminopeptidase N. J Biol Chem 1989; 264:5480-5487.
5. Weller U, Mauler F, Habermann E. Tetanus toxin: Biochemical and pharmacological comparison between its protoxin and some isotoxins obtained by limited proteolysis. Naunyn-Schmiedeberg's Arch Pharmacol 1988; 338:99-106.
6. Whelan SM, Elmore MJ, Bodsworth NJ, Atkinson T, Minton NP. The complete amino acid sequence of the clostridium botulinum type E-neurotoxin, derived by nucleotide-sequence analysis of the encoding gene. Eur J Biochem 1992; 204:657-667.

TETANUS TOXIN AS A TOOL
FOR INVESTIGATING THE CONSEQUENCES
OF EXCESSIVE NEURONAL EXCITATION

Jane Mellanby

Department of Experimental Psychology
University of Oxford
England OX1 3UD

INTRODUCTION

The tetanus toxin model of limbic epilepsy, developed in the late 1970s,[16] has been used to study the evolution and waning of a chronic epileptic focus and its effects on the physiology of the hippocampus and the behaviour of the animal—the integrated output of the nervous system.[2,7,14,18] We have argued that the particular usefulness of this model, complementary to those produced by introduction of directly cytotoxic agents such as kainic acid, alumina cream, iron etc, is that firstly the toxin, in the minute doses which we use, does not apparently kill cells at the injection site, and that secondly the epilepsy wanes within a few weeks and the EEG apparently returns to normal.[17,19] This allows the study of both recovery from epilepsy and the long-term effects of having had epilepsy.

The above statement concerning the lack of direct cytotoxicity of the toxin is based on histological examination of the injection site in more than 100 brains injected with 5-20 mouse LD_{50} of toxin into the ventral hippocampus. It is of interest that where larger doses of toxin are used (>500 MLD) localized cell death does occur,[1] presumably as a result of the excessive excitation produced near the site of injection.[24]

We have recently reported[22,23] the occurrence of cell death at a site remote from the locus of injection in about 1/3 of our injected rats (Table 1). This occurs in *dorsal* CA1 and entails complete loss of the CA1 pyramidal neurones. Jefferys et al.[12] have now found that when they inject similar doses of the toxin into *dorsal* hippocampus, loss of CA1 neurons in dorsal hippocampus (particularly contralaterally) was detected in about 1/10 of the rats. Jefferys *et al.*[12] have also reported that in one strain of rats used, but not in another, many of the pyramidal cells in all divisions of the hippocampus became acidophilic (dark cells), which suggested that genetic and/or environmental factors may be important in determining whether pathology results from the initial toxin injection.

The long-term outcome in human epilepsy is extraordinarily variable. It is well-known that fits in infancy may predispose to the development of temporal lobe epilepsy (TLE) in later life and that loss of pyramidal cells in the hippocampus is commonly found in specimens resected

Botulinum and Tetanus Neurotoxins, Edited by
B.R. DasGupta, Plenum Press, New York, 1993

Table 1. Incidence of pyramidal cell loss in CA1 region of dorsal hippocampus in rats previously injected with tetanus toxin (5-20 mouse LD_{50}).

Time	Total no. of rats	No. with cell loss		% with cell loss
1. Unilateral toxin injection				
6-7 days	22	4	(2*)	18.1
10-20 days	27	10	(2*)	37.0
8 weeks	13	4	(0*)	30.8
2. Bilateral toxin injection				
5 weeks-3 months	10	6		60.0%
9 months	7 ⎫	2 ⎫		39.4%
1 yr	8 ⎬ 23	1 ⎬ 7		30.4%
14 months	8 ⎭	4 ⎭		

* no. with contralateral loss only

at operation on patients with intractable TLE.[3,4] It was therefore of particular interest to investigate the physiology and anatomy of a small minority of our toxin-injected rats which, having apparently recovered from their epilepsy, relapsed at a later date. We have already reported[22] that the injections in ventral hippocampus lead within a few days to activation of microglial cells in the dorsal hippocampus. Since such cells have been implicated in other situations in cell death and the scavenging of debris, and also in synaptic remodelling,[8,10,20] it was an obvious possibility that it might be variability in the early microglial response that contributed to the variability in the occurrence of cell death, and also perhaps to the variability in long-term epileptic outcome.

ACUTE ELECTROPHYSIOLOGICAL CHANGES

We have shown that tetanus toxin (100-200 mouse LD_{50}) injected acutely into the ventral hippocampus of the urethane-anaesthetized rat caused a local block of inhibition within 1-3 hours, followed an hour or so later by a block of excitation also.[24] The early block of inhibition occurs with an overall increase in the size of the population spike and this is usually associated with the occurrence of multiple (2-3) population spikes in response to perforant path stimulation.

EARLY STAGES OF CHRONIC EPILEPSY

Jordan and Jefferys[15] have shown that the lower doses of toxin which are used to produce a chronic epileptic syndrome also block inhibition, and this block lasts for some weeks. An alternative approach, using ex-vivo K^+-stimulated Ca-dependent GABA release from hippocampal slices has shown a block of GABA release 10-14 days after toxin injection (and recovery to normal by 6 weeks).[13] The block of inhibitory processes, which for some reason predominates over the toxin's block also of excitatory transmission, is presumably what leads to the development of an epileptic focus which produces intermittent seizure activity. During the early evolution of the toxin-induced epileptic focus, the seizure activity soon spreads throughout the hippocampus and to connected areas such as nucleus accumbens, hypothalamus and cingulate gyrus. As the frequency and length of the seizure discharges increase over the 2-3 days after injection, so they come to elicit motor fits.[9]

MICROGLIAL CELL RESPONSES

Neurones are not, however, the only cell type which responds to the epileptic challenge. In recent work we have demonstrated that ventral hippocampal injection of toxin leads within a few days to a characteristic activation of microglial cells (the resident macrophages of the CNS) in dorsal CA1 area of the hippocampus. Initially, the microglia are activated in the region of the CA1 dendrites where they align themselves along the dendritic longitudinal axis and adopt the morphology of "rod cells".[21] Preliminary electron microscopy (carried out in collaboration with John Morris) has shown that at this stage the microglia do not appear to be actively phagocytosing. There are then two possible outcomes: either the whole population of CA1 cells dies and at the same time the microglia adopt a 'bushy activated' form in the region of the cell body layer, or the CA1 cells survive and the activated microglia disappear—presumably reverting to their non-activated, ramified morphology. We do not at present know whether the CA1 cells die first, perhaps as a result of the hyperexcitation caused by the toxin-induced epileptic focus in ventral hippocampus, and the debris is then scavenged by the microglia, or whether the activated microglia may play a direct role in killing the cells.

RECOVERY

The finding that the motor fits usually cease within about 3 weeks and the EEG apparently returns to normal within about 6 weeks[9] is likely in part to be associated with the gradual recovery of inhibitory processes.[13] We do not yet know whether this recovery involves sprouting of new terminals from the toxin-blocked synapses (as occurs at neuromuscular junctions after tetanus or botulinum toxin block[5,6]) or whether more radical synaptic reorganization may occur.

LONG-TERM CHANGES IN FUNCTION

However, even though the rats generally appear to have recovered from their epilepsy by 6-8 weeks after injection, there are important long-term changes in physiology at sites remote from the point of toxin injection. Thus, months after injection of toxin into ventral hippocampus, we have found, in *dorsal* hippocampus, decreased excitability of both CA3 and dentate granule cells, and decreased paired pulse inhibition[19] and Jefferys and Empson[11] have found reduced inhibition also in a mirror focus which develops after unilateral hippocampal toxin injection. In parallel, and apparently correlated with the reduction in CA3 cell excitability, we have found that the rats are impaired in learning behavioural tasks which require hippocampal function. Thus it appears that there is considerable adjustment of synaptic efficacy in response to the spread of epileptic activity and such adjustments while being required for the recovery from the epilepsy may also involve a price to be paid in terms of the overall functional efficiency of the brain.

RECURRENT FITS AT LONG SURVIVAL TIMES

Over the years that we have worked on the long-term changes which occur in the tetanus toxin model, we have occasionally observed rats in which there has been a recurrence of fits after they have apparently stopped. This is reminiscent of the clinical situation. In a recent experiment, one rat was found to be having fits again 5 months after toxin injection. The fits, unlike those at the usual early stage, were elicited by handling. This rat was put through our standard electrophysiological procedure for measuring field potentials (excitation and inhibition) in the dentate gyrus in response to stimulation of the perforant path. It was found to have an exceptionally low threshold for the elicitation of a population spike in the dentate gyrus, and the population spike itself appeared double (Figure 1A). Paired pulse inhibition at 15 msec separation was normal, but at 500 msec it appeared weaker than normal. Histological examination showed total bilateral absence of the CA1 layer in dorsal hippocampus. During the next three months, 3 toxin-injected rats from a parallel experiment in which there were originally 20 toxin rats, developed recurring fits. All the surviving rats in that experiment were then kept under surveillance until one year after the original toxin injection and were then examined electrophysiologically (Paru Patel, unpublished) in terminal acute experiments and finally the brains were sectioned and examined histologically (J.A.G. Shaw, unpublished). The 3 with recurring fits had low, but not exceptionally low, thresholds for excitation of the population spike and all had the CA1 cell loss, and so did a further rat where fits did not appear to have recurred. Three out of the 4 rats with CA1 cell loss showed a double spike on stimulating the perforant path (Figure 1B, C, D); none of the rats without cell loss showed this. The double spike echoes the findings usually encountered at earlier times in the syndrome. These were seen both in the acute experiments mentioned above and also in a chronic experiment using implanted electrodes (Yang, Sundstrom and Mellanby, unpublished) where double spikes occurred in most toxin-injected rats up to 8-9 weeks after injection, but the second spike then waned.

The incidence of the late recurrence of fits is of course strategically extraordinarily difficult to assess for the simple reason that continuous monitoring of fits would be needed from an adequate number of rats (say 50) over a period of up to a year. In practice, the experimenter relies on the chance observation of a fit occurring when the rat is being weighed or the cage changed—and we do have the impression that such procedures can elicit fits in susceptible individuals at these long time intervals after toxin injection. The incidence of complete pyramidal cell loss in the CA1 region of dorsal hippocampus is around 30%—and it appears that there is no greater incidence in animals sacrificed after about one year than in those studied after 10-20 days. Table 1 shows the incidence of death of the dorsal hippocampal CA1 pyramidal cell layer in all those rats on which we have completed histological examination of the relevant brain area. This suggests that the cell death in the first weeks is an all-or-none process, after which time the challenge to CA1 cell survival disappears. (The slightly higher incidence of the cell loss in rats sacrificed after 10 days compared with those sacrificed at 6-7 days does not reach statistical significance; chi-squared $p > 0.05$).

In view of the importance of hippocampal pathology in human epilepsy, it is clear that future research must be directed towards determining the factors which decide whether CA1 cells live or die in the tetanus toxin model.

POSSIBLE SEQUENCE OF EVENTS

Thus it can be proposed that the loss of the dorsal CA1 pyramidal cells may, in some indirect manner, lead to the recurrence of chronic epilepsy, although such a causal link is supported by circumstantial evidence only. The reverse causal process i.e. the fits causing the cell loss, is also possible.

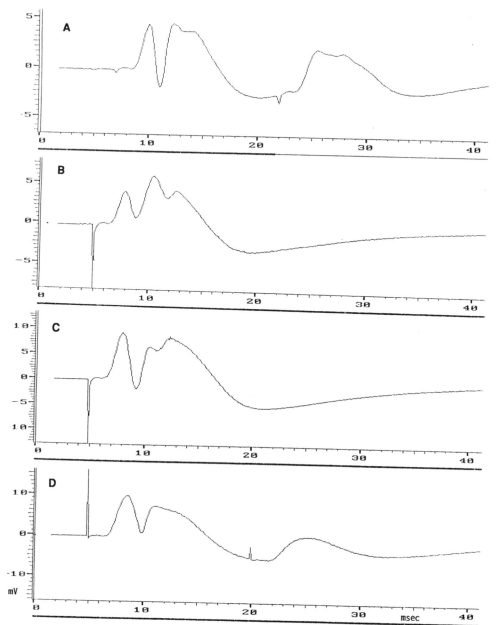

Figure 1. Examples of field potentials recorded in the dentate gyrus of dorsal hippocampus in response to perforant path stimulation at long time intervals after injection of 5-10 LD50 tetanus toxin into ventral hippocampus.

A. A rat investigated 5 months after toxin injection. The rat was still having 2-5 fits per day.

    Paired pulse inhibition. Conditioning stimulus and test stimulus size 2v.

    Histological examination showed complete loss of CA1 pyramidal neurones in dorsal hippocampus.

B, C, D. All investigated around 1 year after toxin injection. Recurrent fits present in all three. All had complete CA1 cell loss as above.

    B,C. Response to single 4v stimulus. In both there appears to be a double spike.

    D. Paired pulse inhibition with 8v stimuli. No double spike.

The sequence of events in dorsal hippocampus might be as follows: excessive activation of all hippocampal circuitry leads to a fall in extracellular calcium concentration in the region of the CA1 dendrites owing to activation of NMDA receptors during seizure activity. This activates astrocytes to move closer to the dendrites (seen closely wrapping the processes in electron micrographs). The activated astrocytes activate the microglia (?via Interleukin-1 secretion) which then elongate and and increase the expression of antigen on their surface. At this stage there is no evidence of phagocytic activity even though in some cases dead neuronal processes and cell bodies are already visible in the electron micrographs. Pyramidal cell death probably occurs if the intraneuronal calcium concentration rises too high as a result of excess stimulation. A few days later, the microglia redifferentiate into the "bushy activated" form (with short, stout processes) and now move in to the cell body layer to phagocytose the debris. What requires further investigation however is whether the microglia have any active role in the death of neurones.

Acknowledgements

I am grateful for the financial support of the British Epilepsy Research Foundation, Ciba-Geigy Pharmaceuticals and St Hilda's College, Oxford.

REFERENCES

1. Bagetta G, Corasaniti MT, Nistico G, Bowery NG. Behavioural and neuropathological effects produced by tetanus toxin injected into the hippocampus of rats. Neuropharmacology 1990; 29:765-770.
2. Brace H, Jefferys J, Mellanby J. Long-term changes in hippocampal physiology and learning ability of rats after intrahippocampal tetanus toxin. J Physiol (Lond) 1985; 368:343-357.
3. Bruton, CJ. The neuropathology of temporal lobe epilepsy. Oxford University Press, 1988.
4. Dam AM. Hippocampal neuron loss in epilepsy and after experimental seizures. Acta Neurologica Scandinavica 1982; 66:601-642.
5. Duchen LW. The effects of tetanus toxin on the motor endplates of the mouse. J Neurol Sci 1973; 19:153-167.
6. Duchen LW, Tonge DA. The effects of tetanus toxin on neuromuscular transmission and on the morphology of motor endplates in slow and fast skeletal muscle of the mouse. J Physiol (Lond) 1973; 228:157-172.
7. George G, Mellanby J. Memory deficits in an experimental hippocampal syndrome in rats. Exp Neurol 1982; 75:678-689.
8. Graeber MB, Steit WJ, Kreutzberg GW. Axotomy of the rat facial nerve leads to increased CR3 complement receptor expression by activated microglial cells. J Neurosci Res 1988; 21:18-24.
9. Hawkins CA, Mellanby JH. Limbic epilepsy induced by tetanus toxin: a longitudinal electro-encephalographic study. Epilepsia 1987; 28:431-444.
10. Hume DA, Perry VH, Gordon S. Immunohistochemical localization of a macrophage-specific antigen in developing mouse retina: phagocytosis of dying neurons and differentiation of microglial cells to form a regular array in the plexiform layers. J Cell Biol 1983; 97:253-257.
11. Jefferys JGR, Empson RM. Development of chronic secondary epileptic foci following intrahippo-campal injection of tetanus toxin in the rat. Experimental Physiology 1990; 75:733-736.
12. Jefferys JGR, Evans BJ, Hughes SA, Williams SF. Neuropathology of the chronic epileptic syndrome induced by intrahippocampal tetanus toxin in rat: preservation of pyramidal cells and incidence of dark cells. Neuropathology and Applied Neurobiology 1992; 18:53-70.
13. Jefferys JGR, Mitchell P, O'Hara L, Tiley C, Hardy J, Jordan SJ, Lynch M, Wadsworth J. Ex vivo release of GABA from tetanus toxin-induced chronic epileptic foci decreased during the active seizure phase. Neurochem Int 1991; 18(3):373-379.
14. Jefferys JGR, Williams SF. Physiological and behavioural consequences of seizures induced in the rat by intrahippocampal tetanus toxin. Brain 1987; 110:517-532.
15. Jordan S, Jefferys JGR. Inhibition is reduced in a chronic model of epilepsy. European Journal of Neuroscience 1989; Suppl 2:274.
16. Mellanby J, George G, Robinson ASA, Thompson PA. Epileptiform syndrome in rats produced by injecting tetanus toxin into the hippocampus. J Neurol Neurosurg Psychiat 1977; 40:404-414.

17. Mellanby J, Mellanby H, Hawkins CA, Rawlins JNP, Impey ME. Tetanus toxin as a tool for studying epilepsy. J Physiol (Paris) 1985; 79:207-215.

18. Mellanby J, Renshaw M, Cracknell H, Rands G, Thompson PA. Long-term impairment of learning ability in rats after an experimental hippocampal epileptiform syndrome. Exp Neurol 1982; 75:690-699.

19. Mellanby J, Sundstrom L. Kindling, memory and behaviour. In: Bolwig TG, Trimble MR, eds. The Clinical Relevance of Kindling. John Wiley and Sons, 1989.

20. Perry VH, Gordon S. Immunohistochemical localization of macrophages and microglia in the adult and developing mouse brain. Neuroscience 1985; 15:313-326.

21. Rio-Hortega P. Microglia. In: Penfield W, ed. Cytology and cellular pathology of the nervous system, Vol. 2. New York: Hocker, 1932:481-584.

22. Shaw JAG, Perry VH, Mellanby J. Tetanus toxin-induced seizures cause microglial activation in rat hippocampus. Neuroscience Letters 1990; 120:66-69.

23. Shaw JAG, Perry VH, Mellanby J. Microglial activation and MHC expression in the tetanus toxin model of epilepsy. Europ J Neuroscience 1991; Suppl 4:100.

24. Sundstrom L, Mellanby J. (1990). Tetanus toxin blocks inhibition of granule cells in the dentate gyrus of the urethane-anaesthetized rat. Neuroscience 1990; 38:621-627.

CLINICAL TETANUS: (SPINAL) DISINHIBITION OR NOT?

Kohsi Takano

Department of Pathoneurophysiology
Faculty of Medicine
University of Göttingen
3400 Göttingen, Germany

INTRODUCTION

The most representative sign of tetanus is hyperactivity of the motor system. Most of the patients die, if not relaxed by adequate methods, due to an apnea, which is caused by "hyperactive" spasm of the respiratory muscles. Tetanus toxin may act not only on the motor system but also on the whole nervous system and therefore tetanus is a multisystemic disease including the disturbance of the autogenic nervous system. However, in this study only the effect of the toxin on the motor system will be discussed.

For the pathogenesis of tetanus intoxication it has been widely believed that tetanus toxin blocks spinal inhibitory transmission without affecting excitatory transmission. This hypothesis of hyperactivity of the motor system is based mostly on the study of Eccles' group (Brooks et al., 1957). They demonstrated that tetanus toxin blocked five types of postsynaptic inhibition of the monosynaptic reflex: tetanus toxin blocks only the inhibitory synapse of the alpha-motoneurone without affecting the excitatory synapse. Since then it has been reported that many kinds of inhibition including the presynaptic type in the central nervous system were blocked by tetanus toxin (e.g. Brooks and Asanuma, 1962; Curtis and de Groat, 1968; Curtis et al., 1973). Some authors who observed hyperactivity during intoxication interpreted their results without observing the inhibition itself, and concluded that the hyperactivity was caused by disinhibition. Thus it was believed that the tetanus was caused by general disinhibition in the central nervous system including the spinal cord (e.g. Kryzhanovsky, 1967; Habermann, 1978; Mellanby and Green, 1981). This selective blocking of inhibition by tetanus toxin has been regarded as the most important aspect.

On the other side several authors reported that tetanus toxin blocks not only inhibitory synapse but also the mammalian endplate (e.g. Davies et al., 1954; Duchen and Tonge, 1973; Kretzschmar et al., 1980) as well as other excitatory

Botulinum and Tetanus Neurotoxins, Edited by
B.R. DasGupta, Plenum Press, New York, 1993

synapses in the CNS (e.g. Sundström and Mellanby, 1989, 1990). Bergey et al. (1987) observed blockade by the toxin of both inhibitory and excitatory transmission to the cultured spinal motor cell (motoneurone).

But Sundström and Mellanby as well as Bergey et al. interpreted their results as follows: Tetanus toxin at high doses can block both inhibitory and excitatory synapses. Inhibition is blocked earlier than excitation by the doses which they used. These authors probably mean that tetanus toxin at low doses blocks only inhibitory transmission and excitatory transmission is also blocked only when the dose of the toxin is high. Thus they may also support the disinhibitory hypothesis.

The intention of this study is to elucidate whether the old disinhibition hypothesis of pathogenesis of tetanus is correct or not. It will be critically reviewed as to whether this hypothesis is based on experiments using relevant methods.

There are several aspects which must be considered before we discuss the pathogenesis of tetanus.

1. **Two Types of Experiment.** There are two types of experiment: clinically relevant and non-relevant. In the former we will try to elucidate the true pathogenesis of the natural disease. In the latter we try to find any effect of tetanus toxin on some types of preparation (from experiments using micropipettes on the patch of the cultured cell membrane, to some in vivo experiments). Some of the authors use tetanus toxin as a tool for the experiment. The results of the latter kinds of experiment must be carefully considered in the discussion about the pathogenesis.

2. **Clinical Picture of General and Local Tetanus.** Tetanus is characterized essentially by generalized convulsions, spasm and/or hypertonia (generalized or general tetanus). Many clinicians have described general tetanus, therefore only one well known point in the management of patients will be emphasized, which is important for discussing the methods of tetanus experiments. Namely, that the nervous system in severe general tetanus shows great lability and therefore any stimulation, photic, acoustic, touch etc. must be kept away from the tetanus patient.

Sometimes the sign of tetanus, rigid hypertonia, is restricted to one muscle and neighbouring muscles (localized or local tetanus), which is (are) near the portal of entry for Clostridium tetani. A pure local tetanus seldom occurs and doctors in practice might be unable to give a diagnosis. Only when local tetanus is contaminated or followed by general tetanus may the local sign be recognized as local tetanus. When a patient with tetanus is delivered to the hospital, the nursing staff fight against the general tetanus. Local tetanus is not fatal (when not in the respiratory muscles) and therefore not important for the doctors.

3. **The Toxin Dose.** In clinical tetanus the mean dose of toxin might be about one minimum lethal dose (MLD) or at most a little more, because the death rate without any care is about 60 %.

Most authors of experimental studies in the second half of this century used tetanus toxin at extremely high doses (see "EXPERIMENTAL GENERAL AND LOCAL TETANUS").

EXPERIMENTAL GENERAL AND LOCAL TETANUS

The animal model of general and local tetanus can be easily produced by intravenous (i.v.) or intramuscular (i.m.) injection of more or less purified toxin.

A current view is that after i.v. injection tetanus toxin is distributed by the blood circulation to the muscles of the whole body, absorbed at the endplate and transported retrogradely via the motor axons (see review e.g. Wellhöner, 1982; for the anterograde transport see Kanda and Takano, 1984) and accumulated in all segments of the spinal cord and in the cranial nerve nuclei (for the distribution of the toxin, see Habermann and Dimpfel, 1973). The distribution occurs at all these sites leading to the disinhibition of the motor system of the whole body. This is the interpretation of general tetanus given by Kryzhanovsky (1981), Habermann (1977) which is accepted by the authors of handbooks and textbooks of neurology (e.g. Struppler, 1987; Hahn and Mielke, 1991). This old hypothesis is confronted with our results.

In the patient as well as in the experimental animal showing severe general tetanus every stimulation must be avoided. Surgical operation or electrical stimulation might cause a violent convulsion leading to death of the animal. Though the experimental general tetanus after i.v. injection of tetanus toxin is a better simulation of the clinical tetanus than the experimental local tetanus there are only a few neurophysiological or electrophysiological experiments of general tetanus including that of Matsuda's (Sugimoto et al. 1982) and our groups (Huck et al., 1981; Takano and Kirchner, 1987; Takano et al., 1991).

On the other hand, tetanus experiments have a peculiar side as pharmacological experiments. Generally speaking, we cannot do, or have only a short time to do, some experiments if drugs at supralethal dose are applied. However, in the case of tetanus toxin high supralethal doses can be injected into the muscle, the nerve, the spinal cord or even into the brain, and because of its relatively slow migration (axonal transport, transsynaptic migration, if any, and binding, internalisation etc.) we have enough time to fulfil some neurophysiological experiments before the death by intoxication of the animal. Thus in almost all neurophysiological experiments of tetanus, animals with local tetanus were used and the results of local tetanus were applied to the explanation of pathogenesis of general tetanus. Notice the enormously high toxin doses in almost all experiments of local tetanus. Notice also the high local concentration of the toxin when it is injected into the spinal cord or brain or high toxin concentration in solutions for cultured cells. Such high local concentration of the toxin will never happen in clinical tetanus. Habermann and Dimpfel (1973) have once assumed that only 0.5 % of the i.v. injected toxin entered the central nervous system (CNS). Much less than 0.01 % of the toxin may reach one spinal segment.

LOCAL TETANUS

Depending on the toxin dose, several hours or days after the i.m. injection of tetanus toxin the symptom of local tetanus appears (Takano et al., 1983). The intoxicated muscle as well as its neighbouring muscles (exactly speaking, muscles which are innervated by motoneurones neighbouring the injected motor unit in the spinal segment and in the neighbouring ones) are shortened. A few days after toxin injection a very high activity in the electromyogram (EMG) can be recorded from these shortened muscles. The muscle shows a spastic paralysis. Several days thereafter, the muscle still remains shortened, but in this later period the muscle shows no EMG activity. There is no difference in the visible sign between active and non-active states. This muscle shortening in the later stage must be called contracture (Takano et al., 1992). There are two types of contracture in local tetanus. When the dose is rather high, the endplate (neuromuscular transmission) is blocked. By the way, attention of readers must be drawn to the fact that the effect of tetanus toxin is stronger on the neuromuscular transmission of fast white muscle like tibialis anterior muscle than that of red slow muscle like soleus muscle (e.g. Kretzschmar et al., 1980). Four times higher doses were necessary for blocking of the neuro-muscular transmission of the soleus muscle than that of tibialis anterior muscle. When the toxin dose is low, the endplate is not blocked but the monosynaptic transmission of the spinal motoneurone is blocked (Takano et al., 1983). Only in the early period of local tetanus was the inhibitory transmission to the spinal motoneurone blocked, as Eccles' group (Brooks et al., 1957) demonstrated.

Brooks et al. (1957) reported virtually no change in the monosynaptic reflex up to 48 h after toxin injection into the spinal cord or into the sciatic nerve at a high dose (7×10^6 mouse MLD, i.e. ca. 4×10^3 cat MLD, into the nerve of the cat). They used such animals with intoxicated spinal segment(s) and found that the postsynaptic inhibition of the motoneurone was blocked but the monosynaptic reflex was left intact *during the time of their experiments*. The following authors who also used high doses confirmed these findings. They concluded that the hyperactivity of the motor system in tetanus resulted from the disinhibition of the spinal motoneurone.

The minimal dose to elicit local tetanus is about 0.001 MLD i.m. in the gastrocnemius (Takano et al., 1983). When we could elicit local tetanus, the excitatory transmission always became blocked. This fact may indicate that the toxin-sensitivity of both types of synapses, the inhibitory as well as the excitatory, is almost identical. This finding could be supported by the experiments of intracellular recording, when we injected tetanus toxin at 0.05 MLD into the gastrocnemius muscle of the cat. Two days after toxin injection the inhibitory postsynaptic potential (IPSP) of the motoneurone disappeared and five days after the injection the excitatory postsynaptic potential (EPSP) was almost totally depressed (Kanda and Takano, 1983).

The following two studies are not local tetanus in the normal sense. They will be discussed here because local effects of tetanus toxin were observed in these experiments. Our findings in local tetanus at low doses could be further supported by the study of Bergey et al. (1987) using cultured mouse motor cells. They concluded: "Inhibitory transmission is reduced by the toxin, however, excitatory transmission is also ultimately reduced and blocked by the concentrations of toxin used in this study". But they interpreted: "Tetanus toxin preferentially affects inhibition" and "this supports the reduction of inhibition as an important mechanism in convulsant action of tetanus toxin". These authors may thus support the disinhibitory hypothesis. Is there really any preference?

Sundström and Mellanby (1989, 1990) observed also that tetanus toxin blocks both excitation and inhibition after the injection of tetanus toxin in a large dose (200 mouse LD_{50}) into the rat's dentate gyrus. They interpreted that they could also block the excitation because they used the toxin at larger doses and/or longer time.

GENERAL TETANUS

Several authors injected tetanus toxin at very high doses i.v. to observe the effect of the toxin on the endplate. They observed flaccid paralysis or "botulinum-like" paralysis (e.g. Davies and Wright, 1954; Matsuda et al., 1982; Laird, 1982; Fedinec et al., 1985). They used large doses (Matsuda et al., 2–1000 µg; Laird, more than 100,000 MLD). Sugimoto et al. (1982) observed that the EMG of the diaphragm disappears earlier than the phrenic nerve activity after the i.v. injection of tetanus toxin at very high doses (10^7 MLD). They have observed also the EMG of the (mixed) gastrocnemius muscle and found that the kinetic (fast, white) component was blocked earlier than the tonic (slow, red) component before the death of animal 1-2 hours after the injection of the toxin. (Their result coincides with our result in experimental local tetanus, Kretzschmar et al., 1980 and that of Huet de la Tour et al., 1979 and not that of Duchen and Tonge, 1973).

In spite of the fact that clinical tetanus can be much better simulated by experimental general tetanus, there is no electrophysiological study on spinal activity in general tetanus after the i.v. injection of the toxin except the studies of our groups. One of the reasons for the lack of these experiments may be the difficulty of the treatment of the animal, which will be easily killed by any surgical operation and electrical stimulation.

We began with a simple experiment with minimal invasion of the animal which is injected by tetanus toxin at 1-5 MLD. After gathering the experience to treat the intoxicated animal carefully we could at last perform a laminectomy in order to get an inhibition curve of the monosynaptic reflex.

Immediately after the appearance of the first sign of general tetanus the animal was anesthetized by a mixture of urethan and chrolarose. Experiments were performed several hours thereafter when the toxin action was anticipated to reach its maximum. Extreme care was taken in the treatment of the animal to avoid any kind of stimulation except those needed. During the time of experiment the animal would have died if it was not anesthetized. The stretch reflexes were elicited manually by the contraction of the antagonistic muscles or by a stretch device. In toxin treated animals the spontaneous EMG activity was inhibited by strong stretching of the tested muscle (autogenic inhibition) or by that of the antagonistic muscle (antagonistic inhibition) (Fig. 3 in Takano and Kirchner, 1987). The stretch reflex of the hind leg extensor muscles elicited by a contraction of the flexor muscles was inhibited by electrical stimulation of the flexor afferent fibres. The stretch reflex elicited by a stretch device as well as the electrically elicited monosynaptic reflex were inhibited by conditioning stimulation of the antagonistic nerve. The inhibition curves were almost the same as those of healthy animals. It is concluded that the spinal postsynaptic inhibitions, such as antagonistic Group Ia, autogenic Group Ib, Groups II and III and also the presynaptic inhibitions, were kept intact in severe general tetanus. For further details see Takano et al. (1991).

DIFFERENCE BETWEEN EXPERIMENTAL LOCAL TETANUS AND GENERAL TETANUS

The results of several authors were presented graphically in Fig. 1. All authors who injected enormously high doses of tetanus toxin intravenously observed flaccid paralysis or weakness of the intoxicated animals or botulinum-like effect of the toxin. None of them could observe the excitation of the motor system which could be introduced by the disinhibitory effect of the toxin (Fig. 1a).

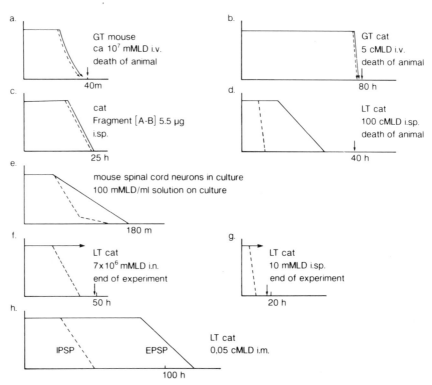

Figure 1. Time course of excitatory (solid line) and inhibitory (dotted line) transmissions of the spinal motoneurone in tetanus (except c). GT, general tetanus; LT, local tetanus; mMLD, mouse minimal lethal dose; cMLD, cat MLD; i.v., i.m., i.sp.; intravenous, intramuscular, intraspinal injection of tetanus toxin respectively. Notice the different time scale. All graphs were drawn by Takano after the discription of different authors. a. Matsuda et al., 1982. b. Takano et al., 1991. c. and d. Takano et al., 1989a. e. Bergey et al., 1987. f. and g . Brooks et al., 1957. h. Kanda and Takano, 1983.

When the toxin at lethal, but near clinical, doses was injected intravenously, the animal showed the typical sign of general tetanus. In these animals not only excitatory but also inhibitory transmission in the spinal cord were intact. They faded away in the same time as the death of the animal (Fig. 1b). Rhythmic activity of the spinal alpha and gamma motoneurones as well as that of the Renshaw cells in lumbar segments in such general tetanus disappeared after the section of the spinal cord at higher segment showing that the origin of the rhythmic activity of the motor system lies not in the spinal cord but in the higher central nervous system (Huck et al., 1981). Rhythmic activity could be produced only when the inhibition is more or less still present.

Fragment [A-B] of tetanus toxin at a very high dose was injected directly into the spinal cord. The animal survived but both excitation and inhibition were blocked at the same time (Fig. 1c). The Fragment [A-B] did not block neuro-muscular transmission when it was intramuscularly injected (Takano et al., 1989a).

When tetanus toxin at a high dose was injected into the spinal cord both inhibitory and excitatory transmission were blocked. The latency of blocking of the inhibition was shorter than that of the excitation (Fig. 1d). In high concentration of the toxin solution both excitatory and inhibitory transmission to cultured spinal motoneurone of the mouse were blocked, first the inhibitory followed by the excitatory (Fig. 1e). Brooks et al. (1957) injected an enormously high dose of tetanus toxin into the sciatic nerve and found the blocking effect of toxin. Their experiment lasted not longer than 50 h (Fig. 1f). In the same study they injected the toxin also into the spinal cord and reported the blockade of inhibition but did not observe blockade of excitation (Fig. 1g). Their experiment lasted only 17 hours after the toxin injection. Just at this time the blocking of the excitation should begin (compare with Fig. 1d).

When tetanus toxin at low doses (above 0.001 MLD) which cause local tetanus was injected, both inhibitory and excitatory transmission were always blocked as shown in Fig. 1h. Assuming that less than 0.01 % of the i.v. injected toxin enters one spinal segment (see "GENERAL TETANUS"), at least the toxin of 10 MLD must be i.v. injected to cause local tetanus. Such high doses may seldom occur in the case of clinical tetanus and are surely impossible in the case of survival with typical symptoms (general tetanus). In other words, the integrated local tetanus cannot be caused by i.v. injection of the toxin.

Spinal Tetanus And Supraspinal Tetanus

As mentioned above the characteristics of the local tetanus and general tetanus are different. Local tetanus is a spinal event including that of the cranial nerves. General tetanus has a supraspinal origin. Therefore:

LOCAL TETANUS = SPINAL TETANUS

GENERAL TETANUS = SUPRASPINAL TETANUS

Presynaptic Inhibition

There are several controversial opinions about the effect of tetanus toxin on presynaptic inhibition in the spinal cord which is transmitted by GABA. Kano and Ishikawa (1972) showed a blocking effect of tetanus toxin on the inhibitory synapses of the crayfish where the transmitter is GABA. Curtis et al. (1973) injected tetanus toxin at doses of 6000 mMLD into the lateral aspect of the ventral horn at the S_1 – S_2 junction of the cat and found that presynaptic inhibition was blocked nine hours after the toxin injection. (Note that the local dose in the spinal segment of this experiment is roughly identical to 10^4 – 10^5 MLD of the i.v. injected toxin). Kryzhanovsky and Lutsenko (1969, 5 rat MLD in the rat i.m.) and our group (Takano et el., 1989b, 0.001-1 MLD in the cat i.m.) observed that the dorsal root potential became greater and lasted longer during the intoxication, showing that the mechanism of the presynaptic inhibition was intact. We could observe presynaptic inhibition of the monosynaptic reflex in doses up to 2000 mMLD (1 cat MLD), provided the monosynaptic reflex was still present.

We conclude that presynaptic inhibition of the spinal motoneurones is kept intact in clinical tetanus, at least in the survival case, and most probably also in the lethal case.

Gamma Motor System

Our group has intensively worked on the effect of tetanus toxin on the gamma motor system (Takano and Kano, 1968; Kano and Takano, 1969; Takano and Kano, 1973; Takano and Henatsch, 1973; Mizote and Takano, 1984) and found that the gamma motor system is highly activated in the early period of local tetanus as well as in general tetanus (Huck et al., 1981). Observations of some authors support our findings (e.g. Liljestrand and Magnus, 1919; Rushworth, 1960). Diazepam is used extensively by clinicians in the treatment of tetanus. We (Takano and Student, 1978) have shown that diazepam acts far more intensively on the gamma system than on the alpha motor system. These facts support the major role of the gamma system in clinical tetanus. However, some reviewers (e.g. Kryzhanovsky, 1981; Mellanby and Green, 1981) have questioned a great role of the gamma motor system. There are some controversial results which are based on the incorrect understanding of the gamma system. For further discussion on this theme, see Takano and Mizote (1985).

SPINAL DISINHIBITION IS NOT THE PATHOGENESIS OF TETANUS

The time course of inhibition and excitation in the local tetanus and that of general tetanus were compared in Fig. 2.

In local tetanus (Fig. 2a) there is a short lasting disinhibitory period followed by a long lasting silent period of the spinal segment(s).

In the severe case of general tetanus both spinal inhibitory and excitatory transmission are almost normal (Fig. 2b).

Thus we can observe only a short disinhibitory period in local tetanus. Therefore we conclude that the hyperactivity of the motor system in tetanus is not caused by spinal disinhibition.

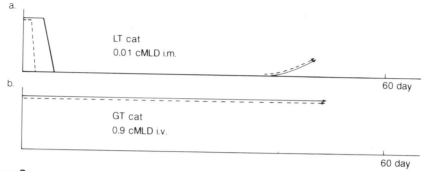

Figure 2. Differential time course of excitatory (solid line) and inhibitory (dotted line) transmissions of the spinal motoneurone in local (a) and general (b) tetanus. a. The same as Fig. 1h but in a long time scale. There is a short disinhibition period followed by a long contracture period in local tetanus. b. There is a blockade of neither inhibitory nor excitatory transmission in general tetanus.

TETANUS: DISINHIBITION OR NOT?

It has been clearly discussed above that the hyperactivity of the motor system during tetanus is not caused by spinal disinhibition. Is it caused by disinhibition in the upper CNS? It might be possible. There are many authors who injected tetanus toxin in the different sites of the CNS (for review see Wellhöner, 1982). Some found hyperactivity in the site of the injection or in its neighbourhood, and they *interpreted* simply that the hyperactivity is caused by disinhibition. The others actually found the disinhibition during the time of their experiments, which did not last enough long. Are these experiments relevant to elucidation at the pathogenesis of tetanus? As already mentioned, our group found that the Fragment [A-B] blocks both inhibitory and excitatory transmission at the same time. If there is any preference of the effect of tetanus toxin on inhibition, the Fragment [C] might play a role. Further investigations have to be done to establish the pathogenesis of tetanus.

REFERENCES

Bergey, G.K., Bigalke, H., and Nelson, P.G., 1987, Differential effects of tetanus toxin on inhibitory and excitatory synaptic transmission in mammalian spinal cord neurons in culture: a presynaptic locus of action for tetanus toxin, J. Neurophysiol. 57:121.

Brooks, V.B., and Asanuma, H., 1962, Action of tetanus toxin in the cerebral cortex, Science 137:674.

Brooks, V.B., Curtis, D.R., and Eccles, J.C., 1957, The action of tetanus toxin on the inhibition of motoneurones, J. Physiol. 135:655.

Curtis, D.R., and de Groat, W.C., 1968, Tetanus toxin and spinal inhibition, Brain Res. 10:208.

Curtis, D.R., Felix, D., Game, C.J.A., and Mc Culloch, R.M., 1973, Tetanus toxin and the synaptic release of GABA, Brain Res. 51:358.

Davies, J.R., Morgan, R.S., Wright, E.A., and Wright, G.P., 1954, Effect of local tetanus intoxication on the hind limb reflexes of the rabbit, Arch. Int. Physiol. 62:248.

Duchen, L.W., and Tonge, D.A., 1973, The effects of tetanus toxin on neuromuscular transmission and on the morphology of motor end-plate in slow and fast skeletal muscle of the mouse, J. Physiol. 228:157.

Fedinec, A.A., Toth, P., and Bizzini, B., 1985, Relation of spastic and flaccid paralysis to retrograde transport ^{125}I-tetanus toxin and its [125] I-Ibc fragment. Modulating effect of F[ab] antibodies directed to specific areas on the toxin molecule, Boll. Ist. Serot. Milan 64:35.

Habermann, E., 1978, Tetanus, in: Infections of the nervous system, Part I, P.J. Vinken and G.W. Bruyn, eds., Handbook of clinical neurology, Vol. 33, North Holland Publ. Amsterdam, New York, Oxford, 491.

Habermann, E., and Dimpfel, W., 1973, Distribution of ^{125}I-tetanus toxin and [125] I-tetanus toxoid in rats with generalized tetanus, as influenced by antitoxin, Naunyn-Schmiedeberg's Arch. Pharmacol. 276:361.

Huet de la Tour, E., Tardieu, C., Tabary, J.C., and Tabary, C., 1979, Decrease of muscle extensibility and reduction of sacomere number in soleus muscle following a local injection of tetanus toxin, J. Neurol. Sci. 40:123.

Hahn, H., and Mielke, M., 1991, Infektion und Abwehr, in: Pathophysiologie des Menschen, K. Hierholzer und R.F. Schmidt, eds., VCH Publ. Weinheim, Wellington, New York, 39.1.

Huck, S., Kirchner, F., and Takano, K., 1981, Rhythmic activity in the cerebellum and spinal cord of rabbits receiving tetanus toxin intravenously, Naunyn-Schmiedeberg's Arch. Pharmacol. 317:51.

Kano, M., and Ishikawa, K., 1972, Effect of tetanus toxin on the inhibitory neuromuscular junction, Exp. Neurol. 37:550.

Kano, M., and Takano, K., 1969, Gamma activity of rigid cat caused by tetanus toxin, Jap. J. Physiol. 19:1.

Kanda, K., and Takano, K., 1983, Effect of tetanus toxin on the excitatory and inhibitory post-synaptic-potentials in the cat motoneurones, J. Physiol. 335:319.

Kanda, K., and Takano, K., 1984, An evidence for anterograde transport of tetanus toxin, in: Seventh Internat. Conf. on Tetanus, G. Nistico, P. Mastroeni, M. Pitzurra, eds., Gangemi Publ. Roma, 79.

Kirchner, F., Rübesamen, V., and Takano, K., Adenosine phosphates concentration in the muscle during local tetanus, in: Ninth Internat. Congr. on Tetanus and Antiinfectious Defense, del Real, ed., (in the press).

Kretzschmar, H., Kirchner, F., and Takano, K., 1980, Relations between the effect of tetanus toxin on the neuromuscular transmission and histological functional properties of various muscles of the rat, Exp. Brain Res. 38:181.

Kryzhanovsky, G.N., 1967, The neural pathway of toxin: its transport to the central nervous system and the state of the spinal reflex apparatus in tetanus intoxication, in: Principles of Tetanus, L. Eckmann, ed., H. Huber, Bern, Stuttgart, 155.

Kryzhanovsky, G.N., 1981, Pathophysiology, in: Tetanus, Important New Concepts, R. Veronesi, ed., Excerpta Medica, Amsterdam, Oxford, Princeton, 109.

Kryzhanovsky, G.N., and Lutsenko, V.K., 1969, Dorsal root potential of the spinal cord in rats with convulsions due to ascending tetanus, Bull. Exp. Biol. Med. 67:16.

Laird, W.J., 1982, Botulinum toxin-like effects of tetanus toxin, in: Sixth Internat. Conf. on Tetanus, Foundation Merieux, Lyon, 33.

Liljestrand, G., and Magnus, R., 1919, Über die Wirkung des Novokains auf den normalens und den tetanusstarren Skelettmuskel und über die Entstehung der lokalen Muskelstarre beim Wundstarrkrampf, Pflügers Arch. 176:168.

Matsuda, M., Sugimoto, N., and Ozutsumi, K., 1982, Acute botulinum-like intoxication by tetanus toxin in mice and the localization of the acute toxicity in the N-terminal papain-fragment of the toxin, in: Sixth Internat. Conf. on Tetanus, Foundation Merieux, Lyon, 21.

Mellanby, J., and Green, J., 1981, Commentary. How does tetanus act? Neurosci. 6:281.

Mizote, M., and Takano, K., 1984, Two kinds of discharge pattern of primary endings to FM muscle vibration in the tetanic cat, in: The Muscle Spindle, I.A. Boyd, M.H. Gladden, eds., The Macmillan Press Lond., 365.

Rushworth, G., 1960, Spasticity and rigidity: an experimental study and review, J. Neurol. Neurosurg. Psychiat. 23:99.

Struppler, A., 1990, Das Nervensystem, in: Pathophysiologie-Pathobiochemie, E. Lang, Ed., Enke, Stuttgart, 193.

Sugimoto, N., Matsuda, M., Ohnuki, Y., Nakayama, T., and Imai, K., 1982, Neuromuscular blocking in acutely tetanus intoxicated mice, Biken J. 25:21.

Sundström, L., and Mellanby, J., 1989, Acute effects of tetanus toxin on granule cells of the dentate gyrus of the rat in vivo, in: Eighth Internat. Conf. on Tetanus, G. Nistico, B. Bizzini. B. Bytchenko, R. Triau, eds., Pythagora Press Rome, Milan, 149.

Sundström, L., and Mellanby, J., 1990, Tetanus toxin blocks inhibition of granule cells in the dentate gyrus of the urethane-anaesthetized rat, Neurosc. 38:621.

Takano, K., and Henatsch, H.D., 1973, Tension extension of the tetanus intoxicated muscle of the cat. Naunyn-Schmiedeberg's Arch. Pharmacol. 276:421.

Takano, K., and Kano, M., 1973, Gamma-bias of the muscle poisoned by tetanus toxin, Naunyn-Schmiedeberg's Arch. Pharmacol. 276:413.

Takano, K., and Kirchner, F., 1987, Pathogenesis of tetanus: clinically relevant new concepts. A mini review, in: Progress in Venom and Toxin Research, P. Gopalakrishnakone and C.K. Tan, eds., Singapore: Singapore National University, 680.

Takano, K., Kirchner, F., Gremmelt, A., Matsuda, M., Ozutsumi, N., and Sugimoto, N., 1989a, Blocking effects of tetanus toxin and its Fragment [A-B] on the excitatory and inhibitory synapses on the spinal motoneurone of the cat, Toxicon 27:385.

Takano, K., Kirchner, F., and Mizote, M., 1991, Intact inhibition of the stretch reflex during general tetanus, Toxicon 29:201.

Takano, K., Kirchner, F., Terhaar, P., and Tiebert, B., 1983, Effect of tetanus toxin on the monosynaptic reflex, Naunyn-Schmiedeberg's Arch. Pharmac. 323:217.

Takano, K., Kirchner, F., Tiebert, B., and Terhaar, P., 1989b, Presynaptic inhibition of the monosynaptic reflex during local tetanus in the cat, Toxicon 27:431.

Takano, K., and Mizote, M., 1985, Central and peripheral effects of tetanus toxin on the gamma motor system, a review, in: Seventh Internat. Conf. on Tetanus, G. Nistico, P. Mastroeni, M. Pitzurra, eds., Gangemi Publ. Roma, 101.

Takano, K., and Student, J.C., 1978, Effect of diazepam on the gamma motor system indicated by the response of the muscle spindle of the triceps surae muscle of the decerebrate cat to the muscle stretch, Naunyn-Schmiedeberg's Arch. Pharmacol. 302:91.

Wellhöner, H.H., 1982, Tetanus neurotoxin, Rev. Physiol. Biochem. Pharmacol. 93:1.

EFFECTS OF BOTULINUM NEUROTOXIN A
ON PROTEIN PHOSPHORYLATION IN SYNAPTOSOMES

Dennis L. Leatherman and John L. Middlebrook

Department of Immunology and Molecular Biology
Toxinology Division
United States Army Medical Research Institute of Infectious Diseases
Frederick MD 21702-5011

ABSTRACT

The phosphorylation of presynaptic proteins was examined in the presence of chlorpromazine using synaptic lysates derived from rat brain synaptosomes that had been preincubated with or without botulinum neurotoxin A. Chlorpromazine increased the phosphorylated state of a number of phosphoproteins with apparent molecular weights (M_rs) between 40 and 60 kilodaltons. The drug's most prominent effect was a 2.5–3.5-fold increase in the incorporation of phosphate into a peptide of M_r~43 kilodaltons (pp43), as determined by sodium dodecyl sulfate polyacrylamide gel electrophoresis. After preincubation with toxin, the chlorpromazine-promoted increase in the phosphorylated state of pp43 was inhibited in a concentration- and temperature-dependent manner.

INTRODUCTION

Botulinum neurotoxin A (BoNT-A) is a potent and highly selective inhibitor of the release of acetylcholine from peripheral motor nerve terminals. The precise mechanism for the toxin's activity on cholinergic neurons has not been completely determined. Nevertheless, evidence from a number of laboratories indicates that the toxin's effects are mediated through "ecto-acceptor" recognition and binding,[1] followed by a temperature-dependent internalization,[1,2] and subsequent inactivation of an intraneuronal component.[3]

Although the toxin's activity *in vivo* is highly selective for cholinergic neurons, its effects *in vitro* have been demonstrated for a variety of neurotransmitters in a number of permeabilized cell preparations.[4,5,6] Moreover, despite the toxin's *in vivo* specificity for peripheral neurons, its effects on neurotransmitter release from rat brain synaptosomes have also been well documented, albeit, at very high toxin concentrations.[5] It has been proposed that the specificity exhibited by the toxin *in vivo* is a result of targeting (via its heavy chain) to specific surface acceptors present in high density only in the peripheral nervous system.[1] However, its intraneuronal target may be a component common to, and essential for, exocytosis of neurotransmitter from neurosecretory cells in general. Hence, the use of synaptosomal preparations to examine the dynamics of toxin binding may be of limited utility, but an elucidation of the toxin's intrasynaptosomal mechanism of action could extrapolate to peripheral motor neurons.

In the present investigation, we used rat brain synaptosomes to search for a potential intrasynaptosomal target for botulinum neurotoxin A. In view of the accumulating evidence for the involvement of phosphoproteins in presynaptic regulatory pathways,[7] we focused our efforts on protein phosphorylation. We report here on a phosphoprotein, present in mammalian nerve terminal preparations, sensitive to botulinum neurotoxin A.

MATERIALS AND METHODS

Toxin

Botulinum neurotoxin type A was the generous gift from Dr. James Schmidt (United States Army Medical Research Institute of Infectious Diseases). The toxin was purified to apparent homogeneity (as determined by sodium dodecyl sulfate polyacrylamide gel electrophoresis) with a specific activity of 2.8×10^8 mouse LD_{50}/mg protein. Prior to use, toxin was dialyzed into 10 mM Tris, pH 7.4 and its concentration was redetermined by the method of Lowry.[8]

Preparation of synaptosomes

Rat brain synaptosomes were prepared from male Sprague-Dawley rats (150-250 g) sacrificed by decapitation. The brain was rapidly removed, trimmed free of primitive brain, rinsed and homogenized in 0.32 M sucrose with a glass–teflon homogenizer. The homogenate was then spun at $1500 \times g$ for 10 min at 4°C. The supernatant was recovered and the P_2 fraction was then prepared as described by Cotman.[9]

Preincubation of intact synaptosomes with toxin

Prior to use, P_2 fractions were washed twice in 0.32 M sucrose, once in Krebs Ringer buffer (KRB) pH 7.4, 0.32 M sucrose (1:1 v/v), and finally resuspended in glucose-supplemented KRB, pH 7.4, containing 0.1 mM EGTA and 1.1 mM Ca^{++}. The synaptosomes were then preincubated with or without toxin as described in the figure legends. After preincubation, the synaptosomes were pelleted for 2 min in an Eppendorf microcentrifuge, washed once in ice-cold KRB, lysed in 10 mM TRIS, pH 7.4, and subsequently sheared by aspiration through a 27 gauge needle. Lysates were then maintained on ice until utilization in phosphorylation reactions.

Chlorpromazine pre-equilibration

Lysates (10.0 μl aliquots containing 12-25 μg protein) were added to 1.5 ml eppendorf tubes and incubated on ice for 30 min with the indicated concentrations of chlorpromazine (Sigma Chemical Co., St. Louis, MO) in a final volume of 40 μl. The reaction mixture (pH 7.4) contained 55 mM Hepes buffer, 12 mM $MgCl_2$, and 1.2 mM EGTA (final concentrations after addition of chlorpromazine).

Phosphorylation of synaptic lysates

After pre-equilibration with or without chlorpromazine, the samples were transferred to a 32°C water bath and thermally equilibrated for 60 sec. Phosphorylation reactions were initiated by the addition of 10 μl of ATP containing 2.5-5.0 μCi ^{32}P-gamma-ATP (sp. act. 3000 Ci/mMole, Amersham Corporation, Arlington Heights, IL) Final concentrations after addition of ATP were 50 mM Hepes buffer, pH 7.4, 10 mM $MgCl_2$, 1.0 mM EGTA and 2–50 μM ATP as indicated in the figure legends. After 30 sec, the reaction was terminated by immersing samples into a temperature block (100°C) with the immediate addition of 10 μl of stop solution containing 5% SDS, 30% glycerol, 5% β-mercaptoethanol, 0.125 M Tris-Cl pH 6.8. Samples were then applied to 8% or 10% polyacrylamide gels with 3% stacking gels and electrophoresed at 30 milliamperes per gel in a BioRad 2001 apparatus. Gels were stained, destained, dried, and ^{32}P-labeled proteins were located by autoradiography.

Acquisition and processing of data

Phosphorylated bands were cut from the dried gels and counted with Liquifluor:Toluene (New England Nuclear, Boston, MA) in a liquid scintillation counter (Beckman 5801, Fullerton, CA). Protein concentrations in synaptic lysates were determined by the method of Lowry after preincubation with or without toxins. Data in graphs were computed as phosphate incorporated per μg of protein lysate and, where indicated, expressed as a percentage of phosphate incorporated into non-toxin controls.

Miscellaneous procedures

Calcium concentrations in stock solutions and free calcium concentrations in synaptic lysates were measured using automated colorimetric procedures. When necessary, total calcium in synaptic lysates was determined by plasma emission and bound/chelated calcium was determined by subtracting free from total.

RESULTS

Preincubation of synaptic lysates with increasing concentrations of chlorpromazine resulted in an increase in the phosphorylated state of pp43 (Figure 1). At a concentration of approximately 1 mM, chlorpromazine produced maximal increases in pp43 phosphorylation, which ranged from 2.5–3.5-fold in different rat brain preparations. The mean increase in pp43 phosphorylation after preincubation with a maximally effective concentration of chlorpromazine was 2.9 ± 0.4 as determined from eight independent experiments.

Incorporation of phosphate into pp43 in the presence of chlorpromazine was linear at 32°C for 45 sec with an ATP concentration as low as 2.0 μM (Figure 2). Chlorpromazine-induced increases in pp43 phosphorylation were also observed at ATP concentrations of 1, 10 and 50 μM (data not shown). Thus, in subsequent experiments, the effects of BoNT-A on pp43 phosphorylation in the presence of chlorpromazine were determined using 30 sec phosphorylations at 32°C with either 2 or 10 μM ATP. These experimental parameters ensured that the determination of the toxin's effects were derived from measurements of initial reaction rates under linear conditions.

The effects of BoNT-A on pp43 phosphorylation are illustrated in Figure 3. At the maximally effective chlorpromazine concentration (0.9 mM), the phosphorylation of pp43 in synaptic lysates was reduced when BoNT-A was preincubated with intact synaptosomes at 0°C, followed by an additional incubation at 37°C. Increasing toxin concentrations produced decreasing levels of pp43 phosphorylation. Based upon single determinations from five independent experiments, pp43 phosphorylation was reduced to $48 \pm 2\%$ of control after preincubation with 100 nM BoNT-A. In contrast, there was little to no effect on pp43 phosphorylation when intact synaptosomes were preincubated with the same concentration of toxin at 0°C with no subsequent incubation at 37°C.

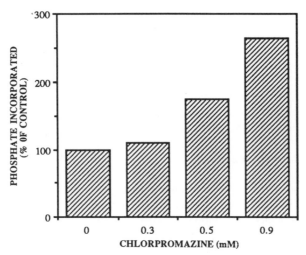

Figure 1. Chlorpromazine-induced increase of pp43 phosphorylation. Synaptosomes were prepared, lysed and pre-equilibrated with the indicated concentrations of chlorpromazine as described in Methods. After thermal equilibration, the lysates (24 μg total protein) were phosphorylated in the presence of 10 μM ATP as described in Methods. The data shown are the averages of single determinations from two independent dose response experiments.

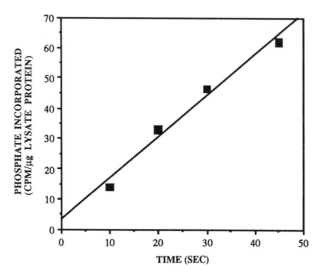

Figure 2. Kinetics of incorporation of phosphate into pp43 in the presence of chlorpromazine. Synaptosomes were prepared, lysed and pre-equilibrated with 0.9 mM chlorpromazine as described in Methods. After thermal equilibration, the lysates (26 μg total protein) were phosphorylated for the times indicated in the presence of 2 μM ATP. The data shown represent single determinations from one experiment repeated twice.

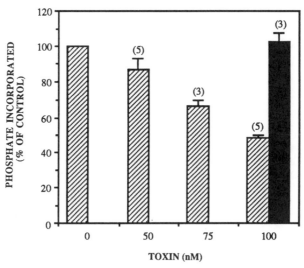

Figure 3. BoNT-A-induced decrease in pp43 phosphorylation. Synaptosomes were prepared as described in Methods, then incubated with the indicated concentrations of toxin at 0°C for 2.5 hr with (cross-hatched bars) or without an additional 45 min incubation at 37°C (solid bar). Lysates (12-26 μg total protein) were prepared and pre-equilibrated with 0.9 mM chlorpromazine and phosphorylated as described in Methods. The data shown were compiled from eight independent experiments. The number of determinations for each toxin concentration are noted on the graph.

DISCUSSION

Inhibition of acetylcholine release from presynaptic nerve terminals by BoNT-A proceeds through a mechanism that has not yet been completely defined. Attempts to identify an intraneuronal target (or target system) have not been successful. In the present investigation, we have utilized chlorpromazine to reveal a toxin-sensitive phosphoprotein (pp43) present in mammalian nerve terminals.

Phosphorylation of pp43 in synaptic lysates was increased by chlorpromazine in a concentration-dependent manner (Figure 1). This effect was significantly attenuated after preincubation of synaptosomes with BoNT-A (Figure 3). Although the toxin's effects could be measured at most chlorpromazine concentrations tested, maximum inhibitory activity of the toxin was most evident at maximally effective chlorpromazine concentrations (data not shown). The physiological significance of pp43 phosphorylation, its potentiation by chlorpromazine, and its apparent inhibition by toxin are not yet known.

It should be noted, that phosphorylation of pp43 in synaptic lysates was inhibited by BoNT-A only after preincubation of synaptosomes with toxin at reduced temperature followed by a subsequent incubation at physiological temperature. Under these conditions, BoNT-A-induced inhibition of pp43 phosphorylation was concentration dependent. In contrast, preincubation with toxin at reduced temperature was insufficient for producing inhibitory effects on pp43 phosphorylation. These data suggest, but do not prove, that the inhibitory effect of the toxin was mediated by an internalization-dependent event. This would be consistent with the data derived from investigations of the toxin's effect on neuromuscular junctions. Investigations by Simpson[2,10] provided evidence that the neuromuscular blockade established by BoNT-A is contingent upon temperature sensitive events similar to, if not identical to, receptor-mediated endocytosis. Similar data were obtained by Ashton and Dolly[5] for the toxin's effect on acetylcholine release from rat brain synaptosomes. To this extent, the effect of BoNT-A reported in this investigation may bear some relevance to the toxin's effect on cholinergic release mechanisms and merits further investigation.

REFERENCES

1. Dolly JO, Black JD, Williams RS, Melling J. Acceptors for botulinum neurotoxin reside on motor nerve terminals and mediate its internalization. Nature 1984; 307:457-460.
2. Simpson LL. Kinetic studies on the interaction between botulinum toxin type A and the cholinergic neuromuscular junction. J Pharmacol Exp Ther 1980; 212:16-21.
3. Poulain B, Tauc L, Mohan PM, Maisey EA, Wadsworth JDF, Dolly JO. Intracellular toxicity of botulinum neurotoxins A and B and their purified subunits studied at central synapses of *Aplysia*. Neuroscience 1987; 22:Suppl 1215p.
4. McInnes C, Dolly JO. Ca^{2+}-dependent noradrenaline release from permeabilized PC12 cells is blocked by botulinum neurotoxin A or its light chain. FEBS Lett 1990; 261:323-326.
5. Ashton AC, Dolly JO. Characterization of the inhibitory action of botulinum neurotoxin type A on the release of several neurotransmitters from rat cerebrocortical synaptosomes. J Neurochem 1988; 50:1808-1816.
6. Bittner MA, Dasgupta BR, Holz RW. Isolated light chains of botulinum neurotoxins inhibit exocytosis. Studies in digitonin-permeabilized chromaffin cells. J Biol Chem 1989; 264:10354-10360.
7. Walass SI, Greengard P. Protein phosphorylation and neuronal function. Pharmacol Rev 1991; 43:299-349.
8. Lowry OH, Rosebrough NJ, Farr AL, Randall RJ. Protein measurement with the Folin phenol reagent. J Biol Chem 1951; 193:265-275.
9. Cotman CW. Isolation of synaptosomal and synaptic plasma membrane fractions. In: Fleischer S, and Packer L, eds. Meth Enzymol 31. New York: Academic Press, 1974:445.
10. Simpson LL. The origin, structure and pharmacological activity of botulinum toxin. Pharmacol Rev 1981; 33:155-188.

INHIBITION OF NOREPINEPHRINE SECRETION FROM DIGITONIN PERMEABILIZED PC12 CELLS BY *C. BOTULINUM* TYPE D TOXIN.

Noriko Yokosawa[1], Kouichi Kimura[1], Kayo Tsuzuki[1], Nobuhiro Fujii[1], Bunei Syuto[2] and Keiji Oguma[1]

[1]Department of Microbiology
Sapporo Medical College

[2]Department of Biochemistry
Faculty of Veterinary Medicine
Hokkaido University

Sapporo, 060 Japan

INTRODUCTION

Toxigenic strains of *Clostridium botulinum* produced seven immunological distinct neurotoxins, type A-G. Neurotoxins are 150-kDa proteins, synthesized as single chain molecules and nicked to the dichain form by endogenous protease(s) in the bacterial culture. Dichain form consists of a heavy chain (approx. 100 kDa) and a light chain (approx. 50 kDa) held together by at least one disulfide bond and noncovalent bonds.

Neurotoxin inhibit the release of acetylcholine from peripheral nerve terminal, producing muscular weakness and paralysis. Electrophysiological studies performed with neuromuscular preparations indicated that the neurotoxins block the quantal acetylcholine release from motor nerve terminals. The toxins apparently do not affect transmitter synthesis and storage (Gundersen, 1980), the propagation of the nerve action potential to the terminal (Harris and Miledi, 1971), the influx of calcium across the terminal membrane (Gundersen et al., 1982; Dreyer et al.,1983), nor intraterminal Ca^{2+} level (Mallart et al., 1989). It was reported that increasing phasic Ca^{2+} entry into terminals enhances evoked synchronized quantal release from terminals poisoned with type A, and E (Molgo et al.,1989a) and induced a period of high frequency asynchronous release from terminals poisoned with type B (Sellin et al.,1983), D (Molgo et al., 1989b), and F (Kauffman et al.,1985). However, intraterminal target and function of toxin are not known.

To study the intracellular action of neurotoxin, we used catecholamine-secreting PC12 cells permeabilized with digitonin, examined the inhibitory action of botulinum type D toxin on the Ca^{2+} evoked norepinephrine secretion and thiophosphorylation dependent norepinephrine secretion.

MATERIALS AND METHODS

Clostridium botulinum type D (strain 1873) was purified as reported previously (Yokosawa et al., 1986). Monoclonal antibodies (ND-1, NDA-122, NDA-75) and polyclonal rabbit antiserum raised against purified type D toxin was prepared previously (Oguma et al., 1984). Preparation and characterization of monoclonal antibody EL161 which was raised against light chain of type E toxin were mentioned previously (Tsuzuki et al.,1988).

The protocol used for measuring Ca^{2+} evoked [^{3}H]NE secretion is based on the method of Pepper and Holtz (1986). The methods for thiophosphorylation dependent norepinephrine secretion and determination of ^{35}S incorporation into the cells were performed as described by Wagner and Vu (1989) and Brook et al. (1984), respectively.

RESULTS AND DISCUSSION

Incubation of permeabilized cells with botulinum neurotoxin from type D strain 1873 prior treated with 10 mM dithiothreitol (DTT) reduce, the [^{3}H]NE secretion evoked with 20 μM free Ca^{2+}. Without DTT pretreatment, type D neurotoxin had no effect on Ca^{2+} evoked secretion. The results obtained by the monoclonal and polyclonal antibodies showed that the inhibition is specific action of the light chain of the toxin. The inhibition was increased by concentration of the toxin (1-30 nM) and incubation time with toxin. The time course of the inhibition were different from the evoked Ca^{2+} concentration.

Calcium added to the evoked media partially counteracted the effects of the toxin; norepinephrine secretion was enhanced despite of the presence of the toxin. With less than 0.7 μM free Ca^{2+}, 70-80 % of inhibition was observed but with more than 2 μM of free Ca^{2+} the inhibition was 30-40 %. Since the action of type D toxin was counteracted by increasing the evoked Ca^{2+} concentration, it was suggested that botulinum type D toxin acted to a certain Ca^{2+} stimulatable process.

Wagner and Vu (1989,1990) showed that Ca^{2+}-stimulated phosphorylation regulate the norepinephrine secretion in rat pheochromocytoma PC12 cells. When ATPγS was used instead of ATP, protein kinase can utilize this analog and thiophosphorylate proteins. As the thiophosphorylated proteins are resistant to phosphatase, the reactions of secretion are separated to the thiophosphorylation and the following secretion. In order to determine function of the toxin to the Ca^{2+}-stimulated phosphorylation, we examined the effect of the toxin to the thiophosphorylation dependent secretion. The toxin inhibited the thiophosphorylation dependent secretion and the inhibition was neutralized by the polyclonal antibodies and monoclonal antibodies which recognize the light chain of the toxin and are able to neutralize the mouse lethality of the toxin.   The concentration of ATPγS affected the inhibition by the toxin; with less than 0.4mM ATPγS, thiophosphorylation dependent secretion was diminished to low level, but with increasing ATPγS concentration, the secretion was enhanced.   Addition of Ca^{2+} enhanced the thiophosphorylation dependent secretion either presence and absence of the toxin.   The toxin inhibited the thiophosphorylation dependent secretion but did not inhibit stimulation of thiophosphorylated dependent secretion by Ca^{2+}.

Finally, we determined the $[^{35}S]$ incorporation from $[^{35}S]ATPγS$ to the permeabilized cells. The toxin inhibited the thiophosphorylation of protein.   The inhibition of thiophosphorylation was antagonized by the polyclonal antibodies.   The thiophosphorylated proteins were analyzed on SDS polyacrylamide gel electrophoresis.   Proteins with wide range of molecular weight were thiophosphorylated in the absence of toxin.   Whereas, in the presence of the toxin thiophosphorylation was decreased in many protein bands.   These results suggested that one of target sites of botulinum type D toxin is a certain proteinkinase.

REFERENCES

Brooks, J.C., Treml, S., and Brook, M., 1984, Thiophosphorylation prevents catecholamine secretion by chemically skinned chromaffin cells, Life Sci. 35:569-574.

Dreyer, F., Mallart, A., and Brigant, J.L., 1983, Botulinum A toxin and tetanus toxin do not affect presynaptic membrane currents in mammalian motor nerve endings, Brain Res. 270:373-375.

Gundersen, C.B.S.,1980, The effect of botulinum toxin on the synthesis, storage and release of acetylcholine, Prog. Neurobiol. 4: 99-119.

Gundersen, C.B.S., Katz, B., and Milede, R., 1982, The antagonism between botulinum toxin and calcium in motor nerve terminals, Proc. R. Soc. Lond. B.216:369-376.

Harris, A.J., and Miledi, R.,1971, Effect of type D botulinum toxin on frog neuromuscular junctions, J. Physiol. ,London, 217: 497-515.

Kauffman, J.A., Way, J.F.Jr., Siegel, L.S., and Sellin, L.C., 1985, Comparison of the action of type A and F botulinum toxin at rat neuromuscular junction, Appl. Pharmacol. 79: 211-217.

Mallart, A., Molgo, j., Angut-Petit, D., and Thesleff, S., 1989, Is the internal calcium regulation altered in type A botulinum toxin poisoned motor endings?, Brain Res.479:167-171.

Molgo, J., DasGupta, B.R., and Thesleff, S., 1989a, Characterization of the actions of botulinum neurotoxin type E at the rat neuromuscular junction, Acta Physiol. Scand. 137:497-501.

Molgo, J., Siegel, L.S., Tabati, N., and Thesleff, S., 1989b, A study of synchronization of quantal transmitter release from mammalian motor ending by the use of botulinum toxin type A and D.

Oguma, K., Murayama, S., Syuto, B., Iida, H., and Kubo, S., 1982, Analysis of antigenicity of *Clostridium botulinum* type C1 and D toxins by polyclonal and monoclonal antibodies, Infect. Immun.43, 584-588.

Peppers, S.C., and Holtz, R.W., 1986, Catecholamine secretion from digitonin-treated PC12 cell , J. Biol. Chem.261:14665-14669.

Sellin, L., Thesleff, S., and DasGupta, B.R., 1983, Different effects of types A and B botulinum toxin on transmitter release at the rat neuromuscular junction, Acta Physiol. Scand. 119:127-133.

Tsuzuki, K., Yokosawa, N., Syuto, B., Ohishi, I., and Fujii, N., Kimura, K., and Oguma,

K., 1988, Establishment of a monoclonal antibody recognizing site common to *Clostridium botulinum* type B, C1, D, and E toxins and tetanus toxin, Infect. Immun. 56:898-902.

Wagner, P.D., and Vu, N.-D.,1989, Thiophosphorylation cause Ca-independent norepinephrine secretion from permeabilized PC12 cells, J.Biol.Chem. 264:19614-19620.

Wagner, P.D., and Vu, N.-D.,1990, Regulation of norepinephrine secretion in permeabilized PC12 cells by Ca-stimulated phosphorylation, J. Biol.Chem. 265: 10352-10357.

Yokosawa, N., Tsuzuki, K., Syuto, B., and Oguma, K., 1986, Activation of *Clostridium botulinum* type E toxin purified by two different procedures, J. Gen. Microbiol. 132:1981-1988.

LONG-TERM EFFECTS OF BOTULINUM TYPE A NEUROTOXIN

ON THE RELEASE OF NORADRENALINE FROM PC12 CELLS

Clifford C. Shone

Division of Biologics
Centre for Applied Microbiology and Research
Porton Down
Salisbury
Wilts SP4 0JG, U.K.

ABSTRACT

Clostridium botulinum type A neurotoxin (BoNTA) slowly inhibits the calcium-dependent release of noradrenaline from PC12 cells in a dose-dependent manner. The effects of BoNTA on PC12 cells are shown to persist for several days and in subsequent cell generations. Examination of the molecular form of cell-associated ^{125}I-labelled BoNTA after 96h incubation within PC12 cells showed virtually no detectable degradation of either the heavy or light chains of the neurotoxin. The data suggest that BoNTA is relatively stable within the PC12 cell with a half-life in the order of several days. More than 50% of the membrane-associated BoNTA within the cell exists in the form of high molecular weight, disulphide-linked aggregates and these appear to be enriched in the light chain of the neurotoxin. A possible role in action of BoNTA for these aggregates is discussed.

INTRODUCTION

Botulinum type A neurotoxin (BoNTA) is one of seven structurally similar protein toxins produced by various strains of the bacterium *Clostridium botulinum*. The neurotoxin acts primarily on the peripheral nervous system where it inhibits the release of the transmitter acetylcholine causing the syndrome botulism which is characterised by a widespread and flaccid paralysis (Simpson, 1981, Shone, 1987). The effects of BoNTA on the neurotransmitter release mechanism often persists for several months. This characteristic of the toxin's action has allowed small doses of the BoNTA to be used clinically in the treatment of a wide range dystonia and cases of muscular spasm.

Data from several studies have shown that BoNTA is capable of inhibiting the calcium-mediated secretion a range of neurotransmitter substances from a variety of neural and non-neural cell types and suggest that the neurotoxin acts on a fundamental step in the process of calcium-mediated secretion. Although the molecular details of the action of BoNTA have yet to be deciphered, the available data suggest a three stage mechanism of toxin action: an initial energy-independent binding stage in which the toxin binds to its acceptor(s) on the presynaptic nerve surface (Black and Dolly, 1986); an energy-dependent internalisation stage in which the toxin enters the nerve cell (Dolly *et al.*, 1984); finally one or several stages in which the toxin disables the transmitter release mechanism by an active site region located on the light chain (Stecher, *et al.* ,1989b; De Paiva and Dolly, 1990). At present none of these stages in the action of BoNTA are well characterised.

Of the various model systems available to study the intracellular action of BoNTA, the rat adrenal pheochromocytoma cell line, PC12 is perhaps the least complex. PC12 cells have been shown to produce and secrete a range of neurotransmitter substances in a calcium-dependent manner (Greene and Rein, 1977; Schubert and Klier, 1977). Although lacking ecto-acceptors for BoNTA, studies in which PC12 cells have been made permeable to the neurotoxin, using either digitonin or streptolysin-O, have shown that BoNTA is able to inhibit the calcium-dependent release process (Stecher *et al.*, 1989a,b; McIness and Dolly, 1990).

In the present study the long-term effects of the BoNTA on neurotransmitter release from PC12 cells have been investigated. It is shown that incubation of intact PC12 cells for several hours with BoNTA induces an inhibition of calcium-dependent noradrenaline release which persists for several days.

MATERIALS AND METHODS

BoNTA was purified and labelled with iodine-125 as described previously (Wadsworth *et al.*, 1990; Williams *et al.*, 1983). PC12 cells were obtained from the American Type Culture Collection and maintained in culture using RPMI 1640 medium containing 5% (v/v) foetal calf serum and 10 % (v/v) doner horse serum, penicillin (100U/ml) and streptomycin (100ug/ml) under an atmosphere of 5% CO_2/air at 37°C (Greene and Tischler, 1976). Gel electrophoresis was carried out essentially by the method of Laemmli (1970) using a 9% polyacrylamide separating gel in conjunction with a 4% polyacrylamide stacking gel.

RESULTS AND DISCUSSION

Inhibition of Noradrenaline Release from PC12 Cells by BoNTA

In the presence of nerve growth factor, PC12 cells differentiate into a sympathetic-like neuron phenotype displaying extensive neurite outgrowths. Both the differentiated and undifferentiated forms of the cells rapidly take up noradrenaline from growth

medium and store the transmitter in the dense-core vesicles (Schubert and Klier, 1977).
Under conditions which depolarise the cell, such as 100mM K$^+$, noradrenaline and a
variety of other neurotransmitters are secreted in a calcium-dependent manner.
Incubation of PC12 cells with various concentrations BoNTA for 20h at 37°C induced a
dose-dependent inhibition of the release of noradrenaline (Fig 1). The concentration of
BoNTA required to inhibit 50% of the calcium-dependent noradrenaline release,
calculated from six such experiments, was found to be 0.12 ± 0.03uM. Pre-treatment of
the toxin at 100°C for 5min or incubation with an excess of BoNTA specific antiserum,
abolished the effects of BoNTA on the PC12 cells.

Binding studies, using 10nM ^{125}I-labelled BoNTA showed no evidence of any
saturable binding of the neurotoxin to the PC12 cells which suggests an absence of high
affinity acceptors for BoNTA on the cells. This apparent lack of acceptors for the toxin

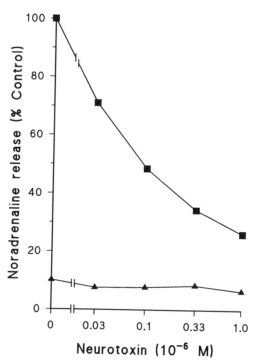

FIG 1. Inhibition of Noradrenaline Release from PC12 Cells By BoNTA.   PC12 cells growing on
collagen-coated 12-well dishes were incubated with a range of concentrations of BoNTA for 16h at 37°C.
[^{3}H]-noradrenaline (1uCi/ml) was then added to the growth medium and the cells incubated for a further
4h. Cells were washed twice with 1ml portions of growth medium and once with 1ml of a low potassium
release buffer (0.05M HEPES pH 7.1, 5mM KCl, 145mM NaCl, 2mM CaCl$_2$, 1mM MgCl$_2$, 10mM glucose).
Cells were then incubated for 15min at 37°C with either 1ml of high potassium release buffer (0.05M HEPES
pH 7.1, 100mM KCl, 50mM NaCl, 2mM CaCl$_2$, 1mM MgCl$_2$, 10mM glucose) or with the low potassium
release buffer. Portions of the buffer were removed, centrifuged at 15000 x g for 2min and the radioactivity
in 0.5ml of the supernatant fluid determined. Released noradrenaline with high potassium release buffer
(■) and low potassium release buffer (▲) were expressed as a percentage of the noradrenaline release in the
absence of BoNTA. Data points are the mean of three determinations; standard errors were less than 10%.

is reflected in the low sensitivity of PC12 cells to the neurotoxin. In comparison with studies using permeablized PC12 cells, in which inhibition of transmitter release was observed with 1-10nM BoNTA, neurotoxin concentrations 10-50 fold higher were required to induce similar levels of inhibition of noradrenaline release in intact cells. The manner in which BoNTA enters the PC12 cells is not clear but in view of the long time course of the inhibition (10-20h) and relatively high toxin concentrations required (0.1-1.0uM), entry into the cell by weak or non-specific interactions with the outer cell membrane seem most likely.

Long-term Effects of BoNTA on Noradrenaline Release from PC12 Cells

In vivo, nerve endings poisoned by BoNTA recover very slowly from the effects of the toxin which frequently persist for a number of weeks. The mechanism by which the toxin exerts its protracted action is presently unknown. In attempt to learn more about this aspect of action of BoNTA, the long-term effects of the toxin on noradrenaline release from PC12 cells were studied. Sister cultures of PC12 cells were either treated with BoNTA for 24h or left untreated. After 24h, both groups of cells were treated with an excess (>10 fold) of BoNTA specific antiserum and then grown for a further 7 days. At daily intervals, noradrenaline release from the toxin-treated cells was compared with that of the control group. Time courses of the inhibition by BoNTA at concentrations of 0.1 and 0.3uM are shown in Fig 2. For both BoNTA concentrations, noradrenaline release was maximally inhibited after 48h and then slowly recovered. Neither toxin concentration appeared to have any significant effect on the cell growth which was identical to the control group in the case of cells treated with 0.1uM BoNTA and only slightly reduced in the case of the 0.3uM BoNTA-treated cells. Noradrenaline release from PC12 cells treated with 0.3uM BoNTA remained at 30% of control values after 5 days despite a more than 4-fold increase in the cell mass. After 8 days, noradrenaline released from the cells was between 60-65% that of the control despite a more than 10-fold increase in the cell mass.

The above data show that the effects of BoNTA persist in the PC12 cells for several days and also in subsequent cell generations. If it is assumed that components of the secretory mechanism are constantly being renewed in the dividing cell, then it must also be assumed that BoNTA has a long half-life within the cell and is able to continue to exert its effects for several days after internalization. It is tempting to speculate that, if BoNTA can exert its effects for several days within a dividing cell line, then the neurotoxin may be able to act for considerably longer in the non-dividing nerve ending. How closely the environment of the neuromuscular junction resembles that of the PC12 cell, however, is unknown.

Fate of BoNTA Within PC12 Cells

PC12 cells intoxicated by prolonged treatment with BoNTA offer a useful model with which to study the action of the neurotoxin. Such cells can be maintained in culture for several days with more than 70% of the calcium-mediated transmitter release mechanism disabled. The above results also suggest that at least a portion of the internalised BoNTA remains active within the cells for several days. To investigate the possibility that the toxin undergoes some form of processing within the cell, such as subunit separation or proteolytic cleavage, the fate of BoNTA within PC12 cells was studied.

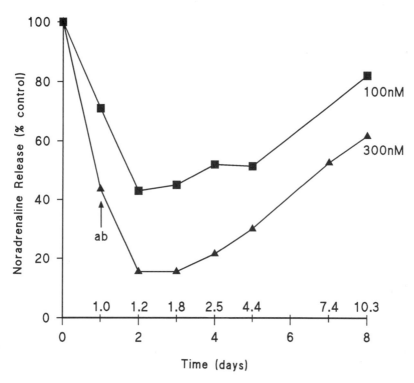

FIG 2. Long-Term Effect of BoNTA on Noradrenaline Release from PC12 Cells. Sister cultures of PC12 cells, grown as in Fig 1, were incubated with 0.1uM BoNTA, 0.3uM BoNTA of left untreated. After 24h the cells were washed once with medium and treated for 20min at 37°C with a > 10 fold excess of BoNTA specific antiserum (ab) and washed three times with 1ml portions of medium. Cells were maintained in culture for eight days and the noradrenaline release from the groups of cells measured at intervals as described in Fig 1. Values obtained from the toxin-treated cells were expressed as a percentage of the release obtained from the control group; (■), 0.1uM; (▲) 0.3uM BoNTA. Data points represent the mean of 6 determinations; standard errors were less than 10%. The numbers above the x-axis represent the relative increase in cell mass of the cells treated with 0.3um BoNTA, assessed by measuring total protein.

PC12 cells were incubated with ^{125}I-labelled BoNTA for 48h, washed and maintained in culture for a further 48h (see legend Fig 3.) After the 96h period, approximately 0.02-0.04% of the BoNTA remained associated with the cells. The distribution of the BoNTA in various cell fractions was then determined. Homogenised cells were centrifuged at 800 x g for 10min to give a pellet containing nuclei and cell debris, the supernatant fluid was removed and then further centrifuged at 20000 x g to give a pellet consisting of the large granules and smaller cell debris (Diliberto et al., 1976, Schubert and Klier, 1977). In three such experiments more than 94% of the cell associated ^{125}I-labelled BoNTA was recovered in the 800g and 20000g pellets. No significant difference in the specific binding of BoNTA to the two membrane fractions was evident; average values of 16 and 20 fmole BoNTA/mg protein were obtained for the 20000g and 800g pellets, respectively. In control experiments, in which ^{125}I-labelled BoNTA was added to cells immediately before homogenisation and incubated for 60min at 0°C, more than

99% of the neurotoxin was recovered in the 20000 x g supernatant fluid. Collectively, the results appear to suggest a slow, non-specific interaction of BoNTA with the various cell membrane components. Although the high proportion of membrane-associated BoNTA found within the cell may favour a membrane bound intracellular target for the toxin, it is not clear as to what proportion of the internalized BoNTA has reached its target and no conclusions may be drawn from the present data.

To examine the form of the intracellular BoNTA, the 20000g pellet containing the large granules was further purified on a linear sucrose density gradient (0.6-1.6M) (Schubert and Klier, 1977) and the ^{125}I-labelled BoNTA, associated with the large granule peak, examined on SDS-polyacrylamide gels (Fig 3). In the presence of thiol, BoNTA associated with the large granule fraction appeared as two bands of molecular weights approximately 97000 and 55000 kDa which were indistinguisable in molecular size from those of the heavy and light chains respectively, of control BoNTA preparations. Only trace bands representing degradation products of the neurotoxin were observed, which suggests that BoNTA is relatively stable within the cell environment. In the absence of thiol, BoNTA associated with the granule fraction appeared as two major bands; one of these bands (approximately 50% of the total) failed to enter the 4% acrylamide stacking gel, while the other band, approximately 150 kDa was indistinguishable in molecular size from that of control BoNTA. These observations suggest that approximately 50% of the BoNTA recovered with the large granule fraction is present in a highly aggregated form. This aggregated form of the BoNTA appears to consist of disulphide-linked toxin molecules, since only a small proportion of the high molecular weight toxin form was visible after thiol treatment. Qualitatively similar results were obtained when the ^{125}I-labelled BoNTA associated with the 800g pellet was examined on SDS polyacrylamide gels. The small proportion of labelled neurotoxin recovered in the 20000g supernatant fluid was found to contain no aggregated forms of the neurotoxin.

Measurement of the light chain to heavy chain ratio of the BoNTA associated with the large granule fraction revealed a small, but consistent increase in the proportion of light chain when compared to control BoNTA preparations. Six determinations from three experiments revealed an increase in the proportion of light chain of 29 $\pm$ 10%. Since approximately 50% of this toxin fraction appeared identical in molecular size to the native toxin, then it must be assumed that the free light chain is present in the aggregated form of the toxin, where it may constitute 60% of the total protein. This observation may be significant since it has been shown that the latter subunit of BoNTA mediates its intracellular action (Stecher, *et al.,* 1989b; De Paiva and Dolly, 1990). It does not seem likely that highly aggregated forms of either BoNTA or its light chain could play a direct role in the intracellular action of the toxin. However, if the protracted action of the BoNTA is due to it having a long-half within the cell, then the toxin must exist in a form which makes it impervious to the cell disposal mechanisms. One may speculate that the aggregated forms of the BoNTA represent protease-resistant pools of toxin within the cell. Slow release of BoNTA or its light chain from these pools could then act to prolong the effects of the toxin within PC12 cells.

Interaction of BoNTA with Liposomes and PC12 Cell Homogenates *In vitro*

The above data suggest that BoNTA slowly aggregates on the membrane cell fractions. To investigate the possibility that other cell proteins may be required for the aggregates of BoNTA to form, the interaction of the neurotoxin with liposomes was studied. ^{125}I-labelled BoNTA (10nM) was incubated with either liposomes or the 800g supernatant fluid of PC12 homogenates for 48h at 37°C. Incubation mixtures were then centrifuged at 100000 x g for 1h and samples of the pellets and supernatant fluids, analyzed by SDS gel electrophoresis (Fig 4).

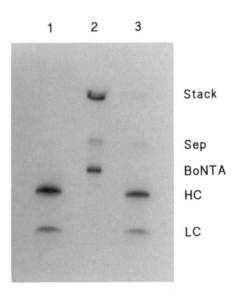

FIG 3. SDS-Polyacrylamide Autoradiogram of BoNTA Associated with the Purified Granule Fraction from PC12 Cells. PC12 cells grown on collagen-coated dishes were incubated at 37°C for 48h with 20uCi/ml ^{125}I-labelled BoNTA, the plates washed with medium and treated with a >10-fold excess of BoNTA specific antiserum for 20min. After a further three washes with medium the cells were grown for a further 48h and then, washed with medium and homogenised in 3mls of HEPES buffer (10mM, pH 7.2) containing 0.1M NaCl and 0.3M sucrose using a glass/teflon homogeniser. Homogenates were centrifuged at 800 x g for 10min, the supernatant fluid removed and centrifuged at 20000 x g for 30min. The pellet was then resuspended in HEPES buffer (5mM, pH 7.2) containing 0.3M sucrose, layered onto a sucrose gradient (0.6-1.8M) and centrifuged at 120000 x g for 2h. Fractions containing ^{125}I-labelled BoNTA were then analyzed on SDS polyacrylamide gels. Track 1, control BoNTA; tracks 2 and 3, cell-associated BoNTA. Tracks 1 and 3 were electrophoresed in the presence of 50mM dithiothreitol. Stack, beginning of 4% stacking gel; Sep, beginning of 9% separating gel; HC, heavy chain band; LC, light chain band.

After the 48h incubation period with liposomes, between 20-35% of the neurotoxin was recovered with the lipid pellet. High molecular weight aggregates of BoNTA which failed to enter the 4% polyacrylamide stacking gel were clearly evident in the BoNTA associated with the liposome pellet while no such forms of the toxin appeared to be present in the supernatant fluid. As with those observed in PC12 cells, the toxin aggregates appeared to consist of disulphide-linked molecules of toxin which dissociated in the presence of thiol (Fig 4). The proportion of the total toxin in an aggregated form

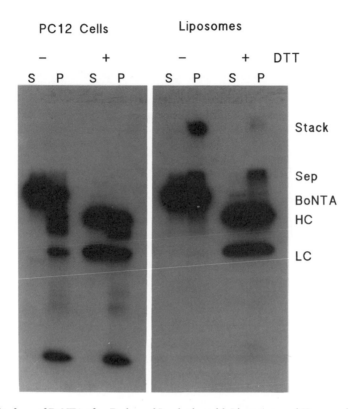

Fig 4. Molecular form of BoNTA after Prolonged Incubation with Liposomes and Homogenised PC12 Cells. Liposomes (phosphatidyl choline:phosphatidyl serine: phosphatidyl glycerol:phosphatidyl ethanolamine; 3:1:1:1) were prepared by the method of Enoch and Strittmatter (1979) and dialysed against KGEP buffer (20mM Pipes pH 7.0, 140mM potassium glutamate, 5mM EGTA, 2mM $MgCl_2$). PC12 cells were homogenised with a glass/teflon homogeniser in KGEP buffer and centrifuged at 800 x g for 10min. The PC12 supernatant fluid and liposomes were incubated with 10nM ^{125}I-labelled BoNTA for 48h at 37°C and then centrifuged at 100000 x g for 1h. The resulting pellets (P) and supernatant fluid (S) were examined on SDS polyacrylamide gels. Stack, beginning of 4% stacking gel; Sep, beginning of 9% separating gel; HC, heavy chain band; LC, light chain band; -/+, absence and presence of 50mM dithiothreitol.

(2-5%) bound to liposomes was, however, considerably less than that observed with PC12 cells. The above observations may indicate that the neurotoxin adopts a different conformation when associated with the lipid bilayer which allows molecules to form disulphide links.

In similar experiments performed with the 800g supernatant fluid of PC12 homogenates, of the 20-30% of the BoNTA recovered in the cell pellet, little or none was observed in an aggregated form. In contrast, a considerable proportion of the BoNTA appeared as protein bands of lower molecular weight indicative of significant proteolytic degradation. Protein bands of similar molecular size to the heavy and light chains were also evident, suggesting some reduction of the inter-chain disulphide bridge. Perhaps, not surprisingly, these data tend to suggest that in homogenised PC12 cells, BoNTA is exposed to greater concentrations of thiols and proteases than the neurotoxin internalised by intact growing cells.

Data from both sets of experiments indicate that BoNTA may adopt a different conformation when associated with lipid bilayers. In the presence of liposomes, BoNTA appears to form high molecular weight disulphide-linked aggregates.

CONCLUSIONS

BoNTA was found to evoke a slow, dose-dependent inhibition of the calcium-dependent release of noradrenaline from growing PC12 cells which persisted for several days and in subsequent cell generations. Thus, PC12 cells offer a useful model with which to study the intracellular mechanism of BoNTA. Toxin-treated cells can be maintained in culture for a number of days which should allow a range of aspects of cell function to be examined.

Analysis of the BoNTA internalised within PC12 cells revealed remarkably little degradation of the toxin molecule after 96h. This observation and the observed prolonged action of the toxin on the transmitter release mechanism, suggest that BoNTA is relatively stable within the PC12 cell.

The bulk of the internalised toxin was recovered in cell membrane fractions, a significant proportion of which was in the form of high molecular weight aggregates of the neurotoxin. The formation of toxin aggregates may well be an artifact of the PC12 cell model system. Alternatively, it is possible that such high molecular weight forms of the toxin could act as protease-resistant pools of BoNTA which serve to prolong the intracellular effects of the neurotoxin.

REFERENCES

Black, J.D. and Dolly, J.O., 1986, Interaction of ^{125}I-labelled botulinum neurotoxin with nerve terminals *J. Cell Biol.*, 103:535-544.

De Paiva, A. and Dolly, J.O., 1990, Light chain of neurotoxin is active in mammalian motor nerve terminals when delivered via liposomes, *FEBS Lett.*, 277:171-174.

Diliberto, Jr., E.J., Viveros, O.H. and Axelrod, J., 1976, Subcellular localisation of protein carboxymethylase and its endogenous substrates in the adrenal medulla: possible role in excitation-secretion coupling, *Proc. Natl. Acad. Sci. USA*, 73:4050-4054.

Dolly, J.O., Black, J.D., Williams, R.S., and Melling, J., 1984, Acceptors for botulinum neurotoxin reside on motor nerve terminals and mediate its internalisation, *Nature*, 307:457-460.

Enoch, H.G, and Strittmatter, P., 1979, Formation and properties of 1000A diameter, single bilayer phospholipid vesicles, *Proc. Natl. Acad. Sci. USA*, 76:145-149.

Greene, L.A. and Rein, G., 1977, Synthesis, storage and release of acetylcholine by a noradrenergic, pheochromocytoma cell line, *Nature*, 268:349-351.

Greene, L.A. and Tischler, A.S., 1976, Establishment of a noradrenergic clonal line of rat adrenal pheochromocytoma cells which respond to nerve growth factor, *Proc. Natl. Acad. Sci. USA*, 73:2424-2428.

Laemmli, U.K., 1970, Cleavage of the structural protein during the assembly of the head of the bacteriophage T4, *Nature*, 227:680-685.

McIness, C. and Dolly, J.O., 1990, Calcium-dependent noradrenaline release from permeabilised PC12 cells is blocked by botulinum neurotoxin A or its light chain, *FEBS Lett.* 261:323-326.

Schubert, D. and Klier, F.G., 1977, Storage and release of acetylcholine by a clonal cell line, *Proc. Natl. Acad. Sci. USA*, 74:5184-5188.

Shone, C.C., 1987, *Clostridium botulinum* neurotoxins: their and modes of action, in:"Natural Toxicants in Foods", D. Watson ed., pp 11-51, Ellis Horwood Ltd., Chichester, U.K.

Shone, C.C., Hambleton, P. and Melling, J., 1985, Inactivation of *Clostridium botulinum* type A neurotoxin and purification of two tryptic fragments, *Eur. J. Biochem.* 151:75-82.

Simpson, L.L., 1981, The origin, structure and pharmacological action of botulinum toxin, *Pharmacol. Rev.* 33:155-188.

Stecher, B., Gratzl, M. and Ahnert-Hilger, G., 1989a, Reductive chain separation of botulinum A toxin - a prerequisite to its inhibitory action on exocytosis in chromaffin cells. *FEBS Lett.*, 248:23-27.

Stecher, B., Weller, U., Habermann, E., Gratzl and Ahnert-Hilger, G., 1989b, The light chain but not the heavy chain of botulinum A toxin inhibits exocytosis from permeabilized adrenal chromaffin cells, *FEBS Lett.* 255:391-394.

Wadsworth, J.D.F., Desai, M., King, H.K., Tranter, H.S., Hambleton, P., Melling, J., Dolly, J.O. and Shone, C.C., 1990, Botulinum type F neurotoxin: purification, radiolabelling and binding to rat brain synaptosomes, *Biochem. J.* 268:123-128.

Williams, R.S., Tse, C.-K., Dolly, J.O., Hambleton, P. and Melling, J., 1983, Radiolabelling of botulinum type A neurotoxin with retention of biological activity and its binding to rat brain synaptosomes, *Eur. J. Biochem.* 131: 437-445.

THOUGHTS ON ACTION OF BOTULINUM TOXIN
SUGGESTED BY REVERSIBILITY OF HEART EFFECTS

Carl Lamanna

Carolina Meadows V267
Chapel Hill, North Carolina 27514

Botulinum neurotoxin poisoning prevents the release of the neurotransmitter acetylcholine into the synaptic cleft. Stimulation of the cholinergic nerve is required for the acetylcholine-containing vesicles to attach to the inner surface at specific points of the nerve cell synaptic membrane. Thereupon at these points of attachment the acetylcholine leaves the vesicles to escape into the synaptic cleft by passage across the synaptic membrane. This event has been called exocytosis and is probably the action frustrated by the presence of toxin.

Classically botulinum neurotoxic activity has been considered to be persistent and almost irreversible. Recently cardiac effects of the toxin have been observed to be readily reversible and thus not persistent.[2,3] I suggest that this paradoxical cardiotoxicity does not challenge the theory of inhibition of cholinergic nerves and that in fact the cardiac phenomena can contribute to productive thinking about the mode of action of botulinum toxin.

Type A hemagglutinin-free toxin causes electrocardiographic changes and a decreased heart rate (bradycardia) within minutes of injection of the toxin in rodents and dogs, the species studied. Within a short time there follows a spontaneous recovery to the normal heart picture. While this spontaneous recovery is not seen *in vitro* with isolated heart preparation a rapid reversal of cardiac effects does follow simple isotonic washing out of the toxin-containing fluid bathing the isolated heart. Another significant finding is repetitious bradycardia following each of successive exposures to toxin of the isolated heart preparation after washing out of the toxin.

The persistence of bradycardia with *in vitro* isolated heart preparations is due to the continuing presence of toxin in the fluid bathing the heart. *In vivo*, after injection of toxin there is a drop in blood level as the toxin is distributed and bound to nerve cells. The heart with its apparent weaker strength of toxin binding loses the toxin to places of greater binding capacity.

Repeated *in vivo* injections of toxin result in repetition of the bradycardia–spontaneous recovery cycle. It can be adduced that spontaneous recovery by *in vivo* exposure to toxin does not mean there is any fundamental difference in nature between *in vivo* heart and *in vitro* isolated heart toxicities.

Ready reversibility of cardiac toxicity and an unimpaired capacity for toxin uptake after prior uptake and washing out of toxin suggest that the intermolecular forces binding toxin to susceptible nerve cells do not lead to profound change in the cellular physicochemical environment for nerve reception of the toxin. This is more consistent with an adsorption type of attraction than chemical bonding, and is not indicative of an enzyme action. With enzymatic activity an exhaustion of substrate would be expected.

The exposure to toxin does not cause any discerned permanent change in the nature of the toxin and membrane receptor. Concomitantly the toxin must be bound to the synaptic membrane for the entire period of poisoning, which can be days, weeks and months long for both experimental and natural botulinum poisoning.

Two possibilities are reasonable to consider for the long-term persistence of nerve cell dysfunction. The cytochemical lesion of botulism is so profound as to resist rapid repair. The alternative I favor is that paralysis persists for as long a time as the toxin remains lodged at the synaptic membrane. The half-life of toxin in the membrane can be a long one since there are no clues to suggest rapid catabolic destruction of the embedded toxin as, for example, by proteolysis. Cytological study of repair of poisoned nerve fibrils is consistent with persistence of membrane-associated toxin. Repair is slow in becoming visible, and significantly the growth of new functional fibrils in the poisoned nerve cell takes place outside the areas of the preexisting poisoned fibrils.[1] Studies should be done for detecting continuous presence of toxin in the poisoned fibrils. This would provide definitive knowledge of the time scale of residence of toxin in the synaptic membrane and suggest if the toxin's presence hinders self-repair of the membrane site toxin occupies.

The following scenario is presented for botulinum neurotoxicity which I believe is consistent with the findings of cardiac studies.

Nerve stimulation is accompanied by two events: attachment of acetylcholine-containing vesicles to specific areas of the interior surface of the membrane and exocytosis of the vesicle. With toxin present nerve stimulation permits membrane uptake of the toxin and vesicle attachment to the synaptic membrane but exocytosis does not take place. Thus poisoned nerve cells do not accumulate acetylcholine in the cytosol. Such intracellular increase of free acetylcholine should happen if acetylcholine was released from vesicles and the released acetylcholine was not able to move across the nerve cell membrane into the synaptic cleft. Toxicity is explained if exocytosis were recognized to be a synaptic membrane–vesicle membrane interaction frustrated by toxin being present at this membrane-to-membrane interface.

Nerve stimulation acts as a valve which opens the nerve membrane to uptake of toxin. It is also accompanied by much studied intracellular and synaptic membrane changes in ionic environment in which calcium is a major participant. Conceivably toxin in the synaptic membrane disturbs access of ions to the trigger spot for exocytosis at the synaptic membrane–vesicle membrane interface. Exocytosis could be prevented by the toxin physically masking the trigger spot. This prevents contact of the proper ions with the trigger spot. Removal of the toxin which yields reversible cardiac toxicity means that the toxin resident in the synaptic membrane causes no irreversible changes in the events associated with initiation of exocytosis and movement of acetylcholine across the synaptic membrane. This is consistent with rapid recovery following removal of the toxin.

An explanation of botulism poisoning must account for differences in sensitivity to toxin among the sympathetic, parasympathetic and central nervous systems in spite of some common possession of cholinergic neurons. There might be differences among these systems in barriers affecting diffusion of toxin from lymph into the synaptic cleft. There may be a limiting width of the synaptic cleft that must be exceeded to allow entrance of the large-sized 150,000 dalton molecular weight toxin, a factor which might operate in the densely packaged brain. The cardiac effects at this stage of knowledge tell us nothing about the possibilities.

A prime factor affecting toxicity would be varying strength of binding forces of synaptic membranes holding toxin. These binding forces could differ among cholinergic neurons associated with different parts of body tissues and organs. If there is a spectrum from high- to low-strength binding forces the heart would appear to represent a marginal- or low-strength binding entity. A task that needs to be undertaken is to identify the sites and nature of toxin-susceptible cardiac innervation. What can be said is that the toxin does not affect the intrinsic automaticity of the beat of isolated heart cells. It can be said that toxin inhibition of heart rates rests on nervous system control at sites located in the heart.

REFERENCES

1. Duchen LW, Strich SJ. The effects of botulinum toxin on the pattern of innervation of skeletal muscle in the mouse. Quart J Exp Physiol Cogn Med Sci 1968; 53:84-89.
2. Lamanna C. Overview of bacterial toxins with a nonreductionist approach to the mode of action of botulinal neurotoxin. In: Pohland AE et al., eds. Microbial toxins in foods and feeds. New York: Plenum Press, 1990.
3. Lamanna C, El-Hage AN, Vick JA. Cardiac effects of botulinal toxin. Arch Int Pharmacodyn Therap 1988; 293:69-83.

PROLIFERATIVE T CELL RESPONSE TO BOTULINUM TOXIN TYPE A IN MICE

Woon Ling Chan, Dorothea Sesardic
and Clifford C. Shone*

National Institute for Biological Standards and Control,
Blanche Lane, South Mimms, Potters Bar, Herts, EN6
3QG, UK and *Centre for Applied Microbiology and
Research. Porton, Salisbury, Wilts, UK

INTRODUCTION

Clostridium botulinum can be divided into seven types, designated A to G. Each type produces an immunologically distinct neurotoxin. For more than 40 years, immunisation with botulinum toxoid has been used to protect laboratory workers at risk for botulism due to contact with the neurotoxins. However, the toxoid preparations used are relatively crude producing undesirable local and systemic reactions. The current pentavalent (ABCDE) toxoid manufactured by the Michigan Department of Public Health using the same methodology as Parke, Davis and Co induces significantly improved response to type B but not to type A and E. Furthermore, a 25-week period rather than 12 weeks is required to achieve a protective level and the range of antibody titres among individuals who received the same number of immunisations is very wide (1). Therefore, it appears that an understanding of the B and T helper cell responses induced by the toxoid would help in the design of improved vaccines. As a first step, we compared the *in vitro* proliferative T cell response induced in mice by botulinum toxoid in the presence or absence of alum. Secondly, we examined sera from immunised mice for neutralising antibody response and showed that increased neutralising antibody titre correlates with the induction of good proliferative T cell response.

PROLIFERATIVE T CELL RESPONSE IN MICE

BALB/c mice were immunised subcutaneously twice at 14 days interval in the flank and footpad with 10μg botulinum toxoid A per mouse in the presence (gp 2) or absence (gp 1) of alum as adjuvant. Control mice were given alum alone (gp 3). Draining lymph node cells were taken 7 days later and assayed for *in vitro* proliferative T cell response as described previously (2). Basically, all assays were performed in triplicates using 96-well flat bottomed tissue culture plates. Lymph node cell suspensions were dispensed at 4×10^5 cell per well and each culture, with or without antigen, was incubated in an atmosphere of 5% CO_2 at 37^0 for 4 days. The cultures were pulsed during the last 18hr with 1μCi per well of [^{3}H]thymidine, harvested with a cell harvester and counted on a ß-counter. The purified BoNT/A toxin fragments L, H and L-H$_N$ used as stimulating antigens were prepared as described previously (3). As shown in figure 1, mice immunised with botulinum toxoid in the presence of alum showed a strong *in vitro* proliferative T cell response to botulinum toxoid and to the BoNT/A toxin fragments L, H and L-H$_N$ but not to tetanus toxoid. In contrast, the T cell response to the same antigens was weak in the group given toxoid without alum. Cells from control mice given alum alone did not respond to any of the antigens *in vitro*.

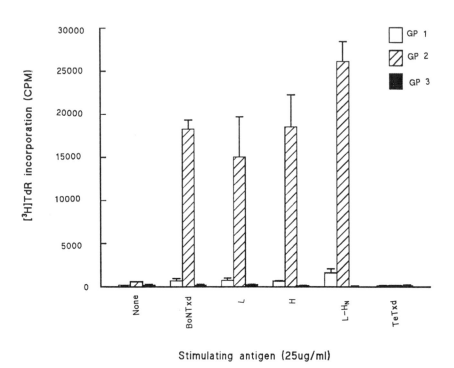

Figure 1. T-cell proliferation with Botulinum toxoid and BoNT/A fragments.

ELISA ANTIBODY RESPONSE

The ELISA assay was performed using 50μl per well of toxoid A, BoNT/A fragments L or H at 2μg/ml to coat 96-well microtitre plates. Antibody binding was detected with 1:2500 dilution of rabbit anti-mouse Ig conjugated with peroxidase and the enzyme substrate used was 2,2'-Azino-bis(3-ethylbenzthiazoline-6-sulfonic acid). O.D. reading were taken at 405nm. All washes were with PBS containing 2% BSA and 0.05% Tween 20. Antibody titres were assayed from sera that were taken from the mice at the same time as the draining lymph node cells. Sera from mice immunised with botulinum toxoid showed a high antibody titre to toxoid and to the purified BoNT/A fragments L and H. While there was no difference in the binding or avidity of the antibody of gp1 and gp2 sera to toxoid, binding to toxin fragment L in particular was greater in sera from gp2 than gp1. There was no significant difference in the antibody titre of sera from gp1 and gp2 to the 3 antigens. Control sera showed no binding to any of the antigens tested.

BOTULINUM TOXIN NEUTRALISATION ASSAY *IN VIVO*

The potency of botulinum antitoxin in the sera of mice immunised with toxoid A was determined by titration against a standard preparation of botulinum antitoxin (WHO International Standard 500 IU/ampoule at IU=0.136 mg protein) to give the same degree of protection with a fixed dose of botulinum toxin ($50LD_{50}$). Sera were diluted 1:100 in 0.05M sodium phosphate buffer pH 6.5 containing 0.2% gelatin. Then graded volumes of the diluted sera or reference antisera (0.5 IU/ml) were mixed with a fixed concentration of botulinum toxin (calculated to deliver a dose of $50LD_{50}$/mouse) to a final volume of 4.5ml per antibody dilution per group of 4 mice. The mixture was incubated at room temperature for 1 hr. Then 1 ml of the mixture was injected intraperitoneally per MF1 mouse. The animals were observed for signs of neurotoxicity over a period of 4 days. Mice immunised with the toxoid produced antibodies that neutralised the toxin. The titre of neutralising antibodies in mice given toxoid with alum was 4-fold higher than those given toxoid alone. Sera from control mice had no toxin neutralising activity.

CONCLUSION

Mice immunised with botulinum toxoid A were able to produce both *in vitro* proliferative T cell and antibody response to purified BoNT/A fragments L and H. While there was no significant difference in the antibody titres (to toxoid, L or H) of sera from mice given toxoid with (gp2) or without (gp1) alum as adjuvant, antibodies from gp2 sera showed higher binding and avidity to BoNT/A fragment L than those from gp1. There was no difference in the binding capacity of sera from both groups to toxoid. However, there was a much stronger *in vitro* antigen specific proliferative T cell response to all antigens tested, with cells from gp2 than from gp1. This difference is also seen in the neutralising antibody titres obtained, with higher antibody titre in gp2 than in gp1. This suggests that the induction of good protective antibody response requires both T and B cells.

REFERENCES

1. Siegel, L. S. *J. Clin. Microbiol.* **26**:2351 (1988).
2. Chan, W. L., Lukic, M. and Liew, F. Y. *J. Exp. Med.* **162**:1304 (1985).
3. Shone, C. C., Hambleton, P. and Melling, J. *Eur. J. Biochem.* **151**:75 (1985).

LIMITED PROTEOLYSIS OF TETANUS TOXIN LIGHT CHAIN BY TRYPSIN AT ITS C-TERMINUS CAUSES A 10–30-FOLD DECREASE OF ACTIVITY AS MEASURED BY INHIBITION OF NORADRENALINE RELEASE FROM PERMEABILIZED CHROMAFFIN CELLS

Dagmar Sanders,[1] Jasminka Godowac-Zimmermann,[2] Ernst Habermann[1] and Ulrich Weller[3]

[1]RBI f. Pharmakologie, JLU, Frankfurter Str. 107, W-6300 Gießen,
  Fed. Rep. of Germany
[2]Inst. f. Physiologische Chemie u. Pathobiochemie, JOGU, Duesbergweg 6,
  W-6500 Mainz, Fed. Rep. of Germany
[3]Inst. f. Med. Mikrobiologie, JOGU, Hochhaus am Augustusplatz,
  W-6500 Mainz, Fed. Rep. of Germany

INTRODUCTION

The light chain (L) of tetanus toxin (TeTx) was shown to be the active principle of the inhibitory action of tetanus toxin in permeabilized cell systems,[1,2] and if injected into neurons from *Aplysia californica*.[3] It was shown to have the same activity as native toxin on a molar basis.[1,3] Different L constructs expressed in *E. coli* and purified to homogeneity turned out to be proteolytically shortened at the C-terminus. Their activity in digitonin permeabilized chromaffin cells (PCC) was reduced to about 10% as compared with L from native toxin. CYS 438 was essential for activity *in vivo* but not in PCC.[4]

Deletion mutants of L expressed by injecting mRNA into neurons of *Aplysia californica* were shown to retain activity up to residue 392.[5,6] In order to investigate structure function relationships at the protein level, L was subjected to limited proteolysis by different proteases.

RESULTS AND DISCUSSION

In pilot experiments, purified L[7,8] was submitted to proteolytic degradation with trypsin (E.C. 3.4.21.4), chymotrypsin (E.C. 3.4.21.1) and carboxypeptidase Y (E.C. 3.4.16.1). Most of L resisted degradation. Proteolysis always resulted mainly in a slight reduction of molecular weight as shown by SDS-PAGE. Due to the higher specificity, trypsin cleaved L was used in the following experiments.

Botulinum and Tetanus Neurotoxins, Edited by
B.R. DasGupta, Plenum Press, New York, 1993

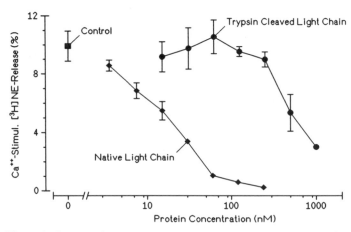

Figure 1. Concentration response curves for inhibition of Ca^{++}-evoked [^{3}H]-noradrenaline ([^{3}H]NE) release from digitonin permeabilized bovine chromaffin cells. Native TeTx light chain (IC_{50} 18,4 nM) was compared with trypsinized light chain (IC_{50} 560 nM). Release in the absence of Ca^{++} (1.5%) was subtracted.

L (6 ml, 1 mg/ml) was dialyzed (0,1 M Na-phosphate pH 7.8, 0.5 mM DTT) and trypsinized (enzyme substrate ratio 1:200, 37°C, 120 min); protease activity was blocked by diisopropylfluorophosphate. SDS-PAGE showed that trypsin cleavage led to a shortening of the chain in two steps (L1, L2). It was not possible to obtain the slightly higher molecular weight form (L1) in high yield.

Edman degradation (Pulsed Liquid Phase ABI 477A sequencer, equipped with ABI 120A PTH analyzer (ABI chemicals) and 610A 1.2.1 Data Analyzer) of cleaved L showed that the N-terminus remains unchanged, similar to recombinant L expressed in *E. coli*.[4] Combination experiments [8,9] with native heavy chain (H) showed that only L1 was able to form toxin, similar to native L. This indicates that CYS 438, which forms the interchain disulfide bridge in native toxin, is still present.[10]

To determine the exact C-terminus of L2, cleavage products were removed by HIC chromatography on Alkyl-Superose 5/5 (Pharmacia). Prior to BrCN cleavage,[10] purified L was dialyzed against distilled water, freeze-dried and alkylated with 4-vinylpyridine. BrCN peptides were separated by C_{18} reversed-phase HPLC chromatography.[10] The resulting chromatograms were compared to native L with identical processing. The peak corresponding to C-terminal BrCN peptide (RVNT) was almost absent. One new peak appeared, and a second increased in height. Both were subjected to Edman degradation. The first peak consisted of a pure peptide which was sequenced 20 steps (initial yield 40%, repetitive yield 92.9%). The sequence RVNTNAFRNVDGSGLVSK (416-433) terminated after 18 steps. The second additional peak contained mainly the peptide 244-260 and about 10% of peptide 416-433 (repetition of analysis from both peaks yielded identical results). This showed that trypsin cleaves L behind LYS 433 and the slightly larger intermediate L1 ends most likely behind LYS 439 or LYS 440. The next trypsin cleavable site C-terminal to Lys 440, is ARG 448. This is only one position away from the C-terminus of native L.[11] Thus separation of chains renders 2 more sites (15 residues) at the C-terminus accessible for trypsin. The N-terminus remains protected from protease attack. The other proteases employed left light chain derivatives of a similar size, indicating that this is a general structural feature.

Removal of 16 residues by limited proteolysis with trypsin reduces inhibitory activity of L on PCC 10-30 times as compared to native L. Since the preparation always contains small amounts of L1, the individual contributions of L1 and L2 to the residual activity remain to be assessed. Reduction of activity due to damage caused by subsequent purification steps can be excluded since activity was measured directly after proteolytic cleavage. The data demonstrate that the C-terminal region of L is important for activity. Reduction of inhibitory potency due to C-terminal proteolysis is in agreement with previous results with recombinant L which was modified by bacterial protease(s) after expression in *E. coli*.[4] These results should be considered in the light of data reported on *Aplysia*, where a construct truncated already after the first 392 residues was still active.[6] The contention has therefore been made that the C-terminal 57 amino acids are not required for activity. Our present data clearly indicate that such an interpretation may be hazardous, and careful quantitative assessment of function is called for. In addition, it may be speculated that species differences (non-vertebrates/mammals) furnish an explanation of this phenomenon.

REFERENCES

1. Ahnert-Hilger G, Weller U, Dauzenroth M-E, Habermann E, Gratzl M. The tetanus toxin light chain inhibits exocytosis. FEBS Lett 1989; 242:245-248.
2. Dayanithi G, Weller U, Ahnert-Hilger G, Link H, Nordmann JJ, Gratzl M. The light chain of tetanus toxin inhibits calcium-dependent vasopressin release from permeabilized nerve endings. Neurosci 1992; 46:489-493.
3. Mochida S, Poulain B, Weller U, Habermann E, Tauc L. Light chain of tetanus toxin intracellularly inhibits acetylcholine release at neuro-neuronal synapses, and its internalization is mediated by heavy chain. FEBS Lett 1989; 253:47-51.
4. Weller U, Fairweather N, Sanders D, Ahnert-Hilger G, Payne M, Habermann E. Expression of tetanus toxin light chain and mutants of CYS 438 in *E. coli:* purification and biological activities of the recombinant proteins. Zbl Bakt Suppl 1992; 23:99-100.
5. Mochida S, Poulain B, Eisel U, Binz T, Kurazono H, Niemann H, Tauc L. Exogenous mRNA encoding tetanus or botulinum neurotoxins expressed in *Aplysia* neurons. Proc Natl Acad Sci USA 1990; 87:7844-7848.
6. Binz T, Grebenstein O, Kurazono H, Eisel U, Wernars K, Popoff M, Mochida S, Poulain B, Tauc L, Kozaki S, Niemann H. Molecular biology of the L chains of clostridial neurotoxins. Zbl Bakt Suppl 1992; 23:56-65.
7. Högy B, Dauzenroth M-E, Hudel M, Weller U, Habermann E. Increase of permeability of synaptosomes and liposomes by the heavy chain of tetanus toxin. Toxicon 1992; 30:63-76.
8. Weller U, Dauzenroth M-E, Meyer zu Heringdorf D, Habermann E. Chains and fragments of tetanus toxin. Separation, reassociation and pharmacological properties. Eur J Biochem 1989; 182:649-656.
9. Kistner A, Habermann E. Reductive cleavage of tetanus toxin and botulinum neurotoxin A by the thioredoxin system from the brain. Naunyn-Schmiedeberg's Arch Pharmacol 1992; 345:227-234.
10. Krieglstein K, Henschen A, Weller U, Habermann E. Arrangement of disulfide bridges and positions of sulfhydrylgroups in tetanus toxin. Eur J Biochem 1990; 188:39-45.
11. Krieglstein K, Henschen A, Weller U, Habermann E. Limited proteolysis of tetanus toxin. Relation to activity and identification of cleavage sites. Eur J Biochem 1991; 202:41-51.

FUNCTIONAL ROLES OF DOMAINS OF CLOSTRIDIAL NEUROTOXINS: THE CONTRIBUTION FROM STUDIES ON *Aplysia*

Bernard Poulain[1], Ulrich Weller[2], Thomas Binz[3], Heiner Niemann[3], Anton de Paiva[4], J. Oliver Dolly[4], Christiane Leprince[1] and Ladislav Tauc[1]

[1]Laboratoire de Neurobiologie Cellulaire et Moléculaire
CNRS
F-91198 Gif-sur-Yvette, France
[2]Institut für Medizinische Mikrobiologie
J.Gutenberg Universität
D-6500 Mainz, GDR
[3]Bundesforschungsanstalt für Viruskrankheiten der Tiere
D-7400 Tübingen, GDR
[4]Department of Biochemistry
Imperial College of Science, Technology and Medicine
London SW7 2AY, U.K.

INTRODUCTION

In recent years there has been increasing interest in the study of the mode of action of botulinum neurotoxin (BoNT) and tetanus toxin (TeTx) at the cellular level (for recent reviews see Simpson, 1986, 1989; Habermann and Dreyer, 1986; Niemann, 1991; Poulain and Molgo, 1992; Wellhöner, 1992, Dolly, 1992, and this volume). This is mainly due to the unique ability of these toxins to effectively block neurotransmitter release in all types of nerve terminals once they gain access inside the nerve endings; BoNTs and TeTx are, thus, useful research tools to study transmitter release mechanisms and tools in medicine. Deciphering their intracellular modes of action still remains a challenge for future research; however, during the last decade continuous progress has been made in the identification of the functional domains of BoNTs and TeTx implicated in the multiple facets of the intoxication. Using neuronal preparations from *Aplysia*, an identification of the functional role of different protein domains of BoNTs and TeTx has been made which contrasts somewhat with those proposed for vertebrates. Thus the aim of the present paper is to review data in *Aplysia* and, also, to stress the similarities and differences into observations made in the various models.

Botulinum and Tetanus Neurotoxins, Edited by
B.R. DasGupta, Plenum Press, New York, 1993

MODELS: FROM VERTEBRATE NEUROMUSCULAR JUNCTION TO NEURONS OF *Aplysia*

The neuroselective action of *Clostridial* toxins has been largely documented and can be grossly depicted as a selective inhibition of cholinergic neurotransmission by BoNT and a preferential blockade by TeTx of non-cholinergic transmitter release. In fact, neither nerve-endings are absolutely resistant to either toxin. Despite its selectivity for "non-cholinergic" synapses, capture of TeTx by the cholinergic motor nerve terminals enables it to reach the central nervous system via the retrograde axonal transport. This had justified the choice of vertebrate skeletal neuromuscular junction to identify the mode(s) and site(s) of action of these two types of *Clostridial* neurotoxins. Notably, the use of such preparations allowed the complex mode of action of the toxins to be split into three steps, namely binding, internalization and intracellular action (BoNT: Simpson, 1981, 1986; TeTx: Schmitt et al., 1981). In addition, nerve-muscle preparations permitted a partial characterization of the blocking events at the elementary level because of an easy quantification of the acetylcholine release process (evoked or spontaneous, quantal or non quantal) using conventional electrophysiological techniques (Habermann and Dreyer, 1986; Molgo et al., 1990). However, due to the very small size of presynaptic nerve terminals, intracellular space is inaccessible except by liposomal delivery (see de Paiva and Dolly, 1990) and too small an amount of material is available from these nerve endings to allow biochemical investigations. This has justified the search for other models, in order i) to establish which protein domains are associated with each step of the toxins' inhibitory action -the aim of this present paper- ii) to indentify the neuronal component serving as the toxins' membrane acceptors (see Schiavo et al., this volume) and iii) finally, to decipher the toxins' intracellular molecular actions (Niemann, 1991; Niemann et al., 1991 and this volume; Dolly et al., 1992 and this volume).

A rapid review of the other available models shows that very few of them are suitable for the elucidation of the functional role of various regions of the toxins. Indeed, studies using intact preparations of the target tissue for TeTx -i.e. the central nervous system (CNS)- are difficult to perform due to the inherent intricacy of the nervous tissue; isolated preparations of nervous tissue (brain sections or slices, brain particles, preparation of nerve-endings or synaptosomes...) have an active life-time of only a few hours in addition to being heterogenous due to the occurence of cholinergic and non-cholinergic nerve endings in varying proportions within the regions of the CNS from which they are dissected (for a review, see Habermann and Dreyer, 1986). Primary cell cultures have similar disadvantages, however they allow prolonged application of lower concentrations of toxins since their life time is not restricted. Though they are feebly sensitive to extracellularly applied *Clostridial* toxins, tumor cell lines (PC12) or secretory cells such as chromaffin cells provide the opportunity of giving access to intracellular space by an injection procedure using a patch pipette (Penner et al., 1986) or by various permeabilization technics (see Bittner et al., 1989; Ahnert-Hilger et al.; 1989a; McInnes and Dolly, 1990; Lomneth et al., 1991). Nevertheless, these preparations do not allow simultaneous access to both intracellular and extracellular spaces. As is evident, no model-preparation is fully satisfactory.

In the quest for an experimental model suitable for the elucidation of the role of the various domains of BoNT and TeTx, ganglionic preparations from the CNS of *Aplysia californica* appeared ideal because they combinate a high susceptibility to the neuroselective actions of TeTx or BoNTs, as well as affording easy access to extra-or/and intracellular spaces. Indeed, in this sea mollusc, ganglia contain easily identifiable cholinergic and non-cholinergic neurons that make synapses (Gardner, 1971; Gaillard and Carpenter, 1986). Transmitter release can be quantified by measurement of the amplitude of postsynaptic responses evoked by presynaptic action potentials or long depolarizations of the presynaptic neuron, using conventional electrophysiological (current- or voltage-clamp) techniques (for further details see Gardner, 1971; Baux et al., 1990). Amongst the numerous advantages of this preparation, is the possibility of simple injection of toxins or fragments inside the presynaptic cell bodies from which they reach (in 5 to 15 minutes) the nearby nerve endings (300-500 μm away). In addition, these preparations have an active life time of several days, or weeks when organotypic culture conditions are used.

TOXINS' DOMAINS IMPLICATED IN THE INTRACELLULAR BLOCKADE OF TRANSMITTER RELEASE

Chains required intracellularly in *Aplysia* and vertebrates

Because of the similar di-chain structure of *Clostridial* toxins with other A-B structured toxins, it was earlier speculated that domain(s) located in the light chain of either BoNTs or TeTx are involved in the intracellular blockade of neurotransmitter release (see Simpson, 1986). To test such a hypothesis one needs highly purified chains (see Figure 1 for nomenclature) of the toxins, coupled to an easy access to intracellular space. For this latter reason, *Aplysia* neurons appeared better candidates than other preparations because their susceptibility to the action of TeTx and BoNT was similar to vertebrates models (Poulain et al., 1990, 1992a; Dolly et al., this volume).

BoNT or TeTx

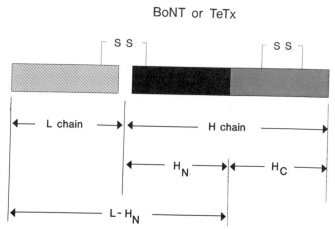

Figure 1. Schematic representation of chains and fragments of BoNT and TeTx. Nomenclature of chains and fragments according to terminology established at the Fifth European Workshop on Bacterial Protein Toxins (Aguilera et al., 1991). H, heavy chain (MW ≈100,000); L light chain (MW ≈50,000); H_C, carboxy-terminus half of heavy chain (MW ≈50,000) and H_N, amino-terminus of heavy chain (MW ≈50,000).

Surprisingly, in *Aplysia*, when applied into cholinergic neurons, neither of the chains of BoNT/A or BoNT/B appeared toxic alone (Poulain et al., 1988; Maisey et al., 1988), even when tested at a very high intracellular concentration (≈100 nM). In a striking contrast, TeTx L chain alone, at an intracellular concentration as low as 1 nM, induced a pronounced blockade of the release process. TeTx H chain had no apparent role to play in the blockade of either acetylcholine (Mochida et al., 1989) or of a non-cholinergic neurotransmitter (Poulain et al., 1991). In the case of BoNT, these findings markedly contrast with those found using mammalian preparations (see Figure 2). In chromaffin cells or PC12 cell line permeabilized in absence of Ca^{2+} with digitonin or streptolysin-O, the release of catecholamine could be triggered by addition of Ca^{2+} to the surrounding medium. When the cells where pre-incubated with the L chain alone before Ca^{2+} addition, release of transmitter was blocked (BoNT/A, /B and /E: Bittner et al., 1989; BoNT/A: Stecher et al., 1989; McInnes and Dolly, 1990). Similar observations were made with permeabilized peptidergic nerve endings (Dayanithi et al., 1990) or with chromaffin cells permeabilized with a "cell cracking" technique (Lomneth et al., 1991). The BoNT H chain alone had no inhibitory action and not was any synergistic effect detected when it was applied together with the L chain (Stecher et al., 1989; Ahnert-Hilger et al., 1990; McInnes and Dolly, 1990). Unlike the situation with BoNT, in the case of TeTx, there is general agreement that only L chain is needed intracellularly for toxic action in vertebrates and in *Aplysia* (see Figure 2). Indeed, *Aplysia* data accords with the observation that transmitter release is blocked after application of L chain of TeTx to permeabilized chromaffin cells, PC12 cells (Ahnert-Hilger et al., 1989b; 1990) or vasopressin releasing nerves endings (Dayanithi et al., 1992). In addition, this aggrees with the earlier observation that blockade of exocytosis ensues injection of the L-H_N fragment of TeTx into bovine chromaffin cells (Penner et al., 1986).

	Intracellularly active moiety			
	Preparation	Method	Moiety	References
BoNT	Aplysia	Injection	H + L	Poulain et al., 1988
	Chromaffin Cells	Permeabilization	L	Bittner et al., 1989 Stecher et al., 1989
	PC12	Permeabilization	L	McInnes & Dolly, 1990
		"Cell cracking"	L	Lomneth et al., 1991
	NMJ	Lipofection	L	DePaiva & Dolly, 1990
	Neurosecretory nerve endings	Permeabilization	L	Dayanithi et al., 1990
TeTx	Aplysia	Injection	L	Mochida et al., 1989
	Chromaffin Cells	Injection	L-H_N *	Penner et al., 1986
		Permeabilization	L	Ahnert-Hilger et al., 1989
	PC12	Permeabilization	L	Ahnert-Hilger et al., 1990
	Neurosecretory nerve endings	Permeabilization	L	Dayanithi et al., 1992
			★	Smaller fragment tested

Figure 2. Identification of BoNT and TeTx moieties intracellular active in various preparations.

The reason for this difference between BoNT and TeTx in *Aplysia* or, in the case of BoNT/A, between *Aplysia* and vertebrates is still unsolved. Apparently, this is not related to the cholinergic/non-cholinergic nature of the released transmitters because introduction of L chain of BoNT/A into cholinergic motor nerve endings via liposomes abolished nerve-evoked twitch mucle contraction (de Paiva and Dolly, 1990; Dolly et al., 1992) and when BoNT chains were injected into non-cholinergic neurons in *Aplysia*, both chains were required (Poulain et al., 1991).

Identification of a domain in the BoNT H_C required within *Aplysia* neurons for inhibitory action

Identification of the protein domain in BoNT H chain required together with the L chain, for intracellular blockade of neurotransmission in *Aplysia* was achieved as follows. Proteolytic deletion of the C-terminal half of the BoNT/A H chain generated a fragment, BoNT/A $L-H_N$, that was inactive when intracellularly applied to *Aplysia* cholinergic neurons (Poulain et al., 1989a) (Figure 3A). Since injection of intact BoNT/A H chain into a neuron pretreated with BoNT/A $L-H_N$ led to restoration of full toxicity, it has been deduced that the H_C region contains a protein domain acting synergistically with BoNT L chain in *Aplysia* neurons. Additional evidence for this observation has been made using intracellular translation of mRNA encoding the BoNT/A H_C fragment together with intra-neuronal administration of BoNT/A L chain; this induced a potent blockade of acetylcholine release (Kurazono et al., 1992) (Figure 3C). Apparently, no functional counterpart to this domain is present in the H chain of TeTx because a mixture of TeTx H chain plus BoNT/A L chain is inactive intracellularly (Poulain et al., 1990, 1991) (Figure 3B). Also, there is no potentiation of TeTx L chain action by BoNT H chain (Poulain et al., 1991).

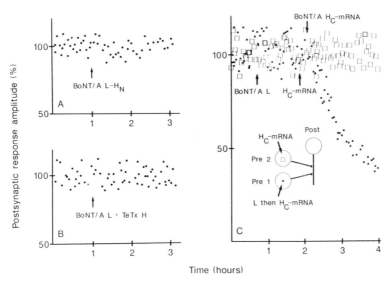

Figure 3. Localisation in the BoNT H_C of the site of BoNT H chain required within Aplysia neurons for toxicity. In all experiments postsynaptic response evoked by a presynaptic action potential was plotted against time, as a % of control amplitude, in various experimental conditions. A) injection (arrow) of BoNT/A $L-H_N$ (final intracellular concentration > 10 nM) into a presynaptic cholinergic neurone produced no inhibition of synaptic transmission, indicating, by inference, that an intracellularly "active" domain of BoNT H chain could be located in its H_C. B) The failure of the injected (arrow, ≈20 nM intracellularly) mixture TeTx H + BoNT/A L to induce blockade of transmission indicates that no site equivalent to that situated in BoNT H_C is present in TeTx H. C) Two different presynaptic neuroneuroninjected with a mRNA encoding the BoNT/A H_C fragment. A depression of transmission was seen only in the neuronneuronjected with BoNT/A L chain (≈50 nM, intracellularly); for further details see Kurazono et al. (1992). This clearly indicated that a protein domain located within the H_C of BoNT must be present inside the neuron for the BoNT L chain act.

The requirement of a H_C domain together with L chain of BoNT inside neurons is not restricted to BoNT/A or BoNT/B; it appears the case of all the BoNTs tested in *Aplysia*. This was deduced indirectly in the case of BoNT/E, /C_1 and /D (see Figure 4) from the following observations:

-i) an intracellular application of H chain (from BoNT/A) with the non-toxic single-chain form of BoNT/E is needed for blockade of transmission (Poulain et al., 1989b).

-ii) application of nicked BoNT/C_1 into cholinergic neurons proved ineffective unless the H chain of BoNT/A was applied. This may indicate that the BoNT/C_1 H_C domain probably does not have the right conformation or structure to act together with its parent L chain.

-iii) injection of a mRNA encoding the L chain of BoNT/D (Binz, Poulain and Niemann, unpublished observation) proved toxic in the presence of BoNT/A H chain (the BoNT/D H chain being not available).

BoNT type	Intracellular	Blockade
BoNT /A or /B	di-chain	+++
	L chain	-
	H + L	+++
BoNT/C_1	di-chain	-
	BoNT/C_1 + BoNT/A H chain	+++
BoNT/D	di-chain	+++
	mRNA BoNT/D L chain	-
	mRNA /D L + BoNT/A H chain	+++
BoNT/E	di-chain	+++
	SC-BoNT/E	-
	SC-BONT/E + BoNT/A H chain	+++

Figure 4. Requirement of both BoNT H and L chains inside Aplysia neurons for blockade of neurotransmission applies to at least BoNTs type A to E.

Speculations on the origin of a different requirement of BoNT chains intracellularly at *Aplysia* and vertebrate preparations

Cytoplasmic versus non-cytoplasmic application? A possibility is that injection into the cell body is not identical to the use of permeabilization or liposomal techniques. In *Aplysia*, due to the large size of the nucleus (80% of the soma diameter) all injections lead to delivery of toxin samples inside the nucleus which is probably broken during this operation. However, this "non-cytoplasmic" delivery could not be the origin of the BoNT H plus L intracellular requirement because, after intra-somatic injection of mRNA encoding the L chain of BoNT/A or of TeTx (and their cytoplasmic translation), a depression of transmission occurred with TeTx L chain but not in the case of BoNT/A L chain unless the complementary BoNT/A H chain was present (Mochida et al., 1990; Niemann et al., 1991 and this volume; Binz et al., 1992).

Intracellular trafficking role for the BoNT H_C? When injected into the neuronal body in *Aplysia*, the chains are assumed to reach the nerve terminal by some unknown mechanisms (cytoplasmic diffusion, axonal transport, migration within the endoplasmic reticulum pathway?) whereas delivery of L chain via permeabilization or liposomes result in an application of the L chain directly to the cytosol, on its site of action. In this regard, it is conceivable that the domain in the BoNT H_C moiety plays some role in the intracellular trafficking of the L chain or mediates its delivery to the final target. If true, why does this not apply to the TeTx L chain?

In conclusion, even if in *Aplysia*, both BoNT H and L chains are required to be present within the neuron to exert toxicity, it may be considered that as in vertebrate preparations (see above quoted references) BoNT L (as TeTx L chain) is the only chain implicated <u>directly</u> into the intracellular of blockade of neurotransmitter release.

TOXINS' DOMAINS IMPLICATED IN THE MEMBRANE STEPS

Heavy chain of *Clostridial* toxins is implicated in the extracellular binding and internalization steps in both vertebrate and *Aplysia* neurons

The first indication of a role for the H chain in binding and internalization of *Clostridial* toxins in *Aplysia* was given in a series of experiments in which the constituent chains were applied separately or together, to the external medium (for a summary see Figure 5). In the case of TeTx, application of high concentrations (up to 5 µM) of L

BoNT/A			TeTx		
Extra-C.	Intra-C.	Blockade	Extra-C.	Intra-C.	Blockade
L + H		−	L		−
H + L		+++		L	+++
L+H		+++	L+H		+++
L-H_N + H		+++	L-H_N + H		+++
L+H_N + H		+++	L+H_N + H		+++

Binding and internalization

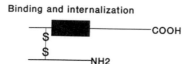

Figure 5. Schematic representation of experiments determining the involvement of the BoNT H_N or TeTx H_N in the internalization of the toxins in Aplysia neurons. Chains or fragments bath applied are denoted in "Extra-C" (extracellular) column and those injected into the presynaptic neuron in the column "Intra-C" (intracellular). The ability of the combination of chains and fragments to block neurotransmission is indicated as +++ (potent as native toxin, on a molar basis); - indicates no effect on transmitter release. It is deduced from this table that internalization of BoNT as well as that of TeTx is mediated by their respective H chain via a domain located in their H_N half (black box on the schematic drawing of toxin).

chain did not lead to blockade of neurotransmission indicating that it was unable to cross the membrane alone. When TeTx H and L chains were mixed together and bath applied, a potent blockade of transmission was observed, similar to that produced by TeTx (Mochida et al., 1989; Poulain et al., 1991, 1992a) (note that it is not possible to know if TeTx H chain enters the neurons, because TeTx L chain is the only chain implicated intracellularly into the blockade of release mechanism and TeTx H chain has no intracellular action).

Likewise, when a mixture of BoNT H + L chains was applied to the extracellular medium, a potent blockade of transmission occurs with similar potency to native BoNT (Poulain et al., 1992a) indicating the efficient internalization of both chains. This data contrasted with the observation that only feeble activity was recovered at the nerve muscle junction when the renatured constituent chains from BoNT were mixed together and bath applied (Maisey et al., 1988) leading to the hypothesis that, at least in mammals, only the di-chain species can mediate uptake (for further details, see Dolly et al., 1988; 1992 and this volume).

A clear demontration that H chain can bind was given by application of complementary chains on opposite sides of the plasma membrane (Figure 5) (this approache can be applied only to BoNTs in *Aplysia* because the two constituents chains of BoNTs are intracellularly required to inhibit transmitter release). Extracellular application of BoNT L chain to a neuron preinjected with BoNT H does not modify acetylcholine release whereas addition to the bath of BoNT H chain to a cell preinjected with BoNT L chain induced the blockade of neurotransmission (Poulain et al., 1988; Maisey et al., 1988). To our knowledge, it is the only experimental evidence demonstrating the internalization of BoNT H chain alone. Moreover this has shown that when BoNT H chain is inside, it cannot mediate the internalization of L chain present extracellularly. These deductions have been confirmed by sequential applications of the chains : when one chain was applied to the external medium, then washed extensively and followed by application of the other, only when the H chain was first did inhibition of transmission occur (Maisey et al., 1988). This clearly demonstrated that BoNT H, but not L chain, binds to the neuronal membrane and that some H chain remains bound to the membrane despite washing because it internalizes the L chain subsequently applied.

In addition, the temperature sensitivity of the internalization step of BoNT provided the opportunity to confirm the identification of the H chain as the binding moiety. At 10°C, a temperature found to prevent internalization but neither the binding nor the intracellular action of BoNT/A (Poulain et al., 1989b; 1992b), preincubation of a cholinergic synapse with a large excess (10 or 20 fold) of BoNT/A H chain for 30 minprior to application of BoNT/A (10 nM) completely prevented the BoNT/A induced decrease of transmission that ensued upon return up to 22°C (Figure 6). Under the same experimental conditions, preincubation with BoNT L chain did not prevent toxicity (Poulain et al., 1989b). Because it is not possible to prevent the inhibitory action of BoNT/A on acetylcholine release by preincubation with an excess of its H chain at 22°C, it was suggested that free acceptors for BoNT continuously recycle at this temperature (Poulain et al., 1989b). Similar deduction was made from labelling studies to vertebrates (Dolly et al., 1984). Similar attempts to study the binding of TeTx H chain by dissociating the binding step from the uptake failed in case of TeTx in *Aplysia* because TeTx uptake cannot be prevented at 10°C (Poulain et al., 1992b). It is not possible to lower the temperature further to achieve a complete blockade of the TeTx uptake process as postsynaptic responses are nearly abolished below 6°C.

The identification in *Aplysia* of the H chain of *Clostridial* toxin as the chain implicated in binding and internalization does not fully agree with observations made from studies in vertebrates. When neurotransmission was measured at the murine neuromuscular junction preincubation with BoNT H chain did not prevent BoNT poisonning (Maisey et al., 1988) leading to the conclusion that in vertebrates it is probable that a more stringent structural requirement (i.e. presence of a di-sulfide linked L chain) is necessary for the productive binding (for a discussion, see Dolly et al., 1988; 1992 and this volume).

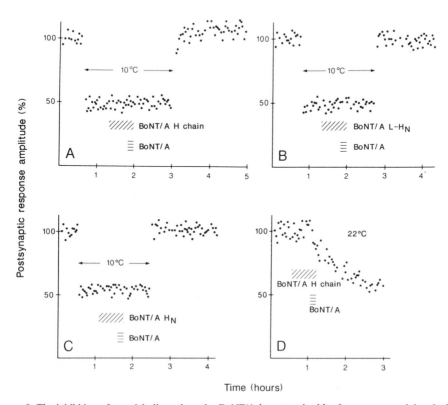

Figure 6. The inhibition of acetylcholine release by BoNT/A is antagonized by fragments containing the H_N fragment only at a reduce temperature. Through the use of reduced temperature to arrest BoNT poisoning at the binding step, the ability of BoNT/A H chain, L-H_N and H_N to compete for interaction with membrane ecto-acceptors was assessed. A, B, C): Following stabilization of the postsynaptic response on lowering the temperature from 22ºC to 10ºC, 200 nM of either toxin fragment was added to the bath (hatched area) for 30 min; BoNT/A (10 nM) was then bath applied for 10 min (horizontal lines), followed by extensive washing. After restoration to 22ºC, no decrease in response amplitude was observed indicating that each of these fragments prevented the binding of BoNT/A at 10ºC. D): When similar experiments were carried out at 22ºC with 200 nM BoNT/A H chain; acetylcholine release decreased upon addition of 10 nM BoNT/A for 10 min indicating that at 22ºC H chain was not able to prevent the toxin's inhibitory action.

Confirmation of the deduction that the H chain contains a binding moiety has been obtained with chimeras consisting of one chain from BoNT and the other from TeTx. In *Aplysia*, the chimera BoNT/A L - TeTx H exhibited a clear non-cholinergic preferential action whereas, that comprising TeTx L - BoNT/A H was most effective in depressing cholinergic transmission at either murine neuromuscular junction or in *Aplysia* (for further details see Poulain et al., 1991; 1992a; Weller et al., 1991). This demonstrated that the H chain was able to internalize the L chain of either type of *Clostridial* neurotoxin and that specificity of the toxins is determined by the H chain.

N-terminal half of H chain mediates both binding and internalization in *Aplysia*

The identification of the domain in the H chain responsible for mediating binding and the subsequent internalization was achieved by studying the consequences on toxicity of the deletion of the C-terminal half of the H chain. In the case of BoNT/A, in *Aplysia*, a prerequisite to studying the internalization properties of the BoNT/A L-H_N fragment was to preinject presynaptic neurons with BoNT/A H chain in order to provide the BoNT H_C required for intracellular toxicity. Subsequent extracellular application of the non-toxic fragment BoNT/A L-H_N or a mixture of its reduced constituents, H_N and L, induced a potent depression in acetylcholine release (Poulain et al., 1989a). This indicated that this fragment can cross the membrane or, at least, that BoNT H_N can mediate the internalization of the L chain (Figure 5). Accordingly, at low temperature in order to selectively prevent internalization, any fragment of BoNT/A containing the H_N domain (H_N-L fragment, H chain, H_N) was able to antagonise productive binding of BoNT/A (Poulain et al., 1989b) (Figure 6). Note that, due to a lack of detectable biological activity, the internalization of BoNT H_N cannot be tested. With TeTx, a similar deduction was made because TeTx L-H_N fragment or a mixture of TeTx L and H_N proved to be as potent as TeTx (Poulain et al., 1992a). These collective findings in *Aplysia* have led to the identification of the H_N domain of the H chain as the prime part of either BoNT or TeTx implicated in productive binding.

Additionally, the ability of the N-terminal moieties of the H chains to mediate internalization appears to be independent of the origin of L chain used because heterologous mixture of H_N plus L chains of TeTx and BoNT was as effective as the parent toxin from where each H_N fragment was derived (Poulain et al., 1991; 1992a). It was thus postulated that this half possesses two functional domains, one being distinct and responsible for the divergent neuronal specificity via an interaction with different acceptors for TeTx and BoNT, whilst the other serves a common role in translocating the L chain of either toxin.

These conclusions contrast strikingly with the situation in vertebrate preparations where the presence of the H_C domain appears a prerequisite for productive binding of the toxins (for review, see Habermann and Dreyer, 1986; Simpson, 1989; Dolly, 1992). Indeed, enzymatic deletion of the C-terminal half of BoNT H chain abolishes both toxicity (Shone et al., 1985; Poulain et al., 1989a) and the ability of BoNT/A ^{125}I-L-H_N fragment to bind to rat cerebrocortical synaptosomes (Shone et al., 1985) or to the neuromuscular junction (Poulain et al., 1989b). Similarly, the L-H_N fragment of TeTx (the so called B fragment) exerts only very low potency when applied to the neuromuscular junction (Simpson and Hoch, 1985). Futhermore, the TeTx H_C was shown to bind alone to nerve membranes and to internalize on its own (Bizzini et al., 1981; Weller et al., 1986; 1989). More recently, binding of TeTx via its H_C to a membrane protein component has been demonstrated (Schiavo et al., 1991 and this volume). In the case of TeTx, by inference, it is deduced that the H_C half was the H chain moiety implicated in binding in vertebrates. However, the experimental conditions may influence the observations (for a discussion, see Bakry et al., 1991) because, when experiments are performed in low ionic strength medium, TeTx, TeTx H chain and TeTx H_C appear to have similar binding properties whereas, in physiological ionic strength medium, binding affinity to brain membranes decreases in the order TeTx > TeTx H chain >> TeTx H_C or TeTx H_N (Weller et al., 1989). It may be that the "binding" seen in these experiments is not related to the productive binding of the toxins. There is no evidence showing that the uptake of TeTx H_C follows the same cellular pathway as the native toxin and an exclusive involvement of the C-terminal H chain domain in the productive uptake of toxin has not been demonstrated. In addition, Takano et al., (1989) observed that TeTx

L-H$_N$ fragment can block transmission at inhibitory and exitatory synapses of the spinal motoneuron of the cat a situation similar to that found in *Aplysia*.

With regard to the internalization process, earlier observations support the identification, using *Aplysia* preparations, of the H$_N$ half as the toxin moiety containing the domain implicated in the internalization process. Indeed, studies on lipid vesicles or on lipid bilayer membranes revealed that this part of the BoNT H chain or of TeTx H chain forms channels at low pH (4-5) (Hoch et al., 1985; Blaustein et al., 1987).Internalization might, thus, result from incorporation of the acceptor toxin complex into endosomes; a decrease in the endosomal pH could allow formation of a "pore" by the H$_N$ fragment and the subsequent delivery of L chain in the cytoplasm through this hydrophobic tunnel (for discussion see Simpson, 1986; 1989; Schiavo et al., 1990; Montecucco et al., 1991). Indeed, several observations provide direct evidence for a TeTx-H chain mediated passage of small molecules and also of protein accross themembrane. TeTx H chain was shown not only to mediate a loss of ^{3}H noradrenaline or ^{3}H-GABA from washed brain homogenates (Weller et al., 1989; Ahnert-Hilger et al., 1990) and a movement to the outside of tetraphenylphosphonium but also to allow efflux of lactate dehydrogenase from rat brain synaptosomes (Högy et al., 1992).

Identification of a putative non-toxigenic binding domain in the TeTx H$_C$ in *Aplysia*

Recent preliminary experiments performed at non-cholinergic synapses in the cerebral ganglion of *Aplysia* provided evidence for the contribution of another protein domain located in the TeTx H$_C$ which probably plays a role in a non-productive binding step (Poulain and Weller, unpublished observations). At room temperature, it was possible to totally prevent 10 nM TeTx-induced inhibition of non-cholinergic transmitter release for at least 3 hours by preincubating the preparation with a 10 fold excess of TeTx H chain (not shown, for a shorter duration see Figure 7B, compare with control inhibition in Figure 7A). This situation contrasts strongly with the fact that BoNT H chain cannot prevent BoNT inhibitory action at room temperature (Figure 5D). Whilst preventing the TeTx inhibitory effect (Figure 7B) and after extensive washing of unbound toxin or H chain, the subsequent addition to the bath of 10 nM TeTx L chain induced a blockade of transmission indicating internalization of TeTx L chain. This indicated that some TeTx H chain remained bound to the membrane as observed with BoNT/A or BoNT/B H chains (Maisey et al., 1988; Poulain et al., 1989a). Importantly, the preventive effect of TeTx H chain could not be reproduced when using its N-terminal half (TeTx H$_N$) (Figure 7C), though this fragment, alone, internalizes TeTx L chain (Mochida et al., 1989; Poulain et al., 1991; 1992a) as well as TeTx H chain does suggesting expected that TeTx L-H$_N$ fragment have similar affinities as TeTx for the productive acceptor.

Since the inability of BoNT H chain to prevent BoNT action at 22ºC (Figure 5D) was thought to result from a recycling of free BoNT acceptors, the comparison of the different data found using TeTx H chain or TeTx H$_N$ suggested, by inference, that a domain, located in the H$_C$ portion, interferes with recycling of acceptors and plays some role in "anchoring" the TeTx H chain to the neuronal plasma membrane so that no free TeTx acceptors were presented. The fact that this situation did not appear when TeTx H$_N$ was used could be interpretated as an indication that during internalization, TeTx H$_N$ dissociates from the TeTx acceptor and allows recycling of free acceptors. To test the hypothesis that a domain located in the H$_C$ portion plays some role in a "non-productive binding step" or "anchoring" of the TeTx H chain to its acceptor, non-cholinergic synapses in *Aplysia* were preincubated with a large excess of TeTx H$_C$ (1000 nM) for 20 min then incubated with 100 nM TeTx H chain for a further 20 min before addition of 10 nM TeTx (see Figure 7D). A potent blockade of neurotransmission exterted upon addition of TeTx indicating that preincubation with TeTx H_C removed the ability of TeTx H chain to prevent TeTx induced blockade of neurotransmission. This experiment showed that saturation of a membrane binding site with TeTx H_C made the TeTx H to behave like H_N in regard to its ability to compete with TeTx. Although preliminary, these results suggest the existence of a domain located in the TeTx H_C that plays a non-toxigenic binding role in *Aplysia*. Nothing is yet known for an equivalent domain in BoNTs and further experiments are needed to know if this domain recognises a common acceptor with BoNT.

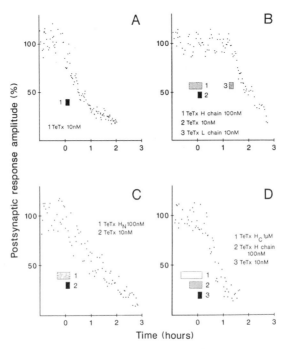

Figure 7. The inhibitory action of TeTx on non-cholinergic transmitter release is antagonized by its H chain but not its H_N fragment at 22ºC. In all experiments (22ºC), after the control (100 %) response was measured, non cholinergic synapses in the cerebral ganglion were incubated in the presence of TeTx 10 nM (A) and after pre-incubation with TeTx H chain (B), TeTx H_N (C) or TeTx H_C followed by TeTx H chain (D) for the duration indicated by hatched areas (concentrations are indicated in each panel). Note that, in B), when TeTx L chain was subsequently added, transmission was inhibited. These collective experiments suggest that the ability of TeTx H chain but not of TeTx H_N to antagonize the TeTx inhibitory action is due to a site located in its H_C half. For further details see text.

If such deductions apply to vertebrates, it is therefore possible that earlier reports on the existence of a binding domain of TeTx in the H_C half have in fact addressed this non-toxigenic domain. The exact role played by this domain is not known. It might participate either in an anchorage of the toxin in the close proximity of its productive acceptor (see Schiavo et al., this volume) or in the intracellular addressing of the internalized molecules toward axonal retrograde transport (see Niemann, 1991). Indeed, an axonal ascent of the TeTx H_C fragment in vertebrates motoneurons has been detected (Bizzini et al., 1981; Weller et al., 1986) and, moreover, TeTx H_C can mediate the uptake of other proteins i.e. glucose-oxydase (Beaude et al., 1990) or peroxidase (Fishmann et al., 1990).

Reassessing the binding model

Since it would seem unlikely that binding and internalization of *Clostridial* toxins follow different patterns in vertebrates and in *Aplysia*, the different data should be reconciliated in a simple model. Reminiscent of the dual model for binding proposed by Montecucco (Montecucco, 1986; Schiavo et al., 1990) and Niemann (1991), we propose that toxin binding involves i) binding of the H_C portion so anchoring the toxin to the membrane, at least in the case of TeTx (non-toxigenic binding) ii) binding of the H_N portion leading to internalization of active moieties (productive binding). In vertebrates preparations, with both BoNT (Dolly et al., 1988; 1992; Poulain et al., 1989b) and TeTx (Weller et al., 1989), it appears that recognition of "productive" high affinity ecto-acceptors is highly stringent and mediated only by the conformation adopted by the intact di-chain toxin species. It is thus plausible that, in vertebrates, the loss of targeting properties of toxin derivatives devoid of the C-terminal region of H chain might result from conformational changes in their N-terminal halves. In *Aplysia* preparations, it is possible that due to a divergent evolution of membrane toxin acceptors, the N-terminal domain of the H chain alone is sufficient for productive binding/uptake.

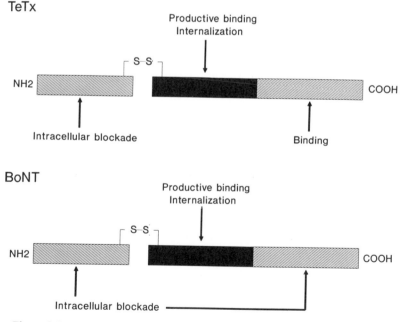

Figure 8. Putative localisation and functional role of domains within BoNT and TeTx.

CONCLUSION

To summarise, using *Aplysia* preparations, a tentative functional structure for BoNT and TeTx has been designed (Figure 8). Similar protein domains located in the N-terminal moieties of the H chain of both BoNT/A and TeTx have been implicated in binding and internalization. Since the H_N portions mediate the neuroselective targetting of the toxin, they recognize different acceptor populations and therefore, these proteins domains must differ in some aspects. An additional protein domain in the H_C has been identified, it seems to anchor the H chain of at least TeTx to the membrane. L chain contains the only protein domain implicated in the intracellular blockade of neurotransmission by acting to targets present in neurons releasing different kind of neurotransmitter release; though an additional site located in the H_C region of BoNT must be present within *Aplysia* neurons for the BoNT L chain acts.

Acknowledgements

This work is supported by grants from *Association Française contre les Myopathies* and *Direction des Recherches et Etudes Techniques* (nº89/145) to L.T., together with a contract (DAMD 17-91-2-1035) from USAMRIID to J.O.D.

REFERENCES

Aguilera, J., Ahnert-Hilger, G., Bigalke, H., DasGupta, B.R., Dolly, J.O., Habermann, E., Halpern, J., Middlebrook, J., Mochida, S., Montecucco, C., Niemann, H., Oguma, K., Popoff, M., Poulain, B., Simpson L., Shone, C.C., Thompson, D.E., Weller, U., Wellhöner, H.H. and Whelan S.M., 1992, Clostridial Neurotoxins - Proposal of a common Nomenclature, *F.E.M.S.* 90:99.

Ahnert-Hilger, G., Bader, M.-F., Bhakdi, S. and Gratzl, M., 1989a, Introduction of macromolecules into bovine adrenal medullary chromaffin cells and rat pheochromocytoma cells (PC12) by permeabilization with streptolysin O: Inhibitory effect of tetanus toxin on catecholaline secretion, *J. Neurochem.* 52:1751.

Ahnert-Hilger, G., Dauzenroth, M.E., Habermann, E., Henschen, A., Krieglstein, K., Mauler, M. and Weller, U., 1990, Chains and fragments of tetanus toxin and their contribution to toxicity, *J. Physiol. (Paris)* 84:229.

Ahnert-Hilger, G., Weller, U., Dauzenroth, M.E., Habermann, E. and Gratzl, M., 1989b, The tetanus toxin light chain inhibit exocytosis, *FEBS Letters* 242:245.

Bakry, N., Kamata, Y. and Simpson, L.L., 1991, Lectins from Triticum vulgaris and Limax flavus are universal antagonists of botulinum neurotoxin and tetanus toxin, *J. Pharmacol. Exp. Ther.* 258:830.

Baux, G., Fossier, P. and Tauc, L., 1990, Histamine and FMRFamideregulate in opposite ways acetycholine release at an identified neuro neuronal synapse in Aplysia, *J. Physiol. (London)* 429:147.

Beaude, P., Delacour, A., Bizzini, B., Domualdo, D. and Remy, M.-H., 1990, Retrograde transport of an exogenous enzyme covalently linked to B-II$_b$ fragment of tetanus toxin, *Biochem. J.* 271:87.

Binz, T., Grebenstein, O., Kurazono, H., Eisel, U., Wernars, K., Popoff, M., Mochida, S., Poulain, B., Tauc, L., Kozaki, L. and H., Niemann, 1992, Molecular biology of the L chains of clostridial neurotoxins, *Zentralblatt für Bakteriologie* Supp.23:56.

Bittner, M.A., DasGupta, B.R. and Holz, R.W., 1989, Isolated light chains ofbotulinum neurotoxins inhibit exocytosis. Studies in digitonin permeabilized chromaffin cells, *J. Biol. Chem.* 264:10354.

Bizzini, B., Grob, P. and Akert, K., 1981, Papain-derived fragment II$_c$ of tetanus toxin: binding to isolated synaptic membrane and retrograde transport, *Brain Research* 210:291.

Blaustein, R.O., Germann, W.J., Finkelstein, A. and DasGupta, B.R., 1987, The N-terminal half of the heavy chain of botulinum type A neurotoxin forms channels in planar phopholipid bilayers, *FEBS Letters* 226:115.

Dayanithi, G., Ahnert-Hilger, G., Weller, U., Nordmann, J.J. and Gratzl, M., 1990, Release of vasopressin from isolated permeabilized neurosecretory nerve terminals is blocked by the light chain of botulinum A toxin, *Neuroscience*, 39:711.

Dayanithi, G., Weller, U., Ahnert-Hilger, G., Link, H., Nordmann, J.J. and Gratzl, M., 1992, The light chain of tetanus toxin inhibits calcium-dependent vasopressin release from permeabilized nerve endings, *Neuroscience,* 46:489.

de Paiva, A. and Dolly, J.O., 1990, Light chain of botulinum neurotoxin is active in mammalian motor-nerve terminal when delivered via liposomes, *FEBS Letters,* 277:171.

Dolly, J.O., 1992, Peptide toxins that alter neurotransmitter release, *in:* "Handbook of Experimental Pharmacology", H. Herkin and F. Hucho, eds, Springer, Berlin.

Dolly, J.O., Black, J.D., Williams, R.S. and Melling, J., 1984, Acceptors for botulinum neurotoxin resisde on motor nerve terminals and mediate its internalization, *Nature* 307:457.

Dolly, J.O., de Paiva, A, Poulain, B., Foran, P., Ashton, A and Tauc, L., 1992, Insights into the neuronal binding, uptake and intracellular activity of botulinum neurotoxins, *Zentralblatt für Bakteriologie,* Supp.23:31.

Dolly, J.O., Poulain, B., Maisey, E.A., Breeze, A.L., Wadsworth, J.D., Ashton, A.C. and Tauc, L., 1988, Neurotransmitter release and K+ channels probed with botulinum neurotoxins and dendrotxin, *in:* "Neurotoxins in Neurochemistry", J.O. Dolly, ed., Ellis Horwood Limited, Chichester.

Fishman, P.S., Savitt, J.M. and Farrand, D.A., 1990, Enhanced CNS uptake of systematically administered proteins through conjugation with tetanus C-fragment, *J. Neurol. Sci.* 98:311.

Gaillard, W.D. and Carpenter, D.O., 1986, On the transmitter at the A-to-B cell in Aplysia californica, *Brain Research* 373:311.

Gardner, D., 1971, Bilateral symmetry and interneuronal organization in the buccal ganglion of Aplysia, *Science,* 173:550.

Habermann, E. and Dreyer, F., 1986, Clostridial Neurotoxins: Handling and action at the cellular and molecular level, *Current Topic Microbiol. Immunol.* 129:93.

Hoch, D.H., Romero-Mira, M., Ehrlich, B.E., Finkelstein, A. DasGupta, B.R. and Simpson, L.L., 1985, Channels formed by botulinum, tetanus and diphteria toxins in planar lipid bilayers. Relevance to translocation of proteins accross membranes, *Proc. Natl. Acad. Sci. USA* 82:1692.

Högy, B., Dauzenroth, M.-E., Hudel, M., Weller, U., and Habermann, E., 1992, Increase of permeability of synaptosomes and liposomes by the heavy chain of tetanus toxin, *Toxicon* 30:63.

Kurazono, H., Mochida, S., Binz, T., Eisel, U., Quanz, M., Grebenstein, O., Poulain, B., Tauc, L., and Niemann, H., 1992, Minimal essential domains specifying toxicity of the light chains of tetanus toxin and botulinum neurotoxin type A, *J. Biol. Chem.,* in-press.

Lomneth, R., Martin, T.F.J. and DasGupta, B.R., 1991, Botulinum neurotoxin light chain inhibits norepinephrine secretion in PC12 cells at an intracellular membranous or cytoskeletal site, *J. Neurochem.* 57:1413.

Maisey, E.A., Wadsworth, J.D.F., Poulain, B., Shone, C.C., Melling, J., Gibbs, P., Tauc, L. and Dolly, J.O., 1988, Involvement of the constituent chains of botulinum neurotoxins A and B in the blockade of neurotransmitter release, *Eur. J. Biochem.,* 177:683.

McInnes, C. and Dolly, J.O., 1990, Ca^{2+}-dependent noradrenaline release from permeabilized PC12 cells is blocked by botulinum neurotoxin A or its light chain, *FEBS Letters* 261:323.

Mochida, S., Poulain, B., Eisel, U., Binz, T., Kurazono, H., Niemann, H. and Tauc, L., 1990, Exogenous mRNA encoding tetanus or botulinum neurotoxins expressed in Aplysia neurons, *Proc. Natl. Acad. Sci.* 87:7844.

Mochida, S., Poulain, B., Weller, U., Habermann, E. and Tauc, L., 1989, Light chain of tetanus toxin intracellularly inhibits acetylcholine release at neuro-neuronal synapses, and its internalization is mediated by heavy chain, *FEBS Letters* 253:47.

Molgo, J., Comella, J.X., Angaut-Petit, D., Pécot-Dechavassine, M., Tabti, N., Faille, L., Mallart, A. and Thesleff, S., 1990, Presynaptic actions of botulinal neurotoxins at vertebrate neuromuscular junctions, *J. Physiol. (Paris)* 84:152.

Montecucco, C., 1986, How do tetanus and botulinum toxins bind to neuronal membranes?, *T.I.N.S.* 11:314.

Montecucco, C., Papini, E. and Schiavo, G., 1991, Molecular models of toxin membrane translocation, *in:* "Sourcebook of Bacterial Protein Toxin", J. Alouf and J. Freer, eds, Academic Press, London.

Niemann, H., 1991, Molecular biology of clostridial toxins, *in:* "Sourcebook of Bacterial Protein Toxin", J. Alouf and J. Freer, eds, Academic Press, London.

Niemann, H., Binz, Th., Grebenstein, O., Kurazono, H., Thierer, J., Mochida, S., Poulain, B. and Tauc, L., 1991, Clostridial Neurotoxins: From Toxins to Therapeutic tools?, *Behring Institute Mitteilungen* 89:153.

Penner, R., Neher, E. and Dreyer, F., 1986, Intracellularly injected tetanus toxin inhibits exocytosis in bovine adrenal chromaffin cells, *Nature* 324:76.

Poulain, B., de Paiva, A., Dolly, J.O., Weller, U. and Tauc, L., 1992b, Differences in temperature dependence of botulinum and tetanus uptake in Aplysia, *Neuroscience Letters* 139:289.

Poulain, B., Mochida, S., Weller, U., Högy, B., Habermann, E., Wadsworth, J.D.F., Dolly, J.O., Shone, C.C. and Tauc, L., 1991 Heterologous combinations of heavy and light chains from botulinum neurotoxin A and tetanus toxin inhibit neuro-transmitter release in Aplysia, *J. Biol. Chem.* 266:2580.

Poulain, B., Mochida, S. Wadsworth, J.D.F., Weller, U., Habermann E., Dolly, J.O. and Tauc, L., 1990, Inhibition of neurotransmitter release by botulinum neurotoxins and tetanus toxin at Aplysia synapses: Role of the constituent chains. *J. Physiol. (Paris)* 84:247.

Poulain, B. and Molgo, J., 1992, Botulinal neurotoxins. Methods used in the study of their mode of action on neuro-transmitter release, *in* "Methods in Neurosciences", P.M. Conn, ed., Academic Press, San Diego.

Poulain, B., Tauc, L., Maisey, E.A., Wadsworth, J.D.F., Mohan, P.M. and Dolly J.O., 1988, Neurotransmitter release is blocked intracellularly by botulinum neurotoxin, and this requires uptake of both toxin polypeptides by a process mediated by the larger chain, *Proc. Natl. Acad. Sci. USA* 85:4090.

Poulain, B., Wadsworth, J.D.F., Maisey, E.A., Shone, C.C., Melling, J., Tauc, L. and Dolly, J.O., 1989a, Inhibition of transmitter release by botulinum neurotoxin A. Contribution of various fragments to the intoxication process, *Eur. J. Biochem.* 185:197.

Poulain, B., Wadsworth, J.D.F., Shone, C.C., Mochida, S., Lande, S., Melling, J., Dolly, J.O. and Tauc, L., 1989b, Multiple domains of botulinum neurotoxin contribute to its inhibition of transmitter release in Aplysia Neurons, *J. Biol. Chem.* 264:21928.

Poulain, B., Wadsworth, J.D.F., Weller, U., Mochida, S., Leprince, C., Dolly, J.O., and Tauc, L., 1992a, Functional studies in *Aplysia californica* on chimeras of botulinum and tetanus toxins, *Zentralblatt für Bakteriologie* Supp.23:46.

Schiavo, G., Bocquet, P., DasGupta, B.R. and Montecucco, C., 1990, Membrane interactions of tetanus and botulinum neurotoxins: a photolabelling study with photoactivable phospholipids, *J. Physiol. (Paris)* 84:180.

Schiavo, G., Ferrari, G., Rossetto, O. and Montecucco, C., 1991, Specific cross-linking of tetanus toxin to a protein on NGF-differentiated PC12 cells, *FEBS Letters* 290:227.

Schmitt, A., Dreyer, F. and John, C., 1981, At least three sequential steps are involved in the tetanus toxin-induced block of neuromuscular transmission, *Naunyn-Schmiedeberg's Arch. Pharmacol.* 317:326.

Shone, C.C., Hambleton, P. and Melling, J., 1985, Inactivation of Clostridium botulinum type A neurotoxin by trypsin and purification of two tryptic fragments. Proteolytic action near the COOH-terminus of the heavy subunit destroys toxin-binding activity, *Eur. J. Biochem.* 151:75.

Simpson, L.L., 1981, The origin, structure, and pharmacological activity of botulinum toxin, *Pharmacological rev.*, 33:155.

Simpson, L.L., 1986, Molecular pharmacology of botulinum toxin and tetanus toxin, *Ann. Rev. Pharmacol. Toxicol.* 26:427.

Simpson, L.L., 1989, Peripheral actions of the botulinum toxins, *in* "Botulinum neurotoxin and tetanus toxin", L. L. Simpson, ed., Academic Press, San Diego.

Simpson, L.L. and Hoch, D.H., 1985, Neuropharmacological characterisation of fragment B from tetanus toxin, *J. Pharmacol. Exp. Ther.* 232:223.

Stecher, B., Weller, U., Habermann, E., Gratzl, M. and Ahnert-Hilger, G., 1989, The light chain but not the heavy chain of botulinum A toxin inhibits exocytosis from permeabilized adrenal chromaffin cells, *FEBS Letters* 255:391.

Takano, K., Kirchner, F., Gremmelt, A., Matsuda, M., Ozutsumi, N. and Sugimoto, N., 1989, Blocking effects of tetanus toxin and its fragment (A-B) on excitatory and inhibitory synapses of the spinal motoneurone of the cat, *Toxicon*, 27:385.

Weller, U., Dauzenroth, M.-E., Gansel, M., and Dreyer, F., 1991, Cooperative action of the light chain of tetanus toxin and the heavy chain of botulinum toxin type A on transmitter release of mammalian motor endplates, *Neuroscience Letters* 122:132.

Weller, U., Dauzenroth, M.-E., Meyer zu Heringdorf, D. and Habermann, E., 1989, Chains and fragments of tetanus toxin: separation, reassociation and pharmacological properties *Eur. J. Biochem.* 182:649.

Weller, U., Taylor, C.F. and Habermann, E., 1986, Quantitative comparison between tetanus toxin, some fragments and toxoid for binding and axonal transport in the rat, *Toxicon* 24:1055.

Wellhöner, H.H., 1992, Tetanus and botulinum Neurotoxins. *in* "Handbook of Experimental Pharmacology", H. Herkin and F. Hucho, eds, Springer, Berlin.

DISSECTING THE L CHAINS OF CLOSTRIDIAL NEUROTOXINS

Heiner Niemann[1], Thomas Binz[1], Oliver Grebenstein[1], Hisao Kurazono[1], Arno Kalkuhl[1], Shinji Yamasaki[1], Ulrich Eisel[1], Johannes Pohlner[2], Germar Schneider[3], Viliam Krivan[3], Shunji Kozaki[4], Sumiko Mochida[5], Ladislav Tauc[5], and Bernard Poulain[5]

[1]Department of Microbiology, Federal Research Centre of Virus Diseases of Animals, P.O. Box 1149; W-74 Tübingen, FRG
[2]Max-Planck-Institute for Biology, Spemannstraße 36, W-74 Tübingen, FRG
[3]Analytik und Höchstreinigung, Universität Ulm, Albert-Einstein-Allee 11, W-79 Ulm, FRG
[4]College of Agriculture, Osaka Prefecture University, Mozuumemachi, Sakai, Osaka, Japan
[5]Laboratoire de Neurobiologie Cellulaire et Moléculaire, CNRS, Gif-sur-Yvette, France

INTRODUCTION

Tetanus toxin (TeTx) and the seven structurally related but serologically distinct botulinal neurotoxins, designated BoNT/A to BoNT/G, bind selectively to nerve terminals where they are internalized and sorted by unknown mechanisms. It is generally accepted that toxification is preceeded by a translocation of the L chains from an acid vesicular environment into the cytosole (Niemann, 1991). Despite such homologies, however, the individual toxins display differences in their primary sites of action, their uptake mechanisms (Poulain et al., 1992) and sensitivity to chemical antagonists (Dreyer et al., 1987; Dolly et al., 1990). We have to assume, therefore, that some of the neurotoxins act on different intracellular targets which appear to play key roles in neurotransmitter release from small translucent as well as large dense-core vesicles. Although tetanus patients immediately receive high doses of human tetanus immune globulin to block any nonbound toxin, the internalized toxin evokes its toxic effects for several weeks (Bleck, 1989). This suggests that the toxins act as enzymes. Several enzymatic activities have been proposed over the past years all of which, however, proved to be wrong later on (Niemann, 1991).

In this paper we present data which suggest that the neurotoxins could constitute zinc-dependent metalloproteases. This hypothesis is based on the one side on the analyses of a large number of TeTx L chain mutants carrying mutations in the putative active center. On the other side, we demonstrate that neurotoxicity of the TeTx L chain requires the presence of zinc. Both the **native** L chain, as isolated from

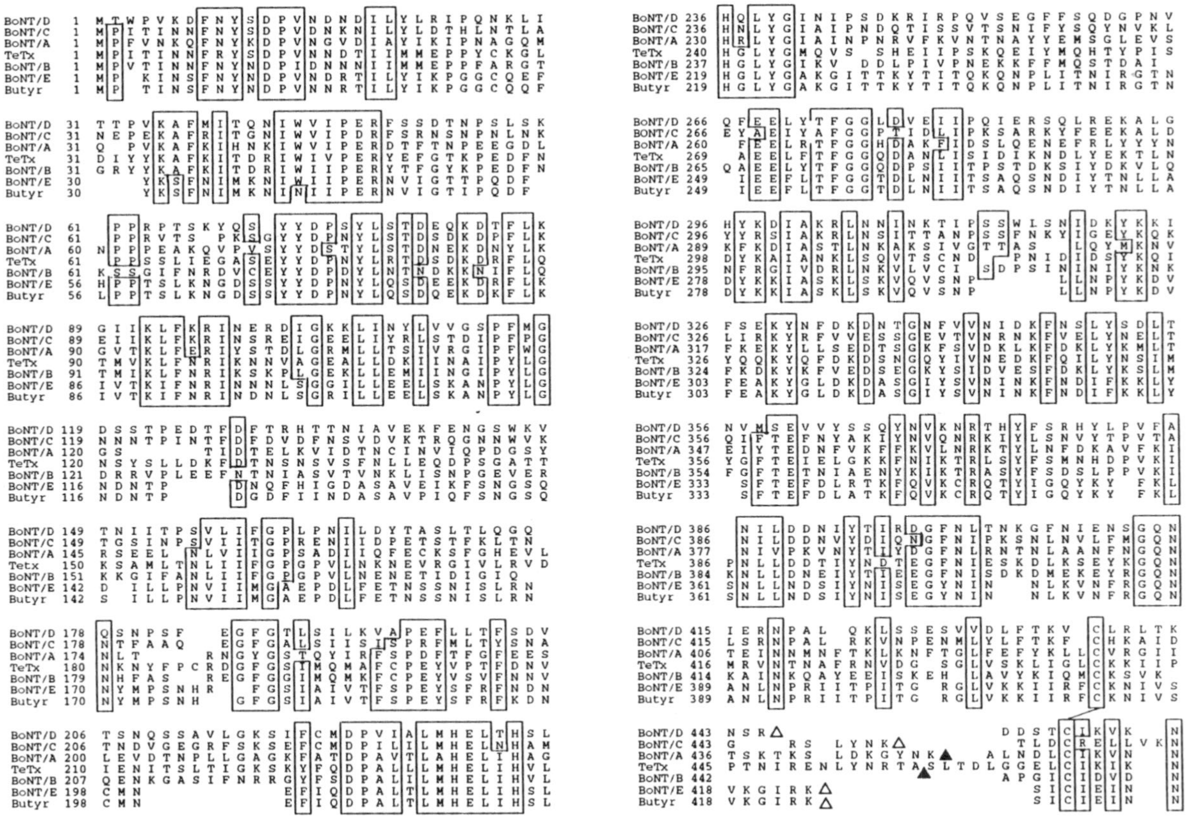

Figure 1. Sequence alignment of the L chains of various clostridial neurotoxins

C. tetani as well as a **recombinant,** biologically active L chain, as expressed in *E. coli,* contained significant amounts of zinc. In contrast, neither the native H chain, nor a biologically inactive L chain mutant generated by site directed mutagenesis contained detectable levels of zinc.

Minimal essential domains of the L chains specifying toxicity

Our group has been involved in the molecular cloning and sequencing of the structural genes encoding TeTx (Eisel et al., 1986), BoNT/A (Binz et al., 1990a), BoNT/B (Kurazono et al., 1992a), BoNT/C (Hauser et al., 1990b), BoNT/D (Binz et al., 1990b), and BoNT/E (Poulet et al., 1992). The sequence alignment of the L chains (Figure 1) shows three domains with increased identity: an amino-terminal domain extending to about residue 120, a central histidine-rich cluster, and a carboxyl-terminal domain involving residues 300 to about 400 (Kurazono et al., 1992a).

We first wanted to know which of the these domains were actually essential in the toxification process. For this purpose, we generated 5′ and 3′ deletions in the genes encoding the L chains of TeTx and BoNT/A. 5′-capped and 3′-polyadenylated mRNA was generated by *in vitro* transcription and microinjected into identified presynaptic neurons B4 or B5 in the buccal ganglia of *Aplysia californica* and the postsynaptic responses of the cells B3 or B6 were recorded as described previously (Mochida et al., 1990).

TeTx L chain

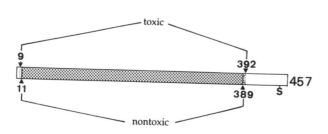

BoNT/A L chain

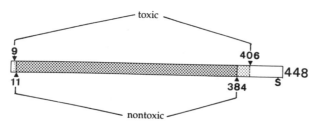

Figure 2.   Minimal essential domains required for neurotoxicity.

Figure 2 depicts the border lines defining toxicity of the L chains of TeTx and of BoNT/A. Two findings are important:

Firstly, C-terminal deletions of 65 (TeTx) or 32 residues (BoNT/A) do not affect toxicity significantly, whereas deletions of 68 (TeTx) or 65 residues (BoNT/A) abolish toxicity completely. This finding implies that the cysteine residues of the L chains that mediate the binding of the corresponding H chain play apparently no role in the toxification process. Theoretically, the sulfhydryl group could constitute a target site for isoprenylation or, in case of BoNT/A, it could be essential for the unexplained, *Aplysia*-specific synergistic interaction with the H chain.

Secondly, the presence of the highly conserved Tyr^{10} in the L chains is essential for the toxifying activity. However, this residue appears not to be part of an active center. This follows from studies involving a series of BoNT/A L chain-specific monoclonal antibodies including the toxin-neutralizing antibody A109 (Kozaki et al., 1989). Each of these antibodies precipitated various C-terminally truncated BoNT/A L chain derivatives that lacked, in addition, the 8 N-terminal residues. However, these antibodies barely precipitated derivatives lacking the 10 N-terminal amino acids indicating that the epitope was non-linear. From these data we concluded that Tyr^{10} interacts with a yet undefined residue located within the N-terminal 188 residues thereby stabilizing the tertiary structure of this N-terminal domain (Kurazono et al., 1992b). The finding that the A109 antibody precipitated an L chain derivative lacking the 265 carboxyl-terminal amino acids further suggests that the N-terminal domain folds independently into tha authentic tertiary stucture that is not influenced by the central and the C-terminal portion of the BoNT/A L chain.

The carboxyl-terminal portion of the BoNT/A L chain interacts with the H_C domain of the H chain

Several mammalian cell systems have been tested with a variety of techniques designed to internalize the L chains into the cytosole and to study the differential effects of the L chains of clostridial neurotoxins (Penner et al., 1986; Bittner et al., 1989; McInnes and Dolly, 1990; Ahnert-Hilger et al., 1989; Stecher et al., 1989). From these data it is clear that in vertebrate systems the L chains of TeTx and botulinal toxins alone are sufficient to block neurotransmitter release. In the *Aplysia* system this is different with respect to the botulinal toxins. Here the botulinal H chains play some as yet unknown synergistic role in activating or stabilizing the L chains (Poulain et al., 1988; 1989). To gain more information about the domains that are involved in this type of interaction, we have microinjected BoNT/A H_C-specific mRNA hoping that the H_C fragment would substitute for the corresponding H chain. As demonstrated in greater detail in the paper by Poulain et al., (this volume), the de novo translated H_C-fragment could indeed replace the H chain and activate the preinjected BoNT/A L chain.

We then asked the question, as to which portion of the BoNT/A L chain interacted with H_C. For this purpose we generated a chimeric L chain gene encoding TeTx-specific sequences from codon 1 to codon 245 (i. e. immediately downstream from the His-rich motif) and then continuing in frame with the BoNT/A-specific codons 236 to 448. As shown in Figure 3, the corresponding gene product exhibited all the features of a botulinal L chain:

(i) the gene product by itself was nontoxic; (ii) no matter whether the purified H chain was preinjected (Figure 3, filled squares) or added after injection of the chimeric mRNA (open squares), toxicity clearly depended on the presence of the H chain; (iii) the shape of the depression curve resembled that commonly induced by botulinal toxins (Poulain et al., 1989).

Together these results allow the following conclusions:

(1). The carboxyl-terminal portion of the BoNT/A L chain interacts with the H_C domain of the H chain.

(2). The carboxyl-terminal portions of the L chains control the substrate specificity of the individual neurotoxins. This conclusion should be further supported by studies involving antagonists of the neurotoxins.

(3). Since activation of the botulinal L chains requires the presence of H_C in the cytosole, this latter domain has to reach this compartment prior to the toxification process. It is tempting to speculate that the H_C domains of the H chains enter the cytosole together with the corresponding L chains in the putative translocation step.

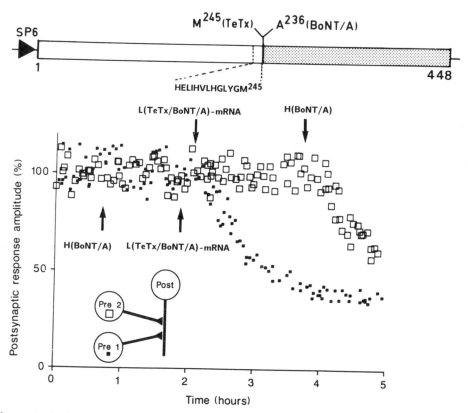

Figure 3. A chimeric L chain consisting of TeTx (residues M^1 to M^{245}) and BoNT/A (hatched area, residues A^{236} to K^{448}) requires the synergistic function of the BoNT/A H chain for toxicity.

Mutations within the histidine-rich motif suggest that at least two of the three histidines are essential for toxicity

The alignment of the L chain sequences reveals a highly conserved core region containing three histidine residues. The conserved motif, hHEnnHnnH in which "h" represents a hydrophobic and "n" noncharged amino acids, is found in numerous

Table 1. Sequence homologies between zinc-dependent metalloproteases and clostridial neurotoxins.

Protease	Sequence
endo-Peptidases	
Thermolysin	VVA **HE** LT **H** AVTDYTAGLIYQNESGAI**NEA**ISDIFGTLVEYANK
B.mesent. MCP	VTA **HE** MT **H** GVTQETANLIYENQPGAL**NE**SFSDVFGYFNDT ED
S.cerev. yscII	VVQ **HE** LA **H** QWFGNLVTMDWWEGLWL **NE**GFATWMSWYSCNE
Mus musc. B cell	VVA **HE** LV **H** QWFGNTVTMDWWDDLWL **NE**GFASFFEFLGVNHAE
Bac. caldolyt.	VVG **HE** LT **H** AVTDYTAGLVYQNESGAI**NE**AMSDIFGTLVEFYANR
Bac. stearotherm.	VVG **HE** LT **H** AVTDYTAGLVYQNESGAI**NE**AMSDIFGTLVEFYANR
Listeria monocyt.	IVG **HE** LT **H** AVIQYSAGLEYEGQSGAL**NE**SFADVFGYFIAPNHWL
Collagenase group	
Serratia emp	TFT **HE** IG **H** ALGLSHPDGYNAGEGNPTYRDV TYA**ED**
prtC gram-	TFT **HE** IG **H** ALGLAHPGEYNAGEGDPSYNDA VYA**ED**
rab stromelysin	VAA **HE** LG **H** SLGLFHSANPEALMYPVYNAFTDLARFRLSQ**DD**VD
por collagen.	VAA **HE** LG **H** SLGLSHSTI GALMYPNYIYTGDVQ LSQ**DD**ID
hum collagen.	VAA **HE** FG **H** ALGLDHSSVPEALMYPMYRFTEGPP LHK**DD**VN
hum neutr. coll.	VAA **HE** FG **H** SLGLAHSSDPGALMYPNYAFRETSNYS LPQ**DD**ID
hum pump1	AAT **HE** LG **H** SLGMGHSSDPNAVMYPTYGNGDPQNFKL SQ**DD**IK
hum strom2	VAA **HE** LG **H** SLGLFHSANTEALMYPLYNSFTELAQFRLSQ**DD**VN
hum strol3	VAA **HE** FG **H** VLGLQHTTAAKALMSAFYTFRYPLS LSP**DD**CR
hum mmp3a	VAA **HE** IG **H** SLGLFHSANTEALMYPLYHSLTDLTRFRLSQ**DD**IN
bov collagen.	VAA **HE** FG **H** SLGLAHSTDIGALMYPSYTFSGDVQ LSQ**DD**ID
rot. atrox ven.	VTM **HE** LG **H** NLGMEHDGKDCLRGASLCIMRPGLTKGRSYEF**SDD**
Snake venom Lepidos.	TMA **HE** LG **H** NLGMKHDENHCHCSASFCIMPPSISEGPSYEF**SDC**
hum bone morph.p.	IVV **HE** LG **H** VVGFWHEHTRPDRDRHVSIVRENIQPGQEYNFLKM
Enteroc. gel	VVG **HE** MT **H** GVTE **H**TAGLEYLGQSGALNESYSDLMGYIISGAS
Various	
hum melanotra.	WLG **HE** YL **H** AMKGLLCDPNRLPPYLRWCVLSTPEIQKCGDMAVA
Angiotensin conv. enzyme	VVH **HE** MG **H** IQYFMQYKDLPVALREGANPGFHEAIGDVLALSVS
Hum Kell blood gr.	IMA **HE** LL **H** IFYQLLLPGGCLACDNHALQEAHLCLKRHYAAFPL
Rat testis mep	TYF **HE** FG **H** VMHQLCSQAEFAMFSG THV**E**RDFVEAPSQMLENW
TeTx	LLM **HE** LI **H** VLHGLYGMQVSS**HE**II PSKQ**E** IYMQHTYPISA**EE**
BoNT/B	ILM **HE** LI **H** VLHGLYGIKVDDLPIVP N**E**KKFFMQSTDAIQA**EE**
BoNT/A	TLA **HE** LI **H** AGHRLYGIAINPNRVFKV NTNAYYEMSGLEVSF**EE**
BoNT/C	ILM **HE** LN **H** AMHNLYGIAIPNDQTISSVTSNIFYSQYNVKLEYA**E**
BoNT/D	ALM **HE** LT **H** SLHQLYGINIPSDKRIRPQVSEGFFSQDGPNVQF**EE**
BoNT/E	TLM **HE** LI **H** SLHGLYGAKGITTKYTITQKQNPLITNIRGT NI**EE**
BoNT/E(butyr)	TLM **HE** LI **H** SLHGLYGAKGITTKYTITQKQNPLITNIRGT NI**EE**
Phytochrome A	VPS **HE** LQ **H** ALHVQQASQQNALTKLKAYSYMRHAINNPLSGMLYS
Rat aminopept.	VIA **HE** LA **H** QSHPLSSPANEVNTPAQISELFDSITYSKGASVLRM
Rat kidney amino-peptidase	VIA **HE** LA **H** QWFGNLVTVDWWNDLWL**NE**GFASYVEFLGADYAEPT
Hum aminopept.N	VIA **HE** LA **H** QWFGNLVTIEWWNDLWL**NE**GFASYVEYLGADYAEPT
E.coli pep N	VIG **HE** YF **H** NWTGNRVTCRDWFQLSLKEGLTVFRDQEFSSDLGSR
Phytochrome	VAS **HE** LQ **H** ALQVQQASEQTSLKRLKAFSYMRHAINNPLSGMLYS
E.coli dcp	TLF **HE** FG **H** GLFARQRYATLSGTNTPRDFVEFPSQINEHWATHP
Oxidoreductase rab.	LCI **HE** LY **H** SALQPLDMTIYADRFTPLTNTDVAAGLWQPYLSDPS
Malate Dehydr. mit.	DKR **HE** LI **H** RIQFGGDEVVKAKNGAGSATLLMAHAGAKFANAVLS
Hydroxybenzoate Dehydrogenase	PVS **HE** LI Y ANHPRGFALCSQRSATRSRYYVQVPLTEKVEDWSDE
Amp.binding prot.	VLL **HE** LY **H** CKRKDMLINYFLCLLKIVYWFNPLVWYLDKEAKTEM

metalloproteases including thermolysin (Table 1). This latter protease has been crystalized and analyzed at 1.6 Å resolution (Holmes and Matthews, 1982).

In thermolysin, a central zinc ion is coordinated in a tetrahedral configuration by the imidazole side chains of His[142], His[146], and by the carboxyl function of Glu[166]. The fourth tetrahedral site is occupied by a water molecule that is stabilized by the zinc ion and is hydrogen-bonded to the carboxyl group of Glu[143] on the other side. That carboxyl group functions as a base to remove a proton and assist attack of the water molecule on the peptide carbonyl. A third histidine residue, His[231], comes close to the active center from a turn structure and acts as a proton donor during the catalytic reaction. As predicted for the histidine-rich motif in the L chains of TeTx and BoNT/A, His[142], Glu[143], and His[146] of thermolysin reside in an α-helix. Apparently this tertiary structural arrangement allows the two imidazole side groups, the carboxyl group together with the water molecule to take the four tetrahedral sites around the central zinc ion (Figure 4).

To assess the biological roles of His^{233}, His^{237}, His^{240}, as well as Glu^{234}, we have generated a number of mutants by site directed mutagenesis and microinjected the corresponding mRNAs into *Aplysia* neurons. The results are summarized in Table 2. To our surprise toxicity was only slightly reduced when His^{233} was converted into Leu^{233} or Val^{233}. It should be noted that in metalloproteases carrying only two histidines in the characteristic motif, a mutation of this first histidine residue results in loss of zinc binding and proteolytic activity (Chang and Lee, 1992). The replacement of His^{233} by leucine or valine is unlikely to destruct the predicted helical

1	ITGTSTVGVG	RGVLGDQKNI	NTTYSTYYYL	QDNTRGDGIF	TYDAKYRTTL
51	PGSLWADADN	QFFASYDAPA	VDAHYYAGVT	YDYYKNVHNR	LSYDGNNAAI
101	RSSVHYSQGY	NNAFWNGSEM	VYGDGDGQTF	IPLSGGIDVV	AHELTHAVTD
151	YTAGLIYQNE	SGAINEAISD	IFGTLVEFYA	NKNPDWEIGE	DVYTPGISGD
201	SLRSMSDPAK	YGDPDHYSKR	YTGTQDNGGV	HINSGIINKA	AYLISQGGTH
251	YGVSVVGIGR	DKLGKIFYRA	LTQYLTPTSN	FSQLRAAAVQ	SATDLYGSTS
301	QEVASVKQAF	DAVGVK			

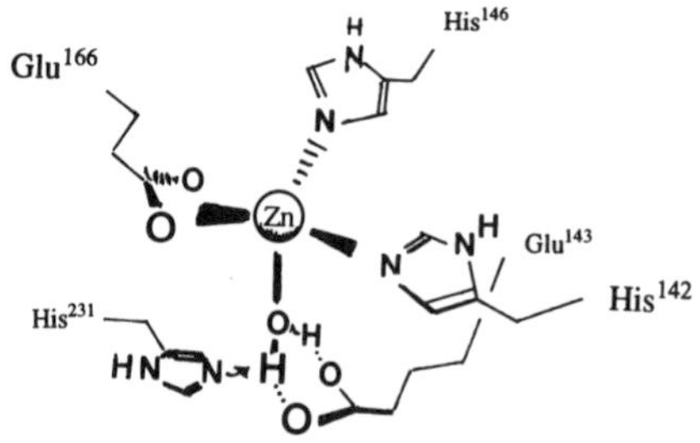

Figure 4. Sequence of thermolysin and configuration around the zinc ion.

structure. However, introduction of a proline residue into this position completely abolished toxicity. Essentially similar results were obtained with respect to His^{237}: again modifications (changes into Asp, Gly, and Val) were tolerated, as long as the overall α-helical structure was not disturbed (mutation into Pro). With regard to His^{240}, apparently any modification seemed to be allowed: basic (Arg^{240}), acidic (Asp^{240}), neutral (Ala^{240}), or even helix-breaking residues (Pro^{240}) had no significant influence on the toxicity of the individual L chains. At a first glance this finding suggests that His^{240} plays absolutely no role in the toxification process. The involvement of His^{240} became apparent, however, when the effects of double mutations were analyzed. The results shown in Table 2 indicate that this amino acid should play a role: when, in addition to a mutation of His^{240}, either His^{233} or His^{237} were replaced by a valine residue (each of these mutations was allowed as single substitutions) toxicity was abolished. The same observation was made for any combination of double mutations.

Table 2. Site-directed mutagenesis of the histidine-rich motif and biological activities of the mutants in *Aplysia* neurons.

Sequence	Helicity	Toxicity
His233-Glu234-Leu- Ile- His237-Val-Leu-His240	h h h h h h h h h h h h h h	+
Leu233	h h h h h h h h h h h h h h	+
Val233	h h h h h h h h h h h h h h	+
Pro233	h h h h h t h h h h h h h	-
Asp237	h h h h h h h h h h h h h h	+
Gly237	h h h h h h h h h h h h h h	+
Val237	h h h h h h h h h h h h h h	+
Pro237	h h h h h h h h h t h h h h	-
Pro233 Pro237	h h h h h t h h h t h h h h	-
Asp240	h h h h h h h h h h h h h h	+
Arg240	h h h h h h h h h h h h h h	+
Ala240	h h h h h h h h h h h h h h	+
Pro240	h h h h h h h h h h h h t h	+
Gln234	h h h h h h h h h h h h h h	-
Lys234	h h h h h h h h h h h h h h	-
Val233 Val237	h h h h h h h h h h h h h h	-
Val233 Ala240	h h h h h h h h h h h h h h	-
Val237 Ala240	h h h h h h h h h h h h h h	-

What is the function of the conserved glutamic acid residue, Glu234, of TeTx? To address tthis question, we have generated two mutants in which this amino acid was either replaced by glutamine or by lysine (Figure 5). In both instances the mutant had lost its neurotoxic activity, indicating that it should indeed play a decisive role.

Taken together our data suggest that at least two of the three histidines and the glutamic acid residue have to remain unaltered within the motif in order to display toxicity. Furthermore, our data suggest that the tertiary structure of the histidine-rich cluster has to be maintained in such a form that the imidazole side chains can interact with a putative metal ion. Whereas these data are compatible with the hypothesis that the clostridial neurotoxins are metalloproteases it remains puzzling that the first of the three histidines can be replaced by other amino acids whithout affecting toxicity.

The L chain of TeTx contains zinc and its biological activity depends on zinc

We then determined by atomic absorption spectroscopy, whether the L chain of TeTx contained significant amounts of zinc. Two different approaches were used:

i) Native L chain, as derived from *C. tetani* and separated from the H chain by isoelectric focussing according to Weller et al., (1989), was analyzed. In this instance the H chain served as a control.

ii) A recombinant wild-type L chain was expressed in *E. coli* and purified to homogeneity. In this instance the double mutant, designated rec. VVH L chain, served as a control.

Our first approach clearly showed that the L chain of TeTx, in contrast to the H chain, contains 0.7 g-atoms zinc per mol of L chain (Figure 6). Since the L and H chains were derived from the same preparation and separated from each other by isoelectric focussing on the same column, it seems unlikely that the L chain just trapped zinc in an unspecific manner.

More direct evidence came from the comparison of the recombinant L chains. To express the TeTx L chain or individual mutants thereof in *E. coli*, we cloned the gene under control of the IPTG-inducible *tac* promoter. We then elongated the C-terminal sequence by introducing a DNA segment that encoded the peptide PPTPGHHHHHH.

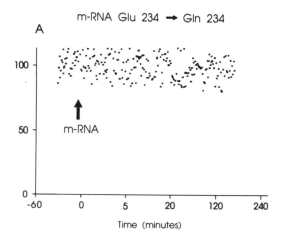

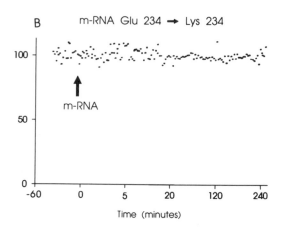

Figure 5. Glutamic acid residue in position 234 of the TeTx L chain is essential for neurotoxicity. Mutations of this residue into glutamine (A) or lysine (B) abolish neurotoxicity in *Aplysia* neurons.

The genetic approach for this modification is outlined in Figure 7. In a first step, a singular NruI blunt end site was generated by the polymerase chain reaction. Into this site we then introduced an oligonucleotide which encoded the above peptide sequence and contained a termination codon for translation. The histidine tag at the C-terminal end of the L chain provided several advantages in the purification procedure of the recombinant L chain.

Firstly, it mediated binding to Ni-agarose columns even after denaturing the L chains by strong detergents such as 6 M guanidinium hydrochloride or 8 M urea (Hochuli et al., 1987).

Secondly, the histidine tag allowed a selection for full-sized L chains, because only those recombinant proteins that contained the tag could bind to the column, whereas prematurely terminated peptides could not.

Thirdly, the tag is preceded by the sequence "PPTP" which represents a cleavage site for IgA-protease (Pohlner et al., 1987). This protease is highly selective, it is resistent against a number of detergents and can cleave the target sequence behind the two prolines even when the corresponding protein is aggregated and presented in form of inclusion bodies (J. Pohlner, unpublished).

A

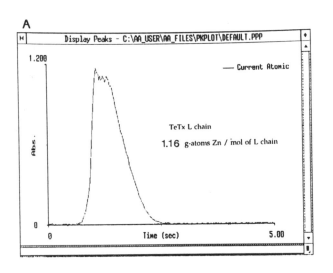

B

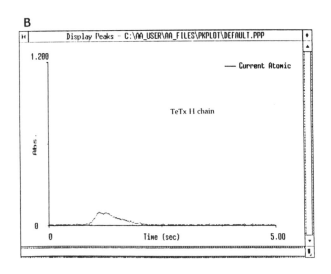

Figure 6. The TeTx L chain is a zinc protein. Solutions containing TeTx L chain (A; 1 mg/ml in artificial sea water) or H chain (B; 0.7 mg/ml artificial sea water) were analyzed by atomic absorption spectroscopy using a Zeeman atomic absorption spectrometer 4100ZL from Perkin Elmer. Mg(NO$_3$)$_2$ (0.05 mg/experiment) was used as a modifier. The amount of zinc detected in the H chain was below back-ground levels.

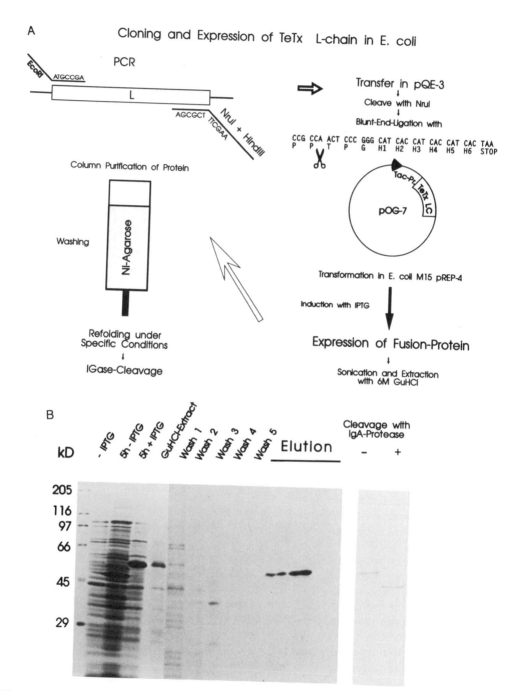

Figure 7. Expression of the TeTx L chain in *E. coli* and its purification from cell lysates by chromatography on Ni-agarose. A. Cloning and purification protocol. B. Analysis of fractions as indicated by SDS-polyacrylamide gel electrophoresis.

Thus purification of the recombinant protein could be achieved in a single step. Even when the protein was denatured by treatment with detergent, a refolding procedure involving washing with a variety of buffers can be attempted, while the protein remains attached to the column. Figure 7B shows the purification of the TeTx L chain and, finally removal of the His-tag. The isolated L chain did not contain any detectable contaminations from E. coli.

Using this approach we also isolated the double mutant, designated TeTx(VVH) in which both His^{233} and His^{237} were replaced by valine residues. Equivalent amounts of protein in artificial sea water (Mochida et al., 1990) were then subjected to atomic absorption spectroscopy. Whereas the wild-type recombinant L chain contained 0.48 g-atoms zinc per mol of L chain, the biologically inactive VVH-mutant (Table 2) lacked any detectable amounts of zinc (Table 3). This is further evidence that indeed the TeTx L chain, and presumably all the botulinal L chains as well are zinc-binding proteins demanding the zinc ion for their biological activity.

Table 3. Zinc content of various TeTx derivatives.

Derivative [Origin]	Protein concentration[a] [µg/ml]	Amount zinc [ppm]	Amount zinc [g-atoms/mol]
L chain [C. tetani]	633,5	0.92	1.16
H chain [C. tetani]	119.8	<0.05	<0.06
rec. wt L chain [E. coli]	476.0	0.29	0.48
rec. VVH L chain [E. coli]	95.0	<0.05	<0.06

[a]Samples were dialyzed against artificial sea water. The original solutions had different protein concentrations due to different solubilities. However, equivalent amounts were subjected to atomic absorption spectroscopy.

Our findings are in keeping with previous reports by DasGupta and Rasmussen (1984) and our own group (Niemann et al., 1987; Binz et al., 1990a) which indicated that a modification of the TeTx or botulinal neurotoxins with diethylpyrocarbonate (to modify the imidazole side chains of histidines) led to inactive toxins. Our findings support previous data by Bhattacharyya and Sugiyama (1989) who have demonstrated convincingly that the L chain of BoNT/A contains significant amounts of zinc and that treatment with various chelating agents including 2,2'-dipyridyl and 1,10-phenanthroline detoxified both TeTx and BoNT/A.

Are the L chains of clostridial neurotoxins metalloproteases?

Let us hypothesize that the L chains are metalloproteases. What additional information can we draw from the neighbouring toxin-specific sequences in comparison with other known metalloproteases containing the His-motif?

Unfortunately, the sequence alignment (Table 1) does not provide any clues with regard to their substrates. What could the natural substrate(s) of the L chain proteases be? The dissimilarities around the histidine-motif could perhaps account for differences in substrate specificity as postulated for TeTx and BoNT/B on the one side and BoNT/A and the other neurotoxins on the other.

It is needless to say that the neurotoxins cannot be simple proteases cleaving any given protein substrate such as casein. In this instance, the toxins would not at all be neurospecific. In fact we have tested a large number of synthetic or natural substrates degraded by other metalloproteases. None of these served as a substrate. Finally, even if we detected that for instance a membrane protein of neuronal synapses was specifically degraded, we would have to prove that such proteo-lytic degradation is sufficient to block neurotransmitter release. There might be more than just one function for the individual L chains which at this moment have yet to be discovered. The search for proteolytic activity and for the natural substrate(s) has begun in many laboratories and, hopefully, it will not take too long until we have deciphered the molecular principles underlying the toxin-induced blockade of neurotransmitter release.

Acknowledgements

We thank O. Dolly for generously providing purified L and H chains of botulinum neurotoxin A. This work was supported by grant Nie 175/5-2 from the Deutsche Forschungsgemeinschaft to H.N. and by a grant from the Association Francaise contre les Myopathies to L.T. H. Kurazono and S. Yamasaki were supported by fellowships from the Alexander-von-Humboldt foundation.

REFERENCES

Ahnert-Hilger, G., Weller, U., Dauzenroth, M.-E., Habermann, E., and Gratzl, M., 1989, The tetanus toxin light chain inhibits exocytosis, *FEBS Lett.* 242:245.

Bhattacharyya, S.D. and Sugiyama, H., 1989, Inactivation of botulinum and tetanus toxin by chelators, *Infect. Immun.* 57:3053.

Binz, T., Kurazono, H., Wille, M., Frevert, J., Wernars, K., and Niemann, H., 1990a, The complete sequence of botulinum toxin type A and comparison with other clostridial neurotoxins, *J. Biol. Chem.* 265:9153.

Binz, T., Kurazono, H., Popoff, M., Eklund, M.W., Sakaguchi, G., Kozaki, S., Krieglstein, K., Henschen, A., Gill, D.M., and Niemann, H., 1990b, Nucleotide sequence of the gene encoding Clostridium botulinum neurotoxin type D, *Nucl. Acids Res.* 18:5556.

Bittner, M.A., Habig, W.H., and Holz, R.W., 1989, Isolated light chains of botulinum neurotoxins inhibit exocytosis, *J. Biol. Chem.* 264:10354.

Bleck, T.P., 1989, Clinical aspects of tetanus, *in:* Botulinum Neurotoxin and Tetanus Toxin, L.L. Simpson, ed., Academic Press, San Diego.

Chang, P.-C., and Lee, Y.-. W., 1992, Extracellular autoprocessing of a metallopro-tease from Streptomyces cacaoi, *J.Biol. Chem.* 267:3952.

DasGupta, B.R. and Rasmussen, S., 1984, Effect of diethylpyrocarbonate on the biological activities of botulinum neurotoxin types A and E, *Arch. Biochem. Biophys.* 232:172.

Dolly, J.O., Ashton, A.C., McInnes, C., Wadsworth, J.D.F. Poulain, B., Tauc, L., Shone, C.C., and Melling, J., 1990, Clues to the multiphase inhibitory action of botulinum neurotoxins on release of transmitters, *J. Physiol.* (Paris) **84**:237.

Dreyer, F., Rosenberg, F., Becker, C., Bigalke, H.,and Penner, R., 1987, Differential effects of various secretagogues on quantal transmitter release from mouse nerve terminals treated with botulinum A and tetanus toxin. *Naunyn Schmiedeberg's Arch. of Pharmacol.* **335**:1.

Eisel, U., Jarausch, W., Goretzki, K., Henschen, A., Engels, J., Weller, U., Hudel, M., Habermann, E., and Niemann, H., 1986, Tetanus toxin: primary structure, expression in E. coli, and homology with botulinum toxins, *EMBO J.* **5**:2495.

Hauser, D., Eklund, M.W., Kurazono, H., Binz, T., Niemann, H., Gill, D.M., Boquet, P., and Popoff, M.R., 1990, Nucleotide sequence of Clostridium botulinum C1 neurotoxin, *Nucl. Acids Res.* **18**:4924.

Hochuli, E., Döbeli, H., and Schacher, A., 1987, New metal chelate adsorbents selective for proteins and peptide containing neighbouring histidine residues. *J. Chromatography* **411**: 177.

Holmes, M.A. and Matthews, B.W., 1982, Structure of thermolysin at 1,6 Å resolution, *J. Mol. Biol.* **160**:623.

Kozaki, S., Miki, A., Kamata, Y., Ogasawara, J., and Sakaguchi, G., 1989, Immunological characterization of papain-induced fragments of clostridium botulinum type A neurotoxin and interaction of the fragments with brain synaptosomes, *Infect. Immun.* **57**:2634.

Kurazono, H., Mochida, S., Binz, T., Eisel, U., Quanz, M., Grebenstein, O., Poulain, B., Tauc, L., and Niemann, H., 1992a, Minimal essential domains specifying toxicity of the light chains of tetanus toxin and botulinum neurotoxin type A, *J. Biol. Chem.*, in press.

Kurazono, S., Kozaki, S., Binz, T., Grebenstein, O., and Niemann, H., 1992b, Monoclonal antibodies as tools to dissect the L chains of clostridial neurotoxins, manuscript n preparation.

McInnes, C. and Dolly, J.O., 1990, Ca^{2+}-dependent noradrenaline release from permeabilized PC12 cells is blocked by botulinum neurotoxin A or its light chain, *FEBS Lett.* **261**:323.

Mochida, S., Poulain, B., Eisel, U., Binz, T., Kurazono, H., Niemann, H., and Tauc, L., 1990, Exogenous mRNA encoding tetanus or botulinum neurotoxins expressed in Aplysia neurons, *Proc. Natl. Acad. Sci. USA* **87**:7844.

Niemann, H., 1991, Molecular biology of clostridial neurotoxins. in: Sourcebook of Bacterial Protein Toxins, Alouf, J. and Freer, J. eds., pp 303-348. Academic Press, New York.

Penner, R., Neher, E., and Dreyer, F., 1986, Intracellularly injected tetanus toxin inhibits exocytosis in bovine adrenal chromaffin cells, *Nature* **324**:76.

Pohlner, J., Halter, R., Beyreuther, K., and Meyer, T., 1987, Gene structure and extracellular secretion of Neisseria gonorrhoeae IgA protease, *Nature,* **325**:458.

Poulain, B., Tauc, L.,Maisey, E.A., Wadsworth, J.D:F., Mohan, P.M., and Dolly, J.O., 1988, Neurotransmitter release is blocked intracellularly by botulinum toxin, and this requires both polypeptides by a process mediated by the larger chain. *Proc. Natl. Acad. Sci. USA* **85**:4090.

Poulain, B., Wadsworth, J.D.F., Shone, C.C., Mochida, S., Lande, S., Melling, J., Dolly, J.O., and Tauc, L., 1989, Multiple domains of botulinum neurotoxin contribute to its inhibition of transmitter release in Aplysia neurons, *J. Biol. Chem.* **264**:21928.

Poulain, B., Paiva, A., Dolly, J.O., Weller, U., and Tauc. L., 1992, Differences in the temperature dependencies of uptake of botulinum and tetanus toxins in *Aplysia* neurons. *Neuroscience Letters* **139**:289.

Poulet, S., Hauser, D., Quanz, M., Niemann, H., and Popoff, M.R., 1992, Sequences of the botulinal neurotoxin E derived from Clostridium botulinum type E (strain Beluga) and Clostridium butyricum (strains ATCC43181 and ATCC43755). *Biochem. Biophys. Res. Commun.* **183**:107.

Stecher, B., Weller, U., Habermann, E., Gratzl, M., and Ahnert-Hilger, G., 1989, The light chain but not the heavy chain of botulinum A toxin inhibits exocytosis from permeabilized adrenal chromaffin cells, *FEBS Lett.* **255**:391.

Weller, U., Dauzenroth, M.-E., Meyer zu Heringdorf, D., and Habermann, E., 1989, Chains and fragments of tetanus toxin, *Eur. J. Biochem.* **182**:649.

STRUCTURE–FUNCTION RELATIONSHIP OF BOTULINUM AND TETANUS NEUROTOXINS

Bal Ram Singh

Department of Chemistry
University of Massachusetts at Dartmouth
N. Dartmouth, MA 02747 U.S.A.

INTRODUCTION

Botulinum and tetanus neurotoxins belong to a general group of bacterial protein toxins with a distinctive characteristics in terms of three structural domains with different complementary functions (Fig. 1). Other toxins of this group include cholera, diphtheria and *Pseudomonas* exotoxin A. These toxins have a polypeptide segment devoted to establish the binding with the target cell, a polypeptide segment that primarily helps translocate the whole or a part of the toxin across the cell membrane, and a third polypeptide segment which possesses an enzymatic or other putative biological activity capable of interfering with certain biochemical reactions within its target cells. Although the third domain elicits the ultimate toxic action, the other two domains are equally important for the biological action of the cellular toxicity of this group of toxins.

Botulinum and tetanus neurotoxins are produced by *Clostridium botulinum* and *C. tetani*, respectively. Both neurotoxins block the release of neurotransmitter from the presynaptic membranes leading to paralytic response. Each neurotoxin consists of a 50 kDa light chain linked to a 100 kDa heavy chain through a disulfide bond (Fig. 2). In their proposed mode of action,[1] the C-terminal half of the heavy chain (H_C) binds to the presynaptic membranes, and the whole neurotoxin is internalized through endocytosis. Inside pH of the endosome is lowered to 5 or below, and the N-terminal half of the heavy chain (H_N) helps form a membrane channel for the translocation of the light chain or the whole neurotoxin. Certain pertinent questions on the existence of structurally separated three functional domains are as follows: Are the domains conformationally interdependent to maintain their active structures? Is there any trend in the general structural characteristics of the three domains? For example, if these domains are preferentially α-helical or β-sheets. What is the relative topography of the three domains? While the last two questions could be answered by a direct structural analysis of the whole neurotoxin and of the three separate domains, the first question can be addressed by determining conformational change upon separation of the domains, and functional activities of the three domains in separated forms as well as in the whole neurotoxin.

Although no systematic study has been conducted to quantitatively compare the binding, membrane translocation and the toxic activities of separated domains and the whole neurotoxin,

Botulinum and Tetanus Neurotoxins, Edited by
B.R. DasGupta, Plenum Press, New York, 1993

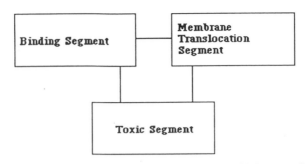

Figure 1. General schematic diagram of protein toxins with three complementary domains.

indirect studies have been done to qualitatively address this issue. For example, approximately 10-fold higher concentration of the botulinum type A heavy chain is required to show a significant antagonistic effect on its parent neurotoxin's toxic effect on the neuromuscular junctions.[2] This could suggest that isolated heavy chain is not as active as the one still associated with light chain in the neurotoxin. At times, even these studies have become more complicated when experiments are performed in a different condition or using a different neuronal system for the assay. The effect of the neurotoxin at a micromolar concentration on neuromuscular junctions of mouse phrenic hemidiaphragm observed by Bandyopadhyay et al.[2] was not seen when nanomolar concentrations were used.[3] Could this be due to the role of the quaternary structure of the neurotoxin that may be different at widely different concentrations? Furthermore, certain observations, such as light chain being sufficient for the toxic action inside the cell lines of mammalian neuronal systems,[4,5] are not supported by the experimental results with *Aplysia* neuronal preparations.[6] In the case of *Aplysia*, it was reported that both light and heavy

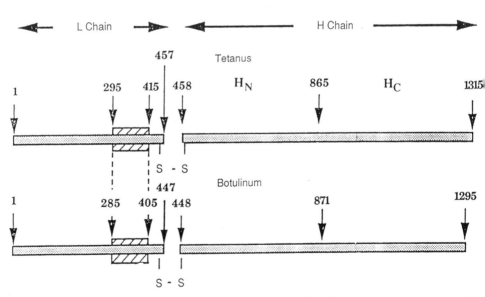

Figure 2. Schematic drawing of light and heavy chain subunits of botulinum and tetanus neurotoxins. The subscripted numbers refer to the residue numbers on the light (L) and heavy (H) chains. Heavy chains are also divided into H_N and H_C to indicate the N-terminal and C-terminal halves. The slanted bar box segments on light chains represent "catalytic" domain.[7]

chains are required together inside the cell to invoke the toxic response. This observation would indicate that light and heavy chains are somehow dependent on each other to maintain an active conformation.

In artificial membrane systems, the channel forming activities of botulinum and tetanus neurotoxins have been investigated. N-terminal halves of both the neurotoxins have been demonstrated to exhibit the membrane channel formation activity.[8,9] Topographical and conformational requirements of the membrane channel forming domain is not clearly understood. For example, the N-terminal half (H_N) of the tetanus neurotoxin when attached to the light chain shows stronger membrane channel forming activity than the heavy chain of the tetanus neurotoxin.[8] This again could indicate a result of specific interaction between light and heavy chains, and between N- and C-terminal domains of the heavy chain.

SIMILARITY BETWEEN BOTULINUM AND TETANUS NEUROTOXINS

Botulinum and tetanus neurotoxins have several common features including over 30% sequence homology,[10] and antagonistic effects of heterologous protein fragments.[11] In addition, both neurotoxins have the so called "catalytic domains" in the C-terminal fragments of their respective light chains.[7] It is also known that tetanus can cause symptoms of botulism at high concentration by blocking the release of acetylcholine from the presynaptic nerve terminals of peripheral nervous system. Therefore, it is tempting to assume a similar mode of action for botulinum and tetanus neurotoxins. Indeed, experimental studies have revealed many similarities between the two neurotoxins including the membrane channel formation by the N-terminal fragments of their respective heavy chains.[8,9]

Despite strong similarities listed above, botulinum and tetanus neurotoxins have significant differences in their physiological functions. The most significant differences include their primary site of action, (botulinum acts primarily on the peripheral nervous system whereas tetanus acts primarily on the central nervous system), and the fact that botulinum neurotoxin is a food poison whereas tetanus is not. The factors responsible for these differences remain to be analyzed. An analysis of the structural basis of the functional similarities and differences between botulinum and tetanus neurotoxins has a potential to provide clues to the structure-function relationship, because the neurotoxins could have mutually inclusive and exclusive domains that can eventually be associated with specific functions involved in their overall toxic actions. A comparative structural analysis could involve primary, secondary, tertiary and quaternary structure.

PRIMARY STRUCTURE

Primary structures of type A botulinum neurotoxin and tetanus neurotoxin have been published.[10,12-14] Nucleotide sequences of type C_1 and D botulinum neurotoxins are also known.[15,16] However, we will not discuss the sequence analysis of types C_1 and D botulinum neurotoxins primarily because a comparison between type A botulinum and tetanus neurotoxin is sufficient to make important points related to the structure-function analysis, and also most other structural studies are not carried out with types C_1 and D.

Alignment of the sequence of light chains of type A botulinum with the corresponding sequence of tetanus neurotoxin reveals a 30.8% homology, whereas the heavy chain shows a sequence homology of 34.8%.[10] However, the sequence homology was 46.4% between the residue number 564-786 of botulinum heavy chain and the corresponding segment of the tetanus heavy chain.[10] Sequence homology itself has not provided any obvious clue to the structural basis of the functions, although even a 30-40% sequence homology could be used to justify closely related functional properties of the two neurotoxins. A clear indication could be derived if any functional motifs are present. For example, hydrophobicity calculations based on a linear sequence revealed two regions in the tetanus neurotoxin with strong hydrophobicity, one in the light chain (residue 223-253) and the other in heavy chain (residue 660-691).[12] In the type A botulinum sequence, three regions (residue 630-648, 652-687, and 773-807) with strong hydrophobicity have been predicted,[14] although the third region (residue 773-807) does not correspond to any significant hydrophobicity in our calculations based on the methods of both Kyte and Doolittle,[17] and Hopp and woods.[18] Despite the presence of 1-2 short stretches of hydrophobic segments in the heavy chains of botulinum and tetanus neurotoxins, it seems unlikely that these segments by themselves may be adequate for forming a membrane channel capable of transporting the light chains of the neurotoxins. There are two possibilities to accommodate this problem: (i) the neurotoxin exists in an oligomeric form, (ii) the neurotoxin has other than just hydrophobic segments, such as amphiphilic segments, compatible for interaction with membranes. It is also possible that a combination of these two factors provides the basis of channel formation by the neurotoxin.

Another structural motif related to the neurotoxin function that could be detected in the primary sequence concerns the toxic activity of the light chains of botulinum and tetanus neurotoxins. There have been wide speculations about ADP-ribosyltransferase activity of botulinum and tetanus neurotoxins[7,19-21] which involves a transfer of an ADP moiety from NAD^+ (a dinucleotide) to an appropriate substrate. One of the characteristic motifs in dinucleotide binding enzymes is considered to be x-h-x-h-G-x-G-x-x-G,[22] where x is any amino acid and h is a hydrophobic amino acid. A search of the light chain of the type A botulinum neurotoxin revealed that such a motif exists as 203-N-P-L-L-G-A-G-K-F-A. It should be noted that Gly(G) can be sometimes replaced by Ala(A) in the dinucleotide binding motif.[22] A similar motif was observed in the tetanus light chain as 156-L-I-I-F-G-P-G-P-V-L-166, except that the last residue is L instead of expected G. This observation is interesting because these motifs occur in the light chain near a domain that is described as "catalytic domain" based on an empirical calculation of Arg/Lys bias ratio.[7]

The Arg/Lys bias ratio (b_{RK}) can be calculated according to London and Luongo:[23]

$$b_{RK} = (n_R - n_K)/n_R \text{ if } n_R < n_K$$

or

$$b_{RK} = (n_R - n_K)/n_K \text{ if } n_R > n_K$$

where n_R and n_K are the number of Arg and Lys, respectively, in a given polypeptide domain.

Although such calculations initially did not reveal any significant bias ratio (b_{RK} of less than -5 or more than +5) when either the whole tetanus neurotoxin was considered or when light and heavy chains were considered individually.[23] A further domain dissection within the light chains of botulinum and tetanus neurotoxins revealed a strong Arg/Lys bias ratio (Table 1). This characteristic was consistent with ADP-ribosyltransferase domains of cholera, diphtheria, *Pseudomonas* exotoxin A, pertussis toxin, etc. Therefore, it is likely that botulinum and tetanus light chains have at least some characteristics of ADP-ribosyltransferase.[7] The problem with this hypothesis is that no true ADP-ribosyltransferase activity has been observed with either pure botulinum or pure tetanus neurotoxin. A partial sequence comparison of this domain with botulinum exoenzyme C_3 (ADP-ribosyltransferase) although showed about 30% sequence homology within its first 22 amino acids,[7] a comparison of the subsequently published C_3 exoenzyme sequence did not reveal any significant homology with either botulinum or tetanus neurotoxin.

One of the major reasons for not observing any ADP-ribosyltransferase activity in botulinum and tetanus neurotoxin could involve a deficiency in its binding with NAD^+, a substrate for the ADP-ribosyltransferase. As mentioned earlier, a peptide motif for dinucleotide binding has been identified in the light chain of type A botulinum neurotoxin, but the motif identified in the tetanus light chain is not entirely consistent with the motif observed in NAD-binding proteins.[22] This could suggest that the presence of such a motif may not play any role in the biological activity of the neurotoxin because both neurotoxins have similar biological activity. Since type A botulinum neurotoxin has the NAD^+ binding motif, we tested its binding with NAD^+ experimentally by monitoring Trp fluorescence of the neurotoxin as a function of NAD^+ concentration (Singh, to be published). We calculated an affinity constant of 4.2×10^4/M, which is about 100-fold lower than the binding constant of NAD^+ to pertussis toxin.[24] The result suggests that despite the presence of NAD^+ binding motif, botulinum neurotoxin does not strongly bind to NAD^+ which may be partly responsible for the lack of ADP-ribosyltransferase activity.

Table 1. Arg/Lys bias ratio in the light chains botulinum and tetanus neurotoxins.

Neurotoxin	Domain	n_R	n_K	b_{RK}
Botulinum	Light chain	14	40	-1.9
	Light chain 285-405	2	18	-8.0
Tetanus	Light chain	14	33	-1.4
	Light chain 295-415	1	15	-12.0

The numbers of Arg(R) and Lys(K) in the botulinum and tetanus neurotoxins were taken from Binz et al.[10] and Eisel et al.,[12] respectively.

SECONDARY STRUCTURE

Secondary structure folding is important for maintaining toxic as well as immunological characteristics of the botulinum and tetanus neurotoxins.[25,26] Secondary structure of proteins is also frequently investigated because of the relative ease and accuracy with which it can be determined for a low concentration of protein solution using a circular dichroism or a FT-IR spectrometer. One major problem with secondary structure determination using these techniques is that only a gross estimation of relative amount of secondary structures can be made (e.g., relative amounts of α-helix, β-sheets, β-turn, and random coil). While a change in the secondary structure content certainly signals a structural change, a non-alternation in the relative contents of the secondary structure may not necessarily mean a no change in the structure because a change in one segment may be compensated by a change in another.

A gross estimation of secondary structure does not allow an identification of a specific segment with a type of secondary structural folding. Sometimes, secondary structure estimation can be compared with predictive methods such as that of Chou and Fasman[27] to identify specific folding of a given peptide segment. A comparison of the secondary structure content of type A botulinum and tetanus neurotoxins based on experimental methods and predictive methods is given in Table 2. Clearly, results from predictive and experimental methods do not match with each other. However, since it is unlikely that the neurotoxins will have 42% β-turn, the predictive method may not be reflecting an accurate estimation. Moreover, the structure of botulinum and tetanus neurotoxins in a physiological condition (pH 7.0) may not reflect its truly functional state because the proposed model for both neurotoxins[1] includes their binding to the membrane, and a decrease in the pH of media surrounding the neurotoxin molecules in their mode of action. Both these steps are known to alter the secondary structure.[28-30]

Table 2. Secondary structure contents of type A botulinum and tetanus neurotoxins based on experimental and predictive methods.

Neurotoxin	Method	α-helix (%)	β-sheets (%)	β-turn (%)	random (%)
Botulinum	CD[a]	21	44	5	30
	Predictive[b]	20	27	41	12
Tetanus	CD[c]	20	50		30
	FT-IR[c]	22	51		27
	Predictive[b]	16	29	42	13

[a]Singh and DasGupta.[31]
[b]Using the method of Chou and Fasman.[27]
[c]Singh et al.[29]

INTERACTION WITH MEMBRANE

Botulinum and tetanus neurotoxins bind to presynaptic membranes, and are translocated across the membrane as part of their action mode. The question is how a water soluble protein, such as these neurotoxins, interacts with the non-polar lipid bilayer? The interaction of the neurotoxins with lipid bilayer in a limited way may still be possible because both neurotoxins have 1-2 stretches of hydrophobic segments (vide supra). However, to form a membrane channel that can accommodate a molecular size of the light chain (50 kDa), the channel may involve many more segments. One possible way to identify other possible segments is to locate amphiphilic peptide segments using hydrophobic moment analysis.[32]

Hydrophobicity (H) and hydrophobic moment (μ) are calculated using a consensus hydrophobicity scale[32] according to the following equation:

$$\mu = \left\{ \left[\sum_n H_n \sin (\Theta_n)^2 \right] + \sum_n H_n \cos (\Theta_n)^2 \right\}^{1/2}$$

where H_n is the hydrophobicity of the nth residue, Θ_n is the angle (in radians) at which successive side chains emerge from the central axis of the structure, Θ for α-helical structures is 100°, whereas for β-sheet structures, it is 160°. When hydrophobic amino acid residues protrude on one side of the structure and hydrophilic amino acid residues on the opposite side, this results in a large hydrophobic moment. Thus, a highly amphiphilic peptide segment will have a large hydrophobic moment. A plot of hydrophobic moment vs. hydrophobicity yields a pattern which can be used, on an empirical basis, to identify amphiphilic (surface-seeking) and trans-membrane segments.[32,33] As an example, one such plot of the tetanus heavy chain is shown in Fig. 3.

Hydrophobic moment analysis of the heavy chains of botulinum and tetanus neurotoxins revealed 13 and 7 amphiphilic/transmembrane peptide segments, respectively (Fig. 4). In addition to the higher number of amphiphilic/transmembrane peptide segments in the type A botulinum heavy chain, other notable points include three transmembrane segments in botulinum heavy chain compared to only one in the tetanus heavy chain and seven amphiphilic segments (7-13; Fig. 4a) in the C-terminal half of the botulinum heavy chain compared to only two (segments 6 and 7; Fig. 4b) in the C-terminal half of the tetanus heavy chain. It may also be noted that transmembrane segments in both neurotoxins are present only in the N-terminal half of their respective chains, and the lone transmembrane segment of the tetanus heavy chain[3] has about 40% sequence homology with the corresponding peptide segment of the botulinum heavy chain (segment 4, Fig. 4a). Another segment of botulinum (segment 11, Fig. 4a) has strong sequence homology (50%) with the corresponding amphiphilic segment of tetanus heavy chain (segment 6, Fig. 4b).

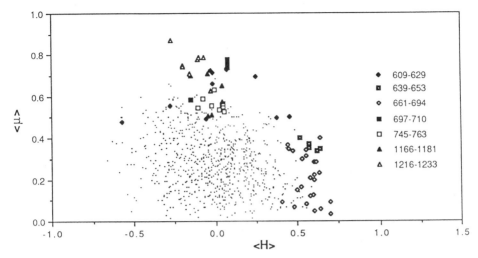

Figure 3. Hydrophobic moment (μ) versus hydrophobicity (H) plot of tetanus heavy chain. The amphiphilic/transmembrane peptide segments are identified by other than dot symbols. Figure derived from Singh and Be.[32]

Role of the amphiphilic/transmembrane domains could be both structural and functional. Structural role could involve complementary interactions among amphiphilic/transmembrane domains to maintain a well-folded neurotoxin structure. A functional role could involve their interaction with membrane bilayers either during initial binding of the neurotoxin or during their translocation across the membrane. The number of segments with strong amphiphilic/transmembrane characteristics appears adequate to form membrane channel even by a monomeric unit of the neurotoxin molecule. Although experimental validity of this hypothesis needs further experiments, such an analysis of the diphtheria toxin and *Pseudomonas* exotoxin A are consistent with their known functional domains.[32,34] One important point that can be derived from these results is that, for two proteins with similar functions, it is not necessary to have very strong sequence homology. The two neurotoxins do not have a very strong sequence homology, but several of the amphiphilic/transmembrane domains exist at similar positions topographically (Fig. 4). There are obviously significant differences in certain domains of the heavy chains of the two neurotoxins, which might explain the differential specificity of botulinum and tetanus neurotoxins vis-a-vis central and peripheral nervous systems, but their implications need further experimental support.

Hydrophobic moment calculations were carried out for the entire sequence of type A botulinum and tetanus neurotoxins (Be and Singh, to be published) assuming α-helical folding, because such an assumption allows detection of strong amphiphilic domains. Conversely, we calculated the α-helical regions by assuming that a segment with strong hydrophobic moment is likely to be in α-helical folding. In this way, type A botulinum and tetanus neurotoxins were estimated to have 19.4% and 20% α-helix, respectively, which is consistent with their α-helical contents determined experimentally (Table 2). This could provide an additional support to the utility of hydrophobic moment in identifying amphiphilic/transmembrane segments.

INTERACTION BETWEEN LIGHT AND HEAVY CHAINS

Secondary structure analysis has been used very effectively to assess structural interaction between light and heavy chains of botulinum and tetanus neurotoxins. Studies with the neurotoxins and their separated light and heavy chains[35,36] have revealed that while secondary

383

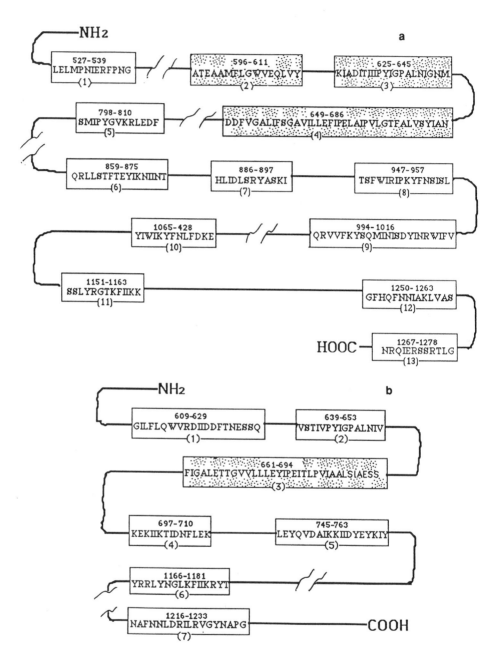

Figure 4. Peptide segments of type A botulinum (a) and tetanus (b) neurotoxin heavy chains with amphiphilic (open boxes) and transmembrane (dot-filled boxes) characteristics as identified from hydrophobic moment versus hydrophobicity plots.

384

structures of light and heavy chains remain virtually intact in type A botulinum neurotoxin (a weighted mean of α-helical contents of separated light and heavy chains as 19.8% vs. 20% observed for the dichain neurotoxin, a difference of 1%), the secondary structural contents of tetanus light and heavy chains changes significantly (a weighted mean of α-helical content of light and heavy chains as 28% vs. 35% observed for the dichain tetanus neurotoxin; a difference of 25%). Differences were more drastic in the β-sheet contents (a 5% change in the case of type A botulinum and 22.6-fold change in tetanus neurotoxin). A larger change in the secondary structures of light and heavy chains upon their separation suggests a stronger binding between the light and heavy chains leading to a tighter folding in tetanus neurotoxin. One preliminary indication of this tight folding could be supported based on the transition temperature for the unfolding of type A botulinum and tetanus neurotoxins. Two transition temperatures for botulinum were observed at 51.8°C and 54.2°C whereas for tetanus neurotoxin, transition temperatures were observed at 55.5°C and 58.4°C (unpublished results). Does this mean that light and heavy chains of tetanus are more interdependent on each other conformationally than that of botulinum neurotoxin? Answer has to be found in experimental results of effectiveness of, for example, heavy chains in antagonizing neurotoxin action. In case of botulinum neurotoxin, at least 4–10-fold higher concentration of type A botulinum heavy chain was required to observe the antagonistic effect.[2] In case of tetanus, studies with the C-terminal half of the heavy chain revealed that at least about 1000-fold higher concentration is needed to observe the antagonistic effect on the toxin activity of tetanus neurotoxin on mouse phrenic hemidiaphragm.[11] How to explain the higher concentration requirement of the botulinum heavy chain even though it retains its secondary structure upon separation? Its answer perhaps lies in the tertiary structural change upon chain separation. Tertiary structural parameters probed in terms of the topography of aromatic amino acid residues indicated significant alternation in the tertiary structure.[35,37]

TERTIARY STRUCTURE

Tertiary structure includes polypeptide folding of a protein, and it is generally determined by x-ray analysis of suitable protein crystals. In the absence of a crystal structure of either botulinum or tetanus neurotoxin, certain parameters related to the polypeptide folding can be obtained for a comparative purpose. As mentioned earlier, we have initiated a protein folding and unfolding study of botulinum and tetanus neurotoxins to derive thermodynamic parameters. Our preliminary studies have indicated that tetanus neurotoxin has a tighter folding than type A botulinum neurotoxin, and each neurotoxin has two independent domains unfolding with transition temperatures of 51.8°C and 54.2°C (type A botulinum neurotoxin), and of 55.5°C and 58.4°C (tetanus neurotoxin, unpublished results).

A further analysis of the polypeptide folding was indirectly conducted by investigating the surface adsorption-induced changes in neurotoxins' secondary structures as determined by the FT-IR spectroscopy. Adsorption of the type A botulinum neurotoxin to ZnSe crystal introduced significant structural changes as indicated by the band position shifts and changes in the curve-fitted area of Amide III IR bands (Table 3).[30] For example, the adsorbed type A botulinum neurotoxin showed a significantly strong extra band at 1299 cm^{-1} (Table 3). A band at 1303 cm^{-1} observed in the neurotoxin solution disappeared but a band at 1309 cm^{-1} appeared with reduced relative band area. These bands correspond to the α-helical regions of a protein. Significant changes were also observed in the random coil regions of the neurotoxin (Table 3).[30] Structural changes were also visible from the Amide I band analysis of the IR spectra, and such changes were largely observed in the bands corresponding to α-helical and random coil regions (Table 3). FT-IR spectral analysis of tetanus neurotoxin (Figures 5 and 6) of Amide I and Amide III regions revealed that the surface adsorption does not induce as much of change in α-helical regions as in the random coil regions (Table 3).

Table 3. FT-IR spectral band positions and relative areas of Amide I and III bands of type A botulinum[30] and tetanus[29] neurotoxins in bulk solution and after adsorption to a ZnSe solid surface.

	Amide I			Amide III		
	Position cm^{-1}	Relative area %	Assignment	Position cm^{-1}	Relative Area %	Assignment
BOTULINUM						
Bulk solution	1678	19	β-sheet	1318	11	α-helix
	1654	46	α + rc	1303	09	α-helix
	1646	05	rc	1295	00	α-helix
	1634	30	β-sheet	1289	09	α-helix
				1279	06	rc
				1261	20	rc
				1245	32	β-sheet
				1227	13	β-sheet
Adsorbed	1678	18	β-sheet	1319	11	α-helix
	1656	36	α + rc	1309	04	α-helix
	1644	13	rc	1299	06	α-helix
	1634	33	β-sheet	1284	11	α-helix
				1267	12	rc
				1253	16	rc
				1241	27	β-sheet
				1226	13	β-sheet
TETANUS						
Bulk solution				1317	10	α-helix
	1678	21	β-sheet			
	1654	44	α + rc	1304	04	α-helix
	1644	06	rc	1294	00	α-helix
	1634	29	β-sheet	1288	08	α-helix
				1271	09	rc
				1258	15	rc
				1248	22	β-sheet
				1237	00	β-sheet
				1233	32	β-sheet
Adsorbed	1678	20	β-sheet	1317	09	α-helix
	1656	50	α + rc	1303	04	α-helix
	1644	01	rc	1293	00	α-helix
	1634	29	β-sheet	1288	09	α-helix
				1272	11	rc
				1259	16	rc
				1249	19	β-sheet
				1241	00	β-sheet
				1233	32	β-sheet

rc = random coil

For example, in Amide I region of tetanus neurotoxin, the band at 1654 cm^{-1} (α-helix + random coil) changes from 44% to 50% upon adsorption, but the change at 1644 cm^{-1} (random coil) is from 6% to 1%. Since bands at both 1644-1646 cm^{-1} and 1654 cm^{-1} could reflect change in unordered segments, these changes may be the result of change in random coil structures. In type A botulinum neurotoxin, corresponding changes were from 46% to 36% (1654 cm^{-1} band) and from 5% to 13% (1644 cm^{-1} band) (Table 3). These changes although opposite in direction are similar in magnitude. A comparison of Amide III bands of botulinum and tetanus neurotoxins (Table 3) indicates much less alteration in tetanus than in botulinum, suggesting a tighter folding in tetanus neurotoxin. It may be noted here that the absolute IR band strengths of botulinum and tetanus neurotoxins are different both in Amide I and Amide III regions. Part of the reason could be the inherent difference between botulinum and tetanus neurotoxin, and a minor contribution could be due to the different pH conditions of botulinum (pH 5.5) and tetanus (pH 7.0) neurotoxins.

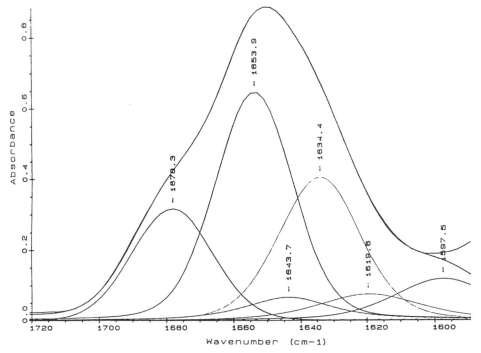

Figure 5. Curve-fitted FT-IR spectral analysis of bulk solution of tetanus neurotoxin in the Amide I and III frequency regions.

Changes in β-sheet bands were minimal both in botulinum and tetanus neurotoxins based on Amide I as well as Amide III bands (Table 3). Structural alterations induced by a solid surface adsorption is likely to involve primarily surface, not the interior, of a protein. This may implicate that β-sheets may be located in the interior of neurotoxins' structure forming the "inner core" surrounded by α-helical and random coil regions. Validity of such a structural model will need further experimental proof. Functional implications of such a model may suggest stronger role of α-helical regions in the neurotoxin interaction to protein receptors and membrane—an aspect consistent with amphiphilic domains identified based on α-helical folding of peptide segments and their suggested role in interactions with membranes.[32] In other words, if amphiphilic α-helical segments are located on the exterior of the neurotoxin protein, their interaction with membranes is more feasible.

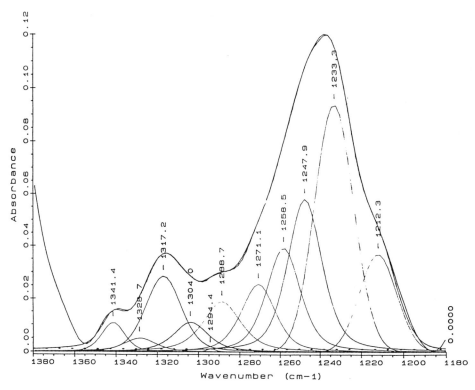

Figure 6. Curve-fitted FT-IR spectral analysis of bulk solution tetanus neurotoxin in the Amide III frequency region.

QUATERNARY STRUCTURE

Quaternary structure of a protein deals with the question of it being a monomer or an oligomer in aqueous solution. Tetanus neurotoxin has been extensively analyzed in the past[38] for its molecular size using analytical ultracentrifugation, and it was reported that purified tetanus neurotoxin exists as a monomer (125-145 kDa), although "aggregation" was observed at certain concentrations. Moreover, to avoid "aggregation," Robinson and co-workers[38] used their toxin preparations "immediately" after purification on a preparative scale electrophoresis which was carried out in 4M urea. Urea is likely to facilitate disaggregation of tetanus neurotoxin into the monomeric form.

Botulinum neurotoxin type A is also reported to form "aggregation" based on native gel electrophoresis band of 450-600 kDa.[39] Whether or not the "aggregated" form is a natural form of the neurotoxins remains to be confirmed. We have carried out native gel electrophoresis of type A botulinum and tetanus neurotoxins (Fig. 7), which indicates that the tetanus neurotoxin exists primarily as a dimer although a small fraction exists as trimer, whereas type A botulinum neurotoxin exists in the form of at least a trimer. Type A botulinum neurotoxin band is spread over a range of 450-600 kDa, but most of it appears as a trimer. This observation is consistent with those of Shone et al.[39] We have further confirmed the existence of oligomeric type A botulinum neurotoxin by chemical cross-linking experiments (to be published). Crystal forms of type A botulinum neurotoxin reported recently are also believed to exist as a dimer.[40]

If botulinum and tetanus neurotoxins exist as dimer or trimer/tetramer in aqueous solutions (Fig. 8), what could be the physiological role of such structures? One likely role could be in the formation of membrane channel during the translocation of the neurotoxin across the

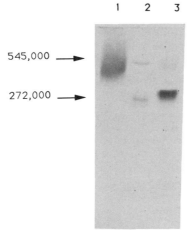

Figure 7. Native polyacrylamide gel electrophoresis bands of type A botulinum (land 1) and tetanus (lane 3) neurotoxins. Lane 2 is the standard protein urease with molecular weight of 545 kDa and 272 kDa.

presynaptic membrane during is mode of action. This could especially be relevant in view of the fact that both the neurotoxins are water-soluble proteins and have only 1-2 hydrophobic stretches of peptide segments which may not be adequate to explain formation of membrane channel capable of translocating at least a 50-kDa light chain. The result could also be relevant to explain the behavior of botulinum neurotoxin with mouse phrenic hemidiaphragm at different concentrations.[2,3] Because both the neurotoxins exist in more than one oligomeric form, it is possible that these oligomeric forms are in equilibrium with each other, and this equilibrium could be altered in different conditions such as in low pH and upon interaction with membranes. We are currently working on a model to further analyze the oligomeric forms of botulinum and tetanus neurotoxins and to rationalize their interaction with membranes.

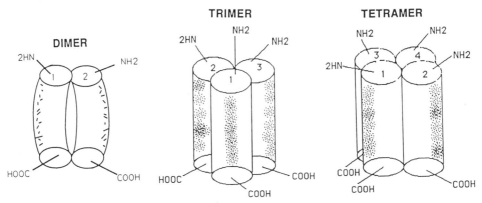

Figure 8. Schematic representation of oligomeric structure of type A botulinum and tetanus neurotoxins. Based on results in Fig. 7, it is assumed that botulinum exists as trimer and tetramer whereas tetanus neurotoxin exists as dimer and trimer. The shaded areas indicate the location of amphiphilic/transmembrane region of the monomeric units.

389

CONCLUDING REMARKS

Botulinum and tetanus neurotoxins are closely related proteins which provide a system to study three different complementary functions in one protein. At the present time, very little is known about the molecular mechanism of their action at all the three functional levels; binding, translocation and the intracellular blockage of the neurotransmitter release. Molecular topography obviously will play a crucial role for the understanding of the physiological function of these neurotoxins.

The discussion in this chapter has only highlighted the potentials of the topographical studies of the neurotoxins to understand their biological function at the molecular level. Further studies are obviously needed to learn not only the molecular topography and polypeptide folding, but also the changes in these parameters in appropriate conditions. Type A botulinum neurotoxin has already been crystallized,[40] and its structure is being worked out. Tetanus neurotoxin's two-dimensional crystal structure is published at 14Å level,[41] and further work is in progress to derive the structure at higher resolution (J. P. Robinson, personal communication). These would be the first major steps in a long journey of solving the structure-function relationship of a protein with three different structural domains corresponding to three complementary functions which ultimately lead to a single physiological action of blocking the neurotransmitter release.

Acknowledgements

This work was in part supported by the UMass Dartmouth Foundation and Nicolet Instruments Corp.

REFERENCES

1. Simpson LL. Molecular pharmacology of botulinum toxin and tetanus toxin. Annu Rev Pharmacol Toxicol 1986; 26:427-453.
2. Bandyopadhyay S, Clark AW, DasGupta BR, Sathyamoorthy V. Role of heavy and light chains of botulinum neurotoxin in neuromuscular paralysis. J Biol Chem 1987; 262:2660-2663.
3. Maisey EA, Wadsworth JDF, Poulain B, Shone CC, Melling J, Gibbs P et al. Involvement of the constituent chains of botulinum neurotoxins A and B in the blockade of neurotransmitter release. Eur J Biochem 1988; 177:683-691.
4. Mochida S, Poulain B, Weller U, Habermann E, Tauc L. Light chain of tetanus toxin intracellularly inhibits acetylcholine release at neuro-neuronal synapses, and its internalization is mediated by the heavy chain. FEBS Lett 1989; 253:47-51.
5. Bittner, M, DasGupta BR, Holz RW. Isolated light chains of botulinum neurotoxins inhibit exocytosis. Studies with digitonin permeabilized cells. J Biol Chem 1989; 264:10354-10360.
6. Poulain B, Tauc L, Maisey EA, Wadsworth JDF, Mohan PM, Dolly JO. Neurotransmitter release is blocked intracellularly by botulinum neurotoxin, and this requires uptake of both toxin polypeptides by a process mediated by the larger chain. Proc Natl Acad Sci USA 1988; 85:4090-4094.
7. Singh BR. Identification of specific domains in botulinum and tetanus neurotoxins. Toxicon 1990; 28:992-996.
8. Boquet P, Ouflot E. Tetanus toxin fragment forms channels in lipid vesicles at low pH. Proc Natl Acad Sci USA 1982; 79:7614-7618.
9. Blaustein RO, Germann WJ, Finkelstein A, DasGupta BR. The N-terminal half of the heavy chain of botulinum type A neurotoxin forms channels in planar lipid bilayers. FEBS Lett 1987; 226:115-120.
10. Binz T, Kurazano H, Wille M, Frevert J, Wernars K, Niemann H. The complete sequence of botulinum neurotoxin type A and comparison with other clostridial neurotoxins. J Biol Chem 1990; 265:9153-9158.
11. Simpson LL. The binding fragment from tetanus toxin antagonizes the neuromuscular blocking actions of botulinum toxin. J Pharmacol Exp Ther 1984; 229:182-187.

12. Eisel U, Jarauseh W, Goretzki K, Henschen A, Engels J, Weller U et al. Tetanus toxin: primary structure, expression in E. coli, and homology with botulinum toxins. EMBO J 1986; 5:2495-2502.

13. Fairweather NF, Lyness VA. The complete nucleotide sequence of tetanus toxin. Nucleic Acids Res 1986; 14:7809-7812.

14. Thompson DE, Brehm JK, Oultram JD, Swinfield TJ, Shone CC, Atkinson T et al. The complete sequence of the Clostridium botulinum type A neurotoxin deduced by nucleotide sequence analysis of encoding gene. Eur J Biochem 1990; 189:73-81.

15. Binz, T, Kurazano H, Popoff MR, Eklund MW, Sakaguchi S, Kozaki S et al. Nucleotide sequence of the gene encoding Clostridium botulinum type D. Nucleic Acids Res 1990; 18:5556.

16. Hauser D, Eklund MW, Kurazano H, Binz T, Niemann H, Michael D et al. Nucleotide sequence of Clostridium botulinum C1 neurotoxin. Nucleic Acids Res 1990; 18:4924.

17. Kyte J, Doolittle RF. A simple method of displaying hydropathic character of a protein. J Mol Biol 1982; 157:105-132.

18. Hopp TP, Woods KR. Prediction of protein antigenic determinants from amino acid sequences. Proc Natl Acad Sci USA 1981; 78:3824-3828.

19. Banga HS, Gupta SK, Feinstein MB. Botulinum toxin D ADP-ribosylates a 22-24 kDa membrane protein in platelets and HL-60 cells that is distinct from $p^{21N-RAS}$. Biochem Biophys Res Commun 1988; 155:263-269.

20. Adam-Vizi V, Rosener S, Aktories K, Knight DF. Botulinum toxin-induced ADP-ribosylation and inhibition of exocytosis are unrelated events. FEBS Lett 1988; 238:277-280.

21. Montecucco C. On the enzymatic activity of tetanus toxin. Toxicon 1987; 25:1255-1257.

22. Moller W, Amos R. Phosphate binding sequences in nucleotide binding proteins. FEBS Lett 1985; 186:1-7.

23. London E, Luongo CL. Domain specific bias in arginine/lysine usage by protein toxins. Biochem Biophys Res Commun 1989; 160:333-339.

24. Lobban D, Moore KJ, van Heyningen S. The interaction of pertussis toxin with NAD^+. In: Rappuoli R et al., eds. Bacterial Protein Toxins. Stuttgart: Gustav Fischer, 1990:95-96.

25. Singh BR, DasGupta BR. Molecular differences between type A botulinum and its toxoid. Toxicon 1989; 27:403-410.

26. Robinson JP, Picklesimer JB, Puett D. Tetanus toxin: Effect of chemical modifications on toxicity, immunogenicity, and conformation. J Biol Chem 1975:256:7435-7442.

27. Chou PY, Fasman GD. Empirical predictions of protein conformation. Annu Rev Biochem 1978; 47:251-276.

28. Lazarovici P, Yanai P, Yavin E. Molecular interactions between micellar polysialogangliosides and affinity purified tetanotoxins in aqueous solution. J Biol Chem 1987; 262:2645-2651.

29. Singh BR, Fuller MP, Schiavo G. Molecular structure of tetanus neurotoxin as revealed by Fourier transform infrared and circular dichroic spectroscopy. Biophys Chem. 1990; 36:155-166.

30. Singh BR, Fuller MP, DasGupta BR. Botulinum neurotoxin type A: structure and interaction with the micellar concentration of SDS determined by FT-IR spectroscopy. J Protein Chem 1991; 10:637-649.

31. Singh BR, DasGupta BR. Molecular topography and secondary structure comparison of botulinum neurotoxin A, B and E. Mol Cell Biochem 1989; 86:87-95.

32. Singh BR, Be X. Use of sequence hydrophobic moment to analyze membrane interacting domains of botulinum, tetanus and other toxins. In: Angeletti RH, ed. Techniques Protein Chemistry. Orlando, FL: Academic Press, 1992:373-383.

33. Eisenberg D, Wilcox W, Eshita S. Hydrophobic moments as tool for the analysis of protein sequences and structures. In: L'Italian JJ, ed. Protein Structure and Function. New York: Plenum Press, 1987:425-436.

34. Eisenberg D, Weiss RM, Terwilliger TC. The hydrophobic moment detects periodicity in protein hydrophobicity. Proc Natl Acad Sci USA 1984; 81:140-144.

35. Singh BR, DasGupta BR. Structure of heavy and light chain subunits of type A botulinum neurotoxin as analyzed by circular dichroism and fluorescence measurements Mol Cell Biochem 1988; 85:67-73.

36. Robinson JP, Holladay LA, Hash JH, Puett D. Conformational and molecular weight studies of tetanus toxin and its major peptides. J Biol Chem 1982; 257:407-411.

37. Singh BR, DasGupta BR. Changes in molecular environments of Trp and Tyr residues of the light and heavy chains of type A botulinum neurotoxins following their separation. Biophysical Chem 1990; 34:259-267.

38. Robinson JP, Hash JH. A review of the molecular structure of tetanus neurotoxin. Mol Cell Biochem 1982; 48:33-44.

39. Shone CC, Hambleton P, Melling J. Inactivation of Clostridium botulinum type A neurotoxin by trypsin and purification of two tryptic fragments. Proteolytic action near the COOH-terminus of the heavy subunit destroys toxin-binding activity. Eur J Biochem. 1985; 151:75-82.
40. Stevens RC, Evenson ML, Tepp W, DasGupta BR. Crystallization and preliminary x-ray analysis of botulinum neurotoxin type A. J Mol Biol 1991; 222:887-880.
41. Robinson JP, Schmid MF, Morgan DG, Chiu W. Three-dimensional structural analysis of tetanus toxin by electron crystallography. J Mol Biol 1988; 200:367-375.

LOW RESOLUTION MODEL OF BOTULINUM
NEUROTOXIN TYPE A

Raymond C. Stevens

Department of Chemistry
Gibbs Chemical Laboratory
Harvard University
Cambridge, MA 02138

The function of botulinum neurotoxin is rapidly being elucidated by biochemical and molecular biology techniques. Although this information is critical to understanding the function of the toxin, the 3-dimensional structure of the neurotoxin is necessary to correlate the structure and function. To better understand the function of botulinum neurotoxin, the three-dimensional structure determination of the neurotoxin is being undertaken using the technique of X-ray diffraction. The technique of dynamical light scattering is being used to aid the crystallographic investigation of the 3-dimensional structure determination of serotype A.

Botulinum neurotoxin serotype A was isolated from liquid culture of *Clostridium botulinum* as the pure 150,000 dalton polypeptide chain.[3] Diffraction quality crystals have been obtained that diffract to 2.6Å.[7] Four different crystal morphologies have been observed for the serotype A (Figure 1). The bipyrimidal shaped crystals (top right) are larger and diffract to a higher resolution than the other crystal forms and is the crystal form of choice in the diffraction analysis. The bipyrimidal shaped crystals crystallize in the space group $P3_121$ (or $P3_221$) with unit cell parameters ($a=b=170.5$Å, $c=161.7$Å). One dimer is present in the asymmetric unit. The observation of the neurotoxin crystallizing as a dimer is consistent with the native state of the enzyme being dimeric based on non-denaturing native gel experiments.

In order to determine the size of the botulinum neurotoxin molecule, the three serotypes that most commonly cause human botulism, obtained from B. R. DasGupta, were examined; dynamical light scattering data were collected on a BIOTAGE dp801 molecular size detector.[5] The neurotoxin types B and E were produced and purified as reported.[2,4] The concentration of neurotoxin was 0.1 mg/ml in 10 mM HEPES, 100mM KCl, 2mM NaN_3 at pH 7.0 for types A, B and E. The diameter of the dimeric molecule of botulinum neurotoxin type A and E is 100 ± 4Å. Interestingly, the diameter of serotype B is 110 ± 4Å. Based on the molecular mass of the dimeric neurotoxin (300,000 daltons) relative to the maximum dimension of the molecule (100–110Å), it appears likely that all three dimeric serotypes are spherical in shape. This is consistent with other proteins with a similar molecular mass and diameter, i.e. the spherical-shaped enzyme aspartate transcarbamylase has a molecular mass of 310,000 daltons and the diameter of the molecule is 110 ± 4Å based on light scattering and crystallographic

Botulinum and Tetanus Neurotoxins, Edited by
B.R. DasGupta, Plenum Press, New York, 1993

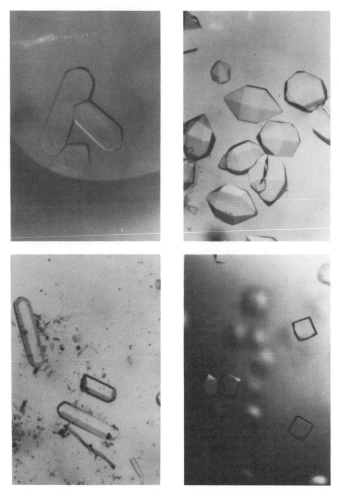

Figure 1. Four different crystal morphologies of *Clostridium botulinum* serotype A.

investigations.[8] Recent results in the structure determination of diphtheria toxin,[1] and in the two-dimensional crystallization of tetanus neurotoxin[6] suggest that a more accurate shape of the botulinum neurotoxin may be "Y"-shaped. This speculation is based on the similarities of the structure and function of botulinum neurotoxin to tetanus and diphtheria toxins and the recent reports of the dimeric nature of botulinum neurotoxin and the diameter/molecular mass of the dimeric botulinum neurotoxin molecule.

REFERENCES

1. Choe S, Bennett MJ, Fujii G, Curmi MG, Kantardjieff KA, Collier RJ, Eisenberg D. The crystal structure of diphtheria toxin. Nature 1992; 357:216-222.
2. DasGupta BR, Rasmussen S. Purification and amino acid composition of *Clostridium botulinum* type E neurotoxin. Toxicon 1983; 535-545.
3. DasGupta BR, Sathyamoorthy V. Purification and amino acid composition of type A botulinum neurotoxin. Toxicon 1984; 21:415-424.

4. DasGupta BR, Woody M. Amino acid composition of *Clostridium botulinum* type B neurotoxin. Toxicon 1984; 22:312-315.
5. Kenney A, Claes P, Boss M. An advanced protein size detector. American Lab 1991; 23:32D-32K.
6. Robinson JP, Schmid MF, Morgan DG, Chiu W. Three-dimensional structural analysis of tetanus toxin by electron crystallography. J Mol Biol 1988; 200:367-375.
7. Stevens RC, Evenson ML, Tepp W, DasGupta, BR. Crystallization and preliminary X-ray analysis of botulinum neurotoxin type A. J Mol Biol 1991; 222:877-880.
8. Warren SG, Edwards BFP, Evans DR, Wiley DC, Lipscomb WN. Aspartate transcarbamylase from *Escherichia coli*: electron density at 5.5A resolution. Proc Natl Acad Sci USA 1973; 70:1117-1121.

STRUCTURE OF BOTULINUM NEUROTOXIN CHANNELS IN PHOSPHOLIPID VESICLES

M. F. Schmid,[1] J. P. Robinson[1] and B. R. DasGupta[2]

[1]Verna and Marrs McLean Department of Biochemistry,
Baylor College of Medicine, Houston, TX 77030
[2]Department of Food Microbiology and Toxicology,
University of Wisconsin-Madison, Madison, WI 53706

Botulinum neurotoxins comprise a class of potent inhibitors of neurotransmitter release, thus acting presynaptically. Their function has been biochemically dissected into distinct stages: i) cell surface attachment; ii) internalization via the endocytotic/lysosomal vesicle pathway; iii) penetration into the cytoplasm from these vesicles through a pH-dependent conformational change; and iv) intracellular poisoning.[1] There are seven distinct serotypes, some of which have been sequenced.[2]

We are interested in the structural expression of the first stages of the intoxication process, namely the attachment of the toxin molecules to the cell surface, and the metamorphosis of this protein from a soluble form to a form that has much in common with an intrinsic membrane channel protein. Several other toxic proteins and some signalling proteins share these properties, so the structure may have wide applicability.[3]

Our approach to studying this membrane-protein complex is the phospholipid monolayer crystallization method. This is a promising and powerful technique of crystalline specimen preparation for electron microscopy (reviewed by Kornberg and Darst[4]). A solution of the 150 kDa neurotoxin was placed in a microtiter well at pH 5.5. Upon it was layered a solution of ganglioside $GT1_b$ and PC (1:20) in chloroform. When the solvent evaporates, a monolayer of phospholipid forms at the air-water interface, with the lipid tails in the air and the head groups in the water. Inevitably, there is an excess of lipid, some of which forms vesicles. Type B botulinum neurotoxin preferentially formed helical arrays by its protein-protein interactions on these surfaces and induced a change in shape of the vesicles from their original spherical shape to closed tubes.

Figure 1 shows an electron micrograph of frozen, hydrated type B botulinum neurotoxin vesicle tubes. In this print, protein is darker than ice for this unstained, unfixed specimen. The micrograph was taken at 30000x magnification. Scale bar is 1000 Å. These tubes form within a few hours. An electron microscope holey grid is applied to the surface and picks up about 3-5 μl of tube suspension. After blotting and quick freezing in liquid ethane, they are transferred to the microscope and examined at low temperature (not exceeding -150°C). These micrographs are taken in the holey grid holes, and thus only protein and vitreous ice are present. The tubes

Botulinum and Tetanus Neurotoxins, Edited by
B.R. DasGupta, Plenum Press, New York, 1993

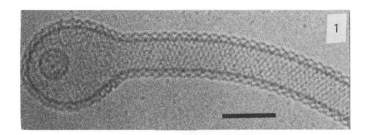

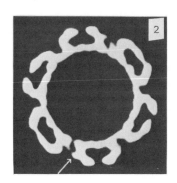

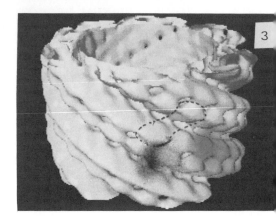

are digitized on a microdensitometer at a resolution equivalent to 6.25 or 8.33 Å/pixel depending on the magnification and straightened using a spline-fitting algorithm.

The computed Fourier transforms were indexed by a program written to define the possible Bessel orders on a layer line, given its spacing from the meridian and the dimensions of the tube followed by a search for selection rules which allow these Bessel orders. Further constraints were imposed by inspection of phases to distinguish odd from even Bessel orders and by making use of the fact that some tubes showed near and far sides in slightly different positions. Finally the absolute sign of a few layer lines was determined from shadowing of the tube surface. The reconstruction was done by methods developed by DeRosier and Moore.[5]

Figure 2 shows a cross-section of the tube taken from a three-dimensional helical reconstruction. The tube is hollow, filled with frozen vitreous ice. The inner radius of the tube is about 240 Å. The thickness of the vesicle wall itself is about 40 Å. The inside of the vesicle wall is not obviously decorated. The protrusions on the outer surface are interpreted to be botulinum neurotoxin molecules. They are connected to the vesicle wall in a localized fashion most of the protein extends away from the surface, eventually to interact with other neurotoxin molecules at high radius (the maximum radius of the object is about 405 Å). Neurotoxin molecules also interact near the vesicle wall, and the channel (see below) is derived from the near-wall interaction of a unit from each protein loop. Note that the molecules also interact short distance above the wall surface, just above the channels (arrow indicates part of the cross bridge).

Figure 3 shows a surface contour map for this three-dimensional reconstruction. It is representation of the surface at which the density map exceeds a threshold. About 450 Å, or little less than 1/2, of the entire helical unit cell distance is shown here, but the features are the same throughout the length.

The most striking feature of the otherwise relatively smooth inside wall is the rows of pits or channels that extend through the entire vesicle wall. There are six such helical rows on the inside wall of the vesicle, one of which is the one visible on the far wall. The outside of the tube is dominated by thick ridges of protein, formed of loops which rise from the surface to meet at high radius. The outside openings of the channels are not the obvious holes where the protein meet at the crests of the ridges, but are almost hidden in the bottoms of the valleys. They are hidden because a bar of density from two ridges crosses just above the channel. This was pointed out in Fig. 2.

The symmetry is such that each channel comes from the interaction of densities from two regions of protein (dotted outline in Fig. 3) related by a two-fold axis centered about the channel. Furthermore, an analysis of the volume of each of these two regions indicates that each is too big to represent a single 150-kDa molecule, but is probably a dimer. Thus, a channel in this map appears to involve the cooperative interaction of 4 150-kDa neurotoxin molecules.

Acknowledgements

This work was partially supported by NIH RR02250 and the W. M. Keck Foundation.

REFERENCES

1. Simpson LL, ed. Botulinum neurotoxin and tetanus toxin. San Diego: Academic Press, 1989.
2. Whelan SM, Elmore MJ, Bodsworth NJ, Brehm JK, Atkinson T, Minton NP. Molecular cloning of the *Clostridium botulinum* structural gene encoding the type B neurotoxin and determination of its entire nucleotide sequence. Appl Environ Microbiol 1992; 58:2345-2354.
3. Simpson LL. Molecular pharmacology of botulinum toxin and tetanus toxin. Ann Rev Pharmacol Toxicol 1986; 26:427-453.
4. Kornberg RD, Darst SA. Two dimensional crystals of proteins on lipid layer. Curr Op Struc Biol 1991; 1:642-646.
5. DeRosier DJ, Moore PB. Reconstruction of three dimensional images from electron micrographs of structures with helical symmetry. J Mol Biol 1970; 52:355-369.

PURIFICATION, CHARACTERIZATION, AND ORAL TOXICITY OF BOTULINUM TYPE G PROGENITOR TOXIN

Masafumi Nukina,[1] Yumi Mochida,[2] Tsutomu Miyata,[1] Sumiko Sakaguchi[2] and Genji Sakaguchi[2]

[1]Public Health Research Institute of Kobe City
Chuo-ku, Kobe, Hyogo 650

[2]College of Agriculture, University of Osaka Prefecture
Sakai-shi, Osaka 591, Japan

PURIFICATION OF BOTULINUM TYPE G PROGENITOR TOXIN[1]

Clostridium botulinum type G (or *C. argentinense*) strain 2470 was grown overnight at 37°C in chopped meat glucose medium and transferred to a medium consisting of proteose peptone 4%, yeast extract 1%, glucose 1%, and L-cysteine 0.2%, pH 7.3, which was incubated for 6 days at 30°C. The 10-L culture with a potential toxicity of 1.3×10^8 mouse i.p. LD_{50} was brought to pH 4.0 with sulfuric acid and kept overnight at 4°C. The precipitate was packed by centrifugation and extracted twice with 0.2 M phosphate buffer, pH 6.0. The extract was treated with 0.5% trypsin (Difco 1:250) at 37°C for 60 min. The extract recovered 1.3×10^8 LD_{50}. The residual cells were sonicated and treated with trypsin in a similar way, which recovered 1.1 $\times 10^8$ LD_{50}. The two extracts were combined, the toxicity of which was set as 100%. Ammonium sulfate was added to the combined extracts to a 0.5 saturation. The precipitate was dissolved in 150 ml of 0.2 M phosphate buffer, pH 6.0, which was clarified by centrifugation. This extract was dialyzed against 0.05 M acetate buffer, pH 4.0, which caused precipitation of the toxin. The precipitate was dissolved in 100 ml of 0.5 M NaCl–0.05 M acetate buffer, pH 4.5. It was clarified by centrifugation, concentrated by salting out, followed by dialysis. The extract of the second salting out recovered 191 mg of protein and 3.3×10^8 LD_{50} (118%). The extract, divided into 20-ml portions, was subjected to gel filtration on Sephadex G-200 (2.5×90 cm). The effluent in the void volume contained 96.3 mg protein and 2.8×10^8 LD_{50} (100%). The fraction was dialyzed against 0.5 M NaCl–0.05 M acetate buffer, pH 4.0, and added were the same volume of 0.5 M NaCl–0.05 M citrate buffer, pH 4.5, and 0.01% protamine. This was clarified by centrifugation and percolated through SP-Sephadex C-50 (2.5×30 cm) equilibrated with 0.5 M NaCl–0.05 M acetate buffer, pH 4.5. The percolate was diluted 2.5-fold with 0.05 M acetate buffer, pH 4.0, applied again to SP-Sephadex C-50 (1.6 $\times 12$ cm) equilibrated with 0.2 M NaCl–0.05 M acetate buffer, pH 4.0, and eluted with NaCl gradient from 0.2 to 0.7 M. The fractions eluted at 0.34 to 0.48 M NaCl contained 48.0 mg of protein and 1.9×10^8 LD_{50} (68%). The toxin fraction was concentrated by salting out and subjected to the second gel filtration on Sephadex G-200 (2.5×90 cm) with the same buffer. A single protein peak eluted contained 22.9 mg of protein and 1.1×10^8 LD_{50} (39%). The specific toxicity was 3.0×10^7 LD_{50}/mg N.

CHARACTERIZATION OF TYPE G PROGENITOR TOXIN

Purity check and molecular-weight determination

PAGE was performed in 5% polyacrylamide gel (pH 4.0) at 2 mA for 4 hr. The gel was stained with 0.2% Coomassie brilliant blue R-250. The purified progenitor toxin formed single band.

Sephadex G-200 gel filtration was performed with 0.2 M NaCl–0.05 M acetate buffer, pH 6.0. Toxins in culture supernatant and sonicated cell suspension and the purified toxin before and after trypsinization had molecular weights between 480K and 520K. Purified type G toxin seems to correspond in molecular weight to L toxin of the other types. No M or L toxin was found. The purified toxin contained a low hemagglutinin activity.

Assay for type G toxin by the time-to-death method

Type G toxin in 0.1-ml doses was injected intravenously into mice, and the time-to-death in minutes was converted to i.p. LD_{50}/ml. A straight line was obtained. The regression coefficient obtained was: $Y = 3.177 - 0.291X$, where Y is $-\log$ death time in minutes and X log dose in mouse i.p. LD_{50}.

Neutral sugar contents of *C. argentinense* 2740 peptidoglycan and its progenitor toxin

Attempts were made to identify the cellular substance that binds type G toxin to the bacterial cells to explain the difficulty of the progenitor toxin to dissociate into the toxic and nontoxic components unlike the toxins of the other types.[2] Total sugar contents were estimated by the phenol–sulfuric acid method with arabinose as a reference. Type G progenitor toxin contained 87 µg of sugar per mg protein, and 1 mg (dry weight) peptidoglycan and lysozyme-treated peptidoglycan contained 290 and 190 µg of sugar/mg dry weight, respectively.

By GLC, type G progenitor toxin was found to contain 100 µg of arabinose and 41 µg of galactose per mg protein. Peptidoglycan contained 130 µg and 210 µg and lysozyme-treated peptidoglycan 170.0 and 49 µg per mg dry weight. The proportion of arabinose to galactose contents of lysozyme-treated peptidoglycan and that of progenitor toxin were close to each other.

ORAL TOXICITIES OF TYPE G PROGENITOR AND GLYCOLYTIC ENZYME-TREATED TOXINS

Each toxin in 0.2-ml doses was administered orally to mice. One oral LD_{50} of type A–L toxin, type G progenitor toxin before and after trypsinization corresponded to 4.0×10^4, 2.0×10^3, and 2.0×10^3 i.p. LD_{50}, respectively; or 6.8×10^3, 1.5×10^4, and 1.5×10^4 oral LD_{50}/mg N, respectively. The oral toxicity of type G progenitor toxin was about twice as high as that of type A–L toxin.

Type G progenitor toxin was treated for 2 h at 37°C with lysozyme (100 µg/ml) in 0.1% EDTA–0.05 M citrate buffer, pH 5.5, with endo-β-galactosidase (1 u/ml) in 0.05 M citrate buffer, pH 5.8, or with N-glycanase (1 u/ml) in the same buffer. One oral LD_{50} value of lysozyme-treated, endo-β-galactosidase-treated, and N-glycanase-treated toxin were 1.6×10^3, 1.2×10^4, and 1.1×10^4 i.p. LD_{50}, respectively. The lysozyme-treated progenitor toxin had an oral toxicity similar to that but those of endo-β-galactosidase- or N-glycanase-treated toxins were 1/6 and 1/5.5 that of untreated progenitor toxin, respectively.

Molecular dissociation of type G progenitor toxin and the glycolytic enzyme-treated progenitor toxin

On Sephadex G-200 gel filtration, the molecular weights of untreated, lysozyme-treated, endo-β-galactosidase-treated, and N-glycanase-treated type G progenitor toxins were estimated at 500K, 280K, 290K, and 290K, respectively. Each sample was dialyzed against 0.05 M Tris–HCl buffer, pH 8.0, and subjected to DEAE–Sephadex chromatography with a linear gradient of NaCl from 0 to 0.3 M in the same buffer to see if separation occurs between toxic and nontoxic components. Untreated, lysozyme-treated and endo-β-galactosidase-treated type G progenitor toxin gave only a single peak, not indicating molecular dissociation, whereas the N-glycanase-treated toxin gave two peaks, the second peak being toxic, indicating molecular dissociation.

The toxic component obtained by N-glycanase treatment and DEAE–Sephadex chromatography of progenitor toxin migrated in one band in SDS–PAGE to the position of a molecular weight of about 150K. Upon DTT treatment, the component migrated in two bands with molecular weights of about 100K and 50K.

DISCUSSION

We have reported that C1-L toxin is bound to peptidoglycan of the cell-wall origin[4] and type E toxin can be obtained in the form of a complex with ribonucleic acid.[5] There was no indication, however, that C1-L toxin purified from the bacterial cells or culture supernatant contained sugars or that type E progenitor toxin purified from culture supernatant contained nucleic acid. Type G toxin is an exception in that the toxin molecule, whether intracellular or extracellular, contains sugars. The sugars may be of the cell-wall origins, and type G strain may lack the system for releasing the progenitor toxin from the sugar moiety.

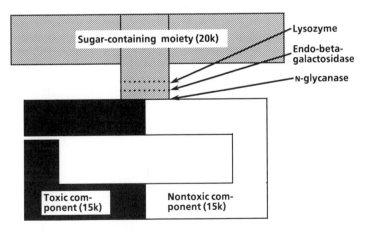

Figure 1. Speculative structure of type G progenitor toxin.

Mouse i.p. LD_{50}/mg N of type G progenitor toxin was about 1/10, whereas its mouse oral LD_{50}/mg N was twice that of type A–L toxin. Type G progenitor toxin may have the potential to cause foodborne botulism. The type G progenitor toxin molecule had a molecular weight of about 500K and the treatment with lysozyme, endo-β-galactosidase, or N-glycanase decreased it to 300K. The oral toxicity of endo-β-galactosidase or N-glycanase-treated toxin was 1/5 that of the untreated or lysozyme-treated toxin. These results indicate that a certain part, being susceptible to endo-β-galactosidase and N-glycanase but not susceptible to lysozyme, of the sugar-containing moiety of type G progenitor toxin is responsible for the higher stability of type G progenitor toxin in the digestive tract.

On DEAE–Sephadex chromatography, the progenitor toxin and the lysozyme- or endo-β-galactosidase-treated toxin were each eluted in a single peak; N-glycanase-treated toxin in two peaks, toxic and nontoxic. These results indicate that a smaller part containing sugar(s) attached by its N-terminus to the toxin prevents molecular dissociation at pH 8.

The type G progenitor toxin we purified may consist of a protein toxin with a molecular weight of about 300K, corresponding to M toxin of the other types and an additional moiety with a molecular weight of about 200K and containing sugar(s) originating from the cell-wall peptidoglycan (Fig. 1).

REFERENCES

1. Nukina M, Mochida Y, Sakaguchi S, Sakaguchi G. Purification of *Clostridium botulinum* type G progenitor toxin. Zbl Bakt Hyg A 1988; 268:220.
2. Nukina M, Miyata T, Sakaguchi S, Sakaguchi G. Detection of neutral sugar in purified type G botulinum progenitor toxin and the effects of some glycolytic enzymes on its molecular dissociation and oral toxicity. FEMS Microbiol Lett 1991; 79:159.
3. Nukina M, Mochida Y, Sakaguchi S, Sakaguchi G. Difficulties of molecular dissociation of *Clostridium botulinum* type G progenitor toxin. FEMS Microbiol Lett 1991; 79:165.
4. Kitamura M, Sakaguchi G. Dissociation and reconstitution of 12-S toxin of *Clostridium botulinum* type E. Biochim Biophys Acta 1969; 194:571.
5. Hyun S-H, Sakaguchi G. Association between the cell-wall peptidoglycan and the progenitor toxin of *Clostridium botulinum* type C. Jpn J Vet Sci 1989; 51:169.

CONSTRUCTION AND EXPRESSION OF THE GENES
FOR NEUROTOXINS AND NON-TOXIC COMPONENTS
IN *C. BOTULINUM* TYPES C AND E.

Nobuhiro Fujii, Kouichi Kimura, Kayo Tsuzuki,
Noriko Yokosawa, and Keiji Oguma

Department of Microbiology
Sapporo Medical Collge
Sapporo 060, Japan

INTRODUCTION

Botulinum toxins are classified into seven groups (type A to G) based on their antigenicity. Type C1 (or C) toxin exists in large molecular sizes of 12S (300 kDa) and 16S (500 kDa) in culture supernatants or in an acid condition, and type E toxin exists in 12S size (Fig. 1). These large (progenitor) toxins are formed by association of the 7S neurotoxin (150 kDa) with a nontoxic component(s). In an alkaline condition, the 12S and 16S toxins dissociate into neurotoxin and nontoxic components. The neurotoxin is produced as a single polypeptide chain and is separated into two fragments, designated as the heavy (H) chain (100 kDa) and the light (L) chain (50 kDa), by reduction of a disulfide bond. On the contrary, the molecular constitution of nontoxic components is not clear. The nontoxic component of the 16S toxin shows hemagglutinating activity, but that of 12S toxin does not. It has been postulated[1] that the nontoxic component of 16S toxin is made up by conjugation of the nontoxic component of 12S toxin (designated as nontoxic-nonHA) with hemagglutinin (HA), and that this conjugation is not separated in an alkaline condition. Molecular mass (Mr) of nontoxic-nonHA component is approximately 120 - 140 kDa in any progenitor toxins, but that of HA is not clear because the preparation of HA alone has not yet been isolated. Suzuki *et al.*[2] reported that Mr. of the nontoxic component of type C 16S toxin was 240 kDa and that it was dissociated into five components (27, 35, 55, 115, and 120 kDa), indicating that HA consisted of several subcomponents. Ohishi *et al.*[3] clarified that the nontoxic components were necessary to maintain the oral toxicity or to cause food poisoning because they prevented the neurotoxin from degradation by gastric juice at a low pH.

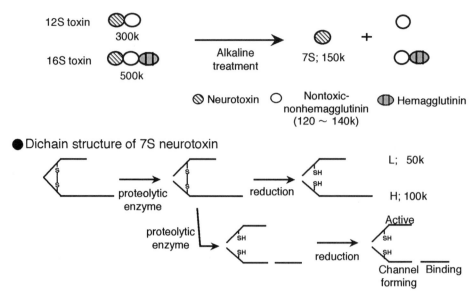

Figure 1. Schematic structure of *C. botulinum* type C1 and E toxin. Type C strain produces both 12S and 16S (progenitor) toxins, and type E strain produces only 12S toxin. These progenitor toxins are dissociated into 7S neurotoxin and a nontoxic component(s) in an alkaline condition. It is suggested that the nontoxic component of 16S toxin is formed by conjugation of the nontoxic component of 12S toxin (nontoxic-nonHA) and hemagglutinin (HA), and that HA consists of several subcomponents such as 53, 33, 17 kDa proteins.

Neurotoxins can be separated into H and L chains by reduction of a disulfide bond, and H chain can be further divided into two parts by mild treatment with proteolytic enzyme. L chain, N-terminus and C-terminus of H chain were presumed to be active, channel-forming, and binding domains, respectively.

We found[4] that the production of type C and D toxin is governed by specific bacteriophages; nontoxic strains are converted to the toxigenic state by infection with phages. HA production is also transmitted by the phages concomitantly with or independently of toxin production[5], indicating that the genes for toxin and HA are located close to each other on a phage DNA. In this report, the construction and the expression of the genes coding for the neurotoxins and the nontoxic components of types C and E are described. Also, the structure and function of type C neurotoxin are discussed.

MATERIALS AND METHODS

Purification of Toxins and Nontoxic Components. The neurotoxin and the 16S toxin were purified from the culture fluid of type C strain, C-Stockholm (C-ST). The organisms were cultured by a cellophane tube procedure. The culture supernatant was centrifuged, and the toxins were precipitated with 50% saturated ammonium sulfate. The neurotoxin was purified by successively applying the precipitate to column chromatography of Sephadex G-75, DEAE-cellulose, and quaternary aminoethyl Sephadex A-50 columns, all of which were equilibrated with 0.064 M borax-phosphate buffer, pH 8.0[6]. Under these conditions, DEAE-cellulose did not absorb the neurotoxin but bound the nontoxic components. The nontoxic components were eluted from this column with 0.3 M NaCl in the same buffer, and then subjected to a Sephacryl S-300 column to get a single peak possessing high HA activity (native nontoxic preparation)[7]. The 16S toxin was purified from the same culture precipitate by column chromatography of Sephadex G-75, SP-Sephadex C-50, and Sephadex G-200 performed under an acid condition[8].

The 12S toxin and its nontoxic component were also purified from type E culture, strain E-Iwanai, according to the procedure reported previously[9]. Young type E cells were collected by centrifugation, washed with 50 mM-sodium acetate buffer (pH 5.0), and then the toxin was extracted with 200 mM phosphate buffer, pH 6.0, at 37°C for 2 hr and then 4 °C overnight. The toxin was precipitated with 60% saturated ammonium sulfate, and purified by sequential column chromatography of CM-Sepharose and Sephadex G-200 in an acid condition. From this purified 12S toxin, nontoxic-nonHA component was separated by applying to a DEAE-Sephadex A-50 column equilibrated with 10 mM phosphate buffer, pH 7.4.

Preparation of Polyclonal and Monoclonal Antibodies. The polyclonal rabbit antibodies against the neurotoxins and nontoxic components of types C and E, and H and L chains of type C neurotoxin were those prepared previously[9,10]. Ten monoclonal antibodies (mAbs) against type C and E neurotoxins were also employed. EL-161-38 and EK-18-2 reacting with L and H chains of type E, respectively, were those previously prepared[11]. The other eight mAbs were newly prepared. They were established as in the same way as reported previously[12] by fusing P3-NS1-1-Ag4-1 myeloma cells with spleen cells obtained from the BALB/c mice immunized with type C toxoid. The recognition sites of these mAbs are shown in Fig. 6 .

Amino Acid Sequencing of N-terminus of Protein. The amino acid sequences of the N-termini of type C and E neurotoxins have been already reported[11,13]. The amino acid sequences of the N-termini of H chain, nontoxic-nonHA, and two subcomponents of HA (HA-33, HA-17) of type C were determined by the direct protein microsequencing method[7,8]. The type C neurotoxin, 16S toxin, and native HA preparation were subjected to 12% SDS-PAGE with 2-ME. The separated proteins were transferred to an Immobilon PVDF membrane. The bands corresponding to each proteins (100, 130, 33, and 17 kDa) were cut off, and their amino acid sequences were determined by a gas phase sequencer coupled to an on-line high-performance liquid chromatograph.

Purification of Phage. The c-st phage was induced from toxigenic strain C-ST, and increased in titer by being passaged through an indicator strain (C)-AO2 in broth medium[4]. The phage particles were precipitated with 8% polyethylene glycol 6000-1M NaCl, and the DNA was extracted by phenol and ethanol treatments[14].

Cloning and DNA Sequencing. Type C and E toxin genes were cloned from c-st phage DNA and cell chromosome, respectively. The phage c-st DNA was digested with

*Eco*RI, cloned into the *Eco*RI site of λgt11 phage DNA by using Packagene (Promega Corp., Madison, Wis.), and then mixed with the *Escherichia coli* (*E. coli*) Y 1090 cells. The DNA library was screened by the PlotBlot Immunoscreening System (Promega) with antisera[15,16]. The positive clones were then recloned into plasmid pUC118, digested with *Pst*I and *Bam*HI, and then deleted by using a TAKARA deletion kit (TAKARA SHUZO Co., Ltd., Japan) containing exonuclease III, Mung bean nuclease, and Klenow enzyme. After transfection into the *E. coli* MV1184 cells, the deleted DNA fragments with the proper sizes were selected by agarose gel electrophoresis. The nucleotide sequence was determined by the dideoxy chain termination method with a T7 sequencing kit (Pharmacia Uppsala, Sweden).

The type E cell chromosome was obtained from a strain, E-Mashike, isolated from a case of fish born botulism named "Izushi" in Hokkaido, Japan[17]. The young cells were cultured in broth medium containing 8% polyethylene glycol 4000, 10 U penicillin G and 10 μg lysozyme per ml. After incubating, the cells collected by centrifugation were lysed with 2-3% SDS. DNA was obtained by phenol and ethanol treatment, digested with *Eco*RI or *Hin*d III, and ligated to λgt11 phage or pUC119 plasmid. The positive clones were immunoscreened, and their nucleotide sequences were determined[18].

Western Blot Analysis. The gene products from the *E. coli* MV1184 cells harbouring the recombinant plasmids were analyzed by Western blot test with polyclonal and monoclonal antibodies as described previously[7].

Toxin Neutralization Test. The toxin of 10 LD50 per ml was mixed with equal volume of diluted antibodies. After incubation at 37℃ for 1 hr, 0.5 ml of the mixture was intraperitoneally injected into two mice, and observed for six days.

Inhibition of Toxin-Binding to NG108 Cells by Antibodies. The toxin-binding test was performed as reported previously[19,20]. The ^{125}I labeled type C neurotoxin was reacted with equal volume of antibodies at room temperature for 30 min, mixed with 2 × 10^5 NG108 cells (hybridoma of mouse neuroblastoma and rat glioma), and then left at 4℃ for 20 min. After the cells were washed by low speed centrifugation at 4℃, the radioactivity was measured in an Autogamma counter.

Hemagglutination Test. The native nontoxic preparation was obtained from the culture supernatant of C-ST as described above. The product of HA-33 gene was partially purified as follows[7]. The *E. coli* cells harbouring the pCL8 were cultured in 300 ml of 2 × TY medium (1.6% tryptone, 1% Yeast extract, 0.5% NaCl, pH 7.6). The cells were collected and sonicated for 5 min at 0℃ to give clear fluids. The gene product was precipitated with 60% saturated ammonium sulfate, suspended in 5 ml of 10mM sodium phosphate buffer (pH 7.4), and separated on a Sephadex G-100 column. Each fraction obtained was analyzed by Western blotting, and the tubes showing a 33-kDa bands were collected and concentrated. Each 0.1 ml of the native- or gene product- preparation was diluted in serial twofold steps and mixed with an equal volume of 1% human type O erythrocytes in phosphate-buffered saline; hemagglutination was observed after 2 h of incubation at 37℃.

RESULTS AND DISCUSSION

Characterization of Toxins and Nontoxic Components.   The purified 16S- and neuro- toxins of type C and 12S type E toxin were subjected to SDS-PAGE with or without reducing agents (Fig. 2-a,b).   The 12S type E toxin demonstrated two bands corresponding to the neurotoxin (Mr.; 140 kDa) and nontoxic-nonHA (Mr.; 120 kDa).   The type C neurotoxin showed one band with Mr. of 150 kDa in the absence of 2 ME, and two bands with 100 kDa (H chain) and 50 kDa (L chain) in the presence of 2 ME.   The (partially) purified native nontoxic preparation demonstrated two major bands (33 and 53 kDa) and several faint bands such as 17 and 25 kDa with or without 2 ME.   The purified 16S type C toxin demonstrated five major bands with Mr. of 130, 100, 53, 50, and 33 kDa, as well as several faint bands in the presence of 2 ME.   Based on previous reports[1,2], it could be concluded that the protein of 130 kDa was the nontoxic-nonHA component, and those of 100 and 50 kDa were the H and L chains, respectively, of type C toxin.   The remaining proteins seemed to be the subcomponents of HA.

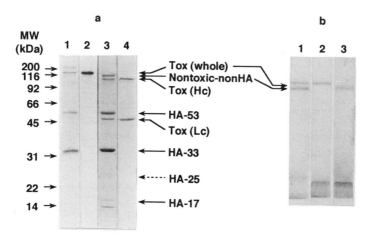

Figure 2. SDS-PAGE of purified toxins and nontoxic components.
(a) 16S toxin (lanes 1 and 3) and neurotoxin (lanes 2 and 4) of type C were subjected in the absence of 2-ME (lane 1 and 2) and in the presence of 2-ME (lane 3 and 4). Neurotoxin (150k) and its H (100k) and L (50k) chains, the subcomponents of HA (53, 33, 25, 17k), and the nontoxic-nonHA component (130k) are indicated by arrows.
(b) 12S toxin (lane 1) purified from type E cell extract, and neurotoxin (lane 2) and nontoxic-nonHA component (lane 3) isolated from this 12S toxin were subjected in the presence of 2-ME. Neurotoxin (140k) and nontoxic-nonHA component (120k) are indicated.

409

Cloning of The Toxin and Nontoxic Component Genes.   Toxin and nontoxic component genes of type C and type E were cloned from phage DNA and cell chromosome, respectively.   These DNAs were digested with *Eco*RI or *Hin*d III and ligated to the polylinker of pUC118 or pUC119.

In type C, the *E. coli* cells transformed with the plasmid containing *Eco*RI-7.8 kb DNA fragment (pCL8) produced two proteins; one was 65 kDa which reacted with the polyclonal and monoclonal antibodies against L chain, the other was 33 kDa which reacted with anti-nontoxic component, and the *E. coli* cells transformed with the plasmid containing *Eco*RI-3 kb DNA fragment (pCH3) produced 74 kDa protein reacted with the antibodies against H chain.   These results indicated that pCL8 contained the structural gene for L chain (and N terminus of H chain) of the toxin and one subcomponent of HA (HA-33, later it was found that the gene for HA-17 subcomponent also existed on pCL8), and that pCH3 contained the structural gene for (almost all of) H chain (Fig. 3).

In an attempt to determine the nucleotide sequence between the DNA fragments cloned in pCL8 and pCH3, polymerase chain reaction with the synthesized 20-mer oligonucleotide primers which were complementary to parts of the 3' and 5' termini of pCL8 and pCH3 insertion fragments were performed[16].   The nucleotide sequence obtained was identical to those of the 3' and 5' termini of the insertion fragments, indicating that those two cloned fragments were directly juxtaposed.

Then, the gene products from a series of plasmids containing various deletions at the 3' end of 7.8 and 3.0 kbp DNA fragments were analyzed by Western blot test.   The results indicated that HA-33 gene was located 4.3 kbp upstream of the neurotoxin gene, and that these two genes were transcribed in the reversed direction[7].

In type E, the *E. coli* cells transformed with the plasmid containing *Hin*d III-6.0 kb fragment (pU9EMH) produced 100 kDa and 120 kDa proteins, the former reacted with mAbs against L and H chains, and the latter reacted with polyclonal anti-nontoxic

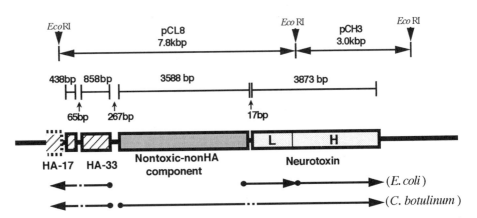

Figure 3. Schematic representation of the hemagglutinin (HA), nontoxic-nonHA, and C 1 neurotoxin genes.   The phage DNAs were digested with *Eco* RI. The fragments (7.8 and 3.0 kb) were cloned into pUC118 plasmids to give pCL8 and pCH3.   The structural genes for HA, nontoxic-nonHA component, and neurotoxin are indicated by boxes. The predicted way of transcription performed in *E. coli* and *C. botulinum* organisms are indicated by arrows.

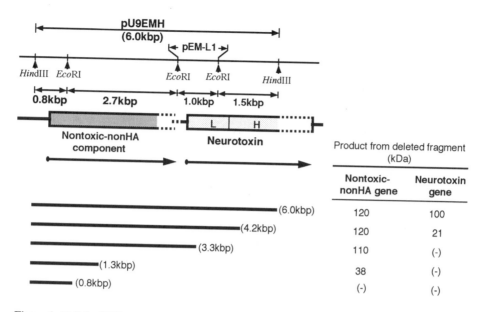

	Product from deleted fragment (kDa)	
	Nontoxic-nonHA gene	Neurotoxin gene
(6.0kbp)	120	100
(4.2kbp)	120	21
(3.3kbp)	110	(-)
(1.3kbp)	38	(-)
(0.8kbp)	(-)	(-)

Figure 4. Cellular DNAs were digested with *Hind*III and *Eco*RI. The *Hind*III 6kb fragment and *Eco*RI 1kb fragment were cloned into pUC119 and pUC118 plasmid to give pU9EMH and pEM-L1, respectively. The 6kb fragment in pUC119 was deleted with exonuclease III, transfected into *E. coli* cells, and then the gene products were analyzed by Western blotting. The structural genes for nontoxic-nonHA component and neurotoxin are indicated by boxes. The predicted way of transcription performed in *E. coli* organisms are indicated by arrows.

component, indicating that pU9EMH contained the structural gene for entire L chain and N-terminus of H chain, and the gene for whole nontoxic component. The location of toxin gene on this 6.0 kb fragment, which was speculated from the results of Western blot test with deleted DNA fragments, and the restriction endonuclease recognition sites are shown in Fig. 4 .

Nucleotide Sequence of The Genes. The nucleotide sequences of these cloned gene or gene-fragments were determined, and their amino acid sequences were deduced (Fig. 5).

The amino acid sequences of N-termini of type C and type E neurotoxins (or L chain) and that of H chain of type E toxin have already been determined by protein analysis [11,13]. Based on these data, the starts of open reading frames (ORF) of type C and E toxins and H chain of type E toxin could be determined[15]. We clarified the amino acid sequences of N-termini of the proteins such as H chain, nontoxic-nonHA component, and two subcomponents (HA-33, HA-17) of HA of type C by direct protein microsequencing technique. Since the amino acid sequence of H chain of type C toxin was Ser-Leu-Tyr-Asn-Lys-Thr-Leu-Asp----, the C terminus of L chain was determined to be Arg-444. The amino acid sequences of HA-33 and HA-17 subcomponents were identical to those deduced from the nucleotide sequences of HA-33 and HA-17 genes. In addition, the gene product

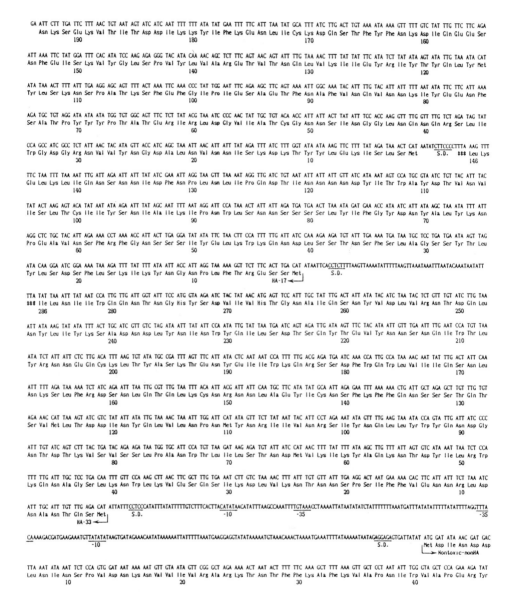

Figure 5. The nucleotide and the deduced amino acid sequences of HA-17, HA-33, nontoxic-nonHA, and neurotoxin. The possible promoter consensus sequences (-10 and -35) and Shine-Dalgarno sequences (S.D.) are underlined.

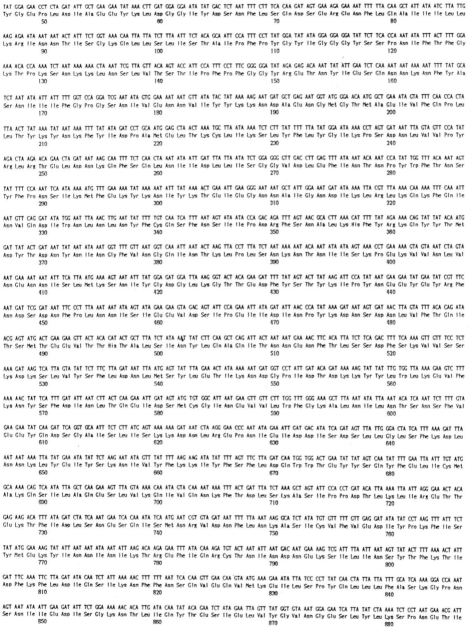

```
TAT GGA GAA CCT CTA GAT ATT CCT GAA GAA TAT AAA CTT GAT GGA GCA ATA TAT GAC TCT AAT TTT CTT TCA CAA GAT AGT GAA AGA GAA AAT TTT TTA CAA GCT ATT ATA ATC TTA TTG
Tyr Gly Glu Pro Leu Asp Ile Ala Glu Glu Tyr Lys Leu Asp Gly Gly Ile Tyr Asp Ser Asn Phe Leu Ser Gln Asp Ser Glu Arg Glu Asn Phe Leu Gln Ala Ile Ile Ile Leu Leu
            50                        60                        70                        80

AAG AGA ATA AAT ACT ATT TCT GGT AAA CAA TTA TTA TCT TTA ATT TCT ACA GCA ATT CCA TTT CCT TAT GGA TAT ATA GGA GGA TAT TCT TCA CCA AAT ATA TTT ACT TTT GGA
Lys Arg Ile Asn Thr Ile Ser Gly Lys Gln Leu Leu Ser Leu Ile Ser Thr Ala Ile Pro Phe Pro Tyr Gly Tyr Ile Gly Gly Gly Tyr Ser Ser Pro Asn Ile Phe Thr Phe Gly
            90                        100                       110                       120

AAA ACA CCA AAA TCT AAT AAA AAA CTA AAT TCG TTA GTT ACA AGT ACC ATT CCA TTT CCT TTC GGG GGA TAT AGA GAG ACA AAT TAT ATT GAA TCT CAA AAT AAT AAA AAT TTT TAT GCA
Lys Thr Pro Lys Ser Asn Lys Lys Leu Asn Ser Leu Val Thr Ser Thr Ile Pro Phe Pro Phe Gly Gly Tyr Arg Glu Thr Asn Tyr Ile Glu Ser Gln Asn Asn Lys Asn Phe Tyr Ala
            130                       140                       150                       160

TCT AAT ATA ATT ATT TTT GGT CCA GGA TCG AAT ATA GTG GAA AAT ATT GTT ATA TAC TAT AAA AAG AAT GAT GCT GAG AAT GGT ATG GGA ACA ATG GCT GAA ATA GTA TTT CAA CCA CTA
Ser Asn Ile Ile Ile Phe Gly Pro Gly Ser Asn Ile Val Glu Asn Ile Val Ile Tyr Tyr Lys Lys Asn Asp Ala Glu Asn Gly Met Gly Thr Met Ala Glu Ile Val Phe Gln Pro Leu
            170                       180                       190                       200

TTA ACT TAT AAA TAT AAT AAA TTT TAT ATA GAT CCT GCA ATG GAG CTA ACT AAA TGC TTA ATA AAA TCT CTT TAT TTT TTA TAT GGA ATA AAA CCT AGT GAT AAT TTA GTA GTT CCA TAT
Leu Thr Tyr Lys Tyr Asn Lys Phe Tyr Ile Asp Pro Ala Met Glu Leu Thr Lys Cys Leu Ile Lys Ser Leu Tyr Phe Leu Tyr Gly Ile Lys Pro Ser Asp Asn Leu Val Val Pro Tyr
            210                       220                       230                       240

AGA CTA AGA ACA GAA CTA GAT AAG CAA TTT TCT CAA CTA AAT ATA ATT GAT TTA TTA ATA TCT GGA GGG GTT GAC CTT GAG TTT ATA AAT ACA AAT CCA TAT TGG TTT ACA AAT AGT
Arg Leu Arg Thr Glu Leu Asp Asn Lys Gln Phe Ser Gln Leu Asn Ile Ile Asp Leu Leu Ile Ser Gly Gly Val Asp Leu Glu Phe Ile Asn Thr Asn Pro Tyr Trp Phe Thr Asn Ser
            250                       260                       270                       280

TAT TTT CCA AAT TCA ATA AAA ATG TTT GAA AAA TAT AAA AAT TAT AAA ACT GAA ATT GAA GGG AAT AAT GCT ATT GGA AAT GAT ATA ATA TTA CGT TTA AAA CAA AAA TTT CAA ATT
Tyr Phe Pro Asn Ser Ile Lys Met Phe Glu Lys Tyr Lys Asn Tyr Lys Thr Glu Ile Glu Gly Asn Asn Ala Ile Gly Asn Asp Ile Ile Leu Arg Leu Lys Gln Lys Phe Gln Ile
            290                       300                       310                       320

AAT GTT CAG GAT ATA TGG AAT TTA AAC TTG AAT TAT TTT TGT CAA TCA TTT AAT AGT ATA ATA CCA GAC AGA TTT AGT AAC GCA CTT AAA CAT TTT TAT AGA AAA CAG TAT TAT ACA ATG
Asn Val Gln Asp Ile Trp Asn Leu Asn Leu Asn Tyr Phe Cys Gln Ser Phe Asn Ser Ile Ile Pro Asp Arg Phe Ser Asn Ala Leu Lys His Phe Tyr Arg Lys Gln Tyr Tyr Thr Met
            330                       340                       350                       360

GAT TAT ACT GAT AAT TAT AAT ATA ACT GGT TTT GTT AAT GGT CAA ATT AAT ACT AAG TTA CCT TTA TCT AAT AAA AAT ACA AAT ATA ATA AGT AAA CCT GCA AAA GTA GTA AAT CTA GTA
Asp Tyr Thr Asp Asn Tyr Asn Ile Thr Gly Phe Val Asn Gly Gln Ile Asn Thr Lys Leu Pro Leu Ser Asn Lys Asn Thr Asn Ile Ile Ser Lys Pro Glu Lys Val Val Asn Leu Val
            370                       380                       390                       400

AAT GAA AAT AAT ATT TCA TTA ATG AAG AAT AAT ATT TAT GGA GAT GGA TTA AAG GGT ACT ACA GAA GAT TTT TAT AGT ACT TAT AAG ATT CCA TAT AAT GAA GAA TAT GAA TAT CGT TTC
Asn Glu Asn Asn Ile Ser Leu Met Lys Asn Asn Ile Tyr Gly Asp Gly Leu Lys Gly Thr Thr Glu Asp Phe Tyr Ser Thr Tyr Lys Ile Pro Tyr Asn Glu Glu Tyr Glu Tyr Arg Phe
            410                       420                       430                       440

AAT GAT TCG GAT AAT TTC CCT TTA AAT AAT ATA AGT ATA GAA GAA GTA GAC AGT ATT CCA GAA ATT ATA GAT ATT AAC CCA TAT AAA GAT AAT AGT GAT AAC TTA GTA TTT ACA CAG ATA
Asn Asp Ser Asp Asn Phe Pro Leu Asn Asn Ile Ser Ile Glu Glu Val Asp Ser Ile Pro Glu Ile Ile Asp Ile Asn Pro Tyr Lys Asp Asn Ser Asp Asn Leu Val Phe Thr Gln Ile
            450                       460                       470                       480

ACG AGT ATG ACT GAA GAA GTT ACT ACA CAT ACT GCT TTA TCT ATA AAT TAT CTT CAA GCT CAG ATT ACT AAT AAT GAA AAC TTC ACA TTA TCT TCA GAC TTT TCA AAA GTT GTT TCC TCT
Thr Ser Met Thr Glu Glu Val Thr Thr His Thr Ala Leu Ser Ile Asn Tyr Leu Gln Ala Gln Ile Thr Asn Asn Glu Asn Phe Thr Leu Ser Ser Asp Phe Ser Lys Val Val Ser Ser
            490                       500                       510                       520

AAA GAT AAG TCA TTA GTA TAT TTC TTA GAT AAT TTA ATG AGT TAT TTA GAA ACT ATA AAA AAT GAT GGT CCT ATT GAT ACA GAT AAA AAG TAT TAT TTG TGG TTA AAA GAA GTC TTT
Lys Asp Lys Ser Leu Val Tyr Ser Phe Leu Asp Asn Leu Met Ser Tyr Leu Glu Thr Ile Lys Asn Asp Gly Pro Ile Asp Thr Asp Lys Tyr Tyr Leu Trp Leu Lys Gly Val Phe
            530                       540                       550                       560

AAA AAC TAT TCA TTT GAT ATT AAT TCT ACT CAA GAA ATT AGT ATG TGT CGC ATT AAT GAA GTT GTT CTT TGG TTT GGG AAA GCT TTA AAT ATT AAT ACA TCA AAT TCT TT GTA
Lys Asn Tyr Ser Phe Asp Ile Asn Leu Thr Gln Glu Ile Ser Met Cys Gly Ile Asn Glu Val Val Leu Trp Phe Gly Lys Ala Leu Asn Ile Leu Asn Thr Ser Asn Ser Phe Val
            570                       580                       590                       600

GAA GAA TAT CAA GAT TCA GGT GCA ATT TCT CTT ATC AGT AAA AAG GAT AAT CTA AGG GAA CCC AAT ATA GAA ATT GAT GAC ATA TCA GAT AGT TTA TTG GGA CTA TCA TTT AAA GAT TTA
Glu Glu Tyr Gln Asp Ser Gly Ala Ile Ser Leu Ile Ser Lys Lys Asp Asn Leu Arg Glu Pro Asn Ile Glu Ile Asp Asp Ile Ser Asp Ser Leu Leu Gly Leu Ser Phe Lys Asp Leu
            610                       620                       630                       640

AAT AAT AAA TTA TAT GAA ATA TAT TCT AAG AAT ATA GTT TAT TTT AAG AAG ATA TAT TTT AGT TTC TTA GAT CAA TGG TCA ACT GAA TAT TAT AGT CAA TAT TTT GAA TTA ATT TGT ATG
Asn Asn Lys Tyr Glu Ile Tyr Ser Lys Asn Ile Val Tyr Phe Lys Lys Ile Tyr Phe Ser Phe Leu Asp Gln Trp Ser Thr Glu Tyr Tyr Ser Gln Tyr Phe Glu Leu Ile Cys Met
            650                       660                       670                       680

GCA AAA CAG TCA ATA TTA GCT CAA GAA AGT TTA GTA AAA CAA GTA CAA AAT AAA TTT GAT TTA TCT AAA GCT AGT ATT CCA CCT GAT ACA TTA AAA TTA ATT AGG GAA ACT ACA
Ala Lys Gln Ser Ile Leu Ala Gln Glu Ser Leu Val Lys Gln Val Gln Asn Lys Phe Asp Leu Ser Lys Ala Ser Ile Pro Pro Asp Thr Leu Lys Leu Ile Arg Glu Thr Thr
            690                       700                       710                       720

GAG AAG ACA TTT ATA GAT CTA TCA AAT GAA TCA CAA ATA TCA ATG AAT CGT GTA GAT AAT TTT TTA AAT AAG GCA TCT ATA TGT GTT TTT GTT GAG GAT ATA TAT CCT AAG TTT ATT TCT
Glu Lys Thr Phe Ile Asp Leu Ser Asn Glu Ser Gln Ile Ser Met Asn Arg Val Asp Asn Phe Leu Asn Lys Ala Ser Ile Cys Val Phe Val Glu Asp Ile Tyr Pro Lys Phe Ile Ser
            730                       740                       750                       760

TAT ATG GAA AAG TAT ATT AAT AAT ATA AAT ATT AAG ACA AGA GAA TTT ATA CAA AGA TGT ACT AAT ATT AAT GAC AAT GAA AAG TCG ATT TTA ATT AAT AGT TAT ACT TTT AAA ACT ATT
Tyr Met Glu Lys Tyr Ile Asn Asn Ile Asn Ile Lys Thr Arg Glu Phe Ile Gln Arg Cys Thr Asn Ile Asn Asp Asn Glu Lys Ser Ile Leu Ile Asn Ser Tyr Thr Phe Lys Thr Ile
            770                       780                       790                       800

GAT TTC AAA TTC TTA GAT ATA CAA TCT ATT AAA AAC TTT TTT AAT TCA CAA GTT GAA GTA ATG AAA GAA ATA TTA TCC CCT TAT CAA CTA TTA TTA TTT GCA TCA AAA GGA CCA AAT
Asp Phe Lys Phe Leu Asp Ile Gln Ser Ile Lys Asn Phe Phe Asn Ser Gln Val Glu Val Met Lys Glu Ile Leu Ser Pro Tyr Gln Leu Leu Leu Phe Ala Ser Lys Gly Pro Asn
            810                       820                       830                       840

AGT AAT ATA ATT GAA GAT ATT TCT GGA AAA AAC ACA TTG TTA CAA TAT ACA GAA TCT ATA GAA TTA GTT TAT GGT GTA AAT GGA GAA TCA TAT CTA AAA TCT CCT AAT GAA ACG ATT
Ser Asn Ile Ile Glu Asp Ile Ser Gly Lys Asn Thr Leu Leu Gln Tyr Thr Glu Ser Ile Glu Leu Val Tyr Gly Val Asn Gly Glu Ser Tyr Leu Lys Ser Pro Asn Glu Thr Ile
            850                       860                       870                       880
```

Figure 5 - CONTINUED

413

```
AAA TTT TCT AAT AAA TTT TTC ACA AAT GGA TTA ACT AAT AAT TTT ACA ATT TGT TTC TGG TTA AGA TTC ACA GGA AAA AAT GAT GAT AAA ACT AGA TTA ATA GCA AAT AAG GTT AAT AAT
Lys Phe Ser Asn Lys Phe Phe Thr Asn Gly Leu Thr Asn Asn Phe Thr Ile Cys Phe Trp Leu Arg Phe Thr Gly Lys Asn Asp Asp Lys Thr Arg Leu Ile Gly Asn Lys Val Asn Asn
            890                           900                           910                           920

TGT GGT TGG GAA ATT TAT TTT GAA GAC AAT GGA TTA GTT TTT GAA ATA ATA GAT TCA AAC GGC AAT CAA GAA AGT GTA TAT TTA TCT AAT ATT ATA AAT GAC AAT TGG TAC TAT ATA TCA
Cys Gly Trp Glu Ile Tyr Phe Glu Asp Asn Gly Leu Val Phe Glu Ile Ile Asp Ser Asn Gly Asn Gln Glu Ser Val Tyr Leu Ser Asn Ile Ile Asn Asp Asn Trp Tyr Tyr Ile Ser
            930                           940                           950                           960

ATA TCC GTT GAT CGT TTA AAA GAT CAA TTA ATA TTT ATT AAT GAT AAA AAT GTT GCA AAT GTA AGT ATC GAT CAA ATA CTA AGT ATT TAT TCT ACC AAT ATA ATA TCT TTA GTT AAT
Ile Ser Val Asp Arg Leu Lys Asp Gln Leu Ile Ile Phe Ile Asn Asp Lys Asn Val Ala Asn Val Ser Ile Asp Gln Ile Leu Ser Ile Tyr Ser Thr Asn Ile Ile Ser Leu Val Asn
            970                           980                           990                           1000

AAG AAT AAT TCA ATC TAT GTA GAA GAA TTG TCA GTT TTA GAT AAT CCT ATT ACA AGC GAA GAA GTT ATA AGA AAT TAC TTT AGT TAT TTA GAT AAT TCA TAT ATA AGA GAT AGT TCC AAA
Lys Asn Asn Ser Ile Tyr Val Glu Glu Leu Ser Val Leu Asp Asn Pro Ile Thr Ser Glu Glu Val Ile Arg Asn Tyr Phe Ser Tyr Leu Asp Asn Ser Tyr Ile Arg Asp Ser Ser Lys
            1010                          1020                          1030                          1040

TCA CTA TTA GAA TAT AAT AAA AAC TAT CAA TTA TAC AAT TAT GTA TTT CCC ACA ACT TCT TTA TAT GAT GTT AAT GAT AAT AAG TCG TAT TTA TCA CTA AAA AAT ACA GAT GGT ATT
Ser Leu Leu Glu Tyr Asn Lys Asn Tyr Gln Leu Tyr Asn Tyr Val Phe Pro Thr Ser Leu Tyr Gly Val Asn Asp Asn Asn Lys Ser Tyr Leu Ser Leu Lys Asn Thr Asp Gly Ile
            1050                          1060                          1070                          1080

AAT ATT TCA AGT GTT AAA TTT AAA TTA ATA AAT GAT GAA AGT AAA GTA TAT GTA CAA AAG TGG GAT GGT ATA ATT TGT GTA TTA GAC GGT ACA GAA AAA TAT TTA GAT ATA TCT
Asn Ile Ser Ser Val Lys Phe Lys Leu Ile Asn Asp Glu Ser Lys Val Tyr Val Gln Lys Trp Asp Gly Cys Ile Ile Cys Val Leu Asp Gly Thr Glu Lys Tyr Leu Asp Ile Ser
            1090                          1100                          1110                          1120

CCT GAA AAT AAT AGA ATA CAA TTA GTA AGT TCC AAA GAT AAT GCA AAA AAG ATT ACA GTT AAT ACT GAT TTA TTT AGA CCT GAT TGT ACA TTT TCA TAT AAT GAT AAA TAT TTT TCT
Pro Glu Asn Asn Arg Ile Gln Leu Val Ser Ser Lys Asp Asn Ala Lys Lys Ile Thr Val Asn Thr Asp Leu Phe Arg Pro Asp Cys Thr Phe Ser Tyr Asn Asp Lys Tyr Phe Ser
            1130                          1140                          1150                          1160

                                                                                                      S.D.
CTA TCA CTT AGA GAT GGA GAT TAT AAT TGG ATG ATA TAT GAT AAT AAC AAG GTG CCT AAA GGT GCA CAT TTG TGG ATA TTA GAA GT TAG GAGATGTTAGTATT ATG CCA ATA ACA ATT
Leu Ser Leu Arg Asp Gly Asp Tyr Asn Trp Met Ile Tyr Asp Asn Asn Lys Val Pro Lys Gly Ala His Leu Trp Ile Leu Glu Ser ***             Met Pro Ile Thr Ile
            1170                          1180                          1190           1196                          → Neurotoxin

AAC AAC TTT AAT TAT TCA GAT CCT GTT GAT AAT AAA AAT ATT TTA TAT TTA GAT ACT CAT TTA ACA CTA GCT AAT GAG CCT GAA AAA GCC TTT CGC ATT ACA GGA AAT ATA TGG GTA
Asn Asn Phe Asn Tyr Ser Asp Pro Val Asp Asn Lys Asn Ile Leu Tyr Leu Asp Thr His Leu Asn Thr Leu Ala Asn Glu Pro Glu Lys Ala Phe Arg Ile Thr Gly Asn Ile Trp Val
            10                            20                            30                            40

ATA CCT GAT AGA TTT TCA AGA AAT TCT AAT CCA AAT TTA AAT AAA CCT CCT CCA GTT ACA AGC CCT AAA AGT GGT TAT TAT CCT AAT TAT TTG AGT GAT TCT GAC AAA GAT ACA
Ile Pro Asp Arg Phe Ser Arg Asn Ser Asn Pro Asn Leu Asn Lys Pro Pro Arg Val Thr Ser Pro Lys Ser Gly Tyr Tyr Asp Pro Asn Tyr Leu Ser Thr Asp Ser Asp Lys Asp Thr
            50                            60                            70                            80

TTT TTA AAA GAA ATT ATA ACG TTA TTT AAA AGA ATT AAT TCT AGA GAA ATA GGA GAA GAA TTA ATA TAT AGA CTT TCG ACA GAT ATA CCC TTT CCT GGG AAT AAC ACT CCA ATT AAT
Phe Leu Lys Glu Ile Ile Thr Leu Phe Lys Arg Ile Asn Ser Arg Glu Ile Gly Glu Glu Leu Ile Tyr Arg Leu Ser Thr Asp Ile Pro Phe Pro Gly Asn Asn Thr Pro Ile Asn
            90                            100                           110                           120

ACT TTT GAT TTT GAT GTA GAT TTT AAC AGT GTT GAT GTT AAA ACT AGA CAA GGT AAC AAC TGG GTT AAA ACT GGT AGC ATA AAT CCT AGT GTT ATA ATA ACT GGA CCT AGA GAA AAC ATT
Thr Phe Asp Phe Asp Val Asp Phe Asn Ser Val Asp Val Lys Thr Arg Gln Gly Asn Asn Trp Val Lys Thr Gly Ser Ile Asn Pro Ser Val Ile Ile Thr Gly Pro Arg Glu Asn Ile
            130                           140                           150                           160

ATA GAT CCA GAA ACT TCT ACG TTT AAA TTA ACT AAC AAT ACT TTT GCG GCA CAA GGA TTT GGT GCT TTA ATA ATT TCA ATA TCC ACT AGA TTT ATG CTA ACA TAT AGT AAT GCA
Ile Asp Pro Glu Thr Ser Thr Phe Lys Leu Thr Asn Asn Thr Phe Ala Ala Gln Gly Phe Gly Ala Leu Ile Ile Ser Ile Ser Pro Arg Phe Met Leu Thr Tyr Ser Asn Ala
            170                           180                           190                           200

ACT AAT GAT GTA GGA GAG GGT AGA TTT TCT AAG TCT GAA TTT TGC ATG GAT CCA ACA CTA ATT TTA ATG GAT ACA CTT AAT CAT GCA ATG CAT AAT TTA TAT GGA ATA GCT ATA CCA AAT
Thr Asn Asp Val Gly Glu Gly Arg Phe Ser Lys Ser Glu Phe Cys Met Asp Pro Thr Leu Ile Leu Met Asp His Gly Leu Asn His Ala Met Asn Leu Tyr Gly Ile Ala Ile Pro Asn
            210                           220                           230                           240

GAT CAA ACA ATT TCA TCT ACA AGT AAT ATT TAT TCT CAA TAT AAT GTG AAA TTA GAG TAT GCA GAA ATA TAT GCA TTT GGA GGT CCA ACT ATA GAC CTT ATT CCT AAA AGT GCA
Asp Gln Thr Ile Ser Ser Thr Ser Asn Ile Tyr Ser Gln Tyr Asn Val Lys Leu Glu Tyr Ala Glu Ile Tyr Ala Phe Gly Gly Pro Thr Ile Asp Leu Ile Pro Lys Ser Ala
            250                           260                           270                           280

AGG AAA TAT TTT GAG GAA AAG GCA TTG GAT TAT TAT AGA TCT ATA GCT AAA GCA CTT AAT AGT ATA ACT GCA AAT CCT TCA AGC TTT AAT AAA TAT ATA GGG GAA TAT AAA CAG AAA
Arg Lys Tyr Phe Glu Glu Lys Ala Leu Asp Tyr Tyr Arg Ser Ile Ala Lys Ala Leu Asn Ser Ile Thr Thr Ala Asn Pro Ser Ser Phe Asn Lys Tyr Ile Gly Glu Tyr Lys Gln Lys
            290                           300                           310                           320

CTT ATT AGA AAG TAT AGA TTC GTA GTA GAA TCT TCA GGT GTT ACA GTA AAT CGT AAT AAG TTT GTT GAG TTA TAT AAT GAA CTT ACA CAA ATA TTT ACA GAA TTT AAC TAC GCT AAA
Leu Ile Arg Lys Tyr Arg Phe Val Val Glu Ser Ser Gly Glu Val Thr Val Asn Arg Asn Lys Phe Val Glu Leu Tyr Asn Glu Leu Thr Gln Ile Phe Thr Glu Phe Asn Tyr Ala Lys
            330                           340                           350                           360

ATA TAT AAT GTA CAA AAT AGG AAA ATA TAT CTT AAT GTA TAT ACT CCG ATT ACG GCG AAT ATA TTA GAC GAT AAT GTT TAT GAT ATA CAA AAT GGA TTT AAT ATA CCT AAA AGT AAT
Ile Tyr Asn Val Gln Asn Arg Lys Ile Tyr Leu Asn Val Tyr Thr Pro Ile Thr Ala Asn Ile Leu Asp Asp Asn Val Tyr Asp Ile Gln Asn Gly Phe Asn Ile Pro Lys Ser Asn
            370                           380                           390                           400

TTA AAT GTA CTA TTT ATG GGT CAA AAT TTA TCT CGA AAT CCA GCA TTA AGA AAA GTC AAT CCT GAA AAT ATG CTT TAT TTA TTT ACA AAA TTT TGT CAT AAA GCA ATA GAT GGT AGA TCA
Leu Asn Val Leu Phe Met Gly Gln Asn Leu Ser Arg Asn Pro Ala Leu Arg Lys Val Asn Pro Glu Asn Met Leu Tyr Leu Phe Thr Lys Phe Cys His Lys Ala Ile Asp Gly Arg Ser
            410                           420                           430                           440

TTA TAT AAT AAA ACA TTA GAT TGT AGA GAG CTT TTA GTT AAA AAT ACT GAC TTA CCC TTT ATA GGT GAT ATT AGT GAT GTT AAA ACT GAT ATA TTT TTA AGA AAA GAT ATT AAT GAA GAA
Leu Tyr Asn Lys Thr Leu Asp Cys Arg Glu Leu Leu Val Lys Asn Thr Asp Leu Pro Phe Ile Gly Asp Ile Ser Asp Val Lys Thr Asp Ile Phe Leu Arg Lys Asp Ile Asn Glu Glu
            450                           460                           470                           480

ACT GAA GTT ATA TAC TAT CCG GAC AAT GTT TCA GTA GAT CAA GTT ATT CTC AGT AAG AAT AAT ACC TCA GAA CAT GGA CTA GAT TTA TTA TAC CCT AGT ATT GAC AGT GGA AGT GAA ATA
Thr Glu Val Ile Tyr Tyr Pro Asp Asn Val Ser Val Asp Gln Val Ile Leu Ser Lys Asn Asn Thr Ser Glu His Gly Leu Asp Leu Leu Tyr Pro Ser Ile Asp Ser Gly Ser Glu Ile
            490                           500                           510                           520
```

Figure 5 - CONTINUED

```
TTA CCA GGG GAG AAT CAA GTC TTT TAT GAT AAT AGA ACT CAA AAT GTT GAT TAT TTG AAT TCT TAT TAT TAC CTA GAA TCT CAA AAA CTA AGT GAT AAT GTT GAA GAT TTT ACT TTT ACG
Leu Pro Gly Glu Asn Gln Val Phe Tyr Asp Asn Arg Thr Gln Asn Val Asp Tyr Leu Asn Ser Tyr Tyr Leu Glu Ser Gln Lys Leu Ser Asp Asn Val Glu Asp Thr Phe Thr
                      530                           540                           550                           560

AGA TCA ATT GAG GAG CCT TTG GAT AAT AGT GCA AAA GTA TAT ACT TAC TTT ACA CTA GCT AAT AAA GTA AAT GCG GGT GTT CAA GGT GGT TTA TTT TTA ATG TGG GCA AAT GAT GTA
Arg Ser Ile Glu Glu Pro Leu Asp Asn Ser Ala Lys Val Tyr Thr Tyr Phe Thr Leu Ala Asn Lys Val Asn Ala Gly Val Gln Gly Gly Leu Phe Leu Met Trp Ala Asn Asp Val
          570                           580                           590                           600

GTT GAA GAT TTT ACT ACA AAT ATT CTA AGA GAA ACA TTA GAT AAA ATA TCA GAT GTA TCA GCT ATT ATT CCC TAT ATA GGA CCC GTA TTA AAT ATA AGT AAT TCT GTA AGA AGA GGA
Val Glu Asp Phe Thr Thr Asn Ile Leu Arg Glu Thr Leu Asp Lys Ile Ser Asp Val Ser Ala Ile Ile Pro Tyr Ile Gly Pro Ala Leu Asn Ile Ser Asn Ser Val Arg Arg Gly
          610                           620                           630                           640

AAT TTT ACT GAA GCA TTT GCA GTT ACT GGT GTA ACT ATT TTA TTA GAA GCA TTT CCT GAA TTT ACA ATA CCT CTT GGT GCA TTT GTG ATT TAT AGT AAG TTT CAA GAA AGA AAC CAG
Asn Phe Thr Glu Ala Phe Ala Val Thr Gly Val Thr Ile Leu Leu Glu Ala Phe Pro Glu Phe Thr Ile Pro Ala Leu Gly Ala Phe Val Ile Tyr Ser Lys Val Gln Glu Arg Asn Glu
          650                           660                           670                           680

ATT ATT AAA ACT ATA GAT AAT TGT TTA GAA CAA AGG ATT AAG AGA TGG AAA GAT TCA TAT GAA TGG ATG ATG GGA ACG TGG TTA TCC AGG ATT ATT ACT CAA TTT AAT AAT ATA AGT TAT
Ile Ile Lys Thr Ile Asp Asn Cys Leu Glu Gln Arg Ile Lys Arg Trp Lys Asp Ser Tyr Glu Trp Met Met Gly Thr Trp Leu Ser Arg Ile Ile Thr Gln Phe Asn Asn Ile Ser Tyr
          690                           700                           710                           720

CAA ATG TAT GAT TCT TTA AAT TAT CAG CGT GCA ATC AAA GCT AAA ATA GAT TTA GAA TAT AAA AAA TAT TCA GGA AGT GAT AAA GAA AAT ATA AAA AGT CAA GTT GAA AAT TTA AAA
Gln Met Tyr Asp Ser Leu Asn Tyr Gln Arg Ala Ile Lys Ala Lys Ile Asp Leu Glu Tyr Lys Lys Tyr Ser Gly Ser Asp Lys Glu Asn Ile Lys Ser Gln Val Glu Asn Leu Lys
          730                           740                           750                           760

AAT AGT TTA GAT GTA AAA ATT TCG GAA GCA ATG AAT AAT ATA AAT AAT TTT ATA CGA GAA TGT TCC GTA ACA TAT TTA TTT AAA AAT ATG TTA CCT AAA GTA TTT GAT GAA TTA AAT GAG
Asn Ser Leu Asp Val Lys Ile Ser Glu Ala Met Asn Asn Ile Asn Asn Phe Ile Arg Glu Cys Ser Val Thr Tyr Leu Phe Lys Asn Met Leu Pro Lys Val Ile Asp Glu Leu Asn Glu
          770                           780                           790                           800

TTT GAT CGA AAT ACT AAA GCA AAA TTA TAT AAT CTT ATA GAT AGT CAT AAT ATT ATT CTA GTT GGT GAA GTA GAT AAA TTA AAA GCA AAA GTA AAT AAT AGC TTT CAA AAT ACA ATA CCC
Phe Asp Arg Asn Thr Lys Ala Lys Leu Tyr Asn Leu Ile Asp Ser His Asn Ile Ile Leu Val Gly Glu Val Asp Lys Leu Lys Ala Lys Val Asn Asn Ser Phe Gln Asn Thr Ile Pro
          810                           820                           830                           840

TTT AAT ATT TTT TCA TAT ACT AAT AAT TCT TTA TTA AAA GAT ATA AAT GAA TAT TTC AAT AAT ATT AAT GAT TCA AAA ATT TTG AGC CTA CAA AAC AGA AAA AAT ACT TTA GTG GAT
Phe Asn Ile Phe Ser Tyr Thr Asn Asn Ser Leu Leu Lys Asp Ile Asn Glu Tyr Phe Asn Asn Ile Asn Asp Ser Lys Ile Leu Ser Leu Gln Asn Arg Lys Asn Thr Leu Val Asp
          850                           860                           870                           880

ACA TCA GGA TAT AAT GCA GAA GTG AGT GAA GAA GGC GAT GTT CAG CTT ATT CCA ATA TTT CCA TTT GAC TTT AAA TTA GGT AGT TCA GGG GAG AGA GGT AAA GTT ATA GTA ACC CAG
Thr Ser Gly Tyr Asn Ala Glu Val Ser Glu Glu Gly Asp Val Gln Leu Ile Pro Ile Phe Pro Phe Asp Phe Lys Leu Gly Ser Ser Gly Glu Arg Gly Lys Val Ile Val Thr Gln
          890                           900                           910                           920

AAT GAA AAT ATT GTA TAT AAT TCT ATG TAT GAA AGT TTT AGC ATT ATT TGG ATT AGA ATA AAT AAA TGG GTA AGT AAT TTA CCT GGA TAT ACT ATT GAT AGT TTT AAA AAT AAC
Asn Glu Asn Ile Val Tyr Asn Ser Met Tyr Glu Ser Phe Ser Ile Ile Trp Ile Arg Ile Asn Lys Trp Val Ser Asn Leu Pro Gly Tyr Thr Ile Ile Asp Ser Val Lys Asn Asn
          930                           940                           950                           960

TCA GGT TGG AGT ATA GGT ATT ATT AGT AAT TTT TTA GTA TTT ACT TTA AAA CAA AAT GAT AGT GAA CAA AGT ATA TTT AGT TAT GAT TCA AAT AAT GCT CCT GGA TAC AAT
Ser Gly Trp Ser Ile Gly Ile Ile Ser Asn Phe Leu Val Phe Thr Leu Lys Gln Asn Asp Ser Glu Gln Ser Ile Asn Phe Ser Tyr Asp Ile Ser Asn Asn Ala Pro Gly Tyr Asn
          970                           980                           990                           1000

AAA TGG TTT TTT GTA ACT GTT ACT AAC AAT ATG ATG GGA AAT ATG AAG ATT TAT ATA AGT GGA AAA TTA ATA GAT ACT AAA GTT AAA GAA CTA ACT GGA ATT AAT TTT AGC AAA ACT
Lys Trp Phe Phe Val Thr Val Thr Asn Asn Met Met Gly Asn Met Lys Ile Tyr Ile Asn Gly Lys Leu Ile Asp Thr Ile Lys Val Lys Glu Leu Thr Gly Ile Asn Phe Ser Lys Thr
          1010                          1020                          1030                          1040

ATA ACA TTT GAA ATA AAT AAA ATT CCA GAT ACC GGT TTG ATT ACT TCA GAT TCT GAT AAC ATC AAT ATG TGG ATA AGA GAT TTT TAT ATA TTT GCT AAA GAA TTA GAT GGT AAA GAT ATT
Ile Thr Phe Glu Ile Asn Lys Ile Pro Asp Thr Gly Leu Ile Thr Ser Asp Ser Asp Asn Ile Asn Met Trp Ile Arg Asp Phe Tyr Ile Phe Ala Lys Glu Leu Asp Gly Lys Asp Ile
          1050                          1060                          1070                          1080

AAT ATA TTA TTT AAT AGC TTG CAA TAT ACT AAT GTT GTA AAA GAT TAT TGG GGA AAT GTT TTA AGA TAT AAT AAA GAA TAT TAT ATG GTT AAT GTA GAT TAT TTA AAT AGA TAT ATG TAT
Asn Ile Leu Phe Asn Ser Leu Gln Tyr Thr Asn Val Val Lys Asp Tyr Trp Gly Asn Val Leu Arg Tyr Asn Lys Glu Tyr Tyr Met Val Asn Val Asp Tyr Leu Asn Arg Tyr Met Tyr
          1090                          1100                          1110                          1120

GCG AAT TCA CGA CAA ATT GTT TTT AAT ACA CGT AGA AAT AAT GAC TTC AAT GAA GGA TAT AAA ATT ATA ATA AAA AGA ATC AGA GGA AAT ACA AAT GAT ACT AGA GTA CGA GGA GGA
Ala Asn Ser Arg Gln Ile Val Phe Asn Thr Arg Arg Asn Asn Asp Phe Asn Glu Gly Tyr Lys Ile Ile Ile Lys Arg Ile Arg Gly Asn Thr Asn Asp Thr Arg Val Arg Gly Gly
          1130                          1140                          1150                          1160

GAT ATT TTA TAT TTT GAT ATG ACA ATT AAT AAC AAA GCA TAT AAT TTG TTT ATG AAG AAT GAA ACT ATG TAT GCA GAT AAT CAT AGT ACT GAA GAT ATA TAT GCT ATA GGT TTA AGA GAA
Asp Ile Leu Tyr Phe Asp Met Thr Ile Asn Asn Lys Ala Tyr Asn Leu Phe Met Lys Asn Glu Thr Met Tyr Ala Asp Asn His Ser Thr Glu Asp Ile Tyr Ala Ile Gly Leu Arg Glu
          1170                          1180                          1190                          1200

CAA ACA AAG GAT ATA AAT GAT AAT ATT ATA TTT CAA ATA CAA CCA ATG AAT AAT ACT TAT TAT TAC CGA TCT CAA ATA TTT AAA TCA AAT TTT AAT GGA GAA AAT TCT GGA ATA TGT
Gln Thr Lys Asp Ile Asn Asp Asn Ile Ile Phe Gln Ile Gln Pro Met Asn Asn Thr Tyr Tyr Tyr Ala Ser Gln Ile Phe Lys Ser Asn Phe Asn Gly Glu Asn Ile Ser Gly Ile Cys
          1210                          1220                          1230                          1240

TCA ATA GGT ACT TAT CGT TTT AGA CTT GGA GGT GAT TGG TAT AGA CAC AAT TAT TTG GTG CCT ACT GTG AAG CAA GGA AAT TAT GCT TCA TTA TTA GAA TCA ACA ACT CAT TGG GGT
Ser Ile Gly Thr Tyr Arg Phe Arg Leu Gly Gly Asp Trp Tyr Arg His Asn Tyr Leu Val Pro Thr Val Lys Gln Gly Asn Tyr Ala Ser Leu Leu Glu Ser Thr Ser Thr His Trp Gly
          1250                          1260                          1270                          1280

TTT GTA CCT GTA AGT GAA TAAATAATGATTAATAATATAAATTATGTTAAATATTTTAATATTAAGTTGAAATATGTTTTGATTAGTCGAACATTAATAAAATAACTAAATAAATTATATTTAAAACTGTTAAAATAGTATATTTATTATCAA
Phe Val Pro Val Ser Glu ***
          1290
```

TTACAGTAGTAAAATACTGATTGCTCTGTCTATAATCTTCCAGAGTTAAAAAAGAACGATGTTTAATTTTTAGAATTTATTAGGTAAAAGAGATACAGTTAGCCGCACTAAAAAATTAAATTAAATTATTTTTAAAAAGTTATTGACATTAGAAAAGTATGG

TAGTATACATTATAAATGTCGAAAGGCAGATAAAGCAAATAAATCTTAGTACAAAATTAAATTAGAAAAATATTAAAGCGAGGATTAAATATGGAACACGATCAAGATT

Figure 5 - CONTINUED

partially purified from the *E. coli* cells harbouring HA-33 gene showed hemagglutinating activity. Therefore, it was concluded that HA of type C 16S toxin consisted of several subcomponents, and that HA-33 subcomponent was important for demonstrating the hemagglutinating activity[7].

The HA-33 gene was located 4.3 kb upstream of the neurotoxin gene. The nucleotide sequence of the DNA between these two genes was determined. It was found that this region contained one ORF coding for 1196 amino acid residues (Fig. 5). The N-terminal amino acid sequence deduced from this nucleotide sequence was identical to that of nontoxic non-HA component determined by the protein microsequencing procedure. The calculated Mr. of this gene product was 138,758, which was in good agreement of 130 kDa[8]. Therefore, it was concluded that this ORF was a gene for the nontoxic-nonHA component. The number of nucleotide bases between the neurotoxin and HA-33 genes, which had previously estimated as 4.3 kb, was found to be 3873.

In type E toxin, almost all of the nucleotide sequences of the both genes coding for the neurotoxin and nontoxic-nonHA component have been determined[18].

Transcription and Translation of The Genes in The Organisms of *C. botulinum* and *E. coli*. In type C, the *E. coli* cells harbouring the recombinant plasmid pCL8 produced N-terminus of the toxin (65 kDa), and the proteins of HA-33 and HA-17 (but did not produce the nontoxic-nonHA component). The *E. coli* cells harbouring the pCH3 produced C-terminus of the toxin (74 kDa). The orientation of the inserted DNA fragment in pCL8 did not affect the expression of the protein. In contrast, the inserted fragment in pCH3 lost its ability to produce the protein when its orientation was changed in plasmid pUC118. These results indicated that the gene encoded in pCL8 was transcribed independently of the *lac* promoter of the plasmid, whereas, the gene in pCH3 was transcribed with the *lacZ* gene[16].

The nontoxic non-HA gene was located 267 bases upstream of the HA-33 gene to 17 bases upstream of the neurotoxin gene. In the 5' noncoding region of the nontoxic non-HA gene, the sequences predicted as a Shine-Dalgarno (SD) and promoter (-10 and -35) could be detected. Also, in the 5' noncoding region of the HA-33 gene, these predicted regulatory sequences were identified. Whereas, in the 17 bases between the nontoxic non-HA and neurotoxin genes, and in the 65 bases between HA-33 and HA-17 genes, only SD sequences could be found. From these results, it was suggested that the nontoxic non-HA and neurotoxin genes, as in the same way as the HA-33 and HA-17 genes, were transcribed by the same mRNA (polycistronic transcription) in the *C. botulinum* organisms (Fig. 3).

In the *E. coli* cells harbouring pCL8 produced both the N-terminus of toxin and the subcomponents of HA, but production of the nontoxic-nonHA component was not detected by a Western blot test as mentioned above. These data suggested that the predicted nontoxic-nonHA promoter gene might not work in the *E. coli* cells but some nucleotide region in the nontoxic-nonHA gene might work as a promoter for the neurotoxin gene; a candidate for such a region is shown in Fig. 5. Downstream of the HA-17 gene, there existed another ORF which was presumed to code for another subcomponent of HA.

In type E case, the nontoxic-nonHA component was produced in *E. coli*, and the results of Western blot test with deleted gene fragments indicated that nontoxic-nonHA and neurotoxin genes were transcribed in the same direction (Fig. 4). Since 1) the sequences predicted as -10 promoter and SD were identified in the 5'-untranslated regions of both nontoxic-nonHA and neurotoxin genes (but no -35 promoter-like sequence was found), and 2) the *Eco*RI 1 kb fragment in pUC 118 plasmid (designated as pEM-L1) was able to be transcribed to produce the protein (N-terminus of the toxin) even if its orientation in the plasmid was changed, it was presumed that these two genes were transcribed independently in the *E. coli* organisms.

Structure and Function of Toxins.   In the nucleotide and amino acid sequences of botulinum type A, C, D, E, and tetanus toxins, there existed many highly homologous regions (Fig. 6); these toxins showed approximately 30 to 50% homology to each other. One of the highly conserved region of L chain was hydrophobic, and all of toxins possessed three His residues at the same positions of this region (Fig. 7).   Binz *et al.* suggested that this region might play the important role either in the internalization of the toxin or in the release of acetylcholine by forming a α-helical structure.   We obtained one mAb (NCL-31) reacting with the site which was unique to type C toxin but very adjacent to this hydrophobic region (Fig. 6).   This mAb did not inhibit the toxin-binding to the synaptosome or NG-108 neuronal cells, but neutralized the toxin-lethal activity in mice.   Therefore, it was suggested that NCL-31 neutralized the type C toxin-lethal activity by changing the three-dimensional structure of this highly homologous region.

We also established four mAbs recognizing the C-terminus (less than 50 amino acid

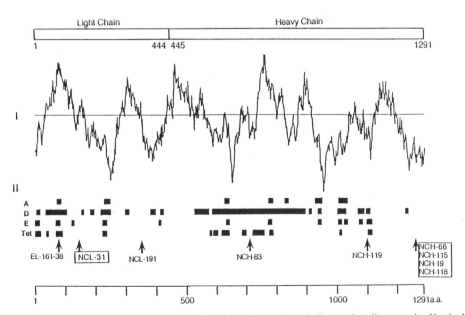

Figure 6. Feature of the botulinum type C toxin. Upper box indicates botulinum toxin. Vertical arrows indicate the recognition site of the monoclonal antibodies. The neutralizing antibodies are boxed. I:Running average of the hydrophilicity of the toxin (window size of 51 amino acids). II:Highly homologous regions (more than 70%) compared with botulinum type A (A)[22,23] type D (D)[24], type E (E)[25,26] and tetanus (Tet)[27,28] toxins.

Type of toxin	Position	Amino acid sequence
		* * *
Butyricum ※ (BL6340)	205-223	D P A L T L M H E L I H S L H G L Y G
Botulinum type E (Mashike)	205-223	D P A L T L M H E L I H S L H G L Y G
botulinum type A (62A)	216-234	D P A V T L M H E L I H A G H R N Y G
botulinum type C1(c-st)	222-240	D P I L I L M H E L N H A M H N L Y G
botulinum type D	222-240	D P V I A L M H E L T H S L H Q L Y G
Tetanus	226-244	D P A L L L M H E L I H V L H G L Y G

Figure 7. One of the highly homologous regions in the L chain of *C. botulinum* type A[22,23], C[15], D[24], E[18] and tetanus[27,28] toxins is shown. All toxins possess three His residues at the same positions of this region.

※ type E toxin[18] produced by a *C. butyricum* strain isolated from infant botulism in Italy.

residues) of type C neurotoxin. These mAbs inhibited the toxin-binding to the target cells and neutralized the toxin-lethal activity in mice. Therefore, it was concluded that C-terminal region which was unique to type C toxin was essential for the binding of type C toxin to the receptor. The low homology in the binding domain is well coincident with the previous studies that the binding efficiency of botulinum toxins was not antagonized by the other types of botulinum toxins [20,21].

REFERENCES

1. G. Sakaguchi, S. Kozaki, and I. Ohishi, Structure and function of botulinum toxins., In: "Bacterial Protein Toxins, "J.E. Alouf, F.J. Fehrenbach, J.H. Freer, and J. Jeliaszewicz, eds., Academic press, London(1984)

2. N. Suzuki, B. Syuto, and S. Kubo, Purification and characterization of hemagglutinin of *Clostridium botulinum* type C strain Stockholm., Jpn.J.Vet.Res.34:269-278 (1986)

3. I. Ohishi, S. Sugii, and G. Sakaguchi, Oral toxicities of *Clostridium botulinum* toxins in response to molecular size., Infect.Immun. 16:107-109(1977)

4. K. Oguma, H. Iida, M. Shiozaki, and K. Inoue, Antigenicity of converting phages obtained from *Clostridium botulinum* type C1 and D., Infect.Immun. 13:855-860(1976)

5. K. Oguma, H. Iida, and M. Shiozaki, Phage conversion to hemagglutinin production in *Clostridium botulinum* types C and D., Infect.Immun. 14:597-602(1976)

6. B. Syuto, and S. Kubo, Isolation and molecular size of *Clostridium botulinum* type C toxin., Appl.Environ.Microbiol. 33:400-405(1977)

7. K. Tsuzuki, K. Kimura, N. Fujii, N. Yokosawa, T. Indoh, T. Murakami, and K. Oguma, Cloning and complete nucleotide sequence of the gene for the main component of hemagglutinin produced by *Clostridium botulinum* type C., Infect.Immun. 58:3173-3177(1990)

8. K. Tsuzuki, K. Kimura, N. Fujii, N. Yokosawa, and K. Oguma, The complete nucleotide sequence of the gene coding for the nontoxic-nonhemagglutinin component of *Clostridium botulinum* type C progenitor toxin., Biochem.Biophys. Res.Commun. 183:1273-1279(1992)

9. N. Yokosawa, K. Tsuzuki, B. Syuto, and K. Oguma, Activation of *Clostridium botulinum* type E toxin purified by two different procedures., J.Gen.Microbiol. 132:1981-1988(1986)

10. K. Oguma, B. Syuto, H. Iida, and S. Kubo, Antigenic similarity of toxins produced by *Clostridium botulinum* types C and D strains., Infect.Immun. 30:656-660(1980)

11. K. Tsuzuki, N. Yokosawa, B. Syuto, I. Ohishi, N. Fujii, K. Kimura, and K. Oguma, Establishment of a monoclonal antibody recognizing an antigenic site common to *Clostridium botulinum* type B, C1, D, and E toxins and tetanus toxin., Infect. Immun. 56:898-902(1988)

12. K. Oguma, S. Murayama, B. Syuto, H. Iida, and S. Kubo, Analysis of antigenicity of *Clostridium botulinum* type C1 and D toxins by polyclonal and monoclonal antibodies., Infect. Immun. 43:584-588(1984)

13. V. Sathyamoorthy, and B.R. DasGupta, Separation, purification, partial characterization and comparison of the heavy and light chains of botulinum neurotoxin types A, B, and E., J.Biol.Chem. 260:10461-10466(1985)

14. N. Fujii, K. Oguma, N. Yokosawa, K. Kimura, and K. Tsuzuki, Characterization of bacteriophage nucleic acids obtained from *Clostridium botulinum* types C and D., Appl.Environ.Microbiol. 54:69-73(1988)

15. K. Kimura, N. Fujii, K. Tsuzuki, T. Murakami, T. Indoh, N. Yokosawa, K. Takeshi, B. Syuto, and K. Oguma, The complete nucleotide sequence of the gene coding for botulinum type C1 toxin in the c-st phage genome., Biochem.Biophys.Res. Commun. 171:1304-1311(1990)

16. K. Kimura, N. Fujii, K. Tsuzuki, T. Murakami, T. Indoh, N. Yokosawa, and K. Oguma, Cloning of the structural gene for *Clostridium botulinum* type C1 toxin and whole nucleotide sequence of its light chain component., Appl. Environ. Microbiol. 57:1168-1172(1991)

17. N. Fujii, K. Kimura, T. Murakami, T. Indoh, T. Yashiki, K. Tsuzuki, N. Yokosawa, and K. Oguma, The nucleotide and deduced amino acid sequences of *Eco*RI fragment containing the 5'-terminal region of *Clostridium botulinum* type E toxin gene cloned from Mashike, Iwanai and Otaru strains., Microbiol.Immunol. 34:1041-1047 (1990)

18. N. Fujii, K. Kimura, T. Yashiki, K. Tsuzuki, K. Moriishi, N. Yokosawa, B. Syuto, and K. Oguma, Cloning and whole nucleotide sequence of the gene for the light chain component of botulinum type E toxin from *Clostridium butyricum* strain BL6340 and *Clostridium botulinum* type E strain Mashike., Microbiol.Immunol. 36:213-220(1992)

19. N. Yokosawa, Y. Kurokawa, K. Tsuzuki, B. Syuto, N. Fujii, K. Kimura, and K. Oguma, Binding of *Clostridium botulinum* type C neurotoxin to different neuroblastoma cell lines., Infect.Immun. 57:272-277(1989)

20. N. Yokosawa, K. Tsuzuki, B. Syuto, N. Fujii, K. Kimura, and K. Oguma, Binding of botulinum type C1, D and E neurotoxins to neuronal cell lines and synaptosomes., Toxicon. 29:261-264(1991)

21. S. Kozaki, Interaction of botulinum type A, B and E derivative toxins with synaptosomes of rat brain., Naunyn-Schmiedergs Arch.Pharmacol. 308:67-70(1979)

22. T. Binz, H. Kurazono, M. Wille, J. Frevert, K. Wernars, and H. Niemann, The complete sequence of botulinum neurotoxin type A and comparison with other Clostridial neurotoxins., J.Biol.Chem. 265:9153-9158(1990)

23. D.E. Thompson, J.K. Brehm, J.D. Oultram, T.J. Swinfield, C.C. Shone, T. Atkinson, J. Melling, and N.P. Minton, The complete amino acid sequence of the Clostridium botulinum type A neurotoxin, deduced by nucleotide sequence analysis of the encoding gene., Eur.J.Biochem. 189:73-81(1990)

24. T. Binz, H. Kurazono, M.R. Popoff, M.W. Eklund, G. Sakaguchi, S. Kozaki, K. Krieglstein, A. Henschen, D.M. Gill, and H. Niemann, Nucleotide sequence of the gene encoding Clostridium botulinum neurotoxin type D., Nucleic Acids Res. 18:5556(1990)

25. S.M. Whelan, M.J. Elmore, N.J. Bodsworth, T. Atkinson, and N.P. Minton, The complete amino acid sequence of the Clostridium botulinum type-E neurotoxin, derived by nucleotide-sequence analysis of the encoding gene., Eur.J.Biochem. 204:657-667(1992)

26. S. Poulet, D. Hauser, M. Quanz, H. Niemann, and M.R. Popoff, Sequences of the botulinal neurotoxin E derived from Clostridium botulinum type E (strain Beluga) and Clostridium butyricum (strains ATCC43181 and ATCC 43755)., Biochem.Biophys.Res.Commun. 183:107-113(1992)

27. U. Eisel, W. Jarausch, K. Goretzki, A. Henschen, J. Engels, U. Weller, M. Hudel, E. Habermann, and H. Niemann, Tetanus toxin: primary structure, expression in E. coli, and homology with botulinum toxins., Embo J. 5:2495-2502(1986)

28. N.F. Fairweather, and V.A. Lyness, The complete nucleotide sequence of tetanus toxin., Nucleic Acids Res. 14:7809-7812(1986)

SEROLOGICAL SUBTYPES OF BOTULINAL NEUROTOXINS

Domingo F. Giménez,[1] Juan A. Giménez[2,3]

[1]Calle 13, Nro 1069, Miramar 7607, Pcia. Bs. As., Argentina
[2]Food Research Institute, University of Wisconsin-Madison
[3]Current address: Biological Technical Services
                    Merck Manufacturing Division
                    Merck & Co., Inc.
                    WP 28B-231
                    West Point, PA 19486

INTRODUCTION

The taxonomy of botulinal bacteria, as responsible for synthesizing botulinal neurotoxin (BoNT), embraces three different but interrelated subjects: (1) the systematic of *Clostridium botulinum* and its physiologic and genetic related bacteria;[1] (2) other clostridial species producing BoNT,[2,3,4,5] and (3) the BoNT. This toxin has a unique pharmacological action, the blockade of the neurotransmitter acetylcholine release from the nerve terminal.[6]

BoNTs are classified in serological types based on neutralization test with antiserum. Although only seven antigenic types of BoNT are known, establishing their classification has been—and it is today as shown by this presentation—a matter of discussion. These matters are primarily concerned to the empirical use of the terms *type* and *subtype*, making these taxa sometimes meaningless. We think the lack of (1) a system of definitions of the different taxa involved and (2) a reliable method for the serological identification and typing, have been the main sources of disagreement. Besides, the advances reported in the last decade on biochemical and antigenic makeup of the BoNTs, and the description of other clostridial species producers of BoNT, make necessary a reordered classification.

The objective of this presentation is to establish the taxon *subtype* in the serological classification of BoNTs. For this purpose, we will examine the behavior of BoNTs in quantitative neutralization tests and redefine the taxa *type*, *subtype* and *intratypic serological variant*.[7] Finally, we will discuss the significance of BoNT subtypes of *C. botulinum* in the fields of systematic, epidemiology, and laboratory diagnosis.

TRACING THE BoNT SUBTYPES

In 1965 the strain 84 of *C. botulinum*, isolated from a vineyard soil sample of Mendoza province, Argentina, was tentatively classified as a subtype of BoNT type A since 20 fifty per cent lethal doses (LD_{50}) but not 200 LD_{50} of its toxin was neutralized by type A antitoxin, whereas its antitoxin neutralized BoNT type A at any of those test doses.[8] Later, it was demonstrated that the aberrant behavior of 84 toxin was due to the presence of a small amount of F toxin showing, for the first time, that a *C. botulinum* cell can produce more than one toxin type.[9]

Botulinum and Tetanus Neurotoxins, Edited by
B.R. DasGupta, Plenum Press, New York, 1993

In 1981 an atypical toxin variant of *C. botulinum* type B (strain 657) associated with infant botulism was reported.[10] Later, we demonstrated that strain B 657 produces a mixture of type B and A BoNTs.[11,12]

The third subtype was isolated in France from a sample of home made meat pie responsible for a case of food borne botulism.[13] Described as type AB, it is yet discussed if it corresponds to a BoNT subtype Ab[13] or subtype Ba.[14]

Finally, the fourth recognized subtype until now was isolated from a feces sample of a case of infant botulism.[15] It corresponds to a complex BoNT—subtype Bf—where type F forms again a toxin mixture as a minor component of a BoNT subtype.

CLOSTRIDIAL SPECIES PRODUCING BoNTs

C. botulinum is a conglomerate of different bacteria where aberrance rather than homology is the rule. The most comprehensive study on cultural and biochemical variations of this and related species was carried out on 58 substrates and metabolic derivative products.[1] These results, together with a precedent proposed grouping,[16] classified these heterogeneous bacteria in three metabolic groups: group I (*C. botulinum* type A, proteolytic strains of *C. botulinum* types B and F, and *C. sporogenes*); group II (*C. botulinum* type E and non-proteolytic strains of *C. botulinum* types B and F, and culturally similar non-toxic organisms), and group III (*C. botulinum* types C and D, and *C. novyi* type A). A fourth group of *C. botulinum* was later proposed for type G.[17]

Table 1. Serological types, subtypes and intratypic variants of BoNTs produced by five clostridial species.

Clostridial species	Physiological groups	Type	Subtype	Intratypic variants (references)
C. botulinum	I	A		(42)
			Af	
			Ab	
		B		(43) (44)
			Ba	
			Bf	
		F		(7)
	II	E, F, B		
	III	C, D		(Discussion in the text)
C. argentinense		G		(45)[a]
C. baratii		F		
C. butyricum		E		
C. novyi type A	(experimental)	C, D		

[a]By fluorescent antibody method

Studies carried out on (1) genetics of toxigenic and non-toxigenic strains of *C. botulinum* [18] and the relatedness of proteolytic strains,[19] (2) cell wall sugar components[20] (3) *r*RNA homology,[21] (4) the interconversion of type C and D strains,[22] (5) the interspecies conversion of *C. botulinum* type C to *C. novyi* type A by bacteriophage,[2] and (6) somatic antigens of proteolytic and non-proteolytic strains[23,24] have widely sustained the physiologic group classification.

The most striking proof of the aberrance of bacteria producing BoNTs was the isolation from a case of infant botulism in the U.S.A. of a type F toxigenic strain that belongs phenotypically to *C. baratii*.[3] The following year, from another clinical case of infant botulism recorded in Italy, a type E toxigenic strain of *C. butyricum* was identified.[4] The genetic confirmation of identity of those strains as *C. baratii* and *C. butyricum* was reported.[25] The molecular parameters of their toxins have been compared with those of known BoNTs.[26,27,28,29,30]

Finally, *C. argentinense* was proposed as the toxigenic organism producing BoNT type G.[5] This new species embraces a genetically homogeneous group composed of all strains producing BoNT type G and some non-toxigenic strains previously identified as *C. subterminale* and *C. hastiforme*.[5]

Because of the description of these non-*C. botulinum* species producing BoNTs, the Smith and Holdeman grouping has a renewed importance. In fact, the current classification by which toxigenic *C. botulinum* is nominated by the BoNT type it produces, seems inadequate. *C. botulinum* should be classified into the respective physiological group whereas the taxon type should be restricted to the serological classification of the BoNTs (Table 1).

SEROLOGICAL BASES FOR IDENTIFICATION AND TYPING OF BoNTs

To summarize the different aspects involved in the identification and typing of BoNTs,[7,31] we will refer to the following definitions and recommendations.

General Principles

(1) Test level: doses of toxin and antitoxin in a neutralization assay.

(2) The serological method selected is the in vivo neutralization test using mice as experimental animals. Once inoculated, animals should be carefully observed. Delayed time-to-death at a given test level indicates either (a) an incomplete neutralization of the homologous toxin of the system or (b) the activity of a small amount of a heterologous toxin, undetectable in the next lower test level.

(3) The serological method is quantitative. That is: (a) the BoNT activity should be estimated by mouse toxicity test and expressed in LD_{50}/ml; (b) specificity, avidity, and potency of antitoxins must be known. Potency of antitoxins should be estimated at a test level of 2,000 LD_{50} and the titers expressed in anti-fifty per cent lethal doses per ml (a-LD_{50}/ml), and (c) neutralization tests should be done at the following three doses of toxin: 20, 200, and 2,000 LD_{50}s.

Antitoxins

(1) The quality of an antitoxin is measured by its avidity or efficiency and is expressed in terms of an index.[31,32] The efficiency index (EI) is determined by the proportional amount of antitoxin required to neutralize preferably no less than three toxin doses (Table 2).

(2) The study of BoNT types A and B in toxin-antitoxin tests showed that antitoxins are progressively less efficient in combining with toxin at higher dilutions,[32] phenomenon also recorded with other toxin-antitoxin systems as types D[31] and F.[7] Only in types C and E systems the neutralization takes place nearly at constant proportions.[31]

(3) Every antitoxin, despite its source or place of preparation as well as indications given on the label on potency, expiration date, etc., should be analyzed for specificity, avidity and potency before its use.

DEFINITION OF EACH TAXON

The complex molecular structure of BoNTs reflects their complex behavior as antigens. The paradox of the BoNTs' specificity for the myoneural end plate versus the diversity of their antigenic types suggests that the homologous peptide(s) responsible for the pathogenesis has a secondary role as immunogen.[33,34,35]

When the antigenic makeup of BoNTs is studied through monoclonal antibodies most of the serotypes, and some other non-botulinal toxins, share common epitopes.[36,37] This is reflected in the neutralization test with polyclonal antibodies in which some degree of cross-neutralization is known between types D and E,[31] and E and F.[38,39] Instead, types A, B, and G are monospecific.[31,40,41] The antigenic relationship between types C and D will be discussed later.

Table 2. Cross-neutralization tests of three type F BoNTs and their antitoxins[7].

Toxin		Antitoxin (a-LD$_{50}$)								
		160			90SL			Langeland		
Strain	TD[1] (LD$_{50}$)	ND[2]	RC[3]	EI[4]	ND	RC	EI	ND	RC	EI
160	2,000	2,000	1	8.5	3,000	1.5	6.3	15,000	7.5	6.6
	200	450	1	3.8	500	2.0	3.8	3,500	17.5	2.8
	20	170	1	1	190	3.8	1	1,000	20.0	1
90SL	2,000	11,200	5.6	6.7	2,000	1	2.5	14,700	7.3	6.8
	200	1,860	4.1	4.0	250	1	2.0	3,000	15.0	3.3
	20	750	4.4	1	50	1	1	1,000	20.0	1
Lange-land	2,000	122,200	61.0	6.4	126,000	63.0	6.1	2,000	1	2.5
	200	26,400	58.6	2.9	26,300	105.2	2.9	200	1	2.5
	20	7,800	46.0	1	7,800	156.0	1	50	1	1

[1]TD: test dose.

[2]ND: neutralizing dose. Amount of antitoxin, expressed in anti-LD$_{50}$, of the highest dilution (set up in 1.2 log progression) that protected a group of six mice during 96 h.

[3]RC: relative consumption of antitoxin. It was calculated as the quotient between the quantity of antitoxin required for the neutralization of the homologous toxin at a given level of test and the quantity required by each other toxin at the same level.

[3]EI: efficiency index. Is defined as the ratio of the quantity of antitoxin required to neutralize 20 LD$_{50}$ to the quantity actually required by 200 or 2,000 LD$_{50}$, corrected by the dilution factor.

Type

(1) This term is the basic taxonomic category from which other subordinate categories will be defined.

(2) The type is monovalent; it has one main distinct antigen borne by one molecule species. The valence is determined by (a) the behavior of each BoNT with its polyclonal antitoxin in the in vivo neutralization test and (b) the EI trend of the homologous antitoxin at test levels of 20, 200, and 2,000 LD$_{50}$s.

(3) The efficiency of a given antitoxin shows the same trend (i.e., EI increases at higher doses of toxin) with any toxin of the type at the same test levels, regardless of the amount required for the neutralization.

(4) Although at low toxin doses some degree of cross-neutralization between some types has been reported, the type is by definition integrated by only one distinct antigen.

Subtype

(1) It is the taxonomic category below the type.

(2) The subtype is bivalent. Its antigenic makeup is borne by two separate molecules. It is a mixture of two toxin types. The subtypes described so far are made up of different proportions of known toxin types. If both toxins were present in equal amount, its classification would be resolved (a) by trypsin activation, by which the subtype would

correspond to the activated toxin as the major antigen, or (b) by convention. Hypothetically, a subtype might be more than bivalent depending on the number of BoNTs of the mixture.

(3) The efficiency of a subtype antitoxin shows the same trend with both its homologous and their corresponding type toxins. On the contrary, the efficiency of the major type antitoxin shows the same trend with the subtype toxin within certain test levels, but abruptly falls at the level where the minor antigen emerges.

Intratypic Serological Variant

(1) Serological variants are within a type[7,42,43,44,45] or, hypothetically, within a subtype.

(2) Variants show quantitative differences in their neutralization with a given antitoxin of a group. The efficiency of the antitoxins shows the same trend with each other toxin of the group (Table 2).

(3) The quantitative serological variations in a group of toxins are expressed in terms of degrees of serological homology.

Degree of Serological Homology

(1) It is defined upon Bulatova's "multiplicity index,"[46] representing the relative similarity among a group of toxins, estimated through the neutralization capacity of every antitoxin against each other toxin of the group.

(2) The degree of homology, expressed in percentage, is calculated as the quotient (multiplied by 100) between the quantity of antitoxin required for the neutralization of the homologous toxin (100%) and the quantity required by each other toxin of the group.

(3) The cross-neutralization tests should be set up preferably at a level of 2,000 LD_{50}.

BoNT Types C and D as Variants of Only One Serological Type

Some discrepancies still exist on the serological classification of types C and D,[47,48,49] and the most important facts are here summarized:

(1) The designation of types Cα and Cβ was proposed to differentiate two toxigenic strains of C. botulinum type C[50,51] for which no quantitative serology was used in the cross-neutralization tests. Yet these results, for the first time, pointed out serological differences within a type.

(2) C1, C2 and D were described as the antigenic components of BoNTs types C and D.[40] Later it was showed that the toxins of strains Cα, Cβ and D were made up of the principal antigen corresponding to such designations plus the other fractions in a lesser amount[46] (Fig. 1).

(3) The toxic fractions of BoNT types C and D were determined by monospecific antisera, showing that type Cα produces factors C1, C2 and D; Cβ produces only factor C2 and type D produces factors C1 and D.[52]

(4) C2 toxin was considered a BoNT until it was proved that it is a vascular, lethal and enterotoxic toxin with a structure different to BoNTs.[49,53] Therefore, the permanence of C2 toxin in the classification may induce the misleading concept that both C1 and C2 toxins are equal in structure, origin, and activity.

(5) For the above reason, and others related to serologic variations[54,55] and genetics[47] of BoNT C and D, C2 toxin as well as the terms Cα and Cβ should not be included in the serological classification of types C and D.

Yet, the Cα and Cβ classification of types C and D[46] was, twenty years ago, useful for a comparative study with serological variants of BoNT type F.[7] When the results obtained in cross-neutralization tests of three type F BoNTs and their antitoxins (Table 2) were translated into degrees of serological homology (Fig. 2) and compared with types C and D (Fig. 1), the similarity of both projections was apparent. On this basis, we hypothesized that BoNT types C and D of those experiments might belong to serological variants of only one type, providing that the principle of efficiency of antitoxins C and D were fulfilled in the cross-neutralization tests.[7] We never had the opportunity to test our hypothesis.

However, recent studies on BoNT types C and D proved overlap of antigenic specificities because of the presence of common epitopes in the molecules of both types.[55,56,57,58,59,60] Therefore, it seems that BoNT types C and D would represent extremes of an intermediate graded series of serological variants.

If this were the case, some BoNT types C and D could be at a critical serological distance between *variant* and *type*. Consequently, *type* should be differentiated from *variant* by establishing the quantitative limit of antitoxin consumption in the cross-neutralization test.

Yet some BoNT types C and D proved to be serological variant of one type, it would be difficult to conciliate differences between both toxins as agents of animal botulism. While type C is the most important pathogen of bird botulism, type D has not been proven as bird pathogen, either spontaneous or experimental.[61,62] Thus, experimental botulism in birds might be a reliable method for classifying some strains producing BoNT types C or D.

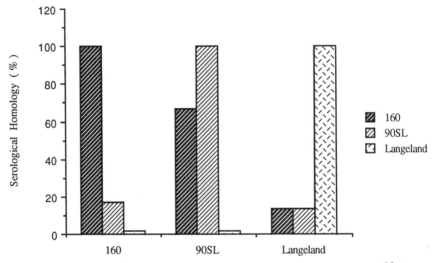

Fig. 1. Degree of serological homology among types Cα, Cβ, and D toxins.[46]

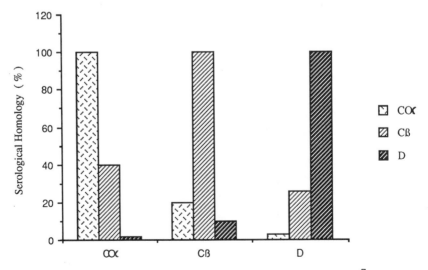

Fig. 2. Degree of serological homology among three type F toxins.[7]

SUBTYPES OF BoNTs

The taxon *subtype* has been used empirically in the serological classification of type C to categorize either strains Cα and Cβ[46,63,64] or two serological variants of C1 toxin.[57,59]

In a previous section we defined the taxon subtype in terms of its bivalent antigenic make up and the behavior of its antitoxin with the homologous and its respective type toxins. Until now four BoNT subtypes have been differentiated (Table 3).

Table 3. Some characteristics of the four known serological subtypes of BoNTs.

Subtype	Identified by	Major/minor toxin type (ratio)	Human botulism	Identified in	Patient outcome	Country
A f	Giménez and Ciccarelli (1966, 1970)	A/F (10:0.1 to 10:1)	Foodborne	Serum[1] Feces Soil	Died	Argentina
B a	Hatheway et al. (1981) Giménez (1984)	B/A (10:0.2 to 10:0.5)	Infant	Feces	Survived	U.S.A.
A b	Poumeyrol et al. (1983)	A/B (10:1)	Foodborne	Serum[1] Food	Died	France
B f	McCroskey and Hatheway (1984)	B/F (10:1)	Infant	Feces	Survived	U.S.A.

[1]Toxin in blood serum was preliminarily identified as type A.

Nomenclature

To name a mixture of two BoNT types at different proportions constituting a unity, it was selected a capital letter followed by a small letter.[7] This notation allowed (a) to indicate a derived antigenic structure from two parental toxin types; consequently, a subordinate taxonomic category, and (b) to emphasize the quantitative differences—major and minor—of both parental toxins. So, the use of the taxon *type* followed by capital letters for nominating a subtype[13,14] is inadequate.

Antigenic makeup of the BoNT subtypes

When the antigenic makeup of Af toxin was first recognized,[9] alternative possibilities of the molecular structure of the toxin were discussed. In fact, according to some observations the toxin behaved as a bimolecular (A plus F) mixture, but other results suggested a double specificity borne by only one molecule. It was demonstrated later that the antigenicity of Af toxin is due to the existence of separated molecules of type A and F specificities.[65]

The most conflicting BoNT described so far belongs to the *C. botulinum* strain B 657, isolated from feces samples of a case of infant botulism.[10] This toxin was described as an atypical variant of BoNT type B, and it was later proved to be a toxin with type B and A antigenicities.[11,12] Studies on the chemical structure of B 657 toxin showed that it behaves as a mixture of single (unnicked) and dichain (nicked) proteins of ~150 kDa, and the N-terminal amino acid sequence of the L (~50 kDa) and H (~100 kDa) chains matched exactly with those of L and H chains of type B, Okra strain.[66] Considering that no serological reactivity to type A toxin was recorded, these results showed indirectly that type B antigenicity of this subtype Ba is borne by one molecule species.

The second puzzling feature of B 657 toxin is the behavior of its homologous antitoxin. Lesser amount of B 657 antitoxin is required to neutralize a heterologous antigen (type B control) than its homologous.[10] This result was confirmed and it was also showed to be equally more specific for type A toxin (control) than for its homologous A fraction.[12]

About 70 times more B and A antitoxins are required for neutralizing B 657 B and A fractions than for A and B type control toxins, and B 657 antitoxin—the prototype of a paradoxical antitoxin—is 20 to 50 times more effective for neutralizing heterologous B and A toxin controls, respectively, than for its homologous toxin.[12] These facts suggest profound differences between B 657 and the standard type A and B BoNTs as antigens and immunogens.

The third subtype studied is the Ab toxin.[14] The purified progenitor toxin was made of both large (L) and medium (M) complexes. Upon trypsinization, the toxicity of M toxin increased mostly due to type B toxin, little to type A toxin. The authors identified four distinct fragments (H and L chains of types A and B) indicating that the activity of Ab toxin comes from two separates molecules of types A and B, but they did not rule out the possibility of a chimera toxin.[14]

There is a disagreement about the A:B ratio in this mixture. By in vivo neutralization test with crude toxin, 90 % of A and 10 % of B was estimated,[13] whereas working with either crude or purified trypsin activated toxin in the in vivo neutralization test and reversed passive latex agglutination, the A:B ratio was estimated at 20 and 80 %, respectively.[14]

Botulism by *C. botulinum* producers of BoNT subtypes

Affecting infant and adults, these bacteria have caused human botulism in three continents (Table 3). The low number of outbreaks and patients involved does not allow generalization on epidemiological or clinical aspects. Furthermore, it is likely that some other outbreaks, identified by its toxin type, might have been produced by subtypes.

The first clinical case recorded involved a 6 weeks old girl.[67] *C. botulinum* and BoNT type B were found in the stool. A sample of blood serum was negative. After 32 days hospitalization, the infant was discharged with a mild residual hypotonia. The organism isolated from the feces was reported as an atypical variant of type B.[10] Three other cases of infant botulism have been described in the U.S.A., two cases produced by subtype Bf and a second by an atypical B.[68,69]

Foodborne outbreaks of botulism involving subtype Ab and Af were recorded in France[13] and Argentina[70], with a sole patient in each outbreak.

In France, a 21 years old woman was affected by a severe botulism after the ingestion of rabbit meat pie. She died after 105 days of evolution because of a complication due to a nosocomial infection. In the first serum sample taken the second day of evolution BoNT type A was identified. Type A antitoxin was immediately given and a second serum sample was obtained 1.5 h later in which it was detected type B. The whole antitoxin treatment—A and B antitoxins—was completed. The strain of *C. botulinum* was isolated from the meat pie and identified as type AB. Feces of the patient were not examined.[13]

Finally, a fatal case of foodborne botulism presumably produced by the ingestion of home prepared pickled trout was recorded in Argentina.[70] The toxin detected in blood serum and feces was preliminarily identified as belonging to type A. The strain of *C. botulinum* isolated from feces produced BoNT subtype Af, a serotype that was isolated from soil samples 16 years before.[8]

DISCUSSION AND CONCLUDING REMARKS

The two cases of foodborne botulism in which *C. botulinum* subtypes Ab and Af were incriminated were characterized by a severe clinical course. In spite of adequate supportive treatments both patients died. In both cases, preliminary laboratory tests on blood serum samples identified only the major toxin of the subtype, an expected result considering the low concentration of toxin in the serum. In the French case, this fact delayed the administration of the second antitoxin. In the Argentine case, *C. botulinum* BoNT subtype Af was identified after the death of the patient, who had not received F antitoxin. Both cases illustrate the precarious aid of the laboratory for the immediate specific antitoxin treatment of the patient in cases of BoNT subtype botulism.

The resemblance of severity in clinical course between the French and the Argentine patients is striking. The very first question is if this fact is connected with the pharmacological mechanism of two associated BoNTs that may block the transmitter release at the end plate by more than one way,[6] potentiating the aggressive action of botulinal toxin. If this is proved, human botulism produced by *C. botulinum* BoNT subtypes may be expected to be of higher severity than that produced by only one BoNT type.

The use of BoNT subtypes as immunogens for medical use deserves a comment. As it has been proved with B 657 toxin-antitoxin systems, the antitoxin produced in rabbits by B 657 toxoid is 20- to 50-fold more active for the standard B toxin[10,12] and A toxin[12] types than their respective B and A antitoxins. If the two-toxin form is truly a better antigen that the individual type toxins, subtype toxoids would be useful for human and animal immunization.

The contradictory results from the study of B 657[10,12,66] imply, at first sight, (1) that the identification of a BoNT subtype can become an arduous task, (2) the apparent low sensitivity of the biochemical methods for detecting small amounts of a heterologous antigen, and (3) the need of an accurate serological method.

Botulinal neurotoxin subtypes appear as a challenging field for research: (a) immunochemical comparison with neurotoxin types; (b) the paradoxical "super antitoxin," that is quantitatively more active for the heterologous than for the homologous toxins; (c) the effect of associated toxic molecules in the pathogenic mechanism of the neurotransmitter release inhibition; (d) the search of subtypes in the environment to recognize distribution, new varieties, ecological matters, etc.; and (e) the genetics of BoNT subtype toxins in the context of the genetic of the parent toxins, are some of the questions to be answered.

REFERENCES

1. L.V. Holdeman and J.B. Brooks, Variation among strains of *Clostridium botulinum* and related clostridia, *in:* "Proceedings of the First U.S. Japan Conference on Toxic Microorganism," M. Herzberg, ed., U.S. Govt. Printing Office, Washington, D. C. (1970).
2. M.W. Eklund, F.T. Poysky, J.A. Meyers and G.A. Pelroy, Interspecies conversion of *Clostridium botulinum* type C to *Clostridium novyi* type A by bacteriophage, *Science.* 186:456 (1974).
3. J.D. Hall, L.M. McCroskey, B.J. Pincomb and C.L. Hatheway, Isolation of an organism resembling *Clostridium barati* which produces type F botulinal toxin from an infant with botulism, *J. Clin. Microbiol.* 21:654 (1985).
4. L.M. McCroskey, C.L. Hatheway, L. Fenicia, B. Pasolini and P. Aureli, Characterization of an organism that produces type E botulinal toxin but which resembles *Clostridium butyricum* from the feces of an infant with type E botulism, *J. Clin. Microbiol.* 23:201 (1986).
5. J.C. Suen, C.L. Hatheway, A.G. Steigerwalt and D.J. Brenner, *Clostridium argentinense* sp. nov.: a genetically homogeneous group composed of all strains of *Clostridium botulinum* toxin type G and some nontoxigenic strains previously identified as *Clostridium subterminale* or *Clostridium hastiforme, Intern. J. Syst. Bacteriol.* 38:375 (1988).
6. L.C. Sellin, Botulinum toxin and the blockade of transmitter release, *Asia Pacific J. Pharmacol.* 2:203 (1987).
7. D.F. Giménez and A.S. Ciccarelli, Variaciones antigénicas en toxinas del tipo F. Ensayo de definiciones para la tipificación serológica y clasificación de *Clostridium botulinum, Medicina (Bs Aires).* 32:596 (1972).
8. D.F. Giménez and A.S. Ciccarelli, A new type of *Cl. botulinum, in:* "Botulism 1966," M. Ingram and T.A. Roberts, ed., Chapman and Hall, London (1967).
9. D.F. Giménez and A.S. Ciccarelli, Studies on strain 84 of *Clostridium botulinum, Zbl. Bakt., I. Abt. Orig.* 215:212 (1970).
10. C.L. Hatheway, L.M. McCroskey, G.L. Lombard and V.R. Dowell Jr., Atypical toxin variant of *Clostridium botulinum* type B associated with infant botulism, *J. Clin. Microbiol.* 14:607 (1981).
11. D.F. Giménez and J.A. Giménez, Identificación de la cepa B 657 de *Clostridium botulinum, Rev. Argent. Microbiol.* 15:51 (1983).

12. D.F. Giménez, *Clostridium botulinum* subtype Ba, *Zbl. Bakt. Hyg.* A 257:68 (1984).
13. M. Poumeyrol, J. Billon, F. Delille, C. Haas, A. Marmonier and M. Sebald, Intoxication botulique mortelle due a une souche de *Clostridium botulinum* de type AB, *Med. Mal. Inf.* 13:750 (1983).
14. G. Sakaguchi, S. Sakaguchi, S. Kozaki and M. Takahashi, Purification and some properties of *Clostridium botulinum* type AB toxin, *FEMS Microbiol. Letters* 33:23 (1986).
15. L.M. McCroskey and C.L. Hatheway, Atypical strains of *Clostridium botulinum* isolated from specimens in infant botulism cases, *ASM Meet. (abs.).* C 159:263 (1984).
16. L.DS. Smith and L.V. Holdeman, "The Pathogenic Anaerobic Bacteria," C. Thomas, Springfield, Illinois (1968).
17. L.DS. Smith and G. Hobbs, *Clostridium, in:* "Bergey's Manual of Determinative Bacteriology," R.B. Buchanan and N.E. Gibbons, ed., Williams and Wilkins, Baltimore (1974).
18. W.H. Lee and H. Riemann, Correlation of toxic and non-toxic strains of *Clostridium botulinum* by DNA composition and homology, *J. Gen. Microbiol.* 60:117 (1970).
19. W.H. Lee and H. Riemann, The genetic relatedness of proteolytic *Clostridium botulinum* strains, *J. Gen. Microbiol.* 64:85 (1970).
20. C.S. Cummins, Cell wall composition in the classification of Gram positive anaerobes, *Intern. J. Syst. Bacteriol.* 20:413 (1970).
21. J.L. Johnson and B.S. Francis, Taxonomy of the clostridia: ribosomal ribonucleic acid homologies among the species, *J. Gen. Microbiol.* 88:229 (1975).
22. M.W. Eklund and F.T. Poysky, Interconversion of type C and D strains of *Clostridium botulinum* by specific bacteriophage, *Appl. Microbiol.* 27:251 (1974).
23. C.S. Gutteridge, B.M. Mackey and J.R. Norris, A pyrolysis gas-liquid chromatography study of *Clostridium botulinum* and related organisms, *J. Appl. Bacteriol.* 49:165 (1980).
24. H.M. Solomon, R.K. Lynt, D.A. Kautter and T. Lilly, Antigenic relationship among the proteolytic and nonproteolytic strains of *Clostridium botulinum*, *Appl. Microbiol.* 21:295 (1971).
25. J.C. Suen, C.L. Hatheway, A.G. Steigerwalt and D.J. Brenner, Genetic confirmation of identities of neurotoxigenic *Clostridium baratii* and *Clostridium butyricum* implicated as agents of infant botulism, *J. Clin. Microbiol.* 26:2191 (1988).
26. J.A. Giménez, M.A. Giménez and B.R. DasGupta, Characterization of the neurotoxin isolated from a *Clostridium baratii* strain implicated in infant botulism, *Infect. Immun.* 60:518 (1992).
27. J.A. Giménez and H. Sugiyama, Comparison of toxins of *Clostridium butyricum* and *Clostridium botulinum* type E, *Infect. Immun.* 56:926 (1988).
28. J.A. Giménez, J. Foley and B.R. DasGupta, Neurotoxin type E from *Clostridium botulinum* and *C. butyricum*: partial sequence and comparison, *FASEB J.* 2: A1750 (1988).
29. B.R. Singh, J.A. Giménez and B.R. DasGupta, Comparative Molecular Topography of Botulinum Neurotoxins from *Clostridium butyricum* and *Clostridium botulinum* type E, *Biochem.Biophys. Acta.* 1077:119 (1991).
30. S. Kozaki, J. Onimaru, Y. Kamata and G. Sakaguchi, Immunological characterization of *Clostridium butyricum* neurotoxin and its trypsin-induced fragment by use of monoclonal antibodies against *Clostridium botulinum* type E neurotoxin, *Infect. Immun.* 59:457 (1991).
31. E.J. Bowmer, Preparation and assay of the international standards for *Clostridium botulinum* types A, B, C, D and E antitoxins, *Bull. Wld Hlth Org.* 29:701 (1963).
32. A.J. Fulthorpe, *Clostridium botulinum* types A and B toxin-antitoxin reactions, *J. Hyg. (Lond.).* 53:180 (1955).
33. H. Sugiyama, *Clostridium botuliunum* neurotoxin, *Microbiol. Rev.* 44:419 (1980).
34. G. Sakaguchi, *Clostridium botulinum* toxins, *Pharmac. Ther.* 19:165 (1983).
35. P. Hambleton, C.C. Shone and J. Melling, Botulinum toxin. Structure, action and clinical uses, *in:* "Neurotoxin and Their Pharmacological Implications," P. Jenner, ed., Raven Press, New York (1987).
36. P. Hambleton, C.C. Shone, P. Wilton-Smith and J. Melling, A possible common antigen on clostridial toxins detected by monoclonal anti-botulinum neurotoxin antybodies, *in:* "Bacterial Protein Toxins," J.E. Alouf, F.J. Fehrenbach, J.H. Freer and J. Jeljaszewic, ed., Academic Press, Inc., London (1984).
37. K. Tsuzuki, N. Yokosawa, B. Syuto, I. Ohishi, N. Fujii, K. Kimura and K. Oguma, Establishment of a monoclonal antibody recognizing an antigenic site common to *Clostridium botulinum* type B, C1, D, and E toxins and tetanus toxin, *Infect. Immun.* 56:898 (1988).
38. C.E. Dolman and L. Murakami, *Clostridium botulinum* type F with recent observations on other types, *J. Infect. Dis.* 109:107 (1961).
39. T.I. Bulatova and E.V. Perova, Antigenic structure of *Cl. botulinum*, types E and F, *Zh. Mikrobiol. Epidemiol. Immuno.* 4:28 (1970).
40. J.H. Mason and E.M. Robinson, The antigenic components of the toxins of *Cl. botulinum* types C and D, *Onderstepoort J. Vet. Sci. Anim. Ind.* 5:65 (1935).
41. D.F. Giménez and A.S. Ciccarelli, Another type of *Clostridium botulinum*, *Zbl. Bakt., I. Abt. Orig.* 215:221 (1970).
42. A.S. Ciccarelli and D.F. Giménez, Estudio serológico de toxinas de *Clostridium botulinum* tipo A y cepa 84, *Rev. lat-amer. Microbiol.* 13:67 (1971).

43. T. Shimizu and H. Kondo, Immunological difference between the toxin of a proteolytic strain and that of a non-proteolytic strain of *Clostridium botulinum* type B, *Jpn. J. Med. Sci. Biol.* 26:269 (1973).
44. D. Rymkiewicz, J. Sawicki and A. Brühl, Study on the immunological heterogeneity of *Clostridium botulinum* B type toxin, *Arch. Immunol. Ther. Exp.* 27:709 (1979).
45. C. Glasby and C.L. Hatheway, Fluorescent-antibody reagents for the identification of *Clostridium botulinum*, *J. Clin. Microbiol.* 18:1378 (1983).
46. T.I. Bulatova and K.I. Matveev, An antigenic structure of *Cl. botulinum*, type C strains, isolated in the USSR, *Zh. Mikrobiol. Epidemiol. Immuno.* 8:79 (1965).
47. M. Eklund, F. Poysky, K. Oguma, H. Iida and K. Inoue, Relationship of bacteriophages to toxin and hemagglutinin production by *Clostridium botulinum* types C and D and its significance in avian botulism outbreaks, *in:* "Avian Botulism: An International Perspective," M.W. Eklund and V.R. Dowell Jr., ed., C. Thomas, Springfield, Illinois (1987).
48. B.C. Jansen, *Clostridium botulinum* type C, its isolation, identification and taxonomic position, *in:* "Avian Botulism: An International Perspective," M.W. Eklund and V.R. Dowell Jr., ed., C. Thomas, Springfield, Illinois (1987).
49. I. Ohishi and B.R. DasGupta, Molecular structure and biological activities of *Clostridium botulinum* C2 toxin, *in:* "Avian Botulism: An International Perspective," M.W. Eklund and V.R. Dowell Jr., ed., C. Thomas, Springfield, Illinois (1987).
50. W. Pfenninger, Toxico-immunologic and serologic relationship of *B. botulinus*, type C, and *B. parabotulinus*, "Seddon". XXII, *J. Infect. Dis.* 35:347 (1924).
51. J.B. Gunnison and K.F. Meyer, Cultural study of an international collection of *Clostridium botulinum* and *parabotulinum*. XXXVIII, *J. Infect. Dis.* 45:119 (1929).
52. B.C. Jansen, The toxic antigenic factors produced by Clostridium botulinum types C and D, *Onderstepoort J. vet. Res.* 38:93 (1971).
53. I. Ohishi, M. Iwasaki and G. Sakaguchi, Vascular permeability activity of botulinum C2 toxin elicited by cooperation of two dissimilar protein components, *Infect. Immun.* 31:890 (1980).
54. B.C. Jansen and P.C. Knoetze, The taxonomic position of *Clostridium botulinum* type C, *Onderstepoort J. vet. Res.* 44:53 (1977).
55. K. Oguma, B. Syuto, T. Agui, H. Iida and S. Kubo, Homogeneity and heterogeneity of toxins produced by *Clostridium botulinum* type C and D strains, *Infect. Immun.* 34:382 (1981).
56. K. Oguma, B. Syuto, H. Iida and S. Kubo, Antigenic similarity of toxins produced by *Clostridium botulinum* type C and D strains, *Infect. Immun.* 30:656 (1980).
57. O.J. Ochanda, B. Syuto, K. Oguma, H. Iida and S. Kubo, Comparison of antigenicity of toxins produced by *Clostridium botulinum* type C and D strains, *Appl. Environ. Microbiol.* 47:1319 (1984).
58. K. Oguma, S. Murayama, B. Syuto, H. Iida and S. Kubo, Analysis of antigenicity of *Clostridium botulinum* type C1 and D toxins by polyclonal and monoclonal antibodies, *Infect. Immun.* 43:584 (1984).
59. J. Terajima, B. Syuto, J.O. Ochanda and S. Kubo, Purification and characterization of neurotoxin produced by *Clostridium botulinum* type C 6813, *Infect. Immun.* 48:312 (1985).
60. K. Moriishi, B. Syuto, S. Kubo and K. Oguma, Molecular diversity of neurotoxins from *Clostridium botulinum* type D strains, *Infect. Immun.* 57:2886 (1989).
61. W.B. Gross and L.DS. Smith, Experimental botulism in gallinaceous birds, *Avian Dis.* 15:716 (1971).
62. J. Haagsma, Laboratory investigation of botulism in wild birds, *in:* "Avian Botulism: An International Perspective," M.W. Eklund and V.R. Dowell Jr., ed., C. Thomas, Springfield, Illinois (1987).
63. A.R. Prévot, A. Turpin and P. Kaiser, "Les Bactéries Anaérobies," Dunod, Paris (1967).
64. A.H.W. Hauschild, *Clostridium botulinum* toxins, *Intern. J. Food Microbiol.* 10:113 (1990).
65. H. Sugiyama, K. Mizutani and K.H. Yang, Basis of type A and F toxicitites of *Clostridium botulinum* strain 84, *Proc. Soc. Exp. Biol. and Med.* 141:1063 (1972).
66. B.R. DasGupta and A. Datta, Botulinum neurotoxin type B (strain 657): partial sequence and similarity with tetanus toxin, *Biochimie* 70:811 (1988).
67. B.J. Edmond, F.A. Guerra, J. Blake and S. Hempler, Case of infant botulism in Texas, *Tex. Med.* 73:85 (1977).
68. C.L. Hatheway and L.M. McCroskey, Examination of feces and serum for diagnosis of infant botulism in 366 patients, *J. Clin. Microbiol.* 25:2334 (1987).
69. C.L. Hatheway and L.M. McCroskey, Unusual neurotoxigenic clostridia recovered from human fecal specimens in the investigation of botulism, *in:* "Recent Advances in Microbial Ecology," T. Hattori, Y. Ishida, Y. Maruyama, R.Y. Morita and A. Uchida, ed., Proceedings of the 5th International Symposium on Microbial Ecology. Japan Scientific Societies Press, Tokyo (1989).
70. R.A. Fernández, A.S. Ciccarelli, G.N. Arenas and D.F. Giménez, Primer brote de botulismo por *Clostridium botulinum* subtipo Af, *Rev. Argent. Microbiol.* 18:29 (1986).

CHARACTERIZATION OF MONOCLONAL ANTIBODIES TO BOTULINUM TYPE A NEUROTOXIN

Bassam Hallis, Sarah Fooks, Clifford Shone, and Peter Hambleton

PHLS-CAMR, Division of Biologics
Porton Down, Salisbury
Wiltshire SP4 0JG, United Kingdom

INTRODUCTION

The botulinum neurotoxin types primarily involved in human illness are types A, B, E and F. Type A neurotoxin is the most intensively studied of the neurotoxins. However the antigenic sites remain unknown. The aim of the current project is to study the structure and determine the location of the antigenic sites of the type A neurotoxin.

In order to study the antigenicity of the type A neurotoxin we have raised a number of monoclonal antibodies against the toxin and produced various fragments of the neurotoxin.

Our results indicate the possibility that formaldehyde-sensitive antigenic sites may be located on the C-terminal half of the heavy subunit.

RESULTS

Botulinum type A neurotoxin was purified using a combination of affinity chromatography and gel filtration techniques. Treatment of the type A toxin with trypsin resulted in the cleavage of the heavy subunit at approximately the middle producing a 100-kDa fragment consisting of the light subunit linked to the N-terminal half of the heavy subunit (LH_N). The toxin was also inactivated with formaldehyde to give a non-toxic immunogenic toxoid.

Mice were immunized by intraperitoneal injections using type A toxoid initially and type A neurotoxin subsequently. Spleen cells were fused with myeloma cells and the resulting hybridomas screened for binding to pure neurotoxin type A using a direct enzyme linked immunosorbent assay (ELISA). Hybridomas responding positively in the ELISA were selected and cloned. Monoclonal antibodies with high affinity to neurotoxin type A were then selected (Table 1) using a screening technique in the presence of high concentration of the chaotropic agent urea. In this test the affinity of the antibody to the solid phase-bound toxin is measured in the presence and absence of 3M urea. Previous studies with this technique have shown that the binding of low affinity antibodies is markedly reduced by the presence of urea. Thus, this simple test appears to provide a means of selecting out low affinity antibodies at an early stage.

Botulinum and Tetanus Neurotoxins, Edited by
B.R. DasGupta, Plenum Press, New York, 1993

433

Table 1. Characteristics of monoclonal antibodies to type A neurotoxin.

Cell line	Antibody subclass	Binding with 3M urea	IgG concentration in ascites (mg/ml)
BA 11	IgG_{2a}	+	2.00
5BA 2.3	IgG_1	+	1.45
5BA 3.3	IgG_1	+	1.15
5BA 4.3	IgG_1	+	3.85
5BA 9.3	IgG_1	+	1.05
5BA 11.3	IgG_{2b}	+	1.45

Table 2. Binding properties of monoclonal antibodies to type A neurotoxin.

Antibody	Toxin	Toxoid	LH_N
BA 11	+	+	+
5BA 2.3	+	-	-
5BA 3.3	+	-	-
5BA 4.3	+	-	-
5BA 9.3	+	-	-
5BA 11.3	+	+	+

Six hybridoma cell lines (Table 1) secreting monoclonal antibodies with high binding activity for type A neurotoxin were selected. Monoclonal antibodies were purified from ascites fluids derived from each cell line by ammonium sulfate precipitation followed by affinity chromatography on protein A superose. Polyclonal antibodies were also produced by immunizing guinea pigs with a combination of type A toxoid and neurotoxin. Purified polyclonal antibodies were conjugated to alkaline phosphatase using the glutaraldehyde method.

The specificity of the type A monoclonal antibodies was tested to both type B and F neurotoxins using the direct ELISA. All six monoclonal antibodies were specific to type A neurotoxin displaying no cross-reactivity with either type B or F neurotoxins. The binding properties of the various monoclonal antibodies against botulinum type A neurotoxin were assessed with the native toxin, its toxoid and a fragment lacking the C-terminal half of the heavy chain (LH_N fragment). Table 2 shows that four out of six antibodies tested bound only to the native toxin and failed to bind both the toxoid and the LH_N fragment which may indicate that these antibodies recognize a formaldehyde-sensitive site located on the C-terminal half of the heavy chain.

Monoclonal antibodies at a fixed concentration of 40 μg/ml were tested both singly and in various combinations for their ability to neutralize a range of concentrations of botulinum type A neurotoxin. Although none of the antibodies were particularly effective at neutralizing toxic activity when used singly, several combinations of antibodies had significant neutralizing activity (Table 3). Combinations of antibodies recognizing different sections of the toxin had greater neutralizing abilities than those that appear to have similar binding sites.

The monoclonal antibodies were used to develop a rapid and sensitive ELISA system for the detection of type A neurotoxin (Table 4). This ELISA was of the single antibody sandwich type which employs a monoclonal antibody as the solid-phase capture antibody with the guinea pig polyclonal antibody alkaline phosphatase conjugate as the second (detecting) antibody. Highly purified type A neurotoxin (10^{-3}-10 ng/ml) was used as a positive reference. The ELISA was amplified enzymatically using the following amplification system. Alkaline phosphatase was reacted with a substrate, NADP. The product of the reaction, NAD, is a catalytic activator of a redox cycle involving alcohol dehydrogenase, ethanol, diaphorase and iodonitrotetrazolium

Table 3. Neutralization of botulinum type A neurotoxin.

Antibody	MDL_{50}/ml
BA 11	10-50
5BA 2.3	100-500
5BA 3.3	50-100
5BA 4.3	5-50
5BA 9.3	10-50
5BA 11.3	5-50
BA 11 + 5BA 2.3	500-1000
BA 11 + 5BA 3.3	500-1000
BA 11 + 5BA 4.3	500*
BA 11 + 5BA 9.3	100*
BA 11 + 5BA 11.3	5-50
5BA 2.3 + 5BA 3.3	50-100
5BA 2.3 + 5BA 4.3	50-100
5BA 2.3 + 5BA 11.3	100-500
5BA 3.3 + 5BA 4.3	50-100
5BA 3.3 + 5BA 11.3	500-1000
5BA 4.3 + 5BA 11.3	1000*

Monoclonal antibodies were added to a final concentration of 40 μg/ml.
*No higher toxin concentrations were used.

violet which becomes reduced to an intensely colored formazan. The developed ELISA was rapid (take less than one hour to perform), sensitive (detection limit 15-50 MDL_{50}/ml) and specific to type A neurotoxin (no cross-reaction with the other botulinum neurotoxins was observed).

Finally, this ELISA was used in parallel with a similar system for type B neurotoxin during the most recent incident of foodborne botulism in the United Kingdom in June 1989 arising from contaminated hazelnut yogurt. A number of hazelnut puree and yogurt samples were analyzed. As a result, we were able to demonstrate the source, nature and the level of contaminant (neurotoxin type B) less than two hours after receiving the food samples. The assay was sensitive enough to detect type B neurotoxin in several of the yogurt (1.50-2.15 ng/g) and hazelnut puree (1000 ng/g) samples. The use of such a detection system, instead of the mouse bioassay which can take up to four days to perform, for the neurotoxins could play a vital role in the future by saving lives by the rapid detection of the source and nature of contaminants.

Table 4. Detection of the type A neurotoxin in the amplified ELISA.

Antibody	Detection limit		Cross-reaction
	MLD_{50}/ml	ng/ml	with types B and F
BA 11	15	0.15	none
5BA 2.3	38	0.38	none
5BA 3.3	20	0.20	none
5BA 4.3	44	0.44	none
5BA 9.3	50	0.50	none
5BA 11.3	33	0.33	none

CONCLUSION

1. Six hybridoma cell lines producing monoclonal antibodies with high affinity for type A neurotoxin were selected. These antibodies are specific for type A neurotoxin and no cross-reactivity with either type B or F neurotoxins was observed.

2. A rapid and sensitive monoclonal antibodies-based ELISA system for the detection of type A neurotoxin was developed. This assay is amplified enzymically and can detect 15-50 MDL_{50}/ml in less than one hour. This assay was used successfully to identify the causative toxin in a botulism outbreak in the U.K. in June 1989.

3. Four out of the six monoclonal antibodies bound only to the native toxin and failed to bind both the toxoid and the LH_N fragment which may indicate the possibility that formaldehyde-sensitive antigenic sites may be located on the C-terminal half of the heavy subunit.

ANTIGENIC STRUCTURE OF BOTULINUM NEUROTOXINS: SIMILARITY AND DISSIMILARITY TO THE TOXIN ASSOCIATED WITH INFANT BOTULISM

Shunji Kozaki, Teiichi Nishiki, Shoji Nakaue, Yoichi Kamata
and Genji Sakaguchi

University of Osaka Prefecture, College of Agriculture
Sakai-shi, Osaka, 593
Japan

INTRODUCTION

Clostridium botulinum toxin has been classified into seven immunological types A through G. The toxin consists of a highly potent neurotoxin and a nontoxic component. The neurotoxin exerts its toxic action by inhibition of acetylcholine release, which results in neuromuscular paralysis.[1, 2] The neurotoxin is produced as a single polypeptide with a molecular weight of about 150 KDa, and is nicked by an endogenous or exogenous protease such as trypsin and other trypsin-like enzymes. The neurotoxin in a nicked form, is made up of two chains, the heavy (ca. 100 KDa) and the light (ca. 50 KDa) chains, which are covalently linked together with at least one disulfide bond.[2-4] The heavy chain is responsible for binding the neurotoxin to the receptor on neural membrane,[5, 6] and the light chain is related to blockade of neurotransmitter release in presynaptic nerve endings.[7, 8] The significance of the nontoxic component in the botulinum toxin molecule as an oral poison has been reviewed by Sakaguchi.[9]

Since infant botulism was first described in 1976[10, 11], more than 800 cases have been reported mainly in the United States.[12] This disease causes neuromuscular paralysis due to the toxin produced by the ingested spores in the intestines. Most cases were caused by types A and B. However types E and F cases were due to the organisms different from the respective types, namely by *C. butyricum* producing type E toxin[13] and *C. baratii* producing type F toxin.[14] Giménez and Sugiyama[15] demonstrated that *C. butyricum* neurotoxin is antigenically similar but not identical to type E neurotoxin. We found, incidentally that the neurotoxin produced by type A strains associated with infant botulism in Japan was not identical to those of classical botulism-related strains.[16]

We prepared the monoclonal antibodies (MAbs) against neurotoxin and analyzed the neurotoxin molecules for their immunological, structural, and functional properties. From these studies, botulinum neurotoxin consists of at least three domains (L, H-1, and H-2) which possess probably functionally different properties. In the current study, we examined also the antigenic properties of neurotoxins associated with infant botulism and clarified their similarity and dissimilarity to already-known neurotoxins.

IMMUNOLOGICAL CHARACTERIZATION OF TYPE A NEUROTOXIN

The neurotoxin of standard and food-borne botulism-associated strain

Type A neurotoxin, produced by type A strain 62A, was treated with papain to obtain fragments that were distinct from either the heavy or the light chain.[17] The papain-digested toxin was loaded onto a column of Protein Pak G-ether and eluted with a linear ammonium sulfate gradient. Eventually, we obtained three fragments of 101 KDa, 45 KDa, and 43 KDa, respectively. On SDS-PAGE, the 101 KDa fragment was separated into two bands (53 KDa and 48 KDa) upon reduction. The mobility of one of the two bands corresponded to that of the light chain (53 KDa); the 101 KDa fragment was found to consist of the light and a part of the heavy chain (H-1 fragment) linked together by a disulfide bond. The other two fragments (43 KDa and 45 KDa) were correlated to the remaining portion of the heavy chain (H-2 fragment).

In order to study the antigenic structure of type A neurotoxin, we prepared MAbs against 62A neurotoxin and examined their reactivities to the neurotoxin and its papain-induced fragments. Of the eleven MAbs, eight reacted to the heavy chain and the remaining three to the light chain. In immunoblotting, six of MAbs reacting to the heavy chain, namely A102, A108, A116, A119, A123, and A124, reacted to the 48 KDa fragment (H-1 fragment). Two antibodies, A101 and A125, bound to the 43 KDa and 45 KDa fragments (H-2 fragment). Of the antibodies reacting to the light chain, A109 only neutralized type A neurotoxin. As for the antibodies reacting to the heavy chain, three, A108, A116, and A124, were capable of neutralizing the neurotoxin, all of which recognized the 48 KDa fragment (Table 1).

When a competitive ELISA was performed with biotin-labeled MAb[18], the epitope recognized by A116 was found to be close to that by A109 (Table 2). Since Niemann[6] demonstrated that A109 recognizes the conformational epitope comprising approximately one-third from the N-terminal of the light chain, the results indicate that the N-terminal region of the light chain is probably adjacent to the H-1 fragment.

The neurotoxin of infant botulism-associated strain

The neurotoxin was purified with an isolate, type A strain 7I03-H, isolated from honey having caused infant botulism.[19] The purified 7I03-H neurotoxin migrated to more slowly than did 62A neurotoxin in SDS-PAGE, suggesting that the former had a molecular weight higher than the latter. Under reducing conditions, 7I03-H neurotoxin was separated also into two bands as was 62A neurotoxin. The light chain of 7I03-H neurotoxin appeared to

Table 1. Properties of monoclonal antibodies against type A neurotoxin strain 62A and their reactivities to 7I03-H neurotoxin

MAb	Fragment*	Neutralizing activity	ELISA value (OD$_{450}$) in wells coated with†	
			62A	7I03-H
A101	H-2	–	> 2.000	> 2.000
A102	H-1	–	1.321	0.009
A108	H-1	+	1.782	> 2.000
A109	L	+	1.439	> 2.000
A115	L	–	1.792	1.743
A116	H-1	+	1.626	> 2.000
A119	H-1	–	1.692	0.192
A122	L	–	1.768	1.549
A123	H-1	–	1.615	1.182
A124	H-1	+	> 2.000	0.624
A125	H-2	–	1.504	0.004

*Reaction to papain-induced fragments in immunoblotting.

†Values were obtained with each MAb at 1 μg/ml. OD$_{450}$, optimal density at 450 nm.

Table 2. Inhibition of the binding of biotin-labeled monoclonal antibodies to 62A neurotoxin with unlabeled monoclonal antibodies

Labeled MAb	Inhibition (%) with unlabeled MAb*		
	A108	A109	A116
A108	99	0	0
A109	0	96.0	43.0
A116	0	97.0	97.0

*Biotin-labeled MAb (100 ng/ml) was mixed with an equal volume of homologous or heterogous MAb (100 μg/ml).

have the same molecular weight as that of 62A neurotoxin. The difference in the molecular weight between the two neurotoxins was accounted for solely by that of the heavy chain.

Agar gel diffusion tests demonstrated that anti-7I03-H neurotoxin gave a spur of 7I03-H neurotoxin over 62A neurotoxin. The converse was also the case; anti-62A neurotoxin recognized both neurotoxins, but the precipitin line of 7I03-H neurotoxin was not identical to that of 62A neurotoxin. The results suggest that 7I03-H neurotoxin was immunological related to 62A neurotoxin but possessed its unique antigenic determinant(s).

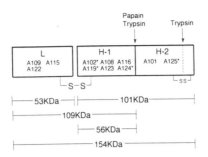

Fig. 1. Probable structure of 62A and 7I03-H neurotoxins. The asterisk shows the MAbs without reactivity to 7I03-H neurotoxin. The arrows show the sites split by papain and trypsin. The site near the C-terminus split by trypsin was described by Shone et al.[20]

In fact, some MAbs against 62A neurotoxin did not react to 7I03-H neurotoxin as shown in Table 1. We therefore attempted to locate the portion accounting for the antigenic differences on the two neurotoxins by use of protease-induced fragments and MAbs against 62A neurotoxin. When 7I03-H neurotoxin was treated with papain in the same way as was 62A neurotoxin, 7I03-H neurotoxin did not appear to have been changed, suggesting that the sensitivities of the two neurotoxin to proteases may be different from each other. 7I03-H neurotoxin was digested with TPCK-treated trypsin by the method of Shone et al.[20] with a slight modification. When analyzed by SDS-PAGE in the absence of a reducing agent, two bands with molecular weights of 109 KDa and 40 KDa emerged. The 109 KDa fragment was separable into further two bands (56 KDa and 53 KDa) under a reducing condition. The 53 KDa fragment appeared to be the light chain because it migrated to the same position as that of the light chain and reacted to the MAbs (A109 and A122) recognizing to the light chain of 62A neurotoxin in immunoblotting analysis. The partial N-terminal amino acid sequences of the 53 KDa and 56 KDa fragments were examined. They were found to be identical to that previously reported for either N-terminal sequence of the heavy or light chain of 62A neurotoxin[21] and of other type A neurotoxins as well.[20, 22, 23] Therefore, the 56 KDa fragment of 7I03-H neurotoxin was the equivalent of H-1 fragment of 62A neurotoxin. In fact, immunoblotting analysis showed that three (A108, A116, and A123) recognizing the H-1 fragment bound to the 56 KDa fragment of 7I03-H neurotoxin. The reactivities of the other MAbs to the fragments of 7I03-H neurotoxin are illustrated in Fig. 1. From these results, the common epitopes in both neurotoxins appeared to be fairly conserved to the corresponding portion in the neurotoxin molecule. The immunological difference between the two neurotoxins is attributable to some regions of the heavy chain.

IMMUNOLOGICAL CHARACTERIZATION OF TYPE B NEUROTOXIN

In type B neurotoxin, chymotrypsin was used to obtain the fragment different from the heavy and light chains.[24] After chymotrypsin treatment by prolonged incubation, the digested material was applied to a CM-Sephadex column and eluted with a linear NaCl gradient. On SDS-PAGE, the eluted protein migrated in a single band in the absence or presence of a reducing agent. This protein (115 KDa) had a little higher molecular weight than that of the heavy chain (105 KDa). The N-terminal residues of this fragment were identical to that of the native neurotoxin, suggesting that the C-terminal portion was digested by chymotrypsin treatment. When this fragment was treated mildly with a lysine-specific protease (e.g. lysyl endopeptidase) and dithiothreitol, it dissociated further into 54 KDa and 59 KDa fragments; the 54 KDa fragment moved to the same position as did the light chain.

The MAbs against type B neurotoxin were tested by ELISA and immunoblotting for their reactivities to the heavy and light chains, and the chymotrypsin-induced fragment. From the results, we confirmed also that the chymotrypsin-induced fragment is composed of the light chain and the N-terminal portion of the heavy chain (H-1 fragment). However, some MAbs recognizing H-1 fragment or the light chain hardly reacted to the native chymotrypsin-induced fragment but to the denatured one. This may suggest that the digestion of the H-2 fragment of the neurotoxin molecule causes conformational change(s) in the remaining fragment.[24]

Table 3. Properties of monoclonal antibodies against type E neurotoxin and their reactivities to butyricum neurotoxin

MAb	Fragment*	Neutralizing activity	ELISA value (OD$_{450}$) in wells coated with†	
			E‡	Buty§
E 1	L	–	1.721	0.065
E 2	L	–	> 2.000	> 2.000
E14	H-1	+	1.194	1.385
E16	H-2	–	1.220	1.037
E20	H-1	–	0.712	0.705
E22	H-1	+	1.007	0.026
E25	L	–	1.353	0.694
E28	H-2	+	1.759	0.058
E32	L	+	1.629	0

*Reaction to chymotrypsin- or trypsin-induced fragments in immunoblotting.

†Values were obtained with each MAb at 1 μg/ml. OD$_{450}$, optimal density at 450 nm.

‡E, type E neurotoxin

§Buty, butyricum neurotoxin

IMMUNOLOGICAL CHARACTERIZATION OF TYPE E NEUROTOXIN AND *C. BUTYRICUM* NEUROTOXIN

We prepared nine MAbs against type E neurotoxin.[25] The recognition site of each MAb on the neurotoxin molecule was determined by immunoblotting. Five MAbs recognized the heavy chain and the other four the light chain (Table 3). Four MAbs neutralized type E neurotoxin. The treatment of type E neurotoxin with trypsin or chymotrypsin induced fragments different from the heavy and light chains. From the

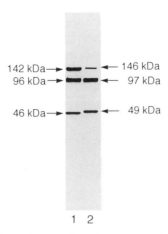

Fig. 2. SDS-PAGE of the mild trypsin-treated butyricum neurotoxin and type E neurotoxin. Lane 1, butyricum; lane 2, type E. Samples were electrophoresed in the presence of dithiothreitol.

experiments with the MAbs, these fragments were found to comprise the light chain and the H-1 fragment. To obtain the remaining portion of the heavy chain (H-2 fragment), the heavy chain was mildly trypsinized in the presence of a reducing agent and urea. The digested material was applied to a QAE-Sephadex column. A single protein peak was eluted with a buffer containing 0.1 M NaCl and urea. It migrated in a single band with a molecular weight of 41 KDa. The MAbs which possess no reactivity to trypsin- or chymotrypsin-induced fragment bound to the 41 KDa fragment. The sum of the molecular weights of the H-1 fragment (56 KDa) and of the 41 KDa fragment is identical to that of the heavy chain (97 KDa).

The neurotoxin was purified from toxigenic *C. butyricum* strain 5262 by the method of Giménez and Sugiyama.[26] Butyricum neurotoxin free from the nontoxic component was isolated by DEAE-Sephacel chromatography. In SDS-PAGE, butyricum neurotoxin (142 KDa) migrated in a single band. After trypsinization and reduction, butyricum neurotoxin was separated into two fragments; the heavy chain (96 KDa) was comparable to that of type E neurotoxin, but the light chain (46 KDa) of the former was slightly smaller than that of the latter (49 KDa). The difference in the molecular size between butyricum and type E neurotoxins may be dependent upon those of the light chains (Fig. 2).

The reactivities of butyricum neurotoxin to the nine MAbs against type E neurotoxin were examined by ELISA. Five (E2, E14, E16, E20, and E25) bound to butyricum neurotoxin, whereas the other four (E1, E22, E28, and E32) did not (Table 3). The latter four did not bind to any band in immunoblotting, while the former five bound to the corresponding heavy or light chain of butyricum neurotoxin. The results suggest that some epitopes different from those of type E neurotoxin are present in both chains of butyricum neurotoxin. The neurotoxin produced by another strain 5520 of *C. butyricum* was also found to show the same reactive pattern to the MAbs, suggesting that the antigenic differences may be held invariably by the butyricum neurotoxin molecule (Table 4). MAb E14, which reacted to butyricum neurotoxin, neutralized type E neurotoxin and butyricum neurotoxin as well. However, the results that E32, neutralizing type E neurotoxin and recognizing the light chain, may suggest that the toxic sites of the two neurotoxins, being probably located on the light chain, are antigenically different from each other.

Table 4. Immunoblotting analysis with monoclonal antibodies of toxins produced by *C. botulinum* type E and *C. butyricum*

MAb	Toxin			
	German sprats*	5545*	5262†	5520†
E 1	+	+	−	−
E 2	+	+	+	+
E14	+	+	+	+
E16	+	+	+	+
E20	+	+	+	+
E22	+	+	−	−
E25	+	+	+	+
E28	+	+	−	−
E32	+	+	−	−

C. botulinum type E

†*C. butyricum*

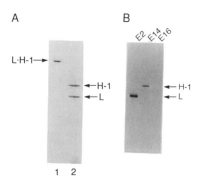

Fig. 3. (A) SDS-PAGE of the trypsin-induced fragment of butyricum neurotoxin. The fragment was electrophoresed in the absence (lane 1) and presence (lane 2) of dithiothreitol. (B) Immunoblotting analysis of the trypsin-induced fragment with the MAbs against type E neurotoxin. The MAb used are indicated above the lanes.

After butyricum neurotoxin was digested at 37°C for 6 h with trypsin at a toxin-to-enzyme ratio of 40:1 at pH 7.5, the mixture was applied to a PD-10 column equilibrated with 20 mM phosphate buffer, pH 5.0. The protein fraction was immediately subjected to a Mono S HR5/5 column and eluted with an NaCl linear gradient.[27] A single peak, containing no toxicity, emerged in 0.1 M NaCl. This protein migrated in SDS-PAGE as a single band (102 KDa) and was further separable into two bands (56 KDa and 46 KDa) under a reducing condition (Fig. 3A). The smaller fragment moved to position identical to that of the light chain of butyricum neurotoxin and reacted to the light chain recognizing antibody E2. Another fragment, being in accordance in the molecular size with the H-1 fragment of type E neurotoxin, also bound to E14. MAb E16, which reacted to

443

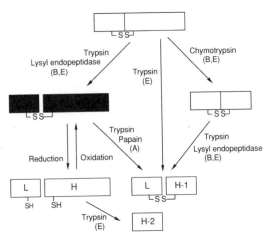

Fig. 4. Illustration of fragmentation of the neurotoxins with proteases. The closed squares indicate to be toxic. The capital letter in parentheses shows the toxin type.

heavy chain of butyricum neurotoxin, did not bind either fragment (Fig. 3B). Since we reported that E16 recognized the H-2 fragment of type E neurotoxin, these results indicate that tryptic digestion results in deletion of the C-terminal portion of butyricum neurotoxin (H-2 fragment) and that the heavy chain of butyricum neurotoxin is split by trypsin at a site comparable to that of type E neurotoxin.

Fragmentation of neurotoxins with proteases is illustrated in Fig. 4.

868   877

Glu·Tyr·Ile·Lys·Asn·Ile·Ile·Asn·Thr·Ser···

Fig. 5. Amino acid sequence of the first ten residues of papain-induced 45 KDa fragment from 62A neurotoxin.

INTERACTION OF NEUROTOXINS AND THEIR FRAGMENTS WITH SYNAPTOSOMES AND GANGLIOSIDES

There are several reports that the neurotoxin binds to brain synaptosomes .[20, 28-32] In inhibitory experiments with the MAbs against type E neurotoxin, E17 recognizing the H-2 fragment completely inhibited the binding of ^{125}I-labeled type E neurotoxin to synaptosomes.[25] Furthermore, ^{125}I-labeled neurotoxin bound to synaptosomes was displaced effectively by E17. MAb E22, recognizing the H-1 fragment, did not compete with the binding of ^{125}I-labeled neurotoxin. We examined also the inhibitory effect of the papain-induced fragments of type A neurotoxin. As we expected, the 43 KDa and 45 KDa fragments antagonized the binding of ^{125}I-labeled type A neurotoxin to the same extent as did unlabeled neurotoxin, but the other 101 KDa fragment had no inhibitory effect.[17] The partial N-terminal sequence of the 45 KDa fragment was different from that of the heavy chain but comparable to residues 868-877 of 62A neurotoxin (Fig. 5).[21] The observations suggest that the binding of neurotoxin to synaptosomes is undertaken mainly by the H-2 fragment of the heavy chain.

The neurotoxin has been found to be inactivated with gangliosides, mainly GT1b, which is a component of neural membrane.[33-35] This finding merely suggests that a particular ganglioside or certain other N-acetylneuraminic acid-containing substances should be regarded as a receptor candidate. However, there remains a question whether gangliosides are the specific binding substance of the neurotoxin, because the neurotoxin does not bind to many non-neural tissues containing gangliosides but to neural membranes. We reported that type E neurotoxin and its heavy chain bound to ganglioside GD1a, GT1b, and GQ1b as assessed by the TLC-immunostaining method in a low ionic strength buffer.[36] The reactivity of type B neurotoxin to gangliosides differs somewhat from those of type A and E neurotoxin. Type B neurotoxin and its heavy chain bound to ganglioside GD1a and GT1b but not GQ1b, which reacts with type A and E neurotoxins. The L·H-1 fragment of the neurotoxins did not bind to gangliosides but to cerebroside, being free of the oligosaccharide portion of gangliosides.[24] The L·H-1 fragment also bound to free fatty acids. These observations, though obtained under low ionic strength conditions, may suggest that the H-2 fragment initially binds to the oligosaccharide portion of gangliosides and then the L·H-1 fragment interacts with cerebrosides and free fatty acids comprising the hydrophobic portion of gangliosides.

Under a physiological ionic-strength condition, the binding of type A and B neurotoxins to synaptosomes was examined after the treatment with lysyl endopeptidase and neuraminidase. The treatment decreased the amounts of both type A and type B neurotoxins bound to the synaptosomes, and the N-acetylneuraminic acid contents of synaptosomes decreased by about 50% (Fig. 6A). The results are consistent with the previous reports.[29,30] Incorporation of exogenous gangliosides into the neuraminidase-treated synaptosomes resulted in effective restoration of the extent of type A and B neurotoxin binding, while that into untreated synaptosomes did not influence the amount of neurotoxin binding. This may suggest that gangliosides alone can not function as the acceptor of botulinum neurotoxin under physiological conditions. Lysyl endopeptidase treatment of synaptosomes decreased also the binding of type B neurotoxin, whereas it retained the ability to bind to type A neurotoxin (Fig. 6B). Incorporation of gangliosides into the lysyl endopeptidase-treated synaptosome did not facilitate the restoration of the binding of type B neurotoxin. This fact suggests that a certain protease-sensitive substance may also play a role in the binding of type B neurotoxin to neural membrane. The different sensitivities of acceptors to lysyl endopeptidase may reflect the difference between the type-specific acceptors to type A and B neurotoxins.

The binding of ^{125}I-labeled butyricum neurotoxin to synaptosomes was effectively inhibited by unlabeled type E neurotoxin and vice versa. The results indicate that both butyricum and type E neurotoxins may share identical or very similar characteristics in respect to the acceptor recognition. However, 7I03-H neurotoxin inhibited the binding of ^{125}I-labeled 62A neurotoxin but to a much lesser extent than 62A neurotoxin, suggesting that the binding site in the neurotoxin molecule and/or the properties of the acceptor on neural membrane are different from each other.

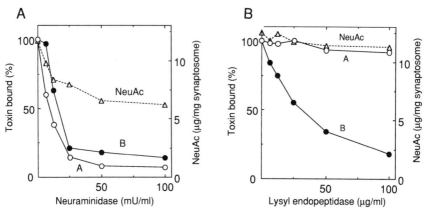

Fig. 6. Effect of neuraminidase (A) or lysyl endopeptidase (B) on the binding of ^{125}I-labeled type A and B neurotoxins to synaptosomes. Synaptosomes were treated for 30 min at 37°C with each enzyme at various concentrations. The ^{125}I-labeled neurotoxins bound to untreated synaptosomes were regarded as 100%. The broken line shows the N-acetylneuraminic acid (NeuAc) contents of synaptosomes.

CONCLUDING REMARKS

During the recent several years, the complete or partial amino acid sequences of botulinum neurotoxins have been deduced from the corresponding nucleotide sequences of toxin genes.[21, 23, 37, 38] The information will lead to understanding of the secondary and tertiary structure of the neurotoxin molecule but it still fail to provide any direct clue in relation to the mechanism of toxic actions. However, biochemical, pharmacological , and immunological studies will help much elevate knowledge obtainable from the molecular biology efforts. From this point of view, MAb is one of very promising reagents. As described herein, we attempted to characterize the protease-induced fragments by use of MAb. We demonstrated that the neurotoxin consists of three different functional domains (L, H-1, and H-2), probably corresponding to those proposed by Simpson.[5] In addition, antigenic similarity and dissimilarity in the neurotoxin associated with infant botulism will contribute to the studies dealing with structure-function relationships of botulinum neurotoxin.

REFERENCES

1. Burgen ASV, Dickens F, Zatman LJ. The action of botulinum toxin on the neuro-muscular junction. J. Physiol. 1949;109:10-24.
2. Sugiyama H. *Clostridium botulinum* neurotoxin. Microbiol. Rev. 1980;44:419-448.
3. Sathyamoorthy V, DasGupta BR. Separation, purification, partical characterization and comparison of the heavy and light chains of botulinum neurotoxin types A, B, and E. J. Biol. Chem. 1985; 260:10461-10466.
4. DasGupta BR, Sugiyama H. A common subunit structure in *Clostridium botulinum* type A, B, and E toxins. Biochem. Biophys. Res. Commun. 1972;48:108-112.

5. Simpson LL. Molecular pharmacology of botulinum toxin and tetanus toxin. Ann. Rev. Pharmacol. Toxicol. 1986;26:427-453.

6. Niemann H. Molecular biology of clostridial neurotoxins. In: Sourcebook of bacterial protein toxins. Academic Press, 1991: 303-348.

7. Bittner MA, DasGupta BR, Holz RW. Isolated light chain of botulinum neurotoxins inhibit exocytosis. J. Biol, Chem. 1989;264:10354-10360.

8. Poulain B, Mochida S, Wadsworth JDF, et al. Inhibition of neurotransmitter release by botulinum neurotoxins and tetanus toxin at *Aplisia* synapses: role of the constituent chains. J. Physiol. (Paris) 1990;84:247-261.

9. Sakaguchi G. *Clostridium botulinum* toxins. Pharmac. Ther. 1983;19:165-194.

10. Pickett J, Berg B, Chaplin E, Shafer MB. Syndrome of botulism in infancy : clinical and electrophysiologic study. New Eng. J. Med. 1976; 295:770-772.

11. Midura TF, Arnon SS. Infant botulism: identification of *Clostridium botulinum* and its toxin in faeces. Lancet 1976;(ii):934-936.

12. Arnon SS. Infant botulism: anticipating the second decade. J. Infect. Dis. 1986;154:201-205.

13. McCroskey LM, Hetheway CL, Fenicia L, Pasolini B, Aureli P. Characterization of an organism that produces type E botulinal toxin but which resembles *Clostridium butyricum* from the feces of an infant with type E botulism. J. Clin. Microbiol. 1986;23:201-202.

14. Hall JD, McCrosky LM, Pincomb BJ, Hatheway CL. Isolation of an organism resembling *Clostridium barati* which produces type F botulinum toxin from an infant with botulism. J. Clin. Microbiol. 1985;21:654-655.

15. Giménez JA, Sugiyama H. Comparison of toxins of *Clostridium butyricum* and *Clostridium botulinum* type E. Infect. Immun. 1988;56:926-929.

16. Sakaguchi G, Sakaguchi S, Kamata Y, Tabita K, Asao T, Kozaki S. Distinct characters of *Clostridium botulinum* type A strains and their toxin associated with infant botulism in Japan. Int. J. Food Microbiol. 1990;11:231-242.

17. Kozaki S, Miki A, Kamata Y, Ogasawara J, Sakaguchi G. Immunological characterization of papain-induced fragments of *Clostridium botulinum* type A neurotoxin and interaction of the fragments with brain synaptosomes. Infect. Immun. 1989;57:2634-2639.

18. Gretch DR, Suter M, Stinski. MF. The use of biotynylated monoclonal antibodies and streptavidin affinity chromatography to isolate herpesvirus hydrophobic proteins or glycoproteins. Anal. Biochem. 1987;163:270-277.

19. Tabita K, Sakaguchi S, Kozaki S, Sakaguchi G. Comparative studies on *Clostridium botulinum* type A strains associated with infant botulism in Japan and in California, USA. Jpn. J. Med. Sci. Biol. 1990;43:219-231.

20. Shone CC, Hambelton P, Melling J. Inactivation of *Clostridium botulinum* type A neurotoxin by trypsin and purification of two tryptic fragments : proteolytic action near the COOH-terminus of the heavy subunits destroys toxin-binding activity. Eur. J. Biochem. 1985;151:75-82.

21. Binz T, Kurazono H, Wille M, Frevert J, Wernars K, Niemann H. The complete sequence of botulinum neurotoxin type A and comparison with other clostridial neurotoxins. J. Biol. Chem. 1990;265:9153-9158.

22. DasGupta BR, Dekleva ML. Botulinum neurotoxin type A: sequence of amino acids at the N-terminus and around the nicking site. Biochimie 1990;72:661-664.

23. Thompson EE, Brehm JK, Outram JO, et al. The complete amino acid sequence of the *Clostridium botulinum* type A neurotoxin, deduced by nucleotide sequence analysis of the encoding gene. Eur. J. Biochem. 1990;189:73-81.

24. Kozaki S, Ogasawara J, Shimote Y, Kamata Y, Sakaguchi G. Antigenic structure of *Clostridium botulinum* type B neurotoxin and its interaction with gangliosides, cerebroside, and free fatty acids. Infect. Immun. 1987;55:3051-3056.

25. Kozaki S, Kamata Y, Nagai T, Ogasawara J, Sakaguchi G. The use of monoclonal antiboies to analyze the structure of *Clostridium botulinum* type E derivative toxin. Infect. Immun. 1986;52:786-791.

26. Giménez JA, Sugiyama H. Simplified purification methods for *Clostridium botulinum* type E toxin. Allp. Environ. Microbiol. 1987;53:2827-2830.

27. Kozaki S, Onimaru J, Kamata Y, Sakaguchi G. Immunological characterization of *Clostridium butyricum* neurotoxin and its trypsin-induced fragments by use of monoclonal antibodies against *Clostridium botulinum* type E neurotoxin. Infect. Immun 1991;59:457-459.

28. Habermann E. ^{125}I-labeled neurotoxin from *Clostridium botulinum* A: preparation, binding to synaptosomes and ascent to the spinal cord. Naunyn-Schmiedeberg's. Arch. Pharmacol. 1974;281:47-56.

29. Evans DM, Williams RS, Shone CC, Hambleton P, Melling J, Dolly JO. Botulinum neurotoxin type B : its purification, radioiodination and interaction with rat-brain synaptosomal membrane. Eur. J. Biochem. 1986;154:409-416.

30. Kitamura M, Sone S. Binding ability of *Clostridium botulinum* neurotoxin to the synaptosomes upon treatment of various kinds of enzymes. Biochem. Biophys. Res. Commun. 1987;143:928-933.

31. Kozaki S. Interaction of botulinum type A, B and E derivative toxins with synaptosomes of rat brain. Naunyn-Schmiedeberg's Arch. Pharmacol. 1979;308:67-70.

32. Murayama S, Syuto B, Oguma K, Iida H, Kudo S. Comparison of *Clostridium botulinum* toxins type D and C1 in molecular property, antigenicity and binding ability to rat-brain synaptosomes. Eur. J. Biochem. 1984; 142:487-492.

33. Kitamura M, Iwamori M, Nagai Y. Interaction between *Clostridium botulinum* neurotoxin and gangliosides. Biochim. Biophys. Acta 1980;628:328-335.

34. Kozaki S, Sakaguchi G, Nishimura M, Iwamori M, Nagai Y. Inhibitory effect of ganglioside GT1b on the activities of *Clostridium botulinum* toxins. FEMS Microbiol. Lett. 1984;21:219-223.

35. Ochanda JO, Syuto B, Ohishi I, Naiki M, Kubo S. Binding of *Clostridium botulinum* neurotoxin to gagnliosides. J. Biochem. (Tokyo) 1986;100:27-33.

36. Kamata Y, Kozaki S, Sakaguchi G, Iwamori M, Nagai Y. Evidence for direct binding of *Clostridium botulinum* type E derivative toxin and its fragments to gangliosides and free fatty acids. Biochem. Biophys. Res. Commun. 1986;140:1015-1019.

37. Binz T, Kurazono H, Popoff M, et al. Nucleotide sequence of the gene encoding *Clostridium botulinum* neurotoxin type D. Nucl. Acids Res. 1990;18:5556-5556.

38. Hauser D, Eklund MW, Kurazono H, et al. Nucleotide sequence of *Clostridium botulinum* C1 neurotoxin. Nucl. Acids Res. 1990;18:4924-4924.

ENZYME-LINKED IMMUNOSORBENT ASSAYS (ELISAs) TO DETECT BOTULINUM TOXINS USING HIGH TITER RABBIT ANTISERA

Gerri M. Ransom, Wei H. Lee, Elisa L. Elliot[1]
and Charles P. Lattuada

United States Department of Agriculture
Food Safety and Inspection Service
Building 322, BARC-East
10300 Baltimore Ave.
Beltsville, MD 20705-2350, USA

INTRODUCTION

Botulinum toxin types A through G are potent neurotoxic proteins produced by heterogeneous clostridia.[1-5] Conventional testing for botulinum toxin is done by mouse bioassay. Toxin type is determined by the mouse protection assay using specific botulinum toxin antisera.[6] The mouse bioassay is a useful test allowing a direct and sensitive determination of LD_{50} and antibody titers. However the test involves tedious manipulations and the use of live animals. Notermans et al.[7] pioneered the development of enzyme-linked immunosorbent assays (ELISAs) for toxins A, B, and E. Since then, numerous radioimmunoassays[8] and ELISAs[9-18] for detecting botulinum toxins have been published. All are immunoassays, and therefore the results are not strictly comparable to the biological activity measured in the mouse bioassay. Since antibodies and toxin standards needed to perform ELISAs are not generally available, these tests cannot readily be duplicated by other laboratories.

The United States Department of Agriculture, Food Safety and Inspection Service (FSIS), Microbiology Division initiated a program to produce high titer rabbit and goat antisera to be used to develop botulinum toxin ELISAs. The FSIS ELISA procedure is a combination of several methods,[7,9] but closely resembles the amplified ELISA developed by Shone et al.[17] The methods for antisera production and the FSIS Amplified ELISA procedures are included in the appendices of this chapter.

Botulinum toxins exist naturally as stable complexes of 2 protein molecules known as the progenitor or bimolecular toxin complex.[19-21] The botulinum toxin complex can be separated by column chromatography to the nontoxic and neurotoxin components,

---

[1]Present address: Food and Drug Administration, 200 C St. S.W., Washington, DC.

Botulinum and Tetanus Neurotoxins, Edited by
B.R. DasGupta, Plenum Press, New York, 1993

which in turn can be reduced to light and heavy chain components.[19] No <u>Clostridium botulinum</u> strains are known to produce the nontoxic protein alone (G. Sakaguchi, Osaka University, Japan, personal communication). Thus, a botulinum toxin ELISA that detects both molecules of the botulinum toxin complex is more sensitive than an ELISA that detects only the neurotoxin protein. Depending on requirements, it is theoretically possible to tailor an ELISA to detect and quantitate the nontoxic protein, specific neurotoxin, specific neurotoxin fragments, neurotoxoid, or even to titer specific botulinum toxin antibody.

BOTULINUM TOXIN SANDWICH ELISA

High Titer Antisera Production

Botulinum toxin ELISAs usually rely upon a combination of monoclonal (MAb) and/or polyclonal antibodies for detection.[10-15] To stimulate polyclonal antibodies against botulinum toxin in animals, toxoid is injected with Freund's adjuvant.[7] (A precipitated alum adjuvant is used with human immunizations.[22]) We produced very high titer rabbit antisera using Ribi R700 adjuvant in booster shots after 2 initial immunizations with Freund's adjuvant (Table 1 and Appendix I). High titer botulinum toxin antisera also were produced in horses using Ribi adjuvant (R. H. Whitlock and C. Buckley, University of Pennsylvania, Kennett Square, PA, personal communication).

Botulinum Toxin ELISA Procedure

The procedure and limitations of the botulinum toxin ELISA have been described in detail.[7] Most double sandwich ELISA procedures use a combination of purified MAb and polyclonal antibodies.[10,11,14,17] The production of MAb to botulinum toxins has been well described.[10,11,14,17] The ELISA procedure of Dezfulian[9] and the FSIS Botulinum Toxin ELISAs rely upon a combination of polyclonal botulinum toxin antisera produced in rabbits, goats or horses as the capture and detection antibodies.[10-11,14,17] T. Rocke, National Wildlife Health Research Center, Madison, WI (personal communication) produced high titer anti-C IgY in chickens and used this with rabbit IgG in an amplified ELISA for detection of type C botulinum toxin. M. Potter, WestReco Inc., Nestles, New Milford, CT (personal communication) used the commercially available Connaught Laboratory goat anti-A, anti-B, or anti-E IgG as the capture antibody and the same goat IgG, biotinylated, as the detection antibody in an amplified ELISA. Her ELISA is completely automated (Beckman Biomek Instrument) to detect botulinum toxin types A, B, E, and F in inoculated food samples. Some examples of amplified botulinum toxin ELISAs are presented in Table 2.

Neurotoxins and the Associated Nontoxic Protein Standards

All ELISAs require standardized sera and reagents to prepare the curves for quantitative assay. In the quantitative FSIS Amplified Botulinum Toxin ELISAs, neurotoxin standards, as well as nontoxic associate protein standards should be used to calibrate the assays. Unfortunately, dilute solutions of these proteins are not available commercially. Aqueous solutions of dilute neurotoxins are unstable even in the frozen state.[8,12,13] B. R. DasGupta, University of Wisconsin, Food Research Institute, Madison, WI (personal communication) preserves concentrated crystalline botulinum neurotoxin proteins in buffered ammonium sulfate solution. We discovered that low concentrations of botulinum toxins and neurotoxins (20-1000 ng/ml) were

Table 1. Botulinum toxin antisera titers.[a]

Antiserum	Mouse Protection IU/ml[bc]	ELISA Titer
I. Antiserum produced using Freund's adjuvant (IM) only		
Goat anti-A	514	1:1000, 1:6000
Goat anti-B	81	Not determined
Goat anti-E	228	1:1000
Rabbit anti-A	86	1:6000
Rabbit anti-B	20	Not determined
II. Antiserum produced using Freund (IM) and Ribi (ID) adjuvants		
Rabbit anti-NTA	775	1:4000-8000
Rabbit anti-NTB	4	Ineffective
Rabbit anti-E	5500, 7100	1:5000-8000
Rabbit anti-F	276	1:5000

[a] Abbreviations: IU, international unit; IM, intramuscular; ID, intradermal; NTA or NTB, botulinum neurotoxin protein A or B (as opposed to botulinum toxin complex A, B, E, or F).
[b] We thank C. Hatheway, the Centers for Disease Control, Atlanta, GA, for performing some of the mouse neutralization titers.
[c] One IU of A, B, or F toxin is 10,000 MLD_{50}. One IU of E toxin is 1000 MLD_{50}.

Table 2. Antisera used in amplified botulinum ELISAs.

Antibodies used	Detects	Reference
MAb[a] + Guinea-pig anti-NTA[b] conjugate	NTA	12,17
MAb + Guinea-pig anti-NTB[c] conjugate	NTB	13,17
Goat anti-A + rabbit anti-NTA	NTA	FSIS
Goat + rabbit anti-A	>Hemagglutinins A/B + <NTA	FSIS
Goat + rabbit anti-E	Nontox E protein = NTE[d]	FSIS
CDC Horse + rabbit anti-F	F	FSIS
Chicken IgY + rabbit IgG anti-C	C	T. Rocke
Connaught[e] goat IgG anti-A, B, E + IgG A, B, E biotin conjugate	A, B, E, F	M. Potter

[a] MAb, monoclonal antibody.
[b] NTA, neurotoxin A protein.
[c] NTB, neurotoxin B protein.
[d] NTE, neurotoxin E protein.
[e] Connaught Laboratory.

stable for several years when stored at -25°C in a 50% glycerol diluent (see Appendix II). Working dilutions of these standards in the 0-100 pg/ml range can be prepared in phosphate casein (PC) buffer (Appendix II) and discarded daily.

Many clinicians have cited the instability of dilute aqueous solutions of botulinum toxins to be a major obstacle to human medical use. Thus, the storage of stable neurotoxin coupled with accurate dispensing as described above may have application in the medical community. However, for human injections, the fetal bovine serum component in the diluent must be replaced with a non-immunogenic protein stabilizer such as sterile human albumin. Sterile neurotoxin can be stabilized by storing at -25°C in a 50% glycerol diluent (concentrated about 100 times the intended use dilution). Just before use, an aliquot of the stable toxin concentrate can be withdrawn with a Hamilton micro-syringe and diluted to the appropriate concentration in sterile saline. The amount of glycerol in the final injection would be minimal (0.005 ml or less).

Other supplies and reagents, such as alkaline phosphatase conjugated anti-rabbit IgG, 96-well plates, and amplified substrates are available commercially (Appendix II). Reagents from different sources may differ in concentration or specificity and thus may have profound effects on ELISA results, so new sera and reagent lots should always be tested.

RESULTS OF FSIS BOTULINUM TOXIN ELISAs

The optimum antibody dilutions used in the FSIS ELISAs were determined by using 100 pg/ml of botulinum neurotoxin standards. The standard curves for the Botulinum Neurotoxin A and Hemagglutinin A/B ELISA (which detects toxin A complex) are presented in Fig. 1. The data show that the neurotoxin A standard curve with concentrations from 0-100 pg/ml is a straight line. This was also true of the FSIS Amplified Botulinum Toxin ELISA standard curves of the other toxin types.

Hemagglutinin A at a concentration of 100 pg/ml was not detected by the FSIS Amplified Botulinum Neurotoxin A ELISA. To accomplish this, the small amount of anti-hemagglutinin A activity in the rabbit anti-neurotoxin A serum was removed by absorption using beads of CH Sepharose 4B-ß-D-thiogalacto-pyranoside-hemagglutinin A.[23] The anti-hemagglutinin activity of the rabbit antiserum was also kept low by administering only small booster doses (150-200 µg) of botulinum neurotoxoid A.

The FSIS Botulinum Neurotoxin A ELISA (Fig. 1) is specific for neurotoxin A, even though the goat anti-A serum, which binds both the hemagglutinin and the neurotoxin A, was used as the capture antibody (first coating). This shows that the rabbit anti-neurotoxin A serum is highly specific, and that only one of the two antisera used in a sandwich botulinum neurotoxin ELISA must be neurotoxin specific. One highly specific antiserum reduces the stringency required of the other antiserum used in a neurotoxin specific ELISA.

Like the Lowry protein assay,[24] the FSIS Amplified Botulinum ELISAs, using linear regression equations to prepare standard curves (Fig. 1), allow determination of unknown quantities of botulinum neurotoxins and hemagglutinins. Hemagglutinin in botulinum toxin A complex can be determined using the Hemagglutinin A/B ELISA by subtracting the absorbance contributed by the neurotoxin A component. This ELISA is 5 times more reactive with hemagglutinin A than with neurotoxin A (Fig. 1). The antibodies prepared for this ELISA also reacted strongly with the hemagglutinin component of type B botulinum toxin complex, but did not react with 100 pg/ml of purified neurotoxin B. The Botulinum Neurotoxin A and the Hemagglutinin A/B ELISAs (Table 2, Fig. 1) were used to determine quantities of neurotoxin A (NTA) and hemagglutinin A or B in various samples (Table 3).

Standard curves for neurotoxin E and the associated nontoxic E protein generated using the Amplified Botulinum Toxin E ELISA are presented in Fig. 2. The data show that the specific activity of the antisera for botulinum neurotoxin E and the type E nontoxic protein were essentially equal. C. botulinum type E cultures produce equal amounts of these 2 proteins, and the molecular size of these 2 proteins is about the same.[25] The standard curves (Fig. 2) were used to determine unknown quantities of botulinum neurotoxin E. The amount of neurotoxin E in crude culture supernatants was estimated by dividing the results by two. However, the test could not be used to

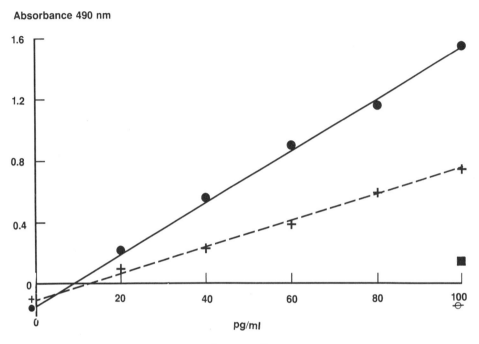

Absorbance 490 nm

Fig. 1. Standard curve for the Amplified Botulinum Neurotoxin A ELISA (●) and its reactivity to 100 pg/ml of hemagglutinin A (⊖). Standard curve for the Amplified Botulinum Hemagglutinin A/B ELISA (+) and its reactivity to 100 pg/ml of neurotoxin A (■).

determine the quantity of neurotoxin E in preparations with unknown proportions of neurotoxin E and nontoxic E proteins. The FSIS Amplified Botulinum Toxin E ELISA (Fig. 2) was able to detect botulinum toxin type E produced by C. butyricum.[26]

One antiserum produced against botulinum toxoid type E, was specific for botulinum toxin E complex (Table 1 and 2). Another antiserum (Table 1), produced with the same toxoid, but using a slightly different immunizing procedure also reacted in the ELISA with culture supernatants of related nontoxic type E and nonproteolytic type B and F strains.[2,3]

We also developed a very sensitive ELISA for the detection of type F botulinum toxin, but lacked purified botulinum neurotoxin F and the associated hemagglutinin F protein standards to prepare standard curves for this ELISA.

Table 3. Neurotoxin and hemagglutinin concentration of some botulinum toxins and culture supernatants as determined by various assays.

Toxin or Culture	NT Bioassay[a]	NTA ELISA	Hem ELISA	Hem/NT Ratio
U. Wisconsin A (ng/vial)	0.5	4.8	40.3	8.3
Wako A[b] (μg/mg)	80	144	1450	10.1
62A Supn (μg/ml)	2	3.8	30.8	8.1
33A Supn (μg/ml)	Not tested	0.45	3.73	8.3
2012A Supn (ng/ml)	2.5	2.9	7.4	1.5
Wako B[b] (μg/mg)	>20	Not tested	474[c]	---
6419B Supn (μg/ml)	0.2	Not tested	11.8[c]	---

Abbreviations: NT, neurotoxin; Hem, hemagglutinin; Supn, supernatant
[a]Mouse Neutralization Test
[b]Wako A and B neurotoxin vials were each labelled 150 μg.
[c]Approximate hemagglutinin B amounts, since the hemagglutinin A standard was used.

BOTULINUM TOXIN ANTIBODY DETECTION USING ELISA

The FSIS Amplified Botulinum Toxin ELISAs can be modified to titer botulinum antisera (R. H. Whitlock, S. A. Crane and C. Buckley, unpublished results). Whitlock et al. used a standard amount of capture antibody with a standard amount of botulinum toxin, and then used the serum to be titered as the detection antibody. They found that the rate of absorbance change is directly proportional to mouse protection titers.

The use of the FSIS Amplified Botulinum Toxin ELISA to determine antisera titers has potential applications, but the development of such assays is greatly hampered by the unavailability of botulinum neurotoxin specific antisera needed as the capture antibody. Currently, C. Hatheway (personal communication) at the Centers for Disease Control is experimenting with this procedure to detect antibodies in the sera of human patients who have received equine toxoid immunization or have been exposed to botulinum toxins.

DISCUSSION

Fig. 1 and 2 show that the FSIS Amplified Botulinum Toxin ELISAs are highly sensitive quantitative protein assays for detecting components of botulinum toxins. In

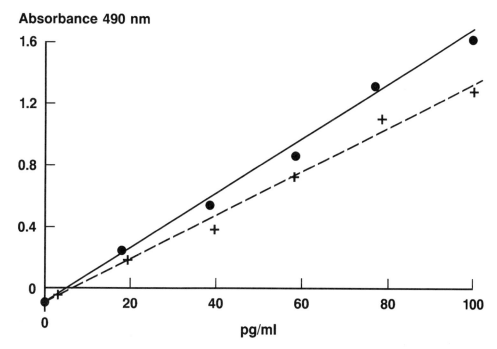

Absorbance 490 nm

Fig. 2. Standard curves for neurotoxin E (●) and the associated nontoxic E protein (+) of the FSIS Amplified Botulinum Toxin E ELISA.

addition to being convenient assays for cultures, the ELISAs are also useful for monitoring toxin production in foods inoculated with C. botulinum. However, a problem which exists with our ELISA, as well as with others which rely on polyclonal antibodies,[7] is cross-reactivity to some nontoxic cultures at low (1:10) dilutions yielding false positive results. Therefore these assays may not be suitable as confirmatory tests for food samples or culture supernatants containing low levels of botulinum toxins. The MAb neurotoxin A and B ELISAs developed by Porton do not appear to exhibit cross-reactivity with nontoxic clostridial cultures.[12,13]

The results in Table 3 show that the FSIS Amplified Botulinum Neurotoxin A ELISA is capable of detecting the biologically active and some biologically inactive forms of neurotoxin A. The Neurotoxin A ELISA also detected neurotoxin A from C. botulinum strain NCTC 2012A, which was not detectable by Porton's earlier MAb-based ELISA.[12] However, our Neurotoxin A ELISA failed to detect formalin treated neurotoxoid A, and reacted poorly with the tetranitromethane treated neurotoxoid A. Tetranitromethane specifically converts botulinum toxin tyrosine residues to nitrotyrosine,[27] whereas formalin, a strong reducing agent, causes a more general denaturation of the protein. Our data (Table 3), as well as that of others,[8,12,13] indicate that prolonged freezing or freeze drying inactivates the biological and immunological activities of botulinum neurotoxins. The relationship between the gradual denaturation of botulinum neurotoxins and their biological or immunological activities is not known. Since the biological and immunological activities of neurotoxin proteins are not always the same,[28] it is best not to equate botulinum toxin ELISA results with the mouse bioassay. However, the assays complement each other.

As future modifications are made to botulinum neurotoxins for the production of therapeutic agents, quantification by mouse bioassay may become no longer possible.[29] Dr. H. Niemann describes elsewhere in this book molecular biology and cloning techniques which allow precise modifications of botulinum neurotoxins. It is reasonable to assume that in the near future it will be possible to create suitable therapeutic agents by modifying botulinum neurotoxins to reduce toxicity and immunological activity. A potential medical use for a liposome encapsulated light chain of botulinum neurotoxin, which has little or no toxicity when measured by the mouse bioassay,[29] has been reported. The FSIS Amplified Botulinum ELISA, using antibodies that react specifically with the light or heavy chain of the botulinum neurotoxin, would facilitate measurement of each of these neurotoxin components. This assay provides a sensitive method for determining the presence of or quantifying the amount of botulinum toxin, or one of its components, in a sample or culture.

REFERENCES

1. Hauschild AHW. Clostridium botulinum toxins. Int. J. Food Microbiol. 1990; 10:113-24.
2. Lee WH, Riemann H. Correlation of toxic and non-toxic strains of Clostridium botulinum by DNA composition and homology. J. Gen. Microbiol. 1970; 60:117-23.
3. Lee WH, Riemann H. The genetic relatedness of Clostridium botulinum strains. J. Gen. Microbiol. 1970; 64:85-90.
4. Suen JC, Hatheway CL, Steigerwalt AG et al. Genetic confirmation of identities of neurotoxigenic Clostridium barati and Clostridium butyricum implicated as agents of infant botulism. J. Clin. Microbiol. 1988; 26:2191-92.
5. Suen JC, Hatheway CL, Steigerwalt AG et al. Clostridium argentinense sp. nov.: a genetically homogenous group composed of all strains of Clostridium botulinum toxin type G and some nontoxigenic strains previously identified as Clostridium subterminale or Clostridium hastiforme. Int. J. System. Bacteriol. 1988; 38:375-81.
6. Kautter DA, Lynt RK. Clostridium botulinum In: Speck ML, ed. Compendium of methods for the microbiological examination of foods. Washington, DC: American Public Health Assoc., 1976:424-36.
7. Notermans S, Hagenaars M, Kozaki S. The enzyme-linked immunosorbent assay (ELISA) for the detection and determination of Clostridium botulinum A, B, and E. Methods Enzymol. 1982; 84:223-38.
8. Ashton AC, Growther JS, Dolly JD. A sensitive and useful radioimmunoassay for neurotoxin and its hemagglutinin complex for Clostridium botulinum. Toxicon. 1985; 23:235-46.
9. Dezfulian M, Hatheway CL, Yolken RH et al. Enzyme-linked immunosorbent assay for detection of Clostridium botulinum type A and B toxins in stool samples of infants with botulism. J. Clin. Microbiol. 1984; 20:379-83.
10. Ferreira JL, Hamdy MK, Herd ZL et al. Monoclonal antibody for the detection of Clostridium botulinum type A toxin. Mol. Cell. Probes. 1987; 1:337-45.
11. Ferreira JL, Hamdy MK, McCay SG et al., Monoclonal antibody to type F Clostridium botulinum toxin. Appl. Environ. Microbiol. 1990; 56:808-11.
12. Gibson AM, Modi NK, Roberts TA et al. Evaluation of a monoclonal antibody-based immunoassay for detecting type A Clostridium botulinum toxin produced in pure culture and an inoculated model cured meat system. J. Appl. Bacteriol. 1987; 63:217-26.

13. Gibson AM, Modi NK, Roberts TA et al. Evaluation of a monoclonal antibody based immunoassay for detecting type B Clostridium botulinum toxin produced in pure culture and an inoculated cured meat system. J. Appl. Bacteriol. 1988; 64:285-91.

14. Henning VK, Meyer H, Cremer J. Schnellnachweis der Clostridium botulinum-Toxine A und B mittels Enzym-Immuno-Assay (ELISA). Arch. Lebensmittelhyg. 1991; 42:49-71.

15. Kamata Y, Kozaki S, Nagai T et al. Evaluation of ELISA techniques for titration of monoclonal antibodies against botulinum toxin. Jpn. J. Vet. Sci. 1986; 48:909-14.

16. Michalik M, Grzybowski J, Ligieza J et al. Enzyme-linked immunosorbent assay (ELISA) for the detection and differentiation of Clostridium botulinum toxins type A and B. J. Immunol. Meth. 1986; 93:225-30.

17. Shone CP, Wilton-Smith P, Appleton N et al. Monoclonal antibody-based immunoassay for type A Clostridium botulinum toxin is comparable to the mouse bioassay. Appl. Environ. Microbiol. 1985; 50:63-7.

18. Wang YC, Sugiyama H. An ELISA for screening and detecting botulinum toxins (A, B, and E). Chin. J. Microbiol. Immunol. 1986; 6:253-56.

19. DasGupta BR. Structure and structure function relation of botulinum neurotoxins. In: Lewis Jr. GE ed. Biomedical aspects of botulism. New York, NY: Academic Press, 1981:1-20.

20. Onishi I, Sugii S, Sakaguchi G. Oral toxicities of Clostridium botulinum toxins in response to molecular size. Infect. Immun. 1977; 16:107-9.

21. Sakaguchi G, Onishi I, Kozaki S. Purification and oral toxicities of Clostridium botulinum progenitor toxins. In: Lewis Jr. GE ed., Biomedical aspects of botulism. Academic Press, New York, NY: 1981:21-34.

22. Anderson Jr. JH, Lewis Jr. GE. Clinical evaluation of botulinum toxoids. In: Lewis Jr. GE ed. Biomedical aspects of botulism, New York, NY: Academic Press, 1981:233-46.

23. Moberg LJ, Sugiyama H. Affinity chromatography purification of type A botulinum neurotoxin from crystalline toxic complex. Appl. Environ. Microbiol. 1978; 35:878-80.

24. Lowry OH, Rosebrough NJ, Farr AL et al. Protein measurement with the Folin phenol reagent. J. Biol. Chem. 1951; 193:265-75.

25. Gimenez JA, Sugiyama H. Simplified purification method for Clostridium botulinum type E toxin. Appl. Environ. Microbiol. 1987; 53:2827-30.

26. Gimenez JA, Sugiyama H. Comparison of toxins of Clostridium butyricum and Clostridium botulinum type E. Infect. Immun. 1988; 56:926-29.

27. Woody M, DasGupta BR. Effect of tetranitromethane on the biological activities of botulinum neurotoxin types A, B and E. Mol. Cell. Biochem. 1989; 85:159-69.

28. Sakaguchi G, Sakaguchi S, Kurazono H et al. Persistence of specific antigenic protein in the serum of chickens given intravenous botulinum toxin type B, C, D, E, or F. FEMS Microbiol. Let. 1987; 43:355-9.

29. De Paiva A, Dolly JO. Light chain of botulinum neurotoxin is active in the mammalian motor nerve terminals when delivered via liposomes. FEBS Let. 1990; 277:171-7.

APPENDIX I. PROCEDURE FOR HIGH TITER RABBIT ANTISERA PRODUCTION

Immunogen Reagents

Formalin or tetranitromethane derived botulinum neurotoxoids were obtained from B. R. Dasgupta, University of Wisconsin, Madison WI. The neurotoxoids were in a stabilized crystalline form in 50% ammonium sulfate solution stored at -25°C. Just before use, the ammonium sulfate solution was removed by centrifugation. The crystalline neurotoxoid pellet was then dissolved in merthiolated saline. Other formalin derived botulinum toxoids were supplied by G. Sakaguchi. They are also available from the Wako Chemical Inc., Richmond, VA. Ribi R700 adjuvants were obtained from Ribi Immunochem Research Inc., Hamilton, MT. Freund's adjuvants were obtained from commercial sources. New Zealand male white rabbits (6 weeks old), obtained from Hazelton Research Products, Inc., Denver, PA, were used to produce the antisera.

Adjuvant-Antigen Preparation

The neurotoxoid precipitate (200 μg), was dissolved in 1.0-1.5 ml merthiolated saline and then combined with an equal volume of Freund's adjuvant. When Ribi R700 lyophilized adjuvant was used, the volume of merthiolated saline was increased to approximately 2.0 ml to provide sufficient volume for immunization. Ribi R730 adjuvant did not improve antibody titer over Ribi R700, and R730 tended to cause severe skin lesions. The entire volume of adjuvant-antigen preparation was administered at each injection time according to the following schedule.

Rabbit Immunization Schedule

Initial injection: Freund's complete adjuvant-antigen preparation, intramuscular (IM).

Second injection: Freund's incomplete adjuvant-antigen preparation, IM, 2 weeks later.

Third injection: Ribi R 700 adjuvant-antigen preparation, intradermal (ID), 2 weeks later.

Fourth and any additional booster injections: Ribi R700 adjuvant-antigen preparation ID every 6 to 8 weeks. Booster immunizations, after the first series of 6 injections can be suspended for extended periods of time (4-9 months) without loss of high serum titer.

Collecting Antisera

Seven days after each booster injection, about 30 ml of blood was collected by ear bleed. Rabbits produced peak titer antisera after the fifth or sixth booster and continued to do so as long as they remained in good health.

Processing and Storing Antisera

After harvest, blood was maintained at room temperature for 2 hours. Then red blood cells were pelleted by centrifugation. Antiserum was decanted and mixed

1:1 (v\v) in sterile glycerol. The 50% glycerol-antisera preparations were stored in cryovials at -25°C. For convenience, antiserum or conjugate was further diluted (1:5 to 1:20) with antiserum diluent. Fresh working dilutions of antisera were prepared in Carbonate (CO_3) buffer or PC Buffer for the ELISA.

APPENDIX II. FSIS AMPLIFIED BOTULINUM TOXIN ELISA PROCEDURES

Reagents and Supplies

Nunc maxisorb 96-well immuno-plates (# 439454)
Nunc minisorb tubes (#468608 & #343306) (for working dilutions of all toxins and antisera)
Amplification substrate (# 9589SA, GIBCO-BRL, Gaithersburg, MD)
Sterile normal goat serum (NGS) (GIBCO-BRL)
Sterile fetal bovine serum (FBS) (GIBCO-BRL)
Anti-rabbit IgG alkaline phosphatase reagent (# 605230, Boehringer Mannheim, Indianapolis, IN)
Casein (# C5890) and other chemicals (Sigma Chemical Co., St. Louis, MO)

Buffers and Diluents

Prepare all buffers with double distilled water and <u>autoclave</u>.

Carbonate (CO_3) Buffer
Add 1.59 g Na_2CO_3 and 2.93 g $NaHCO_3$ per liter of water. Adjust to pH 9.6. Store at 4°C.

5% Casein (10X)
Dissolve 50 g casein in 800 ml of 0.3 $\underline{M}$ KOH by stirring overnight. Then adjust the pH to 8.5 with 5 $\underline{N}$ HCl. Bring volume to 1 liter with distilled water. Store frozen in 40 ml aliquots. Note: Casein solution contains 0.24 $\underline{M}$ KCl. It is not necessary to add KCl to the PC buffer below.

Phosphate Casein (PC) Buffer
Add 8 g NaCl, 1.25 g NaH_2PO_4, 0.2 g KH_2PO_4, and 0.5 g Tween 20 per liter of water. Add 100 of the 5% casein solution per liter and adjust pH to 7.4. Store in 400 ml quantities at-25°C.

Tris (T) Wash Buffer
Add 7.9 g Tris-HCl, 11.7 g NaCl, and 0.224 g KCl per liter of water. Adjust pH to 7.5 with 10 $\underline{N}$ NaOH and store in 400-1000 ml quantities at 4°C.

Antiserum Diluent
Add 50 ml sterile fetal bovine serum (FBS) to 50 ml sterile glycerol. Store at -25°C.

Toxin Stabilizing Diluent
Add 25 ml sterile FBS and 50 ml sterile glycerol to 25 ml sterile 0.1 $\underline{M}$ citric acid, 0.2 $\underline{M}$ K_2HPO_4 buffer, pH 5.5. Store at -25°C.

Botulinum Neurotoxins and Nontoxic Associated Proteins Standards

Dissolve neurotoxin in toxin stabilizing diluent and prepare a known concentration as specified by the weight on the vial label (i.e. 10 μg/vial). Prepare intermediate diluted stock solutions of 20-1000 ng/ml from the original stock solution using toxin stabilizing diluent. Store all toxin standard stock solutions in cryovials at -25°C (toxin stabilizing diluent prevents freezing at this temperature).
[Crystalline botulinum neurotoxins and nontoxic associated proteins were obtained from B. R. DasGupta. Wako botulinum toxins were also used as toxin standards].

Use positive displacement micropipets (Microman) to accurately dispense toxin standards. Wipe the micropipet tip before dispensing into diluent.

Make fresh working dilutions of toxin standards in PC buffer daily. For instance, to prepare 4 ml of a 100 ng/ml diluted stock solution from a 10 μg/ml concentrated stock, make a 1:100 dilution using 40 μl of the concentrate and 3.96 ml of the toxin stabilizing diluent. This solution is kept as an intermediate stock at -25°C. To prepare 5 ml of a working toxin dilution of 100 pg/ml, make a 1:1000 dilution of the 100 ng/ml intermediate stock by adding 5 μl of the stock to 4.995 ml (5.0 ml) of PC buffer. Dilute unknown botulinum toxin samples (i.e., culture supernatants or foods) in PC buffer. Various dilutions, such as 1:10 and 1:100, can be assayed. Use ELISA working dilutions only on the day of preparation.

Antisera Storage and Dilutions

Mix antisera stocks 1:1 with sterile glycerol and store at -25°C in cryovials. For convenience, further dilute antisera or conjugate (1:5 to 1:20) with antiserum diluent. For use in the ELISA, prepare fresh working dilutions of capture antibody in CO_3 buffer or detection antibody in PC buffer. For example, to prepare 10 ml of a 1:1000 dilution of goat anti-A as capture antibody, make a 1:200 dilution of the intermediate 1:5 stock using 50 μl of the stock and 9.95 ml of CO_3 buffer.

Determination of Antisera Dilutions or Titers for the ELISA

The optimum antisera dilutions were determined by a cross titration using 100 pg/ml neurotoxin and a 1:5000 dilution of the Boehringer Mannheim conjugate. The highest antiserum dilution yielding a net absorbance value of 1 at 490 nm was considered an optimum titer. The titers of different lots of antisera and conjugates were determined separately. Examples of antisera working dilutions for the three ELISAs are listed below:

ELISA	Capture Antibody		Detection Antibody
Neurotoxin A:	Goat anti-A 1:1000	+	Rabbit anti-NTA 1:6000
Hem A/B/NTA:	Goat anti-A 1:6000	+	Rabbit anti-A 1:6000
NTE/Nontox E:	Goat anti-E 1:1000	+	Rabbit anti-E 1:4000

Higher dilutions of antisera used for ELISA resulted in lower blank absorbance values. Antiserum coated wells, with PC buffer added in place of toxin or sample, were used to blank or zero the ELISA reader.

FSIS Amplified Botulinum Toxin (Botox) ELISA

Test each sample in duplicate antibody-sensitized wells. As negative controls, add samples to wells coated with NGS diluted to the same concentration as the antiserum. Additionally, to correct for background color, prepare blank wells to zero the ELISA reader: add PC buffer to antiserum coated wells in place of toxin or sample.

For all incubation steps, place the 96-well ELISA plates in a sealed plastic chamber (Rubbermaid) humidified by moistened paper or a sponge.

For each washing step, wash the ELISA plates 5 times with T Wash Buffer (0.35 ml per well) with 1 minute soak intervals between washes. Since the amplification system allows detection of minute quantities of phosphate, it is important to eliminate all traces of phosphate left from the PC diluent before the substrate is added by using T wash buffer for all wash and soaking steps.

Procedure:

1. **Coating of Capture Antibody (Nonspecific Binding of Goat Antiserum or Normal Goat Serum).**
Dilute specific goat antiserum or normal goat serum in CO_3 buffer and add 0.1 ml to appropriate wells. Incubate at 5°C overnight or longer. Do not store plates dry after coating. Wash plate with T wash buffer before starting assay.

2. **Toxin Binding.**
Dilute botulinum toxin standard and unknown samples in PC buffer (prepare fresh). To appropriate wells, add 0.1 ml diluted botulinum toxin standard, diluted unknown sample, or buffer (negative control). Incubate 75 min at 35°C. Wash plate with T wash buffer.

3. **Detection Antibody (Second Antibody) Binding.**
Dilute rabbit antiserum in PC buffer and add 0.1 ml to each well. Incubate 75 min at 35°C. Wash plate with T wash buffer.

4. **Conjugate Binding.**
Dilute anti-rabbit IgG alkaline phosphatase conjugate (1:5000) in PC buffer and add 0.1 ml to each well. Incubate 75 min at 35°C. Wash plate with T wash buffer.

5. **Soaking.**
Soak plate for 10 min using 0.35 ml T wash buffer in each well. Aspirate. Invert the plate and tap on clean lab tissues to remove the last traces of buffer.

6. **Substrate.**
Add 0.05 ml of the GIBCO substrate solution to each well. Incubate 25 min at 35°C. Do not wash.

7. **Amplifier.**
Add 0.05 ml of the GIBCO amplifier solution to each well and mix briefly. Incubate 25 min at 35°C. Do not wash.

8. **Stop.**
Add 0.05 ml of 0.3 $\underline{M}$ sulfuric acid to each well and mix. Do not wash. Read absorbance at A_{490} nm. Note: Some color develops in the negative (buffer) control wells. These wells are used to blank the ELISA reader to zero.[2]

DEVELOPMENT OF A MOLECULAR ENGINEERED VACCINE FOR *C. BOTULINUM* NEUROTOXINS

H. F. LaPenotiere, M.A. Clayton, J. E. Brown and J. L. Middlebrook

Toxinology Division
USAMRIID
Fort Detrick
Frederick, MD 21702

ABSTRACT

Recent work involving recombinant tetanus neurotoxin fragments demonstrate the "C fragment" (H_c domain) is a good vaccine candidate. This report describes initial efforts to determine if botulinum neurotoxin H_c fragments will also elicit a protective immune response. Using clones encoding part of the heavy chain, gene segment constructs were designed to produce a native H_c polypeptide or an H_c polypeptide fused to *E. coli* maltose binding protein. Problems were encountered with both systems. Specifically, the construct for the native protein appeared to be unstable, while the fusion product appeared to be packaged in inclusion bodies that formed insoluble aggregates. Preliminary trials with animals indicated that vaccination with the fusion product conferred protection against native toxin challenge.

INTRODUCTION

Clostridium botulinum produces seven distinct neurotoxins, which are considered among the most potent toxins known to man.[1] The toxins act primarily on the neuromuscular junction, causing paralysis by inhibiting the release of the neurotransmitter, acetylcholine. The botulinum neurotoxins are synthesized as single polypeptide chains, which are subsequently cleaved by endogenous proteases to form two chains of 50 and 100 kd linked by a disulfide bridge. Further proteolysis with trypsin or chymotrypsin results in a second cleavage at an artifactual site at the midpoint of the heavy chain.[2] Previous work with tetanus and botulinum neurotoxins suggested the presence of a protective epitope in the 50 kd carboxyl terminus (Hc) region of the protein.[3,4] In the following work we describe initial efforts to characterize the vaccine potential of the H_c domain of botulinum neurotoxin type A.

Botulinum and Tetanus Neurotoxins, Edited by
B.R. DasGupta, Plenum Press, New York, 1993

MATERIALS AND METHODS

Materials

Restriction endonucleases and DNA modifying enzymes were obtained from GIBCO BRL (Gaithersburg, MD). Polymerase chain reaction (PCR) reagents were purchased from Perkin-Elmer Cetus (Norwalk, CT). SDS-PAGE precast gels and running buffers were acquired from Amersham (Arlington Heights, IL). All oligonucleotides were synthesized by Macromolecular Resources (Ft. Collins, CO). ELISA reagents were obtained in-house or from Sigma (St. Louis, MO) or Kirkegard and Perry Laboratories (Gaithersburg, MD).

Strains, vectors and growth conditions

The *Escherichia coli* host was K12 DH5a, purchased as competent cells from GIBCO BRL. Expression vectors pMAL, from New England BioLabs (Beverly, MA), and pKK233-2, from Pharmacia LKB (Piscataway, NJ), were used according to the manufacturers' standard protocols. The DNA clone coding for the H_c domain of *C. botulinum* toxin serotype A was pCBA3, kindly provided by Nigel Minton.[5]

Methods

Oligonucleotide primers incorporating appropriate terminal restriction enzyme sites were used to PCR amplify the H_c region of the *C. botulinum* clone pCBA3. Gel-purified insert DNA and vector DNA were cleaved with the appropriate restriction enzymes, purified on low melting point agarose, and ligated overnight at room temperature. Competent DH5α host cells were transformed according to suppliers' recommendations and plated on LB plates with 100 μg/ml ampicillin. Protein electrophoresis was run on precast 11-20% SDS PAGE at the manufacturers' recommended parameters. ELISA plates were incubated with capture antibody (horse anti-botulinum A polyclonal serum) overnight, then blocked with skim milk prior to application of various dilutions of test material, signal antibody (rabbit anti-botulinum A polyclonal serum), signal HRP conjugated anti-(rabbit IgG) and ABTS substrate solution. Plates were read on an automated reader at 405 nm.

RESULTS

Cloning Experiment

Supernatants of sonicated, IPTG-induced recombinant pMAL fusion *E. coli* cultures were tested for the presence of the botulinum H_c expression product by ELISA and SDS-PAGE. While ELISA results of the fusion product proved positive (Table 1), attempts to visualize the expression product on gradient SDS-PAGE gels stained with coomassie brilliant blue were unsuccessful. We believe this may be due to formation of an insoluble product. This hypothesis was supported by two observations. First, the induced *E. coli* cells appeared to harbor inclusion bodies not readily observed in non-induced cells. Second, when pellets from lysed, induced cell cultures were resuspended and examined by ELISA, the signal often equaled or exceeded that observed with the lysate supernate. Other laboratories have expressed tetanus neurotoxin gene fragments corresponding to H_c plus additional heavy chain sequence in *E. coli* and, in at least one case, the expressed protein was reported to be insoluble.[6]

Table 1. Representative ELISA data of fusion protein product (OD_{405}).

Dilution factor	Toxin control	pMALc control		BotAC/pMALc		BotAC/MALc		Blank well
		Supe.	Pellet	Supe.#1	Supe#2	Pellet#1	Pellet#2	
102400	1.766	0.205	0.257	0.474	0.444	0.273	0.292	0.237
25600	2.073	0.273	0.258	0.869	0.768	0.429	0.433	0.235
6400	2.295	0.225	0.332	1.095	1.191	0.745	0.799	0.215
1600	2.057	0.243	0.352	1.292	1.196	1.174	1.078	0.168
400	2.437	0.319	0.370	1.227	1.371	1.114	1.345	0.141
100	1.994	0.084	0.342	0.692	1.519	0.997	0.240	0.126

Attempts to express the H_c fragment as a non-fusion product were unsuccessful. Initial characterization of plasmid DNA from putative clones in pKK233-2 demonstrated an insert of the expected size was present. In addition, SDS-PAGE indicated the presence of a protein of approximately 50 kd after induction. However, the recombinants appeared unstable and further preparations of this and other cultures failed to reproduce these results. This approach was subsequently abandoned in favor of the fusion product expression.

Immunization Trials

Although attempts to quantitate expressed H_c fusion products were unsuccessful, limited immunization trials were performed on mice to evaluate the vaccine potential of the product. Initial vaccination employed concentrated, crude *E. coli* lysate with complete Freund's adjuvant. Two weeks later, animals were boosted with amylose column-purified expression product with Freund's incomplete adjuvant. At this time, a second group of five animals received amylose purified product with Freund's incomplete adjuvant as a single vaccination. After two additional weeks, both groups were challenged intraperitoneally with a dose of 3 LD_{50} of toxin. All eleven animals receiving two immunizations with Hc survived while six of the twelve control animals receiving pMAL vector alone died. Likewise, all five animals receiving one H_c vaccination survived, while animals receiving the pMAL vector alone died.

Table 2. Response to botulinum neurotoxin type A challenge in mice (# deaths/total).

Calculated challenge dosage (LD_{50})	One immunization[a]		Two immunizations[b]	
	Control (vector without H_c insert)	Experimental (vector with H_c insert)	Control (vector without H_c insert)	Experimental (vector with H_c insert)
3	3/3	0/5	6/12	0/11
30				1/3[c]
300				1/3[c]
1200				1/3

[a]Animals received 0.5 ml. amylose column-purified product in saline, SC with Freund's incomplete adjuvant followed by challenge in two weeks.
[b]Animals received equivalent of 100 mls. crude sonicated *E. coli* culture lysate resuspended in 0.5 ml. saline with Freund's complete adjuvant at time 0, followed by boost with amylose-purified product as above at week 2 and challenge at week 4.
[c]Deaths were atypical of toxin-associated mortality.

The apparent protection of mice from low doses of toxin encouraged us to use higher challenge toxin amounts. Thus, four weeks after their survival from 3 LD_{50}, nine of the eleven animals receiving two immunizations were exposed to 30, 300, or 1200 LD_{50} (Table 2). The animals succumbing to the toxin challenge of 30 and 300 LD_{50} did not exhibit fatality typical of botulinum toxin poisoning in that they appeared healthy after 18 hours, but were dead a few hours later. In contrast, the animal which died from the 1200 LD_{50} dose appeared moribund when examined at 18 hours and remained so until death, consistent with the symptoms usually observed with botulinum toxin-induced paralysis.

CONCLUSION

The experiments presented here are intended to answer two questions: first, will genetically engineered H_c domain polypeptide of botulinum toxin type A elicit a protective immune response in an animal model, and second, is the genetically engineered product expressed in a stable, functional form in an *E. coli* host? Although preliminary in nature, we believe these data suggest this polypeptide product has potential as a vaccine candidate. Since the inception of this work, several groups have reported problems associated with the cloning and expression of *C. botulinum* genes in *E. coli*.[7] Although it is certainly possible that the problems with stability and solubility we encountered are product specific, we believe the major obstacle to expressing this and other *C. botulinum* gene fragments is related more to the extreme A/T richness of the associated nucleotide sequences. Thus, we are now constructing a synthetic gene encoding the H_c domain of the botulinum toxin A which will incorporate optimal *E. coli* codon usage. In addition, we are investigating the expression of this and other gene fragments in other expression systems.

Acknowledgements

We thank Dr. W. H. Lee of U.S.D.A, Beltsville, MD for kindly providing the rabbit anti-botulinum toxin A serum.

REFERENCES

1. Gill DM. Bacterial toxins: a table of lethal amounts. Microbiological Reviews 1982; 46:86-94.
2. DasGupta BR, Sugiyama H. Limited proteolysis of *Clostridium botulinum* types A and B neurotoxin. Am Soc Microbiol 1978:25.
3. Sheppard AJ, Cussell D, Hughes M. Production and characterization of monoclonal antibodies to tetanus toxin. Infect Immun 1984; 43:710-714.
4. Kozaki S, Kamata Y, Takahashi M, Shimizu T, Sakaguchi G. Antibodies against botulism neurotoxin. In: Simpson LL, ed. Botulism Neurotoxin and Tetanus Neurotoxin, New York: Academic Press, 1989.
5. Thompson DE, Brehm JK, Oultram JD, Swinfield TJ, Shone CC, Atkinson T, Melling J, Minton NP. The complete amino acid sequence of the *Clostridium botulinum* type A neurotoxin, deduced by nucleotide analysis of the encoding gene. Eur J Biochem 1990; 189:73-81.
6. Makoff AJ, Ballantine SP, Smallwood AE, Fairweather NF. Expression of tetanus toxin fragment C in *E. coli*: its purification and potential as a vaccine. Biotechnology 1989; 7:1043-1046.
7. Popoff MR, Hauser D, Boquet P, Eklund MW, Gill DM. Characterization of the C3 gene of *Clostridium botulinum* types C and D and its expression in *E. coli*. Infect Immun 1991; 59:3673-3679.

DEVELOPMENT OF AN AVIAN ANTITOXIN
TO TYPE A BOTULINUM NEUROTOXIN

B.S. Thalley,[1] M.B. van Boldrik,[1] S.B. Carroll,[2] W. Tepp,[3] B.R. DasGupta,[3] and D.C. Stafford[1]

[1]Ophidian Pharmaceuticals, Inc.
[2]Howard Hughes Medical Institute and University of Wisconsin-Madison
[3]University Of Wisconsin-Madison, Food Microbiology and Toxicology

ABSTRACT

Most commercially available antitoxins and antivenoms are raised in horses and purified by bulk fractionation techniques. These preparations frequently elicit deleterious side effects that compromise their efficacy and the treatment of intoxication or envenomation. Alternatively, the use of hyperimmune immunoglobulin from the egg yolks of laying hens has shown considerable promise as a cost-effective and potentially safer source of antitoxins and antivenoms. To investigate the utility of this system in the development of a botulism antitoxin, laying hens were immunized with purified 150 kDa type A botulinum neurotoxin detoxified with formaldehyde. Antibodies present in egg yolk were then isolated by polyethylene glycol (PEG) fractionation and further purified by affinity chromatography using immobilized toxoid. These antibodies exhibited a high titer of toxoid reactivity as measured by enzyme immunoassay. A mouse lethality/protection assay demonstrated the efficacy of these antibodies in neutralizing type A botulinum neurotoxin, where PEG-fractionated and affinity-purified antibodies neutralized 38 and 340 I.U. of type A botulinum neurotoxin per milligram of protein, respectively. Thus, the avian antibody preparations, even if only PEG-fractionated, are significantly more potent (neutralization titer per mg protein) than conventional equine botulism antiserum preparations. In addition to their superior potency, hyperimmune avian antibodies, whether PEG-fractionated or affinity-purified, are expected to significantly increase antitoxin safety.

INTRODUCTION

Botulism, a rare, but potentially fatal disease, can result from ingesting preformed botulinum neurotoxin present in food contaminated by *Clostridium botulinum*. Infant botulism, now more prevalent in the U.S. than foodborne botulism, is toxico-infectious.[1] Botulinum neurotoxin, the most toxic material known, binds tightly to presynaptic membranes of the peripheral nervous system, inhibiting the release of acetylcholine and causing acute flaccid paralysis.[2] Seven antigenically distinct serotypes (A–G) of the neurotoxin have been identified.[1]

Worldwide, the medically accepted standard for the treatment of poisonous intoxication or envenomation is the administration of specific, toxin-neutralizing antibodies. Typically, these antitoxins are preparations of immunoglobulin fractionated from the sera of horses hyperimmunized with one or more toxin types.[3,4] Horses are most commonly used to produce antitoxins because of their large blood volume. Botulism antitoxin (Botulism Antitoxin, Trivalent; Connaught Laboratories, Canada), also produced in horses, is generated by immunization with types A, B and E neurotoxin serotypes. More recently, antibodies derived from immunized human volunteers are being evaluated for their use in treating infant botulism.[5,6]

In an analysis of 132 case histories of type A foodborne botulism in the U.S., patients treated with the equine trivalent antitoxin had lower fatality rates and a shorter clinical course than those not receiving treatment.[7] However, clinical experience has also clearly demonstrated that antitoxins derived from animal serum can produce serious side effects, including serum sickness, anaphylaxis and anticomplement activity.[3,8,9] The relative incidence of these side effects varies according to the kind and amounts of antitoxin or antivenom administered, as well as the methods by which they are manufactured. For example, the currently available commercial equine anti-botulinum antitoxin causes adverse reactions in 9-21% of patients.[8,9]

Methods have recently been developed for harvesting improved animal-derived antitoxins and antivenoms from the eggs of specifically immunized laying hens.[10,11] The specific hen antibodies are highly effective in neutralizing a variety of toxins and venoms, and may offer several advantages over conventional equine antibody technology, including reduced side effects and lower production costs. In this study, avian antibodies to type A botulinum neurotoxin were prepared and their neutralization efficacy evaluated.

MATERIALS AND METHODS

Immunization

White Leghorn strain laying hens, obtained from local breeders, were immunized with purified type A botulinum toxoid, which was a pure (>99%) preparation of 150 kDa type A botulinum neurotoxin[12] detoxified with formaldehyde.[13] One ml of toxoid (1.0 mg/ml in 150 mM NaCl, 10 mM Na phosphate buffer, pH 7.2; or PBS) was emulsified with 1 ml complete Freund's adjuvant and injected at various intramuscular and subcutaneous sites. Booster immunizations of 0.25–0.75 mg toxoid, similarly emulsified in incomplete Freund's adjuvant, were given on days 14 and 21, and thereafter. Eggs were collected prior to and throughout the course of immunization.

Antibody Extraction

Total yolk Ig (IgY) was extracted according to a modification of Polson et al.[14] Preimmune or hyperimmune egg yolks were blended with 4 volumes of extraction buffer (PBS) and polyethylene glycol (PEG) 8000 was added to a concentration of 3.5%. The mixture was centrifuged at 11,000 ×g for 10 min. The supernatant was filtered through cheesecloth to remove any insoluble lipids, and PEG was added to a final concentration of 12% to precipitate total IgY. After centrifugation, the pellet containing crude IgY, was dissolved in the original yolk volume of PBS containing 0.005% thimerosal and stored at 2-8°C. Total protein in the crude immunoglobulin fraction was determined both by absorbance at 260 and 280 nm, and by the Lowry method.[15]

Affinity Purification Of Toxin Specific IgY

Type A botulinum toxoid covalently linked to derivatized agarose (Actigel A, materials and methods supplied by Sterogene Biochemicals, CA) was used to affinity purify of specific IgY. PEG-fractionated antibody was loaded on the column and unbound antibody was removed with successive washes of BBS-Tween (0.1 M boric acid, 0.025 M Na borate, 1 M NaCl, 0.1% (v/v) Tween 20 pH 8.3) and TBS (50 mM Tris, 150 mM NaCl, 0.005% thimerosal, pH 7.5). Antibodies specific for type A botulinum toxoid were removed by applying Actisep elution medium (Sterogene Biochemicals, CA) and immediately dialyzed against several liters of TBS. Affinity-purified antibody concentration was determined based on 1.3 A_{280} = 1 mg/ml IgY.

Enzyme Immunoassay (EIA)

The reactivity of PEG-fractionated and affinity purified IgY antibodies was assessed by EIA. A 96-well immunoassay plate was coated with type A botulinum toxoid (2.5 µg/ml in PBS). After several hours, the toxoid solution was removed and any unoccupied protein binding sites were blocked by filling the wells with bovine serum albumin (3 mg BSA/ml in PBS). The primary antibody (2.5 µg/ml, either PEG-fractionated or affinity-purified IgY) was 5-fold serially diluted in PBS containing 1 mg/ml BSA and placed in the wells for 2 hrs at room temperature. The wells were then washed with BBS-Tween followed by PBS-Tween and PBS to remove any unbound antibodies. The plate was incubated for 2 hours with alkaline phosphatase-conjugated anti-chicken Ig secondary antibody (Sigma, MO), diluted 1:750 in PBS containing 1 mg/ml BSA. The plate was washed as before to remove any unbound secondary antibody, and p-nitrophenyl phosphate (Sigma, MO) at 1 mg/ml in 50 mM Na_2CO_3, 10 mM $MgCl_2$ was added to each well. The color developed after 15-30 minutes at room temperature and the color change was measured at 410 nm using a Dynatech MR 700 microplate reader.

Neurotoxin Neutralization Bioassay

A neurotoxin neutralization assay, based on mouse lethality, was kindly performed by Dr. Charles Hatheway (Centers for Disease Control; Atlanta, GA); 2-fold serially diluted antibody preparations were tested against type A botulinum neurotoxin and the results were compared with the neutralization titer of WHO International Standards for type A botulinum antitoxin.[16] Dilution endpoints were calculated by the Reed and Muench method.[17]

RESULTS

Laying hens were immunized with purified type A botulinum toxoid and immune eggs were collected. IgY was PEG-fractionated from eggs prior to immunization and following various booster immunizations. Immunoreactivity was monitored during the course of immunization using a solid phase EIA. PEG-fractionated antibodies were strongly reactive against immobilized type A botulinum toxoid (Fig. 1, open triangles), as compared to non-reactive pre-immune IgY (Fig. 1, open squares). Purification of this preparation using a type A botulinum toxoid affinity column further enhanced the toxoid-binding activity (Fig. 1, solid squares), and yielded a protein preparation which, judged by SDS-polyacrylamide gel electrophoresis, is greater than 99% immunoglobulin heavy and light chains (data not shown).

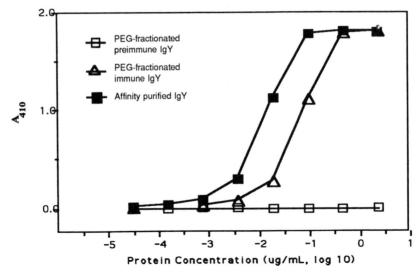

Figure 1. EIA measuring direct binding of avian antibody preparations on immobilized type A botulinum toxoid.

Toxin neutralization assays were performed after antibodies were found to have sufficient titers by EIA. Typically, neutralization results are reported as I.U. (International Unit = 10^4 LD_{50}) neutralized per milliliter of antitoxin solution tested. Using this method, it appears as if the PEG-fractionated IgY is more potent (454 I.U./ ml) than the affinity purified IgY (16 I.U./ ml) (Table 1, middle column). A more precise comparison of antitoxins, however, can be made by analyzing neurotoxin neutralization titer per *milligram* of protein. Such an analysis is presented in Table 1. Note that affinity-purification increased the potency nearly 10 times over the PEG-fractionated antibody (340 I.U./mg vs. 38 I.U./mg, respectively; see right-hand column). While the affinity purified IgY in this study is relatively dilute, these antibodies can be readily concentrated to 10 mg IgY/ml using conventional methods.

Table 1. Neutralization of type A botulinum neurotoxin with avian antibodies.

Sample	Protein concentration (mg/ml)	Titer[1]	I.U./ml	I.U./mg
PEG-fractionated preimmune IgY	12	<1:1	<0.126	>0.01
PEG-fractionated hyperimmune IgY	12	1:36096	454	38
Affinity-purified IgY	0.047	1:1272	16	340

Mouse protection bioassay comparing PEG-fractionated preimmune and hyperimmune IgY, and affinity-purified IgY anti-type A botulinum toxin. Results show I.U. per *milliliter* of test antibody solution and calculations of I.U. per *milligram* based upon protein concentration.
[1]Dilution endpoint at which 50% of mice survive.

DISCUSSION

Conventional antitoxins and antivenoms are typically manufactured by salt fractionation of hyperimmune horse serum. Often these antibodies are then enzymatically digested to remove the Fc fragment which is able to bind complement and trigger inflammatory side effects. While effective in neutralizing venoms and toxins, these concentrated protein products contain both non-immunoglobulin proteins and antibodies that are therapeutically irrelevant to the neutralization of toxins. These contaminating proteins not only significantly reduce the potency (neutralization titer per mg protein) of the antitoxin, but contribute to the onset of serum sickness because of the infusion of large amounts of foreign protein. In this study, we have prepared both PEG-fractionated and affinity purified type A botulinum antitoxin from egg yolks obtained from hyperimmunized laying hens. These methods have yielded a more potent antitoxin, which is anticipated to significantly reduce, if not eliminate, the incidence and severity of serum sickness associated with conventional antitoxin treatment.

The current U.S.P. specifications[18] for botulinum antitoxin (i.e., as manufactured by Connaught Laboratories, Canada) require a minimum neutralization potency against type A neurotoxin of 7500 I.U. per 10 ml vial. The specifications also require that protein levels not exceed 200 mg/ml, however, the protein concentration may be adjusted from lot to lot in order to achieve the appropriate neutralizing activity. Thus, antitoxin meeting the minimum U.S.P. requirements can have a specific activity as low as 3.75 I.U. anti-type A botulinum neurotoxin per mg protein.

For comparison purposes, the protein concentration of a vial of Connaught Botulism Antitoxin was measured and shown to be 171 mg/ml. Assuming that this 10 ml vial contained the minimum specified anti-type A botulinum neurotoxin of 7500 I.U., the specific activity is calculated to be at least 4.4 I.U./mg. Using this value for general comparison, we calculated that both the PEG-fractionated and affinity purified anti-type A botulinum neurotoxin avian antibodies are significantly more potent (as much as 9- and 77-fold, respectively) than the Connaught product. Thus, to achieve the same type A-neutralizing capability, approximately 1/9th to as little as 1/77th the amount of avian antibody need be administered. Such a reduction of heterologous protein should significantly reduce the incidence and severity of serum sickness.

Avian antibodies present additional advantages to the treatment of intoxication because IgY does not fix human complement.[19,20] In contrast, horse,[21] and even processed human immunoglobulin,[22] have been shown to trigger human complement activation. Therefore, the intravenous administration of mammalian antibodies as human therapeutics can result in complement-mediated side effects that can complicate effective treatment and may be confused with true immediate hypersensitivity.[3,23] To avoid this reaction, some equine antivenoms and antitoxins (e.g., botulinum antitoxin) undergo an additional enzymatic processing step to remove the Fc region of the equine immunoglobulin. This processing step is often incomplete, reduces specific antibody yields, and adds to the complexity and cost of manufacturing. The use of non-complement binding avian antibodies would obviate the need for this procedure.

Thus, purified avian antibodies are anticipated to reduce or eliminate side effects common to equine serum products. Furthermore, avian antibodies may be particularly useful for passive immunotherapy against toxins where a suitable human vaccine is not available, when the toxin/ toxoid is rare or unsafe for human use, or when large amounts of antitoxin are required. Clinical studies evaluating human immune globulin from immunized volunteers for the treatment of infant botulism have been reported.[5,6] While this product may be preferable since administration of homologous protein will circumvent serum sickness and anaphylaxis reactions, maintaining a ready pool of hyperimmune human volunteers and adequate plasma stocks may sometimes be difficult.[6] Using a non-human source of antitoxin also reduces the risk for transmittal of agents infectious (i.e., hepatitis, AIDS) to humans. Finally, because avian antibodies are both economical and readily manufactured in large scale, they offer a potentially safer, more effective, and practical alternative to equine-derived antitoxins.

Acknowledgments

We would like to thank Dr. Charles Hatheway for testing the neutralization capability of the avian antitoxin, and Dr. Philip Sobocinski for his advice and encouragement.

REFERENCES

1. Hatheway CL. Toxigenic clostridia. Clinical Microbiol Review 1990; 3:66-98.
2. Middlebrook JL, Dorland RB. Bacterial toxins: Cellular mechanisms of action. Microbiol Rev 1984; 48:199-221.
3. World Health Organization. Progress in the characterization of venoms and standardization of antivenoms. WHO Offset Publication No. 58. Geneva, 1981.
4. Botulism Antitoxin, Trivalent (Equine) Types A, B, and E; Connaught Laboratories package insert.
5. Frankovich TL, Arnon SS. Clinical trial of botulism immune globulin for infant botulism. West J Med 1991; 154:103.
6. Schwartz PJ, Arnon SS. 1992. Botulism immune globulin for infant botulism arrives - one year and a Gulf War later. West J Med 1992; 156:197-198.
7. Tacket CO, Shandera WC, Mann J, Hargrett NT, Blake PA. Equine antitoxin use and other factors that predict outcome in type A foodborne botulism. Am J Med 1984; 76:794-798.
8. Merson MH, Hughes JR, Dowell VR, Taylor A, Barker WH, Gangarosa EJ. Current trends in botulism in the United States. J Am Med Assoc 1974; 229:1305-1308.
9. Black RE, Gunn RA. Hypersensitivity reactions associated with botulinal antitoxin. Am J Med 1980; 69:567-570.
10. Carroll SB. Antivenoms and methods for making antivenoms. U.S. Patent Application (pending) 1989.
11. Thalley BS, Carroll SB. Rattlesnake and scorpion antivenoms from the egg yolks of immunized hens. Biotechnology 1990; 8:934-938.
12. DasGupta BR, Sathyamoorthy V. Purification and amino acid composition of type A botulinum neurotoxin. Toxicon 1984; 22:415-424.
13. Singh BR, DasGupta BR. Molecular differences between type A botulinum neurotoxin and its toxoid. Toxicon 1989; 27:403-410.
14. Polson A, von Wechmar MB, van Regenmortel MHV. Isolation of viral IgY antibodies from yolks of immunized hens. Immunol Comm 1980; 9:475-493.
15. Lowry OH, Rosebrough NJ, Farr AL, Randall RJ. Protein measurement with the folin phenol reagent. J Biol Chem 1951; 193:265-275.
16. Cardella MA. Botulinum toxoids. In: Lewis KH, Cassel K Jr, eds. Botulism: Proceedings of a symposium PHS Publication No. 999-FPI, Washington, D.C.: Government Printing Office 1964:113-130.
17. Reed LJ, Muench H. A simple method of estimating 50% endpoints. Amer J Hyg 1938; 27:493-497.
18. The United States Pharmacopeia, Twenty-second Revision. (United States Pharmacopeial convention, Inc. Rockville, MD.) 16th edition, 1990:186.
19. Benson HN, Brumfield HP, Pomeroy BS. Requirement of avian C1 for fixation of guinea pig complement by avian antibody-antigen complexes. J Immunol 1961; 87:616-622.
20. Rose ME, Orlans E. Fowl antibody IV. The estimation of hemolytic fowl complement. Immunology 1962; 5:642-648.
21. Sutherland SK. Serum reactions: An analysis of commercial antivenoms and the possible role of anticomplementary activity in de-novo reactions to antivenoms and antitoxins. Med J Aust 1977; 1:613-615.
22. Barandun S, Kistler P, Jeunet F, Isliker H. Intravenous administration of human gamma-globulin. Vox Sang 1962; 7:157.
23. Jurkovich GJ, Luterman A, McCullar K, Ramenofsky ML, Curreri PW. 1988. Complications of Crotalidae antivenin therapy. J Trauma 1988; 28(7):1032-1037.

EFFICACY OF PROPHYLACTIC AND THERAPEUTIC ADMINISTRATION OF ANTITOXIN FOR INHALATION BOTULISM

David R. Franz,[1] Louise M. Pitt,[1] Michael A. Clayton,[1] Martha A. Hanes[2] and Kenneth J. Rose[3]

Toxinology[1], Pathology[2] and Veterinary Medicine[3] Divisions
U.S. Army Medical Research Institute of Infectious Diseases
Fort Detrick, Frederick, MD 21702-5011

BACKGROUND

Botulism is caused by intoxication with one or more of the seven neurotoxins produced by *Clostridium botulinum*. Food poisoning is the most common cause of botulism. Accidental inhalation exposure, however, has been reported in the laboratory,[4] and inhalation is the likely route of exposure after the toxin's use as a terrorist weapon or warfare agent. Toxoids are available for immunization, but there are presently no known drugs that can be used to prevent or treat botulinum intoxication. A pentavalent human, hyperimmune globulin product[6] and a heptavalent, equine $F(ab')_2$ (despeciated) antitoxin (unpublished data, G.E. Lewis, Jr. and R.M. Condie) were evaluated in a preliminary study as prophylaxis or therapy for inhalation botulism (serotype A) in rhesus monkeys.

MATERIALS AND METHODS

Monkeys of either sex (4.0-5.8 kg) were anesthetized with Telazol (A.H. Robins, Richmond, VA) and exposed, head-only, for 10 min to aerosolized liquid botulinum toxin, type A in gelatin phosphate buffer. The aerosol concentration of toxin in a 5-min sample collected by using an all-glass impinger was measured by a standard mouse toxicity assay. The inhaled dose was calculated by actual minute volume measured in the anesthetized animals 3-5 minutes before exposure and by using Guyton's formula,[3] which is based on body weight. Mass-median diameter of the aerosol was approximately 1.8 µm. Preliminary studies demonstrated that the aerosol LD_{50} of botulinum toxin A for rhesus monkeys was 300-400 mouse intraperitoneal LD_{50} (MIPLD$_{50}$)/kg inhaled. A lethal, target dose of approximately 2,000 MIPLD$_{50}$/kg was administered for all antitoxin efficacy studies. The human hyperimmune globulin contained an antiserotype A titer of 115 IU/ml and the equine $F(ab')_2$ antitoxin contained 1020 IU/ml. Monkey blood was drawn and antitoxin titers were measured with the mouse-neutralization assay

Botulinum and Tetanus Neurotoxins, Edited by
B.R. DasGupta, Plenum Press, New York, 1993

473

utilizing W.H.O. antitoxin standards. Titers were calculated by Thompson's Moving Average Interpolation with five dose groups, four mice/group and two-fold dilutions.[9] Any animals judged by a veterinary clinician to be terminally ill were killed.

RESULTS

Tables 1a & b describe the effects of administering prophylactic hyperimmune human globulin or prophylactic equine F(ab')$_2$ antitoxin, respectively, on lethal aerosol challenge with botulinum toxin serotype A. The typical untreated control animal died 2-4 days after exposure to toxin. Clinical signs of intoxication were noted 12-18 hours before death; in order of onset, signs included mild muscular weakness, intermittent ptosis, severe weakness of postural muscles of the neck, occasional mouth breathing, serous nasal discharge, salivation, dysphagia, mouth breathing, rales, anorexia, severe weakness, and lateral recumbency.

The animals described in Table 1c were part of a study to evaluate the therapeutic efficacy of equine F(ab')$_2$ antitoxin. At the time of antitoxin administration, animal #19 demonstrated general weakness and paresis of the hind limbs. Animal #20 showed severe muscle weakness, dyspnea and mouth breathing, and a gagging cough. Animal #21 demonstrated muscular weakness, dysphagia, drooling, and serous nasal discharge. Animal #22 showed no signs and #23 minimal signs of intoxication at time of treatment; before recovering fully, #22 experienced difficulty eating on days 3 and 4 and sat quietly in its cage on days 5 and 6 while #23 sat quietly in its cage on days 3-6.

Histopathological change directly attributable to botulinum toxin was not seen. Mild to moderate subacute, submucosal tracheitis, possibly secondary to the intoxication, was observed in three (#10, 17, and 19) of the six animals that died in these studies.

DISCUSSION

Both human hyperimmune globulin and equine F(ab')$_2$ antitoxin, at plasma titers ranging from 0.6 IU/ml to unmeasurable (<0.02 IU/ml), prevented signs of intoxication after exposure to lethal botulinum toxin aerosols. The Centers for Disease Control (Atlanta, GA) recommends that individuals working with the botulinum toxins have at least 0.25 IU of neutralizing antibody per ml of plasma.[2] Our preliminary data suggest that, with regard to aerosol exposure, this benchmark is conservative and should be universally protective; titers 10-fold lower were fully protective in these studies. One would expect actively induced antibody titers to be even more effective. Bear in mind, however, that botulinum toxins are typically 20–80-fold less toxic by aerosol challenge than by intraperitoneal challenge.[1] Therefore, these data should not be used to make assumptions about protection against toxin introduced by ingestion, accidental needle sticks, or other non-inhalation laboratory accidents.

Determination of actual retained dose after inhalation exposure to extremely toxic protein toxins is difficult. Historically, Guyton's minute volume formula has been used to calculate toxin dose inhaled by monkeys after exposure to a known concentration of aerosol for a given period of time. We noted that the received dose calculated by minute volume measured in anesthetized animals immediately before exposure to the toxin averaged 63% of that based on the formula; Guyton's work was based on studies of awake animals.[3] It is possible that Telazol, like many anesthetic agents, significantly reduced minute volume and lowered exposure dose.

Prophylaxis with the equine F(ab')$_2$ antitoxin was more effective than therapy at 48 hr after exposure. Botulinum toxin selectively targets cholinergic nerve terminals by receptors, then is internalized.[8] Internalization of the toxin probably had begun before therapy was administered in the case of animals showing signs of intoxication. Apparently, therapy was unsuccessful when a lethal dose of toxin was internalized, which was thus removed from a compartment that

Table 1. Clinical outcome of rhesus monkeys challenged with botulinum serotype A aerosol and administered antitoxin before or after exposure.

a. Prophylactic Human Hyperimmune Globulin

Animal	Antitoxin	Titer(IU/ml)	Toxin dose[a]	Outcome
1	Untreated Control	N/A	1646/4492	Death/96 hr
2	16 IU/kg 48 hr pre	0.19	1200/4639	No signs
3	"	0.19	2400/4586	No signs
4	"	0.19	2546/4388	No signs
5	8 IU/kg 48 hr pre	0.19	3135/4639	No signs
6	"	0.22	2160/4472	No signs
7	"	0.19	2216/4329	No signs
8	16 IU/kg IgG 6 wk pre	0.11	1520/4586	No signs
9	"	<0.02	2960/4612	No signs

b. Prophylactic F(ab')$_2$

Animal	Antitoxin	Titer(IU/ml)	Toxin dose[a]	Outcome
10	Untreated Control	N/A	3738/4514	Death/72 hr
11	143 IU/kg 48 hr pre	0.38	2997/4612	No signs
12	"	0.45	3103/4368	No signs
13	"	0.6	3057/4348	No signs
14	14 IU/kg 48 hr pre	0.02	1714/4389	No signs
15	"	<0.02	3168/4612	No signs
16	"	<0.02	3038/4368	No signs

c. Therapeutic Equine F(ab')$_2$

Animal	Antitoxin	Titer(IU/ml)	Toxin dose[a]	Outcome
17	Untreated control	N/A	3300/4565	Death/<45 hr
18	Untreated control	N/A	5878/4752	Death/46 hr
19	143 IU/kg 46 hr post	N/A	2800/4586	Minimal signs[b] Death/98 hr
20	"	N/A	3020/4429	Severe signs[b] Death/74 hr
21	"	N/A	3572/4368	Mod. signs[b] Recovery day 10
22	"	N/A	2715/4368	No signs[b] Mod. signs day 4 Recovery day 6
23	"	N/A	3600/4493	Minimal signs[b] Recovery day 6

[a]Dose (MIPLD$_{50}$/kg) calculated from minute volume measured immediately before exposure/Dose calculated by Guyton's formula {minute volume in ml = 2.1(body weight in gm)$^{0.75}$}. Presented as inhaled, not retained, dose; we assume retained dose was approximately 30-40% of this value.

[b]Signs observed immediately before antitoxin therapy.

could be accessed by the antibody. It has been demonstrated previously, with both intraperitoneal and inhalation challenge of guinea pigs, that efficacy of antibody therapy is reduced with time post challenge[5] (unpublished, M.A. Crumrine). These data suggest that, if given at onset of clinical signs or before, the despeciated equine botulinum antitoxin may be effective in protecting individuals from death after exposure to lethal inhaled doses of botulinum toxin. It is likely that supportive care along with antibody therapy could extend the window of opportunity for use of therapeutic antitoxin; this has been demonstrated in rhesus monkeys after intravenous botulinum toxin challenge.[7]

Acknowledgements

The authors thank Dr. Houston Dewey for statistical support.

In conducting the research described in this report, the investigators adhered to the "Guide for the Care and Use of Laboratory Animals," as prepared by the Committee on Care and Use of Laboratory Animals of the Institute of Laboratory Animals Resources Commission of Life Sciences—National Research Council. The facilities are fully accredited by the American Association for Accreditation of Laboratory Animal Care. The views of the authors do not purport to reflect the positions of the Department of the Army or the Department of Defense.

REFERENCES

1. Cardella MS. Botulinum toxoids. In: Lewis KH, Cassel K, eds. Botulism. Cincinnati, OH: Department of Health Education and Welfare, Public Health Service, 1964.
2. Ellis RJ. Immunobiologic agents and drugs available from the Centers for Disease Control: descriptions, recommendations, adverse reactions, and serologic response, Atlanta, GA: Centers for Disease Control, 1982.
3. Guyton AC. Measurement of the respiratory volumes of laboratory animals. Am J Physiol 1947; 150:70.
4. Holzer E. Botulism caused by inhalation. Med Klin 1962; 41:1735.
5. Lewis GE Jr., Metzger JF. Studies on the prophylaxis and treatment of botulism. In: Eaker D, Wadstrom T, eds. Natural Toxins. Oxford: Pergamon Press, 1980.
6. Metzger JF, Lewis GE Jr. Human-derived immune globulins for the treatment of botulism. In: Reviews of Infectious Diseases. Chicago, IL: The University of Chicago, 1979,
7. Oberst FW, Crook JW, Cresthull P, House MJ. Evaluation of botulinum antitoxin, supportive therapy, and artificial respiration in monkeys with experimental botulism. Clin Pharm Ther 1967; 9: 209.
8. Simpson LL. Peripheral Actions of Botulinum Toxins. In: Simpson LL, ed. Botulinum Neurotoxin and Tetanus Toxin. New York, NY: Academic Press, Inc., 1989.
9. Thompson WR, Weil CS. On the construction of tables for moving-average interpolation. Biometrics 1952; 8:51.

CLINICAL TRIAL OF HUMAN BOTULISM IMMUNE GLOBULIN

Stephen S. Arnon

California Department of Health Services
Berkeley, CA 94704

INTRODUCTION

A major theme of this conference is that botulinum toxin, nature's most poisonous poison,[1] has now been "tamed" and has acquired the distinction of becoming, through the work of Scott[2] and Schantz,[3] the first bacterial toxin to serve as a medicine for the betterment of humanity. So in this historic and felicitous context, it appears to be my unwelcome responsibility to remind this conference that botulinum toxin remains very much an adversary for certain select groups of patients, most notably, infants.[4]

As is now generally appreciated, infant botulism is an infectious form of botulism in which ingested spores germinate, colonize the large intestine and produce botulinum toxin in it.[5-7] Infant botulism is the most common form of human botulism in the United States, with approximately 75-100 hospitalized cases recognized annually.[8] Infant botulism thereby fulfills the criterion for an "orphan disease" as defined by the U.S. Food and Drug Administration (FDA), namely, an illness that affects 200,000 or fewer persons in the United States. Approximately half of all United States cases of infant botulism are reported from California, thus making California the logical location in which to investigate this disease and study its treatment. Parenthetically, very rarely and under exceptional circumstances, infant-type botulism can occur in adults,[9-13] in addition to the better-known foodborne and wound forms of the disease.

In the 16 years since infant botulism was first recognized as a distinct disease,[14,15] its treatment has been medically and scientifically unsatisfactory because it consisted only of general supportive (nutritional and respiratory) care. The existing equine botulinum antitoxin could not be used in infants because of its unacceptably high incidence of serious side-effects (anaphylaxis, serum sickness) when used in adults with foodborne botulism.[16] Without specific treatment, patients with infant botulism in 1991 in California were hospitalized for an average of five weeks, often requiring intensive care and mechanical ventilation, at an average cost of approximately $2500 per day and over $85,000 per case.

Botulinum and Tetanus Neurotoxins, Edited by
B.R. DasGupta, Plenum Press, New York, 1993

PRODUCTION OF BIG

For these reasons, the Infant Botulism Prevention Program of the California Department of Health Services undertook to evaluate in the treatment of infant botulism a human-derived neutralizing antibody product to botulinum toxin types A-E, an Investigational New Drug formally known as Botulism Immune Globulin (BIG). The study intended to use human BIG produced by the U.S. Army,[17,18] but was unable to do so because the Army needed to retain its BIG for defensive purposes[19] after the invasion of Kuwait by Iraq in August 1990 and the outbreak of war in January 1991 that followed.

Instead, it was necessary to produce a new supply of BIG, which was done with the support of the Office of Orphan Drug Products of FDA, by plasmapheresing California volunteers who had been previously immunized with botulinum toxoid for occupational (laboratory) safety reasons. The immune plasma was fractionated into immune globulin at the Massachusetts Public Health Biologic Laboratories (MPHBL), the only facility of the seven in Europe and North America licensed by FDA for production of intravenous immune globulin that was able to process the small volume of immune plasma (63 liters) collected for this investigation.

STUDY DESIGN AND LAUNCH

The Clinical Trial of Human Botulism Immune Globulin for the treatment of infant botulism is a randomized, double-blinded, placebo-controlled study (Table 1).

Table 1. The clinical trial of human Botulism Immune Globulin (BIG).

Therapeutic Principle:	Maximize neutralization of toxemia by minimizing the interval between disease onset and BIG administration.
Study Design:	Randomized, placebo-controlled, double-blinded. One IV dose of BIG (50 mg/kg) is expected to provide >4 months of toxemia neutralization (see Table 2).
Placebo:	Commercially-available, FDA-licensed, normal human IGIV (Gammagard®, Baxter-Hyland Laboratories).
Outcome Measures:	1) Length and cost of hospital stay, 2) progression of illness, 3) incidence of complications.
Logistics:	Coded vial hand-delivered to bedside in response to telephone inquiry. Three-year enrollment of approximately 120 patients anticipated. 85 participating hospitals; 59 separate Institutional Review Board approvals obtained.
Cost to patient or hospital:	None

The design of the study is based on an observation derived from foodborne botulism patients treated with the equine botulinum antitoxin, namely, that the shorter the interval between onset and receipt of antitoxin, the better the outcome.[20] Because botulinum toxin is tightly bound to the neuromuscular junction, BIG is not expected to reverse the paralysis that led to the initial hospitalization of the infant. Rather, the therapeutic rationale is to inactivate any toxin in the circulation and extracellular fluid before the toxin can bind to nerve endings, including all toxin subsequently absorbed from its site of production in the large intestine.[7] Because the half-life of human IgG antibody in humans is approximately 20 days, a single intravenous infusion of BIG is expected to provide a protective level of neutralizing antibody against types A and B botulinum toxin for approximately four months (Table 2). The protocols and consent forms for the study were approved by the 59 separate Institutional Review Boards for the protection of human research subjects that serve the 85 participating hospitals and the California Department of Health Services.

Table 2. Expected decay in BIG titer in an infant at a dosage of 50 mg/kg.

Day		Anti-A, IU/ml of ECF[1], $\times 10^{-4}$	Anti-B, IU/ml of ECF[1], $\times 10^{-4}$
0		1250	250
20		650	125
40		312	63
60	(2 months)	156	32
80		78	16
100		39	8
120	(4 months)	20	4

[1]By definition, 1 IU (International Unit) of neutralizing antibody inactivates 10^4 mouse LD_{50} for botulinum toxin types A and B. ECF = Extracellular fluid.

The study is planned to last three years in order to maximize its statistical power; an enrollment of approximately 120 patients (60 BIG, 60 placebo) is expected. One interim data evaluation at 18 months is scheduled to ensure that early efficacy is not overlooked. The primary outcome measure is a reduction in the length of hospital stay; secondary outcome variables are the cost of hospital stay, the progression of illness, and the incidence of complications.

After preparatory efforts and supporting observations that span more than 15 years (Table 3), the Clinical Trial of Human Botulism Immune Globulin formally began on February 24, 1992 with enrollment of its first patient at the Children's Hospital of Oakland. Since then, an additional fourteen patients have been entered without incident or complication.

FUTURE PROSPECTS FOR BIG

If BIG proves to be successful in the treatment of infant botulism, then an application for its licensure may be submitted to FDA. Once licensed, BIG might then be used to treat patients with foodborne botulism and wound botulism, the two other forms of human botulism whose low and sporadic incidence renders impossible a controlled clinical trial of BIG in their treatment. Finally, it is possible that there may be a role for BIG as an adjunct to the therapeutic use of botulinum toxin: clinical outcome may be improved by neutralizing any toxin that escapes from the injected muscle, thereby preventing the induction of host antibody that would render the patient refractory to further treatment with botulinum toxin.

Table 3. Selected milestones in the development of human botulism immune globulin. *(See footnote for explanation of abbreviations.)*

DATE	EVENT
1976	CDC convenes *ad hoc* national expert committee to discuss the newly recognized problem of infant botulism. The committee identifies the need for human Botulism Immune Globulin.
1976	The Director, FRI-UW, writes the Director, CDC, to suggest that immunized FRI-UW staff be considered as a source of immune plasma for BIG.
1976	The Director, Bureau of Laboratories, CDC, replies that the suggestion and offer will be discussed by CDC technical staff.
1977	Schantz (FRI-UW) visits CDHS to discuss further the possibility of using immunized FRI-UW staff as the source of immune plasma for BIG.
1977	The Chief, Infectious Disease Section, CDHS writes the (new) Director, CDC, to urge "...the development of BIG now."
1978	The Assistant Director, CDC, replies that the topic of BIG was referred to the Assistant Secretary for Health, DHEW, with the recommendation that "the National Academy of Sciences be given a contract for...development of...a national plan of action [on infant botulism] involving all concerned Government agencies under the leadership of the U.S. Public Health Service."
1978	The Chief, Infectious Disease Section, CDHS replies to the (new) Director, CDC, to note that the Assistant Secretary for Health did not accept the recommendation and instead requested that CDC convene a committee of federal agencies. "So one year later, we are still at ground zero."
1978	Lewis and Metzger (U.S. Army) publish letter "Botulism Immune Plasma (Human)" (Lancet 2:634-635) that describes Army plans to develop BIG.
1978	Hatheway (CDC) identifies botulinum toxin in serum of infant botulism patient (MMWR 1978, 42:411-412).
1979	Lewis and Metzger demonstrate the efficacy of human BIG in experimental (guinea pig) botulism (*Natural Toxins*, 1980: 601-606).
1982	Condi produces BIG at University of Minnesota under U.S. Army contract and ships BIG to Ft. Detrick; Army retains all BIG for its needs.
1983	Orphan Drug Act enacted (PL 97-414); "orphan disease" defined; Orphan Drug Products Office within FDA and a national Orphan Drug Commission created.
1983	Paton, Lawrence and Stern in Australia report detection of botulinum toxin in serum of infant botulism patient (J Clin Micro 17:13-15).
1983-86	Discussions at IBRCC between the Army, CDC, FDA and CDHS re: BIG for treatment of infant botulism.
1986	Federal Technology Transfer Act (PL 99-502) enacted; Army now authorized to provide BIG to CDHS.
1986	Aureli *et al.* in Italy report detection of botulinum toxin in serum of infant botulism patient (J Infect Dis 154:207-211).
1986	Condi visits CDHS for further discussion of use of Army BIG to treat California infant botulism patients.
1987	Orphan Drug Commission holds its first public hearing (in San Francisco); CDHS notes existence of BIG yet its continued unavailability for use in infant botulism.
1987	Further discussions at IBRCC between the Army, CDC, FDA and CDHS re: BIG treatment for infant botulism.
1987	FDA Orphan Drug Products Office convenes first meeting of Army, CDC and FDA officials "to discuss access to supplies of BIG at Ft. Detrick."
1988	FDA Orphan Drug Products Office holds further discussions with Army, CDC and CDHS to resolve obstacles to BIG's availability.

Table 3 (cont'd)

DATE	EVENT
1988	FDA Orphan Drug Products Office holds further discussions with Army, CDC and CDHS to resolve obstacles to BIG's availability.
1988	More discussions at IBRCC between the Army, CDC, FDA and CDHS re: continued unavailability of BIG for treatment of infant botulism.
1988	Army authorizes CDHS to cross-reference its IND for BIG (BB-IND-1332) in applications to FDA and begins to develop an agreement to provide BIG to CDHS "on a humanitarian basis for use in these [clinical trial] studies."
1989	Funding application, "Clinical Trial of Human Botulism Immune Globulin," submitted to FDA Orphan Drug Products Office.
1989	BIG officially designated an "Orphan Drug."
1989	U.S. Army submits Memorandum of Understanding to CDC and CDHS re: terms of providing BIG.
1989	State of California's Human Subjects Protection Committee gives its final approval to the BIG study (first of 59 IRB approvals obtained for the study as originally designed).
1990	FDA-CDHS Cooperative Agreement funding the "Clinical Trial of Botulism Immune Globulin" begins.
1990	Enrollment of participating hospitals and obtaining IRB approvals nears completion (85 total enrolled).
1990	Iraq invades Kuwait. Army BIG again becomes unavailable. FDA Orphan Drug Products Office counsels patience.
1991	Allied forces invade Iraq and Kuwait. FDA Orphan Drug Products Office encourages feasibility study of replacement source of BIG.
1991	Final agreement on revised Memorandum of Understanding by Army, CDC, and CDHS for eventual supplying of Army's again-unavailable BIG.
1991	Supplementary funds awarded to CDHS by FDA to begin new BIG production.
1991	Plasmapheresis of botulinum toxoid-immunized Northern California volunteers.
1991	Massachusetts Public Health Biologic Laboratories fractionates pooled plasma into BIG.
1991	California BIG assigned its Investigational New Drug number by FDA (BB-IND-4283).
1992	Baxter-Hyland Laboratories (Lessiens, Belgium) bottles placebo (Gammagard®) into Massachusetts vials.
1992	BIG and placebo arrive in California in coded vials of identical appearance.
1992	Completion of a second set of approvals from the 59 IRBs for substitution of California BIG for Army BIG.
1992	Formal opening of patient enrollment into the statewide Clinical Trial of Human Botulism Immune Globulin.

Abbreviations

BIG = human Botulism Immune Globulin
CDC = U.S. Centers for Disease Control
CDHS = California Department of Health Services
DHEW = U.S. Department of Health, Education and Welfare
FDA = U.S. Food and Drug Administration
FRI-UW= Food Research Institute, University of Wisconsin
IBRCC = federal Interagency Botulism Research Coordinating Committee
IRB = Institutional Review Board (i.e. a Human Subjects Research Protection Committee)

ACKNOWLEDGMENTS

The Clinical Trial of Human Botulism Immune Globulin is supported by Cooperative Agreement FD-U-000476 between the U.S. Food and Drug Administration and the State of California and by the California Department of Health Services.

REFERENCES

1. Lammana C. The most poisonous poison. Science 1959; 130:763-772.
2. Scott AB. Botulinum toxin injection into extraocular muscles as an alternative to strabismus surgery. Ophthalmology 1980; 87:1044-1049.
3. Schantz EJ, Johnson EA. Properties and use of botulinum toxin and other microbial neurotoxins in medicine. Microbiol Rev 1992; 56:80-99.
4. Arnon SS. Infant botulism. In: Feigin RD, Cherry JD, eds. Textbook of Pediatric Infectious Diseases, third edition. Philadelphia: WB Saunders, 1992:1095-1102 (ch. 107).
5. Arnon SS, Midura TF, Clay SA, Wood RM, Chin J. Infant botulism: epidemiological, clinical, and laboratory aspects. JAMA 1977; 237:1946-1951.
6. Wilcke BW Jr, Midura TF, Arnon SS. Quantitative evidence of intestinal colonization by *Clostridium botulinum* in four cases of infant botulism. J Infect Dis 1980; 141:419-423.
7. Mills DC, Arnon SS. The large intestine as the site of *Clostridium botulinum* colonization in human infant botulism. J Infect Dis 1987; 156:997-998.
8. Centers for Disease Control. Summary of notifiable diseases, United States, 1990. Morbid Mortal Wkly Rept 1991; 39(53):56.
9. Chia JK, Clark JB, Ryan CA, Pollack M. Botulism in an adult associated with foodborne intestinal infection with *Clostridium botulinum*. N Engl J Med 1986; 315:239-241.
10. Freedman M, Armstrong RM, Killian JM, Boland D. Botulism in a patient with jejunoileal bypass. Ann Neurol 1986; 20:641-643.
11. Sonnabend WF, Sonnabend OA, Grundler P, Ketz E. Intestinal toxicoinfection by *Clostridium botulinum* type F in an adult. Lancet 1987; 1:357-362.
12. McCroskey LM, Hatheway CL. Laboratory findings in four cases of adult botulism suggest colonization of the intestinal tract. J Clin Microbiol 1988; 26:1052-1054.
13. McCroskey LM, Hatheway CL, Woodruff BA, Greenberg JA, Jurgenson P. Type F botulism due to neurotoxigenic *Clostridium baratii* from an unknown source in an adult. J Clin Microbiol 1991; 29:2618-2620.
14. Pickett J, Berg B, Chaplin E, Brunstetter-Shafer M. Syndrome of botulism in infancy: clinical and electrophysiologic study. N Engl J Med 1976; 295:770-772.
15. Midura TF, Arnon SS. Infant botulism: identification of *Clostridium botulinum* and its toxins in faeces. Lancet 1976; 2:934-936.
16. Black RE, Gunn RA. Hypersensitivity reactions associated with botulinal antitoxin. Am J Med 1980; 69:567-569.
17. Lewis GE Jr, Metzger JF. Botulism immune plasma (human). Lancet 1978; 312:634-635.
18. Lewis GE Jr, Metzger JF. Studies on the prophylaxis and treatment of botulism. In: Eaker D, Wadstrom T, eds. Textbook of Natural Toxins. New York: Pergamon Press, 1980:601-606
19. Ember L. Chemical weapons: plan proposed to destroy Iraqi arms. Chem Eng News 1991; 69:6.
20. Tacket CO, Shandera WX, Mann JM, Hargrett NT, Blake PA. Equine antitoxin use and other factors that predict outcome in type A foodborne botulism. Am J Med 1984; 76:794-798.

STUDIES ON THE RELATIONSHIP OF VITAMIN A
DEFICIENCY TO THE ANTIBODY RESPONSE
TO TETANUS TOXOID IN THE RAT

A. Catharine Ross,[1,2] Makiko Kinoshita,[1] and Deepa Arora[1]

[1]Division of Nutrition,
  Department of Physiology and Biochemistry and

[2]Department of Pediatrics
  Medical College of Pennsylvania
  Philadelphia, PA 19129

INTRODUCTION

Morbidity and mortality due to tetanus constitute a very significant public health problem in much of the developing world. Although vaccination with tetanus toxoid has significantly reduced the incidence of tetanus, still nearly 800,000 children per year are estimated to die from this infection.[1] Nutritional deficiencies, including that of vitamin A, are most prevalent in some of the geographic areas where tetanus is also most prevalent. A number of clinical and field-based studies have shown that vitamin A deficiency increases the risk of mortality in young children. In several studies conducted in Indonesia, India, and Nepal, supplementation of pre-school children with vitamin A has resulted in a reduction in mortality.[2] Effects of vitamin A supplementation on morbidity have been more difficult to demonstrate. Evidence consistent with reduced morbidity from respiratory, diarrheal and measles infections has been reported in some studies[3-5] but not in others[6].

To better understand the relationship of vitamin A status to the immune response and resistance to infection, we have conducted studies with the rat as a model in which we have measured antibody production after immunization with various bacterial antigens. In previous publications, we reported that vitamin A deficiency was associated with a marked reduction in the antibody production after immunization with the polysaccharide antigens from *Streptococcus pneumoniae*[7] and *Neiserria meningitidis*, causative agents of bacterial pneumonia and meningitis. Likewise, the primary antibody response to tetanus toxoid, a T cell-dependent protein antigen, was also very low (~18% of normal).[8] For each of these antigens, the antibody response was normal levels in rats which had been repleted with vitamin A (retinol) a few days before they were immunized. In contrast, when similar vitamin A-deficient rats were immunized with the lipopolysaccharides from *Pseudomonas aeruginosa* or *Serratia marcescens* antibody production was normal.[8] Thus, these studies showed that *some*, but *not all*, antibody responses are low during vitamin A deficiency, a finding also supported by the observation that the *total* immunoglobulin G concentration of the vitamin A-deficient rat was significantly greater than that of the controls.[9] Since pneumococcal polysaccharide and meningococcal polysaccharide have been considered T cell independent (TI), *type II*, antigens and the bacterial lipopolysaccharides TI antigens of the *type I* group, these data suggested that vitamin A deficiency may particularly compromise the responses to the TI *type II* and TD antigens.[8]

Two main questions arose from these studies. The first is whether or not immunological memory to tetanus toxoid, on which successful vaccination depends, is also low during vitamin A deficiency. Second, might adjuvants stimulate the immune response, even during vitamin A deficiency? The studies described above raised the possibility that agents like *P. aeruginosa* lipopolysaccharide might activate the immune system in such a way that the response to other antigens, e.g. tetanus toxoid, could be increased even in the vitamin A-deficient state. Similarly, the observation that the immune defect was reversed after rats were repleted with retinol a few days before they were immunized, together with evidence that retinol may act as an adjuvant in some circumstances,[10-13] led us to investigate whether giving LPS or retinol concurrently with immunization with tetanus toxoid would stimulate the tetanus toxoid-specific antibody response.

GENERAL METHODS

The dietary treatments used in these studies have been described in detail previously.[7,8] Briefly, Lewis rats raised in our Animal Facility were weaned onto a purified diet that contained either a nutritionally adequate level of retinol (4 μg/g diet) or no vitamin A. In rats fed the vitamin A-free diet, vitamin A deficiency, as determined by a plasma retinol level of <0.2 μmol/L of plasma, ensued by ~40 days of age in males and ~50 days of age in females. For repletion with vitamin A, rats were given a single oral dose of vitamin A (retinyl palmitate equal to 1.5 mg of retinol).

Rats were immunized with 100 μg of tetanus toxoid (Connaught Labs) diluted in 1 ml of saline and injected intraperitoneally. The timing of immunization and collection of plasma for analysis of anti-tetanus toxoid antibodies is given with the protocol for each study, below. Generally, blood was collected from the tail at intervals to determine the kinetics of the antibody response. At the end of each study, rats were euthanized and blood plasma and tissues (spleen, thymus and liver) were collected to determine vitamin A concentrations by high-performance liquid chromatography.[14]

The antibody response to tetanus toxoid was determined by enzyme-linked immunosorption assays (ELISA) specific for rat anti-tetanus toxoid IgM or anti-tetanus toxoid IgG.[9] The concentrations of total rat plasma IgM and IgG were also determined by ELISA.

SPECIFIC PROTOCOLS AND RESULTS

Protocol for Studies on the Immunological Memory to Tetanus Toxoid

The design of the main study on immunologic memory formation is shown in Figure 1. The primary and secondary antibody responses to tetanus toxoid were studied in four groups of male Lewis rats. The vitamin A-deficient group was fed the vitamin A-free diet throughout the study. The group referred to as the "early repletion" group (R1) was fed the vitamin A-free diet throughout the study but received a single oral dose of vitamin A 1 day after the first immunization with tetanus toxoid. The group referred to as the "late repletion" group (R2) was fed the vitamin A-free diet throughout the study, but received a single oral dose of vitamin A 2 days before re-immunization with tetanus toxoid; thus, these animals were identical with the

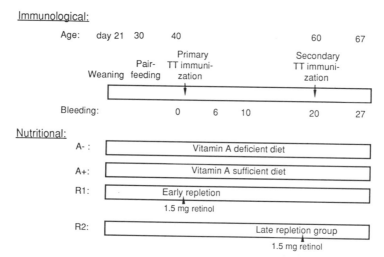

Figure 1. Protocol for immunization and vitamin A repletion.

vitamin A-deficient group from day 0-18 of the primary response and were pooled with them for analysis of the primary response. The control group was fed the vitamin A-sufficient diet throughout the study.

All rats, except a few additional non-immunized controls, were immunized with tetanus toxoid when they were 40 days old. Blood was collected prior to immunization and 4-20 days later to measure the primary antibody response. Rats were reimmunized in the same manner on day 20 and the secondary antibody response was determined in plasma collected up to 11 days later.[9]

Vitamin A Status

The vitamin A-deficient status of the rats fed the vitamin A-free diet was confirmed by a very low plasma retinol concentration (0.12 ± 0.05 μmol/L vs. the normal control group value of 1.09 ± 0.04 μmol/L), liver vitamin A reserves that were nearly exhausted, and spleen and thymus retinol concentrations that were also significantly less than the control group.[9] Repletion with retinol (R1 and R2 groups), even without further dietary vitamin A, resulted in normal plasma retinol concentrations at the end of the study as well as normal liver, spleen and thymus retinol concentrations. In other studies, we have shown that repleting the vitamin A-deficient rats with an oral dose of 1.5 mg of retinol restores the plasma retinol concentration within 2-8 hours[15] and results in significant hepatic storage of vitamin A.

The Primary and Secondary Antibody Responses to Tetanus Toxoid

Both anti-tetanus toxoid IgM and IgG were monitored throughout the study (Figure 2). In the primary response, anti-tetanus toxoid IgM (Fig. 2A) and anti-tetanus toxoid IgG (Fig. 2B) responses of vitamin A-deficient rats were both significantly different from those of the control group. However, the kinetics of response were nearly identical, with peak IgM responses 6-8 days after primary immunization and peak IgG responses 10-12 days after immunization. The previously vitamin A-deficient rats given a single early repletion with retinol (R1) produced on average more anti-tetanus toxoid IgM and IgG than the control group. This effect was consistent

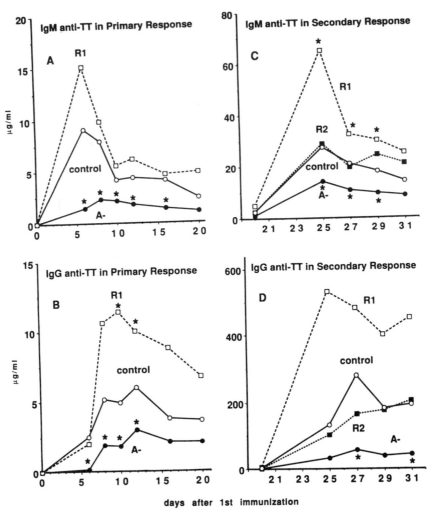

Figure. 2. The primary and secondary IgM and IgG responses to tetanus toxoid. * denotes significant differences from the control group. From reference 9 with permission.

across the time course of the primary response but, due to large variance in the R1 group, the effect was statistically significant only for the peak anti-tetanus toxoid IgG response.

The response of the previously vitamin A-deficient rats repleted with retinol just before re-immunization (R2 group) provided information about the formation and retention of immunologic memory during vitamin A deficiency. If memory has not formed, we might have observed a "primary-type" response, of normal magnitude, following the second immunization in retinol-repleted (R2) rats. However, we observed that the secondary response was of the same magnitude as that of vitamin A-sufficient control rats (Fig. 2C, 2D). Thus, all the characteristics expected of the secondary, recall response—rapid antibody production, greater magnitude of response, and antibodies largely of the IgG class—were present, indicating that the pathway to memory cell formation must have remained intact, and that the memory T and B cells could be activated after vitamin A status was improved.

Ability of Vitamin A-Deficient Rats to Respond to Adjuvants

Previous studies had shown that vitamin A-deficient rats could mount a normal antibody response to bacterial lipopolysaccharides such as those from *P. aeruginosa* and *S. marcescens*. If the response to these TI *type 1* is normal, but the response to TI type 2 and TD antigens is impaired [8], would the factors or signals elicited by the lipopolysaccharide antigens provide "help" to enable a normal response to either TI type 2 or TD antigens? The ability of bacterial lipopolysaccharides to act as adjuvants in the humoral immune response was known from the work of Skidmore et al.[16] and others, and studies of McGhee et al.[17] had shown that this activity was dependent on T cell as well as macrophage functions.

To test whether co-immunization would increase the antibody response to tetanus toxoid, vitamin A-deficient rats were immunized with 100 µg of tetanus toxoid and, simultaneously, with 2-20 µg of *P. aeruginosa* lipopolysaccharide. As indicated in Table 1, the increase in antibody production was very great, far in excess of that produced in the usual anti-tetanus toxoid response of control rats. This preliminary study indicates that the vitamin A-deficient rat does not lack the ability to synthesize and release specific antibodies, but apparently does not generate, transmit, or receive "signals," most likely cytokines and other helper factors, in a normal fashion. In studies of the antibody response to *Streptococcus pneumoniae*, similar amplification of the antibody response specific for pneumococcal polysaccharide was observed in vitamin A-deficient co-immunized rats with *P. aeruginosa* lipopolysaccharide.[24]

Table 1. Effect of co-immunization with *Pseudomonas aeruginosa* lipopolysaccharide (Pa-LPS) on the antibody response to tetanus toxoid in vitamin A-deficient rats.

	Anti-tetanus toxoid IgM, µg/ml		
	Day 4	Day 7	Day 10
Tetanus toxoid	1.2 ± 0.4	1.1 ± 0.5	0.6 ± 0.3
Tetanus toxoid + Pa-LPS[a]	6.3 ± 1.6	10.0 ± 2.4	3.9 ± 0.7
	Anti-tetanus toxoid IgG, µg/ml		
	Day 7	Day 10	Day 12
Tetanus toxoid	0.4 ± 0.4	0.7 ± 0.4	1.5 ± 0.7
Tetanus toxoid + Pa-LPS[a]	28.2 ± 7.0	28.1 ± 7.1	31.9 ± 7.7

[a]Rats were immunized with 100 µg tetanus toxoid and 20 µg of Pa-LPS. The primary anti-tetanus toxoid response was measured by ELISA on the days indicated.

DISCUSSION

The immune response to tetanus toxoid is of special concern for the direct protection of the pregnant woman and for the indirect protection of the neonate who passively receives protection through the transfer of maternal anti-tetanus toxoid IgG across the placenta, as well as for the young child during routine immunization, usually with DPT vaccine.[1] In the young animal or human, the immune response to polysaccharide and protein antigens develops slowly post-natally.[18] Nutritional deficiencies that compromise either the maternal production of antibody, maternal health during pregnancy, the gestational development of the child, or the post-natal development of the immune system could potentially reduce antibody protection. The current studies have focused on the immune response of a rat model of young child vitamin A deficiency in order to determine, first, whether antibody production is significantly reduced and, second, whether the pathway through which memory is formed and maintained is changed in a similar manner, or might remain normal. Since immunologic memory is the foundation on which vaccination rests, this question is of importance to understanding whether anti-tetanus toxoid might still be protective even if the primary antibody response is poor.

From the data described here, we inferred that memory to tetanus toxoid was formed normally. This inference was based on the normal secondary response of previously vitamin A-deficient rats after rehabilitation with retinol. The secondary anti-tetanus toxoid response, both for IgM and IgG was normal in kinetics and magnitude in the R2 group, even though the primary antibody response of these rats had been low. Thus, the amplification of antibody production (ratio of the concentrations of anti-tetanus toxoid IgM or IgG in the secondary response as compared to IgM or IgG the primary response) was significantly greater in the R2 group than in any other group.

Based on these results, specific antibody formation was low during both the primary and the secondary antibody responses when vitamin A status was poor. However, the question could still be raised as to whether this is because the animal's ability to produce and secrete antibodies was diminished, or because intercellular communication and signaling was somehow impaired. Because early repletion with retinol could be delayed at least 1 day after primary immunization with tetanus toxoid, it seemed unlikely that a normal retinol status was essential in the earliest stage of antigen uptake and processing. Indeed, in a limited study, when "early" repletion with retinol was delayed up to 4 days after immunization with tetanus toxoid, the primary and secondary anti-tetanus toxoid responses were nearly normal although slightly delayed (M. Kinoshita, unpublished observations). Other investigators have shown that the T lymphocytes of vitamin A-deficient mice do not stimulate B cells from vitamin A-sufficient mice in a normal manner, and these data suggest that interactions which normally stimulate T cell functions are defective.[19] It is interesting to note another situation in which memory formation but not antibody production was stimulated after immunization with tetanus toxoid. Kraft et al.[20] reported that normal mice immunized with tetanus toxoid complexed with an excess of gamma globulin produced no humoral response to tetanus toxoid, but developed lymph node germinal centers indicative of memory B cell production.

The observation that rats repleted with retinol near the time of the primary immunization had a greater-than-normal primary *and* secondary antibody response was unexpected. This observation suggests that retinol stimulated the development of the pathway leading to plasma cell formation and antibody secretion, but also stimulated the formation of memory cells. The dose of retinol used (1.5 mg) is large as compared to the daily nutritional needs of the rat, but it does not produce any signs of toxicity or intolerance. Following oral dosing, plasma retinol levels increased within hours, and liver, spleen and thymus levels also increased significantly.[15] This dose in the rat is quite similar, when compared on the basis of body weight or metabolic rate, to the doses that have been given to young children in field studies [typically, 200,000 IU (= to 60 mg of retinol)/single dose].[21] The immunostimulation we observed suggests that the provision of high dose retinol, at least in the vitamin A-deficient animal, provides more than simple nutritional rehabilitation. The concept that retinol and other retinoids have adjuvant properties, first shown in studies of Dresser,[10] is now well supported.[11,12,22] However, the route of administration used in other studies has almost always been a non-physiological one, either intraperitoneal[10,22] or intramuscular/subcutaneous[11,12] administration. The observation that orally administered retinol can stimulate antibody production is of interest with regard to vitamin A supplementation programs for young children. Further work is needed to evaluate the effects of large retinoid doses on the immune system, considering both potentially beneficial as well as possibly detrimental consequences of "immunostimulation" in the young animal.

The fact that total plasma IgG concentration was slightly elevated during vitamin A deficiency was an indicator that total antibody production was not impaired, and previous work had shown that vitamin A-deficient rats retained their ability to produce a specific antibody response to bacterial lipopolysaccharides.[8] In the co-immunization study reported here, vitamin

A-deficient rats immunized with tetanus toxoid plus a low dose of *P. aeruginosa* lipopolysaccharide produced very high levels of antibody specific to tetanus toxoid. These results show that the fundamental ability to produce antibodies to T cell-dependent antigens is maintained. Further work to understand the factors involved is in progress. We hypothesize that cytokines released in response to bacterial lipopolysaccharide provide stimulation not only to antigen-specific cells responding to *P. aeruginosa* lipopolysaccharide but also to neighboring cells responding to tetanus toxoid. The response was not simply due to polyclonal B cell activation, as the antibody produced was specific for tetanus toxoid. Immunoglobulin subclass selection is known to be under the regulation of cytokines.[23] Analysis is now underway of the subclasses of anti-tetanus toxoid IgG formed in response to tetanus toxoid immunization during vitamin A deficiency, following repletion with retinol, and following stimulation with lipopolysaccharide. Knowledge of subclass-specific alterations may provide clues as to the types of cytokines that are lacking or overproduced in vivo in these three different situations.

REFERENCES

1. Henderson RH, Keja J, Hayden G, Galazka A, Clements J, Chan, C. Immunizing the children of the world: Progress and prospects. Bulletin of the WHO 1988; 66:535-543.
2. Sommer A. Vitamin A status, resistence to infection, and childhood mortality. Ann NY Acad Sci 1990; 587:17-23.
3. Milton RC, Reddy V, Naidu AN. Mild vitamin A deficiency and childhood morbidity—an Indian experience. Am J Clin Nutr 1987; 46:827-829.
4. Hussey GD, Klein M. A randomized, controlled trial of vitamin A in children with severe measles. N Engl J Med 1990; 323:160-164.
5. Coutsoudis A, Broughton M, Coovadia HM. Vitamin A supplementation reduces measles morbidity in young African children: a randomized, placebo-controlled, double-blind trial. Am J Clin Nutr 1991; 54:890-895.
6. Rahmathullah L, Underwood BA, Thulasiraj RD, Milton RC. Diarrhea, respiratory infections, and growth are not affected by a weekly low-dose vitamin A supplement: A masked, controlled field trial in children in southern India. Am J Clin Nutr 1991; 54:568-577.
7. Pasatiempo AMG, Bowman TA, Taylor CE, Ross AC. Vitamin A depletion and repletion: Effects on antibody response to the capsular polysaccharide of *Streptococcus pneumoniae*, type III (SSS-III). Am J Clin Nutr 1989; 49:501-510.
8. Pasatiempo AMG, Kinoshita M, Taylor CE, Ross AC. Antibody production in vitamin A-depleted rats is impaired after immunization with bacterial polysaccharide or protein antigens. FASEB J 1990; 4:2518-2527.
9. Kinoshita M, Pasatiempo AMG, Taylor CE, Ross AC. Immunological memory to tetanus toxoid is established and maintained in the vitamin A-depleted rat. FASEB J 1991; 5:2473-2481.
10. Dresser DW. Adjuvanticity of vitamin A. Nature 1968; 217:527-529.
11. Friedman A. Induction of immune response to protein antigens by subcutaneous co-injection with water-miscible vitamin A derivatives. Vaccine 1991; 9:122-128.
12. Athanassiades TJ. Adjuvant effect of vitamin A palmitate and analogs on cell-mediated immunity. JNCI 1981; 67:1153-1156.
13. Dennert G. Retinoids and the immune system: Immunostimulation by vitamin A. In: *The Retinoids*, V.2, edited by Sporn, M.B., Roberts, A.B. and Goodman, D.S. Orlando: Academic Press, Inc., 1984, p. 373-390.
14. Ross AC. Separation and quantitation of retinyl esters and retinol by high-performance liquid chromatography. Meth Enz 1986; 123:68-74.
15. Pasatiempo AMG, Abaza M, Taylor CE, Ross AC. Effects of timing and dose of vitamin A on tissue retinol concentrations and antibody production in the previously vitamin A-depleted rat. Am J Clin Nutr 1992; 55:443-451.
16. Skidmore BJ, Chiller JM, Morrison DC, Weigle WO. Immunologic properties of bacterial lipopolysaccharides (LPS): Correlation between the mitogenic, adjuvant and immunogenic activities. J Immunol 1975; 114:770-775.
17. McGhee JR, Farrar JF, Michalek SM, Mergenhagen SE, Rosenstreich DL. Cellular requirements for lipopolysaccharide adjuvanticity. A role for both T lymphocytes and macrophages for in vitro responses to particulate antigens. J Exp Med 1979; 149:793-807.

18. Kimura S, Eldridge JH, Michalek SM, Morisaki I, Hamada S, McGhee JR. Immunoregulation in the rat: Ontogeny of B cell responses to types 1, 2 and T-dependent antigens. J Immunol 1985; 134:2839-2846.

19. Carman JA, Smith SM, Hayes CE. Characterization of a helper T-lymphocyte defect in vitamin A deficient mice. J Immunol 1989; 142:388-393.

20. Kraft R, Buerki H, Schweizer T, Hess MW, Cottier H, Stoner RD. Tetanus toxoid complexed with heterologous antibody can induce germinal center formation and B cell memory in mice without evoking a detectable anti-toxin response. Clin Exp Immunol 1989; 76:138-143.

21. West KP Jr, Sommer A. Delivery of oral doses of vitamin A to prevent vitamin A deficiency and nutritional blindness a state-of-the-art review. United Nations ACC/SCN State-of-the-Art Series 1987; Paper No.2.

22. Bryant RL, Barnett JB. Adjuvant properties of retinol on IgE production in mice. Int Archs Allergy Appl Immun 1979; 59:69-74.

23. Finkelman FD, Holmes J, Katona IM, Urban JF Jr. Beckmann MP, Park LS, Schooley KA, Coffman RL, Mosmann TR, Paul WE. Lymphokine control of in vivo immunoglobulin isotype selection. Annu Rev Immunol 1990; 8:303-332.

24. Pasatiempo AMG, Kinoshita M, Ross AC. Co-immunization with lipopolysaccharide and pneumococcal polysaccharide (SSS-III) restores the antibody response to SSS-III in the vitamin A-deficient rat. FASEB J. 1992; 6:A1659.

BACTERIOLOGY AND PATHOLOGY

OF NEUROTOXIGENIC CLOSTRIDIA

Charles L. Hatheway

Division of Bacterial and Mycotic Diseases
National Center for Infectious Diseases
Centers for Disease Control
Atlanta, Georgia 30333

INTRODUCTION

Neurotoxigenic clostridia are classified in Bergey's Manual[6] as either *Clostridium tetani* or *Clostridium botulinum*. The nomenclature has been strictly determined by the kind of neurotoxin that the organisms produce. For *C. tetani*, the system has been satisfactory because thus far, only one phenotype of organism has been observed to produce the tetanus neurotoxin. It also remains simple in that only one serologic type of tetanus toxin has been identified. With this simplicity, it is possible to refer to organisms which have all of the same characteristics, but lack the ability to produce the neurotoxin, as nontoxigenic *C. tetani*. The system for naming organisms that produce botulinum neurotoxin, however, is laden with a number of problems because of the multiplicity of phenotypes and genotypes of organisms that must be included.

Among the organisms capable of producing botulinum neurotoxin, each phenotypically different organism constitutes the equivalent of an individual species, and DNA comparison studies reveal nontoxigenic organisms that are closely related genetically to the toxigenic organisms at the species level. However, we are forced to make a clear taxonomic distinction based on one single gene. In addition to the taxonomic problems caused by this gene, it also raises speculation on what may be the function of the gene product in the economy of the organism.

Clostridium tetani and *Clostridium botulinum* have come to our attention only because of the diseases they can cause in humans and animals through the action of their neurotoxins. Without the toxins, they would be noticed only as casual or nuisance organisms involved in minor infections of wounds or spoilage of foods. The organisms are widely distributed in nature in soils, and in marine and freshwater sediments. From these sources they can contaminate wounds, foods, and feeds, and in some cases colonize intestinal tracts. There they produce their toxins, and the characteristic neurologic diseases then occur in humans and animals.

Botulinum and Tetanus Neurotoxins, Edited by
B.R. DasGupta, Plenum Press, New York, 1993

DISCOVERY OF NEUROTOXIGENIC CLOSTRIDIA

Both tetanus and botulism were recognized long before they were known to be due to bacteriologic agents. Tetanus has been clearly described by Hippocrates (ca. 400 B.C.).[1] Botulism, known early as sausage poisoning (Wurstvergiftung), had been encountered in Europe at least since the middle ages, but was only first described specifically in the medical literature in 1820 by Justinus Kerner,[42] who detailed an outbreak occurring in Wildbad, Germany in 1793 and reviewed 230 cases. Kitasato first isolated *C. tetani* in pure culture from a patient and showed that the culture as well as the culture filtrate could cause tetanus when injected into animals.[5] In the investigation of a large outbreak of botulism in Ellezelles, Belgium, van Ermengem established that the illness was due to a toxin produced by an anaerobic, sporeforming bacterium.[52] He named the organism *Bacillus botulinus*.

INVESTIGATION OF THE DISEASES

Tetanus

The clinical manifestations of tetanus are so uniquely distinct that a thorough laboratory investigation of individual cases is seldom pursued. Laboratory diagnosis does not play an important clinical role.[5] Positive identification of the toxin requires animal inoculation. Detection of tetanus toxin in the circulating blood has not been reported. It is possible to test a wound discharge for toxin if there is sufficient quantity (more than 1 ml), although it is difficult to cite the documentation of a positive finding by this means. Poor success in isolating *C. tetani* from wounds of patients with tetanus (10-15%) has been attributed to widespread use of antimicrobial agents.[5] Probably most of the information we have on the organism has been from studies of strains isolated from soil or other environmental specimens, which have not actually caused the disease. Reports of tetanus occurring in patients despite "protective levels" of tetanus antitoxin[4,39] suggest the possible existence of variant serologic types of tetanus toxin. This could not be answered in those cases, since neutralization tests were not performed on toxic clinical specimens or on cultures of the organisms responsible for the illness.

Botulism

Since van Ermengem's report[52] on the outbreak of botulism in Ellezelles, the majority of botulism incidents that have come to the attention of medical or public health authorities have been well investigated. This is because 1) the signs and symptoms of botulism are not so self-evident as those of tetanus and may mimic several other conditions (e.g., Guillain-Barré syndrome and myasthenia gravis); 2) clinical specimens of patients often are ready sources of toxin and/or organisms that serve as conclusive evidence for confirming the diagnosis; 3) botulism, unlike tetanus, can affect groups of people in a single incident, and thus identification of the source is often urgent; 4) in foodborne incidents, the suspect food often provides a convenient material for studying the outbreak. After each successful investigation of botulism, another one or more strains of the organism become available.

While tetanus has always been associated with infected wounds, botulism was recognized only as the result of ingesting food in which *C. botulinum* had been growing and producing its toxin, until 1943 with the recognition of the first case of wound botulism,[8] which is indeed analogous to tetanus. Thus, the possibility of growth of *C. botulinum* in the human or mammalian body was established. Colonization of the intestinal tract was then suspected but not confirmed until the discovery of infant botulism and the recognition of its mechanism.[31,40] This form of botulism has become recognized as the one most frequently occurring in the

United States since 1979. Botulism due to intestinal colonization of adults, although very rare, has been confirmed.[7,28]

CHARACTERISTICS OF THE ORGANISMS

Clostridium tetani

The organisms designated by this species name are a very uniform group, just as we would expect for a species. Microscopically, they are gram positive in young cultures, becoming gram negative after about 24-h incubation, and readily develop terminal spores which give the sporulated rods their "drumstick" appearance. The organisms are usually vigorously motile and are noted for their "swarming" growth on agar media. For isolation of pure cultures, advantage can be taken of their sporulation for selection by their heat resistance, and of their motility by picking inoculum from the advanced growth line at the front of the swarming growth. Usually the organisms appear to be negative for almost all the usual biochemical tests. However, some investigators report that glucose fermentation, gelatin liquefaction, and proteolytic activity may be shown if certain media and cultural conditions are met.[43] Acetic, propionic, and butyric acids are produced as metabolic end-products. The outstanding feature of the members of this species is their ability to produce the potent neurotoxin known as *tetanospasmin*, which is the cause of tetanus in humans and in a variety of mammalian species. Organisms are encountered that are similar in all respects to *C. tetani* except for a lack of toxigenicity.[36] This would be expected, since the toxin gene is contained in a plasmid.[13,24] Thus far, only one serologic type of tetanospasmin has been recognized. A highly toxigenic strain, known as the Harvard strain serves as a reference strain for the toxin and has been widely used for the preparation of toxoid for immunization. Besides the neurotoxin, *C. tetani* produces a hemolysin, known as tetanolysin, which appears not to play any role in tetanus.

Nakamura et al.[36] studied the DNA relatedness of a collection of strains of *C. tetani* (including three nontoxigenic strains), *C. tetanomorphum*, *C. cochlearium*, and *C. lentoputrescens*. Using the Harvard strain as the source of the labeled DNA, they found that all 11 of the toxigenic strains had at least 89% homology with one another. Two of the nontoxigenic *C. tetani*-like strains had 89% and 85% homology, respectively, with the Harvard strain. The third nontoxigenic strain (K-130), which was derived from the Harvard strain after heating at 100°C for 30 min, showed only 65% homology. This latter strain was more closely related to a reference strain of *C. cochlearium* (91% homology), and a group of 6 other strains previously identified as members of the other three species. In that study, the type strain of *C. lentoputrescens* was found to be identical with that of *C. cochlearium*, so the former species is no longer recognized; *C. tetanomorphum* is also no longer recognized as a species since available reference strains do not possess the characteristics originally described for the species.[6]

Clostridia that Produce Botulinum Neurotoxin

Cultural and Toxicological Characteristics. Although organisms that were found to cause botulism could be classified into three different groups, each of which could be defined as a separate species, all of them were maintained in a single species known as *Clostridium botulinum*. The first obvious difference among strains was the serologic distinction of different toxin types. Later differences in physiological characteristics of organisms producing different types of toxin became apparent. The American isolates that produce type A or type B toxins (Group I; proteolytic) were different from European strains that produce type B toxin (Group II; nonproteolytic); the latter conform to the description of the organism first isolated by van

Ermengem.[52] Strains that produce type C or type D toxins (Group III) differ from both these groups, and the botulism they cause essentially is restricted to nonhuman species. When the type E toxin strains were discovered,[23] they were noted to resemble the Group II strains. The first recognized type F strain[34] corresponds to Group I, but subsequent isolates of this toxin type recovered from environmental samples have been found that belong to either Group I or Group II. The second incident of type F botulism in humans was caused by a Group II strain.[32]

When Giménez and Ciccarelli reported an organism isolated from soil that produced a new type of botulinum neurotoxin (type G),[16] this marked the only time such a discovery was made in the absence a case or outbreak of botulism in either humans or animals. The organism was asaccharolytic and failed to produce the lipase reaction on egg yolk medium that is characteristic of all previously recognized botulinum toxin-producing organisms. In accord with the practice of including all botulinogenic organisms in the species, *C. botulinum*, Group IV was created to contain it. Thus, there are four groups of organisms known as *C. botulinum* with differential characteristics as listed in Table 1. There are seven serologically distinguishable types of botulinum neurotoxin; with very few exceptions, only one type is produced by a strain. Strains that produce type B or type F neurotoxin may be found in both Group I and Group II. Strains that produce type A belong to Group I, type E to Group II, types C and D to Group III, and type G to Group IV.

Since 1979, two human cases of type E botulism have been caused by neurotoxigenic strains of *C. butyricum*,[2,29] and three of type F botulism by neurotoxigenic *C. baratii*.[20,30] The most recent case caused by the latter organism occurred in January, 1992 (CDC, unpublished investigation). This widens the diversity of organisms we now know can produce botulinum neurotoxin. Both type E cases[2] and one of the type F cases[20] were in infants; the other two cases caused by the type F toxigenic *C. baratii* were in adult males, and had no suspected food source. The adult cases may have been due to intestinal colonization. The cases in which these organisms were involved are shown in Table 2. It may be noted that all of these cases appear to be the result of intestinal colonization.

Another case of type F botulism occurred in 1981 in which I suspect *C. baratii* may have been involved.[19] The serum of this case was toxic for mice, and the toxicity was neutralized by type F antitoxin. Mouse deaths were delayed by type E antitoxin when undiluted sample was tested, and complete protection was observed when the serum was tested at a 1:2 dilution at which the serum was lethal in the presence of type A or B antitoxin. No likely food vehicle was established. The patient was admitted with an initial diagnosis of small bowel obstruction. Only two rectal swabs were available for culture, but no toxigenic organisms were recovered.

DNA Homology Studies. The distinction among Groups I, II, III, and IV based on phenotypic characteristics is confirmed by DNA comparison among representative strains.[25,26,35,37,46] A high degree of DNA relatedness within each group is noted, and genetically related nontoxigenic organisms are found within each group. *Clostridium sporogenes* is related to Group I strains; *C. novyi* as well as nontoxigenic organisms that have apparently lost their phages are related to the type C and D strains in Group III. Nontoxigenic organisms closely related to type E organisms are commonly found, often in the same cultures with toxigenic organisms; thus, there are equivalent nontoxigenic organisms for Group II. One strain each of toxigenic *C. baratii* and *C. butyricum* was tested and found to be closely related to the type strain (nontoxigenic) of its respective species.[47] Noting the variety of species, or species-like groups of organisms that can produce the botulinus neurotoxins, Suen et al. proposed the name, *Clostridium argentinense* for the organisms, both toxigenic and nontoxigenic, in Group IV.[46] Distinguishing the other three groups with appropriate species names would seem acceptable, since we now recognize and accept the fact that botulism has been caused by *C. butyricum* and *C. baratii*.

Table 1. Characteristics of clostridia capable of producing botulinum neurotoxin.

| | *Clostridium botulinum* group | | | | | |
	I	II	III	IV*	*C. baratii*	*C. butyricum*
Toxin type	A,B,F	B,E,F	C,D	G	F	E
Lipase	+	+	+	-	-	-
Lecithinase	-	-	-	-	+	-
Milk digestion	+	-	±	+	-	-
Gelatin	+	+	+	+	-	-
Glucose	+	+	+	-	+	+
Lactose	-	-	-	-	+	+
Mannose	-	+	+	-	+	+
Growth temperature						
Optimum	35-40°	18-25°	40°	37°	30-37°	30-45°
Minimum	12°	3.3°	15°			10°
Spore heat resistance						
Temperature	112°	80°	104°	104°		
D-value	1.23	0.6-1.25	0.1-0.9	0.8-1.2		
Vol acids	A,iB,B,iV	A,B	A,P,B	A,iB,B,iV	A,B	A,B
Non-vol acids	HCA			PP		

*The name *Clostridium argentinense* has been proposed for this group.[46]

Table 2. Characteristics of five cases of botulism caused by *C. baratii* and *C. butyricum*.

Date	Age	Location	Toxin type	Toxin detected	Agent	Isolated from
12/79	2 wk	New Mexico	F	Feces+	*C. baratii*	Feces
7/84	16 wk	Italy	E	Feces+	*C. butyricum*	Feces
2/85	16 wk	Italy	E	Serum+ Feces+	*C. butyricum*	Feces
11/87	54 y	Georgia	F	Serum+ Stool+	*C. baratii*	Enema, feces
1/92	55 y	Maryland	F	Serum+	*C. baratii*	Gastric, feces

NEUROTOXINS AND GENES

The structure and function of tetanus and botulinum neurotoxins are remarkably similar despite the striking contrast in the manifestations of the respective illnesses they cause.[27] The structural gene for tetanus toxin is contained in a plasmid.[9,12,13] Likewise, evidence indicates that the gene for type G botulinum neurotoxin is carried on a plasmid.[11] On the other hand, toxigenicity of group III *C. botulinum* strains has been clearly shown to be phage mediated. Studies reported thus far have failed to show a role for either plasmids or phages in the toxigenicity in Group I or Group II.

The similarity of the serologic features[18] and of the nucleotide sequence of the gene fragment encoding for the N-terminus of the type E toxin genes[14,15] of the neurotoxins produced by both *C. botulinum* type E and neurotoxigenic *C. butyricum* should suggest a common source of the toxin gene for both species. In confirmation of this, Zhou, Sugiyama, and Johnson of the Food Research Institute, University of Wisconsin, have found a bacteriophage which carries the type E toxin gene, capable of interspecies transfer of toxigenicity (personal communication, Dr. Eric Johnson).

The type F toxin produced by the three toxigenic *C. baratii* strains encountered so far appears to differ somewhat serologically from the type F toxin produced by strains of *C. botulinum*. A high ratio of antitoxin to toxin is required for specific neutralization, and there seems to be a more pronounced cross-neutralization by type E antitoxin.[20,30] The toxin was purified and characterized, and the first 32 amino acids of the N-terminus of the molecule have identified.[17] Thus, more is presently known of the type F toxin from this source than from *C. botulinum*, and at present it cannot be compared structurally with any other type F neurotoxin.

INFECTIVE POTENTIAL OF NEUROTOXIGENIC CLOSTRIDIA

The pathogenesis of *C. tetani* appears to be restricted to dissemination of tetanospasmin during growth in an infected tissue. Unlike botulism, there are no reports of tetanus due to consumption of preformed toxin, or by absorption of toxin produced by the organism in the intestine. The difference between the two diseases in this respect may be due to the inability of tetanus toxin to be absorbed, or a greater lability of this toxin and an efficient destruction of its biologic activity in the gastrointestinal tract. The organism is commonly found in human and animal excreta.[3,50,51] It can grow and produce toxin in the intestinal tract of germ-free rats without causing tetanus;[54] in fact, an immune response by the rats to the toxin was observed. The natural presence of *C. tetani* in the intestines of humans and animals may result in a naturally acquired immunity to tetanus.[53]

The rarity of wound botulism despite the human population of the world remaining unimmunized gives evidence to the low potential for growth of *C. botulinum* in wounds. In every case of wound botulism on record (fewer than 60 reported), the toxin type has been either A or B, and whenever the organisms have been isolated and characterized, they have been identified as Group I.

Intestinal colonization, as typified by infant botulism, is also generally restricted to Group I strains.[22] The exceptions for infant botulism have been the two Italian type E cases caused by *C. butyricum*,[2] the one type F case caused by *C. baratii*,[20] and one of type C.[38] Several cases of adult botulism due to intestinal colonization have been documented.[7,28] Surgical modification of the bowel appears to be a contributing factor. It appears that a second isolate of toxigenic *C. baratii* caused this form of botulism in Georgia in 1987 in a patient with previous pyloroplasty and vagotomy.[30] A third toxigenic strain of *C. baratii* has recently been isolated from a patient with botulism in Maryland, with no plausible food vehicle found (CDC, 1992, unpublished laboratory findings). Recently we confirmed botulism in a man with Crohn's disease, whose stools continued to yield *C. botulinum* type A for more than six months (P. M. Griffin and C. L. Hatheway, CDC, unpublished). It is not known if the botulism was a simple colonization of the intestine, or an infection that developed in the intestinal lesions of the underlying disease. The patient developed a protective antitoxin immune response against the type A toxin. This is the only known instance of a naturally acquired protective immune response to botulism. Infection of intestinal lesions by *C. botulinum* has been proposed as the mechanism for "shaker foal" syndrome, a form of botulism in newborn horses.[49]

Thus far, *C. botulinum* type E has not been implicated in botulism that is due to growth of the organism in the body, either in a wound or in the intestine. It has been debated whether this is due to 1) the inability of the organism to grow at body temperature, 2) the inability to produce toxin at this temperature, or 3) the instability of the toxin at body temperature. However, some evidence shows that another Group II organism, the "European type B" *C. botulinum*, has colonized the gut and contributes to the continued absorbance of toxin long after the onset of foodborne botulism. Sebald and Saimot[41] report on type B cases due to consumption of ham in France with toxemias detectable weeks and even months after onset. In one investigation of type B (Group II) botulism in Iceland, one person who was previously asymptomatic and had no serum toxin or organisms detected in the stool developed symptoms and yielded positive serum toxin test results and stool cultures 47 days after onset of botulism in other family members.[28] His onset of symptoms came at a time when awareness of a food hazard was keen, and there was no possibility of his consumption of a food related to the suspected vehicle (sausages) that had caused the illness in the other family members.

Animal Models for Study of Intestinal Colonization

The mechanism of intestinal colonization by Group I strains of C. botulinum has been intensively studied in animal models.[48] Infant mice between the ages of 7 and 13 days are susceptible to colonization after intragastric challenge with 10^5 spores, but do not develop botulism. Adult germ-free mice are susceptible to colonization after orogastric administration of 10 spores, and develop botulism; they quickly become resistant (within 3 days) after exposure to conventional mice. Germ-free mice preinoculated with different mixtures of known flora were partially resistant to colonization, and a group preinoculated with C. perfringens and C. difficile were highly resistant. Cecectomized germ-free mice were resistant to colonization and did not develop botulism. Antibiotic (erythromycin-kanamycin)-treated conventional mice were susceptible to colonization (but did not develop botulism) after challenge with 10^5 spores between 15 and 60 hours after cessation of the treatment.[48] Toxin was detected in the intestines of colonized mice even in the absence of signs of illness. Mice apparently remain asymptomatic if colonization is restricted to the cecum, but develop symptoms of botulism in the toxin is produced in the small intestine where it can be absorbed. Perhaps constipation in infant botulism serves to move the colonizing organisms and toxin up into the small intestine.

In one study on colonization of infant mice after intragastric challenge with spores of C. botulinum type E, colonization and botulism did occur, but required a minimum challenge with 10^7 spores.[33] This suggests that colonization with this organism is possible, but unlikely because of the requirement of such a large inoculum.

Toxin Types C and D

Questions arise on why types C and D botulism occur so rarely in humans. Perhaps it is because of the unique phage relationship and specific cultural requirements for growth and infection of bacterial cells necessary for toxigenicity. These conditions are met with flocks of birds which pick up various combinations of bacterial strains and variants of phage from litter, leading to development of highly toxic combinations within living birds[10]. Carcasses of birds dying from any cause provide a potent source of toxin and toxigenic organisms, and are the means of continuing the outbreak among the birds or spreading it into other scavenging species. The organisms do not seem to be able to grow and produce toxin in foods that are acceptable for human consumption.

Toxin Type G

Although isolation of C. botulinum type G (C. argentinense) from autopsy specimens has been reported,[44,45] it seems doubtful that the organism was the cause of botulism or death in any of the cases. In the second report, seven of the nine alleged cases of botulism yielded unlikely botulinum types: three type G, three type F, and one type C. One should expect the diagnosis of many more cases of botulism of these types in living subjects in the same population that experienced such an incidence of undiagnosed illness. Neither hospitalized nor outpatient cases of type G botulism in humans have been observed, nor have any incidents in animals been reported. I believe it is necessary to raise questions on the occurrence of type G botulism because, although the reports are cited by everyone as conclusive, the evidence is questionable.

ASSAYS FOR NEUROTOXINS

The standard means for detecting and identifying clostridial neurotoxins is the mouse bioassay.[21] Mice injected with lethal amounts of the toxins develop botulism or tetanus and die within 12 to 72 hours. The rigid paralysis of tetanus is clearly recognizable and can hardly be confused with anything else. The flaccid paralysis of botulism accompanied by characteristic respiratory difficulties and development of the "wasp-shaped" body are also quite recognizable to the experienced laboratorian. Specific protection of mice injected with samples mixed with tetanus antitoxin or type-specific botulism antitoxin provides conclusive and specific identification.

When assays are being performed on multiple or replicate samples from known strains of toxigenic organisms, in vitro tests such as enzyme immunoassays that have been shown to detect the toxin from the source employed may well be more desirable.

The only way to positively identify a toxigenic organism has been to show the presence of toxin in the culture. With the current knowledge of nucleotide sequences of toxin genes and molecular methods, gene probes or the polymerase chain reaction (PCR) for detecting the toxin gene may provide an easier means. The PCR approach is under development in a number of laboratories, including our own (G. Franciosa and C.L. Hatheway, unpublished studies).

IMMUNITY AND IMMUNIZATION

Protection of humans and animals against tetanus and botulinum neurotoxins can be provided easily by immunization with preparations of the formalin-inactivated toxoids. For humans, because of the high potential for the terrible disease of tetanus, universal immunization is needed. There is only one known serologic type to tetanus toxin, and the Harvard strain, noted for its high toxigenicity is used as the source of the antigen.[55]

Botulism is so rare in humans that only laboratorians working with the organisms and their toxins are at sufficient risk to justify immunization. The most widely used vaccine for humans is the pentavalent ABCDE toxoid supplied by CDC. Antitoxin titers in toxoid recipients indicate a satisfactory level of protection. The vaccine's shortcomings are that it lacks the type F and G antigens and it is distributed and administered as an "investigational new drug," because its usage has been too limited to qualify for licensing. There have been no reported cases of botulism in any immunized person, many of whom have worked extensively with the toxin. The toxoid is closely monitored, and an adverse reaction rate of about 10% has been reported. If the rate exceeds that of tetanus toxoid, that may be due to a misadvised recommendation of subcutaneous instead of intramuscular inoculation.

For determination of immune status of toxoid recipients (both tetanus and botulism), mouse protection assays are more reliable than in vitro tests at present.

ANTITOXIN THERAPY

Although administration of antitoxin after onset of symptoms is of questionable efficacy, it is very often given in the hopes that it will limit the progression of the paralysis by neutralizing toxin that has not yet bound to receptors on nerve endings. This may be of special interest in cases of botulism due to intestinal colonization, in which there may be a continual absorption of toxin from the gut. Human tetanus immune globulin (TIG) is available for tetanus because of the widespread immunization against tetanus. For treatment against botulism, we are limited to the use of horse antitoxin. However, at present a study under the direction of Dr. Stephen Arnon in California is evaluating the efficacy of treatment of infant botulism with a special lot of human botulism immune globulin (BIG).

Administration of antitoxin before onset of paralysis (prophylactically) would obviously be effective for prevention of both tetanus and botulism. This can be and sometimes is done to prevent tetanus in a nonimmunized person who experiences a contaminated traumatic wound. TIG is readily available and a certain risk of tetanus is predictable. The only reasonable presumption of an exposure to botulism would be in the case of a laboratory accident. Because the scarcity of equine botulism antitoxin and the unavailability of a human product make it impossible to provide every laboratory working in the field with a supply for emergency use, active immunization of laboratorians is highly recommended.

TOPICS FOR FURTHER CONSIDERATION

In conclusion, there remain a number of important topics relating to the bacteriology and pathology of organisms that produce tetanus and botulinum neurotoxins. I will present them here in the form of questions and recommendations.

Questions

1. What do the toxin molecules do when they are not causing tetanus or botulism? If they serve no function, it should make no difference if mutations occur that make them into molecules with no activity. Are there nontoxigenic strains that produce the inactive molecules? Perhaps these questions can be studied by altering toxin genes (see chapter of Dr. Niemann et al.) and comparing the behavior and performance of substrains with variant genes.

2. Are there undiscovered neurotoxigenic species, phenotypes, and toxin types?

3. Why aren't Group II and III organisms involved in wound botulism, and botulism due to intestinal colonization, like Group I organisms?

4. To what extent and for what purposes are in vitro tests reliable? If tests are very specific for certain strains, might they be too specific to detect other strains of the same toxin type? We note "subtype" differences, which Dr. Giménez presented at this meeting (see chapter of Giménez and Giménez). With the pressures to eliminate bioassays, we must beware of being left with only unreliable assay methods. The reports of tetanus occurring in patients with "protective levels" of circulating antitoxin[4,39] may be the result of positive in vitro test results that do not correlate with protective potency.

5. Are existing toxoids adequate for protection of humans? The occurrence of tetanus in presence of "protective levels of antitoxin" might be due to an unrecognized serologic type of tetanospasmin. Are these observations due to a false presumption that an in vitro antitoxin test result is correlated with protection?

6. What features of the organisms will serve as good epidemiologic tools? Differences in physiologic characteristics, antibiotic susceptibilities, plasmid profiles, and isoenzyme electrophoretic mobility patterns have been examined with limited success. Currently under consideration are ribotyping, probing of restriction endonuclease digest fragments for location of the toxin gene, and sequence analysis of PCR-generated portions of toxin genes.

Recommendations

1. Attempt to isolate the strain from every case of tetanus and botulism. This approach to the investigation of botulism has taught us much about the bacteriology of this disease. The same approach should be taken with tetanus.

2. Revise the taxonomy of organisms that produce botulinum toxin to allow a distinctive name for each different group. This seems justified because of the discovery of botulinogenic strains of *C. baratii* and *C. butyricum*. The name *Clostridium argentinense* has been proposed as a first step, to include those that produce type G toxin, as well as genetically related nontoxigenic organisms.

3. Make reliable in vitro tests for neurotoxins available. There is a need for collaborative sharing of protocols and reagents between laboratories, and it seems unlikely that kits and reagents will become commercially available because of the limited market. Assure that bioassays will remain available for justified uses.

4. Assure availability of antitoxin standards. The supply of the International Standards for *Clostridium botulinum* Antitoxins supplied by WHO is critical at this time.

5. Revise the international unit for *C. botulinum* type E antitoxin. The international unit for types A, B, C, D, and F *C. botulinum* antitoxin were originally based of the amount of antitoxin required to neutralize 10,000 mouse LD_{50} of the corresponding toxin, while the type E unit was based on $1000LD_{50}$. One of the problems posed by this discrepancy is that in stating the unitage of a polyvalent antitoxin, it gives a distorted view of the relative potency of the type E component.

6. Assure the availability of adequate toxoids. A new lot of polyvalent botulinum toxoid that will contain types F and G needs to be made at this time. A commitment to this goal needs to be made by some organization.

7. Make human botulism immune globulin available for treatment of botulism patients.

REFERENCES

1. Adams EB, Lawrence DR, Smith JWG. Tetanus. Blackwell Scientific Publications, 1969.
2. Aureli P, Fenicia L, Pasolini B, Gianfranceschi M, McCroskey LM, Hatheway CL. Two cases of type E infant botulism caused by neurotoxigenic *Clostridium butyricum* in Italy. J Infect Dis 1986; 154:201.
3. Bauer JH, Meyer KF. Human intestinal carriers of tetanus spores in California. J Infect Dis 1926; 38:295.
4. Berger SA, Cherubin CE, Nelson S, Levine L. Tetanus despite pre-existing antitetanus antibody. J Am Med Assoc 1978; 240:769.
5. Bytchenko B. Microbiology of tetanus. In: Veronesi R, ed. Tetanus: Important New Concepts. Amsterdam: Excerpta Medica, 1981.
6. Cato EP, George WL, Finegold SM. Genus *Clostridium*. In: Sneath PHA, Mair NS, Sharpe ME, Holt JG, eds. Bergey's Manual of Systematic Bacteriology. Baltimore: Williams and Wilkins, 1986.
7. Chia JK, Clark JB, Ryan CA, Pollack M. Botulism in an adult associated with food-borne intestinal infection with *Clostridium botulinum*. N Engl J Med 1986; 315:239.
8. Davis JB, Mattman LH, Wiley M. *Clostridium botulinum* in a fatal wound infection. J Am Med Assoc 1951; 146:646.
9. Eisel U, Jarausch W, Goretski K, Henschen A, Engels J, Weller U, Hudel M, Habermann E, Niemann H. Tetanus toxin: primary structure, expression in *E. coli* and homology with botulinum toxins. EMBO J 1986; 5:2495.
10. Eklund MW, Poysky F, Oguma K, Iida H, Inoue K. Relationship of bacteriophages to toxin and hemagglutinin production and its significance in avian botulism outbreaks. In: Eklund, MW, Dowell VR Jr, eds. Avian Botulism. Springfield, IL: Charles C. Thomas, 1987.
11. Eklund MW, Poysky FT, Habig WH. Bacteriophages and plasmids in *Clostridium botulinum* and *Clostridium tetani* and their relationship to production of toxins. In: Simpson LL, ed. Botulinum Neurotoxin and Tetanus Toxin. San Diego: Academic Press, Inc., 1989.
12. Fairweather NF, Lyness VA. The complete nucleotide sequence of tetanus toxin. Nucleic Acids Res 1986; 14:7809.
13. Finn CW, Silver RP, Habig WH, Hardegree MC, Zon G, Garon CF. The structural gene for tetanus neurotoxin is on a plasmid. Science 1984; 224:881.

14. Fujii N, Kimura K, Murakami T, Indoh T, Yashiki T, Tsuzuki K, Yokosawa N, Oguma K. The nucleotide and deduced amino acid sequences of *Eco*RI fragment containing the 5'-terminal region of *Clostridium botulinum* type E toxin gene cloned from Mashike, Iwanai and Otaru strains. Microbiol Immunol 1990; 34:1041.

15. Fujii N, Kimura K, Yashiki T, Indoh T, Murakami T, Tsuzuki K, Yokosawa N, Oguma K. Cloning of a DNA fragment encoding the 5'-terminus of the botulinum type E toxin gene from *Clostridium butyricum* strain BL6340. J Gen Microbiol 1991; 137:519.

16. Giménez DF, Ciccarelli AS. Another type of *Clostridium botulinum*. Zbl Bakt I, Abt Orig A 1970; 215:221.

17. Giménez JA, Giménez MA, DasGupta BR. Characterization of the neurotoxin isolated from a *Clostridium baratii* strain implicated in infant botulism. Infect Immun 1992; 60:518.

18. Giménez JA, Sugiyama, H. Comparison of toxins of *Clostridium butyricum* and *Clostridium botulinum* type E. Infect Immun 1988; 56:926.

19. Green J, Spear H, Brinson RR. Human botulism (type F)—a rare type. Am J Med 1983; 75:893.

20. Hall JD, McCroskey LM, Pincomb BJ, Hatheway CL. Isolation of an organism resembling *Clostridium barati* which produces type F botulinal toxin from an infant with botulism. J Clin Microbiol 1985; 21:654.

21. Hatheway CL. Botulism. In: Balows A, Hausler WH Jr, Ohashi J, Turano A, eds. Laboratory Diagnosis of Infectious Diseases. New York: Springer-Verlag, 1988.

22. Hatheway CL, McCroskey LM. Examination of feces and serum for diagnosis of infant botulism in 336 patients. J Clin Microbiol 1987; 25:2334.

23. Hazen EL. A strain of *B. botulinus* not classified as type A, B, or C. J Infect Dis 1937; 60:260.

24. Laird WJ, Aaronson W, Silver RP, Habig WH, Hardegree, MC. Plasmid-associated toxigenicity in *Clostridium tetani*. J Infect Dis 1980; 142:623.

25. Lee WH, Riemann H. Correlation of toxic and nontoxic strains of *Clostridium botulinum* by DNA composition and homology. J Gen Microbiol 1970; 60:117.

26. Lee WH, Riemann, H. The genetic relatedness of proteolytic *Clostridium botulinum* strains. J Gen Microbiol 1970; 64:85.

27. Matsuda M. The structure of tetanus toxin. In: Simpson LL, ed. Botulinum Neurotoxin and Tetanus Toxin. San Diego: Academic Press, Inc., 1989.

28. McCroskey LM, Hatheway CL. Laboratory findings in four cases of adult botulism suggest colonization of the intestinal tract. J Clin Microbiol 1988; 26:1052.

29. McCroskey LM, Hatheway CL, Fenicia L, Pasolini B, Aureli P. Characterization of an organism that produces type E botulinal toxin but which resembles *Clostridium butyricum* from the feces of an infant with type E botulism. J Clin Microbiol 1986; 23:201.

30. McCroskey LM, Hatheway CL, Woodruff BA, Greenberg JA, Jurgenson P. Type F botulism due to neurotoxigenic *Clostridium baratii* from an unknown source in an adult. J Clin Microbiol 1991; 29:2618.

31. Midura TF, Arnon SS. Infant botulism: identification of *Clostridium botulinum* and its toxin in faeces. Lancet 1976; ii:934.

32. Midura TF, Nygaard GS, Wood RM, Bodily HL. *Clostridium botulinum* type F: isolation from venison jerky. Appl Microbiol 1972; 24:165.

33. Mitamura H, Kameyama K, Ando Y. Experimental toxicoinfection in infant mice challenged with spores of *Clostridium botulinum* type E. Jpn J Med Sci Biol 1982; 35:239.

34. Moller V, Scheibel I. Preliminary report on the isolation of an apparently new type of *Cl. botulinum*. Acta Pathol Microbiol Scand 1960; 48:80.

35. Nakamura S, Kimura I, Yamakawa K, Nishida S. Taxonomic relationships among *Clostridium novyi* types A and B, *Clostridium haemolyticum* and *Clostridium botulinum* type C. J Gen Microbiol 1983; 129:1473.

36. Nakamura S, Okado I, Abe T, Nishida S. Taxonomy of *Clostridium tetani* and related species. J Gen Microbiol 1979; 113:29.

37. Nakamura S, Okado I, Nakashio S, Nishida S. *Clostridium sporogenes* isolates and their relationship to *C. botulinum* based on deoxyribonucleic acid reassociation. J Gen Microbiol 1977; 100:395.

38. Oguma K, Yokota K, Hayashi S, Takeshi K, Kumagai M, Itoh N, Tachi N, Chiba S. Infant botulism due to *Clostridium botulinum* type C toxin. Lancet 1990; 336:1449.

39. Passen EL, Andersen BR. Clinical tetanus despite a "protective" level of toxin-neutralizing antibody. J Am Med Assoc 1986; 255:1171.

40. Pickett J, Berg B, Chaplin E, Brunstetter MA. Syndrome of botulism in infancy: clinical and electrophysiologic study. N Engl J Med 1976; 295:770.

41. Sebald, M, Saimot G. Toxémie botulique: intérêt de sa mise en évidence dans le diagnostic du botulisme humain de type B. Ann Microbiol (Inst Pasteur) 1973; 124A:61.

42. Smith LDS, Sugiyama H. Botulism: The Organism, its Toxins, the Disease. Springfield, IL: Charles C. Thomas, 1988.

43. Smith LDS, Williams BL. The Pathogenic Anaerobic Bacteria. Springfield, IL: Charles C. Thomas, 1984.

44. Sonnabend O, Sonnabend W, Heinzle R, Sigrist T, Dirnhofer R, Krech U. Isolation of *Clostridium botulinum* type G and identification of type G botulinal toxin in humans: report of five sudden unexpected deaths. J Infect Dis 1981; 143:22.

45. Sonnabend OA, Sonnabend WF, Krech U, Molz G, Sigrist T. Continuous microbiological study of 70 sudden and unexpected infant deaths: toxigenic intestinal *Clostridium botulinum* infection in 9 cases of sudden infant death syndrome. Lancet 1985; ii:237.

46. Suen JC, Hatheway CL, Steigerwalt AG, Brenner DJ. *Clostridium argentinense*, sp. nov: a genetically homogenous group composed of all strains of *Clostridium botulinum* toxin type G and some nontoxigenic strains previously identified as *Clostridium subterminale* or *Clostridium hastiforme*. Int J Syst Bacteriol 1988; 38:375.

47. Suen, JC, Hatheway CL, Steigerwalt AG, Brenner DJ. Genetic confirmation of identities of neuro-toxigenic *Clostridium baratii* and *Clostridium butyricum* implicated as agents of infant botulism. J Clin Microbiol 1988; 26:2191.

48. Sugiyama H. Mouse models for infant botulism. In: Zak O, Sande MA, eds. Experimental Models in Antimicrobial Chemotherapy. New York: Academic Press, 1986.

49. Swerczek, TW. Toxicoinfectious botulism in foals and adult horses. J Am Vet Med Assoc 1980; 176:217 (1980).

50. Ten Broeck C, Bauer JH. The tetanus bacillus as an intestinal saprophyte in man. J Exp Med 1922; 36:261.

51. Tulloch, WJ. Report of bacteriological investigations of tetanus on behalf of the war office. J Hyg 1919-20; 18:103.

52. Van Ermengem E. Ueber einen neuen anaeroben Bacillus and seine Beziehungen zum Botulismus. Ztschr Hyg Infektionskh 1897; 26:1.

53. Veronesi R, Bizzini B, Focaccia R, Coscina AL, Mazza CC, Focaccia T, Carraro F, Honningman MN. Naturally acquired antibodies to tetanus toxin in humans and animals from the Galapagos Islands. J Infect Dis 1983; 147:308.

54. Wells CL, Balish E. *Clostridium tetani* growth and toxin production in the intestines of germfree rats. Infect Immun 1983; 41:826.

55. Wirz, M, Gentili G, Collotti C. Tetanus vaccine: present status. Bact Vacc 1990; 13:35.

EPIDEMIOLOGICAL ASPECTS OF INFANT BOTULISM IN CALIFORNIA, 1976-1991

Patricia J. Schwarz, Joyce M. Arnon, Stephen S. Arnon

California Department of Health Services
Berkeley, CA 94704

INTRODUCTION

Infant botulism is the most common form of human botulism in the United States and is now known to be global in extent. The disease, an infectious intestinal form of botulism, was first recognized as a distinct medical entity in late 1976 in California, where all suspected cases are diagnosed microbiologically in this laboratory. Although its clinical spectrum ranges from mild outpatient illness to sudden unexpected death, most recognized cases are hospitalized.

METHODS

In these studies a case of infant botulism was defined as 1) an outpatient (N=3) or hospitalized (N=510) illness consistent with the known paralyzing action of botulinum toxin, *and* 2) identification of *Clostridium botulinum* toxin or organisms in feces. In 1976-83, two healthy control infants were matched to each case: one control for age, sex, race and neighborhood; the other for age, sex and county of residence.

RESULTS

For the 15-year period 1977-91 incidence averaged 7.2/100,000 live births (range 2.4-10.7). Age at onset ranged from 2-39 weeks, with 93% of cases occurring by age 6 months in a characteristic age distribution pattern. Unexpectedly for an infectious disease, female cases (51.3%) predominated. Racial distribution of cases generally mirrored the state's birth proportions: White (60.7%), Hispanic (26.5%), Asian (7.6%), Black (2.8%), Native American (0.8%), and other (1.6%). All cases resulted from either type A or type B botulinum toxin (ratio A:B=3:2). For the 320 inpatient cases hospitalized in 1984-91, hospital costs totaled $18,335,987, with a mean cost (1991) of $86,911 per case and a mean stay of 5 weeks. As for risk markers, formula-fed infants were younger at onset (mean 8.8 ± 4.9 weeks) than were breast-fed infants (mean 16.6 ± 7.8 weeks: *t*-test value 13.0, P <0.0001). Slow intestinal transit time predisposed to illness; 50% of cases *vs.* 15% of controls had less than one bowel movement per day (χ^2 = 26.9, P <10^{-6}). Honey remains the one identified, avoidable source of *C. botulinum* spores for susceptible infants; spore-containing honeys linked to cases have been recognized in California, Kentucky, Missouri, Utah, Washington state, and Canada and Japan. Corn syrups, whether of the light or dark variety, were not found to be associated with infant botulism, even when analyzed in precisely the same logistic regression categories reported by the Centers for Disease Control to be associated with infant botulism cases outside of California (Spika *et al.* AJDC 1989; 143:828-832).

CONCLUSIONS

In California infant botulism remains an expensive, protracted illness to which infants of all races appear to be potentially susceptible. Susceptibility may be enhanced by formula feeding, honey feeding and a slow intestinal transit time. The characteristic age distribution of infant botulism is seen in only one other entity, the sudden infant death syndrome (SIDS, crib death). The variety of neurotoxigenic clostridia that can cause infant botulism (*C. botulinum*, *C. baratii*, *C. butyricum*) suggests that this illness may represent the prototype of a class of diseases now termed "the intestinal toxemias of infancy," with other bacteria that can colonize the infant gut and produce toxins in it still awaiting discovery. These clinical illnesses and their severity would be expected to vary according to the target organ(s) of the toxins.

REFLECTIONS ON A HALF-CENTURY
OF FOODBORNE BOTULISM

E. M. Foster

Food Research Institute
University of Wisconsin
Madison, WI 53706

My personal interest in foodborne botulism dates from the first course in food microbiology that I took 55 years ago. The thing that impressed me most, then as well as now, was the spectacular potency of the toxin. It is hard to believe that so little of anything can do so much damage.

My interest in the subject has never waned, and I am pleased to have this opportunity to share some of my thoughts and observations with you after more than 50 years as an interested observer of this fascinating disease.

HISTORICAL PERSPECTIVE

The organism we now know as *Clostridium botulinum* was first associated with an outbreak of foodborne botulism in the Village of Ellezelles, Belgium in 1895, almost 100 years ago.[1,3] Professor E. Van Ermengem of the University of Ghent investigated the incident, which involved 34 musicians who ate some raw salted ham. Thirteen of them developed serious neuroparalytic symptoms, and three died.

Van Ermengem isolated an anaerobic spore-forming bacillus from the ham and from the spleen of one of the victims. Filtrates of the cultures produced fatal paralysis when injected into experimental animals. Van Ermengem noticed the similarity of the symptoms to those of botulism and named the organism *Bacillus botulinus*.

Perhaps I should mention that the disease called botulism, which means sausage poisoning, had been known in Europe for 100 years before Van Ermengem isolated and described the causal organism. Moreover, a similar disease associated with fish—hence named ichythism—had been recognized in Russia for at least 80 years before Van Ermengem appeared on the scene. What Van Ermengem did was isolate the causal organism and demonstrate how it produced the disease; that is, by forming a characteristic neuromuscular toxin.

Nine years after the Van Ermengem report, Landmann isolated *C. botulinum* from a salad made of home-canned beans.[3] This was the first incident involving a vegetable product. Thereafter, many other cultures were isolated from foods incriminated in botulism outbreaks, and it soon became apparent that the cultures showed numerous serological and cultural

Botulinum and Tetanus Neurotoxins, Edited by
B.R. DasGupta, Plenum Press, New York, 1993

differences. This led to the development of our present typing scheme with seven serologically distinct types designated A through G.

The majority of outbreaks in humans have involved types A, B and E. Type A is predominantly proteolytic and produces very heat resistant spores. Type B can be either proteolytic or non-proteolytic and its spores vary widely in heat resistance. Type E is non-proteolytic and its spores are easily killed with heat. Type A is the predominant cause of botulism in the U.S., China and Argentina. Type B is more common in Europe. Type E predominates in the cooler countries of the Northern Hemisphere. These differences are only general and there are many exceptions to each. Their commonality is the super-potent botulinal neurotoxin, which still kills about one-fifth of its victims in the U.S.

I do not propose to discuss the other forms of this disease—fowl botulism, which is usually caused by type C; animal botulism, caused by type D; wound botulism in humans caused by type A or B; and the toxico-infection of very young babies discovered by Arnon and his colleagues in California in 1976. I am referring, of course, to infant botulism, which can be caused by either type A or type B, and which is now the most common form of botulism in the U.S.

Neither do I propose to discuss botulism caused by clostridial species other than C. botulinum. I am aware that a strain of Clostridium barati has been incriminated in an outbreak of botulism and has been shown to produce botulinum toxin type F.[2] I do not wish to belittle the importance of this discovery—in fact, its existence is ominous; but we haven't had enough experience with it yet to evaluate its significance fully, and I have a different assignment anyway.

My primary mission is to consider two very practical questions: (1) Under what conditions does foodborne botulism occur? and (2) How can we protect ourselves against this dread disease? I propose to respond to these questions by frequent reference to the American experience, because I know it best. The general principles of microbial control that I shall mention will apply in any circumstance.

CONDITIONS NECESSARY FOR AN OUTBREAK OF FOODBORNE BOTULISM

Three conditions must be met if foodborne botulism is to occur:

1. Botulinum spores must be present in the food

Clostridium botulinum is widely distributed in soil, mud and surface waters over much of the globe, though usually in very low concentrations. The organism multiplies only in the absence of free oxygen, but it is not a strong competitor even in strictly anaerobic environments like the intestinal tract. Experience has shown, however, that it is prudent to assume the presence of botulinum spores in any food that is exposed to the environment—meat, vegetables, fruits, cereals, milk, eggs; in fact, any raw food product.

2. The spores must be able to germinate and the resulting vegetative cells must be able to grow and produce the neurotoxin

Here is where we have choices. One option is to kill the spores. A dead spore is harmless; it can not reproduce and form toxin. The canning industry is based on this principle, using heat as the lethal agent. Ionizing radiation will also work, but it is not yet applied in most of the world.

Alternatively, we can prevent growth, and thereby achieve safety, in either of two general ways: (1) We can hold the food at a temperature below the point where C. botulinum can multiply (i.e., refrigerate or freeze); or (2) We can prevent growth by changing the composition of the food. We can remove water (dry the food). We can add acid, directly or by fermentation, and produce an inhibitory pH. We can add an inhibitory chemical (salt, nitrite); or we can do a combination of adjustments, no one of which is sufficient by itself.[7] All of these approaches are now used in food preservation. In one way or another, but usually by simple trial and error, we have learned what works. Traditional procedures have proved their effectiveness, or they are discarded. But new ideas are constantly arising, usually from people who do not understand the principles of microbial control. Some of these ideas have led to disaster, as I shall point out later.

3. The food must be eaten without cooking

Fortunately, botulinum neurotoxin is easily destroyed by heating to 80°C for 10 minutes; so if there is any question about safety one can simply bring the product to a boil and eat it without danger. This is the basis of the common recommendation to boil home canned vegetables before serving. Boiling is not appropriate, however, for some foods.

FOODS USUALLY INVOLVED IN OUTBREAKS OF BOTULISM

As might be expected, the method of preparation and the conditions of storage determine the likelihood that any particular food product will develop botulinal toxin.

(1) "Fermented" fish and parts of sea mammals have been responsible for many outbreaks in the cooler northern countries.[1,3] Prominent among these foods have been Kirikomi (fermented fish) and Izushi (fermented fish with rice) in Northern Japan. Raw or parboiled meat of sea mammals, Urraq (seal flippers fermented in oil), Muktuk (whale blubber with skin and meat), and fermented salmon eggs have been responsible for numerous outbreaks in Alaska and Northern Canada. Lightly salted, partly dried or fermented fish and fish eggs have been the vehicles of outbreaks in Sweden, Iran and parts of the USSR. Practically all of the incidents attributed to seafood have been caused by *C. botulinum* type E.

(2) Preserved vegetable products have been the primary vehicles of botulism in China, the United States, Argentina, Italy, Spain and parts of the USSR.[3] In China, particularly Sinkiang Province, the predominant vehicles are fermented bean products such as Tofu, Natto and Miso.[6] Type A is predominant in Chinese, American and Argentine outbreaks, whereas Type B is more common in Italy, Spain and the USSR.

(3) Meats are the primary food vehicles in Germany, France and Poland, with Type B predominant.[3] Canned meats have been the main problem in Poland, and home-cured hams the chief offender in France.

THE AMERICAN EXPERIENCE

I propose to spend the rest of my time discussing some of our problems and experiences here in the United States. They have taught me a great deal about this wily and resourceful organism. I call it that because, no matter what we do to stop it, *Clostridium botulinum* always seems to find a way to evade our protective measures and do the job it is equipped to do; that is, poison us.

Home-canned foods

I have already mentioned that home-canned low-acid foods, primarily vegetables, are the chief vehicles of foodborne botulism in the United States today. That was equally true when I was a child on a small farm in Texas. I clearly remember how my mother spent all summer canning the food that enabled us to survive those long dreary Texas winters. Her pantry bulged with quart and half-gallon jars of the many fruits and vegetables that grew in our garden.

Did we ever have botulism? No, because my mother had access to a pressure cooker and she followed directions. I don't think she knew what botulism was, but she knew that the products would spoil if she did not heat them enough. So, in preventing spoilage she prevented botulism.

Not everybody uses a pressure cooker correctly, however. One can find directions for canning with a water bath or even in an oven. In these the temperature does not exceed 100°C, and that will not kill botulinum spores in a reasonable period of time. So most years we still have a few incidents of botulism from home-canned food in this country. When there is an outbreak it usually involves a single family, but sometimes home-canned products are sold in restaurants. Then many more people will be at risk. The largest botulism outbreak on record in this country, 59 cases, resulted from serving home-canned hot sauce at a Mexican-type restaurant in Michigan.[6]

I still remember the story of a family of four from Illinois who ate some delicious spaghetti sauce at a restaurant in Pennsylvania while they were on a vacation trip. The family purchased several jars of the sauce to take home. Sometime later all four members of the family checked into our University Hospital with botulism. One of the jars they took home with them was toxic. Fortunately, nobody died; but the father was in the hospital for three months.

Education seems to be the only solution to this problem. No commercial interests are involved, so the only way we can protect against botulism from home-canned food is to teach every home canner how to do the job properly. This is a difficult thing to do; so I see no prospect of total elimination of the botulinum hazard from home-canned food in the United States.

Commercially canned foods

The canned food industry in America started in the 1800s but achieved its most rapid growth with the development of agriculture in California. Heat processes were developed by each operator, largely on the basis of trial and error. Spoilage was a common problem, but the industry continued to grow.

The first recorded outbreak of botulism from a commercially canned food in the U.S. occurred in 1906.[4] It involved pork and beans canned in California, with 7 cases and 3 deaths. Between 1910 and 1914 commercially canned foods were responsible for five more outbreaks with 18 cases and 14 deaths.

Matters got rapidly worse in 1915 (Table 1), and still worse by 1921. Botulism was giving canned foods a bad name in this country, and something clearly had to be done.

Table 1. Botulism from commercially canned foods in the U.S. and Canada.[4]

Year	Outbreaks	Cases	Deaths
1915	3	9	7
1916	0	0	0
1917	0	0	0
1918	3	4	3
1919	5	54	29
1920	10	36	23
1921	5	42	10
1922	2	11	6
1923	0	0	0
1924	3	22	8
1925	4	13	9
1926	0	0	0
1927	0	0	0
1928	0	0	0
1929-1940	0	0	0

The canning industry, through its National Canners Association, enlisted the aid of top scientists, who developed safe, scientifically based heat processes for all kinds of canned foods. They used *C. botulinum* Type A spores as the test organism and developed process times and temperatures that would kill one thousand billion spores—the famous 12 D heat process. This is equivalent to heating for 2.4 minutes at 121°C.[3]

The new processes became available in the mid-1920s and the botulinum crisis was over. Table 2 shows the foods involved in the outbreaks from 1919 to 1925. It also shows there was not a single outbreak attributable to U.S.-produced commercially canned foods from 1926 to 1941.

Table 2. Products involved in botulism outbreaks from commercially canned foods.[4]

Year	Food	Outbreaks	Cases	Deaths
1919	Beets[a]	1	23	12
	Olives	3	28	17
	Sausage	1	3	0
1920	Beets	1	5	5
	Spinach	2	8	5
	Olives	5	15	9
	Milk	1	4	0
	Ham	1	4	4
1921	Spinach	4	37	7
	Olives	1	5	3
1922	Spinach	2	11	6
1923	—	0	0	0
1924	Olives	3	22	8
1925	Sardines	2	4	4
	Spinach	1	5	1
	Potted meat	1	4	4
1926-1940	—	0	0	0

[a]Canada

The record since that time continues to be remarkably good. Table 3 lists the recognized outbreaks of botulism attributable to canned foods produced commercially in the United States since 1941. You will notice that four of the seven incidents were caused by *C. botulinum* Type E. In view of this organism's very low heat resistance it seems likely that these outbreaks resulted from container leakage after processing.

The Type A botulism outbreak in 1961 from vichyssoise soup had a critical impact on regulation of the canning industry in the U.S. Prior to that time canning plants were regulated largely by state authorities, and there was considerable variation from state to state. The 1971 outbreak appeared to be a clear case of underprocessing, so our Food and Drug Administration developed and adopted new nationwide regulations that involved all commercial canneries in the country.

Table 3. Outbreaks of botulism from commercially canned foods in the U.S. 1926-1986.[5]

Year	Product	Toxin type	Cases	Deaths	Probable cause
1941	Mushroom sauce[a]	E	3	1	Leakage
1963	Tuna fish	E	3	2	Leakage
1971	Vichyssoise soup	A	2	1	Underprocess
1974	Beef stew[a]	A	2	1	Unknown
1978	Salmon[a]	E	4	2	Leakage
1982	Salmon[a]	E	2	1	Leakage
1982	Tomatoes[a]	A	1	0	Unknown

[a]Single can

The agency's action also was influenced by several reports of botulinum spores or toxin in canned foods that were not involved in disease outbreaks (Table 4). In these instances the contamination was discovered during routine examination of cans of spoiled food.

Table 4. Presence of *C. botulinum* or its toxin in U.S. commercially canned foods—not involved in disease outbreaks.[5]

Year	Product	Toxin type	Cause of spoilage
1965	Spinach	A	Underprocess
1970	Mushrooms	B	Underprocess
1971	Chicken vegetable soup	A,B	Underprocess
1973	Mushrooms[a]	B	Underprocess
1974	Tuna fish[b]	C	Leakage
1980	Mushrooms	B	Underprocess
1981	Mushrooms[c]	B	Underprocess

[a]Involved 7 packers
[b]Single can
[c]Malfunctioning flame sterilizer

It has been estimated that the U.S. industry produces about 30,000,000,000 cans of low-acid food each year.[5] In the 42 years between 1942 and 1984 it has been calculated that we Americans consumed approximately 1,300,000,000,000 containers of low-acid canned food.[5] When we consider that only seven incidents of botulism have been traced to U.S. commercially canned foods in all that time (Table 3) it is clear that the risk of botulism is very small indeed. The canning industry and its scientists deserve a great deal of credit for rescuing this important preservation system from the danger it faced 70 years ago.

Smoked fish from the Great Lakes

The Great Lakes of the United States and Canada have long been an important fishery resource for people who live along the shores of the lakes. Many years ago a smoked fish industry developed to utilize some of the abundant small fish, such as chubs, ciscoes and whitefish. The smoking process was very simple. Cleaned fish were soaked in mild salt brine and then hung by the tail over an open fire inside a smokehouse. The result was lightly salted cooked fish with a smoke flavor. They were delicious and most were consumed by local people in nearby bars and restaurants within 2 or 3 days.

Then about 35 years ago a packaging entrepreneur proposed to distribute the smoked fish over a larger area by packaging them under vacuum to prevent mold growth, dehydration and rancidity. Several processors actually started distributing vacuum-packaged smoked fish at that time.

About 1960 a woman in Minneapolis, Minnesota, bought a package of smoked ciscoes from Lake Superior. The weather was warm and she kept the smoked fish in her car for several hours. That evening she served them to guests at dinner, and two people died from botulism.

This was a problem we didn't know we had, and my research group, with support from our Food and Drug Administration, undertook a survey of the Great Lakes for *C. botulinum* Type E. Not surprising in retrospect, the spores were found in all the lakes.

Meanwhile, on a beautiful warm weekend in September, 1963, a man and wife took a drive along the Lake Superior shore to see the fall colors. Along the way they purchased a large smoked whitefish, which they kept in their car and sampled from time to time. Twenty-four hours after they returned home both were dead of botulinum poisoning. The remaining part of the fish contained Type E toxin.

About the same time in 1963 a processor in Michigan shipped a container of vacuum-packaged smoked chubs by truck to a major supermarket chain in another state. The product was packed in dry ice, but it was missent to Chicago and then transshipped on another truck to its destination in Tennessee. All of this took about five days. By the time the product arrived the dry ice had long since evaporated. The smoked fish was distributed over parts of three states and 17 people contracted botulism. Five of them died.

This is a good example of what can happen if we rely totally on refrigeration for protection against botulinum growth. Refrigeration sometimes fails, and when it does disaster can strike.

I want to emphasize that vacuum packaging *per se* was not the reason why the anaerobic *C. botulinum* Type E was able to grow in Great Lakes smoked fish. We and others have shown that the spores can germinate, grow and produce toxin equally well whether the smoked fish was vacuum packaged or not. Vacuum packaging merely makes it possible to hold the product longer and abuse it more without signs of mold growth, rancidity or dehydration.

We still find Great Lakes smoked fish in our food stores, but the processing and distribution systems are much better regulated and controlled than they used to be.

Foil-wrapped baked potatoes

In recent years we have had at least two outbreaks, one involving more than 30 cases, that were traced to ordinary potato salad. This familiar product is made from diced cooked potatoes mixed with chopped pickles, onion, celery and other raw vegetables, then treated with a liberal quantity of mayonnaise or other acidic salad dressing.

Our outbreaks were the result of a packaging innovation intended to save labor. Beefsteak with a baked potato is a popular item in many U.S. restaurants. Someone started a business of supplying restaurants with cleaned raw potatoes of uniform size wrapped in metal foil for baking. All the chef had to do was estimate the number he would need for a meal and put them in the oven.

Frequently, there are left-over baked potatoes. Rather than waste them the chef will save the foil-wrapped cooked vegetables and make potato salad of them at a later time. At first the potatoes were simply held at room temperature until needed.

After potato salad was incriminated in botulism outbreaks, two laboratories showed that *C. botulinum* can grow and produce toxin in foil-wrapped baked potatoes at room temperature. This happened whether the potatoes were injected with botulinum spores or simply held with their natural inoculum from the soil. If they had not been wrapped in metal foil the potatoes would have dried out somewhat and might not have become toxic.

As a result of these experiences our Food and Drug Administration has classified baked potatoes as perishable food that must be refrigerated.

Chopped garlic in oil

Another labor-saving innovation that brought disaster was chopped garlic in oil. Garlic toast is a popular item in North American restaurants. It is simply toasted bread coated with garlic-flavored butter. Preparing garlic-flavored butter by a chef is tedious and time-consuming. As an alternative an American company suspended finely chopped raw garlic in edible oil and sold it to restaurants. The chef could easily mix the garlic in oil with softened butter to make garlic toast. The chopped garlic in oil was prominently labelled "KEEP UNDER REFRIGERATION."

Nonetheless, a Canadian restaurant was involved in an outbreak of botulism with at least 37 cases attributable to chopped garlic in oil made by the American company. It is generally believed that the product was not kept refrigerated at the restaurant as directed on the label.

We have had a subsequent small outbreak from chopped garlic in oil in the United States. In response, the Food and Drug Administration has banned this product from the market unless it is acidified.

Sauteed onions

Several customers of a restaurant in the central United States developed botulism after eating patty-melt sandwiches containing sauteed chopped onions. The onions were cooked in the usual way and then kept warm in a pan of melted margarine. The pan was not emptied and cleaned on a frequent basis; the chef simply added more sauteed onions as needed. *C. botulinum* actually grew in the onion-margarine mixture.

Fried lotus root

Not everything wrong happens in the U.S. Several years ago a commercial "vacuum packed deep-fried lotus root" product caused an outbreak of Type A botulism in Japan, with 36 cases and 11 deaths.[6] We have comparable products in this country, but they are usually distributed frozen. There is nothing to stop someone from refrigerated distribution, however, and if refrigeration failed we could suffer the Japanese experience.

Other possibilities

The list of unexpected botulism vehicles continues to grow. Raw mushrooms are being packaged in plastic film and sold at room temperature. Although I am not aware of any actual botulism outbreaks, research has shown that *C. botulinum* can grow in these mushrooms at room temperature. Two small holes in the film will admit air and prevent botulinal growth.

There are many other examples, but these should be enough to convince us of the versatility and resourcefulness of *C. botulinum*. Over the years we have learned in one way or another how to control the organism in our traditional foods; but when we change the way we produce, transport and store foods we often do things that affect their stability without realizing how the change may affect their safety. Right now, for example, we are being urged to reduce our consumption of salt, nitrite and other preservative chemicals. Unknowing consumers respond positively to this advice and food marketers try to give consumers what they want. The result is constant pressure on manufacturers to reduce the salt content of their products. At the same time there is intense pressure for longer and longer shelf life, which gives *C. botulinum* more time to grow.

Few people, even in the food industry, understand how important salt is to the stability of cheeses and cured meats. If present trends continue I predict that we shall see more outbreaks of botulism, especially in those products that depend primarily on constant refrigeration for their safety.

CLOSING COMMENTS

I should like to close with a personal observation on the foodborne botulism hazard. Earlier in my career as a food microbiologist I used to wonder why we don't have more botulism in those vast areas of the world where there is no refrigeration. Then I remembered my own experience on my family's farm and the answer came to me.

Everything we ate was either dried, canned, pickled or fresh. When we got up in the morning my mother cooked a good hot breakfast—oatmeal, biscuits, and eggs with home-cured ham or bacon. Then we went out into the fields to work. When we came in at noon we had another hot meal to carry us through the afternoon. At night, however, we ate the left-overs from the noon meal. What we did not eat went out to the pigs and chickens. We had no refrigeration and each day was just as I have described. I am told that millions of the world's people who have no refrigeration follow essentially the plan my family followed—they prepare their food, they cook it thoroughly, and they eat it right away. They do not attempt to keep ready-to-eat cooked foods for very long.

Why does this system work? It simply does not give *C. botulinum* spores enough time to germinate, grow and produce their deadly toxin. Time is a vital element in the botulinum hazard equation. If we give the organism enough time it will manage somehow to bite us.

No one can make light of the botulism problem. Botulism is a terrible disease if you have it, and it is a disaster to a food manufacturer whose product is identified as the vehicle in an outbreak. Even so, I am encouraged by the fact that we don't have even more foodborne botulism today. In the United States, at least, we can credit any successes we have achieved to better education of the public, better use of scientific information by the food industry, and more effective regulation by the government.

I believe I can safely say that more people in this country now are receiving their botulinum toxin by injection than by ingestion.

REFERENCES

1. Dolman CE. Botulism as a world health problem. In: Lewis KH, Cassel K Jr, eds. Botulism, proceedings of a symposium. Cincinnati, OH: U.S. Public Health Service, 1964.
2. Hall JD, McCroskey LM, Pincomb BJ, Hatheway CL. Isolation of an organism resembling *Clostridium barati* which produces type F botulinal toxin from an infant with botulism. J Clin Microbiol 1985; 21:654-655.
3. Hauschild AHW. *Clostridium botulinum*. In: Doyle MP, ed. Foodborne bacterial pathogens. New York: Marcel Dekker, Inc., 1989:111-189.
4. Meyer KF, Eddie B. Sixty-five years of human botulism in the United States and Canada. San Francisco: George Williams Hooper Foundation, University of California, 1965.
5. NFPA/CMI Container Integrity Task Force, Microbiological Assessment Group Report. Botulism risk from post-processing contamination of commercially canned foods in metal containers. J Food Prot 1984; 47:801-816.
6. Sakaguchi G. Botulism. In: Cliver DO, Cochrane BA, eds. Progress in food safety. Madison, WI: Food Research Institute, University of Wisconsin, 1986:18-34.
7. Tanaka N, Traisman E, Plantinga P, Finn L, Flom W, Meeks L, Guggisberg J. Evaluation of factors involved in antibotulinal properties of pasteurized process cheese spreads. J Food Prot 1986; 49:526-531.

CONTRIBUTIONS OF THE U.S. ARMY
TO BOTULINUM TOXIN RESEARCH

John L. Middlebrook

Toxinology Division
U. S. Army Medical Research Institute for Infectious Diseases
Frederick, MD 21702

This conference on botulinum and tetanus neurotoxins has covered a wide range of scientific topics, from basic toxin characterization studies to use of the toxins in modern clinical settings. At the request of the conference organizers, I will briefly summarize the U.S. Army's past and present programs on botulinum toxin, highlighting several of the very practical accomplishments from which this field of research has benefited. Some of this information is not highly technical or scientific in nature, but represents contributions which have gone unrecognized by many of those working in the field. I should hasten to say that the views expressed here are mine alone and should not be construed to be official positions of my Institute or the U.S. Army.

While this is not intended to be a political or historic document, it is necessary at the outset to consider why the U.S. Army has a program on botulinum toxin research and to briefly describe its genesis and development. Simply stated, the Army works with botulinum toxin because of the potential that this extremely potent toxin might be used as a biological warfare agent against the United States or its allies. Although actual employment of biological warfare can be traced to the 6th century B.C.,[1] little or no systematic concern was expressed at governmental levels until World War II. During the war, the Allies learned of several efforts by the Axis powers to develop certain infectious agents as biological weapons. In some cases, there were apparently cruel and inhumane experiments carried out on prisoners of war.[2] It was recognized that medical science and technology had advanced to the point where biological warfare could be a strategic threat and the U.S. made a decision to develop a program to deal with the problem. Since biological warfare was considered a military threat, the problem was handed over to the military, specifically to the U.S. Army Chemical Corps. Initially, the research program was established at then Camp Detrick, in Frederick, Maryland. Largely, I believe, because of the post WW II and pre-cold war mentalities, it was decided to embark on both offensive and defensive programs. That is to say, the U.S. Army sought to develop biological weapons as well as to develop protective measures from biological weapons. Parenthetically, I would note that this was the approach of several other nations who developed biological warfare programs at the same time. Be that as it may, this stance was maintained by the U.S. until 1969, when, then President Richard Nixon, unilaterally renounced the development of biological warfare weapons, disestablished the offensive program and destroyed the U.S. arsenal of biological weapons. Because the programs were so intricately entwined, these moves also eliminated the medical defensive efforts of the labs, and this mission fell to a small and little-known unit, also at Ft. Detrick, called the U.S. Army Medical Unit. This Unit was later reorganized and designated the U.S. Army Research Institute of Infectious Diseases (USAMRIID). Because both the staff and the physical plant of USAMRIID were far too small to handle the program, a new facility was constructed and the staff increased.

Simultaneously, a important development took place in the security classification. During the days of weapon development, much of the work was done at the "secret" level. Publication was difficult for the scientists involved and there was an element of distrust from the outside scientific community. With the changeover to solely medical defensive work, the program became open and unclassified; publication of the scientists' work was encouraged and research was open to public scrutiny. These policies continue today; we receive visiting scientists from the U.S. and foreign countries alike. Numerous collaborations have taken place, including work with scientists and public health officials in countries such as the People's Republic of China and the former Soviet Union. While work on defenses against the potential use of biological warfare continues at USAMRIID and other Army and Navy laboratories, it is my purpose here to highlight some of the results of work which have already brought benefits to the area of botulinum toxin research and clinical care of intoxicated patients.

BOTULINUM TOXIN VACCINE

One of the most important contributions of the early Ft. Detrick workers was the research and development effort behind the botulinum toxoid vaccine we all use today. It was obvious at the outset of the Army program that protection of laboratory workers from botulinum toxin was imperative. The classical work of Roux with diphtheria toxin[3] suggested that a toxoid would be a good candidate for eliciting protective immunity in humans, and the Army began such an effort. There had been a botulinum toxoid produced in the 1930s by a Russian researcher, Velikanov;[4] his was apparently the first work which involved immunization of humans against the toxin.[4]

The Army's first efforts with botulinum toxin were production and testing of a toxoid against serotypes A and B.[5] The toxoid was a filtered autolysate of a three-day culture to which formalin was added. The material lost toxicity after 2-3 weeks, and was refiltered and bottled. This material was referred to as a "fluid toxoid". With slight variations, another lot of fluid toxoid was prepared and adsorbed to alum to make an "alum-precipitated toxoid". Several hundred human volunteers were then immunized with these two types of toxoids, and their sero-responses followed as a function of time. Each serum sample was evaluated by a mouse neutralization test, wherein various dilutions of the test serum were incubated with botulinum toxin, and then injected into mice.

The fluid toxoid was administered in the following regimens: (A) 0, 3 and 6 weeks; (B) 0, 2, 4 and 6 weeks; (C) 0, 3 and 9 weeks; (D) 0, 1, 2 and 3 weeks. Initially, regimens A and B were compared, and the results were virtually the same. Only a few individuals responded with a measurable neutralizing titer before 8 weeks. At that time, 70-80% had titers of 0.02 unit/ml serum or greater, and these titers were maintained in the absence of additional boosters for up to 10 months. Compared to regimens A and B, regimens C and D were inferior from two standpoints. The titers took more time to develop and the fraction of individuals exhibiting ≥ 0.02 unit/ml serum titers was lower.

Because of work demonstrating the increased efficacy of toxoids adsorbed to a particulate carrier,[3] immunization with alum-precipitated toxoid was compared to immunization with fluid toxoid. Only two injections of precipitated toxoid at 0 and 8 weeks elicited titers of ≥ 0.1 units in 40-60% of subjects, compared to only about 15% achieving that level with fluid toxoid and regimen B above. From the data presented, one cannot determine how long those higher titers were maintained in precipitated-toxoid recipients, since the subjects were apparently not followed after 3 months. However, additional regimens with precipitated-toxoid were carried out, and individuals monitored for more extended time periods. For example, precipitated-toxoid was given 2 or 3 times, spaced 3-4 weeks apart. A much higher percentage (approximately 60-70%) of those receiving 3 doses exhibited titers than did those receiving only 2 doses (<10%). Moreover, the titers of the 3-dose recipients were still well elevated after 4 months.

However, the number of immunizations was not the only important factor behind attaining good responses. Thus, 4 injections at 2-week intervals elicited an early response of titers, but immunization twice, at 0 and 8 weeks, produced equal or better responses by 10 weeks and recipients retained their titers at 4 months. Only one immunization or two immunizations 3 weeks apart produced poor responses. Thus, this study seemed to indicate that a minimum of 2 immunizations was required to obtain a high percentage of responders, and that at least 4 weeks should elapse (perhaps as many as 8 weeks) before the second injection is administered.

Because there was a significant number of undesirable local and systemic reactions to the crude toxin-based toxoid, some effort was spent to devise a simple, large scale purification scheme for both serotypes A[6] and B.[7] The schemes were very similar and consisted of acid precipitation of toxin from the culture supernatants, $CaCl_2$ extraction of the precipitate, a second acid precipitation and a final alcohol precipitation. Since this work was done before the availability of such analytical techniques as gel electrophoresis, analytical ultracentrifugation was the only technique used to determine purity. By that criterion, the material appeared to be pure and has been referred to many times as "crystalline toxin." Today, we know that the toxin is only about 10-15% pure 150,000 kD neurotoxin at this stage of purification. However, it is clear by the data provided that the scheme did remove a considerable number of extraneous materials. The conditions necessary to detoxify both crystalline A and B toxins were investigated in detail.[8] Basically 2-3 weeks in 0.6% formalin made a stable toxoid that elicited protective immunity in mice and guinea pigs. This toxoid appeared to be stable for at least 36 months. Therefore, it was decided to test these improved toxoids in humans.[9]

Purified toxin-based toxoids were prepared in two large lots by Parke, Davis and Company. The toxoids contained approximately the same amount of protein toxoid and were both adsorbed to aluminum phosphate, then mixed together to produce a bivalent vaccine. Four vaccination schemes were undertaken; 0-2-4-6, 0-8, 0-2-10 and 0-10 weeks.[9] The results for serotype A indicated that the 0-2-10 week schedule was markedly superior to the others in producing rapid immunity. Titers of approximately 0.18 units/ml of serum were achieved by 12 weeks and these titers held to the level of approximately 0.08 units at 52 weeks. The 0-2-10 week schedule was also the best for serotype B, but only marginally. Immunization at 0-2-4-6 weeks achieved almost the same results. Moreover, it was clear that higher titers could be achieved against serotype A than B, despite there being the same amount of antigen in the preparations. However, if rapid immunity was the goal, the 0-2-10 week schedules were best for both serotypes.

In contrast, if single boosts were given at 1 and 2 years after the above initial series, the 0-2-4-6 scheme produced a better long-term immunity than did the initial series of 0-2-10. Thus, the average titer against both serotypes was four-fold higher at 1 year, post-boost after the series 0-2-4-6 in comparison with the series 0-2-10 and 4 to 10 times higher at two years, post-boost. Nevertheless, 100% of the recipients were judged to have a satisfactory response 8 weeks after their one year booster regardless of the initial regimen.

Since the bivalent toxoid seemed to work well in humans but there were additional toxin serotypes of concern, there was a subsequent effort to produce and test a pentavalent toxoid.[10] Using partially purified toxins prepared like the bivalent preparations above, toxin serotypes A-E were toxoided and mixed to make a pentavalent vaccine. Based on the data provided, it is difficult to determine the exact weight ratios of the different serotypes. However, those values are probably close to 1.7 parts A, 0.5 parts B, unknown parts C, 4 parts D and 0.5 parts E. The immunization schedule selected was 0-2-10 weeks with a booster at 52 weeks, and comparison was made with monovalent toxoid for each serotype. Sera were taken at 12, 52 and 60 weeks for titers by the standard mouse neutralization test. Results were qualitatively the same as with the monovalent and bivalent toxoids. Namely, there was a clear immune response to each serotype at 12 weeks. By 52 weeks, titers were very low or undetectable, but, after the boost, individuals exhibited a robust increase in titers to levels approximately ten-fold those observed at 12 weeks. In almost all instances, individuals receiving monovalent toxoid exhibited higher titers than those receiving the same amount of antigen mixed as the pentavalent product. By most measures, the poorest response was to serotype B. Immunization with this antigen elicited the lowest percentage of individuals exhibiting measurable titers and, in those who responded, titers were usually lower than with the other serotypes. Serotype E toxoid produced slightly better results than serotype B. Responses to serotypes A, C and D were good. However, since there was as much variation between lots of the vaccine as there was between these three serotypes, one cannot determine which was the best immunogen.

Following this work, large amounts of components for the pentavalent vaccine were prepared by the Michigan State Department of Public Health using technology developed by the Ft. Detrick scientists. These preparations have served as the U.S. supply of botulinum toxin vaccine until the present time.

BOTULINUM TOXIN ANTISERA

In addition to vaccine research, scientists at Ft. Detrick played a role in the development of antisera products designed to protect from botulinum toxin. Since there are significant adverse side effects to the use of equine antitoxin,[11] it was deemed advisable to develop a human antitoxin product. An effort was undertaken to boost and plasmapherese humans who had been previously immunized with the pentavalent vaccine.[12] More than 2000 liters of plasma were collected, most of which was processed for clinical use under a contract with Dr. Richard Condie at the University of Minnesota. A small amount of product is kept on hand at USAMRIID available in the event of a laboratory accident. While this plasma has proven very effective at protecting mice from experimental challenges with the toxin, we have (fortunately) not had the need to use it in humans.

More recently, USAMRIID acquired a large number of horses and immunized them in a shortened regimen with unusually high doses of monovalent toxoid. The animals were later boosted with toxin. This approach proved very effective and the animals developed extraordinarily high titers in only a few weeks. Thousands of liters of plasma have been obtained and are being processed to produce a "despeciated" or $F(Ab)_2$ product suitable for human use.

FINANCIAL SUPPORT OF BASIC RESEARCH AND CONFERENCES

A final contribution to the area of botulinum toxin research has been financial support in the form of grants and research contracts to academic and private laboratories. By and large, this has taken the form of support for basic scientific studies and several of the scientists attending and contributing to this conference have received such support. These contracts are awarded on a competitive basis and involve first a preproposal letter broadly stating the proposed work. If there is a programmatic need for the work, a full proposal is then requested and it receives both a scientific and budget review. Frequently there are negotiations on the nature of the work scope or budget before a funding decision is made. Final decisions concerning the contract are based on the reviews and availability of funds. In most cases rejections are made not because of the scientific quality of the proposal, but because of a lack of need for the proposed research within the mission of the overall program. Scientific conference organizers apply for support by essentially the same process. Once again, decisions about support are based on relevance to the mission and availability of funds.

FUTURE DIRECTIONS

In spite of the decreasing world tension and apparent end of the cold war, there is still a concern about the possible use of biological warfare against the U.S. Many references have been made to biological warfare as the "poor man's atomic bomb". It is much easier and cheaper to acquire the means and raw materials necessary to make biological than it is to make conventional or atomic weapons. Thus, there are more concerns than ever that nations may undertake such efforts. Because of these concerns and because of the extreme toxicity of botulinum toxin, our medical protection program will continue. During the next few years, we intend to develop an entirely new and more efficacious vaccine against botulinum toxin. There is a short paper elsewhere in these proceedings which describes our preliminary efforts in this regard. In addition, we are attempting to develop new immunotherapeutic products in the form of molecularly engineered monoclonal antibodies. Some of the exciting data presented at this conference suggest new approaches to drug therapy and these, too, will be pursued. Scientists who are involved in this program hope that the materials which we strive to develop never have to be used to defend against a biological warfare attack. However, there is little doubt that these vaccines and drugs find their way into public health use, as did some of the products described above.

REFERENCES

1. Huxsoll DL, Patrick WC III, Parrott CD. Veterinary services in biological disasters. J Amer Vet Med Assoc 1987; 190:714-722.
2. Tsuneishi K. The germ warfare unit which disappeared: the Kwangtung Army's 731st unit. Tokyo: Kai-mei-sha 1982; 140 (translated by SCITRAN).
3. Roux E, Yersin A. Contribution a l'étude de la diphterie. Am Inst Pasteur 1889; 3:273-288.
4. Velikanov IM. Experimental immunization of man against botulism. Klin Med 1934; 12:1802-1806.
5. Reames HR, Kadull PJ, Housewright RD, Wilson JB. Studies on botulinum toxoids, types A and B III. Immunization in man 1947; 55:309-324.
6. Duff JT, Wright GG, Klerer J, Moore DE, Bibler RH. Studies on immunity to toxins of *Clostridium botulinum* I. A simplified procedure for isolation of type A toxin. J Bacteriol 1957; 73:42-47.
7. Duff JT, Klerer J, Bibler RH, Moore DE, Gottfried C, Wright GG. Studies on immunity to toxins of *Clostridium botulinum* II. Production and purification of type B toxin for toxoid. J Bacteriol 1957; 73:597-601.
8. Wright GG, Duff JT, Fiock MA, Devlin HB, Soderstrom RL. Studies on immunity to toxins of *Clostridium botulinum* V. Detoxification of purified type A and B toxins, and the antigenicity of univalent and bivalent aluminum phosphate adsorbed toxoids. J Immunol 1960; 84:384-389.
9. Fiock MA, Devione LF, Gearinger NF, Duff JT, Wright GG, Kadull PJ. Studies on immunity to toxins of *Clostridium botulinum* VIII. Immunological response of man to purified bivalent AB botulinum toxoid. J Immunol 1962; 88:277-283.
10. Fiock MA, Cardella MA, Gearinger NF. Studies on immunity to toxins of *Clostridium botulinum* IX. Immunological response of man to purified pentavalent ABCDE botulinum toxoid. J Immunol 1963; 90:697-702.
11. Black RE, Gunn RA. Hypersensitivity reactions associated with botulinal antitoxin. Am J Med 1980; 69:567-70.
12. Lewis GE Jr. Approaches to the prophylaxis, immunotherapy and chemotherapy of botulism. In Lewis GE, ed. Biomedical aspects of botulism. New York: Academic Press, 1981.

ALTERED SENSITIVITY OF RECOGNITION SITES FOR A NEUROTRANSMITTER IN THE ABSENCE OF CHANGES IN RECEPTOR BINDING PARAMETERS: CO-SENSITIZATION OF AN ALTERNATE SYSTEM

Richard M. Kostrzewa

Department of Pharmacology
James H. Quillen College of Medicine
East Tennessee State University
Johnson City, TN 37614-0577, U.S.A.

INTRODUCTION

During a period of about 5 years we have attempted to permanently modify the organization of different components of the dopamine (DA) neurotransmitter system in the brain. As part of this approach rats have been treated either as neonates or adults with DA receptor agonists, DA receptor antagonists or overt DA neurotoxins. In several instances we successfully attenuated the neuroteratogenic effects of these agents. And as a continuum of these studies we have ascertained that another neurotransmitter system, serotoninergic, actually mediates supersensitization of DA systems. While the study of DA and other neurotransmitter systems may seem remote from the symposium objectives, there are common themes with the neural adaptation and neural mediation that can be expected when botulinum and tetanus neurotoxins exert their effects. As important, is the message that simple measurements of presynaptic and postsynaptic elements (e.g., receptor site status and second messenger changes) are simply inadequate for determining the functional changes that are consequent to the injury produced by these neurotoxins.

HISTORICAL PERSPECTIVE

For catecholaminergic systems such as DAergic and noradrenergic neurons, the effects of specific neurotoxins have been extensively studied and described during the past 25 years. It was in 1967 that Thoenen and Tranzer discovered that 6-hydroxydopamine (6-OHDA) is a specific neurotoxin for noradrenergic fibers. Literally thousands of papers have now been published on this and related neurotoxins. The reader is referred to recent reviews (Kostrzewa, 1988, 1989). More insidious, is the change that is produced by receptor agonists or antagonists, which are not considered to be neurotoxins at all. In adult animals a classical receptor proliferation ensues following chronic administration of DA receptor antagonists

Botulinum and Tetanus Neurotoxins, Edited by
B.R. DasGupta, Plenum Press, New York, 1993

(Burt, Creese and Snyder, 1977; Creese and Chen, 1985; Porceddu et al., 1985; Muller and Seeman, 1977). In contrast, a receptor sensitization has been found, in some instances, following chronic administration of DA receptor agonists (Morelli and Di Chiara, 1990; Morelli et al., 1990).

In the above studies the changes in receptor status are either temporary or dependent on previous administration of a neurotoxin. In the studies to be described, the changes reported are generally long-lived and possibly permanent. The nature of this change relates to the fact that the assortment of agents is administered during ontogenetic development, a period in which the developing presynaptic and postsynaptic components of the DA nervous system are still immature. Insult by agonists, antagonists and neurotoxins produce neural adaptive responses which are often different from that induced by similar treatment of adults. This is the focus of the paper.

ONTOGENETIC TREATMENTS WITH DA RECEPTOR ANTAGONISTS

DA D_1 Receptor Antagonist Treatment

To determine whether repeated treatments with the DA D_1 receptor antagonist, R-(+)-7-chloro-8-hydroxy-3-methyl-1-phenyl-2,3,4,5-tetrahydro-1-H-3-benzazepine (SCH 23390; 0.30 mg/kg per day, i.p.), could alter the development of DA receptors, rats were treated with this agent once a day for the first 32 days after birth. This period corresponds to the most rapid phase of development of DA fibers (Breese and Traylor, 1972; Coyle and Axelrod, 1972; Smith et al., 1973), D_1 receptors (Giorgi et al., 1987) and D_2 receptors (Pardo et al., 1977).

Between 5 and 14 weeks after birth there was an approximate 75% reduction in DA D_1 receptor binding in the striatum (Fig. 1), a region with a dense DA fiber network and abundance of D_1 and D_2 binding sites (Saleh and Kostrzewa, 1988). This change correlated with an approximate 75% decrease in the B_{max} for D_1 binding, while the K_d remained unaltered (Kostrzewa and Saleh, 1989b). Despite the reduction in numbers of D_1 binding sites, these sites retained their adaptive response to repeated daily treatments with SCH 23390 in adulthood (Fig. 2). Ontogenetic treatments with SCH 23390 did not affect the binding to D_2 sites (Saleh and Kostrzewa, 1988; Fig. 3).

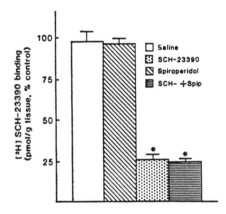

FIG. 1. Total binding of the D_1 radioligand, [^{3}H]SCH 23390 (300 pM), to striatal homogenates of 12 week old rats that were treated daily in postnatal ontogeny with SCH 23390 HCl(0.30 mg/kg per day x 32 d) and/or spiperone HCl (1.0 mg/kg per day x 32d), respective D_1 and D_2 receptor antagonists. (*)P < 0.001 vs. the saline control group. (From Saleh and Kostrzewa, 1988.)

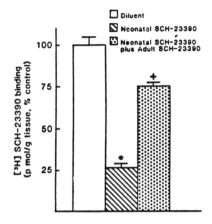

FIG. 2. Total binding of [³H]SCH 23390 (300 pM) to striatal homogenates of 15 week old rats that were treated in ontogeny as per fig. 1 and challenged at 12 weeks by SCH 23390 HCl (0.30 mg/kg per day x 17 d) or saline (x 17 d). (*)$P< 0.001$ vs. the diluent control group; (+)$P< 0.001$ vs. the other 2 groups. (From Saleh and Kostrzewa, 1988.)

DA D_2 Receptor Antagonist Treatment

To determine whether repeated treatments with the DA D_2 receptor antagonist, spiperone (1.0 mg/kg per day, i.p.), could alter the development of DA receptors, rats were treated with this agent once a day for the first 32 days after birth.

Between 5 and 14 weeks after birth there was an approximate 75% reduction in DA D_2 receptor binding in the striatum (Fig. 3). This change correlated with an approximate 75% decrease in the B_{max} for D_2 binding, while the K_d remained unaltered (Kostrzewa and Saleh, 1989b). In contrast to the proliferative response of D_1 sites after repeated antagonist treatments in adulthood, D_2 sites did not increase in number when spiperone was given repeatedly for 2 weeks in adulthood (Fig. 4). Ontogenetic treatments with spiperone did not affect the binding to D_1 sites (Saleh and Kostrzewa, 1988).

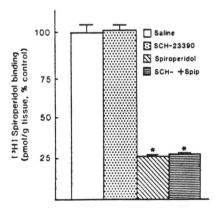

FIG. 3. Total binding of the D_2 radioligand, [³H]spiperone (300 pM), to striatal homogenates of 12 week old rats that were treated in ontogeny as per fig. 1. (*)$P< 0.001$ vs. the saline control group. (From Saleh and Kostrzewa, 1988.)

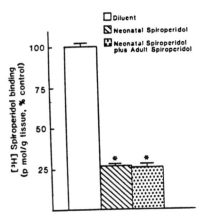

FIG. 4. Total binding of [³H]spiperone, to striatal homogenates of 15 week old rats that were treated in ontogeny as per fig. 1 and challenged at 12 weeks by spiperone HCl (1.0 mg/kg per day x 17 d) or saline (x 17 d). (*)P< 0.001 vs. the diluent control group. (From Saleh and Kostrzewa, 1988.)

Attenuation of Receptor Changes by L-Prolyl-L-Leucyl-Glycinamide

A tripeptide, L-prolyl-L-leucyl-glycinamide (PLG), also known as melanocyte stimulating hormone-release inhibitory factor (MIF-1), attenuates a variety of behavioral and receptor adaptative responses to DA agonists and antagonists (Bhargava, 1984a,b; Chiu et al., 1981a,b; Chiu et al., 1985; Kostrzewa et al., 1978; Spirtes et al., 1976). This occurs in the absence of effect of PLG on DA turnover (Kostrzewa et al., 1979a), or change in activity of DA synthesizing enzymes (Kostrzewa et al., 1979b) or change in DA uptake inactivation (Kostrzewa et al., 1976).

When PLG was co-administered with SCH 23390 during postnatal development, the tripeptide attenuated the SCH 23390-induced alteration of striatal D₁ receptor development (Kostrzewa and Saleh, 1989a; Fig. 5). Similarly, when PLG was co-administered with spiperone during postnatal development, the tripeptide attenuated the spiperone-induced alteration of D₂ receptor development (Saleh and Kostrzewa, 1989; Fig. 6). These effects occurred despite the inability of PLG to interact *in vitro* with D₁ (Kostrzewa and Saleh, 1989a) or D₂ sites (Kostrzewa et al., 1979b).

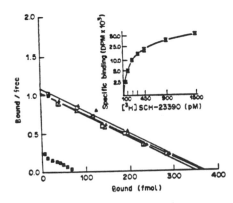

FIG. 5. Scatchard plot of the binding of [³H]SCH 23390 to striatal homogenates of 5 week old rats that were treated in ontogeny with saline (x 32 d)(△), SCH 23390 HCl (0.30 mg/kg per day x 32 d) (■), PLG (1.0 mg/kg per day x 32 d)(▲) or SCH 23390 plus PLG (□). (From Kostrzewa and Saleh, 1989a.)

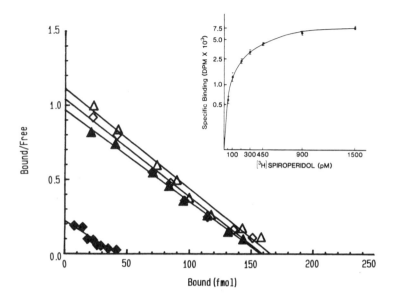

FIG. 6. Scatchard plot of the binding of [³H]spiperone to striatal homogenates of 5 week old rats that were treated in ontogeny with saline (x 32 d)(△), spiperone HCl (1.0 mg/kg per day x 32 d) (♦), PLG (1.0 mg/kg per day x 32 d)(▲) or spiperone plus PLG (◊). (From Saleh and Kostrzewa, 1989.)

In addition to PLG modifying receptor binding per se, this peptide apparently modifies the sensitivity of the receptors, as shown by the altered behavioral responses of rats to DA agonists (Kostrzewa et al., 1989).

Summary of Changes Produced by DA Receptor Antagonists

Antagonists to the DA D_1 and D_2 receptor, when administered for prolonged periods during postnatal ontogeny, are able to prevent the development of a full complement of D_1 and D_2 binding sites in the striatum. This alteration is prevented by PLG co-treatments during ontogeny.

ONTOGENETIC TREATMENTS WITH DA RECEPTOR AGONISTS

DA D_1 Receptor Agonist Treatment

To determine whether repeated treatments with the DA D_1 receptor agonist, 1-phenyl-2,3,4,5-tetrahydro-[1H]-3-benzazepine-7,8-diol (SKF 38393), could alter the development of DA D_1 receptors, rats were treated with this agent once a day for the first 33 days after birth. When striata were assessed at 7 weeks for DA D_1 receptor site binding, there was no change in either the B_{max} or K_d for D_1 binding. Also, there was no behavioral supersensitivity to adult challenge doses of SKF 38393 (Hamdi and Kostrzewa, 1991; Table 1).

DA D_2 Receptor Agonist Treatment

To determine whether repeated treatments with the DA D_2 receptor agonist, quinpirole, could alter the development of DA D_2 receptors, rats were treated with this agent once a day for the first 28 to 33 days from birth. When striata were assessed at 2 months for binding status, there was no change in either the B_{max} or K_d for *in vitro* [^{3}H]spiperone binding (Kostrzewa and Brus, 1991b; Table 2). Since this radioligand will bind to D_2, D_3 and D_4 sites, it cannot be stated with certainty that there is no change in any of these subclasses of binding sites.

TABLE 1

EFFECTS OF POSTNATAL SKF 38393 AND/OR 6-OHDA TREATMENTS ON DOPAMINE D_1 RECEPTOR BINDING PARAMETERS IN RAT STRIATUM.

Developmental treatment	D_1 receptor binding	
	B_{max} (fmol/mg tissue)	K_d (pM)
Vehicle	109.5 ± 3.8	363 ± 13
SKF 38393	122.9 ± 1.7	399 ± 13
6-OHDA	103.2 ± 6.2	359 ± 15
SKF 38393+6-OHDA	109.6 ± 6.7	399 ± 24

SKF 38393 HC1 (3.0 mg/kg per day i.p.) or vehicle was administered once a day for the first 33 days after birth. 6-OHDA HBr (200 μg, half in ea. lateral ventricle) or vehicle was injected at 3 days after birth. Each group is the mean of five rats. (From Hamdi and Kostrzewa, 1991.)

However, despite the lack of an overt change in receptor binding parameters, there was a marked behavioral supersensitization of rats to challenge with low doses of quinpirole. This was shown as an increase in potency and efficacy for quinpirole-induced yawning (Kostrzewa and Brus, 1991b; Fig. 7). Although this is generally considered to be a DA D_2-mediated effect, it is suggested that this may represent a D_3-mediated response (Kostrzewa and Brus, 1991a). A change in receptor responsiveness is also seen with higher doses of quinpirole. This is reflected by an enhanced eating response to acute quinpirole treatment (Kostrzewa et al., 1990) and greater antinociceptive effect (Fig. 8). This latter effect was attenuated by spiperone pretreatment immediately prior to hot plate testing. Also, the quinpirole antinociceptive action was additive to that of morphine (Kostrzewa et al., 1991).

TABLE 2

EFFECT OF ONTOGENIC QUINPIROLE TREATMENTS ON DOPAMINE D$_2$ RECEPTOR BINDING IN RAT STRIATUM*

Group	D$_2$ Receptor Binding	
	B$_{max}$(fmol/mg tissue)	K$_d$(pM)
Vehicle	40.8 ± 1.9	93.5 ± 4.2
Quinpirole	43.2 ± 2.9	98.8 ± 14.3

*Each group is the mean of 5 or 6 rats.
†Quinpirole HC1 (3.0 mg/kg/day, IP) was administered to male rats once a day for the first 28 days from birth. Rats received no other treatments and were sacrificed at 2 months of age. (From Kostrzewa and Brus, 1991b.)

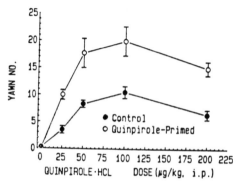

FIG. 7. Quinpirole-induced yawning in male rats that were ontogenetically primed with quinpirole HCl (3.0 mg/kg per day x 28 days) or saline, and studied at 8 to 10 weeks from birth. Each rat was observed for 1 hr., starting 10 min. after each dose of quinpirole or saline. (From Kostrzewa and Brus, 1991b.)

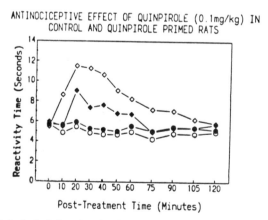

FIG. 8. Quinpirole-induced antinociception in the hot plate test of analgesia in rats. One group of rats was treated in development with quinpirole HCl (3.0 mg/kg per day x 28 d) and challenged at 4 months with saline (o) or quinpirole HCl (100 μg/kg)(◇). The other group was treated similarly in development with vehicle and also challenged at 4 months with saline (●) or quinpirole HCl (100 μg/kg)(◆). (From Kostrzewa et al., 1991.)

Summary of Changes Produced by DA Receptor Agonists

An agonist of the DA D_1 receptor, administered during postnatal ontogeny, does not produce evident changes in either D_1 binding parameters or behavioral sensitivity to D_1 agonist challenge treatment.

An agonist of the DA D_2 receptor, administered during postnatal ontogeny, does not produce evident changes in D_2 binding parameters. However, because of the similar affinity of D_2, D_3 and D_4 receptors for the radioligand, [^{3}H]spiperone, it is possible that one or more of these subclasses of receptors is altered. Also, there is behavioral sensitization to challenge doses of quinpirole in adulthood.

ONTOGENETIC TREATMENT WITH THE DA NEUROTOXIN, 6-HYDROXYDOPAMINE

Neonatal 6-Hydroxydopamine Treatment Alone

Neonatal treatment of rats with a high dose of 6-hydroxydopamine (6-OHDA) results in the marked destruction of DA fibers that innervate the striatum (Breese and Traylor, 1972; Cooper et al., 1975; Breese et al., 1984, 1985, 1987). This effect alone does not markedly alter the behavioral response to challenge doses of D_1 or D_2 agonists in adulthood (Breese et al., 1987; Hamdi and Kostrzewa, 1991). Also, the B_{max} and K_d for DA D_1 receptors remains unchanged (Breese et al., 1985, 1987; Hamdi and Kostrzewa, 1991; Table 1). When D_2 receptors are assessed in the striatum there is a notable proliferation of these sites in rostral striatum (Dewar et al., 1990). These rats appear to be subsensitive to challenge doses of D_2 agonists in adulthood (Bruno et al., 1985; Duncan et al., 1987).

Adult Priming of Neonatal 6-OHDA Treated Rats

When neonatal 6-OHDA-lesioned rats are treated as adults with L-dihydroxyphenylalanine (L-DOPA) or D_1 agonists, there is a priming of D_1 receptors. That is, repeated D_1 agonist treatments sensitize the D_1 receptor (Breese et al., 1984, 1985, 1987; Criswell et al., 1989). This behavioral supersensitivity is exemplified by markedly enhanced stereotyped and locomotor responses of rats to D_1 agonist treatment.

Neonatal Priming of Neonatal 6-OHDA Treated Rats

Neonatal 6-OHDA-lesioned rats which are treated daily in postnatal ontogeny with the D_1 agonist, SKF 38393, display enhanced behavioral sensitivity to SKF 38393 in adulthood. This is shown as increased stereotyped responses, in the absence of changes in striatal B_{max} and K_d for D_1 receptors (Hamdi and Kostrzewa, 1991; Fig. 9).

When the neonatal 6-OHDA-lesioned rats are treated daily in postnatal ontogeny with the D_2 agonist, quinpirole, there is a development of behavioral sensitivity to quinpirole in adulthood (Kostrzewa et al., 1990). However, this effect is not as dramatic as the neonatal priming induced by a D_1 agonist.

Non-Primed Behavioral Supersensitivity of Neonatal 6-OHDA Treated Rats

As noted above, behavioral supersensitivity of neonatal 6-OHDA-lesioned rats appears to be latent, being induced by DA agonist treatments. However, one exaggerated behavior that is evident without a priming event is D_1 agonist-induced oral activity. When neonatal 6-OHDA-lesioned rats are challenged in adulthood

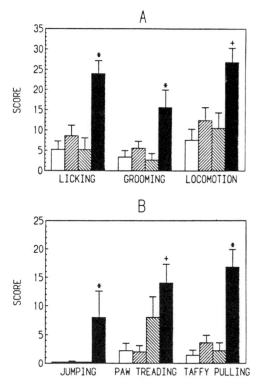

FIG. 9. Effect of acute SKF 38393 HCl (3.0 mg/kg) on stereotyped and locomotor behaviors of rats at 6 to 7 weeks from birth. Rats were treated in ontogeny with either saline (x 33 d)(□), SKF 38393 HCl (3.0 mg/kg per day x 33 d)(▨), 6-OHDA HBr (200 μg, intracerebroventricularly [icv] at 3 d after birth)(▧) or combined SKF 38393 and 6-OHDA treatments (■). Rats were challenged at 6 to 7 weeks from birth with a single dose of SKF 38393 HCl (3.0 mg/kg i.p.), 10 min. before the first observation. (*)P< 0.05 vs. other treatment groups; $^+$ P< 0.05 vs. saline group. (From Hamdi and Kostrzewa, 1991.)

with either a D_1 receptor agonist or D_2 receptor antagonist, there is an enhanced oral activity response (Kostrzewa and Hamdi, 1991; Kostrzewa and Gong, 1991; Fig. 10). This is reflective of both an increase in potency and efficacy (Fig. 11). Since the D_1 receptor antagonist, SCH 23390 attenuates the oral activity response of both the D_1 agonist and D_2 antagonist, it appears that the DA D_1 receptor is most closely involved in mediating the response (Figs. 12 and 13).

SUPERSENSITIZATION OF SEROTONIN SYSTEMS BY THE DA NEUROTOXIN, 6-OHDA

Serotonin (5-HT) Fiber Response to Neonatal 6-OHDA Treatment

When DA fibers are destroyed in rat striatum by neonatal 6-OHDA treatment, 5-HT afferents to this brain region sprout and subsequently hyperinnervate the striatum (Breese et al., 1984; Stachowiak et al., 1984; Berger et al., 1985; Luthman et al., 1987; Towle et al., 1989). However, until recently, there was no dramatic behavioral event that could be associated with these 5-HT fibers (Luthman et al., 1987; Towle et al., 1989).

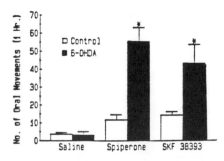

FIG. 10. Effect of neonatal 6-OHDA HBr treatment (200 μg icv) on SKF 38393-induced (3.0 mg/kg) or spiperone-induced (80 μg/kg) oral activity in rats. Numbers of oral movements were determined for one minute every 10 minutes over a 60-minute period, beginning 10 minutes after SKF 38393 hydrochloride (3.0 mg/kg, IP) or 60 minutes after spiperone hydrochloride (80 μg/kg, IP). (*)$P < 0.005$ vs. respective vehicle control groups. (From Kostrzewa and Gong, 1991.)

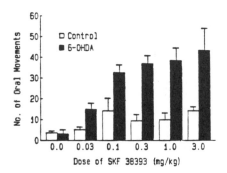

FIG. 11. Dose-response relationship for SKF 38393 HCl-induced oral activity in control and neonatal 6-OHDA HBr-lesioned (200 μg icv) rats. Numbers of oral movements were determined as in fig. 10. The number of oral movements in 6-OHDA vs. control rats is different at each respective dose of SKF 38393 ($P < 0.01$). (From Kostrzewa and Gong, 1991.)

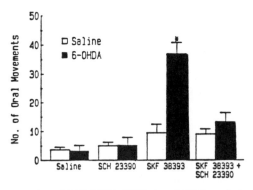

FIG. 12. Attenuation of SKF 38393 HCl-induced (0.30 mg/kg) oral activity in control and neonatal 6-OHDA HBr-lesioned (200 μg icv) rats by SCH 23390 HCl (0.30 mg/kg). Numbers of oral movements were determined as in fig. 10. (*)$P < 0.005$ vs. other 6-OHDA groups. (From Kostrzewa and Gong, 1991.)

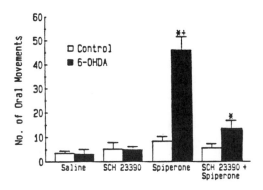

FIG. 13. Attenuation of spiperone HCl-induced (80 μg/kg) oral activity in control and neonatal 6-OHDA HBr-lesioned (200 μg icv) rats by SCH 23390 HCl (0.30 mg/kg). Numbers of oral movements were determined as in fig. 10. (*)P< 0.001 vs. respective saline control group. (+)P< 0.005 vs. 6-OHDA groups treated with SCH 23390 alone or with SCH 23390 plus spiperone. (From Kostrzewa and Gong, 1991.)

Enhanced 5-HT Agonist-Induced Oral Activity Response in Neonatal 6-OHDA-Lesioned Rats

Recently, the 5-HT receptor agonist, *m*-chlorophenylpiperazine (*m*-cpp), was found to induce oral activity responses in intact rats (Stewart et al., 1989). Since 5-HT sprouting occurs in neonatal 6-OHDA-lesioned rats, the effect of *m*-cpp was determined in this animal model. As shown in Fig. 14, *m*-cpp produced a markedly enhanced oral activity response in adult rats that were lesioned neonatally with 6-OHDA. The effect was greater than the maximal response that was observed after D$_1$ agonist treatment (Gong and Kostrzewa, 1992). This finding indicates that damage to a DA neurotransmitter system can sensitize another monoaminergic system, 5-HTergic.

Apparent Supersensitization of 5-HT$_{1C}$ Receptors in Neonatal 6-OHDA-Lesioned Rats

Since a full range of specific and selective agonists and antagonists for each of

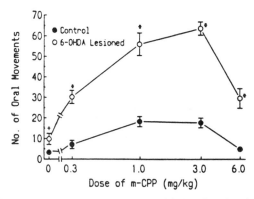

FIG. 14. Dose-response curve to *m*-cpp 2HCl-induced oral activity in adult rats that were treated at 3 days after birth with vehicle or 6-OHDA HBr (200 μg). Numbers of oral movements were determined as in fig. 10, starting 10 min. after *m*-cpp or saline. P< 0.001 vs. vehicle group challenged with the same dose of *m*-cpp. (+)P = 0.024 vs. saline group. (From Gong and Kostrzewa, 1992.)

the serotonin receptor subtypes has not been developed, a variety of 5-HT receptor agonists and antagonists were used to study 5-HT receptor involvement in the oral activity response of D_1 supersensitized rats. The respective 5-HT$_{1A}$ and 5-HT$_{1B}$ receptor agonists, (±)8-hydroxydipropylaminotetralin (8-OH-DPAT) and 7-trifluoromethyl-4(4-ethyl-1-piperazinyl)-pyrrolo[1,2-alquinoxaline] (CGS 12066B), did not increase oral activity. The beta-adrenoceptor antagonist, pindolol, which also is an antagonist for 5-HT$_{1A}$ and 5-HT$_{1B}$ receptors, failed to attenuate the enhanced oral activity response to m-cpp in D_1 supersensitized rats.

The respective 5-HT$_2$ and 5-HT$_3$ receptor antagonists, ketanserin and 3-tropanyl-3,5-dichlorobenzoate (MDL-72222), likewise failed to attenuate the oral activity response to m-cpp. Only mianserin, an antagonist at both 5-HT$_{1C}$ and 5-HT$_2$ sites, attenuated the oral activity response to m-cpp (Fig. 15). Since ketanserin, a 5-HT$_2$ receptor antagonist did not attenuate the effect of m-cpp, it then seems that the induction of oral activity by m-cpp occurs because of its activity at 5-HT$_{1C}$ sites.

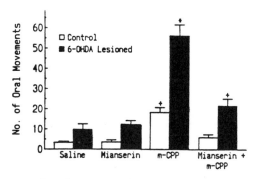

FIG. 15. Attenuation of m-cpp 2HCl-induced (3.0 mg/kg) oral activity in control and neonatal 6-OHDA HBr-lesioned (200 µg icv) rats by mianserin HCl (1.0 mg/kg). Numbers of oral movements were determined as in fig. 14. (*)$P < 0.001$ vs. other groups; (+)$P < 0.05$ vs. respective control group treated with both m-cpp and mianserin. (From Gong and Kostrzewa, 1992.)

MEDIATION OF DA RECEPTOR SUPERSENSITIVITY BY 5-HT$_{1C}$ RECEPTORS

Although it is interesting that there is coincident supersensitization of DA D_1 and 5-HT$_{1C}$ receptors in neonatal 6-OHDA-lesioned rats, the 2 events could be either independent or interdependent phenomena. That is, DA D_1 agonists could act through one nerve pathway to produce a response, while 5-HT$_{1C}$ agonists could act through a separate pathway. Conversely, DA and 5-HT pathways could be 'in series', with DA fibers acting on 5-HT fibers or vice versa.

The following study was done to determine the most likely arrangement. One group of neonatal 6-OHDA-lesioned rats was injected with the DA D_1 receptor antagonist, SCH 23390, before m-cpp treatment. As shown in fig. 16, the DA receptor antagonist failed to modify the m-cpp effect, thereby suggesting that 5-HT events occur either independent of DA events or subsequent to DA events. Another group of neonatal 6-OHDA-lesioned rats was injected with the 5-HT receptor antagonist, mianserin, before treatment with the DA D_1 agonist, SKF 38393. As shown in fig. 17, the 5-HT receptor antagonist attenuated the response to SKF 38393. This finding suggests that DA fibers act on 5-HT fibers which then induce oral activity effects (Fig. 18). Therefore, it seems that ontogenetic destruction of DA nerves produces supersensitization of both DA and 5-HT receptors, with the DA nerves being dependent on 5-HT processes.

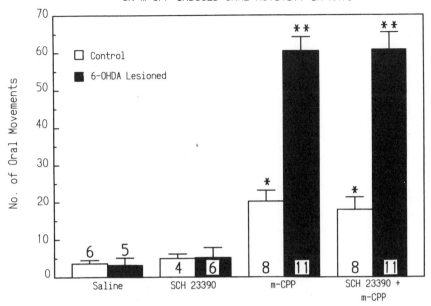

FIG. 16. Failure of SCH 23390 HCl (0.30 mg/kg) to attenuate *m*-cpp 2HCl-induced (1.0 mg/kg) oral activity in control and neonatal 6-OHDA HBr-lesioned (200 μg icv) rats. Numbers of oral movements were determined as in fig. 14. (*)P< 0.01; (**)P< 0.001 vs. respective group challenged with saline. (From Gong et al., 1992.)

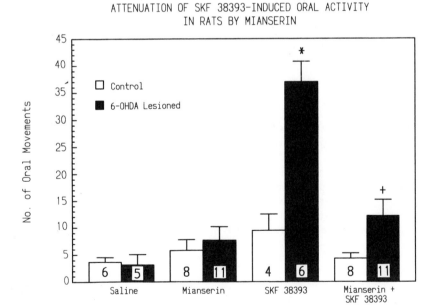

FIG. 17. Attenuation of SKF 38393 HCl-induced (1.0 mg/kg) oral activity in control and neonatal 6-OHDA HBr-lesioned (200 μg icv) rats by mianserin HCl (1.0 mg/kg). Numbers of oral movements were determined as in fig. 14. (*)P< 0.001 vs. other groups; (+)P< 0.001 vs. SKF 38393 effect in the 6-OHDA-lesioned group of rats. (From Gong et al., 1992.)

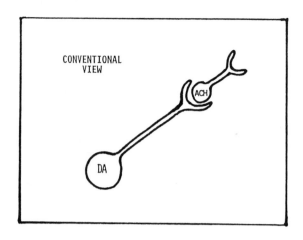

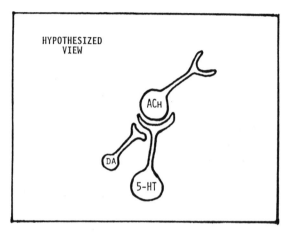

FIG. 18. Conventional and hypothesized view of the interaction between DAergic, 5-HTergic and cholinergic neurons that are involved in the genesis of oral activity in rats.

SUMMARY

1. A variety of neurochemicals can produce alterations in the development of neurotransmitter receptors. This altered development may be manifest as a change in receptor sensitivity.
2. Supersensitization of neurotransmitter receptors can occur in the absence of changes in receptor number or affinity.
3. Neuroteratogenic effects are sometimes produced by seemingly non-toxic agents.
4. Supersensitization of one type of neurochemical receptor may be the consequence of changes in a different neurochemical system.

CONCLUSIONS

Several related neurochemical systems must be studied when attempting to determine the effects of toxins that are reportedly selective for one of the systems. Otherwise, one may be confronted with unexplained lack of changes in receptor binding characteristics or lack of changes in second messengers.

534

Acknowledgements

The author acknowledges research support from the Scottish Rite Schizophrenia Research Program, N.M.J., U.S.A., John E. Fogarty International Center Scientist Exchange Program, and National Institute of Neurological Disorders and Stroke NS29505. The illustration in figure 18 was prepared by Ms. Ditty Hagardorn. The manuscript was expertly prepared by Ms. Lottie Winters.

REFERENCES

Berger, T. W., Kaul, S., Stricker, E. M. and Zigmond, M. J., 1985, Hyperinnervation of the striatum by dorsal raphe afferents after dopamine-depleting brain lesions in neonatal rats, Brain Res. 366:354-358.

Bhargava, H. N., 1984, Effects of prolyl-leucyl-glycinamide and cyclo (leucylglycine) on the supersensitivity of dopamine receptors in brain induced by chronic administration of haloperidol, Neuropharmacology 23:439-444.

Bhargava, H. N., 1984, Enhanced striatal [^{3}H] spiroperidol binding induced by chronic haloperidol treatment inhibited by peptides administered during the withdrawal phase, Life Sci. 34:873-879.

Breese, G. R. and Traylor, T.D., 1972, Developmental characteristics of brain catecholamines and tyrosine hydroxylase in the rat: effects of 6-hydroxydopamine, Br. J. Pharmacol. 44:210-222.

Breese, G. R., Baumeister, A. A., McCown, T. J., Emerick, S. G., Frye, G. D., Crotty, K and Mueller, R. A., 1984, Behavioral differences between neonatal and adult 6-hydroxydopamine-treated rats to dopamine agonists: relevance to neurological symptoms in clinical syndromes with reduced brain dopamine, J. Pharmacol. Exp. Ther. 231:343-354.

Breese, G. R., Baumeister, A., Napier, T. C., Frye, G. D. and Mueller, R. A., 1985a, Evidence that D-1 dopamine receptors contribute to the supersensitive behavioral responses induced by L-dihydroxy-phenylalanine in rats treated neonatally with 6-hydroxydopamine, J. Pharmacol. Exp. Ther. 235:287-295.

Breese, G. R., Napier, T. C. and Mueller, R. A., 1985b, Dopamine agonist-induced locomotor activity in rats treated with 6-hydroxydopamine at differing ages: functional supersensitivity of D-1 dopamine receptors in neonatally-lesioned rats, J. Pharmacol. Exp. Ther. 234:447-455.

Breese, G. R., Duncan, G. E., Napier, T. C., Bondy, S. C., Iorio, L. C. and Mueller, R. A., 1987, 6-hydroxy-dopamine treatments enhance behavioral responses to intracerebral microinjection of D_1 and D_2-dopamine agonists into nucleus accumbens and striatum without changing dopamine antagonist binding, J. Pharmacol. Exp. Ther. 240:167-176.

Bruno, J. P., Stricker, E. M. and Zigmond, M. J., 1985, Rats given dopamine-depleting brain lesions as neonates are subsensitive to dopaminergic antagonists as adults, Behav. Neurosci. 99:771-775.

Burt, D. R., Creese I. and Snyder, S.H., 1977, Antischizophrenic drugs: chronic treatment elevates dopamine receptor binding in brain, Science 196:326-328.

Chiu, S.C., Paulose, C.S., and Mishra, R. K., 1981a, Effect of L-prolyl-L-leucyl-glycinamide (PLG) on neuroleptic-induced catalepsy and dopamine neuroleptic receptor binding, Peptides 2:105-111.

Chiu, S.C., Paulose, C. S., and Mishra, R. K., 1981b, Neuroleptic drug-induced dopamine receptor supersensitivity: antagonism by L-prolyl-L-leucyl-glycinamide, Science 214:1261-1262.

Chiu, S. C., Rajakumar, G., Chiu, S., Johnson, R.L., and Mishra, R. K., 1985, Mesolimbic and striatal dopamine receptor supersensitivity: prophylactic and reversal effects of L-prolyl-L-leucyl-glycinamide (PLG), Peptides 6:179-183.

Cooper, B.R., Smith, R. D., Konkol, R. J. and Breese, G. R., 1975, Alteration of growth and development after neonatal treatments with 6-hydroxydopamine. In Chemical Tools in Catecholamine Research, ed. by G. Johsson, T. Malormfors and C. H. Sachs, North Holland Publishing Co., New York, vol. 1, 197-210.

Coyle, J. T. and Axelrod, J., 1972, Tyrosine hydroxylase in rat brain: developmental characteristics, J. Neurochem. 19:117-123.

Creese, I. and Chen, A., 1985, Selective D_1 dopamine receptor increase following chronic treatment with SCH 23390, Eur. J. Pharmacol. 109:127-128.

Criswell, H., Mueller, R. A. and Breese, G. R., 1989, Priming of D_1-dopamine receptor responses: long-lasting behavioral supersensitivity to a D_1-dopamine agonist following repeated administration to neonatal 6-OHDA-lesioned rats, J. Neurosci. 9:125-133.

Dewar, K. M., Soghomonian, J. J., Bruno, J. P., Descarries, L. and Reader, T.A., 1990, Elevation of dopamine D_2 but not D_1 receptors in adult rat neostriatum after neonatal 6-hydroxydopamine denervation, Brain Res. 536:287-296.

Duncan, G. E., Criswell, H. E., McCown, T. J., Paul, I. A., Mueller, R. A. and Breese, G. R., 1987, Behavioral and neurochemical responses to haloperidol and SCH 23390 in rats treated neonatally or as adults with 6-hydroxydopamine, J. Pharmacol. Exp. Ther. 243:1027-1034.

Giorgi, O., De Montis, G., Porceddu, M. L., Mele, S., Calderini, G., Tofano, G., and Biggio, G., 1987, Developmental and age-related changes in D_1-dopamine receptors and dopamine content in the rat striatum, Dev. Brain Res. 35:283-290.

Gong, L. and Kostrzewa, R. M., 1992, Supersensitized oral responses to a serotonin agonist in neonatal 6-OHDA-treated rats, Pharmacol. Biochem. Behav. 41:621-623.

Gong, L., Kostrzewa, R. M., Fuller, R. W. and Perry, K., 1992, Supersensitization of the oral response to SKF 38393 in neonatal 6-OHDA-lesioned rats is mediated through a serotonin system, J. Pharmacol. Exp. Ther. 261:1000-1007.

Hamdi, A. and Kostrzewa, R. M., 1991, Ontogenic homologous supersensitization of dopamine D_1 receptors, Eur. J. Pharmacol. 203:115-120.

Kostrzewa, R. M. and Garey, R. E., 1976, Effects of 6-hydroxydopa on noradrenergic neurons in developing rat brain, J. Pharmacol. Exp. Ther. 197:105-118.

Kostrzewa, R. M., Kastin, A. J., and Sobrian, S. K., 1978, Potentiation of apomorphine action in rats by l-prolyl-l-leucyl-glycine amide, Pharmacol. Biochem. Behav. 9:375-378.

Kostrzewa, R. M., Fukushima, H., Harston, C. T., Perry, K. W., Fuller, R. W. and Kastin, A. J. , 1979, Striatal dopamine turnover and MIF-1, Brain Res. Bull. 4:799-802.

Kostrzewa, R. M., Hardin, J. C., Snell, R. L., Kastin, A. J., and Bymaster, F., 1979, MIF-1 and dopamine postsynaptic receptor sites, Brain Res. Bull. 4:657-662.

Kostrzewa, R. M., 1988, Reorganization of noradrenergic neuronal systems following neonatal chemical and surgical injury, Prog. Brain Res. 73:405-423.

Kostrzewa, R. M., 1989, Neurotoxins that affect central and peripheral catecholamine neurons, in: Neuromethods Vol. 12, eds. A. A. Boulton, G. B. Baker and J. M. Baker (Humana Press, Clinton, NJ) 1-48.

Kostrzewa, R. M., White, T. G., Zadina, J. E. and Kastin, A. J., 1989, MIF-1 attenuates apomorphine stereotypies in adult rats after neonatal 6-hydroxydopamine, Eur. J. Pharmacol. 163:33-42.

Kostrzewa, R. M. and Saleh, M. I., 1989a, Attenuation of SCH 23390-inducedalteration of striatal dopamine D_1 receptor ontogeny by prolyl-leucyl-glycinamide in the rat, Neuropharmacol. 28:805-810.

Kostrzewa, R. M. and Saleh, M. I., 1989b, Impaired ontogeny of striatal dopamine D_1 and D_2 binding sites after postnatal treatment of rats with SCH-23390 and spiroperidol, Dev. Brain Res. 45:95-101.

Kostrzewa, R. M., Hamdi, A. and Kostrzewa, F. P., 1990, Production of prolonged supersensitization of dopamine D2 receptors, Eur. J. Pharmacol. 183:1411-1412.

Kostrzewa, R. M. and Brus, R., 1991a, Is dopamine-agonist induced yawning behavior a D3 mediated event, Life Sci. 48:PL-129.

Kostrzewa, R. M. and Brus, R., 1991b, Ontogenic homologous supersensitization of quinpirole-induced yawning in rats, Pharmacol. Biochem. Behav. 39:517-519.

Kostrzewa, R.M. and Gong, L., 1991, Supersensitized D1 receptors mediate enhanced oral activity after neonatal 6-OHDA, Pharmacol. Biochem. Behav. 39:677-682.

Kostrzewa, R. M. and Hamdi, A., 1991, Potentiation of spiperone-induced oral activity in rats after neonatal 6-hydroxydopamine, Pharmacol. Biochem. Behav. 38:215-218.

Kostrzewa, R. M., Brus, R. and Kalbfleisch, J., 1991, Ontogenic homologous sensitization to the antinociceptive action of quinpirole in rats, Eur. J. Pharmacol. 209:157-161.

Luthman, J., Bolioli, B., Tustsumi, T. Verhofstad, A. and Jonsson, G., 1987, Sprouting of striatal serotonin nerve terminals following selective lesions of nigrostriatal dopamine neurons in neonatal rat., Brain Res. Bull. 19:269-274.

Morelli, M. and Di Chiara, G., 1990, Stereospecific blockade of N-Methyl-D-aspartate transmission by MK 801 prevents priming of SKF 38393-induced turning, Psychopharmacology (Berlin) 101:287-288.

Morelli, M., De Montis, G., and Di Chiara, G., 1990, Changes in the D_1 receptor-adenylate cyclase complex after priming, Eur. J. Pharmacol. 180:365-367.

Muller, P. and Seeman, P., 1977, Brain neurotransmitter receptors after long-term haloperidol: dopamine, acetylcholine, serotonin, -noradrenergic and naloxone receptors, Life Sci. 21:1751-1758.

Pardo, J. V., Creese, I., Burt, D. R., and Snyder, S.H., 1977, Ontogenesis of dopamine receptor binding in corpus striatum of the rat, Brain Res. 125:376-382.

Porceddu, M.L., Ongini, E. and Biggio, G., 1985, ^{3}H-SCH 23390 binding sites increase after chronic blockade of D_1 dopamine receptors. Eur. J. Pharmacol. 118:367-370.

Saleh, M. I. and Kostrzewa, R. M., 1988, Impaired striatal dopamine receptor development: differential D-1 regulation in adults, Eur. J. Pharmacol. 154:305-311.

Saleh, M. I. and Kostrzewa, R. M., 1989, MIF-1 attenuates spiroperidol alteration of striatal dopamine D_2 receptor ontogeny, Peptides, 10:35-39.

Smith, R. D., Cooper, B. R., and Breese, G. R., 1973, Growth and behavioral changes in developing rats treated intracisternally with 6-hydroxydopamine: evidence for involvement of brain dopamine, J. Pharmacol. Exp. Ther. 185:609-619.

Spirtes, M. A., Plotnikoff, N. P., Kostrzewa, R. M., Harston, C. T., Kastin, A. J., and Christensen, C.W., 1976, Possible association of increased rat behavioral effects and increased striatal dopamine and norepinephrine levels during the DOPA-potentiation test, Pharmacol. Biochem. Behav. 5(Suppl. 1):121-124.

Stachowiak, M. K., Bruno, J. P., Snyder, A. M., Stricker, E. M. and Zigmond, M. J., 1984, Apparent sprouting of striatal serotonergic terminals after dopamine-depleting brain lesions in neonatal rats, Brain Res. 291:164-167.

Stewart, B. R., Jenner, P. and Marsden, C. D., 1989, Induction of purposeless chewing behavior in rats by 5-HT agonist drugs., Eur. J. Pharmacol. 162:101-107.

Towle, A. C., Criswell, H. E., Maynard, E. H., Lauder, J. M., Joh, T. H., Mueller, R.A. and Breese, G. R., 1989, Serotonergic innervation of the rat caudate following a neonatal 6-hydroxydopamine lesion: an anatomical, biochemical and pharmacological study, Pharmacol. Biochem. Behav. 34:367-374.

NEUROTOXINS THAT AFFECT CENTRAL SEROTONINERGIC SYSTEMS

Tomás A. Reader, and Karen M. Dewar

Centre de recherche en sciences neurologiques, and
Centre de recherche psychiatrique Fernand-Séguin
Départements de Physiologie et de Psychiatrie
Faculté de Médecine, Université de Montréal
Montréal (Québec) H3C3J7 Canada

INTRODUCTION

The last 30 years have witnessed an explosive increase in our knowledge of neuronal communication in the central nervous system (CNS), mainly due to the methodological progress which has made it possible to identify neurons and neuronal networks by the chemical substance they use as their transmitter. Historically, the first chemically-identified neurotransmitter systems were found to utilize aromatic monoamines; either the catecholamines noradrenaline (NA), adrenaline (AD) and dopamine (DA) or the indoleamine serotonin (5-hydroxytryptamine, 5-HT). The first comprehensive descriptions of central 5-HT were based on biochemical determinations of this aromatic monoamine in different brain regions [1-4] and very soon followed by the use of histofluorescent microscopy [5-9], measurements of tryptophan hydroxylase [10], immunocytochemical surveys using antibodies against 5-HT itself [11] as well as autoradiographic tracing procedures based on uptake properties of the 5-HT system. [12-15] A strategy in the study of a particular nerve function that has been widely accepted has been to remove surgically either a nerve or the ganglion from which the nerve originates, and then establish the characteristics, magnitude and extent of the functional loss. In the particular case of aromatic monoamine neurotransmitters, this approach can be exemplified by the work pioneered by Cannon [16] in the sympathetic nervous system, and by the use of immunosympathectomies using antibodies againts nerve growth factor [17]. The research on central catecholamine and 5-HT systems has benefited from the use of different drugs and agents used as chemical scalpels to dissect out the different projections and at the same time this strategy provided interesting models for the study of changes in sensitivity, uptake mechanisms, behavioural modifications, axonal sprouting and plasticity. The first compound used as a selective neurotoxin for the catecholamines was 2,4,5-trihydroxyphenylethylamine or 6-hydroxydopamine (6-OHDA)

Botulinum and Tetanus Neurotoxins, Edited by
B.R. DasGupta, Plenum Press, New York, 1993

that produces a permanent reduction of NA and DA in the CNS [18]. Other compounds that more or less selectively destroy catecholamines include DSP-4 [19] and MPTP [20]. The most frequently used compound to produce catecholamine denervations has been 6-OHDA , and it has been proposed that it undergoes spontaneous autooxidation within neurons to intermediate *p*-quinones, which then affect membrane integrity or interfere with energy production by interrupting the mitochondrial respiratory chain [21]. Compounds that affect 5-HT neurons were developed later, and similar mechanisms of action have been proposed for the most frequently used , i.e.: 5,6-dihydroxytryptamine [22-24] and 5,7-dihydroxytryptamine [25-27]. In addition, other neurotoxins for central 5-HT neurons include synthesis inhibitors and amphetamine derivatives and will also be discussed in this review.

DIHYDROXYTRYPTAMINES

Mechanism of Action of the Dihydroxytryptamines

The neurotoxins 5,6-dihydroxytryptamine (5,6-DHT) and 5,7-dihydroxytryptamine (5,7-DHT) are preferentially taken up by the 5-HT transport mechanism of serotoninergic neurons, and this first step constitutes a requisite for their effects (Figure 1). Once taken up, both dihydroxytryptamines form quinone-like compounds of the indolequinone group. The intraneuronal autooxidation is the initial and essential step of a chain of chemical events, that involves oxidation and formation of hydrogen peroxyde as well as covalent binding to intracellular proteins, leading to the destruction of 5-HT nerve terminals [28,29].

The 5-HT uptake carrier recognizes dihydroxytryptamines only in their reduced state and will not transport oxidized compounds; therefore to be used as intracellular neurotoxins they must be protected from premature oxidation, usually by adding ascorbic acid as antioxidant. After the dihydroxytryptamines are taken up by the neurons, only those that become autooxidized at a biological pH and at a moderate rate will be able to act as neurotoxins. In the case of 5,6,7-trihydroxytryptamine (5,6,7-THT), it was soon established that this compound is rapidly converted into a quinone in either the cerebrospinal fluid and/or the extracellular (interneuronal) compartment and can not be used as a substrate of the 5-HT carrier. Therefore, 5,6,7-THT could not replace the first neurotoxic indoleamine employed, namely 5,6-DHT [22]. Although 5,6-DHT is taken up with relatively high affinity, its oxidation velocity within neurons, determined by oxygen consumption rates, is slower than the 5,7-derivative. In addition, 5,6-DHT can be oxidized quite rapidly in the extracellular environment before being subjected to uptake. The neurotoxin 5,7-DHT is also taken up quite efficiently and within neurons is oxidized at a more rapid rate than 5,6-DHT. These two characteristics, i.e.: slower oxidation in the extracellular compartment and rapid oxidation within neurons explains the greater potency of 5,7-DHT. On the other hand, 5,7-DHT is a *meta*-substituted dihydroxytryptamine; an aqueous solution will be composed at equilibrium of a mixture of different tautomeric keto-enol forms, which retain their affinity for the 5-HT uptake mechanisms. Even if a fraction of 5,7-DHT is oxidized rapidly, there may still be conservation of its substrate properties since one of its hydroxyl groups will remain functional. Furthermore, brain tissue apparently provides a more protective environment for 5,7-DHT than for 5,6-DHT; the latter compound is in fact autooxidized catalitically by the oxygen peroxide it generates. Although the initial oxidation of 5,6-DHT is slow [30], it is followed by an accelerated reaction with O_2 , and this autocatalytic oxidation is caused by the formation of hydrogen peroxyde (Figure 2). The dihydroxytryptamines can also be metabolized by monoamine oxidase (MAO) or by mitochondrial aldehyde dehydrogenases, resulting in dihydroxyindoleacetic acids and nondeaminated dihydroxytryptamines; these metabolites may oxidize further, in the presence of cytochrome c oxidase [31,32].

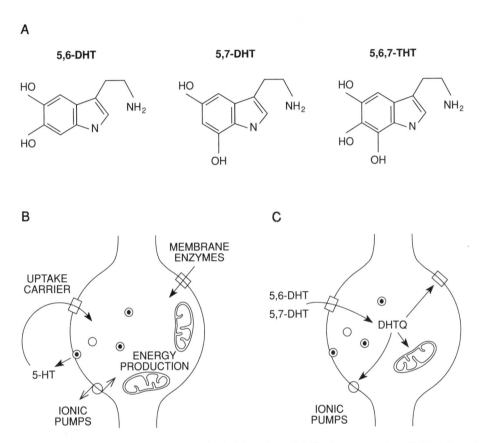

Figure 1. A) Structures of the neurotoxic indoletamines 5,6-dihydroxytryptamine (5,6-DHT) and 5,7-dihydroxytryptamine (5,7-DHT). The trihydroxylated derivative 5,6,7-trihydroxytryptamine (5,6,7-THT) is rapidly oxidized, mainly in the extraneuronal compartment and can not be taken up by the uptake carrier mechanism. B) Serotonin nerve terminals, i.e.: varicosities, have an uptake carrier that constitutes the main physiological mechanism of 5-HT inactivation. In addition, terminals abound with mitochondria for energy production, ionic pumps, membrane transporters and other essential enzymes. C) The neurotoxic dihydroxytryptamines 5,6-DHT and 5,7-DHT are taken up by the uptake carrier, oxidize within the terminals to dihydroxytryptamine-quinones (DHTQ) and affect mitochondrial enzymes and other proteins required to maintain normal cellular metabolism.

The neurotoxic dihydroxytryptamines can covalently bind to certain nucleophilic protein groups, in particular to sulfhydryls, and thus exibit denaturation effects. In fact, the quinones produced in the first intraneuronal step can alkylate sulfhydryl groups of proteins, which are extremely reactive (Figure 2). The alkylation of sulfhydryl bonds can cause protein denaturation and thus inactivate essential components for cell regulation, including enzymes, carriers and transporters in membrane pump mechanisms, and even energy producing enzymes of the mitochondrial respiratory chain. A further step involves the alkylation of secondary sulfhydryl groups, thus enabling the polimerization by crosslinking of homologous or heterologous proteins and/or peptides [33]. This is particularly important for the quinone derived from 5,6-DHT, that rapidly reacts with nucleophiles and cross-link proteins and peptides leading to the production and accumulation of melanoid polymers of high molecular weight, that can be found in combination with non-oxidized 5,6-DHT. On the other hand, 5,7-DHT does not easily form high molecular weight polymers.

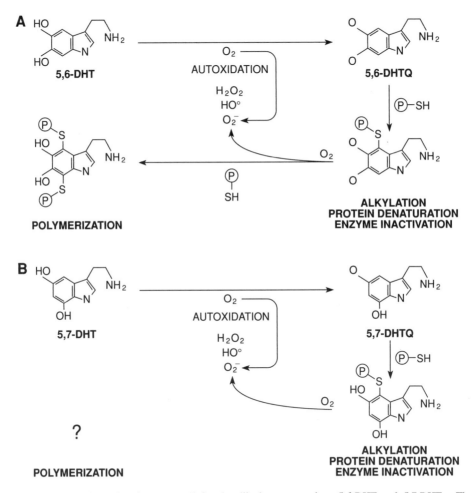

Figure 2. Mechanism of action proposed for the dihydroxytryptamines 5,6-DHT and 5,7-DHT. The compounds once transported into 5-HT neurons by the high-affinity uptake carrier are autooxidized into the quinones 5,6-DHTQ and 5,7-DHTQ. This process is autocatalytic, favored and maintained by hydrogen peroxyde. The quinones formed can alkylate sulfhydryl groups (-SH) of proteins, which are extremely reactive. The alkylation of -SH bonds leads to protein denaturation and essential cellular enzymes will be inactivated. Alkylation of a second -SH bond may cause polymerization by crosslinking proteins and peptides; the polymerization is seemingly more frequent with 5,6-DHT than with 5,7-DHT.

An important aspect in the toxicity of dihydroxytryptamines is their interaction with monoamine oxidase or MAO [30-32]. The α-methylated analogs of tryptamines are resistant to oxidation by MAO and have only very slight effects on both 5-HT and NA neurons [34]. In addition, inhibitors of MAO seemingly antagonize some of the non-specific toxic effects of 5,7-DHT on NA neurons [28]. Although it has been proposed that metabolism by MAO may be required for neurotoxicity of dihydroxytryptamines[35], it seems to play no critical role for the *in vivo* neurotoxicity of 5,6-DHT and only a minor role for the *in vivo* effects of 5,7-DHT [27,32,36]. It has also been shown that dyhydroxytryptamines can be oxidized intracellularly by cytochrome *c* oxidase, and this increases their toxicity [37]. It should be pointed out that the important intracellular effect is on energy production, and this takes place because the indolequinones replace ubiquinones in the mitochondrial respiratory chain with a consequent blockade of adenosine triphosphate (ATP) production. In addition, the dihydroxytryptamines and/or their degradation products may uncouple oxidation and phosphorylation mechanisms, leading to a further reduction in ATP formation.

Specificity of the dihydroxytryptamines

The specificity of effects produced by the neurotoxic dihydroxytryptamines depends on the mode of administration, including the site of injection and volume, the type and amounts of antioxidant used to protect the dihydroxytryptamines from premature extraneuronal oxidation, the type and dosage of the anesthetic employed, as well as the use of a catecholamine uptake blocking agent to prevent the transport of the neurotoxin into noradrenergic, adrenergic, or dopaminergic neurons (for reviews see refs. 27-29). In order to obtain selective destructions of 5-HT neurons, the most important single factor is the choice of a catecholamine uptake blocking agent that will prevent the transport into NA and DA neurons and thus limit the neurotoxic effects to the serotoninergic systems [34,38,39]. In our studies [29,40-43], the animals were pretreated with desmethylimipramine (desipramine; 25 mg/kg i.p.) 1 hour before the intraventricular administration of the 5,7-DHT. As shown in Table 1, the neurotoxin 5,7-DHT produced important decreases in endogenous 5-HT contents, that varied in magnitude according to the region examined. In cerebral cortex, hypothalamus, cerebellum and spinal cord, the depletions were greater than 70%, but in pons-medulla and mesencephalon, the effect was of a lesser degree; of 64 % and 57 %, respectively. The two latter regions contain 5-HT neuronal cell bodies, that may be more refractory to 5,7-DHT than nerve terminals and thus could explain why the decreases in 5-HT levels were not as important as in the terminal fields of innervation.

In most of the regions examined in which 5-HT deafferentations were obtained, there were no significant changes in endogenous NA contents, thus confirming that inhibition of the uptake carrier for catecholamines of NA fibers prevents their nonselective destruction by 5,7-DHT. However, in some regions such as pons-medulla and cerebellum, there were significant reductions in NA levels. In these two regions, NA content decreased to 51% and 47 % of control values, in pons-medulla and cerebellum, respectively. Most likely, this nonspecific effect can be related to the relative high dose employed, i.e.: 200 µg free base versus 150 µg free base, classically employed and recommended [37] and to the greater susceptibility to 5,7-DHT of the noradrenergic cell bodies of the *Locus coeruleus* (pons-medulla) and neighboring fibers projecting to the cerebellum. There were also slight increases in NA levels in hippocampus and hypothalamus, probably due to the activation of tyrosine hydroxylase that is under the control of a tonic 5-HT inhibition [44].

Table 1. Endogenous serotonin (5-HT), dopamine (DA) and noradrenaline (NA) in rat brain regions following administration of 5,7-dihydroxytryptamine (5,7-DHT).

Region		5-HT ng/mg p.	DA ng/mg p.	NA ng/mg p.
Cerebral cortex	Control	2.29 ± 0.39	0.78 ± 0.24	2.26 ± 0.32
	5,7-DHT	0.60 ± 0.10***	0.42 ± 0.12	1.69 ± 0.28
Hypothalamus	Control	4.32 ± 0.40	3.00 ± 0.30	7.33 ± 0.96
	5,7-DHT	0.82 ± 0.30***	1.80 ± 0.24**	8.96 ± 0.82
Hippocampus	Control	2.72 ± 0.12	0.69 ± 0.15	2.14 ± 0.36
	5,7-DHT	0.48 ± 0.13**	0.48 ± 0.14	2.28 ± 0.25
Mesencephalon	Control	3.23 ± 0.49	2.21 ± 0.30	2.96 ± 0.66
	5,7-DHT	1.39 ± 0.30**	1.62 ± 0.26	2.39 ± 0.47
Cerebellum	Control	1.95 ± 0.67	0.21 ± 0.07	1.69 ± 0.32
	5,7-DHT	0.50 ± 0.26*	0.27 ± 0.08	0.80 ± 0.18*
Pons-medulla	Control	5.70 ± 0.31	0.77 ± 0.22	5.60 ± 1.07
	5,7-DHT	2.06 ± 0.59***	0.40 ± 0.10	2.88 ± 1.25**
Spinal cord	Control	2.87 ± 0.25	0.43 ± 0.10	2.57 ± 0.35
	5,7-DHT	0.62 ± 0.27***	0.38 ± 0.08	2.27 ± 0.23

The 5,7-DHT-treated animals received a single intraventricular injection of 5,7-dihydroxytryptamine (200 µg free base). One hour before, they were pretreated with desipramine (25 mg/kg; i.p.) to protect catecholamine neurons. The values in nanograms per milligram of protein (ng/mg p.) are the means ± SEM (n=6). Statistical significance was calculated by unpaired t Student's test: * $p < 0.05$, **$p < 0.01$ and *** $p < 0.001$. (Data modified from refs. 29, 42 and 43).

However, the overall effect of 5,7-DHT on catecholamines was a slight reduction in endogenous DA contents in hypothalamus (40%) as well as of NA levels in cerebellum (53%) and pons-medulla (49%), indicating that in this study the protection provided by desipramine alone was not sufficient. A further degree of specificity could have been obtained with a better blockade of the catecholamine uptake transporters by using a combination of desipramine and nomifensine [36,37], or with desipramine plus the MAO inhibitor pargyline [45]. At present, it can not be ruled out that other non-catecholaminergic and non-serotoninergic neurons may also uptake dihydroxytryptamines; this may be mediated by other types of carriers, i.e.: amino acid transporters. Although there is no evidence indicating this theoretical possibility, this issue warrants consideration when studying the pharmacological properties, genetic expression and regulation of membrane carriers of neurons and glial cells.

OTHER COMPOUNDS THAT AFFECT SEROTONIN SYSTEMS

Synthesis inhibitors.

The compound *p*-chlorophenylalanine (PCPA) was originally reported to inhibit tryptophan hydroxylase, the rate-limiting enzyme in 5-HT synthesis (Figure 3) and thus depletes transmitter levels without affecting the integrity of nerve terminals or cell bodies [46]. It was very soon found out that it could decrease NA levels as well, but this depended on the dose of PCPA, the duration of treatment and the brain region examined [47,48].

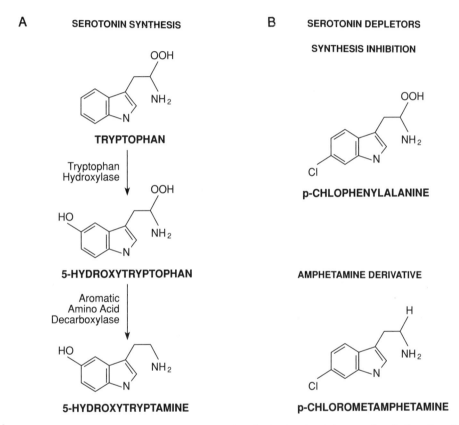

Figure 3. A) The precursor amino acid tryptophan is hydroxylated by tyrosine hydroxylase into 5-hydroxytryptophan (5-HTP), which can be then decarboxylated to 5-hydroxytryptamine (5-HT) by an aromatic amino acid decarboxylase. B) Structures of the synthesis inhibitor *p*-chlorophenylalanine (PCPA) and the amphetamine derivative *p*-chlorometamphetamine (PCMA).

In Table 2 we compare, for rat frontoparietal cortex, the effects of destroying 5-HT nerve terminals using 5,7-DHT with the depletions obtained by inhibiting synthesis with PCPA. In addition, electrolytic lesioning of the nuclei of origin of the cortical 5-HT projections, i.e.: the *Raphe nuclei* complex, provided us with a non-chemical model of deafferentation to compare 5-HT depletions. It can be seen that PCPA reduced endogenous 5-HT levels to the same extent as 5,7-DHT. However, PCPA also affected NA content; this was not the case for 5,7-DHT since it was given under the protection of the specific uptake blocker desipramine.

The dopaminergic system however, was not protected and this explains the reduction in DA levels after 5,7-DHT (see above). This lack of specificity of PCPA has been interpreted as either a direct effect on tyrosine hydroxylase or as the result of interacting neuronal mechanisms. Although the effect of PCPA on 5-HT levels can be attributed to a direct inhibition of tryptophan hydroxylase, blockade of amino acid uptake could also play a role. It has also been shown that 5-HT depletions by PCPA are of greater magnitude in the terminal fields of innervation than in the nuclei of origin [49]. However, using doses between 300-400 mg/kg/day for 2-3 days, important and significant 5-HT depletions can be obtained and documented [50] even in the *Raphe nuclei* complex (Table 3). Due to its mode of action, the 5-HT depletions induced by PCPA are more generalized than those caused by the neurotoxic dihydroxytryptamines.

Table 2. Endogenous levels of serotonin (5-HT), dopamine (DA) and noradrenaline (NA) in rat frontoparietal cortex following administration of 5,7-dihydroxytryptamine, *p*-chloroamphetamine (PCPA) or after electrolytic lesion of the dorsal Raphe nuclei.

		5-HT ng/mg p.	DA ng/mg p.	NA ng/mg p.
Control	(n=10)	3.60 ± 0.62	0.16 ± 0.03	3.30 ± 0.21
5,7-DHT	(n=10)	0.32 ± 0.08***	0.05 ± 0.01**	3.09 ± 0.17
PCPA	(n=10)	0.32 ± 0.02***	0.14 ± 0.04	2.07 ± 0.30*
Lesion	(n=10)	0.55 ± 0.19***	0.05 ± 0.01**	3.27 ± 0.18

The 5,7-DHT injections and the electrolytic lesions of the Raphe nuclei complex were made under chloral hydrate anesthesia (400 mg/kg, i.p.). For the electrolytic lesions, a cathodal current of 2.5 nA was passed for 10 s through a stainless steel electrode implanted in the Raphe nuclei. The PCPA-treated rats received the tryptophan hydroxylase inhibitor *p*-chlorophenylalanine methyl esther hydrochloride daily (400 mg/kg i.p.) for two days. The 5,7-DHT-treated animals received a single intraventricular injection of 5,7-dihydroxytryptamine (200 µg free base). One hour before, they were pretreated with desipramine (25 mg/kg; i.p.) to protect catecholamine neurons. The values in nanograms per milligram of protein (ng/mg p.) are the means ± SEM (n=10). Statistical significance was calculated by unpaired *t* Student's test: * $p < 0.05$, **$p < 0.01$ and *** $p < 0.001$. (Data modified from ref. 29).

For example, a 80-90 % decrease in 5-HT content caused by PCPA measured in the terminal fields of innervation implies that synthesis is probably equally blocked by 80-90 % in all 5-HT terminals. In contrast, a 80-90 % depletion of 5-HT caused by neurotoxic dihydroxytryptamines is the result of the loss of 80-90 % of nerve terminals, and the 10-20 % of terminals present may actually contain normal transmitter amounts [29,40]. These considerations are important when dealing with functional and trophic properties such as hypersensitivity phenomena [40], regeneration and axonal sprouting, as well as when considering the use of other indoleamine neurotoxins related to amphetamines [50].

Amphetamine derivatives.

Besides the neurotoxic dihydroxytryptamines and the synthesis inhibitor PCPA, several compounds derived from amphetamines have been used to decrease 5-HT levels within the brain, and they include p-chloroamphetamine, p-chloro-N-methylamphetamine [47,51-54], N-hydroxy-4-chloroamphetamine [55] and methylenedioxyamphetamine [56-58] as well as fenfluramine [59,60].

Table 3. Endogenous serotonin (5-HT), dopamine (DA) and noradrenaline (NA) in rat brain after p-chloroamphetamine (PCA) or p-chlorophenylalanine (PCPA) treatments.

Region		5-HT ng/mg p.	DA ng/mg p.	NA ng/mg p.
Cerebral cortex	Control	2.58 ± 0.13	0.35 ± 0.03	2.46 ± 0.19
	PCA	0.23 ± 0.06^c	0.41 ± 0.05	2.52 ± 0.11
	PCPA	0.32 ± 0.08^c	$0.14 \pm 0.06^{a,e}$	$1.70 \pm 0.11^{b,e}$
Neostriatum	Control	2.77 ± 0.29	67.21 ± 5.12	5.26 ± 0.82
	PCA	0.60 ± 0.21^c	49.48 ± 3.13	2.73 ± 0.45
	PCPA	0.66 ± 0.12^c	$48.73 \pm 4.25^{a,b}$	$13.47 \pm 0.11^{b,e}$
Hippocampus	Control	2.50 ± 0.18	0.009 ± 0.002	3.37 ± 0.20
	PCA	0.50 ± 0.09^c	0.029 ± 0.014	3.35 ± 0.45
	PCPA	0.15 ± 0.06^c	$0.126 \pm 0.047^{a,b}$	$1.55 \pm 0.14^{c,e}$
Raphe nuclei	Control	7.02 ± 0.19	0.42 ± 0.06	7.47 ± 0.93
	PCA	4.55 ± 0.32^c	0.18 ± 0.05^b	6.98 ± 0.92
	PCPA	$0.73 \pm 0.11^{c,f}$	0.07 ± 0.02^c	4.82 ± 0.81^a
Substantia nigra	Control	6.90 ± 0.83	2.52 ± 0.49	2.82 ± 0.61
	PCA	2.22 ± 0.69^c	1.97 ± 0.43	1.98 ± 0.39
	PCPA	$0.11 \pm 0.08^{c,d}$	1.37 ± 0.43	1.78 ± 0.43

The PCA-treated animals received daily injections of DL-p-chloroamphetamine hydrochloride (300 mg/kg, i.p.) for three days, and were sacrificed 7 days later. The PCPA-treated rats were treated with p-chloro-phenylalanine ethyl esther hydrochloride for two days (300 mg/kg, i.p.) and sacrificed at the third day. The values in nanograms per milligram of protein (ng/mg p.) are the means ± SEM (n=6-8). Statistical significance was determined by one-way analysis of variance (ANOVA) followed by post-hoc t test: $^a p < 0.05$, $^b p < 0.01$ and $^c p < 0.001$ between PCA or PCPA treatments and control, and $^d p < 0.05$, $^e p < 0.01$ and $^f p < 0.001$ between PCA or PCPA treatments.

These agents have an initial reserpine-like effect, i.e.: after releasing 5-HT they further decrease endogenous 5-HT pools, and this more sustained effect may involve toxic metabolites of serotonin and/or the inhibition of tryptophan hydroxylase. The results summarized in Table 3 compare the effects of the neurotoxin p-chloroamphetamine (PCA) with those of the synthesis inhibitor PCPA. In all in five brain regions examined, both PCA and PCPA produced important 5-HT depletions. In the case of PCA, the depletions were almost complete in cerebral cortex (91%) and very important in hippocampus (80%) and neostriatum (79%). However, in regions were 5-HT neurons are found, such as the *Raphe nuclei* complex, the depletion was of only 35%. Interestingly, the samples of the mesencephalic *Substantia nigra* were also quite resistant to PCA treatment and depleted by only 68%, suggesting that part of the 5-HT of this area is contained in neurons. In contrast, after PCPA, the depletion of 5-HT was generalized and throughout all brain regions, indicating that the blockade of synthesis affected not only the terminal fields of innervation but also the 5-HT neurons of the *Raphe nuclei* and *Substantia nigra*.

An important aspect in the utilization of 5-HT neurotoxins and synthesis inhibitors is the evaluation of the degree, extent and magnitude of the depletions, and constitutes an important methodological aspect in the experimental use of these compounds. The most widely employed methods for the assessment of 5-HT depletions have been to estimate the amounts of serotonin contained in the terminals and neurons either by direct visualization through histofluorescence microscopy, or to rely on biochemical assays of tissue 5-HT levels, of tryptophan hydroxylase, or to quantify 5-HT uptake sites. The last approach is probably the most reliable procedure to determine the numbers of 5-HT nerve terminals in the CNS. However, until recently this was not easy to apply since the only available marker was labelled serotonin which for quantification purposes could only be used in *in vitro* or in limited *ex vivo* experimentation following very precise incubation conditions [15].

Table 4. Distribution of [^{3}H]paroxetine binding sites in rat brain following administration of p-chloroamphetamine (PCA) or p-chlorophenylalanine (PCPA).

Region	Control fmol/mg p.	PCA fmol/mg p.	PCPA fmol/mg p.
Cerebral cortex	420 ± 98	77 ± 24[a,b]	315 ± 67
Neostriatum	363 ± 79	53 ± 10[a,b]	351 ± 45
Hippocampus	324 ± 25	113 ± 29[a,b]	329 ± 31
Raphe nuclei	491 ± 70	390 ± 101	459 ± 62
Substantia nigra	407 ± 70	213 ± 34[a,b]	526 ± 73

The PCA-treated animals received daily injections of DL-p-chloroamphetamine hydrochloride (300 mg/kg, i.p.) for three days, and were sacrificed 7 days later. The PCPA-treated rats were administered p-chlorophenylalanine ethyl ester hydrochloride for two days (300 mg/kg, i.p.) and were sacrificed at the third day. The receptor density (BMAX) values in femtomoles per milligram of protein (fmol/mg p.) are the means ± SEM (n=4). Statistical significance was determined by one-way analysis of variance (ANOVA) followed by post-hoc t test: [a]$p < 0.05$, significantly different from controls, and [b]$p < 0.05$, significantly different from PCPA treatment.

Alternatively, other methods sometimes used to estimate 5-HT deafferentations have called upon a combination of histofluorescence microscopy and biochemical assays and/or the use of antibodies raised against either 5-HT or tryptophan hydroxylase. With the recent development of new ligands that label specific uptake sites, located on 5-HT varicosities or nerve terminals and on 5-HT cell bodies, it is possible to quantifify this innervation in both the control situation as well as following treatments with neurotoxins or synthesis inhibitors. The data given in Table 4 obtained by [3H]paroxetine binding [50] shows the densities of 5-HT uptake sites following destruction by PCA or synthesis inhibition by PCPA, and belongs to the same experimental study previously illustrated in Table 3.

The novel antidepressant paroxetine is a good marker of the 5-HT uptake recognition site in the mammalian CNS, and we found that [3H]paroxetine binding was closely correlated with endogenous 5-HT concentrations in various brain regions [50,61]. Following PCA treatment, [3H]paroxetine binding was dramatically reduced and this reduction was accompanied by a substantial decrease in the concentration of cortical 5-HT and 5-HIAA content [50]. Furthermore, [3H]paroxetine binding and tissue concentrations of 5-HT and the metabolite 5-hydroxyindole-3-acetic acid (5-HIAA) measured in different regions of the rat brain: cingulate, parietal, and visual cortical areas; dorsal and ventral hippocampus; rostral and caudal halves of neostriatum; ventral mesencephalic tegmentum and midbrain raphe nuclei region, were also significantly reduced. The correlation between regional 5-HT levels and specific [3H]paroxetine binding was found to be conserved following PCA treatment, indicating that the relationship between 5-HT innervation and [3H]paroxetine binding was maintained despite a substantial loss of serotonin neurons. The psychotropic amphetamine derivatives produce a differential loss of 5-HT neurons within the CNS with a greater loss of 5-HT terminal fields than neuronal cell bodies. This heterogenous destruction of 5-HT neurons closely paralled the reduction in [3H]paroxetine as well as the depletions in both 5-HT and 5-HIAA. Systemic administration of PCPA produced a sustained depletion in central 5-HT and 5-HIAA concentrations without any subsequent change in the binding parameters of cortical [3H]paroxetine binding. Under these conditions where the 5-HT innervation remains unchanged while the concentrations of 5-HT and 5-HIAA are greatly reduced, the correlation between [3H]paroxetine binding and endogenous 5-HT content was lost. The 5-HT uptake recognition site as labeled by [3H]paroxetine appears to be refractory to regulation following prolonged inhibition of 5-HT synthesis, while loss of 5-HT innervation following PCA administration was reflected by a reduction in [3H]paroxetine binding. Thus, it would appear that [3H]paroxetine provides a reliable marker of 5-HT innervation density within the mammalian CNS.

It has been proposed that the neurotoxicity of PCA is dependent on a releasable pool of serotonin [54]. When rats were treated with PCA alone or with reserpine and PCA, they exhibited a profound loss of 5-HT innervation in cerebral cortex, after a 2-week survival period. However, if a 5-HT depletion was induced by combined treatment with PCPA and reserpine, there was substantial protection against the neurotoxic effects of PCA. These results suggest that the release of 5-HT is necessary in the neurotoxicity of PCA, and that probably a peripheral source of 5-HT is involved. Based on this evidence, it has been proposed that 5-HT released from platelets into the peripheral circulation may result in the formation of a neurotoxic 5-HT metabolite. Although the nature and source of such toxic metabolite(s) has not yet been established, oxidized derivatives of tryptamines may exhibit deleterious effects on neurons, and the compound tryptamine-4,5-dione could be a potential candidate [55] to be considered as an *endogenous* neurotoxin for 5-HT neurons (see below).

Another amphetamine derivative of interest is methylenedioxyamphetamine (MDMA) that has been shown to produce serotoninergic deafferentations in several species [56]. The *in vivo* administration of MDMA to both rats and Rhesus monkeys produces widespread and long-lasting degeneration of 5-HT neurons in brain, apparently without any major or consistent effects on catecholamine neurons. From both neurochemical and neuroanatomical studies, including the detailed examination of the parameters involved in the neurotoxic and neurodegenerative effects of MDMA on central 5-HT neurons, it is now clear that the severity of the lesion depends on the dose of drug administered; interestingly the drug was more potent in Rhesus monkeys than in rats. Furthermore, the neuro-degenerative effects of MDMA are long-lasting (up to one year) with respect to neuronal regeneration evaluated by the absence of recovery of 5-HT uptake sites. Therefore, an interesting feature of MDMA is that functional recovery may be permanently impaired since 5-HT content remains markedly (40-50%) below levels, in age-matched controls, for as long as one year after drug administration. In addition, the neurochemical and autoradiographic data suggest that there are some regional differences and morphological specificity to the neurodegenerative effects of MDMA as demonstrated by greater reductions in 5-HT uptake sites in brain regions containing primarily terminals, while regions containing axons of passage and cell bodies are relatively unaffected. In this respect MDMA is very similar to PCA; the latter compound affects primarily the terminal fields of innervation [50], as shown in Tables 3 and 4.

Besides the destruction of 5-HT nerve endings, treatments with MDMA appear to perturb other brain function, including energy utilization. In rats, after a four day treatment with MDMA (20 mg/kg twice daily, s.c.) glucose utilization was altered in several CNS region; this perturbation overlasting MDMA administration [57]. In this study, local cerebral glucose utilization (LCGU) was measured between six and nine weeks after MDMA treatment using [^{14}C]2-deoxyglucose quantitative autoradiography. To evaluate the effect on the number of 5-HT nerve endings, samples of frontal cortex taken from these animals were quantified by *in vitro* [^{3}H]paroxetine binding and revealed a 64% reduction in the number of 5-HT uptake sites. In the majority of the 31 functionally diverse brain areas examined, no significant changes in LCGU were measured, however, there were significant increases in glucose utilization in the anterior cingulate cortex (+16%) and in the sensorimotor cortex (+21%). Furthermore, the most profound increases were found in *Globus pallidus* (+30%) and in the molecular layer of the hippocampus (+34%). It would thus appear, that treatment with MDMA results in long-lasting alterations in cerebral functional activity, as reflected by glucose utilization, but only in restricted CNS regions. This aspect of MDMA toxicity on energy metabolism may be also relevant to its mode and/or mechanism of action. In fact, MDMA has been reported to produce a significant hyperthermia in rats, which could be blocked in a competitive manner by the selective 5-HT2 receptor antagonist MDL-11939. Interestingly, the 5-HT2 antagonist also blocked MDMA-induced neurotoxicity, determined by the decline in regional 5-HT concentrations observed 1 week later [58]. These two effects of MDL-11939 could be dissociated at higher doses of MDMA; the antagonist still provided virtually complete protection against the neurochemical deficits but only partially attenuated the hyperthermic response. In contrast to the effect of the 5-HT2 antagonist MDL-11939, the classic DA antagonist haloperidol did not alter MDMA-induced hyperthermia but did antagonize its long-term neurochemical effects. In addition, when a both MDL-11939 and the selective 5-HT uptake inhibitor MDL-27777 were administered together, the hyperthermia produced by a high dose of MDMA was still present, but this combined treatment with a 5-HT uptake blocker and a

5-HT2 antagonist completely prevented the depletion of 5-HT. In addition, if the MDMA-induced hyperthermia was prevented by temporarily maintaining animals at reduced ambient temperature, the neurochemical changes normally observed 1 week later were also blocked. Although these results demonstrate that the drugs tested do not antagonize MDMA-induced neurotoxicity by interfering with its effect on body temperature, they seem to imply that MDMA-induced hyperthermia may contribute to the development of the drug's long-term neurochemical effects.

As above mentioned, the amphetamine-derived indoleamine neurotoxins have an initial acute depleting effect acting in a *hit-an-run* manner like reserpine, but in addition will further decrease more permanently endogenous 5-HT pools. In this way they have been used to examine the consequences of a lack of serotoninergic neurotransmission on numerous CNS functions. In addition, they have also allowed for a better understanding of some of the basic structural and biochemical properties of central 5-HT systems. In fact, 5-HT neuronal cell bodies have a different susceptibility to PCA than nerve endings (Tables 3 and 4). Moreover, it seems that there are different types of 5-HT nerve terminals that can be distinguished by the selective use of particular neurotoxins. One good example is the denervation caused by fenfluramine [59,60] that has been recently characterized at the structural level [62]. Using immunocytochemistry to visualize and quantify 5-HT innervation, it has been shown that this compound causes structural damage to 5-HT neurons, but that it is exerted only on the fine terminals while beaded axons were seemingly spared. In addition, this study revealed the presence of markedly swollen and fragmented 5-HT axons at a very early time interval, i.e.: at 36 h following fenfluramine administration. This anatomical observations were interpreted as signs of axonal degeneration, and provides morphologic evidence for a neurotoxic effect of fenfluramine upon 5-HT axon terminals.

Other indoleamine neurotoxins

Besides dihydroxytryptamines, amphetamine derivatives and 5-HT synthesis inhibitors, other compounds and pharmacological manipulations can result in decreases in 5-HT contents and/or 5-HT deafferentations. It should be recalled that reserpine was one of the first drugs used to modify aromatic monoamine content in the CNS. This alkaloid, extracted from *Rawolfia serpentiana* has an inmediate destructive action on synaptic vesicles, and this amphetamine-like effect is to release neurotransmitters, including 5-HT and NA. Because of this *hit-and-run* effect on the storage vesicles, reserpine very soon leads to a decrease in endogenous 5-HT levels; both uptaken 5-HT as well as newly synthesized 5-HT molecules can not be stored in synaptic vesicles so that this non-vesicular pool is now susceptible to oxidation by MAO. It is only when new vesicles are formed, that these two 5-HT pools can be protected, and reconstitute the vesicular 5-HT compartment. The effects of reserpine are long-lasting but reversible, since neither the structure of the 5-HT nerve terminals nor the enzymes of synthesis are affected. However, this compounds also decreases catecholamines, and because of this lack of selectivity it was very early abandoned in the study of central 5-HT systems.

The fact that several compounds such as amphetamine-derived neurotoxins and reserpine induce a release of 5-HT, could presently explain why the neurotoxicity of these compounds may be exerted and/or enhanced. During the early phase of 5-HT release, there may be a production of very high amounts of oxidized metabolites that exceed the normal levels and these metabolites could result in neurotoxicity. For example, it as recently been shown that tryptamine-4,5-dione, a novel oxidation product of serotonin increases 5-HT efflux from the neostriatum and hippocampus, and can cause selective neuronal death [55]. Interestingly, following incubations of synaptosomes or the brain guanine nucleotide-binding regulatory proteins G_i and G_o with [^{3}H]tryptamine-4,5-dione, a major band was isolated, with an apparent molecular mass of about 40,000 daltons, equivalent to that of the α subunits of G_i and G_o. In addition, tryptamine-4,5-dione inhibited in a dose-dependent manner the binding of [^{35}S]GTP-γ-S to G_i and G_o proteins as well as the pertussis toxin-catalyzed [^{32}P]ADP-ribosylation of the G protein α subunits. Thus, the possibility exists that naturally-occurring indoleamines derived from the oxidation of endogenous 5-HT such as tryptamine-4,5-dione, may exert their neurotoxic effects through specific interactions with G proteins and could therefore be considered as endogenous neurotoxins.

Role of 5-HT neurons in other types of toxicity

The pyridine derivative 1-methyl-4-phenyl-1,2,3,6-tetrahydropyridine (MPTP) is known to produce the destruction of DA-containing neurons in certain species, and in humans produces the symptoms of Parkinson's disease [20]. An essential step in its neurotoxicity is the enzymatic conversion [63] of MPTP to 1-methyl-4-phenylpyridinium ion or MPP$^+$ by monoamine oxidase type B or MAO-B [64-66]. Since MAO-B is located primarily in 5-HT neurons and astrocytes, the production of MPP$^+$ is thought to occur in extra-dopaminergic loci. Interestingly, after pretreating mice with the 5-HT uptake inhibitor fluoxetine, the dopaminergic neurotoxicity of MPTP was greatly attenuated [67]. This effect was not the result of a nonspecific inhibition of dopaminergic uptake, as fluoxetine pretreatment did not attenuate the dopaminergic neurotoxicity resulting from the intrastriatal administration of MPP$^+$. In addition, further localization of the primary site of MPP$^+$ production as being astrocytes was provided by the failure of 5,7-DHT-induced lesions of serotonergic neurons to attenuate the neurotoxicity produced by administration of MPTP. Moreover, in similarly 5,7-DHT-lesioned animals, fluoxetine pretreatment was still effective to attenuate MPTP-induced dopaminergic neurotoxicity. These results are consistent with the hypothesis that astrocytes are a principle site of conversion of MPTP to the active dopaminergic neurotoxin MPP$^+$. It has also been recently shown, using the technique of microdialysis [68], that MPP$^+$ causes a cumulatively dose-dependent increase in the release of 5-HT while concomitantly producing a decrease in the efflux of 5-HIAA. It has been proposed that this MPP$^+$-induced increase in 5-HT release, may play an important role in the acute 5-HT syndrome seen in animals after MPTP administration.

CONCLUDING REMARKS

The research on chemically-identified neurotransmitter systems has benefited from the use of compounds that either destroy neurons and axons terminals or that deplete endogenous stores by synthesis inhibition or increased release. In the case of the dihydroxytryptamines, these drugs have been proven to be authentic neurotoxins, and will produce chemical denervations once inside 5-HT nerve terminals and cell bodies. The possibility exists that non-serotoninergic cells may be susceptible to 5,6-DHT and 5,7-DHT if membrane transporters are made available to uptake the dihydroxytryptamines. This implies a membrane mechanism composed of a *recognition site* coupled to an appropriate *uptake carrier* that is normally quiescent in non-serotoninergic cells, but that could be activated through genetic or pharmacological manipulations. On the other hand, compounds that inhibit synthesis, such as PCPA, or drugs that affect the vesicular compartment, i.e.: reserpine, have reversible effects and can not be truly considered neurotoxins for indoleamine neurons; they are less specific for 5-HT neurons, although they have wide applications to decrease endogenous neurotransmitter levels. The amphetamine derivatives will also release 5-HT, but their main action is to produce serotoninergic deafferentations, although sometimes restricted to certain types of 5-HT nerve fibers and terminals. It remains to be determined whether the enhanced release, or an increase in indoleamine metabolite(s) production due to the greater 5-HT overflow caused by amphetamine derivatives contributes to their neurotoxicity.

Acknowledgements

The original studies refered to in this chapter were funded by grants from the Medical Research Council of Canada (MT-6967) to T.A. Reader, and from the Banting Research Foundation to Karen M. Dewar. Personal support was from the Fonds de la recherche en santé du Québec. The technical assistance of Ms. Louise Grondin and the secretarial help of Ms. Helène Dussault are greatly appreciated. Credit for the graphic artwork goes to Messrs. Claude Gauthier and Daniel Cyr.

REFERENCES

1. Twarog BM, Page IH. Serotonin contents of some mammalian tissues and urine and a method for its determination. J Physiol (Lond) 1953; 175:157-61.
2. Amin AH, Crawford TBB, Gaddum JH. The distribution of substance P and 5-hydroxytryptamine in the central nervous system of the dog. J Physiol (Lond) 1954; 126:596-618.
3. Bogdanski DF, Pletscher A, Brodie BB, Udenfriend S. Identification and assay of serotonin in brain. J Pharmacol Exp Ther 1956; 117:82-8.
4. Zieher LM, De Robertis E. Subcellular localization of 5-hydroxytryptamine in rat brain. Biochem Pharmacol 1963; 12:596-8.
5. Dahlström A, Fuxe K. Evidence for the existence of monoamine containing neurons in the central nervous system. I. Demonstration of monoamines in the cell bodies of brainstem neurons. Acta Physiol Scand 1964; 62(suppl):1-55.
6. Anden N-E, Dahlström A, Fuxe K, Larsson K. Mapping out of catecholamine and 5-hydroxy-tryptamine neurons innervating the telencephalon and diencephalon. Life Sci 1965; 4:1275-9.
7. Anden N-E, Fuxe K, Ungerstedt U. Monoamine pathways to cerebellum and cerebral cortex. Experientia 1967; 23:838-42.
8. Ungerstedt U. Stereotaxic mapping of the monoamine pathways in the rat brain. Acta Physiol Scand 1971; 82(suppl 367):1-48.
9. Moore RY, Halaris AE, Jones BE. Serotonin neurons of the midbrain raphe: Ascending projections. J Comp Neurol 1978; 180:417-38.
10. Kuhar MJ, Aghajanian GK, Roth RH. Tryptophan hydroxylase activity and synaptosomal uptake of serotonin in discrete brain regions after midbrain raphe lesions: correlations with serotonin levels and histochemical fluorescence. Brain Res 1972; 44:165-76.
11. Steinbusch HWM. Distribution of serotonin-immunoreactivity in the central nervous system of the rat: cell bodies and terminals. Neurosci 1981; 6:557-618.

12. Azmitia EF, Segal M. An autoradiographic analysis of the differential projections of the dorsal and median raphe nuclei in the rat. J Comp Neurol 1978; 179:641-68.

13. Beaudet A, Descarries L. The fine structure of central serotonin neurons. J Physiol (Paris) 1981; 77:193-203

14. Parent A, Descarries L, Beaudet A. Organization of ascending serotonin systems in the adult rat brain. A radioautographic study after intraventricular administration of [^{3}H]-5-hydroxytryptamine. Neurosci 1981; 6:115-38.

15. Descarries L, Audet MA, Doucet G, Garcia S, Oleskevich S, Séguéla P, Soghomonian J-J, Watkins KC. Morphology of central serotonin neurons. Brief review of quantified aspects of their distribution and ultrastructural relationships. Ann NY Acad Sci 1990; 600:81-92.

16. Cannon WB. A law of denervation. Amer J Med Sci 1939: 198:737-50.

17. Levi-Montalcini R, Booker B. Destruction of the sympathetic ganglia in mammals by an antiserum to the nerve growth promoting factor. Proc Nat Acad Sci USA 1960; 46:384-91.

18. Thoenen H, Tranzer JP. The pharmacology of 6-hydroxydopamine. Ann Rev Pharmacol 1973; 13: 169-80.

19. Jonsson G, Wiesel F-A, Hallmam H. Developmental plasticity of central noradrenaline neurons after damage-changes in transmitter functions. J Neurobiol 1979; 10:337-53.

20. Langston JW, Ballard P, Tetrud JW, Irwin I. Chronic parkinsonism in humans due to a product of meperidine analog synthesis. Science 1983; 219:979-80.

21. Wagner K and Trendelenburg U. Effect of 6-hydroxydopamine on oxidative phosphorylation and on monoamine oxidase activity. Naunyn-Schmiedeberg's Arch Pharmacol 1971; 269:110-6.

22. Baumgarten HG, Björklund A, Lachenmayer L, Nobin A, Stenevi U. Long-lasting selective depletion of brain serotonin by 5,6-dihydroxytryptamine. Acta Physiol Scand 1971; 373:1-16.

23. Costa E, Daly J, Lefevre H, Meck J, Revuelta A, Sparro F, Strada, S, Daly J. Serotonin and catecholamine concentrations in brains of rats injected intracerebrally with 5,6-dihydroxytryptamine. Brain Res 1972; 44:304-8.

24. Daly J, Fuxe K, Jonsson G. Effects of intracerebral injection of 5,6-dihydroxytryptamine on cerebral monoamine neurons: evidence for selective degeneration of central 5-hydroxytryptamine neurons. Brain Res 1973; 49:476-82.

25. Baumgarten HG, Lachenmayer L. 5,7-Dihydroxytryptamine: improvements in chemical lesioning of indoleamine neurons in the mammalian brain. Z. Zellforsh 1972; 135:399-414.

26. Daly J, Fuxe K, Jonsson G. 5,7-Dihydroxytryptamine as a tool for the morphological and functional analysis of central 5-hydroxytryptamine neurons. Res Commun Chem Path Pharmacol 1974; 7:175-87.

27. Baumgarten HG, Klemm HP, Lachenmayer L, Björklund A, Lovenberg W, Schlossberger HG. Mode and mechanism of action of neurotoxic indoleamines. A review and progress report. Ann NY Acad Sci 1978; 305:3-24.

28. Baumgarten HG, Björklund A. Neurotoxin indoleamines and monoamine neurons. Ann Rev Pharmacol Toxicol 1976; 16:101-11.

29. Reader TA. Neurotoxins that affect central indoleamine neurons. In: Boulton AA, Baker GB, Juorio AV, eds. Drugs as Tools in Neurotransmitter Research. Clifton, New Jersey, USA: Humana Press, 1989: 49-102.

30. Klemm HP, Baumgarten HG, Schlossberger HG. Polarographic measurements of spontaneous and mitochondria-promoted oxydation of 5,6- and 5,7-dihydroxytryptamine. J Neurochem 1980; 35:1400-8.

31. Klemm HP, Baumgarten HG, Schlossberger HG. Interaction of 5,6- and 5,7-dyhydroxytryptamine with tissue monoamine oxidase. Ann NY Acad Sci 1978; 305:36-56.

32. Klemm HP, Baumgarten HG, Schlossberger HG. In vitro studies on the interaction of brain monoamine oxidase with 5,6- and 5,7-dihydroxytryptamine. J Neurochem 1979; 32:111-9.

33. Sinhababu AK, Ghosh AK, Borchardt RT. Molecular mechanism of action of 5,6-dihydroxy-tryptamine. Synthesis and biological evaluation of 4-methyl-, 7-methyl- , and 4,7-dimethyl-5,6-dihydroxytryptamines. J Med Chem 1985; 28:1273-9.

34. Björklund A, Baumgarten HG, Rensch A. 5,7-dihydroxytryptamine: Improvement of its selectivity for serotonin neurons in the CNS by pretreatment with desipramine. J Neurochem 1975; 24:833-5.

35. Creveling CR, Lundstrom J, McNeal ET, Tice L, Daly JW. Dihydroxytryptamines: effects on noradrenergic function in mouse heart in vivo. Med Pharmacol 1975; 11:211-22.

554

36. Baumgarten HG, Jenner S, Klemm HP. Serotonin neurotoxins: recent advances in the mode of administration and molecular mechanism of action. J Physiol (Paris) 1981; 77:309-14.

37. Baumgarten HG, Klemm HP, Sievers J, Schlossberger HG. Dihydroxytryptamines as tools to study the neurobiology of serotonin. Brain Res Bull 1982; 9:131-50.

38. Gerson S, Baldessarin RJ. Selective destruction of serotonin terminals in rat forebrain by high doses of 5,7-dihydroxytryptamine. Brain Res 1975; 85:140-5.

39. Breese GR, Cooper BR. Behavioral and biochemical interactions of 5,7-dihydroxytryptamine with various drugs when administered intracisternally to adult and developing rats. Brain Res 1975; 98:517-27.

40. Ferron A, Reader TA, Descarries L. Responsiveness of cortical neurons to serotonin after 5,7-DHT denervation of PCPA depletion. J Physiol (Paris) 1981; 77:381-4.

41. Gauthier P, Reader TA. Adrenomedullary secretory response to midbrain stimulation in rat: effects of depletion on brain catecholamines or serotonin. Can J Physiol Pharmacol 1982; 60:1464-74.

42. Ferron A, Descarries L, Reader TA. Altered neuronal responsiveness to biogenic amines in rat cerebral cortex after serotonin denervation or depletion. Brain Res 1982; 231:93-108.

43. Reader TA, Gauthier P. Catecholamines and serotonin in the rat central nervous system after 6-OHDA, 5,7-DHT and p-CPA. J Neural Transm 1984; 59:207-27.

44. Mc Rae Deguerce A, Berod A, Keller A, Chouvet G, Joh TH, Pujol J-F. Alterations in tyrosine hydroxylase elicited by raphe nuclei lesions in the rat locus coeruleus: evidence for the involvement of serotonin afferents. Brain Res 1982; 235:285-301.

45. Lyness WH, Demarest KT, Moore KE. Effects of d-amphetamine and disruption of 5-hydroxy-tryptaminergic neuronal systems on the synthesis of dopamine in selected regions of the rat brain Neuropharmacol 1980; 19:883-9.

46. Koe BK, Weissman A. p-Chlorophenylalanine: a specific depletor of brain serotonin. J Pharmacol Exp Ther 1966; 154:499-516.

47. Miller FP, Cox RHJr, Snodgrass WR, Maickel RP. Comparative effects of p-chlorophenylalanine, p-chloroamphetamine and p-chloro-N-methylamphetamine on rat brain norepinephrine, serotonin and 5-hydroxyindole-3-acetic acid. Biochem Pharmacol 1970; 19:435-42.

48. Reader TA. Catecholamines and serotonin in rat frontal cortex after PCPA and 6-OHDA: absolute amounts and ratios. Brain Res Bull 1982; 8:527-34.

49. Aghajanian GK, Kuhar MJ, Roth RH. Serotonin-containing neuronal perikarya and terminals: differential effects of p-chlorophenylalanine. Brain Res 1973; 54:85-101.

50. Dewar KM, Grondin L, Carli M, Lima L, Reader TA. [^{3}H]Paroxetine binding and serotonin content of rat cortical areas, hippocampus, neostriatum, ventral mesencephalic tegmentum, and midbrain raphe nuclei region following p-chlorophenylalanine and p-chloroamphetamine treatment. J Neurochem 1992; 58:250-7

51. Fuller RW, Hines CW, Mills J. Lowering of brain serotonin by chloroamphetamines. Biochem Pharmacol 1965; 14:483-8.

52. Pletscher A, Burkard WP, Bruderer H, Gey KF. Decrease of cerebral 5-hydroxytryptamine and 5-hydroxyindoleacetic acid by an arylalkylamine. Life Sci 1963; 2:828-33.

53. Sanders-Bush E, Bushibg IA, Sulser F. Long-term effects of p-chloroamphetamine on tryptophan hydroxylase activity and on the levels of 5-hydroxytryptophan and 5-hydroxyindoleacetic acid in brain. Eur J Pharmacol 1972; 20:385-90.

54. Berger UV, Grzanna R, Molliver ME. Depletion of serotonin using p-chlorophenylalanine (PCPA) and reserpine protects against the neurotoxic effects of p-chloroamphetamine (PCA) in the brain. Exper Neurol 1989; 103:111-5.

55. Fishman JB, Rubins JB, Chen JC, Dickey BF, Volicer L. Modification of brain guanine nucleotide-binding regulatory proteins by tryptamine-4,5-dione, a neurotoxic derivative of serotonin. J Neurochem 1991; 56:1851-4.

56. De Souza EB, Battaglia G, Insel TR. Neurotoxic effect of MDMA on brain serotonin neurons: evidence from neurochemical and radioligand binding studies. Ann NY Acad Sci 1990; 600: 682-97.

57. McBean DE, Sharkey J, Ritchie IM, Kelly PA. Chronic effects of the selective serotoninergic neurotoxin, methylenedioxyamphetamine, upon cerebral function. Neurosci 1990; 38:271-5.

58. Schmidt CJ, Black CK, Abbate GM, Taylor VL. Methylenedioxymethamphetamine-induced hyperthermia and neurotoxicity are independently mediated by 5-HT2 receptors. Brain Res 1990; 529:85-90.

59. Trulson ME, Jacobs BL. Behavioral evidence for the rapid release of CNS serotonin by PCA and fenfluramine. Eur J Pharmacol 1976; 36:149-54.

60. Fuller RW. Pharmacology of central serotonin neurons. Ann Rev Pharmacol Toxicol 1980; 20:111-127.

61. Dewar KM, Reader TA, Grondin L, Descarries L. [³H]Paroxetine binding and serotonin content of rat and rabbit cortical areas, hippocampus, neostriatum, ventral mesencephalic tegmentum and midbrain raphe nuclei region. Synapse 1991; 9:14-26.

62. Molliver DC, Molliver ME. Anatomic evidence for a neurotoxic effect of (±)-fenfluramine upon serotonergic projections in the rat. Brain Res 1990; 511:165-8.

63. Markey SP, Johannessen JN, Chiueh CC, Burns RS, Herkenham MA. Intraneuronal generation of a pyridinium metabolite may cause drug-induced Parkinsonism. Nature 1984; 311:464-7.

64. Chiba K, Trevor A, Castagnoli NJr. Metabolism of the neurotoxic tertiary amine, MPTP, by brain monoamine oxidase. Biochem Biophys Res Commun 1984; 120:574-8.

65. Heikkila RE, Hess A, Duvoisin RC. Dopaminergic neurotoxicity of 1-methyl-4-phenyl-1,2,5,6-tetrahydropyridine (MPTP) in mice. Science 1984; 224:1451-3.

66. Heikkila RE, Sonsalla PK, Duvoisin RC. Biochemical models of Parkinson's disease. In: Boulton AA, Baker GB, Juorio AV, eds. Drugs as Tools in Neurotransmitter Research. Clifton, New Jersey, USA: Humana Press, 1989: 351-84.

67. Brooks WJ, Jarvis MF, Wagner GC. Astrocytes as a primary locus for the conversion MPTP into MPP⁺. J Neural Trans 1989; 76:1-12.

68. Miyake H, Chiueh CC. Effects of MPP⁺ on the release of serotonin and 5-hydroxyindoleacetic acid from rat striatum in vivo. Eur J Pharmacol 1989; 166:49-55.

A CLINICAL PREFACE

Alan B. Scott

Smith-Kettlewell Eye Research Institute
San Francisco, CA 94115

In 1970-71 three things came together: First, strabismus surgery had reoperative rates of 40% in some categories, and we were systematically looking for alternatives. Second, we had developed electromyographic techniques and equipment to localize injections of anesthetic in human eye muscles for physiologic studies. We applied this in animals to test the effects of locally injected alcohol, enzymes, enzyme blockers, long-acting anesthetics and snake neurotoxin. Third, the chapter by Drachman in Simpson's 1971 book, *Neuropoisons*, showed that he could induce local paralysis with botulinum toxin. The idea of using botulinum toxin clinically had been floating around for several years (Bach-y-Rita, Crone, Jampolsky and Maumanee probably all had it independently), but it had seemed too wild to consider until then. Schantz supplied Drachman and nearly the whole experimental world with toxin, using at the Food Research Institute the crystalline type A technique developed by Lamanna in 1946 and later modified by Duff in 1957 at Fort Detrick. Schantz generously supplied toxin for our experiments, published in 1973, in animals (later, humans) on dosage, duration of effect, pharmaceutical methods of freeze-drying and so on. The long reluctance of FDA to issue an IND for clinical trial was overcome by the intercession of David Cogan, whose awesome reputation and logic availed, and clinical studies began in 1977. By 1982 we had injected the eye muscles for strabismus and nystagmus, the lid muscles for retraction and blepharospasm, and the limbs and neck for dystonia, as predicted in 1973. At that point safety and efficacy seemed substantial. The subsequent contributions of over 200 clinical investigators then expanded the knowledge base rapidly. The groups at Columbia University, Houston, NINCDS and Vancouver deserve special mention, and excellent work from Moorfields (London) was the basis for the Porton development.

The valuable clinical attributes of botulinum toxin are its specificity for motor nerve terminals and its long duration of action. In appropriate doses one can block most muscle groups without any other effect. The immune response problem will be overcome by more potent preparations (less protein), and other serotypes of botulinum toxin are on the way as alternatives for the few patients (less than 40) documented to produce antibodies to botulinum toxin type A.

Weakness of adjacent muscles due to drug diffusion can be reduced by protective antitoxin injection of those muscles. Prolonging clinical action is the major challenge and will be very valuable to patients. Loading the nerve terminal with a mixture of more than one

Botulinum and Tetanus Neurotoxins, Edited by
B.R. DasGupta, Plenum Press, New York, 1993

serotype of toxin to take advantage of several types of nerve terminal receptors or of separate enzymatic effects in the nerve terminal, and conjugating the heavy chain receptor portion to other toxin molecules are some of the ways this is being approached.

Clinicians and their patients are grateful to the basic investigators and their colleagues whose work underlies the valuable addition of botulinum toxin to treatment of neurological disease. This symposium brings together these clinical and basic science groups, whose continued interdependent work should extend this beginning. We are grateful to Bibhuti DasGupta for such a success.

DISORDERS WITH EXCESSIVE MUSCLE CONTRACTION: CANDIDATES FOR TREATMENT WITH INTRAMUSCULAR BOTULINUM TOXIN ("BOTOX")

Mitchell F. Brin,[1] Andrew Blitzer,[2]
Celia Stewart,[1,2] Zachary Pine,[3] Joanne Borg-Stein,[4]
James Miller,[1] Nagalapura S. Viswanath,[5] and
David B. Rosenfield[5]

[1]Department of Neurology, [2]Otolaryngology, and [3]Rehabilitation Medicine,
Columbia University College of Physicians and Surgeons,
Columbia Presbyterian Medical Center, New York, NY
[4]Department of Rehabilitation Medicine, Tufts University, Boston, MA
[5]Stuttering Center Speech Motor-Control Laboratory,
Department of Neurology, Baylor College of Medicine
& the Methodist Hospital, Houston, Texas

INTRODUCTION AND GENERAL CONCEPTS

The initial clinical use of local injections of type A botulinum toxin (botox) was for the treatment of strabismus.[1] The intended goal was to block acetylcholinergic neuromuscular junctions and rebalance neural input to the extraocular rectus muscles; this enhanced convergence.[2] Several drugs, including alpha-bungarotoxin, had been considered prior to botox, but each had limitations. These included lack of selectivity, undesired side effects, short duration of action, and substantial antigenicity. Botox has the advantage of being a potent neuromuscular blocking agent while not suffering as many limitations as other putative therapeutic agents.

Botox has provided a dramatically effective therapy for many focal dystonias. Because botox temporarily relieves muscle contractions, regardless of etiology, its potential applications are numerous (Table 1). The list of areas treated has expanded to include dystonic and non-dystonic involuntary movements, including those excessive muscle contractions accompanying stroke, demyelinating disease, tremor and cosmetic conditions.

Botox injections have several advantages over surgical therapy in the management of intractable disease. The patient is awake and there is no risk of anesthesia. Graded degrees of weakening can be achieved by varying the dose injected. Most adverse effects are transient and are due to an extension of the pharmacology of the toxin. If the patient has a strong response to therapy and too much weakness occurs, strength gradually returns. The patient's acceptance is high, and in most cases, botox therapy is preferred to alternative pharmacotherapy.

Botulinum and Tetanus Neurotoxins, Edited by
B.R. DasGupta, Plenum Press, New York, 1993

Table 1. Botox: wherever muscles overcontract.*

Common feature: *excessive muscle contraction*

I. Dystonic spasms
   Blepharospasm
   Cervical dystonia
   Laryngeal dystonia
   Oromandibular dystonia
   Occupational cramps
   Limb dystonia
   Dystonic tremor

II. Non-dystonic excessive muscle contraction
   Back pain
   Bladder: detrusor-sphincter dyssynergia
   Bruxism
   Cosmetic: "brow furrows", "frown lines", "crows feet", platysma lines
   Gastrointestinal: anismus (constipation), cricopharyngeal spasm, lower esophageal sphincter spasms, rectal spasms
   Eyelid spasms: hemifacial spasm & synkinesis, benign eyelid fasciculation
   Headache
   Myokimia
   Spasticity: stroke, cerebral palsy, head injury, paraplegia, multiple sclerosis
   Sports medicine injuries
   Stuttering
   Tics
   Temporomandibular joint associated muscle spasm
   Tremor: Parkinson's disease, essential tremor

*This list includes both body regions treated with proven efficacy, and regions considered experimental (from Brin, 1991[1]).

In order to treat a patient with botox, one needs toxin, a standard freezer, saline without preservative, syringes with small-gauge needles, alcohol swabs, gauze, and in many circumstances an electromyography (EMG) machine. During our clinical trials for the past decade, we have successfully utilized the botulinum toxin manufactured by Allergan Pharmaceuticals, (Irvine, California), marketed as Oculinum[R]/Botox[TM]. Rather than using a measurement unit of weight (nanograms), the unit of measurement for this product is a unit of bioactivity or potency, the mouse unit. This uniform unit of measurement, when determined according to published specifications,[2] has advantages in that it is standardized.

The European preparation of botulinum toxin-A, Dysport[R] is distributed by Porton Products Limited, England. This preparation has been used clinically with equivalent success in blepharospasm and torticollis, and is licensed for distribution by the Ministry of Health in England. By review of the literature, it is very difficult to calculate the relationship between Oculinum/Botox and Dysport from published reports. (1) In Quinn's report,[3,4] it is implied that the Dysport nanogram is 16 times more potent than the Oculinum/Botox nanogram. (2) The prescribing information that accompanies Dysport recommends a standard dilution of 2.5 ml saline in a 500 Unit vial, or 200 Units/1.0 ml solution. The standard injection of Dysport for blepharospasm is 0.1 ml-0.2 ml, or 20 Units-40 Units per site. The standard injection of Oculinum/Botox is 2.5 Units-5.0 Units per site. This implies that the Dysport Unit is less potent than the Oculinum/Botox Unit. Nevertheless, this would appear to be a contradiction because the Unit is defined as a biological unit. There have been no published guidelines or standards for converting between the two products. From a discussion with our European colleagues, we suspect that one Oculinum/Botox Unit is approximately equivalent to 4-5 Dysport Units. Oculinum/Botox units are used in the subsequent discussion.

Chiba Serum Ltd, in Japan produces botulinum toxin type A[5] with which we have no experience nor information about relative potency.

There is a paucity of data regarding use of botox during pregnancy. In one report, of nine patients treated during pregnancy (dose unspecified), one gave birth prematurely; "this was thought not to be related to the drug".[6] We have treated 2 patients with small doses (5-50 units) of botox during pregnancy with normal outcomes. Currently, we recommend not injecting patients who are pregnant or lactating.

Although we have treated some patients with preexisting disorders affecting neuromuscular junction function, we recommend proceeding with caution in treating patients with conditions such as myasthenia gravis, Eaton-Lambert syndrome, and motor neuron disease, particularly when large doses are required, such as in the treatment of cervical dystonia. Patients receiving aminoglycosides should not be treated with botox unless there is an emergent extenuating circumstance that justifies the potential risk.

In all conditions discussed, appropriate diagnostic studies to identify the etiology of symptoms should be carried out prior to institution of therapy. The treating physician should be knowledgeable in the diagnosis, differential diagnosis and medical management of the disorder of the body region being treated. The relevant anatomy should be studied and basic knowledge in the use of the EMG machine may be required.

In each situation, whether treating with botox or other therapies, *informed consent* must be given: disclosure of the major risks of treatment or procedure being contemplated, an accurate assessment of the benefits that can be reasonably expected, and a discussion of alternative forms of treatment.[7] Treatment of most movement disorders is challenging, and it is wise that the treating physician and associated staff nurture a responsive relationship with the patient. In most situations, we counsel the patient that both the doctor and the patient "hold hands" in proceeding through the treatment options. Dose of medication is carefully monitored and adjusted according to the response benefits vs. adverse effects. Patients build a portfolio of "response to therapy"; this portfolio is consulted frequently as each new strategy is considered. So that the patient profile is easily accessible, a "drug list" is kept in each chart.

When treating patients, we recommend maintaining a database of each patient's treatment profile and response to therapy. This is helpful in tailoring therapy to each individual patient's needs. In addition to storing data on the dose delivered to each muscle, we use clinical rating scales to follow the response to therapy. Some scales are global and generic (Table 2), and some are specific to the region being treated. A patient diary is recommended.

Table 2. Global rating scales (from Brin, 1991[1]).

Ask the patient to use the following 2 scales: "For the area of the body being treated, how would you rate your current condition?"

Disability Rating Scale:

0	-	Normal
1	-	Mild discomfort or functional impairment
2	-	Mild to moderate discomfort or functional impairment
3	-	Moderate discomfort or functional impairment
4	-	Moderate to severe discomfort or functional impairment
5	-	Severe discomfort or functional impairment
6	-	Completely disabled or incapacitated

Percent of Normal Function:

0%	-	Fully disabled
100%	-	Normal

Table 3. Botox: typical adverse effects.

Around eyes (blepharospasm, cosmetic): ptosis

Face (hemifacial spasm, cosmetic): droopy lip

Jaw (oromandibular): difficulty chewing

Lingual: dysphagia

Neck (torticollis or cervical dystonia): generalized fatigue, dysarthria, dysphagia, antibody formation

Laryngeal: mild choking, without true aspiration; breathy hypophonia

Limb: generalized fatigue

Any region: excessive local weakness

This report reviews the clinical applicability of botox for some of the disorders that have been studied in detail, and some of the disorders that have recently been explored. The information reviewed below applies primarily to utilization of the toxin in adults; although efficacy in children parallels that in adults, safety has not been studied in great detail. Preliminary studies in children have not revealed a more extensive adverse effect profile (Table 3); side effects are typically reversible, and represent an extension of the pharmacology of the drug. It is remarkable that in over a decade of therapeutic application of the toxin, no persistent side effects nor deaths have occurred. Treatment has been reviewed for the American Academy of Neurology,[8] American Academy of Ophthalmology,[9,10] American Academy of Otolaryngology[11] and National Institutes of Health;[12] all organizations have documented the positive response to therapy in selected situations. Additional reviews of the conditions and their management with botox are available.[1,13,14] Detailed reviews of the disorders that follow can be found in major review books and articles.[15-21] In our communication, we will try not to overlap with other reports in this volume.

DYSTONIA

Dystonia is a syndrome dominated by sustained muscle contractions frequently causing twisting and repetitive movements, or abnormal postures that may be sustained or intermittent. Dystonia can involve any voluntary muscle. Because the movements and resulting postures are often unusual, and the condition is rare, it is one of the most frequently misdiagnosed neurological conditions.[22] The prevalence of the condition is unknown, but we estimate at least 50,000-100,000 cases of idiopathic dystonia in this country.[23]

As a clinical syndrome, we can classify patients according to clinical distribution and symptomatology, age at onset, and etiology (Table 4). Classification may be important as it can give us clues about prognosis and also an approach to management. According to the etiologic classification, patients with idiopathic disease have no evidence by history, examination or laboratory studies of any identifiable cause for the dystonic symptoms. Therefore, there must be a normal perinatal and early developmental history, no prior history of neurologic illness or exposure to drugs known to cause acquired dystonia (e.g., phenothiazines), normal intellectual, pyramidal, cerebellar and sensory examinations, and diagnostic studies are normal. Patients who have abnormalities noted above are classified as having secondary dystonia.

The clinical phenomenology will often be a clue as to etiology. Primary dystonia is typically action-induced; symptoms are enhanced with use of the affected body part and the region may appear normal at rest. Secondary dystonia frequently results in fixed dystonic postures. The presence of extensive dystonia limited to one side of the body (hemidystonia) suggests a secondary etiology.

Table 4. Classification of dystonia.

I. Age at onset
  A. Infantile (<2 yr)
  B. Childhood (2-26 yr)
  C. Adult (>26 yr)
II. Etiology
  A. Primary
    1. With hereditary pattern
      a. Autosomal dominant
          Classical types
            Childhood-onset dystonia
            Focal dystonia
          Variant types
            Dopa-responsive dystonia
            Myoclonic dystonia
      b. X-linked recessive
    2. Sporadic (without a documented hereditary pattern)
          Classical types
          Variant types
  B. Secondary
    1. Associated with other hereditary neurologic disorders: e.g., Wilson's disease, Huntington's disease, ceroid lipofuscinosis
    2. Environmental: e.g., post-traumatic post-infectious, vascular, tumor, toxic, phenothiazines (tardive).
    3. Dystonia associated with parkinsonism
    4. Psychogenic
III. Distribution
  A. Focal
    1. Blepharospasm (forced, involuntary eye-closure)
    2. Oromandibular dystonia (face, jaw, or tongue)
    3. Torticollis (neck)
    4. Writer's cramp (action-induced dystonic contraction of hand muscles)
    5. Spasmodic dysphonia (vocal cords)
  B. Segmental (cranial/axial/crural)
  C. Multifocal
  D. Generalized (ambulatory, non-ambulatory)

When classified by distribution, patients are categorized as having focal, segmental or generalized symptoms. Focal dystonia symptoms involve one small group of muscles in one body part, segmental disease involves a contiguous group of muscles, and generalized dystonia is widespread. Common examples of focal dystonia include blepharospasm (forced, involuntary eye-closure), oromandibular dystonia (face, jaw, or tongue), torticollis (neck, Figure 1), writer's cramp (action-induced dystonic contraction of hand muscles, Figure 1), and spasmodic dysphonia (vocal cords). Patients with dystonia limited to one region of the body (focal or segmental dystonia) are the best candidates for treatment with botox. However, patients with severe generalized dystonia (Figure 2) may benefit when selected regions of the body are treated.

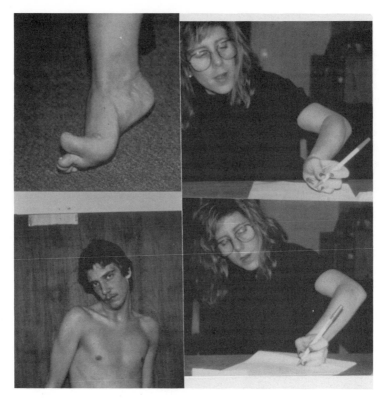

Figure 1. Dystonia is characterized by abnormal involuntary excessive muscle contraction, frequently causing abnormal postures. Although some postures are present at rest, many examples are action-induced, as in writers cramp.
Top left, foot dystonia; right, writer's cramp; bottom left, cervical dystonia (torticollis).

In families with dystonia, various members may have generalized, segmental, or focal dystonia, suggesting that those with milder manifestations are related to the more generalized cases. This gives rise to the concept that the less involved individuals are *formes frustes* of generalized dystonia. The presence of dystonic symptoms in other family members supports the concept that the focal involvement is dystonic and genetic in these cases.

We have recently identified a marker linked to a gene for childhood onset dystonia on chromosome 9.[24,25] There is genetic heterogeneity among idiopathic patients with dystonic symptomatology, with an X-chromosome form with parkinsonism,[26-30] a DOPA responsive form[31] in addition to other clinical types that are on chromosome 9.

Laboratory investigations are typically normal in patients with idiopathic dystonia. Patients with secondary dystonia will have laboratory findings consistent with the underlying disorder.

There are no consistent brain pathological findings in patients with idiopathic dystonia. Among the various reviews[32-36] of primarily idiopathic dystonia, the most frequently cited lesions are in the basal ganglia, including the putamena, head of caudate, and upper brainstem. Hedreen[37] proposed that the putamen and the striatopallidal-thalamo-cortical circuit appear to be the most likely sites in which to search for the unknown defect in primary dystonia.

There is also little known about the precise biochemistry of dystonia. Some patients derive benefit from pharmacotherapy with anticholinergics, benzodiazepines, baclofen, dopamine

Figure 2. Severe generalized dystonia in an ambulatory (above) and non-ambulatory (below) patient.

depletors, or dopamine blocking agents. However, from the few autopsies,[38] there is normal choline acetyltransferase in the cortex and striatum. In examples of secondary dystonia,[39] there is a marked and consistent elevation of norepinephrine in several regions of brainstem. In dopa-responsive dystonia, a rare form of childhood-onset dystonia, there is reduced biological half life of dopamine, suggesting impaired storage is responsible for diurnal fluctuations. Studies with positron emission tomography (PET) have suggested abnormalities in the sensory[40,41] or motor[42] cortex in classical idiopathic disease, and motor cortex[43] in X-linked dystonia-parkinsonism ("Lubag"[26,27]). However, there are no consistent findings across all forms of dystonia.

Therapy of dystonia

Various options are available for treating patients with dystonia; aside from the case of DRD,[44,45] no drug has emerged as uniformly effective and the surgical approaches are not uniformly beneficial or without risk. One must choose a treatment strategy keeping the risk to a minimum. In order to avoid potential harm when searching for effective therapy, surgical therapies and drugs that may cause irreversible harm should be avoided. Very few systematic drug trials have been conducted in dystonia;[46] much of what we know about pharmacotherapy has resulted from empiric observations. Greene reported on 358 patients treated at the Neurological Institute.[47]

As our genetic research advances, our concern about ablative CNS surgery has grown. The goal of genetic research is to find out more about the etiologic pathophysiology with the intent to develop specific pharmacotherapeutic intervention. With that goal in mind, we do not want to ablate regions of the brain that may hold important receptors to future interventional pharmacotherapy.

Specific pharmacotherapy directed at the underlying identified biochemical defect is available for only a limited number of symptomatic dystonias.[48] The most notable is Wilson's disease. For tardive dystonia, the best treatment is avoidance of offending medications when possible, and providing the patient with a list of these medications to avoid.

Levodopa-responsive dystonia (DRD) usually starts in childhood with gait disturbance and is often familial. Most patients respond dramatically to less than 500 mg/day levodopa (with carbidopa). For this reason, treatment with levodopa is our first-line pharmaceutical approach.

Patients with focal dystonia

Most of the focal dystonias are now effectively treated with local injections of botulinum toxin (botox). We have employed this modality of therapy since 1984 in over 1500 patients, and in many cases, the response is dramatic. For many indications, botox has replaced pharmacotherapy or surgical therapy for this patient population. Although at the current time, botox is Food and Drug Administration approved and labeled for blepharospasm and related facial dystonia, and hemifacial spasm, it has been demonstrated as safe and effective appropriate therapy for torticollis (cervical dystonia), spasmodic dysphonia, and many cases of oromandibular dystonia, and writers cramp.[1,8-14] Active investigations are being conducted in other regions of the body.

Patients with more than focal dystonia

Patients with more than focal dystonia (segmental, multifocal, generalized) are usually treated with pharmacotherapy. However, many of our patients have benefited with botox therapy directed towards one or many discrete regions of the body. Pharmacotherapy is usually initiated with either an anticholinergic, benzodiazepine or baclofen. The choice of drug to initiate usually depends on the age of the patient, prior exposure to medications and other concurrent medications or medical problems. Drug dosage is low initially, and gradually increased as tolerated. If the medication is of no benefit at a dose necessary to cause adverse effects, then it is gradually tapered and discontinued. If a medication is documented helpful, then it can be continued, and the next medication is added. In difficult cases, dopamine depleting and receptor blocking agents may be added. Surgery is considered in only the most refractory cases that are unresponsive to botox and/or extensive pharmacological trials.

HEMIFACIAL SPASM (HFS)

Hemifacial spasm is a chronic distressing and embarrassing movement disorder of the face characterized by twitching, tonic spasm and synkinesis of the muscles innervated by the facial nerve.[49,50] Physical methods (rubbing tonics into the face, electrical shocks, irritation therapy) are of no proven therapeutic value. Alternative pharmacotherapy with membrane stabilizing anticonvulsants (phenytoin, carbamazepine), and clonazepam result in variable relief in some patients, but no large series with adequate follow-up are available. Surgical therapy with facial nerve injections with alcohol or phenol, or physical block have been reported to have a good response in over 500 patients but the procedure is painful and because the goal is to replace spasm with paresis, the patient often has severe facial weakness. Gentle heating of the facial nerve (percutaneous thermolysis) is reported to be efficacious with fewer adverse effects. In selected patients, intracranial microvascular decompression of the seventh nerve can bring about a cure of HFS in a high percentage of patients, rather than just preventing its manifestations peripherally. However, serious surgical complications may occur. The procedures that have been commonly used for blepharospasm, myectomy and differential section of the seventh nerve, are also affective for treating HFS.

Botox injections into facial muscles have been widely used to manage HFS.[51-69] Botox treatment is safe and effective and may be offered as primary therapy. There is no evidence of antibody production by patients who have received repeated doses of botox for the treatment of HFS, nor any systemic complications during over 8 years of continued therapy.

TREMORS

Tremor is a rhythmical, oscillatory movement of a body part.[20,70] Small amplitude tremors may only be detectable by sensitive recording devices. Tremor may occur at rest, with action, or with the maintenance of posture. A kinetic tremor is one that occurs during any form of movement. Most tremors are due to either synchronous or alternating contractions of antagonist muscles, and may be primary (essential tremor) or associated with another disorder (Parkinson's disease, dystonia).

Mild tremors may be responsive to pharmacotherapy;[71] large amplitude tremors may cause substantial disability.[72] Thalamotomy may be efficacious,[73,74] but may carry a significant neurosurgical risk. We and others[75] have found botox useful in the management of voice, neck and limb tremor due to essential tremor, dystonia, and Parkinson's disease. Although no large series are published, the preliminary results appear to be comparable to that seen in dystonia; we await the results of double-blind studies for confirmation. In most cases of essential limb tremor, benefit is accompanied by mild limb weakness.

SPASTICITY

Spasticity is a major problem in the management of many neurologic disorders. It is a complex process in which the loss of upper motor neuron inhibition of the spinal reflex arcs plays a major role.[17,18,76,77] Spasticity impacts on all spheres of function, especially including mobility, self care, and hygiene. It can cause disabling pain as well. Maintaining some spasticity is clearly of benefit in many individuals, as it may compensate for loss of voluntary motor control. A frequently seen example of this phenomenon is spasticity of the hip and knee extensors in hemiplegia, where the spasticity allows the hemiplegic person to bear weight and ambulate despite poor voluntary control of the hemiplegic leg.

Existing treatments for spasticity may be categorized as systemic or locally acting. Systemic medications commonly used are baclofen, diazepam and other benzodiazepines, and sodium dantrolene.[78-81] Tinzanidine is used in Europe but is not yet available in the United States.[82-84] With the exception of sodium dantrolene which acts directly on the muscles[81], these medications act on the central nervous system, and sedation is a limiting side effect. Dantrolene and baclofen may also cause hepatotoxicity. All of these agents may "unmask" weakness in other muscles, and as a result may decrease the patient's functional abilities. Because they are nonselective they frequently improve function in one muscle group while severely compromising others. In patients with severe spasticity and painful spasticity, systemic antispasticity drugs are rarely effective in doses which can be tolerated.

Locally acting treatments include motor point blocks, surgical or chemical rhizotomy, and intrathecal baclofen. Local injection of motor points with phenol has been used to provide temporary relief in some individuals with spasticity,[85,86] but the caustic nature of phenol limits the dosage which can be safely administered, and it is irreversibly neurotoxic. In non-ambulatory patients, chemicals into the subarachnoid space (intrathecal) have been administered with some success,[87,88] but these are irreversible and highly invasive techniques.

Intrathecal phenol has the potential for producing devastating weakness should the zone of instillation become wider than intended. Despite the destructive nature of these procedures, spasticity frequently returns, often in muscle groups less amenable to treatment. Intrathecal baclofen appears to be of benefit in selected severe cases of spasticity in the lower extremities and back.[89-91] Delivery of intrathecal baclofen by subcutaneous implanted pump has recently been approved for use in the United States for these situations; this treatment modality is still under investigation for other indications.

Advantages of intramuscular injections of Botox for treatment of spasticity include the ability to target specific muscle groups, the ability to weaken these muscles in a graded fashion, the absence of caustic chemical effects, and a sustained but reversible effect.[92-94]

Botox for spasticity

There are only a limited number of studies using botox to treat spasticity. Das[93,94] treated spastic limbs in 8 patients with hemiplegic stroke with similar benefit. Snow[92] treated the thigh adductor muscles in 9 patients with multiple sclerosis and found a significant reduction in spasticity and a significant improvement in the ease of nursing care; our experience has been similar.[95]

We have seen similar benefit in our series.[95] For instance, two subjects with chronic progressive multiple sclerosis achieved gains in function or ease of hygiene following botox injections. In both cases, treatment effects were seen in non-injected muscles. One patient developed knee buckling while standing, following adductor injections. Another patient manifested reduced spasticity throughout both lower extremities following unilateral hamstring injection. In this case, the widespread effect was beneficial, enabling improved sitting and transfers. The distant effects observed may be explained by reduced input to spinal cord pathways, by axonal transport of toxin to the spinal cord and subsequent spread within the cord,[96] or by an effect of circulating toxin (see discussion below). These potential mechanisms of action deserve further exploration.

STUTTERING

Stuttering is the oldest known, and possibly least understood communication disorder. Stuttering was first described around 2000 BC during the Middle Egyptian Dynasty in hieroglyphics that refer to a speech disorder as "to walk haltingly with a tongue that is sad".[97] Since then there has been considerable debate about the psychological versus learned versus organic etiology of the condition.

Over the years many theories about the cause of stuttering have emerged. Few theories have been proved or disproved. From the 1930s to the 1940s most of the dominant theories came from a medical model that focussed on the organic cause of stuttering. These theories are shaped by research in perseveration,[98] blood sugars,[99] cerebral dominance,[100] and delayed myelination of the nerve fibers.[101] Ingham[102] described how, in the 1940s, Freud and Rogers change the favored perspective from a medical to a psychological approach. Ingham also stated that Skinner's theories on learning helped to change the orientation back towards a medical model. The behaviorist approach became popular, experimental methods were used, and programmed therapy became popular. Webster[103,104] theorized that impaired auditory feedback causes stuttering. Aberrant interaction of air-conduction and bone-conduction components of the feedback system were believed to distort the feedback signal from speech and cause the stutterer to compensate for the speech perception distortion by stuttering.

With the availability of new technology, organic theories became popular in the 1970s and 1980s. Using electromyography, Freeman[105] measured muscle activity during fluent and nonfluent utterances. Analysis revealed that stuttering is accompanied by high levels of

Table 5. Stuttering: Clinical measures.

1. Stuttering Severity Index (SSI):
   a. Percentage of dysfluencies
   b. Length of 3 longest blocks in seconds*
   c. 4 Physical components* (Distracting sounds, noisy breathing, facial grimaces, head movements, limb movements)
2. Perceptions of Stuttering Inventory (PSI):
   Self rating of behaviors: struggle, avoidance, anticipation
3. Percent of syllables stuttered during reading of a standard passage

*Subsection analyzed separately.

laryngeal muscle activity and disruption of normal reciprocity between abductor and adductor muscle groups. Results have been interpreted to demonstrate a strong correlation between abnormal laryngeal muscle activity and stuttering. Recently, Nudelman et al.[106] model stuttering as a momentary instability in a complex multiloop control system.

Over the years, speech therapists have employed a number of therapies including temporal therapy (metronome), gentle onset therapy, slow rate of speech (pacing), cancellation therapy, biofeedback, masking therapy (white noise in patient's ears), breathing techniques, hypnosis, acupuncture, psychotherapy, psychoanalysis and combination therapies (total emersion therapy). Occasional drug therapeutic trials have been performed with limited success. Surgical removal of portions of the tongue have resulted in death from aspiration. Most traditional therapies have limited long term benefit;[21] however, there are very few follow-up studies.

Local injections of botulinum toxin are effective in treating the inappropriate laryngeal muscle spasms of spasmodic dysphonia.[14,107-109] Subsequent to our first laryngeal injection in 1984,[14] 3 patients referred with a diagnosis of laryngeal dystonia displayed the clinical features of idiopathic stuttering. Since their stuttering was associated with severe laryngeal blocks, they were treated, and the outcome was excellent.

We therefore undertook a more detailed study of 7 adult stutterers. The clinical measures analyzed are outlined in Table 5. Patients were treated with injections of 1.25 U botulinum toxin into each vocal cord using our established technique.[14] Results are summarized in Table 6. Adverse effects are similar to those observed in our larger series.[14]

We hypothesized that relief of laryngeal blocks would modify the stuttering phenomenon and increase fluency. Local injections of botulinum toxin are effective in relieving the laryngeal spasms, decreasing the struggle behavior and duration of the stuttered blocks. Further study with a placebo controlled clinical trial is required to evaluate long-term efficacy.

HYPERFUNCTIONAL FACIAL LINES: COSMETIC USES

Hyperfunctional lines are a result of pull on the skin of the underlying facial mimetic musculature.[110] These include the corrugator supercilii (glabellar lines), frontalis (for deep forehead lines or an overly active brow with a Mephistophelean appearance), lateral aspect of the orbicularis oculi (crow's feet), and the zygomaticus minor, levator superioris, levator superioris alaeque nasi, and well as the orbicularis oris (nasolabial fold). Excessively prominent facial lines are often interpreted as anger, anxiety, fatigue, fear or melancholia.

In the past, patients with these hyperfunctional lines have been treated with surgical excision, leaving unsightly scars and/or abnormal facial motion.[111-116] Other patients have had implants of silicone, collagen, suture material, fibrin and others under the furrows in an attempt to lessen the cosmetic deformity, by ballooning the skin underlying the furrow.[116-121] These

Table 6. Results of treatment of stuttering in 7 adult stutterers.

Measure	Paired t-test vs. Initial Baseline		
	2 weeks	6 weeks	12 weeks
PSI	ns	0.004	0.05
SSI	ns	0.09	ns
Duration Blocks	ns	0.04	ns
Physical Mvts	ns	0.02	ns

treatments however, do not appropriately address the fact that these lines are functional, related to the attachment of the underlying mimetic facial musculature.

Over the past several years we have observed that patients who have a Bell's palsy, have minimal or no wrinkles or hyperfunctional facial lines. We have also treated patients with botox who have blepharospasm, Meige syndrome involving the upper and/or lower face, hemifacial spasm, and post-Bell's palsy facial synkinesis. Many of these patients, after treatment also displayed a loss of wrinkles or hyperfunctional lines. We have therefore extended the use of botox to selected patients who have hyperfunctional muscle pull creating deep wrinkle deformities, to assess the efficacy of botox.

Twenty six patients were treated: 24 with dystonia involving other regions of the body, and 2 for purely cosmetic indications. The regions injected included 18 frontalis/corrugator muscles; 14 lateral canthus; 9 nasolabial fold/upper lip muscles; and 1 platysma muscle. Patients experienced benefit from the injections within 24-72 hours with partial or total resolution of painful contractions and/or unsightly hyperfunctional lines and spasms Figures 3). Benefit ranged from 3 to 6 months, whereupon patients returned and were reinjected.[122]

Clark[123] described safely using botox for the treatment of facial asymmetry caused by facial nerve paralysis. The Carruthers'[124] treated 18 patients with glabellar frown lines showing improvement in 17 for 3 to 11 months. Our expanded series supports the use of botox to treat hyperfunctional facial lines. A double blind trial will be necessary to confirm these open trials.

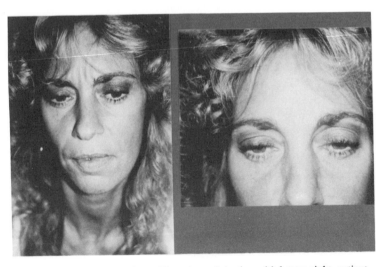

Figure 3. Left, glabellar frown line prior to injection with botox; right, patient attempting to frown 6 weeks after treatment.

OTHER APPLICATIONS

Chemodenervation with local injections of botox have been demonstrated as safe and effective in relieving the disabling spasms for a variety of disorders. Recognizing that chemodenervation can be effective in relieving unwanted muscle contractions, it is not surprising that it has been useful in managing selected cases (many of these are by personal communication) of tics, detrusor-sphincter dyssynergia,[125,126] constipation,[127] back spasm pain, and cricopharyngeal spasm (personal experience). We anticipate that additional creative approaches to a variety of disorders will emerge.

CONCLUSION, AND THE TASK AT HAND

Patient acceptance of his form of therapy is high. Compared to alternative methods of therapy, treatment with botox is preferred by most patients with inappropriate focal muscle contraction, and is currently offered as primary therapy for many conditions.

There are a many clinically relevant issues that need to be resolved; some are addressed in this volume:

(1)   Why do we observe a decrease in clinical muscle activation in muscles neighboring those injected? Ludlow[128] and Woodson[109] proposed that reduced muscle-action feedback to motoneuron pools may be the mechanism of action of botox. We[14] offered the possibility that toxin might block intrafusal fibers, resulting in decreased activation of muscle spindles. This would effectively change the sensory afferent system by reducing the Ia traffic. Alternative hypotheses include a central effect of the toxin due to retrograde axonal transport of toxin, and subsequent spread within the cord[96] or brainstem, or by an effect of circulating toxin. These and other hypotheses need to be explored.

(2)   What are the fundamental differences between the American (Allergan), United Kingdom (Porton) and Japanese (Chiba serum) toxins, and how do these differences account for clinical activity?

(3)   Is there any direct central effect of the toxin?

(4)   To what extent does local diffusion occur, and how can it be harnessed or controlled for varying clinical situations? Is there an advantage to giving one injection into the middle of the muscle vs. multiple small injections that may afford greater dispersion throughout the muscle? Would the latter approach potentially present toxin closer to adjacent muscles which would not be desired?

(5)   What is responsible for heterogeneity of effect in different patients? Is there toxin-specific enzymatic substrate specificity? What factors are responsible for modifying the proteolytic effect within a patient or between patients?

(6)   Why do different serotypes have different clinical effects; is this due to binding, internalization, enzymatically mediated blockade or substrate saturation?

(7)   Whereas the majority of patients do not develop antibodies, why does the occasional patient develop antibodies or "clinical resistance to therapy"? What are the factors that control the development of antibodies, and how can we control them?

(8)   What are the long-term histological and neurochemical effects of repeated exposure to toxin?

(9)   Why is the United Kingdom Dysport Unit different from the United States Oculinum/Botox™ Unit? Is it possible to have a standard Unit that can be used worldwide in training physicians and treating patients?

(10)   What is the most appropriate model to assess stability and also potency in human muscle? How stable is the toxin when stored at room temperature, in the refrigerator or freezer?

Can a toxin be formulated which will maintain potency when stored?

(11) What is the preferred nomenclature? We used the term "botox" in our original communications;[129,130] and we[131] and other clinicians[132] have also used "btx". Our basic science colleagues have used BoNT, Botx, BoTX and BoTx as abbreviations (see other chapters in this book). In 1992, Allergan Pharmaceuticals trademarked their product as "Botox™" and then registered the name, and as a result the word is no longer a "generic." We propose a standard terminology, either BoNT or BTX as a generic, but in all cases, specify the source of the toxin in professional communications. Serotypes can be indicated as a modifier, such as BTX-A, BTX-F, or BoNT-A, BoNT-F.

This is an exciting time in interventional neurology.[1]

Acknowledgments

Research conducted by Dr. Brin has been supported, in part, by the United States Public Health Service (NS-26656, DC-01140, NS-24778, MO1-RR00645), The Dystonia Medical Research Foundation and Allergan Pharmaceuticals. Dr. Rosenfield acknowledges support from the Ariel-Benjamin-Jeremiah-Gideon-Abigail Maida Lowin and the M.R. Bauer Medical Research Foundations.

REFERENCES

1. Brin MF. Interventional neurology: treatment of neurological conditions with local injections of botulinum toxin. Arch Neurobiol 1991; 54 (Suppl 3):7-23.
2. Sellin LC. The action of botulinum toxin at the neuromuscular junction. Med Biol 1981; 59:11-20.
3. Quinn N, Hallett M. Dose standardisation of botulinum toxin [letter]. Lancet 1989; 1:964.
4. Quinn NP, Hallett M. Dose standardisation of botulinum toxin—published erratum correction. Lancet 1989; 1:1989.
5. Nagamine T, Kaji R, Hamano T, Kimura J. Treatment of focal dystonia with botulinum toxin. Clin Neurol 1991; 31:32-37.
6. Scott AB. Clostridial toxins as therapeutic agents. In: Simpson LL, ed. Botulinum neurotoxin and tetanus toxin. New York: Academic Press, 1989:399-412.
7. Klawans HL. Taking a risk. In: Trials of an expert witness: tales of clinical neurology and the law. Boston: Little Brown, 1991:93-94.
8. AAN. Assessment: the clinical usefulness of botulinum toxin-A in treating neurologic disorders. Report of the Therapeutics and Technology Assessment Subcommittee of the American Academy of Neurology. Neurology 1990; 40:1332-1336.
9. AAO Statement. Botulinum-A toxin for ocular muscle disorders. Lancet 1986; 1:76-77.
10. AAO Statement. Botulinum toxin therapy of eye muscle disorders. Safety and effectiveness. American Academy of Ophthalmology. Ophthalmology 1989; Part 2 (Sept):37-41.
11. AAO-HNS. American Academy of Otolaryngology-Head and Neck Surgery Policy Statement: Botox for spasmodic dysphonia. AAO-HNS Bulletin 1990; 9(12/December):8.
12. NIH Censensus Development Conference Consensus Statement. Clinical use of botulinum toxin. N I H Cons Dev Conf 1990; 8:12-14.
13. Jankovic J; Brin M. Therapeutic uses of botulinum toxin. N Engl J Med 1991; 324:1186-1194.
14. Brin MF, Blitzer A, Stewart C, Fahn S. Treatment of spasmodic dysphonia (laryngeal dystonia) with local injections of botulinum toxin: review and technical aspects. In: Blitzer A, Brin MF, Sasaki CT, Fahn S, Harris KS, eds. Neurological disorders of the larynx. New York: Thieme, 1992:214-228.
15. Parkinson's Disease and Movement Disorders. Baltimore:Urban & Schwarzenberg, 1988:
16. Weiner WJ, Lang AE. Movement disorders: a comprehensive survey. Mount Kisco:Futura, 1989:
17. Garrison SJ, Rolak LA, Dodaro RR, O'Callighan AJ. Rehabilitation of the stroke patient. In: DeLisa JA, ed. Rehabilitation Medicine: Principles and Practice. Philadelphia: Lippincott, 1988:565.
18. Whyte J, Rosenthal M. Rehabilitation of the patient with head injury. In: DeLisa JA, ed. Rehabilitation medicine: Principles and Practice. Philadelphia: Lippincott, 1988:585.
19. Cleeves LB, Findley LJ. Tremors. Med Clin North Am 1989; 73:1307-1319.

20. Capildeo R, Findley LJ. Classification of tremor. In: Findley LJ, Capildeo R, eds. Movement disorders: tremor. New York: Oxford University Press, 1984:3-13.
21. Bloodstein O. A Handbook on Stuttering. Chicago:National Easter Seal Society, 1987:
22. Fahn S. The varied clinical expressions of dystonia. Neurol Clin 1984; 2:541-554.
23. Nutt JG, Muenter MD, Aronson A, Kurland LT, Melton LJ, III. Epidemiology of focal and generalized dystonia in Rochester, Minnesota. Mov Disord 1988; 3:188-194.
24. Ozelius L, Kramer PL, Moskowitz CB, et al. Human gene for torsion dystonia located on chromosome 9q32-q34. Neuron 1989; 2:1427-1434.
25. Kramer PL, de Leon D, Ozelius L, et al. Dystonia gene in Ashkenazi Jewish population is located on chromosome 9q32-34. Ann Neurol 1990; 27:114-120.
26. Wilhelmsen KC, Weeks DE, Nygaard TG, et al. Genetic mapping of the "Lubag" (X-linked dystonia-parkinsonism) in a Filipino kindred to the pericentromeric region of the X chromosome. Ann Neurol 1991; 29:124-131.
27. Fahn S, Moskowitz C. X-Linked recessive dystonia and parkinsonism in Filipino males. Ann Neurol 1988; 24:179.
28. Lee LV, Kupke KG, Caballar Gonzaga F, Hebron Ortiz M, Muller U. The phenotype of the X-linked dystonia-parkinsonism syndrome. An assessment of 42 cases in the Philippines. Medicine (Balt) 1991; 70:179-187.
29. Kupke KG, Lee LV, Muller U. Assignment of the X-linked torsion dystonia gene to Xq21 by linkage analysis. Neurology 1990; 40:1438-1442.
30. Kupke KG, Lee LV, Viterbo GH, Arancillo J, Donlon T, Muller U. X-linked recessive torsion dystonia in the Philippines. Am J Med Genet 1990; 36:237-242.
31. Kwiatkowski DJ, Nygaard TG, Schuback DE, et al. Identification of a highly polymorphic microsatellite VNTR within the argininosuccinate synthetase locus: exclusion of the dystonia gene on 9q32-34 as the cause of dopa-responsive dystonia in a large kindred. Am J Hum Genet 1991; 48:121-128.
32. Obeso JA, Gimenez Roldan S. Clinicopathological correlation in symptomatic dystonia. Adv Neurol 1988; 50:113-122.
33. Marsden CD, Obeso JA, Zarranz JJ, Lang AE. The anatomical basis of symptomatic hemidystonia. Brain 1985; 108:463-483.
34. Narbona J, Obeso JA, Tunon T, Martinez Lage JM, Marsden CD. Hemidystonia secondary to localized basal ganglia tumour. J Neurol Neurosurg Psychiatry 1984; 47:704-709.
35. Jankovic J, Patel SC. Blepharospasm associated with brainstem lesions. Neurology 1983; 33:1237-1240.
36. Zweig RM, Hedreen JC, Jankel WR, Casanova MF, Whitehouse PJ, Price DL. Pathology in brainstem regions of individuals with primary dystonia. Neurology 1988; 38:702-706.
37. Hedreen JC, Zweig RM, DeLong MR, Whitehouse PJ, Price DL. Primary dystonias: a review of the pathology and suggestions for new directions of study. Adv Neurol 1988; 50:123-132.
38. Jankovic J, Svendsen CN, Bird ED. Brain neurotransmitters in dystonia. N Engl J Med 1987; 316:278-279.
39. de Yebenes JG, Brin MF, Mena MA, et al. Neurochemical findings in neuroacanthocytosis. Mov Disord 1988; 3:300-312.
40. Tempel LW, Perlmutter JS. Abnormal vibration-induced cerebral blood flow responses in idiopathic dystonia. Brain 1990; 113:691-707.
41. Perlmutter JS, Raichle ME. Pure hemidystonia with basal ganglion abnormalities on positron emission tomography. Ann Neurol 1984; 15:228-233.
42. Playford ED, Rassingham RE, Marsden CD, Brooks DJ. Abnormal activation of striatum and dorsolateral prefontal cortex in dystonia. Neurology 1992; 42 (Suppl 3):377.(Abstract)
43. Takahashi H, Snow B, Waters C, et al. Evidence for nigrostriatal lesions in Lubag (X-linked dystonia-parkinsonism in the Philippines). Neurology 1992; 42 (Suppl 3):441.(Abstract)
44. Nygaard TG, Trugman JM, de Yebenes JG, Fahn S. Dopa-responsive dystonia: the spectrum of clinical manifestations in a large North American family. Neurology 1990; 40:66-69.
45. Fletcher NA, Holt IJ, Harding AE, Nygaard TG, Mallet J, Marsden CD. Tyrosine hydroxylase and levodopa responsive dystonia. J Neurol Neurosurg Psychiatry 1989; 52:112-114.
46. Burke RE, Fahn S, Marsden CD. Torsion dystonia: a double-blind, prospective trial of high-dosage trihexyphenidyl. Neurology 1986; 36:160-164.
47. Greene P, Shale H, Fahn S. Analysis of open-label trials in torsion dystonia using high dosages of anticholinergics and other drugs. Mov Disord 1988; 3:46-60.
48. Bressman SB, Greene PE. Treatment of hyperkinetic movement disorders. Neurol Clin 1990; 8:51-75.
49. Digre K, Corbett JJ. Hemifacial spasm: differential diagnosis, mechanism, and treatment. Adv Neurol 1988; 49:151-176.

50. Digre KB, Corbett JJ, Smoker WR, McKusker S. CT and hemifacial spasm. Neurology 1988; 38:1111-1113.

51. Brin MF, Fahn S, Moskowitz C, et al. Localized injections of botulinum toxin for the treatment of focal dystonia and hemifacial spasm. Mov Disord 1987; 2:237-254.

52. Biglan AW, May M, Bowers RA. Management of facial spasm with Clostridium botulinum toxin, type A (Oculinum). Arch Otolaryngol Head Neck Surg 1988; 114:1407-1412.

53. Carruthers J, Stubbs HA. Botulinum toxin for benign essential blepharospasm, hemifacial spasm and age-related lower eyelid entropion. Can J Neurol Sci 1987; 14:42-45.

54. Elston JS. Botulinum toxin therapy for involuntary facial movement. Eye 1988; 2:12-15.

55. Kraft SP, Lang AE. Cranial dystonia, blepharospasm and hemifacial spasm: clinical features and treatment, including the use of botulinum toxin. Can Med Assoc J 1988; 139:837-844.

56. Newman NM. Chemical denervation for hemifacial spasm. West J Med 1986; 145:520.

57. Savino PJ, Sergott RC, Bosley TM, Schatz NJ. Hemifacial spasm treated with botulinum A toxin injection. Arch Ophthalmol 1985; 103:1305-1306.

58. Roggenkamper P, Laskawi R, Damenz W, Schroder M, Nubgens Z. Botulinus-toxin-behandlung bei synkinesien nach fazialisparese. HNO 1990; 38:295-297.

59. Iwashige H, Maruo T. [Treatment of blepharospasm and hemifacial spasm with botulinum A toxin (Oculinum)]. Nippon Ganka Gakkai Zasshi 1988; 92:1637-43.

60. Kraft SP, Lang AE. Botulinum toxin injections in the treatment of blepharospasm, hemifacial spasm, and eyelid fasciculations. Can J Neurol Sci 1988; 15:276-280.

61. Saraux H. [Treatment of blepharospasm and facial hemispasm by injection of botulinum toxin]. J Fr Ophthalmol 1988; 11:237-40.

62. Tolosa E, Marti MJ, Kulisevsky J. Botulinum toxin injection therapy for hemifacial spasm. Adv Neurol 1988; 49:479-491.

63. Marti MJ, Kulisevsky J, Tolosa E. [Treatment of dystonic blepharospasm and hemifacial spasm with botulinum toxin: preliminary studies]. Med Clin (Barc) 1987; 89:321-323.

64. Elston JS. Botulinum toxin treatment of hemifacial spasm. J Neurol Neurosurg Psychiatry 1986; 49:825-829.

65. Arthurs B, Flanders M, Codere F, Gauthier S, Dresner S, Stone L. Treatment of blepharospasm with medication, surgery and type A botulinum toxin. Can J Ophthalmol 1987; 22:24-28.

66. Carta A, Stocchi F, Berardelli A, Bramante L, Viselli F. Botulinum toxin therapy for focal dystonias. New Trends in Ophthalmology 1989; 3:68.

67. Erbguth F, Claus D, Neundorfer B. Treatment of facial-cervical dystonia and hemifacial spasm with botulinum-toxin A. New Trends in Ophthalmology 1989; 3:69.

68. Geller BD, Hallett M, Ravits J. Botulinum toxin therapy in hemifacial spasm: Clinical and electrophysiologic studies. Muscle Nerve 1989; 12:716-722.

69. Jankovic J, Schwartz K, Donovan DT. Botulinum toxin treatment of cranial-cervical dystonia, spasmodic dysphonia, other focal dystonias and hemifacial spasm. J Neurol Neurosurg Psychiatry 1990; 53:633-639.

70. Gresty MA, Findley LJ. Definition, analysis and genesis of tremor. In: Findley LJ, Capildeo R, eds. Movement disorders: tremor. New York: Oxford University Press, 1984:15-26.

71. Hubble JP, Busenbark KL, Koller WC. Essential tremor. Clin Neuropharm 1989; 12:453-482.

72. Busenbark KL, Nash J, Nash S, Hubble JP, Koller WC. Is essential tremor benign? Neurology 1991; 41:1982-1983.

73. Hariz MI. Correlation between clinical outcome and size and site of the lesion in computed tomography guided thalamotomy and pallidotomy. Stereotact Funct Neurosurg 1990; 54-55:172-185.

74. Kelly PJ. Applications and methodology for contemporary stereotactic surgery. Neurol Res 1986; 8:2-12.

75. Jankovic J, Schwartz K. Botulinum toxin treatment of tremors. Neurology 1991; 41:1185-1188.

76. Zager EL. Neurosurgical management of spasticity, rigidity, and tremor. Neurol Clin 1987; 5:631-647.

77. Davidoff RA. Skeletal muscle tone and the misunderstood stretch reflex. Neurology 1992; 42:951.

78. Cartlidge NEF, Hudgson P, Weightman D. A comparison of baclogen and diazepam in the treatment of spasticity. J Neurol Sci 1974; 23:17.

79. Feldman RG, Kelly Hayes M, Conomy JP. Baclofen for spasticity in multiple sclerosis; double blind crossover and three year study. Neurology 1978; 28:1094.

80. Sawa GM, Paty DW. The use of baclofen in treatment of spasticity in multiple sclerosis. J Neurol Sci 1979; 6:351.

81. Schmidt RT, Lee RH, Spehlemann R. Comparison of dantrolene sodium and diazepam in the treatment of spasticity. J Neurol Neurosurg Psychiatry 1976; 39:350.

82. Rinne UK. Tizanidine treatment of spasticity in multiple sclerosis and chronic myelopathy. Curr Ther Res 1980; 28:827.

83. Stein R, Nordal HJ, Offedal SI, Slettebo M. The treatment of spasticity in multiple sclerosis; a double blind clinical trial of a new anti spastic drug tizanidine compared with baclofen. Acta Neurol Scand 1987; 75:190.

84. Bass B, Weinshenker B, Rice GP, et al. Tizanidine versus baclofen in the treatment of spasticity in patients with multiple sclerosis. Can J Neurol Sci 1988; 15:15-19.

85. Keenan MAE, Todderud EP, Henderson R, Botte M. Management of intrinsic spasticity in the hand with phenol injection or neurectomy of the motor branch of the ulnar nerve. Journal of Hand Surgery 1987; 12A:734.

86. Garland DE, Lilling M, Keenan MA. Percutaneous phenol blocks to motor points of spastic forearm muscles in head-injured patients. Arch Phys Med Rehabil 1984; 65:243.

87. Nathan PW. Intrathecal phenol to relieve spasticity in paraplegia. Lancet 1959; 2:1099.

88. Kelly RE, Gautier-Smith PC. Intrathecal phenol in the treatment of reflex spasms and spasticity. Lancet 1959; 2:1102.

89. Penn RD, Kroin JS. Continuous intrathecal baclofen for severe spasticity. Lancet 1985; 2:125.

90. Penn RD, Kroin JS. Long-term intrathecal baclofen infusion for treatment of spasticity. J Neurosurg 1987; 66:181.

91. Penn RD, Savoy SM, Corcos DM. Intrathecal baclofen for severe spinal spasticity. N Engl J Med 1989; 320:1517.

92. Snow BJ, Tsui JK, Bhatt MH, Varelas M, Hashimoto SA, Calne DB. Treatment of spasticity with botulinum toxin: A double-blind study. Ann Neurol 1990; 28:512-515.

93. Das TK, Park DM. Botulinum toxin in treating spasticity. Br J Clin Pract 1989; 43:401-403.

94. Das TK, Park DM. Effect of treatment with botulinum toxin on spasticity. Postgrad Med J 1989; 65:208-210.

95. Borg-Stein J, Pine ZM, Miller J, Brin MF. Botulinum toxin for treatment of spasticity in multiple sclerosis. Association of Academic Physiatrist, Orlando, Florida, 1992; (Abstract)

96. Wiegand H, Erdmann G, Wellhoener HH. 125I labelled botulinum A neurotoxin: pharmacokinetics in cats after intramuscular injection. Naunyn Schmiedebergs Arch Pharmacol 1976; 292:161-165.

97. Eldridge M. A History of the Treatment of Speech Disorders. London:Livingston, 1968:

98. Schuell H. Sex differences in relation to stuttering: part 1. J Speech Hear Disord 1946; 11:297-298.

99. Kopp GA. Metabolic studies of stutterers: I. Biochemical study of blood composition. Speech Monogr 1934; 1:117-132.

100. Orton S, Travis LE. Studies in stuttering: IV. Studies of action currents in stutterers. Arch Neurol 1929; 21:61-68.

101. Karlin IW, Strazzulla M. Speech and Language problems of mentally deficient children. J Speech Hear Disord 1952; 17:286-294.

102. Ingham R. Stuttering and Behavior Therapy: Current Status and Experimental Foundations. San Diego:College Hill Press, 1984:

103. Webster RL, Lubker BB. Interrelationships among fluency producing variables in stuttered speech. J Speech Hear Res 1968; 11:754-66.

104. Webster RL, Lubker BB. Masking of auditory feedback in stutterers' speech. J Speech Hear Res 1968; 11:221-3.

105. Freeman FJ, Ushijima T. Laryngeal muscle activity during stuttering. J Speech Hear Res 1978; 21:538-562.

106. Nudelman HB, Herbrick KE, Hoyt BD, Rosenfield DB. Phonatory response times of stutterers and fluent speakers to frequency-modulated tones. J Acoust Soc Am 1992; (In Press)

107. Ludlow CL, Naunton RF, Fujita M, Sedory SE. Spasmodic dysphonia: botulinum toxin injection after recurrent nerve surgery. Otolaryngol Head Neck Surg 1990; 102:122-131.

108. Ludlow CL, Naunton RF, Sedory SE, Schulz GM, Hallett M. Effects of botulinum toxin injections on speech in adductor spasmodic dysphonia. Neurology 1988; 38:1220-1225.

109. Zwirner P, Murry T, Swenson M, Woodson G. Acoustic changes in spasmodic dysphonia after botulinum toxin injection. J Voice 1991; 5:78-84.

110. Pierard GE, Lapiere CM. The microanatomical basis of facial frown lines. Arch Dermatol 1991; 125:1090-1092.

111. Ellenbogen R, Wethe J, Jankauskas S, Collini F. Curette fat sculpture in rhytidectomy: improving the nasolabial and labiomandibular folds. Plast Reconstr Surg 1991; 88:433-442.

112. Ezrokhin VM. Removal of wrinkles and redundant skin in the region of the forehead. Acta Chir Plast 1991; 33:1-7.

113. Horibe EK, Horibe K, Yamaguchi CT. Pronounced nasolabial fold: A surgical correction. Aesth Plast Surg 1989; 13:99-103.

114. Rafaty FM, Cochran J. A technique of nasalabioplasty. Laryngoscope 1978; 88:95-99.
115. Riefkohl R. The nasolabial fold lift. Ann Plast Surg 1985; 15:1-6.
116. Webster R, Kattner MD, Smith RC. Injectable collagen for augmentation of facial areas. Arch Otolaryngol 1984; 110:652-656.
117. Kaplan EN, Flaces E, Tolleth H. Clinical utilization of injectable collagen. Ann Plast Surg 1983; 10:437.
118. Matti BA, Nicolle FV. Clinical use of zyplast in correction of age and disease related contour deficiencies of the face. Aesth Plast Surg 1990; 14:227-234.
119. Milliken C, Rosen T, Monheit G. Treatment of depressed cutaneous scars with gelatin matrix implant: A multicenter study. J Am Acad Dermatol 1989; 16:1155-1162.
120. McKinney P, Cook JQ. Liposuction and the treatment of nasolabial folds. Aesth Plast Surg 1989; 13:167-171.
121. Pollack SV. Silicone, fibrel and collagen implantation for facial lines and wrinkles. J Dermatol Surg Oncol 1990; 16:957-961.
122. Blitzer A, Brin MF, Keen MS. Botulinum toxin for the treatment of hyperfunctional lines of the face. Arch Otolaryngol Head Neck Surg 1992; (In Press)
123. Clark RP, Berris CE. Botulinum toxin: a treatment for facial asymmetry caused by facial nerve paralysis. Plast Reconstr Surg 1989; 84:353-355.
124. Carruthers JD, Carruthers JA. Treatment of glabellar frown lines with botulinum A exotoxin. J Dermatol Surg Oncol 1992; 18:17-21.
125. Dykstra DD, Sidi AA, Scott AB, Pagel JM, Goldish GD. Effects of botulinum A toxin on detrusor-sphincter dyssynergia in spinal cord injury patients. J Urol 1988; 139:919-922.
126. Dykstra DD, Sidi AA. Treatment of detrusor-sphincter dyssynergia with botulinum A toxin: a double-blind study. Arch Phys Med Rehabil 1990; 71:24-26.
127. Hallan RI, Williams NS, Melling J, Waldron DJ, Womack NR, Morrison JF. Treatment of anismus in intractable constipation with botulinum A toxin. Lancet 1988; 2:714-717.
128. Ludlow CL, Hallett M, Sedory SE, Fujita M, Naunton RF. The pathophysiology of spasmodic dysphonia and its modification by botulinum toxin. In: Berardelli A, Benecke R, Manfredi M, Marsden CM, eds. Motor Disturbances II. New York: Academic Press, 1990:273-288.
129. Brin MF, Fahn S, Moskowitz CB, et al. Injections of botulinum toxin for the treatment of focal dystonia. Neurology 1986; 36 (Suppl 1):120.(Abstract)
130. Blitzer A, Brin MF, Fahn S, Lange D, Lovelace RE. Botulinum toxin (BOTOX) for the treatment of "spastic dysphonia" as part of a trial of toxin injections for the treatment of other cranial dystonias [letter]. Laryngoscope 1986; 96:1300-1301.
131. Lange DJ, Warner C, Brin MF, List T, Fahn S, Lovelace R. Botulinum toxin therapy: Distant effects on neuromuscular transmission [abstract]. Muscle Nerve 1985; 8:624.
132. Sanders DB, Massey EW, Buckley EC. EMG monitoring of botulinum toxin in blepharospasm. Neurology 1985; 35 (Suppl 1):272.(Abstract)

IMPROVED OUTCOME AFTER REPEATED INJECTIONS OF BOTULINUM TOXIN FOR TREATMENT OF SPASMODIC TORTICOLLIS AND AXIAL DYSTONIA: EXPERIENCE WITH JAPANESE BOTULINUM TOXIN A

Ryuji Kaji[1] Takahiro Mezaki[1], Nobuo Kohara[1],
Takashi Nagamine[2], Tomoko Shimizu[3], and Jun Kimura[1]

[1]Department of Neurology
[2]Department of Brain Pathophysiology
 Kyoto University Hospital, Kyoto
[3]Chiba Serum Laboratory , Chiba Japan

INTRODUCTION

Although the efficacy of botulinum toxin in the treatment of cervical dystonia has been demonstrated[1-3], significant number of patients fail to respond to this therapy. These "non-responders" show little clinical improvement from the beginning (primary failure), or become resistant to the therapy after repeated injections (secondary failure)[3]. The latter is ascribed to development of antibodies against the toxin, while factors underlying the primary failure are yet to be determined.

Through clinical trials with a new crystallized type A-toxin product (Oculin A®, Chiba Serum Laboratory) in patients with spasmodic torticollis or axial dystonia for over 28 months, we found that repeated injections of this toxin virtually eliminated non-responders without eliciting antibodies or tachyphylaxis.

SUBJECTS AND METHODS

Twenty patients with spasmodic torticollis and 2 with axial dystonia, aged 18-62, 15 men and 7 women, were enrolled. Duration of illness ranged from 6 months to 18 years. Two patients with spasmodic torticollis and a patient with axial dystonia were diagnosed as secondary tardive dystonia, the rest being idiopathic. We excluded those patients with phasic neck movement from analysis, in whom objective video-rating was not feasible, but anti-toxin antibody was screened in 36 patients with spasmodic torticollis who had similar treatment.

Patients were injected with 300 units of botulinum toxin A (Oculin A®, Chiba Serum Laboratory) at an interval of 1 month, divided into a fraction of 100-150 unit, each injected in 2-3 muscles with increased activity evidenced by surface EMG or inspection and palpation. After the clinical plateau, at which additional injection gave no more improvement, reinjection was made by the patient's request with intervals ranging 2 - 9 months. The total follow-up periods were 9-28 months.

Clinical symptoms were rated before the initial injection and two weeks after every injection, both by subjectively and objectively. Subjective score ranged from 0 to 100, with 0 being no improvement and 100 being a complete cure. Objective score (Table) was obtained with the aid of videotaping rated by an examiner, blinded as to the injection schedule.

Anti-toxin antibody titer was measured using mouse-bioassay.

Table. Objective Score: total sum of the scores listed below. Degrees are measured from the neutral position of the head.

	degrees	score		degrees	score
Rotation of the head	0	0	Lateral Tilt	0	0
	0-15	1		0-15	1
	15-30	2		15-30	2
	30-45	3		30-	3
	45-	4	Shoulder Elevation		
				0	0
				0-7	1
Antero-Posterior Flexion	0	0		7-15	2
	0-15	1		15-	3
	15-30	2			
	30-45	3			
	45-	4	Scoliosis	0	0
				0-15	1
				15-30	2
				30-	3

RESULTS

In 10 out of 22 patients, the most favorable outcome with regard to the objective score was attained after 3 injections. None had the best response after single injection. As a whole, the objective score did not differ significantly after single injection (p=0.07; one-way ANOVA), but did significantly after 3 injections (p=0.001) (Fig.1).

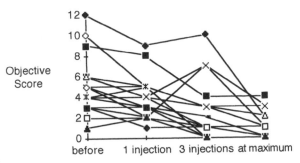

Fig.1. Changes in objective score in patients before injection, after single injection, after 3 injections and at the maximum effect reached after 3 or more injections.

All patients showed improvement in the objective scale after repeated injections, which ranged from 1 to 10 (mean 3.83). At the peak of this objective improvement, 17 out of 22 patients had subjective score more than 50 points, and 9 patients marked more than 80 points. Representative cases are described as below.

Case 1

A 32-year-old man with 8-year history of psychosis and major tranquilizer administration developed spasmodic toricollis 3 years ago. Over the following 2 years, the dystonia extended to the trunk. He had been bed-ridden for over a year. On examination, he had prominent laterocollis and scoliosis to the right, which made him unable to sit up without support. He had injections of 300 units of the toxin every month into 2 or 3 muscles chosen from right sternocleidomastoid, trapezius, splenius capitis and paraspinals. Initial 7 injections did not result in any substantial improvement in subjective as well as objective scores, although his neck pain was relieved. After 8-11 injections, the head and trunk position greatly improved so that he was able to sit up without support, and he eventually became ambulatory after 12 injections. The cumulative dose is 4500 unit. No anti-toxin antibody was detected.

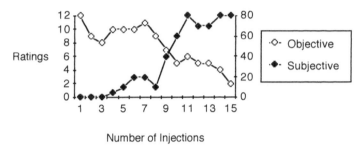

Fig.2. Ratings in case 1, evaluated after each number of injection shown below.

Case 2

A 49-year-old woman had spasmodic torticollis for 2 years. Before injection, she had neck rotation towards left, initially mediated by increased activity in right sternocleido-mastoid. Left trapezius was then silent. After single injection, left trapezius was activated in such a way as to maintain the original rotation. The patient recovered normal head position only after the second injection, which included left trapezius in addition to left splenius capitis and right sternocleidomastoid.

In 15 patients in this series, one muscle was activated after another to resume the original abnormal head posture as in case 2. We termed this as "scratch-where-it-itches" phenomenon. After injecting into all the muscles involved in maintaining a specific posture, the patient recovered normal head position.

Anti-toxin antibodies

Anti-toxin antibody was detected in no patients.

DISCUSSION

All the patients treated in this study showed objective as well as subjective improvement. Previous studies[1-3] demostrated beneficial effect of the toxin injection, but not all the patients were associated with objective improvements. Some of our patients showed a substantial improvement only after repeated injections (e.g. case 1). This raises a possibility that a few number of injections are not sufficient to overcome the tendency of another uninjected muscle to be activated to maintain the original head position[2] (scratch-where-it-itches phenomenon), best illustrated in case 2. An important implication of this observation is that clinical picture of dystonia is preprogrammed by abnormal proprioceptive body images, rather than uncontrolled activities set in each muscle[2]. This phenomenon and insufficient delivery to each muscle may result in primary failure, which may be overcome by repeated injections over a short period (booster injections)[3].

Major concern of repeated injection is development of antibodies. The present study did not show any antibodies to the toxin nor failure of response in the course of the treatment (secondary failure). Duration and dose of the treatment are comparable to those in the previous studies[1-3]. Although there still remains a chance that some of the patients eventually develop antibodies, it is possible that this newly developed toxin has less immunogenicity than the other toxins.

When treating patients with dystonia involving large muscles, failure of treatment may often result from insufficient delivery to the target. Since the total amount of toxin injected in a single session is limited up to 300-600 unit by the possible toxicity, repeated injections are superior to single injection, if using a toxin which has much less immunogenicity than those used previously.

CONCLUSION

In order to obtain the maximum efficacy, we injected botulinum toxin repeatedly at an interval of 1 month for treating 22 patients with spasmodic torticollis or axial dystonia. Single injection consisted of 300 unit of the toxin, divided into 100-150 unit per each muscle. The effect was rated both subjectively and objectively with the aid of videotaping. Single injection was never sufficient, and 3 injections were most frequently required to attain the maximum effect of the treatment. After injecting the toxin in a muscle with increased activity, another previously "silent" muscle was activated as if it reproduced the original abnormal posture ("scratch-where-it-itches phenomenon"). Thus, abnormal proprioceptive body image, rather than abnormal activity of individual muscle, is preprogrammed in these dystonias. Since these patients did not develop an antibody to the toxin thus far, repeated injections were superior to single injection at this dosage.

REFERENCES

1. Jankovic J, Schwartz K.Botulinum toxin injections for cervical dystonia. Neurology 1990;41:277-80.
2. Gelb DJ, Yoshimura DM, Olney RK Lowenstein DH, Aminoff MJ. Change in pattern of muscle activity following botulinum toxin injections for torticollis. Ann Neurol 1991;29:370-6.
3. Green P, Kang U, Fahn S, Brin M, Moskowitz C, Flaster E. Double-blind, placebo-controlled trial of botulinum toxin injections for the treatment of spasmodic torticollis. Neurology 1990;40:1213-8.

THE USE OF BOTULINUM TOXIN
IN THE MANAGEMENT OF CEREBRAL PALSY
IN PEDIATRIC PATIENTS

L. Andrew Koman, James F. Mooney, III, and Beth P. Smith

Department of Orthopaedic Surgery
The Bowman Gray School of Medicine
Wake Forest University
Winston-Salem, North Carolina 27157-1070

INTRODUCTION

The intramuscular use of botulinum toxin (Botox) for the purpose of neuromuscular blockade is accepted therapy for certain ocular and facial muscular imbalances, and its mechanism of action has been well described.[1-15] The therapeutic use of Botox in larger skeletal muscle groups is not as well established. Direct intramuscular injections of Botox into the sternocleidomastoid, trapezius, and splenii muscles have resulted in clinical improvement in spasmodic torticollis;[16,17] preliminary studies using Botox to treat focal foot and hand dystonias, upper limb spasticity in stroke patients, and spastic adductor muscles in patients with multiple sclerosis have been reported.[4,5,18] Until recently, however, there have been no reports of Botox being used to manage the spasticity of cerebral palsy. Before discussing the use of Botox in this population, it is important to present a general overview of the pathophysiology of cerebral palsy and to point out the large number of patients with this disorder.

Cerebral palsy was originally described in 1843 by Sir William Little as a nonprogressive disorder of motion and posture caused by a brain insult occurring in the early period of brain growth.[19] This growth period was considered to include the first two years of life. Historically, the majority of patients with cerebral palsy have developed central nervous system (CNS) problems resulting from cerebral anoxia and kernicterus secondary to maternal/fetal Rh incompatibility. Although the current lower incidence of Rh abnormalities and improvements in prenatal care have led to increased survival of premature and low-birth-weight neonates, brain injury and anoxia in these infants are not considered by some to be the leading causes of new cases of cerebral palsy. As such, the most common cause of new cases is still a matter of great discussion. Currently, cerebral palsy is an imprecise term attached to a large number of patients who have varying CNS injuries but tend to exhibit somewhat similar spastic symptoms.[19] There are approximately 500,000 to 700,000 patients with cerebral palsy in the United States of America, and approximately 5,000 to 7,000 new cases appear each year.[20]

Botulinum and Tetanus Neurotoxins, Edited by
B.R. DasGupta, Plenum Press, New York, 1993

In patients with cerebral palsy, all extremity muscle groups will exhibit some level of symptomatic spasticity. This spasticity, with or without rhythmic motion disorders such as athetosis, is the primary dynamic deformity that interferes with motion or function and thus results in problems in patient positioning, hygiene, functional upper extremity use, and ambulation. Spastic muscle groups undergo exaggerated and excessive contractions due to an abnormal stretch reflex, which is combined with a central lesion within the pyramidal or extrapyramidal systems.[19,21] In time, these sustained repetitive and abnormal contractions increase the resting length of the antagonist muscle groups, and those groups become weakened or nonfunctional. The results of such uncorrected muscular imbalances are fixed contractures (i.e., deformity that with maximal passive force cannot be corrected to neutral) as well as bony deformities. Dynamic contractures are, by definition, those deformities that can be corrected to normal position with maximal, or less than maximal, passive force.

Historically, multiple methods and interventions have been recommended and used with varying rates of success to reduce the spasticity and contractures seen in cerebral palsy patients. These include various forms of physical therapy; neuropharmacologic therapy, neurosurgical procedures, including local neurectomies and recently selective posterior rhizotomies; orthotics; and orthopaedic surgical procedures.[19, 21-25] Recent reports question the long-term benefit of physical therapy on spasticity and extremity function.[26,27] Oral neuropharmacologic treatment with diazepam and baclofen has proved unsuccessful and carries significant risks as well as problems with oversedation. Local injections of intramuscular blocking agents such as 45% alcohol or phenol are useful, but are painful and require general anesthesia during their administration.[28-30] Use of these agents also may result in long-term muscle weakness or denervation.

Neurectomies, previously highly regarded, are no longer recommended in patients with cerebral palsy, because of the increased risk of contracture secondary to the denervation. The most common example of this complication has been seen in patients who have undergone obturator neurectomy. A significant number of these patients have developed some degree of abduction contracture due to complete adductor denervation. Recently, reports of selective posterior rhizotomy with adequate followup have appeared.[24] This procedure seems to benefit a carefully selected group of younger cerebral palsy patients, who, for the most part, have mild contractures. However, the procedure is not widely available, has strict preoperative criteria, and is not recommended in profoundly involved children or in patients with athetosis.[24]

Orthotic devices such as braces and shoe inserts have been the mainstays in the day-to-day care of patients with cerebral palsy. Orthoses improve the energy efficiency of gait, are of great benefit in controlling and maintaining joint alignment, and are capable of maintaining correction of some mild dynamic deformities.[23] However, extremity deformities in some patients may be too severe or rigid to allow comfortable or effective bracing.

Orthopaedic surgical intervention in spasticity is the treatment of choice if the deformity is progressive despite the use of all nonoperative techniques. The most common surgical procedures involve lengthening the tendons of the contracted muscle groups. However, this procedure does not lengthen the muscle fibers and thereby does not increase musculotendinous excursion.[25,31] The primary value of tendon lengthening is to allow more functional use of the extremity and to facilitate bracing, which will improve positioning of all patients as well as gait pattern in patients who are ambulatory. Despite usually predictable and successful results, one of the goals of the orthopaedic management of cerebral palsy patients is to postpone any needed surgery until the patient is close to maturity.[19] In addition, it is advisable to complete all necessary procedures during one single course of anesthesia, so as to allow the patient to recuperate from all procedures at once and to expose the patient to the inherent surgical and anesthetic risks as infrequently as possible. These risks include the potential for respiratory and anesthetic complications as well as postoperative infections. Timing of surgery is critical because, in some patients—particularly those requiring surgical intervention early in childhood—an operation may be required after each growth spurt. Due to these risks, orthopaedic surgical management of patients with cerebral palsy is indicated only after all other options have been exhausted.

The previous use of Botox in other anatomic locations for the management of conditions involving involuntary movement disorders makes it seem an ideal candidate for use in the management and control of spasticity in pediatric patients with cerebral palsy. The senior author (LAK) undertook the first clinical trials of Botox in this patient population in the United States beginning in 1987. Initial study goals were to determine any potential efficacy in this specific patient population, any possible side effects, the maximal duration of action, and the optimal dosage. For the patients enrolled in the original study, as well as for those in all subsequent studies, surgery or intramuscular blocks requiring general anesthesia were the only remaining realistic treatment options.

PRELIMINARY STUDY METHODS

From March 1, 1988, through May 31, 1990, 39 cerebral palsy patients with dynamic deformities secondary to spastic muscles were entered into the study and assigned to one of four groups according to their specific type of muscular imbalance. Indications for patient participation included dynamic deformities that had been unresponsive to other treatments, with surgery being the only other realistic alternative.

All three patients placed in Group I had severe spastic or mixed quadriparesis. All were nonambulatory and severely debilitated, and all had painful paraspinal spasticity that interfered with sitting balance. In all patients trial injections of lidocaine in the paraspinal muscles had been found to relieve pain and to aid in positioning.

The eight Group II patients were children with lower extremity spasticity that interfered with positioning and hygiene. Their diagnoses included spastic diplegia, spastic quadriparesis with athetosis, and spastic quadriparesis without athetosis. These patients were given Botox injections in the muscles of the lower extremities to decrease spasticity, increase range of motion, improve positioning, and possibly inhibit further hip subluxation.

The sixteen Group III patients were ambulatory children with cerebral palsy who were injected with Botox to improve their gait. All had progressive dynamic equinovarus or equinovalgus foot and ankle deformities that had been unresponsive to bracing or other modalities. In the ten patients with equinovalgus deformities, doses of Botox were divided between the medial gastrocnemius and the lateral gastrocnemius muscles. In the six patients with equinovarus deformities, Botox was injected into the posterior tibialis muscle as well as into the medial gastrocnemius and lateral gastrocnemius muscles.

Group IV consisted of 12 ambulatory patients with cerebral palsy, ages 11 to 14 years, who had been included in a randomized, double-blind, controlled preliminary evaluation. Their enrollment requirements included: 1) no significant health problems other than the spasticity; and 2) equinovarus or equinovalgus foot deformities associated with dynamic joint contractures that had not responded to standard physical therapy, orthotics, or nonoperative management, including the use of neuropharmacologic agents. Four of the 12 patients were hemiplegic; the remaining eight were diplegic. Six of the patients were randomized to a placebo group, and six received Botox. The goal of this portion of the trial was to determine the effects of Botox injections on the lower extremity during a four- to six-week treatment period.

Evaluation Methods

All patients were assessed by physical therapy evaluations, gait analysis (of those who were ambulatory), a physician rating scale developed by the senior author (Table 1), and a parental, or caregiver, evaluation of response. The physical therapy evaluation was developed to assess global function both before and after injection, and was conducted after any gait analyses were done. Video recordings were made of all gait analyses. The physician rating scale and caregiver evaluations were assessed for each patient. The patients were evaluated clinically before injection; at approximately two- to four-week intervals after injection until a significant clinical result had been attained; and then every two to three months thereafter. Second injections were made at two to three weeks after the initial injection if clinical improvement was not apparent.

Table 1. Physician rating scale (PRS) for gait analysis (determined for each limb injected).

Dynamic function (range of motion)	Score
1. Crouch	
Severe (>20° hip, knee, ankle)	0
Moderate (5° to 20° hip, knee, ankle)	1
Mild (<5° hip, knee, ankle)	2
None	3
2. Equinus Foot	
Constant (fixed contracture)	0
Constant (dynamic contracture)	1
Occasional heel contact	2
Heel-to-toe gait	3
3. Hind Foot	
Varus at foot strike	0
Valgus at foot strike	1
Occasionally neutral at foot strike	2
Neutral at foot strike	3
4. Knee	
Recurvatum >5°	0
Recurvatum 0° to 5°	1
Neutral (no recurvatum)	2
5. Speed of Gait	
Only slow	0
Variable (slow-fast)	1
6. Gait	
Toe-to-toe	0
Occasional heel-to-toe	1
Heel-to-toe	2
Total _____	

Safety Considerations

Possible side effects, complications, and effects of increasing dosage were monitored closely throughout the study. According to a rigid protocol, the patient's caregivers were called each day during the first two weeks after each injection and questioned about possible problems. The caregivers also were instructed to call the clinical nurse coordinator if any undesirable side effects were noted.

RESULTS

In all, 99 toxin injections were given to the 39 children. Target threshold was reached at doses of approximately one to four units of toxin per kilogram of body weight for each major muscle group injected.

The effect of an injection was not evident immediately in most cases, but the patients or caregivers noted significantly decreased spasticity within the first 12 to 72 hours after injection. The positive response lasted from three to six months and was repeatable. Interestingly, the duration of action of the toxin appeared to be longest in the patients with an element of athetosis and in the larger, more active patients, possibly because those patients were able to cooperate with physical therapists to strengthen their antagonist muscles during the course of treatment.

Side Effects and Complications

The injections were described as causing a sensation of "cold" as the toxin was instilled. The most common side effects were soreness at the injection site, which lasted one to two days, and a feeling of generalized fatigue in a small number of patients. There were no generalized or clinically significant systemic complications in this series.

Efficacy and Clinical Assessment of Effect

In the Group I patients with paraspinal muscle spasticity, caregivers reported that painful paraspinal spasticity was reduced in all patients. The children rested more comfortably and could be positioned more easily following the Botox injections.

Group II patients (non-ambulatory lower extremity spasticity) exhibited decreased spasticity in the larger muscle groups and improved positioning. Injections in one patient reduced spasticity to the extent that bracing, which previously had not been tolerated, was possible. Hip subluxation was noted radiographically to be reduced in two patients. Botox injections in the hip adductors of two patients resulted in increased range of motion, allowing for more comfortable positioning. Scissoring and crouching with attempted standing were reduced following injections in the hamstrings of one patient. Knee extension was also improved in that patient.

In Group III, 15 of the 16 patients demonstrated changes in the tone of injected muscles without global weakness or systemic toxicity. The single patient exhibiting minimal clinical improvement was found to have a fixed contracture, the components of which had not been evident until unmasked by the Botox injections. The physician's rating scale values were significantly improved post-injection, with a p value of less than 0.05. Finally, in the randomized double-blind protocol group (Group IV), five of the six patients in the toxin group (83%) showed an improvement in gait pattern and on the physician's rating scale, whereas only two of the six patients in the placebo group (33%) showed an improvement.

CONCLUSION

Information generated in these preliminary studies suggests the efficacy of Botox as an alternative treatment modality in cerebral palsy patients with dynamic contractures, although not in such patients with fixed contractures. The effective dose used appeared safe in all patients, and the side effects or complications of such injections were relatively minor and short lived. After the target threshold was achieved, most of the patients exhibited marked relief of symptoms and significant improvement in function for an average of three to six months. Treatment with Botox allowed noticeable improvements in patient care, positioning, and function without the risks of surgery or general anesthesia. Because of the apparent positive results in the preliminary study, a large multi-center, double-blind clinical trial employing more strict controls and conforming to a rigorous design is being initiated for a more objective assessment of the results of Botox injections in a pediatric cerebral palsy population. The larger study will include formal computerized gait analysis and electromyographic evaluation in addition to the evaluation scales and assessments used in this preliminary study.

Overall, we believe that our preliminary work provides compelling evidence that Botox (1) is a potentially valuable adjunctive treatment in the management of dynamic deformity in pediatric patients with cerebral palsy, and (2) is a nonsurgical mechanism with which to manage dynamic extremity and truncal deformity in the absence of fixed contractures. We found the Botox injections to be safe and without significant complications in this pediatric population. Clinical improvement was noted in patients receiving these injections at a dose of one to four units per kilogram body weight for each muscle group; the duration of improvement appeared to be relatively constant for three to six months and was repeatable with a second injection. The best results were in those patients able to strengthen antagonistic muscles through aggressive physical therapy after the injections. It is clear that the benefits and advantages of Botox injections in spastic muscles in children with cerebral palsy merit the prospective and objective evaluations that are being undertaken at this time.

REFERENCES

1. Alpar AJ. Botulinum toxin and its uses in the treatment of ocular disorders. Am J Optom Physiol Optics 1987; 64:79.
2. Arthurs B, Flanders M, Codere F, Gauthier S, Dresner S, Stone L. Treatment of blepharospasm with medication, surgery and Type A botulinum toxin. Can J Ophthalmol 1987; 22:24.
3. Carruthers J, Stubbs HA. Botulinum toxin for benign essential blepharospasm, hemifacial spasm and age-related lower eyelid entropion. Can J Neurol Sci 1987; 14:42.
4. Cohen LG, Hallett M, Geller BD, Hochberg F. Treatment of focal dystonias of the hand with botulinum toxin injections. J Neurol Neurosurg Psychiatry 1989; 52:355.
5. Das TK, Park DM. Botulinum toxin in treating spasticity. Br J Clin Pract 1989; 43:401.
6. Dutton JJ, Buckley EG. Botulinum toxin in the management of blepharospasm. Arch Neurol 1986; 43:380.
7. Elston JS. Botulinum toxin therapy for involuntary facial movement. Eye 1988; 2:12.
8. Kraft SP, Lang AE. Cranial dystonia, blepharospasm and hemifacial spasm: Clinical features and treatment, including the use of botulinum toxin. Can Med Assoc J 1988; 139:837.
9. Ludlow CL, Naunton RF, Sedory SE, Schulz GM, Hallett M. Effects of botulinum toxin injections on speech in adductor spasmodic dysphonia. Neurology 1988; 38:1220.
10. Miller RH, Woodson GE, Jankovic J. Botulinum toxin injection of the vocal fold for spasmodic dysphonia. A preliminary report. Arch Otolaryngol Head Neck Surg 1987; 113:603.
11. Tsui JKC, Eisen A, Calne DB. Botulinum toxin in spasmodic torticollis. Adv Neurol 1988; 50:593.
12. Tsui JK, Eisen A, Mak E, Carruthers J, Scott A, Calne DB. A pilot study on the use of botulinum toxin in spasmodic torticollis. Can J Neurol Sci 1985; 12:314.
13. Tsui JKC, Fross RD, Calne S, Calne DB. Local treatment of spasmodic torticollis with botulinum toxin. Can J Neurol Sci 1987; 14:533.
14. Tsui JKC, Eisen A, Stoessl AJ, Calne S, Calne DB. Double-blind study of botulinum toxin in spasmodic torticollis. Lancet 1986; 2:245.
15. Brin MF, Fahn S, Moskowitz C, Friedman A, Shale HM, Greene PE, Blitzer A, List T, Lange D, Lovelace, RE, McMahon D. Localized injections of botulinum toxin for the treatment of focal dystonia and hemifacial spasm. Mov Disord 1987; 2:237.
16. Gelb DJ, Lowenstein DH, Aminoff MJ. Controlled trial of botulinum toxin injections in the treatment of spasmodic torticollis. Neurology 1989; 39:80.
17. Stell R, Thompson PD, Marsden DC. Botulinum toxin in spasmodic torticollis. J Neurol Neurosurg Psychiatry 1988; 51:920.
18. Tsui JKC, Snow B, Bhatt M, Varelas M, Hashimoto S, Calne DB. New applications of botulinum toxin in the lower limbs (Abstract). Neurology 1990; 40(Suppl 1):382.
19. Bleck EE. Orthopaedic Management in Cerebral Palsy. Philadelphia: J.B. Lippincott (1987).

20. Cerebral Palsy—Facts and Figures (Fact Sheet). United Cerebral Palsy Association, Inc., New York, August 9 (1986).
21. Barabas G, Taft LT. The early signs and differential diagnosis of cerebral palsy. Pediatr Ann 1986; 15:203.
22. Albright AL, Cervi A, Singletary J. Intrathecal baclofen for spasticity in cerebral palsy, JAMA 1991; 265:1418.
23. Diamond M. Rehabilitation strategies for the child with cerebral palsy. Pediatr Ann 1986; 15:230.
24. Peacock WJ, Arens LJ, Berman B. Cerebral palsy spasticity. Selective posterior rhizotomy. Pediatr Neurosci 1987; 13:61.
25. Silver RL, de la Garza J, Rang M. The myth of muscle balance: A study of relative strengths and excursions of normal muscles about the foot and ankle. J Bone Joint Surg 1985; 67-B:432.
26. Piper MC, Kunos VI, Willis DM, Mazer BL, Ramsay M, Silver KM. Early physical therapy effects on the high-risk infant: A randomized controlled trial. Pediatrics 1986; 78:216.
27. Wright T, Nicholson J. Physiotherapy for the spastic child: An evaluation. Dev Med Child Neurol 1973; 15:146.
28. Carpenter EB. Role of nerve blocks in the foot and ankle in cerebral palsy: Therapeutic and diagnostic. Foot Ankle 1983; 4:164.
29. Carpenter EB, Seitz DG. Intramuscular alcohol as an aid in management of spastic cerebral palsy. Dev Med Child Neurol 1980; 22:497.
30. Tardieu G, Tardieu C, Hariga J, Gagnard L. Treatment of spasticity by injection of dilute alcohol at the motor point or by epidural route. Clinical extension of an experiment on the decerebrate cat. Dev Med Child Neurol 1968; 10:555.
31. Silver RL, Rang M, Chan J, de la Garza J. Adductor release of nonambulant children with cerebral palsy. J Pediatr Orthop 1985; 5:672.

USE OF BOTULINUM A TOXIN
IN THE URETHRAL SPHINCTER

John B. Nanninga

Department of Urology
Northwestern University Medical School
Chicago, IL 60611

INTRODUCTION

The initial use of botulinum A toxin in the extraocular eye muscles to correct strabismus demonstrated effectiveness in reducing or eliminating unwanted muscle activity.[1] In recent years, the toxin has been used to correct muscle hyperactivity in blepharospasm, hemifacial spasm, torticollis, leg muscle spasticity associated with cerebral palsy or multiple sclerosis, and sphincter muscle (anal or urethral) hyperactivity.[2,3]

Specifically, the reason for blocking the urethral sphincter is to reduce the muscle activity which is causing a relative obstruction to the flow of urine during urination (sometimes called "sphincter dyssynergy". In this study patients who were treated suffered from spinal cord injury, multiple sclerosis, and idiopathic urinary retention. The diagnosis of urethral sphincter dyssynergy was based on sphincter EMG performed during attempted urination and/or x-ray study during attempted urination.

METHODS

Treatment consisted of injection of the sphincter through an endoscope (cystoscope) with a 22g injection needle (male) or insertion of the needle alongside the urethra to a depth of 12-15 mm (female). The sphincter muscle EMG activity was monitored via a needle (monopolar). The dose used initially was 1 unit per kg in 5 ml normal saline (no preservative); more recently 100 units have been injected in four quadrants of the sphincter (1.25 ml per quadrant). EMG activity appears to subside after 15-20 minutes.

RESULTS

Five patients have been treated. In four patients the effect was as expected, providing improved urination and bladder emptying. In one patient incontinence proved more trouble-

Botulinum and Tetanus Neurotoxins, Edited by
B.R. DasGupta, Plenum Press, New York, 1993

some than expected and the patient returned to an internal catheter for drainage. One patient has been injected three times and one patient twice. The effect seems to have lasted about six months, based on the reappearance of sphincter hyperactivity. There were no major complications. Transient urethral bleeding was noted in two of the patients.

The use of botulinum A toxin would seem to be an effective though temporary means of controlling urethral sphincter dyssynergy. Dykstra found the drug to be valuable in this regard and was the first to describe its effectiveness for this condition.[3] The fact that the effect is not permanent is of value in a disease such as multiple sclerosis which fluctuates over time. For patients with permanent disability such as spinal cord injury the frequency of injection is not disabling or particularly troublesome, considering the advantage of being catheter free.

Acknowledgments

My thanks to Smith-Kettlewell Eye Institute for initially supplying the botulinal A toxin and to Dr. Dennis Dykstra for showing me his technique of injection.

REFERENCES

1. Scott A. Botulinum toxin injection of eye muscles to correct strabismus. Tr Am Ophth Soc 1991; LXXIX:736-770.
2. Jankovic J, Brin MF. Therapeutic uses of botulinum toxin. N Engl J Med 1991; 324:1186.
3. Dykstra D, Sidi A. Treatment of detrusor-sphincter dyssynergia with botulinum A toxin; a double blind study. Arch Phys Med Rehabil 1990; 71:24-26.

BOTULINUM TOXIN TREATMENT OF SPASMODIC TORTICOLLIS:

EFFECTS ON PSYCHOSOCIAL FUNCTION

Marjan Jahanshahi and C. David Marsden

Medical Research Council Human Movement & Balance Unit &
Department of Clinical Neurology, Institute of Neurology
The National Hospital for Neurology & Neurosurgery
Queen Square, London WC1N 3BG

INTRODUCTION

In torticollis, depression can be either secondary to the onset and experience of living with the illness or result from the primary central neurotransmitter dysfunction. The results of a number of previous studies have suggested that in torticollis depression may develop as a reaction to the postural abnormality of the head. One hundred patients with torticollis were more depressed than an equally chronic group of cervical spondylosis sufferers (Jahanshahi & Marsden, 1988). In torticollis, depression centered around a negative view of the self and disfigurement accounted for 14% of its variance (Jahanshahi & Marsden, 1990a). In a sample of 67 torticollis patients, a number of psychosocial variables (self-deprecation, satisfaction with available social support, maladaptive coping) together with disability and extent of control over head position accounted for 75% of the variance in depression. The self-deprecation associated with the negative body concept resulting from the postural abnormality emerged as the major predictor of depression (Jahanshahi, 1991). Longitudinal follow-up of these patients over 2.5 years showed that change in the clinical status of torticollis had a significant effect on depression, disability, and body concept (Jahanshahi & Marsden, 1990b).

These studies provide support for the proposal that in torticollis, depression is a reaction to the disorder. A further approach to addressing this issue would be to assess the changes in psychosocial function before and after treatment of torticollis with botulinum toxin injections.

Injections of botulinum toxin into the superficial neck muscles have become the treatment of first choice for torticollis. The efficacy has been evaluated in a number of studies (Tsui et al, 1986; Brin et al, 1988; Stell et al, 1988; Blackie & Lees, 1990). In most studies, substantial normalization of head posture with the injections was achieved. Despite the success of botulinum toxin, no study has examined the psychosocial changes that may be associated with the symptomatic improvement resulting from the injections. The aim of this study was to do so and to determine whether symptomatic improvement of torticollis with botulinum toxin injection is accompanied by improvement in psychosocial function.

Botulinum and Tetanus Neurotoxins, Edited by
B.R. DasGupta, Plenum Press, New York, 1993

METHOD

Sample

The sample consisted of 26 cases (5 male, 21 female) with a clinical diagnosis of idiopathic torticollis. Rotational turning or lateral tilting of the head were the major abnormal posture in respectively 10 (38.5%) and 4 (15.4%) patients. Forward flexion and backward extension of the head were respectively present in 2 (7.7%) and 5 (19.2%) cases. In a further 5 (19.2%) cases, the abnormal head posture was a combination of turn/tilt and flexion or extension of the head. The mean age was 51.5 years (SD=10.8). The mean age of onset was 44.3 years (SD=10.4, range 24 to 72 years), with an average duration of illness of 7.2 years (SD=6.6, median=5.3 years). All patients had undergone extensive pharmacotherapy including treatment with anticholinergics which had not been fully successful in controlling torticollis.

Material

The following questionnaires were completed by the patients before starting treatment and after one or more toxin injections.

The Beck Depression Inventory The 21 items cover cognitive, somatic and behavioural symptoms of depression. Patients indicate how they have generally felt during the previous week. Total scores range from 0 to 63.

The Functional Disability Questionnaire A 27-item scale was developed to measure disability in daily functioning. Each item is rated on a 4 point scale (1: "not at all affected", 4: "severely affected"). Details of the items and data showing it to have high reliability and validity have been presented in a previous publication (Jahanshahi & Marsden, 1990a).

The Body Concept Scale This scale consisted of 22 pairs of opposite body-related adjectives rated on 7 point semantic differential scales. "My body" was the concept rated by the patients. Details of the scale and data showing it to be reliable and valid have been provided elsewhere (Jahanshahi & Marsden, 1990a).

Rosenberg's Self-esteem Scale The 10 items are rated on a 4 point agree-disagree format and assess an individual's feelings of self-worth and self-deprecation.

Botulinum Toxin Questionnaire At follow-up, the patients indicated the direction and extent of change in their torticollis with the injections by selecting one of 8 categories (much worse, worse, slightly worse, unchanged, slightly better, better, much better, no head deviation). The most important benefits of the injections were indicated by rank ordering the following four statements: "Relief of neck pain", "Straightening of head", "Reduction/cessation of involuntary head movements", "Improved ability to move head in different directions". The patients also described any undesirable side effects and provided the following toxin-related information: the number of injections, date of the first and last injections, interval between injections, delay in benefit after injections, duration of benefit.

Rating of toxin-related change by a relative At follow-up, a spouse/relative indicated how the patient's torticollis was compared to before the injections using the same 8 categories as the patients.

RESULTS

Symptomatic Change In Torticollis With Botulinum Toxin Injection

The mean interval since the first injection was 29.9 weeks (SD=26.2, range=8 weeks to 21 months). The mean number of injections was 3, (range=1 to 11). The mean delay in the onset of benefit was 8.9 days (median 5 days, range 1 to 28 days). The average duration of benefit was 8 weeks (SD=3.2, minimum=1, maximum=14 weeks). The average interval between successive sets of injections was 11.6 weeks (SD=2.0).

Twenty-two of the 26 patients (84.6%) considered their torticollis to be better following the injections. Torticollis was reported to be unchanged in 2 (7.7%) and worse by 2 patients (7.7%). The ratings of change by the relatives were similar: with improvement, no change and deterioration respectively reported in 88%, 8%, and 4% of the patients.

Straightening of the head was rank ordered as the first or second most important benefit of the injections by 14 (54%) patients. Relief of neck pain was rank ordered as the first or second most important benefit by 11 cases (42%). Nine patients (35%) considered reduction of involuntary movements of the head as the first or second most beneficial result. Improved ability to move the head in different directions was ranked as the first or second most important benefit derived from the injections by 7 patients (27%). Eight cases (31%) had no side-effects from the injections. Neck stiffness or floppiness of the head were reported as side effects by 10 patients (39%). Difficulty swallowing and lack of energy were respecively reported by 7 (27%) and 6 (23%). Five patients (19%) considered their voice or speech to be in some way altered.

Changes In Psychosocial Function With Botulinum Toxin

The pre-treatment psychosocial scores of the improved (n=22) and unimproved (n=4) cases did not differ (p>.05). However, the 4 patients who had not benefitted from the injections showed less improvement in depression (Mann-Whitney z=2.3, p=.02) and body concept (Mann-Whitney z=2.2, p=.02) than those whose torticollis had improved. There were no differences between the improved and unimproved cases in terms of disability, self-worth, and self-deprecation (p>.05).

Table 1. Mean and standard deviation scores of the 22 improved cases on the measures of psychosocial function before and after botulinum toxin injections

	Before Injections	After Injections	
Depression	14.4 (10.9)	11.3 (9.9)	* p<.05
Disability	52.9 (14.6)	46.4 (15.7)	* p<.05
Body concept	79.5 (28.9)	69.2 (30.7)	
Self-esteem			
Self-worth	13.3 (3.4)	14.5 (2.6)	
Self-deprecation	14.6 (3.4)	13.9 (3.8)	

To examine whether symptomatic improvement in torticollis was associated with improved psychosocial function, the 4 unimproved cases were excluded, as their inclusion may make the sample less homogenous and mask the beneficial effects of the injections on psychosocial function. For the remaining 22 cases, the scores on the psychosocial measures before and after treatment with botulinum toxin are shown in the above Table. There was a significant improvement in depression (t=2.7, df=21, p<.05), and disability (t=2.6, df=21, p<.05). The improvement in body concept was not significant (t=1.8, df=18, p=.08). Self-worth and self-deprecation were also not significantly altered from before to after the injections (p>.05).

CONCLUSIONS

1. Botulinum toxin injection was an effective treatment for the majority of cases. Improvement was reported by 85% of the patients and 88% of the relatives.
2. Straightening of head posture and relief of neck pain were the most common symptomatic benefits.
3. Neck stiffness or floppiness of the head and difficulty swallowing were the most common side effects.
4. Significant reduction of depression and disability were the major psychosocial benefits of the injections.
5. After the injections body concept had shifted towards a more positive view of the body, but the changes were not significant. The beneficial effects of the injections wear off over time and the head deviation and involuntary movements recur in the interval between successive injections. Therefore, as freedom from the postural abnormality of the head is neither complete nor permanent, it is possible that "normalization" of body concept will not be achieved with botulinum toxin treatment of torticollis. It is also possible that the lack of effect of the injections on body concept and self-esteem results from the fact that in the course of living with torticollis, alterations of body concept and low self-esteem have become partially dissociated from the postural abnormality and maintained by other independent factors.

REFERENCES

Beck, A.T., Ward, C.H., Mendelson, M., Mock ,J.E. and Erbaugh, J.K., 1961, An inventory for measuring depression, *Archives of General Psychiat.* 4:561-571.

Blackie, J.D. and Lees, A.J., 1990, Botulinum toxin treatment in spasmodic torticollis, *J Neurol Neurosurg & Psychiat*, 53:640-643.

Brin, M.F. et al 1988, Localised injections of botulinum toxin for the treatment of focal dystonia and hemifacial spasm, In Fahn S. et al (eds) *Advances in Neurology,* 50, Raven Press: New York, pp 599-608.

Jahanshahi, M. and Marsden, C.D., 1988, Depression in torticollis: A controlled study, *Psych Med.* 18:925-933.

Jahanshahi, M. and Marsden, C.D., 1990a, Body concept, disability and depression in torticollis, *Behav Neurol,* 3:117-131.

Jahanshahi, M. and Marsden, C.D., 1990b, A longitudinal study of depression, disability and body concept in torticollis, *Behav Neurol,* 3:233-246.

Jahanshahi, M., 1991, Psychosocial factors and depression in torticollis, *J Psychosom Res.* 35: 493-507.

Rosenberg, M.I. 1965 *Society and the adolescent self-image.* Princeton University Press: Princeton.

Stell, R., Thompson, P.D. and Marsden, C.D. 1988, Botulinum toxin in spasmodic torticollis, *J Neurol Neurosurg & Psychiat,* 51:920-923.

Tsui, J.K., Eisen, A., Stoessl, A.J., Calne, S. and Calne, D.B. 1986, Double-blind study of botulinum toxin in spasmodic torticollis, *Lancet,* August:245-247.

The work reported in this manscript was previously reported in an article that appeared in the Journal of Neurology, Neurosurgery, & Psychiatry (1992), 55, 229-231.

PHYSIOLOGICAL CHANGES FOLLOWING TREATMENT OF SPEECH AND VOICE DISORDERS WITH BOTULINUM TOXIN

Christy L. Ludlow and Geralyn M. Schulz

Voice and Speech Section
National Institute on Deafness and Other Communication Disorders
Bethesda, MD 20892

INTRODUCTION

The symptoms of adductor spasmodic dysphonia (SD) include uncontrolled voice breaks during vowels in sentences, intermittent hoarseness and a slower than normal speech rate. Each of these symptoms is improved following botulinum toxin injections into selected laryngeal muscles.[1-3] Similarly, in oral mandibular dystonia, patients' sentence and word durations are increased relative to normal prior to treatment,[4] demonstrating a slow speech rate. Following botulinum toxin injections into selected lingual and/or mandibular muscles, the word and sentence durations become reduced to within the normal range.[4] Both studies, then, suggest that botulinum toxin injections improved the speed of speech movements in patients following treatment. Both disorders are thought to be due to abnormalities of muscle tone. In spasmodic dysphonia, for example, spasmodic bursts in the thyroarytenoid muscles, contained in the vocal folds, are associated with interruptions in speech.[5] Because botulinum toxin impairs neuromuscular transmission producing a muscle weakness, these improvements in speech and voice function were thought to be due to reduced spasm amplitude in the injected muscles.

When botulinum toxin injection is used in the treatment of these disorders, some unexpected changes have been noted clinically. In SD, prior to injection, intermittent spasmodic movements are seen in both vocal folds during speech.[6,7] When the thyroarytenoid muscle is injected only on one side, most SD patients have an initial breathiness for about 2 weeks after which the voice returns but without symptoms lasting for about 3 months following injection.[1,8,9] On viewing the larynx during this period of improved voice, movement is reduced in the vocal fold on the side of injection. On the opposite non-injected side, no such reduction occurs in movement but the previous spasmodic movements are absent. This suggests some physiological changes are seen in non-injected muscles that may be responsible for symptom improvement.

Another unexpected clinical finding has been the number of patients with spasmodic dysphonia or oral mandibular dystonia who have a longer period of benefit than would be expected. Following intramuscular injection of botulinum toxin, recovery of function begins as early as 14 days as a result of sprouting of new nerve terminals.[10] In SD, vocal fold movement restitution usually occurs within 4 months after injection.[1] Close to 14% of our patients with adductor spasmodic dysphonia, however, have remained symptom-free for over a year following injection[8] despite at least partial if not total return of movement. In orolingual mandibular dystonia, 4 of 25 such patients, 16%, have had a marked reduction in symptoms and have not required further treatment for over 4 years. These findings may be due either to psychological factors in our patients or an alteration in the physiological factors affecting these disorders.

Because of these observations, we have been conducting physiological studies in the voice and speech musculature in patients with spasmodic dysphonia or orolingual mandibular dystonia before and after injection to address the following issues:

1. What changes occur in the resting and volitional activation patterns for speech and voice production in injected and non-injected muscles in patients with spasmodic dysphonia and orolingual mandibular dystonia?

2. Are there changes in the occurrence of spasmodic bursts in both injected and non-injected muscles in these patients?

3. Are there long-term effects on motor unit physiology in the laryngeal muscles following injection?

CHANGES IN INJECTED AND NON-INJECTED MUSCLE ACTIVATION PATTERNS

Spasmodic Dysphonia

Improvements in SD patients' speech have been demonstrated following botulinum toxin injection into either one or both thyroarytenoid muscles.[8,11,12] For unilateral injections, dosages of between 15 and 30 Mouse Units are injected into one muscle[1] which result in a unilateral vocal fold paresis, usually lasting up to 4 months. In bilateral injections, smaller amounts are injected, usually between 2 and 4 Units on each side with no marked reductions apparent in vocal fold motion on either side although the speech symptoms are improved.[11] We evaluated the physiological effects of these two injection types in 5 patients with adductor spasmodic dysphonia; three received unilateral injections and two received bilateral injections.[13] The patients' muscle activation levels were also compared with 5 normal controls of similar age and sex characteristics before treatment.

Electromyographic recordings were made from the right and left thyroarytenoid and cricothyroid muscles prior to injection. The thyroarytenoid (TA) muscles are contained within the vocal folds and shorten and close the vocal folds. The cricothyroid (CT) muscles act to lengthen and tense the vocal folds.[14] Each of the patients was also recorded between 2 and 3 weeks following treatment with botulinum toxin in either one or both TA muscles. During the EMG recording, non-speech muscle verification gestures were used to assure correct electrode placement,[14] and then subjects were recorded while resting quietly. Respiratory movements were recorded simultaneously with the EMG to examine changes in muscle activation during inspiration and expiration. Subjects were then asked to repeat several speech items including prolonged vowels and the sentences, "We mow our lawn all year", and "A dog dug a new bone". Both sentences contain many voiced segments and are difficult for patients with adductor spasmodic dysphonia to produce without voice interruptions.[1]

Prior to injection, we compared the amplitude of EMG recordings in the SD patients with normal controls. Significantly higher levels of muscle activation in microvolts were found during quiet respiration in all 4 muscles (both TAs and both CTs) in the patients.[15] This increase in resting levels was interpreted as muscle hyperactivity. In another study,[13] we computed the percent increase in muscle activation during the inhalation and exhalation phases of respiration over mean minimum resting levels to compare muscle activation for respiration in patients and controls. During expiration, when the TA muscles are normally activated,[16] the percent increase in muscle activation was greater in the patients than in the controls. On the other hand, during inspiration, when the CT muscles are normally active,[17] the controls had a higher percent increase over resting levels than the patients (Figure 1). Both these differences were statistically significant.[13]

Injected and non-injected muscles were recorded following treatment to determine whether changes could be found in muscle activation levels during respiration. In the unilateral injection group, both the injected left TA and the non-injected right TA muscles and both CT muscles decreased in percent activation levels following treatment. Similarly, in the bilaterally injected patients, activation levels were decreased in both the injected TA muscles and the non-injected CT muscles (Figure 1).

Significant reductions relative to baseline levels, therefore, were found in the percent increase in muscle levels during quiet respiration in the non-injected muscles. In the unilaterally injected patients, the non-injected muscles included the ipsilateral CT and the contralateral TA and CT. Reductions in these three muscles are unlikely due to diffusion, only the ipsilateral CT is proximate to the injected TA where the CT attaches to the lower medial surface of the thyroid cartilage. However, the bulk of the CT muscle, and that portion studied with electromyography would be the rectus portion covering the outer surface of the cricoid cartilage. In bilaterally injected patients, the other muscles showing a decrease in activation level were the two CTs. It is unlikely that these effects are due to diffusion because bilaterally injected patients only received a total of 2.5 Units in each TA which does not produce an observable change in vocal fold movement range, speed or lengthening.

EMG measures were also made during prolonged vowels and during sentence production before and after treatment. The mean muscle activity was measured for the entire sentence, from the time of muscle activation onset prior to speech, to muscle activation offset after

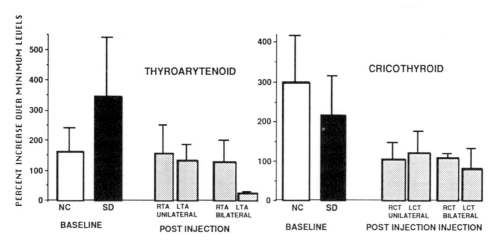

Figure 1. Mean and one standard deviation in the percentage increase in muscle activation during quiet respiration for normal controls (NC) and for patients with adductor spasmodic dysphonia (SD). The post injection patient groups include those who received a unilateral left thyroarytenoid injection and those who received bilateral thyroarytenoid injections. Values are presented for the right thyroarytenoid (RTA), the left thyroarytenoid (LTA), the right cricothyroid (RCT) and the left cricothyroid (LCT). Adapted from Ludlow et al.[13]

speech. During speech, the patients had higher mean percent increases in muscle activation over resting levels than the controls in both the TA and CT muscles (Figure 2). A large amount of variation in the mean activation levels was seen during speech within and between patients in both the TA and CT muscles (Figure 2). This is probably due to the muscle spasms which occurred in the patients' muscles during speech (Figure 3). In contrast, the normal control's muscles shown in Figure 4 exhibit smooth slow tonic activity during speech.

After treatment, decreases in mean muscle activation during speech were seen during speech in the unilateral group in the TA and CT muscles only on the injection side. High levels of muscle activation continued to be seen on the side opposite the injection possibly because increased activation was required on the non-paralyzed side to close the larynx for phonation. This suggests that changes in muscle activation on the opposite side were not because of diffusion. For example, Figure 5 shows muscle recordings from the same patient following injection as shown in Figure 3 prior to injection. The right non-injected TA activation level, in microvolts, is similar to pre-treatment levels. The pattern of muscle activation is more consistent, however and the other three muscles show reduced levels of activation. In the bilaterally injected group, decreases were found in all four muscles (Figure 2). Again, it is unlikely that such changes are due to diffusion because of the very small dosages used in the TA muscles.

Orolingual Mandibular Dystonia

Although botulinum toxin injections have been shown helpful for patients with orolingual mandibular dystonia, the response is much less consistent than in patients with adductor spasmodic dysphonia.[18] In an effort to better understand these differences we have completed analyses on four patients with orolingual mandibular dystonia. In each patient a combination of jaw closing muscles, the medial pterygoid and masseter, and jaw opening muscles, the lateral

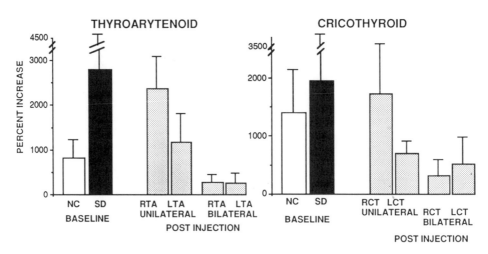

Figure 2. Mean and one standard deviation in the percentage increase in muscle activation during phonation and sentence production for normal controls (NC) and for patients with adductor spasmodic dysphonia (SD). The post injection patient groups include those who received a unilateral left thyroarytenoid injection and those who received a bilateral thyroarytenoid injection. Values are presented for the right thyroarytenoid (RTA), the left thyroarytenoid (LTA), the right cricothyroid (RCT) and the left cricothyroid (LCT). Adapted from Ludlow et al.[13]

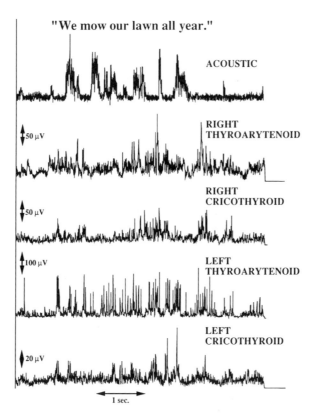

"We mow our lawn all year."

ACOUSTIC

50 μV

RIGHT
THYROARYTENOID

50 μV

RIGHT
CRICOTHYROID

100 μV

LEFT
THYROARYTENOID

20 μV

LEFT
CRICOTHYROID

1 sec.

Figure 3. Simultaneous recordings of the acoustic signal of speech production and electromyographic (EMG) recordings in a patient with adductor spasmodic dysphonia (SD) while saying the sentence, "We mow our lawn all year" during baseline recording prior to injection with botulinum toxin. The acoustic and EMG signals have been rectified to display signal amplitude over time. Signals are presented for the right thyroarytenoid, the left thyroarytenoid, the right cricothyroid and the left cricothyroid. Adapted from Ludlow et al.[13]

pterygoid and anterior digastric, were studied. In two of the patients, excessive tongue elevation was also affected and the genioglossus and styloglossus were also recorded. All recordings were made with bipolar hooked wire electrodes, while repeating syllables ("pa", "ta", "ka"), words ("tomato", "potato", and "baggage"), and sentences ("I have the baggage" and "I thought about my sister"). Measures of the mean muscle activation made during each speech token were normalized by computing the percent of maximum mean activation levels measured during resisted isometric activation. The same EMG recording procedures were repeated within 2 to 3 weeks following a series of injections into one or two muscles at dosages of between 20 and 40 Units per muscle.

Figure 6 shows the rectified muscle recordings from one patient before and after injection. This patient had uncontrolled jaw opening movements which were elicited each time she started speaking. The involuntary jaw opening also occurred during eating. These problems developed after several years of extensive and painful dental work. Speech was effortful, slow and distorted. The muscle recordings (Figure 6) depict muscle activation patterns pre- and post-treatment during repetition of the syllable "pa" which requires jaw elevation for lip closure for each "p", (the silent portion of the acoustic signal). The second and third recordings show the

"We mow our lawn all year."

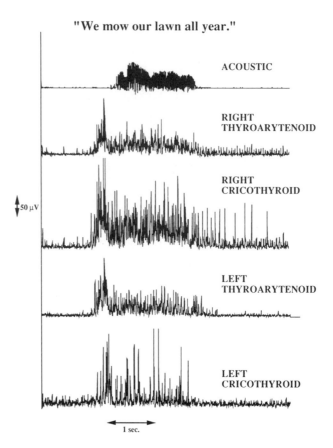

Figure 4. Simultaneous recordings of the acoustic signal of speech production and electromyographic (EMG) recordings in a normal control while saying the sentence, "We mow our lawn all year." The acoustic and EMG signals have been rectified to display signal amplitude over time. Signals are presented for the right thyroarytenoid, the left thyroarytenoid, the right cricothyroid and the left cricothyroid. Adapted from Ludlow et al.[13]

masseter and medial pterygoid, two jaw closing muscles, and the right and left digastric muscles, jaw openers, in the fourth and fifth tracings. The signal amplitudes have been converted into percent of maximum activation levels. Before treatment, abnormally high levels of muscle activation for speech production can be seen in both the opening and closing muscles. This excessive contraction could be the direct result of the motor control disorder. Alternatively, the high activity in the medial pterygoid, a jaw closing muscle, could be the result of the patient attempting to close the jaw for speech and overcome excessively high activation levels in the anterior digastrics. Most of the EMG signals show sustained high levels of tonic activity. Only the bottom recording from the anterior belly of the digastric shows a normal pattern of muscle recruitment with activation being greatest during jaw opening for the vowel sound following the voiceless silent "p" consonant.

The anterior digastric muscles were injected bilaterally and the patient improved her ease of movement, jaw control and speech clarity. Some of her chewing difficulties remained, however. In the post treatment muscle recording (Figure 6), the levels of activity in all four

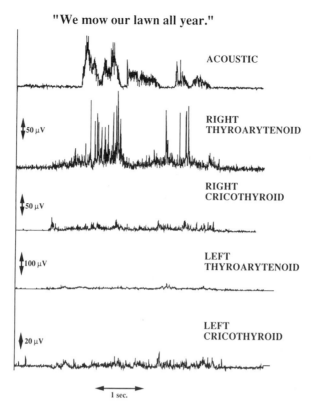

"We mow our lawn all year."

ACOUSTIC

RIGHT
THYROARYTENOID

50 μV

RIGHT
CRICOTHYROID

50 μV

LEFT
THYROARYTENOID

100 μV

LEFT
CRICOTHYROID

20 μV

1 sec.

Figure 5. Simultaneous recordings of the acoustic signal of speech production and electromyographic (EMG) recordings in a patient with adductor spasmodic dysphonia (SD) while saying the sentence, "We mow our lawn all year" during recording two weeks following injection with botulinum toxin. The acoustic and EMG signals have been rectified to display signal amplitude over time. Signals are presented for the right thyroarytenoid, the left thyroarytenoid, the right cricothyroid and the left cricothyroid. Adapted from Ludlow et al.[13]

muscles relative to maximum activation are plotted during the same speech task. Activation was reduced in both digastrics and also in the medial pterygoid, a jaw closer, which was the most active muscle prior to treatment. In addition, the masseter was active at normal levels, suggesting that by decreasing some muscles, this muscle may now have become more appropriately modulated. Overall, normal phasic activation patterns were more evident in all four muscles post-treatment (Figure 6).

The mean percent activation levels for injected and non-injected muscle before and after treatment are plotted for all 4 patients in Figure 7. All but one of the injected muscles decreased in mean percent activation levels following treatment. Note that the scale is normalized into percent of maximum activation, showing that post treatment the patients used proportionally less muscle activation in the injected muscles. No similar trend was found in the non-injected muscles. As shown with the patient in Figure 6, some muscles tended to increase in mean percent activation levels while others tended to decrease.

Spasmodic dysphonia and orolingual mandibular dystonia seem to differ then, in the type of abnormalities present. In orolingual mandibular dystonia, high levels of sustained tonic activity were found in several muscles. This may be the result of antagonistic cocontraction

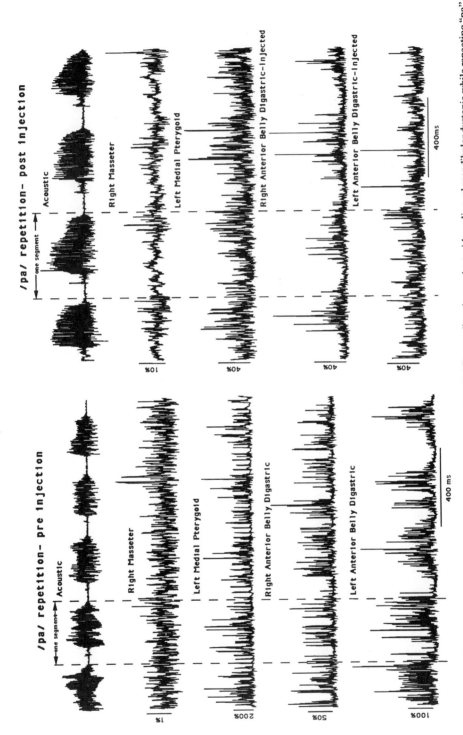

Figure 6. Simultaneous recordings of the acoustic speech signal and electromyographic (EMG) recordings in a patient with orolingual mandibular dystonia while repeating "pa", "pa", "pa" during recording prior to injection and two weeks following injection of botulinum toxin into the anterior belly of the digastric bilaterally. The acoustic and EMG signals have been rectified to display signal amplitude over time in percent of maximum activation during isometric resisted maneuvers. Signals are presented for the right masseter, the left medial pterygoid, the right and left anterior digastrics.

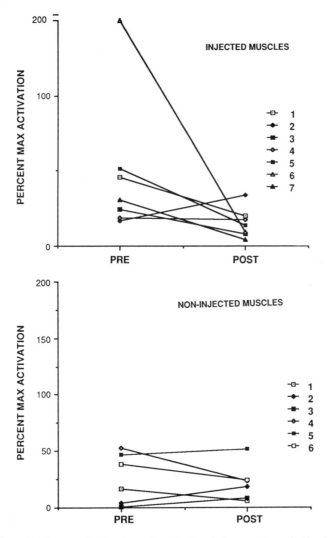

Figure 7. The mean levels of muscle activation during speech production in 7 injected and 6 non-injected muscles recorded and measured in 4 patients before and after injection with botulinum toxin. The mean values are in percent of maximum activity recorded during isometric resisted maneuvers.

being used in an attempt to override the abnormal activation levels in the affected muscle groups. In spasmodic dysphonia, on the other hand, the difficulty was one of involuntary spasms occurring at random in all the muscles (Figure 3). Successful treatment in this disorder, then, would require the elimination of the spasmodic bursts and the production of a more sustained tonic activation pattern.

CHANGES IN PATHOPHYSIOLOGY IN SPASMODIC DYSPHONIA

To determine whether botulinum treatment alters the pathophysiology of spasmodic dysphonia, the frequency of spasmodic bursts in both injected and non-injected muscles was examined.[13] The authors read oscillographic tracings of the electromyographic recordings without identifying information and counted the number of spasmodic bursts in both injected and non-injected muscles. Speech items requiring sustained vocal fold positions were studied such as prolonged vowels and all voiced sentences. Those spasmodic bursts which occurred after speech onset and exceeded the vocal fold closure onset burst amplitude, were counted.[13] For example, in the all voiced sentence, "We mow our lawn all year", following the initial closure, the vocal folds remain in the same position throughout the sentence. In the normal tracing, after the initial burst for vocal fold closure, the activation pattern is continuously tonic (Figure 4). In the SD patient, however (Figure 3), several bursts of activity, greater than or equal to the onset burst amplitude, are seen in each of the 4 muscles throughout sentence production. The authors counted all such bursts from unidentified tracings for the controls and the patients before and after treatment (Figure 8). The mean number of bursts per muscle during speech are shown for the controls, and for the five spasmodic dysphonic patients before injection, and after unilateral and bilateral injections. Significant differences were found between the patients and the controls at baseline for both the TA and CT muscles. Following treatment, decreases in the numbers of bursts were found in both injected TA muscles as well as non-injected CT muscles. Therefore, changes in the abnormalities present were found in all the muscles studied following treatment in spasmodic dysphonia.

CHANGES IN MOTOR UNIT PHYSIOLOGY OVER TIME FOLLOWING INJECTION

To evaluate the time post injection required for speech symptoms to return in spasmodic dysphonia, we have injected patients only after a return of their speech symptoms.[8] In injections followed by symptom return (86% of our patients), the mean time of return is 125 days. However, in 10% of these, a significant benefit continued past 6 months and in 3% a significant benefit continued past 10 months. Because of these reports, we have studied seven patients who all received only unilateral TA injections. Motor unit characteristics were

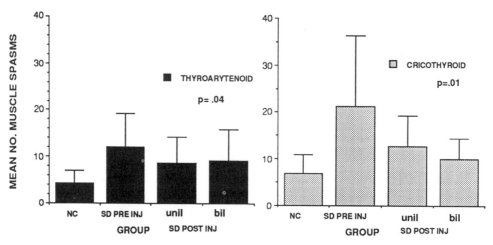

Figure 8. Mean and one standard deviation in the number of spasmodic bursts during phonation and speech in normal controls (NC) and patients with adductor spasmodic dysphonia (SD). The post injection patient groups (SD POST INJ) include those who received a unilateral left thyroarytenoid injection (UNIL) and those who received a bilateral thyroarytenoid injection (BIL). Values are presented for the right and left thyroarytenoid muscles combined (THYROARYTENOID) and the right and left cricothyroid muscles combined (CRICOTHYROIDS). The p values are for ANOVA comparisons between the 4 group means. Adapted from Ludlow et al.[13]

measured in injected and non-injected TA muscles for up to 38 months following the last treatment with botulinum toxin.[19] All had received more than one injection and some as many as 7 treatments with a combined total dosage of up to 142 Mouse Units. Motor units in the TA were measured on the non-injected side for comparison in each patient. Our purpose was to determine when and if the motor unit characteristics become similar on the injected and non-injected sides in SD patients.

Motor unit durations were studied as an indication of motor nerve terminal sprouting, which occurs during recovery from botulinum toxin injection.[20,21] Because small numbers of fibers are innervated by a laryngeal motor neuron, motor unit durations in the TA and CT muscles are extremely short.[22] It has been suggested that in some instances a motor neuron may innervate only a few muscle fibers, and that there may be multiple innervation of some laryngeal muscle fibers.[23,24] With botulinum toxin treatment resulting in nerve terminal sprouting, the laryngeal motor unit durations may increase if collateral fibers become innervated by the same motor neuron.[25,26] The preliminary data were reported by Davidson and Ludlow last year.[19]

Bipolar or concentric needle electrodes were used to sample motor unit firings from previously injected and non-injected TA muscles in three to five locations during quiet breathing. The recordings were reviewed visually on a computer screen after digitizing at 10 kHz. Motor units with a minimum of 10 identical firings, identified by visual inspection, were included. Duration measures were made automatically from the initial to the third zero crossing. Significant increases in duration were found in motor units recorded from injected muscles in comparison with units recorded from non-injected muscles. However, in patients who had not received an injection for over a year, unit durations recorded from injected muscles were more similar in duration to unit durations of non-injected muscles than in the patients who were less than a year since their last botulinum injection. The results, then, suggest that motor unit durations were increased, as was expected, following recovery from botulinum toxin injection, probably due to nerve terminal sprouting. What was surprising, however, was the finding that unit duration became further reduced after a year following injection. This suggests that some normalization process continued after the initial reinnervation process and that this further recovery process may be responsible for the restitution of normal motor unit physiology in the long term. Therefore, when patients were not re-injected until the reinnervation process was at least partially complete, fewer differences were found in motor unit characteristics.

Deficits were also found in the recovery of symmetric vocal fold motion in four of the seven patients despite a 6 to 13 month hiatus since botulinum toxin injection.[19] The presence of vocal fold asymmetry, however, did not correlate with measures of motor unit durations or numbers of turns in the EMG interference pattern, an estimate of motor unit firings. Movement asymmetries may have been the result of injury during repeated needle insertions during injections. Alternatively, muscle contractures may have developed during the paralysis that occurred following each injection. Further evaluations of patients receiving injections over several years are needed to determine if these movement changes are due to factors associated with repeated muscle injections other than botulinum toxin. The motor unit results are encouraging, however, in that they suggest that when 6 to 12 months elapse between treatments, most changes are reversible in the laryngeal muscles. These studies need to continue, however, to evaluate the changes in vocal fold movements seen on some patients.

DISCUSSION

Our studies this far have suggested that changes take place in the central control of movement following treatment with botulinum toxin. We have seen that muscle activation levels tend to be reduced during volitional movement and as a result speech is more fluent and rapid. Similar changes in eyelid movement were reported by Manning et al[27] following treatment of blepharospasm with obicularis oculi injection. Thus, volitional movement may increase after injection, presumably because of reduced hyperactivity in muscles which previously interfered with fine adjustments in vocal fold length and tension during voicing control for speech. Some of these improvements, however, may not be the result of selective denervation of injected laryngeal muscles. We have found that spasmodic bursts are no longer present in non-injected muscles on the opposite side of the larynx, although these muscles are not reduced in activation amplitude during speech. Although effects such as jitter, have been demonstrated distant from locally injected botulinum toxin in patients,[28] such effects would not account for the changes in muscle activation patterning. Further, because laryngeal dosages are low, changes in spasmodic bursts in non-injected muscles may not be the result of diffusion. A recent study of muscle activation pattern changes in torticollis patients demonstrated changes in the muscle activation patterns in both injected and non-injected muscles.[29] Several of the non-injected muscle changes reported by these authors were increases in activation during head movements. Therefore, the authors concluded that these were due to changes in central patterning of muscle activation and not due to the peripheral effects of the botulinum toxin injections. The authors were not more specific, however, on what mechanisms might be responsible for these patterning changes.

Others have investigated the possibility of central effects of botulinum toxin injection on inhibitory mechanisms. They failed to find effects on Renshaw cells in animal studies[30] or on blink reflex recovery curves[31] in patients with blepharospasm successfully treated with botulinum toxin injection. Hagenah et al[30] suggested that with axonal transport, the toxin may be altered and no longer able to have an effect on central transmission. Another possible explanation for our results may be that these changes are due to central reorganization following peripheral changes in neuromuscular transmission. Recently it has been demonstrated that changes can occur in the motor cortex within hours of a peripheral change due to unmasking of latent connections not usually functional except after peripheral alteration.[32] The role of central nervous system plasticity in changes in motor control following botulinum toxin injection must be considered in future studies.

Several investigations are needed to explore our physiological results further. First, these effects need to be further investigated in larger numbers of patients. Second, the possibility of diffusion from the TA to the CT needs to be studied. The TA and CT muscles are adjacent to each other in the subglottic region at the medial inferior edge of the thyroid cartilage. Portions of the CT run in a superior direction between the two cartilages and insert on the medial inferior surface of the thyroid cartilage (Kahane, personal communication). Possibly injections may have been in this region, either directly affecting CT function, or diffusing into the CT. Changes in muscle spasms on the opposite side, however, are less likely to have been due to diffusion.

Another possibility may be that reductions in muscle activation in injected muscles may reduce strain on the other laryngeal muscles and thus the level of activation needed to achieve vocal fold movement. Because the TA muscle has a vocal fold shortening action and the CT has a lengthening action, reduced levels of TA activation as a result of injection, may result in less CT activation being required to achieve the same vocal fold length tension characteristics. The mechanism for eliciting spasms in spasmodic dysphonia is unknown. Because many of these disorders are action induced, the elicitation of abnormalities in muscle activation levels may be dependent upon the level of motor neuron pool drive. That is, they may be triggered by high levels of motor neuron pool activation. After injection of the TA, high levels of CT activation may no longer be required to produce vocal fold adduction for speech. Thus, spasmodic bursts may no longer be elicited in the CT muscles as lower levels of CT motor neuron firing were needed.

This would not account for the reduced spasms found in the TA and CT on the opposite side, however, in the unilaterally injected patients. As shown in Figure 2, the same levels of activation were found in the non-injected muscles on the opposite side during speech, probably because the non-paralyzed side had to increase its activation to achieve vocal fold closure for speech. This can also be seen in the SD patient's muscle recordings pre- and post-treatment during sentence production (Figures 3 and 5).

In a previous paper[13] we suggested that these changes may be due to reductions in feedback to the motor neuron pool from injected muscles which are no longer activated because of neuromuscular blockade, thus reducing the stimulation of the fusal afferents for injected muscles. This seems unlikely, however, as the number of muscle spindles are few in the human larynx and the physiological significance of a stretch reflex in the human laryngeal muscles has not yet been demonstrated.

Others have suggested that with botulinum toxin injections, the metabolic requirements decrease in the injected muscles[33]. This may have a modulating effect on blood flow to the region and could cause secondary effects on the metabolic function in the adjacent laryngeal muscles. Studies evaluating the effects of blood flow reductions on adjacent muscle activation patterning are needed to explore this possibility.

CONCLUSIONS

A great deal more information is needed on the physiological effects, both peripherally and centrally, which may be responsible for improved functioning in spasmodic dysphonia and orolingual mandibular dystonia after botulinum toxin injection. We currently have little understanding of the physiological bases for these disorders. The dramatic improvements with botulinum toxin injection, particularly in spasmodic dysphonia, should be further investigated physiologically to improve our understanding of the genesis of the pathophysiology of these disorders.

REFERENCES

1. Ludlow CL, Naunton RF, Sedory SE, Schulz GM, Hallett M. Effects of botulinum toxin injection on speech in spasmodic dysphonia. Neurology 1989; 30:123-456.
2. Ludlow CL, Naunton RF, Terada S, Anderson BJ. Successful treatment of selected cases of abductor spasmodic dysphonia using botulinum toxin injection. Oto-Laryngol Head Neck Surg 1991; 104:849-855.
3. Ludlow CL, Naunton RF, Fujita M, Sedory SE. Spasmodic dysphonia: Botulinum toxin injection after recurrent nerve surgery. Oto-Laryngol Head Neck Surg 1990; 102:122-131.
4. Schulz GM, Ludlow CL. Botulinum treatment for orolingual-mandibular dystonia. In: Mooore CA, Yorkston KM, Beukelman DR, eds. Dysarthria and apraxia of speech: Perspectives on management. Baltimore: Brooks,P.H. Publishing Co., Inc., 1991:227-241.
5. Shipp T, Izdebski K, Reed C, Morrissey P. Intrinsic laryngeal muscle activity in a spastic dysphonic patient. J Speech Hearing Dis 1985; 50:54-59.
6. Parnes SM, Lavorato AS, Myers EN. Study of spastic dysphonia using videofiberoptic laryngoscopy. Ann Otol 1978; 87:322-326.
7. Ludlow CL, Nauton RF, Bassich CJ. Procedures for the selection of spastic dysphonic patients for recurrent laryngeal nerve section. Otolaryngology-Head Neck Surg 1984; 92:24-31.
8. Ludlow CL, Bagley JA, Yin SG, Koda J, Rhew K. A comparison of different injection techniques in the treatment of spasmodic dysphonia with botulinum toxin. J Voice 1991; (in press)
9. Ludlow CL. Treatment of speech and voice disorders with botulinum toxin. JAMA 1990; 264:2671-2675.
10. Capra NF, Bernanke JM, Porter JD. Ultrastructural changes in the masseter muscle of Macaca fascicularis resulting from intramuscular injections of botulinum toxin type A. Archs oral Biol 1991; 36:827-836.
11. Blitzer A, Brin MF. Laryngeal dystonia: A series with botulinum toxin therapy. Ann Otol Rhinol Laryngol 1991; 100:85-89.
12. Truong DD, Rontal M, Rolnick M, Aronson AE, Mistura K. Double-blind controlled study of botulinum toxin in adductor spasmodic dysphonia. Laryngoscope 1991; 101:630-634.

13. Ludlow CL, Hallett M, Sedory SE, Fujita M, Naunton RF. The pathophysiology of spasmodic dysphonia and its modification by botulinum toxin. In: Berardelli A, Benecke R, Manfredi M, Marsden CD, eds. Motor Disturbances. 2nd ed. Orlando: Academic Press, 1990:274-288.

14. Hirano M, Ohala J. Use of hooked-wire electrodes for electromyography of the intrinsic laryngeal muscles. J Speech Hear Res 1969; 12:362-373.

15. Ludlow CL, Baker M, Naunton RF, Hallett M. Intrinsic laryngeal muscle activation in spasmodic dysphonia. In: Benecke R, Conrad B, Marsden CD, eds. Motor Disturbances. Orlando: Academic Press, 1987:119-130.

16. Kuna ST, Insalaco G, Woodson GE. Thyroarytenoid muscle activity during wakefulness and sleep in normal adults. J Appl Physiol 1988; 65:1332-1339.

17. Woodson GE. Respiratory activity of the cricothyroid muscle in conscious humans. Laryngoscope 1990; 100:49-53.

18. Jankovic J, Schwartz K, Donovan DT. Botulinum toxin treatment of cranial-cervical dystonia, spasmodic dysphonia, other focal dystonias and hemifacial spasm. J Neurol Neurosurg Psychiat 1990; 53:633-639.

19. Davidson BT, Ludlow CL. Long term effects of botulinum toxin injections in spasmodic dysphonia. Otolaryngology-Head Neck Surg 1991; 105:237.

20. Holds JB, Alderson K, Fogg SG, Anderson RL. Motor nerve sprouting in human orbicularis muscle after botulinum A injection. Invest Ophthalmol Vis Sci 1990; 31:964-967.

21. Wojno T, Campbell P, Wright J. Orbicularis muscle pathology after botulinum toxin injection. Ophthal Plast Reconstr Surg 1986; 2:7174.

22. Chanaud CM, Ludlow CL. Single motor unit activity of human intrinsic laryngeal muscles during respiration. Ann Otol Rhinol Laryngol 1991 (submitted).

23. Bendiksen FS, Dahl HA, Teig E. Innervation pattern of different types of muscle fibers in the human thyroarytenoid muscle. Acta Otolaryngol 1981; 91:391-397.

24. Rossi G, Cortesina G. Morphological study of the laryngeal muscles in man. Acta Otolaryngol (Stockh) 1965; 59:575-592.

25. Mukuno K, Scott AB, Ishikawa S. Histopathological study of the monkey extraocular muscles under botulinum toxin injection. Proc Meet Int Strabismol Assoc 1982; 4:707-710.

26. Spencer RF, McNeer KW. Morphological basis of junctional or muscle adaptation in pharmacological denervation of monkey extraocular muscles. In: Keller EL, Zee DS, eds. Adaptive processes in visual and oculomotor systems. New York: Pergamon, 1986:13-20.

27. Manning KA, Evinger C, Sibony PA. Eyelid movements before and after botulinum therapy in patients with lid spasm. Ann Neurol 1990; 28:653-660.

28. Lange DJ, Rubin M, Greene PE, et al. Distant effects of locally injected botulinum toxin: a double-blind study of single fiber EMG changes. Muscle and Nerve 1991; 14:672-675.

29. Gelb DJ, Yoshimura DM, Olney RK, Lowenstein DH, Aminoff MJ. Change in pattern of muscle activity following botulinum toxin injections for torticollis. Ann Neurol 1991; 29:370-376.

30. Hagenah R, Benecke R, Wiegand H. Effects of type A botulinum toxin on the cholinergic transmission at spinal Renshaw cells and on the inhibitory action at IA inhibitory interneurons. Naunyn Schmiedeberg's Arch Pharmacol 1977; 299:267-272.

31. Valls-Sole J, Tolosa ES, Ribera G. Neurophysiological observations on the effects of botulinum toxin treatment in patients with dystonia blepharospasm. J Neurol Neurosurg Psychiat 1991; 54:310-313.

32. Jacobs KM, Donohue JP. Reshaping the cortical motor map by unmasking latent intracortical connections. Science 1991; 251:944-947.

33. Spencer RF, McNeer KW. Botulinum toxin paralysis of adult monkey extraocular muscle. Structural alterations in orbital, singly innervated muscle fibers. Arch Ophthalmol 1987; 105:1703-1711.

TREATMENT OF SPASMODIC DYSPHONIA
WITH BOTULINUM TOXIN: CLINICAL EXPERIENCES
AND RESEARCH ISSUES

Gayle E. Woodson,[1] Thomas Murry,[1] Petra Zwirner,[2] and Michael Swenson[3]

[1]Department of Surgery/Division of Otolaryngology
University of California, San Diego
School of Medicine and VA Medical Center, San Diego
225 Dickinson Street, 9112C
San Diego, CA 92103

[2]HNO Klinik, Klinikum Grosshadern
Marchiainistr. 15
8000 Munich 70
Germany

[3]Department of Neurosciences
University of California, San Diego
School of Medicine
225 Dickinson Street
San Diego, CA 92103

INTRODUCTION

Spasmodic Dysphonia is a disabling speech disorder characterized by excessive contractions of laryngeal muscles. Botulinum toxin has been demonstrated in several clinical trials to be very effective in improving speech in most patients and its clinical use is becoming more widespread. Although the benefits of botulinum toxin treatment are considerable, few patients achieve a totally normal voice and minor side effects are common. Clinical research is needed to determine the factors limiting improvement, to determine the role of adjunctive speech therapy, and to develop treatment protocols which will maximize benefits and diminish side effects.

PATHOPHYSIOLOGY

Two major types of spasmodic dysphonia have been described, adductor and abductor.[6] Spasm of adductor muscles results in tight closure of the larynx, with vocal strain and frequent

Botulinum and Tetanus Neurotoxins, Edited by
B.R. DasGupta, Plenum Press, New York, 1993

voice breaks. In abductor dysphonia, the voice is impaired by air escape, due to abduction of the vocal folds. In the former, the goal of therapy is to decrease the strength of laryngeal closure, to prevent voice breaks and decrease the effort of speaking. In the latter, laryngeal opening during speech must be prevented or reduced to diminish vocal breathiness. Adductor dysphonia is the most common form, occurring in more than 90% of the patients treated by the authors. Although pure abductor dysphonia is much less common, it can coexist with adductor spasms.[9]

Spasmodic dysphonia primarily affects the process of phonation, or production of sound by the larynx. Normal phonation requires adequate breath support, normal vibratory capacity of the vocal folds (AKA vocal cords), and control of pitch by changes in length and tension of the vocal folds. A final requirement is that the vocal folds must be precisely approximated to produce a slit like opening in the larynx, so that exhalation can easily induce vibration. If the vocal folds are pressed together too tightly, as occurs in adductor dysphonia, excessive expiratory pressure is required, and the voice sounds strained or strangled. If the vocal folds are not close enough together, excitation of vibration is inefficient, resulting in diminished vocal power, and audible escape of air. The latter situation occurs in abductor dysphonia.

Phonation is only one component of the process of speech, which also involves resonance (filtering and amplification of sound by the induction of vibration in other body structures) and articulation (the shaping of sounds into words) Although spasmodic dysphonia primarily impairs phonation, there are secondary effects on resonance and phonation. An analogous situation to speaking in the presence of unpredictable spasms of the larynx would be trying to walk barefoot across a dark room with tacks scattered on the floor. Speech becomes effortful, with irregular speaking rate, and may be less intelligible. Various compensatory maneuvers are adapted to reduce the sequela of laryngeal spasm, and such adaptations impact on both voice quality and speech patterns. Abnormalities such as substitutions, omissions, distortions, and prolonged segment duration have been detected by spectrographic analysis of the voice.[9,12] Other upper airway muscles are also involved by spasms. Flexible laryngoscopy has documented some motions responsible for disturbances in articulation, such as spasms and tremor in the velum or the posterior tongue.[19]

Patients with spasmodic dysphonia commonly experience considerable fluctuations in symptoms, worsening with vocal demand or stress, and sometimes improving with singing, reciting, or other non-speech vocalizations. Some patients can speak normally much of the time, with only occasional voice breaks. Because of this variability in symptoms, as well as the difficulty in observing pathologic function inside the vocal tract, spasmodic dysphonia was until recent years believed to be a psychiatric problem.[2] Diagnosis was established by the presence of a "strained and strangled" voice in the absence of any morphologic defect of the larynx. Through the years, it has become apparent that in many, if not the vast majority of patients with this condition, the cause is neurologic.[3-5,16,17,19]

The prevailing opinion today is that the most patients with spasmodic dysphonia have a medical disease: a focal dystonia of the larynx.[1] It seems, however, that spasmodic dysphonia is not a single homogenous entity. The clinical presentation and response to therapy varies, at least in part due to the inconsistent use of compensatory mechanisms, such as whispering and pitch modulation, to mask symptoms. It is also likely that pathophysiology differs among patients. Many patients with spasmodic dysphonia have tremor.[4] Rosenfield et al.[18] found essential tremor in 71 of 100 subjects with SD, involving the larynx and/or pharynx in 59. We noted tremor during laryngoscopy in 29 of 38 patients.[21] Other authors have also noted a high incidence of tremor in association with SD.[5,12] Blitzer et al.[7] noted extralaryngeal dystonias in 66 of 110 patients with laryngeal dystonias.

Head trauma, stress, viral illness, and toxic exposure have all been implicated as causes of spasmodic dysphonia, and many cases are familial. Because the etiology is obscure and may be multifactorial it should be considered as a syndrome and not a disease.

Although the vast majority of spasmodic voice problems can be attributed to an organic cause, there are cases which are undeniably psychogenic, and some patients develop substitution symptoms when the laryngeal spasms are controlled. Further, because a vocal handicap is very disruptive, some degree of emotional response is expected in all patients, and psychogenic overlay could make a substantial contribution to symptoms. Detecting the presence and significance of psychogenic factors is quite difficult. Among dystonias, spasmodic dysphonia is unique in that the muscle spasms are not readily evaluated by physical examination. The larynx can be inspected by laryngoscopy, but the muscles cannot be directly palpated. Currently there is a lack of clinical criteria for differentiating functional or psychiatric disorders from organic spasmodic dysphonia.

TREATMENT

Until about twenty years ago, the only available treatments for spasmodic dysphonia were speech therapy and psychotherapy. Treatment was prolonged and labor intensive, usually with disappointing results. Then surgical induction of unilateral laryngeal paralysis, by transection of the recurrent laryngeal nerve, was demonstrated to produce a dramatic decrease in symptoms.[10] Although this treatment is satisfactory in some patients, others have unacceptable levels of breathiness, or develop recurrent spasms.

Botulinum toxin was developed as an alternative therapeutic modality, based on the success in treating spasms in other parts of the body. It was reasoned that the toxin could produce a graded paresis of specific muscles, rather than global hemiparalysis of the larynx. Injection of the thyroarytenoid muscle permits specific weakness of adduction. With weakness, rather than flaccid paralysis, the breathiness and swallowing problems due to glottal incompetence can be minimized or avoided. Injection of the posterior cricoarytenoid muscle limits abduction. In either case, the paresis would be temporary, so that if the treatment produced an unsatisfactory result, it would dissipate with time.

Injection is most often percutaneous, with EMG guidance.[7,13,15,20] An alternative approach is transoral injection.[11] Our preference is for a percutaneous approach with EMG guidance, because our experience allows reliable placement, and because this approach causes minimal patient discomfort. A 27 gauge hypodermic needle, coated with Teflon except at the point and the hub, serves as both the electrode and the port of injection. To access the thyroarytenoid muscle, for adductor dysphonia, the needle is passed through the cricothyroid membrane, just off the midline, and angled upwards and slightly laterally. The posterior cricoarytenoid muscle is reached by rotating the larynx manually to bring the posterior larynx nearer to the lateral neck skin. In either case, electrode position is confirmed by obtaining crisp action potentials on activation of the target muscle. The muscles can be approximately located without EMG guidance, but toxin effect could be limited by injection near, rather than in muscle, and there is also a possibility that the needle tip could be in the air space.

Assessment of the results of such treatment indicates that botulinum toxin is extremely effective in decreasing spasmodic vocal fold adduction.[21] The mechanism of this improvement is not entirely clear. Ludlow et al.[14] have recently found increased levels of thyroarytenoid muscle activity during both respiration and phonation, as well as spasmodic bursts in the cricothyroid muscle. Botulinum toxin injection decreased spasms and respiratory activation not only in the injected muscle, but also in the cricothyroid muscles and the opposite thyroarytenoid muscle. They suggest that reduced feedback to motoneuron pool could be the mechanism of action of botulinum toxin, and postulated that the pathophysiology of spasmodic dysphonia could be an increased sensitivity of motoneuron pools to muscle action feedback.

At UCSD, we have treated 125 patients with laryngeal injections of botulinum toxin. 117 had spasmodic dysphonia, while the remainder had various other neurologic causes of laryngeal spasm. 90 SD patients were female, and 27 were male. 8 had abductor dysphonia, 4 of each gender.

All patients with adductor dysphonia have had a remarkable improvement in vocal function, although few have achieved a totally normal voice. We initially used only unilateral injections, but have since begun to use bilateral injections in some cases. 37 adductor patients were eventually determined to be managed best by bilateral injection. The average effective dose in patients with bilateral treatment is 1.5 units per vocal fold, while the average effective unilateral dose is 10 units. Thus, a larger dose is required for effectiveness with unilateral injections. After one month, both techniques result in comparable elimination of voice breaks and improvement in airflow. However, at one week, there are substantial differences, with increase in phonatory airflow to abnormally higher levels and a greater instability of pitch in patients treated bilaterally. Side effects of breathiness of mild aspiration and vocal weakness and breathiness during the first two weeks are also more common in patients with bilateral injections.

We have found that many patients differ in overall response to the two approaches, some doing best with unilateral injections and others responding best to bilateral injections. Although no definite criteria have been developed to predict which approach will produce the best response, we have found in general that patients with the most severe spasms are best managed by bilateral injections, because a greater effect is produced with less toxin. In contrast, patients with mild to moderate, or intermittent symptoms often have unacceptable levels of breathiness and aspiration with bilateral injection. In these patients, a greater control of spasm, with less glottal incompetence, can be achieved with a larger unilateral injection. Controlled studies are needed to enable accurate prediction of optimum dose and placement for each patient.

Duration of action is most commonly around 4 months, although some patients have relief of symptoms for as long as one year, or as briefly as two weeks. Recently, our clinical impression has been that voice therapy prolongs the duration of benefits from botulinum treatment for spasmodic dysphonia. Further studies are needed to test the role of adjunctive voice therapy. Some patients who have had stable voice improvement for several months often note a sudden relapse of symptoms after an upper respiratory infection or some stressful events. These occurrences suggest that some other factors in addition to the biochemical effects on motor nerve endings are responsible for the benefits of treatment with botulinum toxin.

Our group has defined some of the specific voice effects of botulinum toxin treatment for adductor dysphonia: elimination of voice breaks, stabilization of the fundamental frequency, and increase in phonatory airflow.[23] The effects are quite different at one week and one month after treatment. At one week, airflows are often elevated above the normal range and the pitch is still unstable. At one month, however, airflow decreases to near normal, and the pitch (as measured by standard deviation of the fundamental frequency) stabilizes.[25] There are still substantial abnormalities of the voice, as determined by acoustic and perceptual analyses.

To seek the mechanisms of the persisting vocal dysfunction, we performed detailed analysis of video laryngoscopy of 17 patients before and after botulinum toxin. Video segments were randomly recorded on a cassette, and then blindly evaluated by three independent observers. We evaluated excessive glottic closure and vocal fold rigidity as indicators of intrinsic laryngeal muscle spasm, and measured the glottal posture to determine the activity of extrinsic laryngeal muscles. Tremor was also detected and scored. After treatment, intrinsic laryngeal muscle hyperfunction was abolished or diminished in all cases. Mean tremor score was reduced, but not all patients were improved. Extrinsic muscle hyperfunction was not significantly improved by therapy, and was judged to become worse in 3 patients. This suggests that treatment results could be improved, by treating tremor with medication, and by controlling extrinsic laryngeal activity. Tremor is very common physical finding in patients with spasmodic dysphonia. In some patients, this represents essential tremor, which worsens with intention, and the head and upper extremities may also be involved. In other cases tremor appears to be a feature of the dystonia. Although tremor usually responds to Botulinum treatment, the rate of satisfaction is slightly lower in patients with tremor. Extrinsic activity may represent functional adaptation, and if so, it should be amenable to speech therapy. However, if these muscles are involved in the dystonia, additional injection sites should be added to the specific muscles involved.

There are other factors which limit the success of botulinum toxin. The weakening by toxin mitigates spasm at the expense of vocal power. In some subjects the fluency they acquire may not be worth the loss of vocal volume and projection. Pitch range can also be limited. Another possible contributing factor is failure to recognize and treat the coexisting abductor spasms in patients with primarily adductor spasms.

Results of botulinum toxin treatment for abductor spasmodic dysphonia are less encouraging than for adductor dysphonia, owing both to the location of the abductor muscles and to apparent differences in the diseases process. The abductor muscle, the posterior cricoarytenoid muscle, is located between the larynx and pharynx and is more difficult to inject. The treatment also has a greater potential for complications. Dysphagia or aspiration are more likely to occur and be significant, because the pharyngeal constrictors are very nearby, and therefore more susceptible to migration of toxin. If the abductor muscles are weakened too much, the airway would be impaired. Injection of the cricothyroid muscle, located in the anterior larynx, is less risky, and has been reported to be beneficial in some patients with abductor dysphonia (Ludlow, pers. comm.).

A recent paper reported the Columbia experience in treating thirty-two patients with the abductor form of the dysphonia.[8] Most patients required bilateral posterior cricoarytenoid muscle injection, and improved to an average of 70% of normal function, which is less than the 80 to 100% normal function observed in patients treated for adductor dysphonia.[7] Ten patients also received cricothyroid muscle injection and/or type I thyroplasty, resulting in further vocal improvement. At UCSD, we have treated 8 patients with abductor spasmodic dysphonia. All have required frequent, bilateral injection, with the duration of benefit lasting only 4 to 8 weeks. In one patient, despite adequate control of abductor spasms, dysphonia persisted due to diaphragmatic spasms. In another, breathy voice breaks occurred despite apparently adequate control of arytenoid abductor spasm. An anterior glottal gap was observed during these breaks. Further research is needed to elucidate the reasons for difficulties in treating patients with abductor spasmodic dysphonia.

ASSESSMENT

The task of assessing function in patients with spasmodic dysphonia is difficult. Clinical evaluation of patients with spasmodic dysphonia differs significantly from that of patients with other dystonias, because the involved muscles are not superficial enough to be palpated or to permit direct observation of action. On the other hand, the voice produced by the patient is an objective physical phenomenon which can be quantitatively measured.

Although experienced clinicians can easily recognize the vocal characteristics of spasmodic dysphonia, objective and quantitative assessment is difficult. Objective measures, such as acoustic and aerodynamic analysis are only relevant and valid if they correlate with perceptually recognized features of the disorder. The parameters measured in evaluating vocal function may not reflect the perceived speech defect in spasmodic dysphonia. The situation is further complicated because symptoms vary considerably over time and in different situations. Many patients are more symptomatic when under stress or during telephone conversations, and the testing environment can strongly influence results. The testing environment is artificial, and relevance to everyday life situations may be minimal. The objective parameters used to measure vocal function may not actually reflect the handicap experienced by the patient, as many symptoms may be more related to the degree of effort required to speak than to the quality of the voice produced. Moreover, two factors which contribute significantly to overall severity are the difficulty, both physical and emotional, experienced by the patient in his attempts to communicate. This cannot be directly measured.

Despite these difficulties, assessment is vital to the management of patients with spasmodic dysphonia. It is important to confirm the diagnosis and rule out other diseases. Information on specific parameters can be used to plan therapy and predict response to treatment.

Several methods can be used to assess patients with spasmodic dysphonia. Optimal assessment should include a battery of tests, including both subjective perceptual ratings and direct physical measurements.[22] Most methods of diagnosing spasmodic dysphonia and assessing treatment are based on the sound of the voice, rather than the measurement of laryngeal function during speech. Acoustic measures are relatively easy to measure and quantify, but give little insight into the pathophysiology of the disorder. The speech of patients with SD has been described as strained and strangled, effortful, choked, laborious, and jerky, with overpressure, hard glottal attacks, and voice stoppages. But the characteristic strained and strangled voice of spasmodic dysphonia which is easily recognized by most clinicians is very similar to the voice of many patients with neurologic diseases such as Parkinsonism, cerebellar ataxia, and essential tremor.[3] Because acoustic changes are not sufficiently specific to separate spasmodic dysphonia from other neurological voice disorders, they should not be regarded as diagnostic tests, but as indicators of function.[5,24]

Aerodynamic measures are very useful. The glottal spasms in spasmodic dysphonia produces significant changes in phonatory airflow and subglottic pressure. In adductor SD, subglottic pressure is increased, mean phonatory airflow is decreased, and sudden drops in airflow can be observed. Phonatory airflow is extremely useful in monitoring the effects of botulinum therapy in adductor SD. Maximal benefit is attained when a dose of Botox is administered which is just sufficient to increase air flow to a normal level. Further Botox results in excessive vocal fold weakness and, subsequently, breathiness. In contrast, airflow is above normal in most cases of abductor spasmodic dysphonia, and bursts of airflow can indicate a component of abductor spasm in a patient who clinically appears to have only adductor spasm. Data regarding the utility of airflow measurements in abductor dysphonia is insufficient to draw conclusions at the present time, but preliminary data suggests that decreased airflow correlates with improved function.

Functional status is extremely difficult to determine objectively, as it depends not only on the patients physical handicap, but also on the patient's functional adjustment and emotional response to the physical limitation. And spasmodic dysphonia symptoms vary greatly depending only on the vocal task, as well as levels of stress

We have used a functional status rating scale to indicate the severity of the disorder, based on its impact on employment and social interaction.[21] Class 0 indicates mild symptoms. Class I patients are mildly annoyed by the disorder. Class II patients have significant problems, but can carry on in their work and social activities. Class III patients have employment problems, career changes, or restriction of social activities. Blitzer and Brin have also used a functional rating scale to follow patients with spasmodic dysphonia, with clinicians rating the patients performance as a percentage of normal.[8] The patient's perception of his problem, which is of course the "bottom line" in determining patient satisfaction with treatment, has also been used to evaluate the results of their treatment.[20] Such ratings are essential to validate the relevance of objective measures, but are not per se amenable to statistical evaluation. A standardized rating system utilizing a questionnaire, could be very useful in permitting comparison between centers.

GOALS IN RESEARCH

It is clear that botulinum toxin has beneficial effects in patients with spasmodic dysphonia, but that the goal of a totally normal voice is rarely achieved. Clinical research is needed to identify factors which limit success and identify the means to overcome these. A major limitation to progress toward this goal, as in the treatment of dystonias in general, is the difficulty in objectively assessing and quantifying symptoms and signs, so that comparisons can be made between subjects and among treatment centers. The following questions should be addressed:

1. How can the optimal dose for each individual patient be clinically predicted?

2. How do the effects of bilateral and unilateral injection differ, and how can the best mode be predicted?

3. What are the factors which limit successful treatment of abductor dysphonia?

4. What role do extrinsic laryngeal muscles play in spasmodic dysphonia? Can injection of these muscles improve results?

5. What is the role of speech therapy as an adjunct to botulinum toxin therapy? What specific techniques should be used?

6. How can psychogenic spasmodic dysphonia be diagnosed and treated? Can it also respond to toxin?

7. How do psychogenic factors modulate the course of the disease and response to treatment of patients with organically based spasmodic dysphonia?

8. Can therapy be improved by the use of an alternative toxin with a longer duration of action?

REFERENCES

1. Adams RD, Victor M. Principles of Neurology, 4th ed. New York: McGraw-Hill, 1989:62-67.
2. Arnold GE. Spastic dysphonia: I. Changing interpretations of a persistent affliction. Logos 1959; 2:3-14.
3. Aronson AE, Brown JR, Litin EM, et al. Spastic dysphonia. I. Voice, neurologic, and psychiatric aspects. J Speech Hear Disord 1968; 33:203-218.
4. Aronson AE, Brown JR, Litin EM, et al. Spastic dysphonia II. Comparison with essential (voice) tremor and other neurologic and psychogenic dysphonias. J Speech Hear Disord 1968; 33:219-231.
5. Aronson AE, Hartman DE. Adductor spastic dysphonia as a sign of essential (voice) tremor. J Speech Hear Disord 1981; 46:52-56.
6. Aronson AE. Psychogenic voice disorders: an interdisciplinary approach to detection, diagnosis and therapy. In: Audio Seminars in Speech Therapy. Philadelphia: Saunders, 1973.
7. Blitzer A, Brin M, Fahn S, Lovelace RE. Clinical characteristics of focal laryngeal dystonia: Study of 110 cases. Laryngoscope 1988; 98:636-640.
8. Blitzer A, Brin MF, Stewart C, Aviv J, Fahn, S. Abductor laryngeal dystonia: a series treated with botulinum toxin. Laryngoscope 1992; 102:163-167.
9. Cannito M, Johnson J. Spastic dysphonia: A continuum disorder. J. Commun Dis 1981; 14:215-223.
10. Dedo HH. Recurrent laryngeal nerve section for spastic dysphonia. Ann Otol Rhinol Laryngol 1976; 85:451-59.
11. Ford CN. A multipurpose laryngeal injector device. Otolaryngol Head Neck Surg 1990; 103:135-137.
12. Freeman F, Cannito MP, Finitzo-Hieber T. Classification of Spasmodic Dysphonia by Perceptual-Acoustic-Visual Means. Spastic dysphonia: State of the Art 1984. New York: The Voice Foundation, 1984:5-13.
13. Ludlow CL, Naunton RF, Sedory MA, Schulz MA, Hallett M. Effects of botulinum toxin injections on speech in adductor spasmodic dysphonia. Neurology 1988; 38:1220-122?
14. Ludlow CL, Hallett M, Sedory SE et al. The Pathophysiology of spasmodic dysphonia and its modification by botulinum toxin. In: Beradelli A, Benecke R, eds. Motor Disturbances II. London: Academic Press, 1990.
15. Miller RH, Woodson GE, Jancovic J. Botulinum toxin injection of the vocal for spasmodic dysphonia. Arch Otol Laryngol Head Neck Surg 1987; 113:603-605.
16. Rabuzzi DD, McCall GN. Spasmodic dysphonia: A clinical perspective. Trans Am Acad Ophthalmol Otolaryngol 1972; 76:724-728.
17. Robe E, Brulik J, Moore, P. A study of spastic dysphonia. Neurologic and electroencephalographic abnormalities. Laryngoscope 1960; 50:219-245.
18. Rosenfield DB, Donovan DT, Sulek M, Viswanath NS, Inbody GP, Nudelman HB. Neurologic aspects of spasmodic dysphonia. J Otolaryngology 1990; 19:231-236.

19. Schaefer S. Neuropathology of spasmodic dysphonia. Laryngoscope 1983; 93:1183-1204.
20. Truong EE, Rontal M, Rolnick M et al. Double-blind controlled study of botulinum toxin in adductor spasmodic dysphonia. Laryngoscope 1991; 101:630-634.
21. Woodson GE, Zwirner P, Murry T, Swenson M. Use of flexible laryngoscopy to classify patients with spasmodic dysphonia. J Voice 1991; 5:85-91.
22. Woodson GE, Murry T, Zwirner P, Swenson M. Assessment of patients with spasmodic dysphonia. J Voice 1992, in press.
23. Zwirner P, Murry T, Swenson M, Woodson G. Acoustic changes in spasmodic dysphonia after botulinum toxin injection. J Voice 1991; 5:78-84.
24. Zwirner P, Murry T, Woodson G. Phonatory function of neurologically impaired patients. J Commun Dis 1991; 24:287-300.
25. Zwirner P, Murry T, Woodson G. Effects of botulinum toxin therapy in patients with spasmodic dysphonia: Acoustic, aerodynamic, and videoendoscopic findings. Laryngoscope 1992; 102:400-406.

PRELIMINARY OBSERVATIONS ON THE DIFFUSION OF BOTULINUM TOXIN FROM THE SITE OF INJECTION IN LARYNGEAL MUSCLES

K. Linnea Peterson,[1,2] Erich S. Luschei,[2,3] and Mark T. Madsen[4]

[1]Department of Otolaryngology-Head and Neck Surgery
[2]The National Center for Voice and Speech
[3]Department of Speech Pathology and Audiology
[4]Department of Radiology, Division of Nuclear Medicine
The University of Iowa
Iowa City, Iowa 52242

INTRODUCTION

Botulinum neurotoxin A is a very potent neurotoxin which causes a flaccid paralysis upon ingestion or injection. Although it has been a feared food poisoning for years, clinical uses in medicine have been developed and refined so that localized injection is now an accepted form of treatment of various neuromuscular disorders. The clinical use of botulinum toxin was pioneered by Scott[1], who developed it as a treatment modality for strabismus and blepharospasm. It has been used in ophthalmology since about 1980. The implications for otolaryngology and neurology have also been noted and it has since been used for such disorders as hemifacial spasm, torticollis, and spasmodic dysphonia[2,3,4,5,6,7]. Despite the known systemic effects from ingestion of the toxin, very few systemic effects from local injection have been reported. Ingestion is known to cause symptoms which suggest CNS involvement (e.g. dizziness, lethargy, general locomotor dysfunction), however the toxin has not been shown unequivocally to cross the blood brain barrier[8]. Antibody production has been noted in patients who received the larger doses (200-300 Units) used in the treatment of torticollis[3]. In addition, a patient with parkinsonism was noted in one study[3] to develop fatigue after two series of injections (81.7 total units). Those studies that have explored the systemic effects physiologically have involved single fiber EMG studies at sites distant from the injection[9,10]. These studies have shown increased "jitter" in distant muscles, but no clinically apparent effects. A recent study looking at spread of the toxin after intramuscular injection looked at whether or not the toxin crossed fascial planes[11]. The results showed that botulinum toxin passed through muscle fascia even at subclinical doses, with fascia reducing the spread of toxin by only 23%.

It has seemed possible to us that a significant fraction of the botulinum toxin injected into a muscle for treatment of muscle spasm is picked up in the blood circulation at the injection site before it is bound to cholinergic nerve terminals. Whether or not this occurs must depend upon the ability of toxin molecules, in areas of high concentration, to enter the muscle's microcirculation. The purpose of our study was to determine whether this occurs in laryngeal muscles, and if so, to estimate the degree and time course of this path of diffusion from the site of injection. To do this, we have injected radiolabelled botulinum toxin into the cricothyroid muscles of anesthetized dogs and quantitatively measured the amount of radioactivity in this region at frequent intervals. If toxin molecules do not enter the systemic circulation, then the amount of radioactivity in the area of the larynx should not change substantially over time. On the other hand, if they do enter the circulation, we would expect the amount of radioactivity to decrease substantially over the first few hours of observation.

MATERIALS AND METHODS

Radiolabelling

Botulinum Neurotoxin A (Sigma Chemical Company, St. Louis) was labelled with [131]I (New England Nuclear) according to the Chloramine T method as described by Williams[12], with reference to Greenwood, et al.[13] [131]Iodine was chosen due to its energy level, which allows both *in vivo* imaging with a standard gamma camera as well as scintillation counting *in vitro*. It has a half life of 8.5 days, and a high energy decay which contributes to protein degradation.

Once the toxin had been labelled, it was important to verify what has been labelled. Since radiolabelling could result in breakdown of the protein, it was critical to know what was being measured in order to be able to draw valid conclusions. Thus a purification step was used to eliminate components of varying activity, including free iodine. We used SDS-PAGE as described by Tse et al.[14], in order to evaluate the labelling efficacy.

Experimental Procedures

Three mongrel dogs, obtained from the Animal Care Unit at the University of Iowa, were initially anesthetized with an intramuscular injection of ketamine and zylazine. The femoral vein was then cannulated for infusion of supplemental anesthetic and normal saline, and for withdrawal of blood samples. Nembutal was given in a dose of approximately 25 mg/kg, or as necessary to produce and maintain deep surgical anesthesia, and then the animal's head was placed in a head-holding frame in a supine position. A ventral incision in the neck region over the larynx was used, with minimal dissection, to expose the cricothyroid muscles. During the course of the experiment, the depth of the anesthesia was checked frequently and supplemental nembutal given as needed.

Serial images of the radiolabelled botulinum toxin were acquired by a gamma camera interfaced to a computer. After the gamma camera was centered over at the larynx of the dog, at a distance of about 25 cm, 10 μC (experiment 1) or 5 μC (experiment 2 and 3) of purified radiolabelled botulinum toxin was injected into the cricothyroid muscle on each side. The radiolabelled toxin had been diluted to obtain a solution of approximately 10 μC/0.1 ml. The actual injection required about 15 seconds. The syringe containing the labelled toxin was immediately withdrawal from the field of the camera, and records of the radioactivity were immediately started. Scanning took place at time intervals of every five seconds for the first minute, then approximately every minute for the next five minutes, approximately every five minutes out to thirty minutes, then once every ten minutes. The clearance of the radioactivity was monitored from the counts in each computer image in a region of interest covering the injected muscle (Figure 1). The collimation used with this camera was not very precise, but was sufficient to ensure that radiolabelled material located in the animal's body would not register in the region that was analyzed.

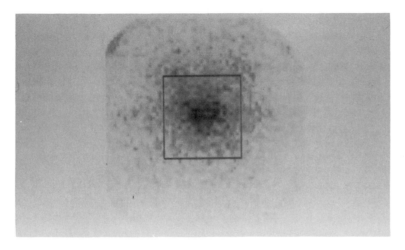

Figure 1. Typical computer-generated image of the region covered by the gamma camera (experiment 2). The computer provided total counts in the area defined by the square. Two slightly brighter "peaks" can be seen in the middle of the bright area; they correspond to the two cricothyroid muscles.

In addition to imaging the site of injection, serum radioactivity was measured by drawing one milliliter blood samples at various intervals corresponding to the imaging protocol (except during the first minute following injection). Once the blood samples were all collected for a given experiment, they were analyzed by running them through a scintillation counter, using a ten minute counting period.

Additional Studies of the Radiolabelled Botulinum Toxin

A Lowry protein assay was done to correlate the radioactivity of the labelled toxin with its protein concentration. Also, an LD-50 (amount of toxin which would kill 50 percent of a group of mice injected with this same dose, within four days) was conducted using Swiss-Webster mice as described by Williams, et. al., (12) with reference to Reed and Muench (15). This procedure was necessary to correlate the biological activity of the labelled toxin with its specific radioactivity.

RESULTS

Results of the sodium dodecyl sulfate-polyacrylamide gel electrophoresis (SDS-PAGE) under reducing conditions showed two bands consistent with the results of other investigators[14], one at 97 kilo Daltons, and one at 55 kilo Daltons. These two

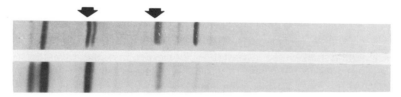

Figure 2. SDS-PAGE gel of radiolabelled botulinum toxin. The stained gel (top) shows prominent bands corresponding to 55 kilo Daltons (left arrow) and 97 kilo Daltons (right arrow). These bands are radioactive, as shown on the film exposure of the gel (below). Two bands corresponding to lighter-weight fractions, are also radioactive.

bands are consistently labelled with the [131]I (Figure 2). In addition there are two bands of much smaller molecular weight, 29 kilo Daltons and 34 kilo Daltons, which are also labelled.

The Lowry protein assay of each batch of radiolabelled toxin showed that the amount of toxin recovered from the labelling procedure was quite variable. The protein concentration of the labelled product ranged from 0.02 mg/ml to 0.09 mg/ml, as compared with the unlabelled toxin concentration of 1.0 mg/ml to 1.2 mg/ml. Obviously some of the toxin was lost in the labelling and purification steps, with dilution all playing a role. This observation draws attention to the need for assay of all batches of labelled toxin.

The results of the experiment to establish the biological activity of the labelled toxin were very disappointing. Only the first batch of labelled material resulted in death of the mice injected with higher doses. In all the subsequent experiments, none of the mice died even though the doses given were large enough to compensate for the fact that the protein concentrations of these latter batches was lower than the first batch. Nevertheless, the first batch did exhibit biological activity, which represented about 300 ng per mouse Unit.

The time course of the radioactivity in the neck region of the three dogs during the first two hours after injection is shown in Figure 3. The results are reasonably consistent in showing a rapid initial decay followed by a much slower loss of radioactivity. The most rapid decay is within the first twenty minutes for each trial. The higher levels of radioactivity for the first experiment are a result of using twice the amount of radioactivity (20 μC) as compared with subsequent trials (10 μC). Despite the different scale, the pattern is the same. Expressed as a percentage, the loss of radioactivity was about 25%, 46% and 43% in experiments 1-3 respectively.

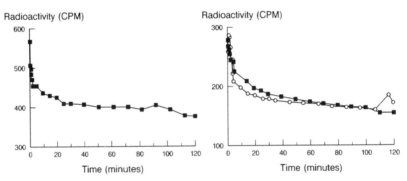

Figure 3. Change in the radioactivity in the neck region of the dogs over two hours. Results of experiment 1 are shown on the left, and experiments 2 (solid squares) and experiment 3 (open circles) on the right.

None of the blood samples collected from the dogs were found to contain significant levels of radioactivity. Radiation was detected from control samples of diluted labelled toxin, however, suggesting there was nothing fundamentally wrong with our procedure. Calculations taking into account the sensitivity of the scintillation counter and dilution within the blood volume of the dog suggest that our procedure should have been able to detect the diffusion of less than 1% of our injected dose. Such a calculation makes the assumption, however, that the toxin lost from the injection site remains in the blood. Our failure to detect what appears to be much larger losses of radioactivity from the injection site may reflect the fact that toxin molecules entering the blood system at the injection site quickly leave the circulation in areas of lower toxin concentration, i.e. other muscles in the body, and become bound to receptors.

DISCUSSION

Our results suggest, however tentatively, that a substantial fraction of a dose of botulinum toxin injected into a well-perfused muscle disappears from that site within the first two hours. This result is not too surprising if one supposes that the toxin molecule can cross the epithelium of muscle capillaries, which it must be able to do if ingestion of the toxin is able to paralyze muscles throughout the body. Yet our experiment is seriously flawed in several respects, so even this "common sense" interpretation must be regarded with suspicion.

Determining the systemic distribution of botulinum toxin emerges as a more difficult problem than we had initially supposed it to be. Part of the problem is the toxin itself; it is very potent, having a median lethal dose of 5-50 ng/kg body weight[16], thus making detection very difficult. There are various labelling techniques that can be considered, each with advantages and disadvantages. We chose to use ^{131}I for several reasons. Radioiodination of botulinum toxin has previously been described in detail[12], with good preservation of biological activity. Previous descriptions of the technique have described the use of ^{125}I. ^{131}Iodine, however, has a higher energy level than ^{125}I, making dynamic detection more feasible, and standard equipment used for clinical imaging of radioisotopes is scaled to the energy levels of ^{131}I. ^{131}I has drawbacks of a shorter half-life and greater risk of protein degradation, however.

The SDS-PAGE gel appears to be a necessary check on labelling procedure since free label present in the injected solution would obviously give false results. Our gels showed the label attached to heavy fragments of the toxin, but smaller fragments were also labelled. This fact needs to be kept in mind when interpreting our findings.

Our experience also indicates that it is necessary to determine the biological activity of the labelled material. Inactive toxin may fail to bind to receptors locally, and thus falsely suggest diffusion from the injection site. Unless the biological activity is known with some precision the results will never be suitable for extrapolating back to the doses that are typical of clinical use.

In presenting these very preliminary results we hope to draw attention to what we regard as an important question, and to address a few of the problems that one may encounter in trying to use radiolabelling methods to answer the question. It is important to refine a labelling technique for ^{131}I until it provides a reasonably high and consistent concentration of protein, has the label attached to the known heavy fragments of botulinum toxin, and has biological activity suggesting that most of the labelled toxin molecules are intact. These factors must be controlled in order to draw valid conclusions.

ACKNOWLEDGEMENTS

This research was supported by grant P60 DC00976 from the National Institutes on Deafness and Other Communication Disorders.

REFERENCES

1. A.B. Scott, Botulinum toxin injection of eye muscles to correct strabismus, *Trans. Am. Ophthalmol. Soc.* 79:734-770 (1981).
2. C.L. Ludlow, R.F. Naunton, S.E. Sedory, G.M. Schultz, and M. Hallett, Effects of botulinum toxin injections on speech in adductor spasmodic dysphonia, *Neurology* 38:1220-1225 (1988).

3. M.F. Brin, S. Fahn, C. Moskowitz, A. Friedman, H.M. Shale, P.E. Greene, A. Blitzer, T. List, D. Lange, R.E. Lovelace, and D. McMahon, Localized Injections of botulinum toxin for the treatment of focal dystonia and hemifacial spasm, *Movement Disorders* 2:237-254 (1987).

4. A. Blitzer, M.F. Brin, P.E. Greene, and S. Fahn, Botulinum toxin injection for the treatment of oromandibular dystonia, *Ann. of Otol., Rhin. and Laryngol* 98: 93-97 (1989).

5. J. Jankovic and J. Orman, Botulinum A toxin for cranial-cervical dystonia: A double-blind, placebo-controlled study, *Neurology* 37:616-623 (1987).

6. C.L. Ludlow, R.F. Naunton, S. Terada, and B.J. Anderson, Successful treatment of selected cases of abductor spasmodic dysphonia using botulinum toxin injection, *Otolaryngol. Head Neck Surg.* 104: 849-855 (1991).

7. C.L. Ludlow, M. Hallett, S.E. Sedory, M. Fujita, and R.F. Naunton, The pathophysiology of spasmodic dysphonia and its modification by botulinum toxin, *in*: "Motor Disturbances II," A. Berardelli, R. Benecke, M. Manfredi, and C.D. Marsden, eds., Academic Press Limited, London (1989).

8. D.A. Boroff, and G.S. Chen, On the question of permeability of the blood-brain barrier to botulinum toxin, *Int. Archs. Allergy Appl. Immun.* 48:495-504 (1975).

9. H. Lundh, H.H. Schiller, and D. Elmqvist, Correlation between single fibre jitter and endplate potentials studied in mild experimental botulinum poisoning, *Acta Neurol. Scandinav.* 56:141-152 (1977).

10. D.B. Sanders, E.W. Massey, aand E.G. Buckley, Botulinum toxin for blepharospasm: Single-fiber EMG studies, *Neurology* 36:545-547 (1986).

11. C.M. Shaari, E. George, B-L Wu, H.F. Biller, and I. Sanders, Quantifying the spread of botulinum toxin through muscle fascia, *Laryngoscope* 101: 960-964 (1991).

12. R.S. Williams, C-K Tse, J.O. Dolly, P. Hambleton, and J. Melling, Radioiodination of botulinum neurotoxin type A with retention of biological activity and its binding to brain synaptosomes, *Eur. J. Biochem.* 131:437-445 (1983).

13. F.C. Greenwood, W.M. Hunter, and J.S. Glover, The preparation of ^{131}I-labeled human growth hormone of high specific radioactivity, *Biochem J.* 89:114-123 (1963).

14. C-K Tse, J.O. Dolly, P. Hambleton D. Wray, and J. Melling, Preparation and characterisation of homogenous neurotoxin type A from Clostridium botulinum, *Eur. J. Biochem.* 122:493-500(1982).

15. L.J. Reed and H. Muench, A simple method of estimating fifty percent endpoints, *Amer. J. Hygiene.* 27:493-497 (1938).

16. L.C. Sellin, The action of botulinum toxin at the neuromuscular junction, *Med. Biol.* 59:11-20 (1981).

THERAPEUTIC BOTULINUM TOXIN: HISTOLOGIC EFFECTS AND DIFFUSION PROPERTIES

Gary E. Borodic[1], L. Bruce Pearce[2], and Robert Ferrante[3]

[1]Department of Ophthalmology, Massachusetts Eye and Ear Infirmary
[2]Department of Pharmacology, Boston University School of Medicine
[3]Department of Neuropathology, Massachusetts General Hospital, Harvard Medical School

INTRODUCTION

Since the introduction of botulism toxin into therapeutic medicine in 1978,[1,2] the use of this drug has been expanded to other indications including blepharospasm, adult onset spasmodic torticollis, spasmodic dysphonia, occupational hand disease and jaw dystonia. Application of this therapy to other disorders is on the horizon and is further contributing to the driving force for expansion of clinical and basic research. However, despite the success obtained with botulinum toxin for treatment of blepharospasm and other focal and segmental movement disorders, its application is limited by the following properties of the therapy:

1. Repeated injections are required indefinitely when treating chronic disease.
2. Untoward spread of toxin to other muscles not targeted for injection.
3. Antibody formation with resistance to the therapeutic action of the toxin subsequent to repeated injections.
4. The consistency of biologic activity contained within the labeled vials.
5. Lack of standardization of the injections sites for the treatment of each syndrome.
6. Placement of the therapeutic toxin preparation into the correct anatomic position when access to the muscles is difficult requiring teflon coated electromyographic assistance (particularly for treatment of occupational hand disorders).
7. Adequate understanding of long term effects of repeated treatment with the therapeutic preparation.

This chapter will address the clinical importance of diffusion model to the efficacious administration of this drug and histologic changes which occur with therapeutic injections. Issues of efficacy as well as safety can be addressed by the diffusion model. The rabbit longissimus dorsi muscle model provides a method to study the diffusion properties of intramuscular injections of therapeutic botulinum toxin.

Botulinum and Tetanus Neurotoxins, Edited by
B.R. DasGupta, Plenum Press, New York, 1993

Strabismus

Strabismus is an ophthalmic term which describes the pathologic misalignment of the eye. The term is very general and is inclusive of paralytic disorders of the extraocular muscles from muscle and cranial nerve diseases, congenital defects within the extraocular muscles, or muscle imbalances associated with optical errors, visual sensory deficits or disorders of the image fusion coordinating centers within the central nervous system. In the past, therapy has included spectacles lens, and surgical procedures used to tighten and loosen various muscles to achieve better alignment. Botulinum toxin had been advocated by Scott[1,2] as an alternative to conventional extraocular muscle surgery because of the increased simplicity of the procedure avoiding a more complex surgical procedure. Unfortunately therapy offered by the toxin injection is temporary, and this injection needs to be repeated in a majority of cases to maintain ocular alignment. The ability of the clinician to exactly target the desired muscle is limited by botulinum toxin diffusion away from the site of injection. Treatment usually involves injection of the medial rectus or lateral rectus muscles (horizontal rectus muscles) to treat horizontal deviations of the eyes. Diffusion of botulinum toxin into the vertical rectus muscles resulting in the complications is common.[1-4] Induced vertical deviation of the eye when horizontal deviations are being treated is clearly a limiting factor occurring in approximately 15% of cases.[4]

Blepharospasm and Meige Syndrome

Over the course of the past 12 years the therapy has expanded to treatment of focal and segmental dystonias.[5-11] These conditions involve involuntary contractions of a group of muscles in a region of body causing debilitating symptoms. Neurologic imaging and autopsy usually do not reveal structural lesions within the brain. There is general agreement that these conditions occur as a result of a disorder within the central nervous system. The chapter in this volume by Dr. Mitch Brin will more comprehensively review the clinical details of these syndromes.

The first dystonia that botulinum toxin therapeutic technology had been applied was benign essential blepharospasm and Meige Disease.[12] This form of neurologic blepharospasm is associated with blinding involuntary eyelid closure which leads to debilitation and desperation. Driving an automobile becomes impossible as the condition progresses and in worse cases situations, the patients can become unable to independently ambulate. In the past, therapy had included the use of neuroleptic medications which were usually ineffective.[13] Surgical therapy involving transecting the facial nerve or removing the orbicularis oculi in surgical stripping procedures[14,15] was tried, but in many patients this approach is only partially effective and occasionally associated with undesirable complications.[14,15]

A distinctive form of blepharospasm occurs with hemifacial spasm and aberrant regeneration of the seventh cranial nerve. In each of these conditions, there is an involuntary movement in muscles only on one side of the face.[16] Hemifacial spasm is characterized by intermittent synchronous involuntary contractions in all muscles supplied by one facial nerve. This disorder is thought to result from damage to the intracranial portion of the facial nerve caused by tortuous blood vessels at the base of the brain. Aberrant regeneration of the facial nerve causes constant involuntary eyelid closure during active facial expressions.

Botulinum toxin has become the only consistent therapy for the treatment of neurologic forms of blepharospasm. The eyelids are injected with small quantities of botulinum toxin (15-75 IU) producing an effective weakening of the protagonist muscle of eyelid closure, the orbicularis oculi. This muscle must be injected every three months in the case of essential blepharospasm and Meige syndrome and every 5 months in blepharospasm associated with hemifacial spasm and aberrant regeneration of the seventh cranial nerve.[5,7] As there were very few alternative therapies for these conditions, the considerable degree of efficacy obtained for these diseases was a therapeutic advance.

As botulinum toxin treatment for this disease has proven to be effective,[5,7,17,18] it has also been limited by several complications, including ptosis, exposure keratopathy, diplopia (double vision) and epiphora (tearing). Ptosis is defined as a drooping of the upper eyelid causing encroachment of the upper lid on the visual axis and effectively decreasing vision. This complication is generally transient and disappears as the denervative effect of the botulinum toxin wears off. It occurs approximately in 10% of patients treated and is particularly prone to occur when the injections of botulinum toxin are made close to the superior sulcus (Figure 1). In that patients already have eyelid disease causing obstruction of vision, the ptosis further aggravates their visual function. Understanding the cause of this complication is vital to the effective application of the therapy. The occurrence of ptosis is thought to relate to diffusion of the toxin from injected orbicularis oculi into the superior orbit effectively weakening the retractor of the upper eyelid, the *levator palpebrae superioris* muscle (Figure 2). Weakening of the muscular portion of the levator causes dropping of the upper lid margin. Given that the control of the upper eyelid movement is primarily a balance between the *orbicularis muscle* and the *levator palpebrae superioris muscle*, effective application of botulinum toxin for this condition involves containing biologic effect of the toxin only to the *orbicularis* muscle, the protagonist muscle causing the abnormal movement. The anatomic configuration of the muscles of the upper eyelid allow selective weakening of the orbicularis muscle and therefore effective therapy with botulinum. The long tendon of the upper eyelid retractor, the *levator aponeurosis*, extends along the undersurface of the orbicularis muscle into the tarsal plate. In that the muscular portions of the levator palpebrae superioris muscle is remote to the antagonist (orbicularis), there is little opportunity for the botulinum toxin to diffuse into the levator muscle of the eyelid. This anatomic distance between eyelid pretarsal eyelid orbicularis

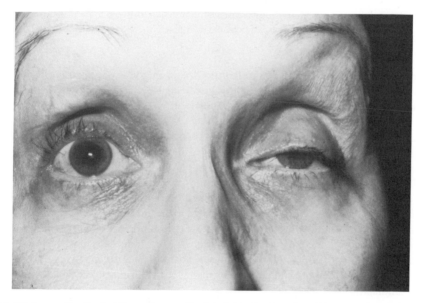

Figure 1. Ptosis complicating botulinum therapy. Ptosis is defined as drooping of the upper eyelids. The patient shown here suffered ptosis for a period of 3 weeks following botulinum toxin injection for the treatment of hemifacial spasm. The injection sites in this patient were close to the superior sulcus and specifically not close to the lash line along the lateral and medial extent. This complication results from diffusion of toxin into the orbit causing weakening of the lid retractor musculature.

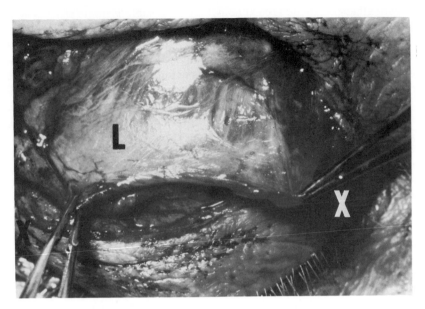

Figure 2. Importance of eyelid anatomy to the beneficial clinical effect of injectable botulinum toxin. This photograph represents a surgical field demonstrating the levator palpebrae superioris muscle. Note that the upper lid retractor extends only as a tendon underneath the orbicularis muscle throughout the upper lid. The actual muscular portion of the upper lid retractor is deep in the orbit, distant from superficial upper lid structures. Such an anatomical arrangement allows botulinum toxin to specifically effect the protagonist of eyelid closure, the orbicularis muscle, which is targeted in the treatment of blepharospasm. The anatomic space between the upper lid retractor and the orbicularis muscle allows for a selective denervation of the orbicularis muscle without involvement of the antagonist levator muscle. Anatomic distances between muscles are important in the administration of botulinum toxin because its potential to diffuse.

(pretarsal orbicularis) and levator muscle provides an anatomic explanation for therapeutic success in selectively targeting the orbicularis in treating blepharospasm. This explanation also provides insight into the cause of this complication and a method to mitigate against its occurrence (Figure 3).

Another complication associated with therapeutic injections of botulinum toxin into the eyelids is diplopia. This occurs less commonly (<5% patients) and is transient, lasting several days to several weeks. Nelson and her co-workers[19] have linked this complication to injections in the lower lid, particularly the inner aspect of the lower lid. The reason for this complication appears to relate to the anatomic proximity of the *inferior oblique muscle* to the inner portion of the lower lid. The inferior oblique muscle arises in the very anterior portions of the medial orbit and penetrates the major fascia of the lower lid (capsulopalpebral fascia) very close to the cutaneous surface of the medial lower lid. Injections into the inner aspects of the lower lid bring the toxin within several millimeters of this muscle so that the toxin can readily diffuse into this region resulting in this complication. Avoiding medial lower lid injections reduces the incidence of this complication.[19]

With respect to the issue of efficacy in the treatment of blepharospasm, it appears that diffusion may again play a role. The location of injection sites used today was empirically derived and based on apparent efficacy as well as a strategy that limited complications. Because multiple injection points are needed, this has added to the discomfort of the application procedure. We initially studied the effects of injecting only electrically determined motor points of the orbicularis oculi.[20] The motor points of orbicularis oculi are essentially in two

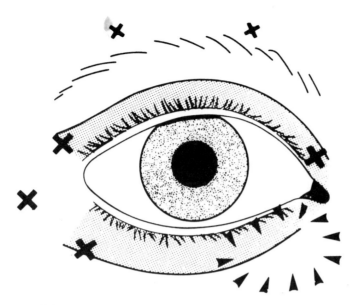

Figure 3. Anatomy and injection sites. The usual injection sites for the treatment of blepharospasm are outlined by cross marks. Note that upper lid injections should be close to the lash line and at the lateral and medial extreme positions. These anatomic positions maximize the distance between injected orbicularis and the antagonistic muscular portion of the levator palpebrae superioris. Maximizing this distance allows a more specific effect on the orbicularis muscle and avoids the complication of ptosis as seen in Figure 1. Also, note that the inner aspects of the lower lid are avoided (circle of arrows). Avoiding the area marked helps prevent diplopia which can result from diffusion of the biologic effects of the toxin into the inferior oblique muscle which is directly deep to this area of the lid.

locations. The first is in the upper outer portion of the eyelid and the second motor point is in the medial portion of the eyelid. The motor point is defined as the area within muscle which has the lowest threshold for contraction to external electrical stimulation. Generally the motor point corresponds to points of major motor nerve branch penetration into the muscle proper.[20] Comparisons of single motor injection to multiple point injections were done on a series of ten patients using the multiple point injection strategy on one eye and the motor point injection strategy on the other, each group receiving the same dose. Eight of the 10 patients found that the multiple point injected eye had a substantially better effect than the motor point injected. Two of the 10 patients noted no difference between the two eyes.[20] The results of this clinical study suggested several points:

1. Dose independent variables are involved in determining efficacy in botulinum toxin treatment.

2. Spreading the toxin throughout the muscle may have increased the toxin diffusion throughout most of its innervation zone.

3. The results tend to negate the prevailing opinion that motor points could provide a superior injection location.

In an effort to explore the *innervation zone*, that is, *the topographic distribution neuromuscular junctions within a muscle* , strips of orbicularis oculi muscle taken from the pretarsal portions of the muscle taken during routine ptosis surgery were analyzed for concentration and distribution of the neuromuscular junctions using a standard acetylcholinesterase stain. Although only the pretarsal portions of the muscle were analyzed in this study, however, this portion of the muscle is exactly over the lateral motor point of the upper lid.

The results indicated that the neuromuscular junctions were diffusely distributed throughout the entire upper portion of the orbicularis muscle. This finding clearly correlates with the large degree of facial nerve projection and ramification into the orbicularis muscle from temporal, zygomatic and buccal branches. This observation lead to the hypothesis *that multiple injection points were preferable because this injection technique tended to cover more of the innervation zone of the orbicularis muscle* as compared to a single large dose point injection. For orbicularis muscle, this concept appears to be a plausible explanation for the superiority of the multiple point injection technique. It appears that diffusing the toxin along the muscle proper was preferable in this particular anatomic location. This study also *disproved the notion that motor point of the muscle corresponds to the innervation zone (areas of concentrated motor end plates).*

ADULT ONSET SPASMODIC TORTICOLLIS

Adult onset spasmodic torticollis is a segmental dystonia involving cervical muscles. The condition can occasionally be associated with Meiges Syndrome or other forms of head and neck movement disorders. A thickened protruding sternomastoid muscle is the most characteristic appearance of these patients on physical examination (Figure 4A). Torticollis has variable expression with the most common being a distorted posture with head *rotated to a side of shoulder elevation* (type 1, type 2).[6] Another variation involves the head *being tilted toward the side of shoulder elevation* (type 3). Another variation includes the head being tilted backwards involuntary (retrocollis, type 4a) or being flexed toward the chest (antecollis, type 4b). Patients often develop pain early in the course of the disease which often becomes progressive. The disease is often associated with involuntary jerking movements of varying frequency and amplitude. The patients in addition to pain, develop substantial stiffness and eventually obvious hypertrophied muscles.

This disease is chronic, and often unrelenting and progressive. Functional disabilities including driving, maintaining gainful employment often detract from quality of life. Often the patients develop depression and futile attitude towards relief because of the constant nature of the symptoms and prior experience with suboptimal therapy.

Past therapy has included the use of neuroleptic medications, use of various forms of myectomy and denervating surgery and occasionally biofeedback.[21,22] Unfortunately previous medical therapies have not been satisfactory. Surgical procedures are often associated with inconsistent results and occasionally disfiguring scarring and further impairment in posture.

The implementation of botulinum toxin for the treatment of this segmental disorder has been a great contribution to neurologic medicine. The application has proven to be the most efficacious therapy for torticollis and can be maintained over a period of years.[6,23]

The major complication associated with the use of botulinum toxin for spasmodic torticollis has been dysphagia.[24] Dysphagia is defined as difficulty swallowing which can occasionally lead to the misdirection of food into the upper airway. Such a misdirection can occasionally cause complete upper airway obstruction which is a medical emergency possibly leading to death. Upper airway obstruction has occurred in at least one patient involved in the North American clinical studies. This patient was immediately treated by the Heimleich maneuver successfully. Other complications have included weakness within the cervical muscle.

The diffusion model has proven to be important in finding a solution to the dysphagia problem. In retrospective studies, dysphagia appeared to be linked to the dose of botulinum toxin dose injected to the sternomastoid muscle (see Figure 4B). This retrospective data

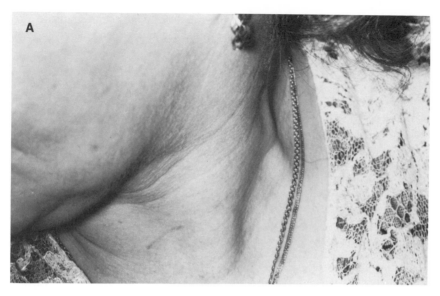

1 Sternocleidomastoid
2 Trapezius
3 Sternohyoid
4 Sternothyroid
5 Omohyoid
6 Scalenus Anterior
7 Scalenus Medius
8 Longus Capitis
9 Longus Cervicis
10 Transverse Arytenoid
11 Lateral Cricoarytenoid
12 Inferior Constrictor of
   Pharynx

13 Laryngopharynx
14 Thyroid Cartilage
15 Internal Carotid Artery
16 Internal Jugular Vein
17 Cervical Vertebra
18 Levator Scapulae
19 Splenius Capitis and Cervicis
20 Longissimus Capitis
21 Semispinalis Capitis
22 Semispinalis Cervicis
23 Multifidis

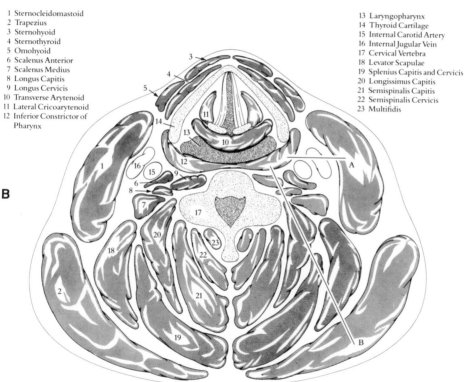

Figure 4. A. Dystonic sternomastoid muscle seen in adult onset spasmodic torticollis. Adult onset spasmodic torticollis is characterized by the presence of a large thickened indurated muscle as seen in this patient. The involuntary impulses coming from the central nervous system causes increased resting tone, frequent spasmodic contractions which leads to hypertrophy of this muscle. **B.** Dysphagia and the treatment of torticollis. Dysphagia (difficulty swallowing) can result from diffusion of botulinum toxin into the peripharyngeal musculature. This complication particularly occurs when high doses of botulinum toxin is given to the sternomastoid muscle which directly overlays the peripharyngeal muscles. This complication occurs as a result of direct spread of biologic effect of from injected sternomastoid muscle into the deeper structures of the neck. Table 1 and Table 2 outline the results of the clinical study linking dysphagia to the dose of botulinum toxin given over the sternomastoid muscle.

Table 1. Comparison of botulinum toxin dose and injection strategies in patients who later experienced dysphagia and those who did not.

	Dysphagia	No dysphagia
Sternocleidomastoid Dose		
No. Injection	7	42
Median dose	150	100
Interquartile Range	10	150
Wilcoxon test	Z=2.22	p=0.026
Total Dose		
No. Injection	7	42
Median dose	160	150
Interquartile range	25	100
Wilcoxon test	Z=0.75	p=0.45

Table 2. Influence of changing the injection protocol limiting the dose to the sternocleidomastoid muscle to 100 IU or less per injection: Prospective analysis.

	Dose not limited	Dose limited
Dysphagia	7	0
No Dysphagia	42	31
Total	49	31

Numbers refer to injections. Fisher's exact test (1-tailed), p = 0.027.

indicated that doses in excess of 100 IU injected to the sternomastoid muscle was associated with the complication (see Table 1). Prospective data analysis limiting the dose to the sternomastoid muscle was shown to markedly decrease the incidence of this complication (see Table 2). The complication rate reported initially in our studies was comparable to those noted in other studies, approximately 15%.[8,11,25] Limiting the sternomastoid dose to less than 100 IU during an injection session reduced the incidence of this complication to less than 2% (see Table 2). Plausible explanation for these findings was that diffusion of toxin from the sternomastoid into the peripharyngeal musculature resulted in weakening the muscles involved in the swallowing reflex. The sternomastoid muscle lies directly over the peripharyngeal musculature while the other muscles usually injected in the treatment of spasmodic torticollis (levator scapulae, posterior scalene, trapezius, splenius capitis, splenius cervicis as well as others) are more posterior located and remote to the pharyngeal muscles.

As it appears that dysphagia was secondary to toxin jump, that is, *toxin spread and diffusion from the sternomastoid muscle, limiting this complication can be accomplished by reducing the dose over this muscle* (see Figure 4B). Such an explanation for the results of these clinical studies suggests a *dose dependent diffusion phenomenon* over the sternomastoid muscle.

Efficacy in the treatment of adult onset spasmodic torticollis may also be dependent on diffusion of the biologic effects of the toxin from the injection sites. As this disease is definitely associated with multiple muscle group involvement in remote areas of the neck, injection of the entire complex of muscles involved with the posture disfigurement is necessary to achieve the most beneficial result.

Table 3. Response rates comparing multiple-single point injections.

	Single point	Multiple point	Mean dose of favorable response (IU)	Mean dose No response (IU)
Pain	15/31	27/31[a]	165.71	147[b]
Posture deformity	13/42	33/44[a]	162.7	148.3[b]
Range of motion	15/39	33/44[a]	156.4	144.3[b]
Activity	13/39	29/38[a]	187.9	145.6[c]
Hypertrophy	27/39	34/44	161.4	154.6[b]
Tremor	4/17	9/17	163.5	146.9[b]

[a]$p < 0.002$ by Chi square.
[b]not statistically significant (Wilcoxan).
[c]$p = 0.04$.

Given the superior clinical results obtained with multiple point injections used to treat blepharospasm, a study group was designed to test multiple point versus single point injection per muscle for the treatment of torticollis patients. Using typical efficacy criteria for spasmodic torticollis, that is, the effectiveness with respect to pain, posture deformity, range of motion of the cervical spine, hypertrophy, activity limitation, there was clearly better clinical results achieved within the multiple point injection group (see Table 3, Figure 5). There was no significant differences in the total dose given to each of these groups. These findings suggested that the administration technique within individual large muscles was important to efficacy. It is unknown whether the innervation zone to these muscles are diffusely spread throughout the muscle proper rather than focally distributed. However, the results of the clinical study tend to suggest a diffuse innervation of most muscles involved with the syndrome.

A DEPICTION OF THE DIFFUSION DENERVATION FIELD IN A PATIENT

An excellent example of the denervation field produced by a point injection of botulinum toxin is the regional depression of the vertical furrowing lines produced by the contraction of the corrugator muscles. These lines are associated with aging. Another application demonstrating the denervation field is to directly inject the frontalis muscle in patients with expressionistic overcorrection after ptosis surgery by frontalis sling procedure.[26] For instance, the patient shown in Figure 6 had undergone a frontalis sling to correct a total ptosis using a tendon graft taken from her leg. During the period of high amplitude facial movements naturally occurring during expression, the upper lid would become overcorrected. The contractility of her frontalis muscle was reduced on both sides for symmetry with point injections of botulinum toxin (see Figure 6). This resulted in the blunting of the transverse creases of the forehead over a circular area. The transverse creases are generally produced by the insertion of the frontalis muscle into the dermis of the forehead skin. With high amplitude contractions and high tone in the muscles, these creases are accentuated. The circular area of blunting of creases indicates decreased tone in the frontalis muscle over a defined region within this muscle. Of note, this patient was injected with 10 IU to each location and has produced a denervation field of approximately 15 mm radius.

The denervation field may be a phenomena that may vary within various muscle fiber arrangements, after repetitive injections and different preparations of the toxin. Further evaluations with respect to these variables are currently under study.

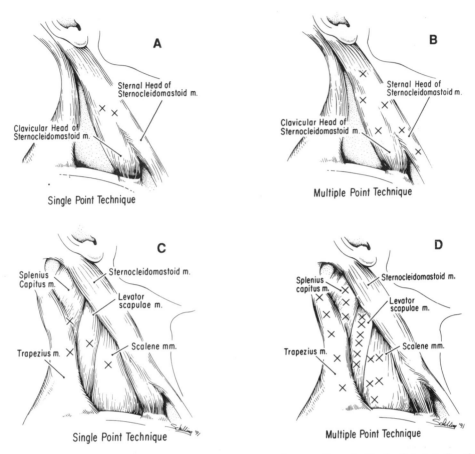

Figure 5A-D. Efficacy and injection technique in the treatment of spasmodic torticollis. In clinical studies, the method of administration is as important to the beneficial results as is the dose of botulinum administered. Figures 5 A-D depicts two types of injection strategies to large dystonic muscles along the anterior and posterior cervical muscles. The multiple point injection strategy per muscle produces a superior clinical benefit compared to single injection points per muscle or injection points just along a single area. The reason for these clinical may be that a multiple point injection strategy allows a more homogeneous diffusion of the biologic effect of the toxin along the targeted muscles (see Table 3).

HISTOLOGIC CHANGES AFTER BOTULINUM TOXIN INJECTION

The first apparent histologic effects appear to be clustering of vesicles at the presynaptic membrane indicating an impairment of release of acetylcholine.[27] Shortly within 7 to 10 days of injection, collateral axonal sprouting appears to occur at the terminal axon or occasionally from the distal node of Ranvier.[28,29] Within 10 to 14 days, the muscle fibers appear to undergo substantial atrophy. The atrophy continues to develop over a 4 to 6 week period. Figure 7 shows the fiber atrophy achieved in an albino rabbit longissimus dorsi 4 weeks after injection of 10 IU of botulinum toxin compared to a saline injected control specimen. The collateral axonal sprouts reestablish proximity to neuromuscular junction as demonstrated by Aldersen.[30]

During a period of 3 to 4 weeks after injection, there is considerable spread of acetylcholinesterase staining activity on muscle fibers. This diffuse acetylcholinesterase staining characteristic persists until the 12th to 16th week after injection after which there is considerable

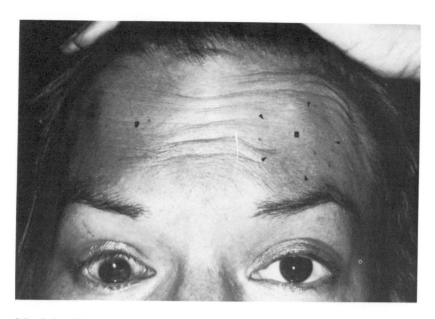

Figure 6. Depiction of denervation field in a patient. The forehead creases are generated by attachments to the frontalis muscle to the skin dermis. This patient was injected along the forehead. Note that the creases are blunted over a circular area after a point injection of botulinum toxin. The blunting of these creases indicates the field of effect of botulinum toxin in this clinical situation.

contraction into the neuromuscular junction. The fiber atrophy measured as fiber size diameter variability on cross sectional analysis is also a reversible phenomenon with recovery over a 4 to 6 month period. Spread of acetylcholinesterase on human muscle fibers 5 weeks after botulinum injection to the orbicularis oculi muscle is seen on Figure 8A. Figure 8B demonstrates the correlation between fiber size (diameter) variability and cholinesterase staining pattern 5 weeks after injection in albino rabbit muscle.

The cycle of histologic changes at the neuromuscular junction was first described by Duchen and co-workers.[28,29] Similar spread of cholinesterase staining activity and collateral axonal sprouting has been demonstrated on human muscle fibers.[24,30,31] Seventeen orbicularis oculi muscle specimens taken from patients during ptosis and myectomy surgery were evaluated for cholinesterase staining characteristics and fiber variability.[31] It appears that cholinesterase staining characteristics generally started at weeks 3 to 4 and maintained a diffuse staining pattern through 3 to 4 months after injection. In all patients studied after 6 months, the cholinesterase staining pattern could not be distinguished from controlled specimens. It is also of note that fiber size variability appeared to correlate temporally with cholinesterase staining pattern. Fiber size variability appeared to be transient lasting 3 to 12 weeks after injection (see Figure 9). These temporal relationships between these histologic changes closely correlates with the duration of action achieved with therapeutic doses of botulinum toxin.[31]

Another important goal in evaluating human muscles specimens after the injection of botulinum toxin is assessing long term effects. If human specimens were not injected within 6 months of biopsy, there appeared to be no difference in fiber morphometric analysis or cholinesterase staining pattern compared to the controls.[31] As there was no substantial differences in fiber size variability, cholinesterase staining pattern compared to controls in patients who had received multiple injections over several years, it appears that chronic denervation does not occur with repetitive use of the toxin. However, other authors have

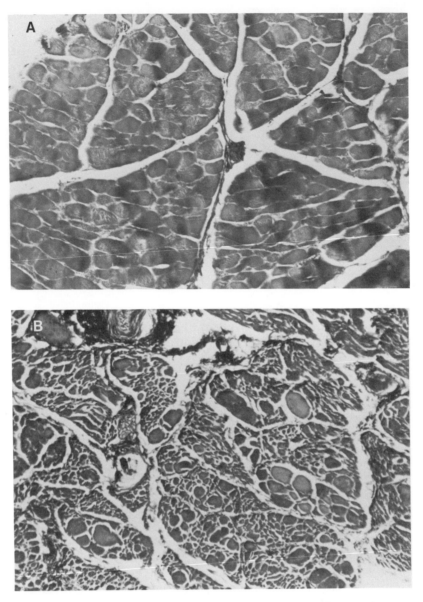

Figure 7 A,B. Muscle fiber response after botulinum toxin injection. Intense fiber atrophy occurs several weeks after botulinum toxin is administered. **B.** This photograph demonstrates large degree of fiber atrophy after 10 IU of botulinum toxin injected into the longissimus dorsi muscle of an albino rabbit back. A is a control injected with saline.

demonstrated possible permanent changes within the myoneural junction such as increased number of preterminal axon sprouts and projections into the myoneural junctions.[30] Although changes may be present at the neuromusclar junction with respect to sprouting, there was no substantial residual atrophy after multiple injections of botulism toxin. Such data appears to indicate that the renervation process is nearly complete after the botulinum toxin is no longer administered and there are no residual atrophic effects.

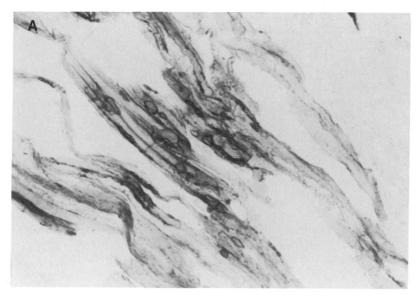

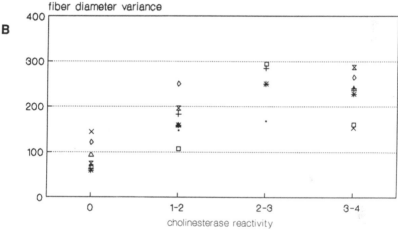

Figure 8. A. Acetylcholinesterase staining in human orbicularis muscle. Spread of acetylcholinesterase is seen on human muscle fibers 5 weeks after injection of botulinum toxin. This is a similar effect that was described in animal muscle fibers in the past. **B.** Correlation between fiber size variability and cholinesterase spreading characteristic. A direct correlation between fiber size variability and acetylcholinesterase spread characteristics on muscle biopsies evaluated from the animal study.

ATP-ase staining characteristics were also evaluated in animal muscle tissues injected with botulism toxin. There was a slight increase in the proportion of type I fibers at a pH of 9.4. However, type I muscle fiber grouping appears greater after the injection and was the most significant finding with this stain (see Figure 10). The grouping of Type I fibers is also associated with some fiber atrophy and increased fiber size variability. These changes appear to be quite subtle compared to the massive spread of cholinesterase seen in animal tissues 3 to 5 weeks after injection.

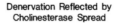

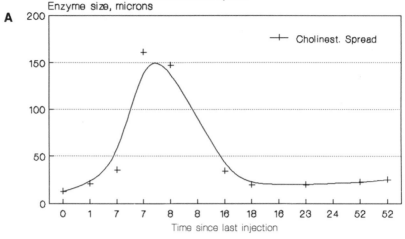

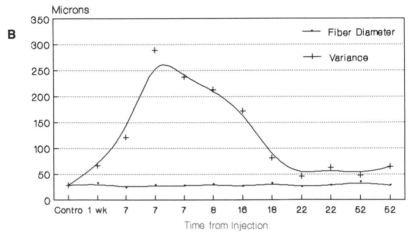

Figure 9. Cyclic histologic changes after botulinum toxin in human muscle. The duration of action of botulinum toxin appears to correlate with fiber variability and cholinesterase spread characteristics seen in human orbicularis muscle specimens. The results of these specimens were evaluated with respect to the last injection of botulinum toxin. The specimens were developed at the time of ptosis surgery and myectomy surgery in patients being treated by this technique for their diseases.

Histologic Determination of the Denervation Field

Muscle fiber size, fiber size variability, and acetylcholinesterase staining characteristics were used to assess muscle fiber response and diffusion of biologic activity from a point injection within a long muscle in an attempt to further define the denervation field scientifically. The albino rabbit longissimus dorsi muscle was chosen because of its length, generally parallel fiber orientation, and easy accessibility for muscle biopsy and injection. Longissimus dorsi in 2 to 3 kgs albino rabbits were injected at a point along the mid-dorsal spine.[24] At the injection point, a tattoo over the skin and muscle was applied using India ink. In addition to the tattoo, the injection point was anatomically placed at the dorsal eminence of the right scapulae to insure reproducible localization of the injection site even if the tattoo faded or tissue plane sliding made identification of the injection point difficult.

636

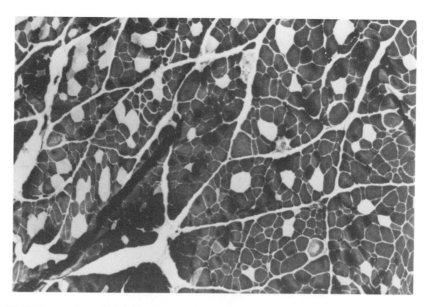

Figure 10. ATP-ase stain at pH 9.4. The amount of Type I fiber grouping appears to be greater after injection of botulinum toxin compared to controls. However, the total number of Type I and Type II fibers did not seem to be substantially different over controls. This appearance is consistent with denervation.

Botulinum A toxin was obtained from Oculinum Incorporated (Oculinum R). It is prepared in a lyophilized form and stabilized with human serum albumin. It is reconstituted in sodium chloride .9% for injection and was diluted to 1 IU per .1 ml, 2.5 IU per .1 ml, 5 IU per .1 ml in 10 IU per .1 ml. These dilutions were chosen because these concentrations are often used in clinical practice. Control injections were made with .9% sodium chloride diluent.

After injection, five weeks were allowed to elapse in order to provide an adequate interval for optimum muscle fiber atrophy and acetylcholinesterase spread.[24,28,29] After five weeks, the animals are sacrificed by lethal injection of Nembutol.

Dissection was taken over the dorsal spine removing the latissimus dorsi muscle and exposing the longissimus dorsi muscle down its entire length to the caudal end of the lumbar spine.

Biopsies were taken at 15mm intervals and processed with liquid nitrogen cooled isopentane -140°C (Figure 11A). Tissue specimens were subsequently stored at -80°C until stained using enzyme histochemistry for acetylcholinesterase and hematoxylin and eosin. Cut sections were made at 15 microns on the cryostat.

Histologic measurements were made with the bioquant II computer system for data collation and statistical analysis with fiber size variation comparisons were conducted using standard deviation and variance from measuring at least 200 fibers. Statistical variations in fiber diameter were compared with F ratio analysis.

The hematoxylin and eosin stained specimens were used to analyze fiber size and fiber size variability because of higher contrast obtained with this stain. Fiber diameter and diameter variance analysis were done in each specimen beginning at the injection site, and at 15mm intervals to 45mm from the injection point.

Intensity of acetylcholinesterase staining was estimated using reference photographs representing gradations of spread and intensity of staining (see Figure 11B–E). These gradations were rated 0-4. The spread and intensity of acetylcholinesterase staining activity for each biopsy was matched to the closest reference photograph. Four biopsies were developed from each

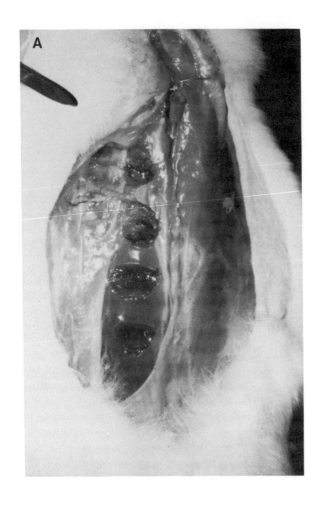

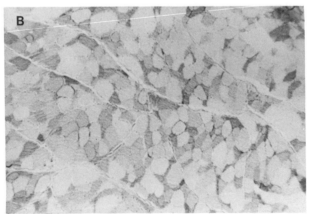

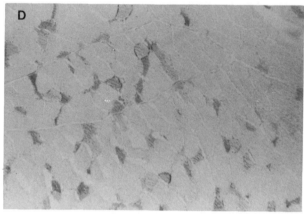

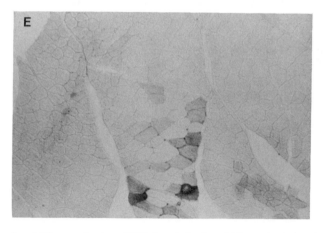

Figure 11. A. Animal model for quantization of diffusion. Animal model for evaluating diffusion of botulinum toxin down longissimus dorsi involves taking multiple biopsies along this muscle. The contralateral muscle is also used for assessment of extramuscular spread. Other contiguous muscles can also be used. **B–E.** Diffusion down the longissimus dorsi muscle can be monitored using the Acetyl- cholinesterase staining characteristic. Note that at the injection site after 1.25 IU of botulinum toxin, there is diffuse spread of cholinesterase. Over 45 mm, there is gradual decrease in cholinesterase spreading until the stain become concentrated only at the neuromuscular junctions. The photographs used in this illustration provided reference standards by which tissue at varying distances were used to evaluate diffusion of biologic effect at varying doses (see Figures 12,16).

muscle including controls and 2-3 animals were used for each dose determination in the histochemical analysis as well as fiber diameter variability and size analysis.

The muscle fiber average diameter was determined from summation of counts on four biopsies taken at 15mm intervals over a linear distance of 4.5 cm from the injection site. The average muscle fiber diameter appeared to correlate to the dose of botulinum toxin administered (Figure 12). The average muscle fiber diameter after a 10 IU injection was 26.7 microns (s=14.8) (n=1600), average diameter at 5 IU was 31.7 microns (s=14.6) (n=1600), the average diameter at 2.5 IU was 30.4 microns (s=14.0) (n=1600), the average diameter at 1 IU was 30.7 microns (s=11.1) (n=2400). Control fiber diameter was 35.4 microns (s=9.2) (n=1600).

Fiber size variability also correlated directly with the dose administered at the injection site (Figure 13) and throughout the entire muscle.

The field of biologic activity within the injected muscle was assessed using fiber size variability and acetyl cholinesterase staining characteristic. The diffusion of biologic activity within the injected muscle correlated with dose administered. Fiber size variability at 1 IU became insignificant compared to controls at 15 mm (F ratio <1.4 based on 200 fiber counts per specimen). At 2.5, 5, 10 IU the biologic effect reflected by fiber size variability was sustained throughout a 45 mm length along the muscle strip (see Figure 14). Fiber size variation was significantly different from controls at all higher doses down the entire muscle strip (F ratio > 1.4). Spread and intensity of acetylcholinesterase staining confirmed the biologic effect substantially diminished at 15 mm for the 1 IU dose. The acetylcholinesterase staining characteristic suggested higher doses (2.5-10 IU) produced a biologic effect throughout the 45 mm length of the muscle strip (see Figure 15).

In order to assess extramuscular diffusion properties of botulinum A toxin, fiber diameter variations and fiber diameter size were determined on the *contralateral longissimus dorsi muscle at 45 mm from the injection site at each dose*. Fiber size variation was significantly greater in the injected muscle at 45 mm than 45 mm at the extramuscular location for 10 IU (F=2.5, p<0.01),and 5 IU (F=1.7, p<0.01). For 2.5 IU and 1 IU the fiber variation comparing the intramuscular injection and the extramuscular injection was not significant. This data indicated that linear spread of biologic effect may be greater within the injected muscle than in a remote muscle at an equivalent distance from the point injection of botulinum toxin although the biologic effect did spread to contiguous muscles when larger doses were used.

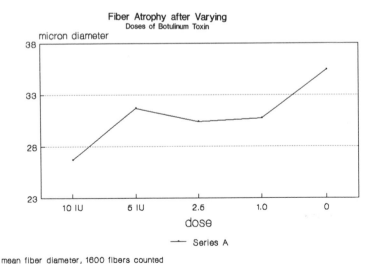

Figure 12. Dose-dependent muscle fiber responses. Fiber atrophy is seen after botulinum toxin injections and the muscle response appears to be dose related.

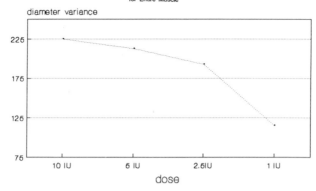

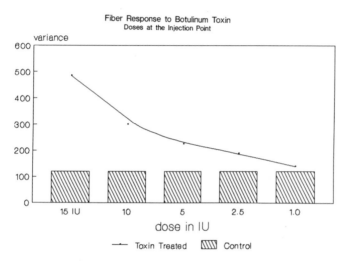

Figure 13. Dose-dependent muscle fiber responses. **A**, at injection point; **B**, for entire muscle. Fiber size variability correlated directly with the dose at the injection site, as well as the average fiber variability throughout the entire muscle.

The acetylcholinesterase staining characteristic and the fiber variability pattern confirmed the presence of a dose dependent field of action for botulinum A toxin.

As diffusion of toxin away from targeted muscles appears to cause complications ("toxin jump"),[24] therefore it is useful to attempt to quantify diffusion of biologic activity from a point injection at various doses used in clinical practice. The findings indicate that the degree of fiber atrophy, fiber size variability as well as intensity of acetylcholinesterase staining at the point of injection are directly related to dose administered. Furthermore, the diffusion of biologic effects from the point of injection within the longissimus dorsi muscle is dose dependent. Animals given 2.5 to 10 IU showed substantial diffusion of the toxin's biologic effects over a linear distance of 45 mm within this individual large muscle. In contrast, animals given 1 IU demonstrated a graduated biologic effect inversely related to the distance from the point injection. There appeared to be collapse of biologic effect using fiber size variation and acetylcholinesterase staining characteristics between 15-30 mm from the point injection of 1 IU within the injected muscle.

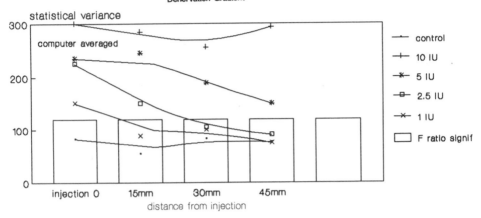

Diffusion Gradient after Botulinum A
Denervation Gradient

Figure 14. Dose-dependent diffusion, denervation gradient after botulinum A. The biologic effect within longissimus dorsi muscle appears to be dose related. The larger the dose the more homogeneous the effect is throughout the muscle strip evaluated. Fiber variations were compared using F-ratios.

Contiguous muscles were evaluated for biologic effect using primarily the longissimus dorsi muscle contralateral to the injection. At lower doses, the toxin did not disseminate to 45 mm on a muscle remote from the injected muscle. In a previous study,[24] the biologic activity of botulinum toxin has been shown to cross fascial planes and cause histochemical changes within non-injected muscles within lessor distances (<2 cm.). The explanation for the extension of biologic effect at a greater distance within the injected longissimus muscle than extramuscular locations may be that the toxin's activity diffuses to a greater degree within an individual muscle. Alternatively, the linear distance over which chemodenervation occurs may be exaggerated in muscles with very long parallel muscle fibers such as the longissimus

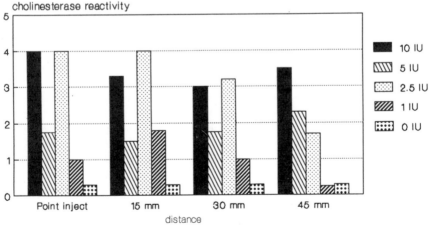

Figure 15. Dose-dependent diffusion, denervation field determined by cholinesterase stain. Acetylcholinesterase staining characteristic suggested that higher dose produced a biologic effect throughout the entire strip. Whereas, the smaller dose produced a gradient down the length of the strip studied.

dorsi. At higher doses there appeared to be dissemination of biologic activity down the contralateral muscle, although not to the same degree as the injected muscle.

In that containment of biologic activity within a targeted area of the body represents a desirable goal for botulinum toxin injection therapy, this scientific data offers insight into the intensity and diffusion of biologic activity within the injected muscle and at muscles remote from the injection site. Diffusion of biologic activity within the muscle appears to be a function of dose and can be graduated. The *denervation field* can be defined as a linear distance from the point injection over which botulinum toxin causes a denervation effect. The degree of denervation as indicated by fiber size atrophy and fiber size variation is at any distance from the point of injection a function of dose. The size of the denervation field is also a function of dose as indicated by the homogeneous effects the larger injection doses produced down the long muscle strip. As denervation field size and degree of neurogenic fiber atrophy are both dose dependent, *larger doses of botulinum toxin can be expected to produce more weakness and fiber atrophy but with greater spread of the toxin from the injection sites* (Figure 14). As complications in clinical practice are often related to undesirable toxin spread ("toxin jump"), the field of denervation must be considered in the clinical use of botulinum toxin.

As muscle atrophy has been noted to qualitatively occur after injection in the clinical setting on gross inspection, it is of interest to quantitate the degree of muscle fiber atrophy in this experimental model. Using cross sectional muscle fiber diameter changes after botulinum injection, it appears that a fiber atrophy of 25% was possible. This analysis is consistent with clinical observations in the sternomastoid muscle in patients treated for torticollis.

In that the diffusion of biologic activity away from a point injection is dose related and measurable, it may be possible to calculate reasonable diffusion fields away from a site of a given dose of botulinum toxin in clinical protocols. Diffusion fields can be established within injected muscles, within contiguous muscles, and within muscles of various fiber orientations. The results of this study underscore the importance of the clinician being knowledgeable of the anatomic distances between important muscles within the area being injected as well as the action of muscle groups in which the botulinum toxin is administered. Such information may provide a scientific approach to determining distances between injection sites at various doses. Furthermore, the minimum dose necessary to produce a homogeneous denervation effect down a long muscle would be ideal if just that muscle was being targeted for injection. Doses in excess of the minimum dose for homogeneous denervation would be more prone to spread outside the fascial planes of the muscle and into contiguous muscle groups potentially causing complications.

SUMMARY

1.   Diffusion of therapeutic botulinum toxin from points of intramuscular injection appears to be a dose dependent phenomena.

2.   Diffusion of the denervative effect outside the muscles targeted for injection is responsible for side effects associated with the clinical use of botulinum toxin ("toxin jump").

3.   Limiting the dose of botulinum toxin in critical anatomic areas can be helpful in preventing complications (e.g. limiting the sternomastoid dose to prevent dysphagia in spasmodic torticollis patients).

4.   Multiple point injections within targeted muscles have produced the most desirable clinical effects in patients with blepharospasm and adult onset spasmodic torticollis. This injection approach may be more beneficial because it diffuses the biologic effect of the toxin more evenly throughout the innervation zone of the muscle.

5.   Muscle fiber size variability and spread of acetylcholinesterase on muscle fibers are consistent histologic markers for therapeutic botulinum toxin effect. A dose dependent gradient

of biologic activity can be demonstrated both within muscles injected and within adjacent muscles.

6. Long term and repetitive therapeutic injections doses not seem to produce permanent denervative effects using fiber size variation and cholinesterase staining characteristics in the evaluation of human muscle specimens.

REFERENCES

1. Scott AB. Botulinum toxin injections to eye muscles to correct strabismus. J Am Ophthalmol Soc 1981; 79:734-770.
2. Scott AB, Magoon EH, McNeer KW, Stager DR. Botulinum treatment of strabismus in children. Trans Am Ophthalmol Soc 1990; 87:174-180 (discussion 180-184).
3. Scott AB, Kennedy, EG, Stubbs HA. Botulinum A toxin injection as a treatment for blepharospasm. Arch Ophthalmol 1985; 103:347-350.
4. Scott AB. Strabismus injection treatment. NIH Consensus Development Conference on Clinical use of botulinum toxin. 1990; Nov 12-14:117-118.
5. Borodic GE, Cozzolino D. Blepharospasm and its treatment, with emphasis on the use of botulinum toxin. Plast Reconst Surg 1989; 83(3):546-554.
6. Borodic GE, Mills L, Joseph M. Botulinum A toxin for the treatment of adult-onset spasmodic torticollis. Plast Reconst Surg 1991; 87(2):285-289.
7. Borodic GE, Cozzolino D, Townsend, DJ. Dose-response relationships in patients treated with botulinum toxin for more than three years (abstract). Sixth International Meeting of the Benign Essential Blepharospasm Research Foundation: 1988 Aug 25-27; Cambridge, MA. Ear Nose Throat J 1988; 67(12):914.
8. Jankovic J, Orman J. Botulinum toxin for cranial cervical dystonia: A double blind placebo controlled study. Neurology 1987; 37:616-623.
9. Blitzer A, Brin MF, Greene PE, Fahn S. Botulinum toxin injection for the treatment of oromandibular dystonia. Ann Oto Rhin Largyngol 1989; 98:93-97.
10. Fletcher NA, Quinn N. Dystonic syndromes. Curr Opin Neurol Neurosurg 1989; 2:330-333.
11. Gelb DJ, Lowenstein DH, Arminoff MJ. Controlled trial of botulinum toxin injections in the treatment of spasmodic torticollis. Neurology 1989; 39:80-84.
12. Scott AB, Kennedy RA, Stubbs HA. Botulinum toxin injection as a treatment for blepharospasm. Arch Ophthalmol 1985; 103:347-350.
13. Jankovic J, Ford J. Blepharospasm and oro-facial dystonia. Pharmacologic findings in 100 patients. Ann Neurol 1979:36:635.
14. Frueh BR, Callahan A, Dortzbach RR et al. The effects of differential section of the seventh nerve on patient with blepharospasm. Trans Am Acad Ophthalmol Otolaryngol 1976; 81:595.
15. Callahan A. Surgical correction of intractable blepharospasm, technical improvement. Am J Ophthalmol 1965; 60:788.
16. Janetta PJ, Abbasy M, Maroon JC, Ramos FM, Albin MS. Etiology and differential micro-surgical treatment of hemifacial spasm. J Neurosurg 1977; 47:321.
17. Borodic GE, Pearce LB, Johnson EJ, Schantz E. Clinical and scientific aspects of botulinum toxin. Ophthalmol Clin N Am 1991; 4(3):491-503.
18. Borodic GE, Metson R, Townsend D, McKenna M, Pearce LB. Botulinum toxin for aberrant facial nerve regeneration. Dose response relationships. Plast Reconst Surg, December 1991 (in press).
19. Frueh BR, Nelson CC, Kapustiak JF, Musch DC. The effects of omitting the lower eyelid in blepharospasm treatment. Am J Ophthalmol 1988; 106:45-47.
20. Borodic GE, Weigner A, Ferrante R, Young R. Orbicularis oculi innervation zone and implications for Botulinum A toxin therapy for blepharospasm. Ophthalmol Plast Reconst Surg 1991; 7(1):54-59.
21. Bertrand C, Molina-Negro P, Martinez SN. Technical aspects of selective peripheral denervation for spasmodic torticollis. Appl Neurophysiol 1982; 45:326-330.
22. Bertrand C, Molina-Negro P, Bouvier G, Gorczyca W. Observations and results in 131 cases of spasmodic torticollis after selective denervation. Appl Neurophysiol 1987; 50:319-323.
23. Borodic GE, Pearce LB, Smith, K, Joseph M. Botulinum A toxin for spasmodic torticollis: Multiple vs single injection points per muscle. Head Neck 1992; 14:33-37.
24. Borodic GE, Joseph M, Fay L, Cozzolino D, Ferrante R. Botulinum A toxin for the treatment of spasmodic torticollis. Dysphagia and regional toxin spread. Head Neck 1990; 12:392-398.

25. Tsui JK, Eisen A, Mak E, Carruthers J et al. A pilot study on the use of botulinum toxin in spasmodic torticollis. Can J Neurol Sci 1985; 12:314-316.

26. Borodic, GE. Botulinum A toxin (expressionistic) ptosis overcorrection after frontalis sling. Ophthalmic Plast Reconst Surg 1991; 2:137-142.

27. Kao I, Drachman D, Price DL. Botulinum toxin: mechanism of presynaptic blockade. Science 1976; 193:1256.

28. Duchen LW. Changes in motor innervation and cholinesterase localization induced by botulinum toxin in skeletal muscle of mouse: Differences between fast and slow muscles. J Neurol Neurosurg Psychiatry 1970; 33:40-54.

29. Duchen LW. Histologic differences between soleus and gastrocnemius muscles in the mouse after local injection of botulinum toxin. J Physiol (Lond) 1969;204:17-18.

30. Aldersen K, Holds JB, Andersen RL Botulinum induced alteration of nerve muscle interactions in human orbicularis oculi following treatment for blepharospasm. Neurology 1991;41:1800-1805.

31. Borodic GE, Ferrante R. Effects of repeated botulinum toxin injections on orbicularis oculi muscle. J Clin Neuro-Ophthalmol 1992; 12(2):121-127.

THE USEFULNESS OF ELECTROMYOGRAPHY (EMG)
IN BOTULINUM TOXIN TREATMENT OF CERVICAL DYSTONIA

Cynthia L. Comella, Aron S. Buchman,
and Glenn T. Stebbins

Department of Neurological Sciences
Rush-Presbyterian-St. Luke's Medical Center
Chicago, Illinois

INTRODUCTION

Electromyography (EMG) serves as a supplement to the clinical examination. Traditional EMG techniques, by displaying the electrical activity of muscle, permit precise localization and identification of muscles not easily palpated from the surface. A more recently developed EMG technique, Turns analysis, allows a quantitative assessment of muscle activity.

Both the traditional and more recently developed EMG techniques may be useful in the botulinum toxin treatment of cervical dystonia (CD).

TRADITIONAL EMG

The treatment of focal dystonia with botulinum toxin relies on the accurate injection of toxin into abnormally overactive muscles. For this reason, EMG is used widely in the botulinum treatment of hand and foot dystonia, jaw-opening oromandibular dystonia and spasmodic dysphonia.[1-4] In each of these disorders, the abnormally overactive muscles are not accessible to palpation by clinical examination, but are accurately localized and targeted for injection with the assistance of EMG.

The application of EMG techniques to botulinum toxin treatment of CD has not been clearly established. Some investigators maintain that EMG assisted injections enhance the clinical response by facilitating the accuracy of injection.[5] Others report greater efficacy and reduced adverse effects with EMG assistance in comparison to centers injecting without EMG.[5-7] Direct comparisons of results among centers, however, has been precluded by differences in treatment protocol, dose and preparation of botulinum toxin, injection technique (single versus multiple injection sites), and ST rating scales.[8-10] In the absence of compelling data, many experienced investigators feel that the additional time and expense required for EMG-assisted botulinum toxin injections make the routine use of this technique impractical. In some centers, EMG is reserved for those patients who have had minimal or no benefit from previous clinically guided injections, or those in whom surface anatomy is obscured.

We conducted a prospective, randomized, blinded study of 52 ST patients comparing EMG-assisted botulinum toxin injections to injections based entirely on clinical examinations alone.[11] Patients were randomized into two groups: EMGRX (muscle selection and botulinum toxin injection based on combined clinical and EMG, N=28) and CLINRX (muscle selection and injection based only on clinical examination, N=24). All patients in both groups were treated by the same physician. Outcome was assessed in two ways: 1) Randomized videotapes of all patients before and following botulinum toxin treatment were evaluated using the modified Toronto Western Spasmodic Torticollis Scale (TWSTRS)[12] by a movement disorders specialist blinded to the patient treatment group; 2) A visual analogue scale of CD severity was completed by the patients before and following botulinum toxin treatment. Overall, the number of patients improving did not differ between groups according to either assessment technique (71% improved by TWSTRS and 86% improved by patient report). However, in the EMGRX group, the magnitude of improvement was significantly greater for overall TWSTRS score and patient self-assessment score. In particular, head tilt, retrocollis and shoulder elevation were more improved in the EMGRX group. We conclude that EMG assistance provides for a greater magnitude of improvement at the same Botulinum toxin dose. The total time for EMG-assisted injection was approximately 15-20 minutes. The additional cost of the specialized needle electrode ($25-30) and the use of the EMG machine were economic considerations.[11]

QUANTITATIVE EMG TECHNIQUES

Turns analysis has been applied to other disorders affecting the neuromuscular junction.[13-15] A recent study at this center has evaluated the usefulness of a quantitative EMG technique, Turns Analysis, for assessment of the effect of botulinum toxin treatment in CD. We studied 19 patients with CD using Turns analysis to determine if quantitative EMG measures of muscle activity changed following botulinum toxin injections.[16] Before and after botulinum toxin injections, six muscles were evaluated bilaterally. Quantitative EMG analysis of active muscles injected with botulinum toxin showed a significant decline in muscle activity after botulinum toxin ($F=[1,41]$ 55.0; $p < 0.001$). Significant reductions in quantitative EMG parameters were also noted in non-injected active muscles after botulinum toxin treatment ($F=[1,51]$ 59.15; $p < 0.001$). The sum of EMG activity for all active muscles was calculated for each subject and showed a significant reduction after botulinum toxin injection (Manova:($F=[1,5]$ 5.69; $p < 0.05$). Quantitative EMG assessment provides an objective measure of response to botulinum toxin. Decreased muscle activity in non-injected muscles may result from toxin diffusion or reflect relaxation of muscles only secondarily involved in cervical dystonia.

FUTURE APPLICATIONS OF EMG

1. The formulation of a botulinum toxin dose-effect curve using quantitative EMG techniques.
2. Quantitative description of cervical dystonia and its response to a specific intervention by the integration of quantitative EMG data and three dimensional measurements of head position.

REFERENCES

1. Report of the Therapeutics and Technology Assessment Subcommittee of the American Academy of Neurology. Assessment: The clinical usefulness of botulinum toxin-A in treating neurologic disorders. Neurology 1990; 40:1332-1336.

2. National Institutes of Health Consensus Development Conference: Clinical use of botulinum toxin. Arch Neurol 1991; 48:1294-1298.
3. Cohen LG, Hallett M, Geller BD, Hochberg F. Treatment of focal dystonia of the hand with botulinum toxin injections. J Neurol Neurosurg Psychiatry 1989; 52:355-363.
4. Jankovic J, Brin M. Therapeutic applications of botulinum toxin. N Engl J Med 1991; 324:1186-1194.
5. Stell R, Thompson PD, Marsden CD. Botulinum toxin in torticollis [letter]. Neurology 1989; 39:1403-1404.
6. Dubinsky RM, Grey CS, Vetere-Overfield B, Koller WC. Electromyographic guidance of botulinum toxin treatment in cervical dystonia. Clin Neuropharmacol 1991; 14:262-267.
7. Shannon KM, Comella C, Tanner CM. EMG-guided botulinum toxin injections in torticollis: objective clinical improvements [abstract]. Neurology 1989; 39 (suppl 1):352.
8. Jankovic J, Schwartz K, Donovan DT. Botulinum toxin treatment of cranial-cervical dystonia, other focal dystonia and hemifacial spasm. J Neurol Neurosurg Psychiatr 1990; 53:633-639.
9. Brin MF, Fahn S, Moskowitz C, Friedman A, Shale HM, Greene PE, Blitzer A, List T, Lange D, Lovelace RE, McMahon D. Localized injection of botulinum toxin or the treatment of focal dystonia and hemifacial spasm. Adv Neurol 1988; 50:599-608.
10. Tsui JKC, Fross RD, Calne S, Calne DB. Local treatment of spasmodic torticollis with botulinum toxin. Can J Neurol Sci 1987; 14:533-535.
11. Comella C, Buchman AS, Tanner CM, Brown-Toms NC, Goetz CG. Botulinum toxin injection for spasmodic torticollis: increased magnitude of benefit with electromyographic assistance. Neurology 1992; 42:878-882.
12. Consky E, Basinski A, Belle L, Ranawaya R, Lang AE. The Toronto Western Spasmodic Torticollis Rating Scale (TWSTRS): Assessment of validity and inter-rater reliability. Neurology 1990; 40 (suppl 1): 445.
13. Fuglsand-Frederissen A, Ronager Jesper R. EMG power spectrum, turns-amplitude and motor unit potential duration in neuromuscular disease. J Neurol Sci 1990; 97:81-91.
14. McGill KC, Lau K, Dorfman LJ. A comparison of turns analysis and motor unit analysis in electromyography. EEG & Clinical Neurophys 1991; 81:8-17.
15. Dorfman LJ, McGill KC. AAEE minimonograph #29: Automatic quantitative electromyography. Muscle Nerve 1988; 11:804-818.
16. Buchman AS, Comella CL, Garratt MJ, Tanner CM. Quantitative electromyographic analysis of response to botulinum toxin therapy for spasmodic torticollis [abstract]. Ann Neurol 1990; 28:234.

DEVELOPMENT OF ANTIBODIES TO BOTULINUM TOXIN TYPE A IN PATIENTS WITH TORTICOLLIS TREATED WITH INJECTIONS OF BOTULINUM TOXIN TYPE A

Paul Greene, and Stanley Fahn

Columbia-Presbyterian Medical Center
New York, NY 10032

INTRODUCTION

Botulinum toxin (botox) injections are now recognized as a major means of treating focal dystonia, e.g. blepharospasm,[1,2] torticollis,[3-9] and spastic dysphonia.[10] Botox has also been used with success in treating patients with dystonia of jaw and other lower facial muscles,[11] brachial dystonia,[12,13] and a variety of other hyperkinetic movement disorders. Although local side effects, such as ptosis after blepharospasm injections and dysphagia after neck or tongue injections, may complicate treatment in a minority of patients, systemic side effects after therapeutic botox injections are uncommon. In most centers, early experience with the use of botox for blepharospasm, in doses of 25–100 units every three months, indicated that loss of benefit did not occur (a unit of botox is the mouse LD_{50} by intraperitoneal injection). Treatment of torticollis requires higher doses of botox, about 150u–300u every 3 months. At these doses, patients have been recognized with antibodies to botox in sufficient titers to block all clinical effect.[14,15,16,17] In addition, some patients may have loss of response to botox, and failure to develop muscle atrophy after botox injections, without the presence of detectable serum antibodies. The prevalence of these problems in patients receiving botox therapy for torticollis is not known. Nonetheless, they are a major therapeutic problem for some patients. We have reviewed our experience in treating torticollis with botox to determine the prevalence of antibody mediated resistance to botox, and to identify possible risk factors for the development of loss of response to botox.

METHODS

Between 1984 and the beginning of 1992, over 550 patients with torticollis were injected with over 450,000 units of botox at the Dystonia Clinical Research Center at Columbia-Presbyterian Medical Center. For their first injection series, most patients received a moderate dose of botox and returned in 2–3 weeks for supplemental injections to maximize improvement (mean total dose for both sessions about 240 u). Some patients continued to have supplemental injection series after subsequent treatments. For purposes of analysis, when an injection treatment occurred within 3 weeks of the prior treatment, we labeled such treatment as a "booster". With the exception of a small number of patients, patients were injected with botulinum toxin from lot 79-11 as prepared by Alan Scott and Allergan, Inc. After dilution in saline, unused toxin is frozen and may be thawed and reused days to several weeks later. Toxin undergoes at most 1 freeze-thaw cycle and is not reused more than once.

Some patients with loss of clinical efficacy had serum assayed by mouse neutralization assay[18] in the laboratory of Dr. Charles Hatheway at the Centers for Disease Control in Atlanta, Georgia, or in the laboratory of Dr. Alan Scott at Smith-Kettlewell Eye Institute in San Francisco, California. The results of these assays allow us to provide a minimum estimate of the prevalence of serologically documented antibodies to botox in our population.

We also wanted to identify risk factors that predict loss of response in the absence of muscle atrophy after botox injections, whether or not antibodies are detected in serum. Muscle atrophy was usually judged after injection of 60-80 units into the sternocleidomastoid muscle (50 units results in complete atrophy of the sternocleidomastoid in most patients). Since 1987, in part because of the recognition of the problem of antibodies, the frequency of injections, the total dose per injection session, and the number of injection sites per muscle have all been reduced, and the concentration of toxin has increased. We currently use a concentration of 5-10 units/0.1ml, 2-5 sites per muscle and an average of 250 units per treatment. Except for a "booster" session after the first injection, we encourage patients to wait at least 3 months between treatments, and not to have "booster" sessions. Because of this change in technique, which may affect the rate of antibody formation, we have studied a cohort of patients who were injected with similar technique: all patients first injected in 1988. To identify possible risk factors for botox resistance in this group, 4 control patients without botox resistance from the 1988 cohort were randomly selected for each patient with botox resistance. For each resistant patient, data from control patients was considered only for the same duration of therapy as the index patient.

RESULTS

Twenty-four patients had serological evidence of antibodies to botox since 1984. All of these failed to develop muscle atrophy after botox injection. There were 17 women and 7 men (not significantly different by Chi-square from the sex ratio of all torticollis patients treated with botox). Patients with antibodies had developed torticollis at a significantly younger age than those without antibodies (Table 1). The mean duration of treatment at the time response was lost was 15.6 months, but seropositivity was not documented until a mean of 23.5 months of treatment. The longest duration of treatment before loss of efficacy was 40 months; the earliest loss of efficacy was after 6 treatments over 3 months (Table 1).

In 1988, 102 patients had their first botox injection at our Center. Only 76/102 had multiple injection sessions with sufficient follow-up to detect the loss of efficacy. Of these, 3 had serological evidence of antibodies and 5 more lost benefit and no longer had muscle atrophy after botox injections. These 8 botox resistant patients had more frequent injections, a larger number of "booster" injections, and a higher dose every 3 months (quarter), than did the other 68 patients first injected in 1988 (Table 2). In addition to more frequent injections, patients with botox resistance had more units of toxin injected at each non-"booster" session, while there was no significant difference in dose at "booster" sessions (Table 2). This suggests that high dose is a risk factor for developing resistance to botox, independent of frequency of injection.

Table 1. Antibodies to botox A: patient characteristics. 24 patients with serological evidence of antibodies.

	Antibodies	No antibodies		
Sex ratio (F/M)	17/7	341/218	N.S.	(Chi-square)
Age at onset	30.9±16.7 years	39.5±14.6 years	p<.008	(t-test)
Duration of symptoms at time of injection	12.2±9.3 years	10.3±11.5 years	N.S.	(t-test)
Duration of treatment prior to seropositivity	23.5 months (range 6–46 months)			
Duration of therapy prior to loss of efficacy	15.6 months (range 3–40 months)			

Table 2. Physiological resistance to botox in 1988 patients.

	Resistant patients	Non-resistant patients		
Time between injections	8.8±5.7 (weeks)	11.4±7.4 (weeks)	p=.0083	(t-test)
Number of "booster" sessions	24/81	45/250	p<.05	(Chi-square)
Dose per quarter	295±120 u	220±129 u	p=.00025	(t-test)
Dose per non-"booster"	230±54 u	208±67	p=.02	(t-test)
Dose per "booster"	167±50 u	148±62 u	N.S.	(t-test)

CONCLUSION

Patients who do not improve after initial botox injections in adequate doses, and some patients who do improve but lose response, continue to have muscle atrophy after injection. In these patients, lack of response is due to some property of the dystonia (such as contraction of uninjected muscles, or shift in the pattern of muscle activity after injection). Patients with serological evidence of antibodies, however, fail to develop muscle atrophy after large doses of botox. We have documented the presence of antibodies to botox in 24 patients. This provides us with a crude estimate of the prevalence of antibodies to botox in our population. Patients who develop antibodies after an injection series will not have clinical evidence of immunity until the next treatment, so the 489 patients having at least 2 treatments were at risk. The minimum prevalence is, therefore, 24/489 = 4.9%. In fact, we know of no patients developing antibodies after the first treatment, so the 336 patients having at least 3 treatments are a more realistic at risk population, yielding a minimum prevalence estimate of 24/336 = 7.1%. At least 12 other patients have failed to develop muscle atrophy after repeated injection series, but have not had serological studies, or have had negative serological studies. We suspect that these patients have antibodies to botox which we have not detected, perhaps because of insensitivity of the mouse assay, so the actual prevalence may be considerably higher than 7%. It is possible, however, that there may be another mechanism underlying their lack of response. For this reason, we propose to designate patients who do not have muscle atrophy after botox injection as having "physiological resistance" to botox.

We retrospectively evaluated the risk factors for developing resistance to botox in a cohort of patients first injected at our center in 1988. Patients developing botox resistance tended to have more frequent injections, especially injections 1–3 weeks apart ("boosters"), and higher doses of botox per treatment. Other factors, such as concentration and volume of toxin, number of injection sites, and purity of toxin which might have influenced the development of resistance, could not be evaluated in our population.

REFERENCES

1. Jankovic J, Orman J. Botulinum A toxin for cranial-cervical dystonia: A double-blind, placebo-controlled study. Neurology 1987; 37:616-623.
2. Fahn S, List T, Moskowitz C, Brin MF, Bressman S, Burke R, Scott A. Double-blind controlled study of botulinum toxin for blepharospasm. Neurology 1985; 35(Suppl1):271.
3. Tsui JKC, Eisen A, Stoessl AJ, Calne S, Calne DB. Double-blind study of botulinum toxin in spasmodic torticollis. Lancet 1986; 2:245-247.
4. Jankovic J, Orman J. Botulinum A toxin for cranial-cervical dystonia: A double-blind, placebo-controlled study. Neurology 1987; 37:616-623.
5. Gelb DJ, Lowenstein DH, Aminoff MJ. Controlled trial of botulinum toxin injections in the treatment of spasmodic torticollis. Neurology 1989; 39:80-84.
6. Greene P, Kang U, Fahn S, Brin M, Moskowitz C, Flaster E. Double-blind, placebo-controlled trial of botulinum toxin injections for the treatment of spasmodic torticollis. Neurology 1990; 40:1213-1218.
7. Blackie JD, Lees AJ. Botulinum toxin treatment in spasmodic torticollis. J Neurol Neurosurg Psychiatry 1990; 53:640-643.
8. Lorentz IT, Subramaniam SS, Yiannikas C. Treatment of idiopathic spasmodic torticollis with botulinum toxin A double-blind study on twenty-three patients. Mov Disord 1991; 6:145-150.
9. Moore AP, Blumhardt LD. A double-blind trial of botulinum toxin-A in torticollis, with one year follow up. J Neurol Neurosurg Psychiatry 1991; 54:813-816.
10. Troung DD, Rontal M, Rolnick M, Aronson AE, Mistura K. Double-blind controlled study of botulinum toxin in adductor spasmodic dysphonia. Laryngoscope 1991; 101:630-634.
11. Blitzer A, Brin MF, Greene PE, Fahn S. Botulinum toxin injection for the treatment of oromandibular dystonia. Ann Otol Rhinol Laryngol 1989; 98:93-97.
12. Cohen LG, Hallett M, Geller BD, Hochberg F. Treatment of focal dystonias of the hand with botulinum toxin injections. J Neurol Neurosurg Psychiatry 1989; 52:355-363.
13. Rivest J, Lees AJ, Marsden CD. Writer's cramp: Treatment with botulinum toxin injections. Mov Disord 1991; 6:55-59.
14. Brin MF, Fahn S, Moskowitz C, Friedman A, Shale HM, Greene PE, Blitzer A, List T, Lange D, Lovelace RE, McMahon D. Localized injections of botulinum toxin for the treatment of focal dystonia and hemifacial spasm. Mov Disord 1987; 2:237-254.
15. Tsui JK, Wong NLM, Wong E, Calne DB. Production of circulating antibodies to botulinum-A toxin in patients receiving repeated injections for dystonia. Ann Neurol 1988; 23:181.
16. Hambleton P, Cohen HE, Palmer BJ, Melling J. Antitoxins and botulinum toxin treatment. BMJ 1992; 304:959-960.
17. Jankovic J, Schwartz KS. Clinical correlates of response to botulinum toxin injections. Arch Neurol 1991; 48:1253-1256.
18. Sakaguchi B. Clostridium botulinum toxins. Pharmac Ther 1983; 19:165-194.

EFFECTS OF INTRAMUSCULAR INJECTION OF BOTULINUM TOXIN TYPE B IN NONHUMAN PRIMATES

E. D. Moyer[1], E. A. Stephens[2], F. Esch[1], M. Litwak[3], and J. Arezzo[3]

[1]Athena Neurosciences, South San Francisco, CA
[2]Michigan Department of Public Health, Lansing MI
[3]Albert Einstein College of Medicine, Bronx, NY

Botulinum Toxin Type B (Bot B) is proposed as a therapeutic agent for the treatment of uncontrolled contractions of specific muscles in the shoulder and neck, associated with the disease, cervical dystonia. When injected into specific muscles, Botulinum Toxin Type B is expected to produce a localized pre-synaptic neuromuscular block by preventing release of acetylcholine from nerve endings, reducing the excessive muscle contractions characteristic of cervical dystonia.

The paralytic efficacy of Botulinum Toxin Type B (Bot B) was evaluated in female cynomolgus monkeys, one animal per dose group. The animals received the toxin as a single dose: half of the total amount was injected into the trapezius, and the other half was injected into the abductor pollicis brevis (APB). Doses tested were 0.75, 1.5, 3, 12, and 24 U/kg body weight per muscle, or a total dose per animal of 1.5, 3, 6, 24, and 48 U/kg.

After injection, the animals were observed for 4 months. Electrophysiological measurements were made of injected and contralateral muscles as well as of a distal, uninjected muscle (extensor digitorum brevis, EDB), evaluating the latency and peak amplitude of the evoked compound muscle action potential. Somatosensory evoked potentials were also recorded over the cauda equina, cervical spinal cord and sensorimotor cortex following median nerve stimulation, to evaluate potential central nervous system (CNS) dysfunction. These measurements were made before, and at 2, 4, 8, 12, and 16 weeks after injection. At the end of the study, a gross necropsy was performed. There were no deaths prior to the necropsy.

During the observation period, the animals in all dose groups maintained normal activity and posture. No unusual behavior or change in eating or elimination patterns was observed in any animal. However, Bot B significantly reduced the evoked compound motor response of the injected muscles in a dose-responsive manner. There also appeared to be an interactive effect of muscle mass: a significant decrease in response was found at a dose of 1.5 U/kg body weight for the APB, but not until 6

U/kg body weight for the trapezius. The alteration in muscle response after injection of Bot B was seen in all animals by 2 weeks postinjection (the earliest time point assessed). The duration of recovery after Bot B injection also appeared to be dose- and muscle mass-related, with lower doses to the trapezius wearing off faster than higher doses into the APB. Bot B was not associated with any signs of systemic peripheral or CNS toxicity at any of the doses tested. There were no electrophysiologic alterations of distal or contralateral muscles, even at the highest dose tested (48 U/kg total dose).

QUALITY OF BOTULINUM TOXIN FOR HUMAN TREATMENT

Edward J. Schantz and Eric A. Johnson

Food Research Institute
University of Wisconsin
1925 Willow Drive
Madison, WI 53706

INTRODUCTION

The crystallized botulinum toxin type A now used in treatment, Batch 79-11 prepared in November 1979 and licensed as an Orphan drug by the FDA in December 1989, has served well for treatment of many human dystonias and other involuntary muscle movement disorders. The continued production of high quality botulinum toxin presents some important problems that must be investigated. In this presentation we wish to make some proposals that should improve the quality of the toxin as a pharmaceutical.

The quality of the type A toxin prepared and purified for the human volunteer program in 1979 had to be specific as to production, purity, toxicity and stability if it was to be approved by the FDA for injection.[1] The following criteria were proposed as a means of defining the quality of the toxin for use (Table 1): (1) production of 10^6 mouse LD_{50} per ml with the Hall strain of *C. botulinum* type A, (2) purification by a procedure not exposing the toxin to synthetic resins, solvents, or antigenic substances that might be carried over into the final crystalline toxin, (3) a maximum UV absorption at 278 nm, (4) a 260/278 nm absorption ratio of 0.6 or less, and (5) a specific toxicity of $3 \pm 20\% \times 10^7$ mouse LD_{50} per mg. Analytical gel electrophoresis was used later as a criterion of quality. The 2x crystalline toxin was best preserved by storing it in the mother liquor of the 2nd crystallization.

CONSISTENCY OF TOXIN LOTS

Through the years a number of batches of toxin have been prepared using as near as possible identical culturing and purification procedures. All batches appeared identical in absorption characteristics and toxicity, and very similar on gel electrophoresis. In certain batches small differences are observed on electrophoresis. Also differences among batches in stability to heat, drying, and other processes involved in the preparation of toxin for distribution were observed by Dr. Alan B. Scott. These differences we feel may be due to genetic changes in the organism or undetectable variations in culturing or purification.

Botulinum and Tetanus Neurotoxins, Edited by
B.R. DasGupta, Plenum Press, New York, 1993

657

Table 1. Criteria proposed as a means of defining toxin quality.

1.	Production in 12 or 16 liter deep culture containing hydrolyzed casein, yeast extract, and dextrose (no animal meat products) with the Hall strain of *C. botulinum* type A yielding at least 10^6 MLD_{50} per ml.
2.	Purification by a procedure not exposing the toxin to synthetic resins, solvents, or antigenic substances that might be carried over in trace amounts to the final (2x) crystalline toxin.
3.	A maximum absorbance at 278 nm.
4.	A 260/278 nm absorption ratio of 0.6 or less.
5.	A specific toxicity of $3 \pm 20\% \times 10^7$ MLD_{50} per mg.
6.	Analytical gel electrophoresis was used later as a criterion of quality.

Consistent production of high quality botulinum toxin is paramount. The continual genetic changes that take place in the organism used to produce the toxin (at present type A) makes it imperative that we carry on a program of selecting and maintaining a stock culture of the organism that consistently produces 10^6 mouse LD_{50} or more per ml in deep cultures of 12 or 16 liters. A more detailed understanding of toxin gene expression should result in consistently high production of type A toxin complex, and could possibly lead to varying ratios of nontoxic and toxic proteins in the complex that may impart beneficial properties such as increased stability. Our laboratory is currently studying the genetics of toxin production in *C. botulinum*.

Nutritional and culturing conditions affect the quantity and quality of toxin produced. Botulinum toxin is currently produced in a complex medium composed of dextrose, casein hydrolysate, and yeast extract. Our laboratory has shown that *C. botulinum* grows well in a chemically defined medium but that toxin production is very low in this medium.[2] Production of type A botulinum toxin in a chemically defined medium would make it possible to obtain a uniform product free from any antigenic material other than toxin.

PROPERTIES OF BOTULINUM TOXIN

Crystalline type A toxin is composed of neurotoxin molecules noncovalently bound to nontoxic proteins that have an important role in the stability of the toxin during dilution, drying, reconstitution, and injection into patients. Compared to the neurotoxin, little is known of the structures and functions of these proteins in the complex except that one causes hemagglutination. The nontoxic proteins including the hemagglutinin in the complex contribute to stability and rate of diffusion in vitro in agarose gels. It would be valuable to determine the relation of these molecules to stability and rate of diffusion in human tissues, and to select strains with high production of these toxin-associated proteins.

Similar to other biologically active proteins, botulinum toxin type A loses stability when diluted to nanogram concentrations per ml, which is necessary for dispensing. This instability is probably due to a stretching of the molecule out of its toxic shape, but is greatly reduced by the presence of a small amount of another protein in solution with the toxin, such as gelatin or albumin used at two grams per liter of solution. At the present time human serum albumin is used for stabilization. However, some risk may be associated with its use due to stable viruses being carried over from donor. Another means of stabilization should be sought.

It is important to develop International standards set down by FDA or WHO for comparison of different preparations of toxin. A reference standard toxin preparation has been used by the FDA[3] and could be made available to different producers of toxin.

Stability of the toxic properties or acetylcholine blocking power is a vital factor in the dilution and preparation of the toxin for distribution to the medical profession who for treatment must depend on the potency stated on the container. A very important concern in the stability of the toxin during preparation for dispensing is that the inactivated toxin may act as a toxoid and evoke antibodies in patients being treated. At the present time, Dr. Scott has observed that 80 to 90% of the toxin is inactivated during drying and further handling at pH 7.3. This inactivated toxin may function as a toxoid and physicians should be made aware of this fact. Antibodies to the toxin have been found in a few (>20) patients who received large doses of toxin for treatment of spasmodic torticollis. Presently, dilution and lyophilization of toxin causes some inactivation of toxic activity and variation between toxin lots. To help preserve activity during lyophilization, sodium chloride should be omitted from the solution for drying. Preparation of the toxin for dispensing, and even for injection into the patient, should be carried out at lower pH, for example 6.4 instead of 7.3 now used, because of the higher stability of the toxin complex to heat and drying at the lower pH. An important advantage to injecting the toxin at a lower pH is that the entire large molecule (neurotoxin plus nontoxic proteins) remains intact and should reduce the diffusion rate of the toxin to adjoining muscles.

COMBINED USES OF MICROBIAL TOXINS

The uses of serotypes of botulinum toxin other than type A are seeing increased use for treatment of muscle disorders, particularly in patients refractory to treatment with A. It is likely that neurotoxins other than botulinum including tetanus, alpha-bungarotoxin, saxitoxin and others may have application in medicine.

There are reasons to speculate on the use of combined toxins for treatment. Saxitoxin which produces paralysis by blockage of the sodium channel in concentrations of one part in 10,000 parts of a local anaesthetic like procaine increases the effectiveness of procaine several fold. There may be a new field of exploration on the pharmacology of botulinum toxin combined with other substances that affect nerve function.

REFERENCES

1. Schantz EJ, Johnson EA. Properties and use of botulinum toxin and other microbial neurotoxins in medicine. Microbiol Rev 1992; 56:80-99.
2. Patterson-Curtis S, Johnson EA. Regulation of neurotoxin and protease formation in *Clostridium botulinum* Okra B and Hall A by arginine. Appl Environ Microbiol 1989; 55:1544-1548.
3. Schantz EJ, Kautter DA. Standardized assay for *Clostridium botulinum* toxins. J Assoc Off Anal 1978; 61:96-99.

STABILITY OF BOTULINUM TOXIN IN CLINICAL USE

Henry T. Hoffman+* and Michael G. Gartlan.*

*Department of Otolaryngology--Head and Neck Surgery
University of Iowa Hospitals and Clinics
Iowa City, Iowa 52242
+The National Center for Voice and Speech
Department of Speech Pathology and Audiology
The University of Iowa
Iowa City, Iowa 52242

The Food Research Institute currently produces all the Botulinum neurotoxin (Botox) used in the United States which is marketed as Oculinum (Allergan, Inc., Irvine, CA).[1] Oculinum is packaged in freeze dried form in vials of 100 mouse units (MU) that require reconstitution with preservative free normal saline for clinical use. Although the FDA recommends discarding toxin that is not used within 4 hours of reconstituting,[2] there is little data describing Oculinum activity past this point.[3]   Information is also lacking regarding the effect of refreezing Botox that has been reconstituted for use at a later time.

Botulinum toxin (Botox) injection has become accepted as a safe and effective method of treating focal dystonias and other conditions associated with involuntary muscle movements.[4] Extensive experience has accumulated employing a wide range of doses with injections into muscles ranging in size from minute intrinsic laryngeal muscles to large shoulder girdle muscles such as the trapezius. Total doses administered to patients range from as little as 1 MU for laryngeal dystonia up to more than 250 MU in the treatment of cervical dystonia. Repeated injections are generally required at 3 to 4 month intervals as the Botox effect wears off.

Although clinicians generally maximize the use of the Botox available in each 100 MU vial by treating several patients within a four hour period, the logistics of this approach may be difficult. As a result, a significant amount of Botox is commonly discarded to comply with FDA guidelines. The high cost of Botox as well as limited availability until the past year have prompted some clinicians to preserve Botox for use at a later time by refreezing unused reconstituted toxin.

Freezer storage of Botox in clinical practice is generally accomplished by either of two methods. First, the Botox may be reconstituted with saline, aliquoted into smaller vials, and then frozen for use at a later time. Alternatively, freezer storage of Botox may be done following completion of a treatment session in which unused Botox remains. Most investigators maintain reconstituted Botox either in a refrigerator or on ice during a treatment session to slow the degradation process.

To date several questions exist regarding the preparation and storage of Botox in the clinical setting. These are:

1. How long after reconstituting can Botox be used without an appreciable drop in potency when maintained at refrigerator temperature?

2. In anticipation of the need for repeated treatments of small amounts of Botox over time, is it possible to divide reconstituted Botox into small aliquots that can be frozen and then thawed later without loss of activity?

Botulinum and Tetanus Neurotoxins, Edited by
B.R. DasGupta, Plenum Press, New York, 1993

3. Does freezing Botox which is left over after a treatment session result in a drop in potency that would prohibit its use at a later date?

These questions were answered in part from research reported at the First Annual Conference on Botulinum and Tetanus Neurotoxins.[5] It was noted that prolonged refrigerator storage and freezing of reconstituted Botox resulted in loss of activity as determined by the mouse LD50 acute toxicity study. This study is similar to that used by Allergan, Inc to determine the potency of Oculinum.[6] Refrigerator storage of reconstituted Oculinum for 6 hours resulted in a small drop in activity that was not statistically significant. Refrigerator storage for 12 hours resulted in a more substantial loss of activity that did reach statistical significance.

Oculinum tested for activity after two weeks of refrigerator storage, which began with freezing immediately upon reconstituting, demonstrated a more severe drop in potency. The greatest loss of activity was seen with Oculinum tested following refrigeration for 6 hours and freezing for one week.[6]

Despite these findings, a drop in potency following use of Botox after refreezing has not been reported from clinical observations. This failure to identify loss of activity may be due to the normal variation that exists in potency of Botox from vial to vial. Botox is aliquotted into vials with an accepted error of 30%. Although a vial is generally expected to contain 100 MU, it may actually contain anywhere between 70 and 130 MUs. Differences in response to treatments that had been attributed to this variability may mask diminished responses that actually occurred as a result of loss in Botox activity after freezing. Additionally, the clinical effect of Botox is not easily quantified. The effect of an attenuated treatment may therefore not be noted unless rigorously searched for.

A drop in potency of Botox has been noted following prolonged refrigeration. When reconstituted the evening before treatment, refrigerated, and used the next morning, clinical response appeared to be diminished when compared to reconstituting Botox immediately before injecting.[7]

Oculinum, the only currently FDA approved toxin for human use, is freeze-dried in vials containing 500 ug of human serum albumin, and 900 ug of sodium chloride at pH 7.3. Although storage in this buffer may make Botox more compatible with body fluids, it also makes the Botox relatively unstable. This less stable method of storage may increase the safety of Oculinum which, if spilled, will rapidly lose its potency at room temperature.[5] However, since the amount of Botox in a 100 MU vial is well below the dose at which systemic toxicity would occur, this safety factor may not be important.

Schantz and Kautter reported that Botox can be frozen and thawed without appreciable loss of toxicity when buffered in a sodium phosphate solution of pH 6.2.[8] They also demonstrated that activity can be maintained at room temperature for several days when Botox is stored in a pH 6.2 phosphate buffer. Although this lower pH may cause some increased pain upon injection, it would be expected to be less than that of lidocaine with epinephrine which is marketed within a pH range of 3.3 to 5.5 to enhance solubility and to prolong shelf-life.[9]

Although methods to increase the stability of Botox may permit multiple uses from a single 100 MU vial over several treatment sessions, an alterative approach to maximize the use of Botox is to distribute vials containing smaller amounts. The full spectrum of patient requirements for Botox may well be served with an assortment of vials containing 5, 20 and 100 MU per vial.

We feel that although refreezing of unused Botox has been done with the patient's best interest in mind -- to diminish cost and increase availability of treatment -- it is not a recommended practice. Inconsistent responses to treatment using attenuated Botox may introduce dangerous irregular dosing patterns. Other concerns exist regarding alteration of the molecular structure of Botox by freezing. Although the development of antibodies to Botulinum neurotoxin A is rare, it is possible that the loss of activity associated with refreezing and thawing may reflect structural changes that may also increase its immunogenicity.

ACKNOWLEDGEMENTS

This work was supported in part by Grant No. P60 DC00976-0X from the National Institutes on Deafness and Other Communication Disorders.

REFERENCES

1. E.J. Schantz and E.A. Johnson, Dose standardization of botulinum toxin, *Lancet*. 335:421 (1990).
2. Oculinum (Botulinum Toxin Type A) package insert, Irvine, CA: Allergan; Inc., (December, 1989).
3. A.B. Scott, Botulinum toxin injection of eye muscles to correct strabismus, *Trans Am Ophthalmol Soc.* 129:735-751 (1981).
4. NIH Consensus Development Conference on Clinical Use of Botulinum Toxin. National Institute of Health, Bethesda Maryland (November 12-14, 1990).
5. H.T. Hoffman, and M.G. Gartlan, Stability of botulinum toxin in clinical use, Presentation at "Botulinum Tetanus Neurotoxins; Neurotransmission and Biomedical Aspects" International Conference. Madison, Wisconsin. (May 12, 1992).
6. M.G. Gartlan and H.T. Hoffman, Botulinum neurotoxin A: degradation in potency with storage, Submitted to *Otolaryngol Head Neck Surg*. (June, 1992).
7. C.L. Ludlow, Personal Communication. (May 12, 1992).
8. E.J. Schantz and D.A. Kautter, Standardized assay for clostridium botulinum toxins, *J Assoc Off Anal Chem*. 61:1, (1978).
9. G.K. McCvoy and K. Litrak, eds, American hospital formulary service drug information. Bethesda MD: American Society of Hospital Pharmacists, Inc., 1951-2 (1991).

AN OVERVIEW OF SOME ISSUES
IN THE LICENSING OF BOTULINUM TOXINS

Jane L. Halpern and William H. Habig

Division of Bacterial Products
Center for Biologics Evaluation and Research
Food and Drug Administration
8800 Rockville Pike
Bethesda, MD 20892

Botulinum toxin is a protein neurotoxin produced by *Clostridium botulinum*. Botulinum toxin blocks neurotransmission at the neuromuscular junction by inhibiting the presynaptic release of acetylcholine (for a recent review, see ref. 1). The ability of botulinum toxin to block neurotransmission and paralyze or weaken muscles has been useful in the treatment of different neurologic disorders. A number of focal dystonias and other conditions with involuntary muscle movements currently are being treated with botulinum toxin, and it appears to be a potentially valuable therapy for many of these conditions.[2] A preparation of botulinum toxin type A (trade name Oculinum, manufactured by Allergan) was approved for human use by the Food and Drug Administration (FDA) in 1989. Oculinum is approved for the treatment of blepharospasm and strabismus, and also has been used experimentally for a number of other disorders. Oculinum has also been designated as an orphan drug under the Orphan Drug Act (ODA) of 1983, which entitles the manufacturer to a seven year period of exclusive marketing for the specific indication for which the product has been approved. This exclusive marketing granted to the manufacturer of Oculinum raises questions of how additional botulinum products can be licensed, for either the same approved indications as Oculinum or for other indications. This article will review general requirements for approval of botulinum toxin by the FDA, and discuss how regulations in the Orphan Drug Act might affect approval of additional botulinum toxin products.

REQUIREMENTS FOR LICENSING BIOLOGIC PRODUCTS

Botulinum toxin is a biologic product, and as such is regulated by the Center for Biologics Evaluation and Research (CBER) of the FDA. The regulations for licensing biologic products are described in the Code of Federal Regulations (21 CFR 601). Both an establishment and a product license application are required for biologic products. The manufacturer must submit adequate data to the FDA demonstrating that the product meets prescribed standards of safety,

Botulinum and Tetanus Neurotoxins, Edited by
B.R. DasGupta, Plenum Press, New York, 1993

purity, and potency. The results from both non-clinical laboratory and clinical studies are included in a license application. Clinical trials generally are required in order to demonstrate the effectiveness of the product. Studies demonstrating that the product is stable throughout the dating period and a full description of the manufacturing method are also required.

Oculinum has been determined to meet the standards required for licensing. This product consists of a single lot of type A toxin produced at the Food Research Institute of the University of Wisconsin. The toxin was purified by a series of precipitations to yield a crystalline complex which comprises the active neurotoxin and an associated hemagglutinin protein. The potency of Oculinum is quantitated in units (U); one unit corresponds to the calculated median lethal intraperitoneal dose (LD_{50}) in mice.

Except for considerations arising from the Orphan Drug Act, new botulinum toxin products are not subject to requirements other than those for other licensed biologic products, nor are there specific recommended procedures for its production. The policy of CBER is to evaluate each product on a case-by-case basis. Different methods for the purification of botulinum toxin have been described in the literature, and many methods could conceivably be adapted to produce botulinum toxin for clinical use. The considerations for producing botulinum toxin are similar to those for other biologic products. The final product should be free of extraneous materials and contaminants and pass tests for potency, general safety, sterility and pyrogenicity.

CONSIDERATIONS FOR ORPHAN DRUG DESIGNATION

The Orphan Drug Act was enacted by Congress in 1983 to promote the development of drugs that are needed by people in the United States with rare diseases or conditions. These drugs were felt to have such limited commercial potential that they would not be profitable and would be unlikely to be developed without incentives. This act provides the manufacturer of an orphan drug product with a seven year period of exclusive marketing, and tax credits for money spent on clinical testing of the product. A rare disease has been defined by the FDA as one affecting fewer than 200,000 people at the time of the request for orphan drug designation. In order to be designated as an orphan drug, the sponsor must meet certain conditions. It must be demonstrated that:

1. The drug is being or will be investigated for a specified rare disease.
2. The drug is subject to approval under the appropriate federal licensing regulations.
3. The marketing approval would be for the stated use or condition.

The sponsor also must provide a rationale for why the drug will be used for the stated indication. Orphan drug designation may be requested at any time prior to filing a license application. Until a designated orphan drug has been licensed by the FDA, there is no limit on the number of manufacturers who may receive orphan drug designation for the same product intended for the same disease. The exclusive marketing rights are granted to the manufacturer whose product is approved by the FDA first. Designation as an orphan drug does not circumvent other aspects of the drug approval process discussed above. An application for a drug that has received orphan drug designation is reviewed similarly to other license applications to ensure that the manufacturer has provided data demonstrating that the product is both safe and effective.

One important point that has been a source of confusion is that orphan drug designation specifies both the drug, and the disease that the drug is approved for. This does not mean that other manufacturers cannot seek approval for and market a drug that has been designated as an orphan drug, it means that they cannot receive approval or market this drug for the same indication. A second manufacturer could market the same orphan drug for a different disease, if the drug is shown to be effective for that disease.

Table 1. Indications for treatment with botulinum toxin.

Blepharospasm
Strabismus
Cervical dystonia (torticollis)
Laryngeal dystonia (spastic dysphonia)
    adductor
    abductor
Oromandibular dystonia
Pediatric cerebral palsy
Hemifacial spasm
Hand dystonia

DESIGNATION OF BOTULINUM TOXIN AS AN ORPHAN DRUG

The provisions in the ODA are very relevant to the development of new botulinum toxin products, since most of the diseases that are being treated with the toxin fit the FDA definition of a rare disease. Table 1 lists a number of diseases for which botulinum toxin is being evaluated as a therapy. This list is not intended to be comprehensive, but to give an indication of the number of different rare diseases that may qualify for orphan designation.

In response to the recognition of botulinum toxin as a useful product for a number of neurologic disorders, several different sponsors have applied for and received orphan drug designation for botulinum toxin products. These products and the indication for which each one has been designated as an orphan drug are listed in Table 2. This list includes products from four different sponsors, which represent three different botulinum toxin serotypes. Four different diseases are included in this table.

LICENSING OF NEW BOTULINUM TOXIN PRODUCTS

Because Oculinum has been licensed and received exclusive marketing rights, additional botulinum toxin products may be barred from receiving approval for the same disease unless they are determined to be a different drug than Oculinum. The question of whether two drugs are the same or different can be complex in the case of macromolecules such as proteins; therefore it is important to establish scientifically reasonable criteria for such decisions.

Drugs that are relatively small molecules whose chemical structure can be completely characterized are considered to be different compounds if there is any difference in the chemical structure of the active moiety. Parts of the molecule that make it a salt or ester are not considered to be the active moiety. This can make the decision of whether two drugs are the same or different quite simple. Macromolecules such as proteins, nucleic acids, and polysaccharides, which are becoming more important as therapeutics, present a more complicated problem, since many, and perhaps most preparations can be shown to be heterogeneous by some criteria. In the case of a protein such as botulinum toxin, this means that one manufacturer could probably always show that his product was different from another, even if these differences had no demonstrable effect on pharmacologic activities. Evaluating macromolecules by the same criteria used for small, well-defined molecules could result in all of them qualifying as orphan products, making the exclusive marketing provision meaningless. Therefore, the FDA has developed criteria for determining if two drugs are the same or different that reflect the intent of the ODA. These criteria have been proposed to be used as guidelines and contain specific criteria for proteins, polysaccharides and nucleic acids.[3] For proteins, the FDA proposed that two protein drugs would be considered the same if the only differences in structure between them were due to:

1. Post-translational events,
2. Infidelity of transcription or translation, or
3. Minor differences in amino acid sequences.

Potentially important differences, such as different glycosylation patterns or different tertiary or quaternary structure, would not cause the drugs to be considered different unless the differences were shown to result in a clinically superior product.

Botulinum toxins are not known to be modified by the addition of carbohydrate or lipid, and there is no evidence that heterogeneity exists in the primary sequence for a neurotoxin of any given serotype. So, how might these criteria be applied in the approval of new botulinum toxin products? Several different situations concerning new botulinum toxin products can be envisaged, including:

1. A new product consisting of a botulinum toxin serotype different from an already approved toxin product.
2. A new product consisting of the same serotype as an already approved product, but with differences in the composition of associated proteins.
3. A new or already approved product for a rare disease different from those already approved.

The first case addresses products with different serotypes. Botulinum toxin types A, B, and F have received orphan drug status (Table 2). Each neurotoxin serotype is generally regarded as a different protein. While there is significant amino acid sequence homology among the neurotoxins, the proteins are immunologically very distinct, and each clearly represents a distinct gene product. Thus it seems likely that botulinum toxin products of distinct serotypes will be considered to be different drugs under current orphan drug regulations.

Table 2. Orphan product designations.

Name Generic/Chemical TN=Trade name	Indication designated	Sponsor DD=Date designated MA=Marketing approval
Botulinum Toxin Type A TN=Oculinum	Treatment of blepharospasm associated with dystonia in adults (Patients 12 years of age and above)	Allergan, Inc. DD 03/22/84 MA 12/29/89
Botulinum Toxin Type A TN=Oculinum	Treatment of strabismus associated with dystonia in adults (Patients 12 years of age and above)	Allergan, Inc. DD 03/22/84 MA 12/29/89
Botulinum Toxin Type A TN=Oculinum	Treatment of cervical dystonia	Allergan, Inc. DD 08/20/86
Botulinum Toxin Type A TN=Dysport	Treatment of essential blepharospasm	Porton International, Inc. DD 03/23/89
Botulinum Toxin Type A TN=Oculinum	Treatment of dynamic muscle contracture in pediatric cerebral palsy patients	Allergan, Inc. DD 12/06/91
Botulinum Toxin Type B TN=no trade name	Treatment of cervical dystonia	Athena Neurosciences, Inc. DD 01/16/92
Botulinum Toxin Type F TN=no trade name	Treatment of essential blepharospasm	Porton International, Inc. DD 12/03/91
Botulinum Toxin Type F TN=no trade name	Treatment of cervical dystonia	Porton International, Inc. DD 10/24/91

Products approved for orphan designation as of July 1, 1992

A recent publication demonstrated the value of having multiple serotypes of botulinum toxin available for clinical use.[4] Ludlow et al. reported on the efficacy of botulinum toxin type F in four patients. Two of these patients had torticollis, one had oromandibular dystonia, and one stuttered. The patients had become clinically resistant to type A botulinum toxin, and an in vivo mouse bioassay confirmed that they had developed antibodies against the toxin. Each patient had an improvement in symptoms after injections of type F toxin.

The botulinum neurotoxins frequently are purified in a non-covalent complex with hemagglutinin and other non-toxic proteins. These complexes can be heterogeneous, leading to the second case described above, that of two products of the same serotype with different compositions. A new product and an already approved orphan product of the same serotype with differences in associated proteins might be considered to be the same drug if these differences were not thought to be significant. However, if the second drug were shown to be clinically superior, this could provide a basis for granting it orphan drug status. A new drug is considered to be clinically superior by one of the following three criteria:

1. If it is shown to have greater effectiveness than the already approved orphan drug as measured in a well controlled clinical trial,
2. If it is shown to have greater safety in a substantial portion of the target population,
3. If the drug otherwise makes a major contribution to patient care.

A sponsor requesting orphan drug designation for a drug that is the same as an already approved orphan drug must provide an explanation of why the proposed variation may be clinically superior to the first drug. Then, prior to licensing, data demonstrating the increased clinical superiority must be supplied. This interpretation seems to preserve the intent of the exclusive marketing provision of the ODA, without unnecessarily blocking the approval of improved products.

The case of a new botulinum toxin product being approved for a novel indication is fairly straightforward. Orphan drugs are designated for a specific rare disease, rather than for a broad indication. For example, Oculinum has received orphan drug status for blepharospasm and strabismus, rather than for dystonia in general. The same drug or a very similar drug made by two different manufacturers can each receive orphan designation and be approved for a different rare disease. The criteria in this case is that the disease subset being defined for orphan drug designation must be medically plausible, and not just an arbitrary subset. The sponsor of the drug needs to submit data supporting the claim that the requested patient population is medically plausible, and the appropriate FDA center makes a decision.

The points discussed in this paper are intended to clarify the guidelines that FDA uses in decisions on licensing of biologic products, specifically the botulinum neurotoxins. It is important to remember that the nature of biologic products makes it impossible to write very specific guidelines that anticipate every situation. Thus, questions concerning whether two products are the same, or whether one product may be clinically superior to another are very difficult to answer except on a case-by-case basis, using scientifically based criteria. The guidelines adopted by the FDA hopefully will further the intent of the Orphan Drug Act and promote the development of treatments for rare diseases.

REFERENCES

1. Wellhoner HH. Tetanus and botulinum neurotoxins. In: Herken H, Hucho F, eds. Handbook of Experimental Pharmacology, Volume 102. Berlin: Springer-Verlag, 1992:357-417.
2. Jankovic J, Brin MF. Therapeutic uses of botulinum toxin. N Engl J Med 1991; 324:1186-1194.
3. Federal Register, January 29, 1991, Notice of Proposed Rulemaking Orphan Drug Regulations.
4. Ludlow CL, Hallett M, Rhew K, Cole R, Shimizu T, Sakaguchi G, Bagley JA, Schulz GM, Yin SG, Koda J. Therapeutic use of type F botulinum toxin. N Engl J Med 1992; 326:349-350.

Appendix

Attendees to the International Conference on Botulinum and Tetanus Neurotoxins

Kathy Alderson
Salt Lake City V.A. Medical Center
Dept. of Neurology
University of Utah
50 North Medical Drive
Salt Lake City, UT 84132

Stephen Arnon
Infant Botulism Prevention Program
California Dept. of Health Services
2151 Berkeley Way, Room 716
Berkeley, CA 94704

Roger Aoki
Allergan, Inc., Mail Code LS-OB
2525 Dupont Dr.
Irvine, CA 92715

Michael Balady
US Army, USAMMDA
Ft. Detrick
Frederick, MC 21702

Abhijit Banerjee
Dept. Food Microbiology & Toxicology
University of Wisconsin-Madison
1925 Willow Dr.
Madison, WI 53706

Julie M. Barkmeier
Otolaryngology, Speech and Hearing
University of Iowa Hospitals & Clinics
Iowa City, Iowa 52242

Marsha J. Betley
Dept. of Bacteriology
University of Wisconsin-Madison
1550 Linden Dr.
Madison, WI 53706

Hans Bigalke
Med. Hochschule Hannover
Institute of Toxicology
3000 Hannover 61
GERMANY

Steve Binion
Allergan, Inc.
2525 Dupont Dr.
P. O. Box 19534
Irvine, CA 92713-9543

Torsten Binscheck
Med. Hochschule Hannover
Institute of Toxicology
3000 Hannover 61
GERMANY

Mary A. Bittner
Dept. Pharmacology
M6322 Med Sci Bldg I
University of Michigan Medical School
Ann Arbor, MI 48104-0626

Bernard Bizzini
Dept. of Molecular Toxinology
Pasteur Institute
218 Rue du Dr Roux
75724 Paris Cedex 15
FRANCE

Gary E. Borodic
Beyer, Townsend & Borodic Ophthalmol.
  Assoc., P.C.
100 Charles River Plaza
Boston, MA 02114

Mitchell F. Brin
Dystonia Clinical Research Center
Neurological Institute
710 West 168th St.
New York, NY 10032

Douglas R. Brown
Toxinology Division
US Army Med. Res. Inst. Infect. Diseases
Fort Detrick
Frederick, MD 21702

J. Edward Brown
Toxicology Division
US Army Med. Res. Inst. Infect. Diseases
Fort Detrick
Frederick, MD 21702

George H. Burgoyne
Division of Biologic Products
Michigan Dept. Public Health
3500 N. Logan, P. O. Box 30035
Lansing, MI 48909

Frank J. Cann
US Army Med. Res. Inst. Chem. Defenses
SGRD-UV-YN
Aberdeen Proving Ground, MD 21010

Moses V. Chao
Dept. of Cell Biology and Anatomy
Cornell University Medical College
1300 York Avenue
New York, NY 10021

Milton P. Charlton
Dept. Physiology
University of Toronto
Toronto, Ontario M561A8
CANADA

Julie Coffield
School of Veterinary Medicine
University of Wisconsin-Madison
2015 Linden Dr.
Madison, WI 53706

Cynthia Comella
Dept. of Neurological Sciences
Rush-Presbyterian-St. Luke's Med. Center
1725 West Harrison
Chicago, IL 60612

Richard M. Condie
MN ALG Program
University of Minnesota
1900 Fitch Ave.
St. Paul, MN 55108

Robert V. Considine
Dept. of Physiology
408 JAH, Thomas Jefferson Univ.
1020 Locust St.
Philadelphia, PA 19107

Richard D. Crosland
Toxicology Division
US Army Med. Res. Inst. Infect. Diseases
Fort Detrick
Frederick, MD 21702

Bibhuti R. DasGupta
Dept. of Food Microbiol. & Toxicology
University of Wisconsin
1925 Willow Dr.
Madison, WI 53706

Pietro De Camilli
Dept. of Cell Biology
Yale University School of Medicine
Boyer Center for Molecular Medicine
295 Congress Ave.
New Haven, CT 06536-0812

Anton DePaiva
Dept. Biochemistry
Imperial College of Science & Technol.
London SW7 2AY
UNITED KINGDOM

Sharad S. Deshpande
US Army Med. Res. Inst. Chem. Defense
Aberdeen Proving Ground, MD 21010

J. Oliver Dolly
Dept. of Biochemistry
Imperial Coll. Science, Technol. & Med.
London SW7 2AY
ENGLAND

Mary L. Evenson
Dept.of Food Microbiol. & Toxicology
University of Wisconsin-Madison
1925 Willow Dr.
Madison, WI 53706

Quinn Farnes
Allergan, Inc.
2525 Dupont Dr.
Irvine, CA 92713-9534

Alexander A. Fedinec
Dept. Anatomy & Neurobiology
College of Medicine
University of Tennessee
875 Monroe Avenue
Memphis, TN 38163

Margaret G. Filbert
US Army Med. Res. Inst. Chem. Defense
Aberdeen Proving Ground, MD 21010

Paula F. Flicker
Dept. Molecular Biology
Vanderbilt University
Box 1820, Station B
Nashville, TN 37235

E. Michael Foster
Dept. of Food Microbiol. & Toxicology
University of Wisconsin
1925 Willow Dr.
Madison, WI 53706

David R. Franz
Chief, Toxinology Division
US Army Med. Res. Inst. Infect. Diseases
Fort Detrick
Frederick, MD 21702

Juergen Frevert
Immuno & Cell Biology
Battelle Europe
Am Roemerhof 35
D-6000 Frankfurt am Main 90
GERMANY

Daniel Gardner
Dept. of Physiology and Biophysics C-533
Cornell University Medical College
1300 York Avenue
New York, NY 10021

Klaus Geldsetzer
Allergan Europe
High Wycombe HP1235H
UNITED KINGDOM

Domingo F. Gimenez
Calle 13, No. 1069
7607 Miramar
Pcia, Buenos Aires
ARGENTINA

Mike Goodnough
Dept.of Food Microbiol. & Toxicology
University of Wisconsin-Madison
1925 Willow Dr.
Madison, WI 53706

Michael Grayston
Manager, Oculinum Technology
Allergan, Inc.
2525 Dupont Dr.
Irvine, CA 92715

Paul Greene
Neurological Institute
Columbia Presbyterian Medical Center
710 West 168th St.
New York, NY 10032

Ernst Habermann
Rudolf-Buchheim-Institut für Pharmakol.
Der Justus-Liebig-Universitat
Frankfurter Strasse 107
D-6300 Giessen
GERMANY

Dallas C. Hack
US Army Med. Res. Inst. Infect. Diseases
Fort Detrick
Frederick, MD 21702-5011

Jane L. Halpern
Division of Bacterial Products
Center for Biologics Evaluation & Res.
FDA, Bldg. 29, Room 103
8800 Rockville Pike
Bethesda, MD 20892

Peter Hambleton
PHLS Center for Appl. Microbiol. & Res.
Division of Biologics
Porton Down, Salisbury
Wiltshire SP4 OJG
UNITED KINGDOM

Charles L. Hatheway
Chief, Botulism Laboratory
Div. of Bacterial & Mycotic Diseases
National Center for Infectious Diseases
Centers for Disease Control
Atlanta, GA 30333

Kazuko Ceci Hirakubo
Porton Products Ltd.
1 Bath Road
Maidenhead
Berkshire
ENGLAND

Henry Hoffman
Dept. of Otolaryngology
University of Iowa Hospital
Iowa City, IA 52242

I. Maged Ismail
Allergan, Inc.
2525 Dupont Dr.
Irvine, CA 92713-9534

Marjan Jahanshahi
MRC Human Movement and Balance Unit
The Institute of Neurology
Queen Square
London WC1N 3BG
UNITED KINGDOM

Robert A. Jassmond
Biopure Corporation
68 Harrison Avenue
Boston, MA 02111

Wu Jigao
Sanitation & Antiepidemic Station of
    Tacheng Area
Tachen City, Xinjiang
CHINA

Eric A. Johnson
Dept. Food Microbiology & Toxicology
University of Wisconsin-Madison
1925 Willow Dr.
Madison, WI 53706

Abhay Joshi
Allergan, Inc.
2525 Dupont Dr.
Irvine, CA 92715

Ryuji Kaji
Dept. of Neurology
Kyoto University Hospital
54 Shogoin-Kawaharacho
Sakyoku
Kyoto 606-01
JAPAN

Richard M. Kostrzewa
East Tennessee State University
James H. Quillen College of Medicine
Dept. of Pharmacology
Box 70577
Johnson City, TN 37614-0577

Shunji Kozaki
Dept. Veterinary Science
College of Agriculture
University of Osaka Prefecture
Sakai-shi, Osaka 593
JAPAN

*Carl Lamanna
Carolina Meadows V267
Chapel Hill, NC 27514

Hugh F. LaPenotiere
Toxinology Division
USAMRIID
Fort Detrick
Frederick, MD 21702-5011

Dennis L. Leatherman
Toxinology Division
USAMRIID
Fort Detrick
Frederick, MD 21702-5011

---

*Contributor; did not attend the conference.

Wei Hwa Lee
United States Dept. of Agriculture
Food Safety and Inspection Service
Microbiology
Bldg. 322, ARC-East
1300 Baltimore Ave.
Beltsville, MD 20705-2350

J. Leon
Allergan, Inc.
2525 Dupont Dr.
Irvine, CA 92715-1599

Richard Lomneth
Dept. Food Microbiology & Toxicology
University of Wisconsin-Madison
1925 Willow Dr.
Madison, WI 53706

Christy L. Ludlow
Chief, Voice and Speech Unit
Division of Intramural Research, NIDCD
Building 10, Room 5D38
9000 Rockville Pike
Bethesda, MD 20892

Eric S. Luschei
Dept. of Speech Pathology & Audiology
Room 122B-SHC
University of Iowa
Iowa City, IA 52242

Hiromi Maezono
Allergan-Japan
6, Sanbancho
Chiyoda-Ku
Tokyo 102
JAPAN

T. F. J. Martin
Zoology Research Bldg., Room 121
1117 W. Johnson St.
University of Wisconsin-Madison
Madison, WI 53706

Morihiro Matsuda
Dept. of Tuberculosis Res. (Bact.
  Toxinology)
Res. Inst. Microbial Dis.
Osaka University
3-1, Yamadaoka, Suita
Osaka 565
JAPAN

James McManaman
Associate Director, Neuroscience
Synergen
1885 33rd St.
Boulder, CO 80301-2546

Jane Mellanby
Dept. Experimental Psychology
University of Oxford
South Parks Road
Oxford, OX1 3UD
ENGLAND

John L. Middlebrook
Toxinology Division
US Army Med. Res. Inst. Infect. Diseases
Fort Detrick
Frederick, MD 21702-5011

Ricardo Miledi
Laboratory of Cell/Molecular Neurobiol.
Dept. Psychobiology
Steinhaus Hall
University of California
Irvine, CA 92717

Narendra K. Modi
Chemical & Biological Defense Estab.
Porton Down, Salisbury
Wiltshire SP4 OJQ
ENGLAND

Cesare Montecucco
Dept. Biomedical Sciences
University of Padova
Via Triaste 75
35121 Padova
ITALY

James Mooney
Dept. Orthopaedic Surgery
Bowman Gray School of Medicine
Medical Center Blvd.
Winston-Salem, NC 27157-1070

Alan Morgan
Dept. of Physiology
University of Liverpool
P.O. Box 147
Liverpool L69 3BX
UNITED KINGDOM

Elizabeth Moyer
Athena Neurosciences Inc.
800F Gateway Blvd.
South San Francisco, CA 94080

Hiroyuki Nakano
Food Research Institute
University of Wisconsin-Madison
1925 Willow Dr.
Madison, WI 53706

John B. Nanninga
Dept. of Urology
Northwestern University Medical School
251 East Chicago Avenue, #629
Chicago, IL 60611

Heiner Niemann
Federal Res. Center Virus Dis. Animals
Dept. of Microbiology
P.O. Box 1149
Paul-Ehrlich-Str. 28
D-7400 Tübingen
GERMANY

Keiji Oguma
Professor, Dept. of Bacteriology
Okayama University Medical School
2-5-1, Shikata-cho
Okayama 700
JAPAN

Julio Barrera Oro
Director of Special Projects
The Salk Institute
Government Services Division
Maryland

L. Bruce Pearce
Boston University
80 E. Concord St.
Boston, MA 02118

Adrienne L. Perlman
VA Medical Center
Iowa City, IA 52246

K. Linnea Peterson
Dept. Otolaryngology–Head & Neck Surg.
University of California–Los Angeles
10833 Le Conte Ave.
Los Angeles, CA 90024

A. M. Pickett
Porton Products Ltd.
Porton Down, Salisbury
Wiltshire 5P4 OJS
ENGLAND

Bernard Poulain
Lab de Neurobiology Cellulaire et
  Moleculaire
CNRS
91198 Gif-sur-Yvette Cedex
FRANCE

Tomas A. Reader
University of Montreal
Dept. of Physiology
Ctr. Rech. Sci. Neurol.
CP 6128 Succ. A
Montreal, Quebec
CANADA H3C 3J7

Richard Robitaille
Dept. Physiology
Medical Science Bldg.
University of Toronto
Toronto, Ontario M5S 1A8
CANADA

Tonie Rocke
Natl. Wildlife Health Research Center
6006 Schroeder Road
Madison, WI 53711

*A. Catharine Ross
Dept. of Physiology & Biochemistry
Division of Nutrition
The Medical College of Pennsylvania
3300 Henry Avenue
Philadelphia, PA 19129

Genji Sakaguchi
Japan Food Research Laboratories
3-1 Toyotsu-cho
Suita-shi
Osaka 564
JAPAN

---

*Contributor; did not attend the conference.

Renee J. Sandler
Food Research Institute
University of Wisconsin-Madison
1925 Willow Dr.
Madison, WI 53706

V. Sathyamoorthy
Virulence Assessment Branch
Division of Microbiology
FDA; HFF236
200 C St., S.W.
Washington, D.C. 20204

Edward J. Schantz
Dept. Food Microbiology & Toxicology
University of Wisconsin-Madison
1925 Willow Dr.
Madison, WI 53706

Cara-Lynne Schengrund
Dept. of Biological Chemistry
Hershey Medical Center
Hershey, PA 17033

Giampietro Schiavo
Dpt. Scienze Biomediche
Universita degli studi di Padova
Via Trieste 75
35121 Padova
ITALY

Michael F. Schmid
Dept. of Biochemistry
Baylor College of Medicine
1 Baylor Plaza
Houston, TX 77030

E. A. Schwartz
Dept. of Pharmacological & Physiological
    Sciences
The University of Chicago
947 East 58th St.
Chicago, IL 60637

Alan B. Scott
Smith-Kettlewell Inst. Visual Sciences
Institute of Medical Sciences
2232 Webster St.
San Francisco, CA 94115

Dorothea Sesardic
National Institute for Biological Standards
    & Control
Blanche Lane, South Mimms
Potters Bar, Hertfordshire EN6 3QG
UNITED KINGDOM

Paulette E. Setler
Executive Vice President, Research
Athena Neurosciences
800F Gateway Blvd.
South San Francisco, CA 94080

Linda F. Shoer
President
List Biological Lab., Inc.
501-B Vandell Way
Campbell, CA 95008

Clifford C. Shone
PHLS Centre for Applied Microbiology &
    Research
Division of Biologics
Porton Down
Salisbury
Wiltshire SP4 0JG
UNITED KINGDOM

Lance L. Simpson
Dept. of Medicine
Jefferson Medical College
1020 Locust St.
Philadelphia, PA 19107

Bal Ram Singh
Dept. of Chemistry
University of Massachusetts-Dartmouth
Old Westport Road
North Dartmouth, MA 02747

Leonard A. Smith
Division of Toxicology
US Army Med. Res. Inst. Infect. Diseases
Fort Detrick
Frederick, MD 21701

Carles Solsona
Dept. de Biologia Cellular i Anatomia
  Patologica
Facultat de Medicina
Universitat de Barcelona
143 Casanova St.
E-08036 Barcelona
SPAIN

Douglas Stafford
Ophidian Pharmaceuticals, Inc.
2800 South Fish Hatchery Road
Madison, WI 53711

Philip D. Stahl
Head, Dept. Cell Biology and Physiology
Washington University
School of Medicine
660 South Euclid Avenue (Box 8228)
St. Louis, MO 63110

E. Ann Stephens
Division of Biologic Products
Michigan Dept. of Public Health
3500 N. Logan
Lansing, MI 48909

Raymond C. Stevens
Chemistry Dept.
12 Oxford St.
Harvard University
Cambridge, MA 02138

Dusan Suput
Institute of Pathophysiology
School of Medicine
University of Ljubljana
Zaloska 4, P.O. Box 11
61105 Ljubljana
SLOVENIA

Lauren Swenarchuk
Dept. Physiology
Medical Science Bldg.
University of Toronto
Toronto, Ontario M5S 1A8
CANADA

Kohsi Takano
Abt. Pathoneurophysiologie
der Universität Göttingen
Humboldtalle 23
D-3400 Göttingen
GERMANY

Diane Tang-Liu
Allergan
2525 Dupont Dr.
Irvine, CA 92715

Ladislav Tauc
Lab de Neurobiology Cellulaire et
  Moleculaire
CNRS
91198 Gif-sur-Yvette Cedex
FRANCE

William Tepp
Dept. Food Microbiology & Toxicology
University of Wisconsin
1925 Willow Dr.
Madison, WI 53706

Stephen Thesleff
Dept. of Pharmacology
University of Lund
Sölvegatan 10
S-223 62 Lund
SWEDEN

Lutz Thilo
Dept. Medical Biochemistry
University of Cape Town
Medical School
Observatory, Cape Town
7925 SOUTH AFRICA

H. S. Tranter
Public Health Laboratory Service
Center for Applied Microbiol. & Research
Division of Biologics
Porton Down, Salisbury
Wiltshire
ENGLAND

Margaret Van Boldrik
Ophidian Pharmaceuticals
2800 S. Fish Hatchery Road
Madison, WI 53711

678

Paul Wagner
Bldg. 37, Rm. 4C24
Laboratory of Biochemistry
National Cancer Institute
Bethesda, MD 20892

Ulrich Weller
Inst. f. Med. Mikrobiologie
Johannes-Gutenberg-Universität Mainz
Hochhaus am Augustusplatz
W-6500 Mainz
GERMANY

Robert Wellner
Toxicology Division
US Army Med. Res. Inst. Infect. Diseases
Fort Detrick
Frederick, MD 21702

James R. Weston
Biopure Corporation
68 Harrison Avenue
Boston, MA 02111

Judy Williamson
Dept. Veterinary Science
University of Wisconsin-Madison
1655 Linden Dr.
Madison, WI 53706

Lura C. Williamson
NICHD, LDN
Bldg. 36/2A21
National Institute of Health
Bethesda, MD 20892

Gayle Woodson
Dept. Surgery/Div. Otolaryngology
Univ. of California, San Diego
School of Medicine and VA Medical
  Center
225 Dickinson, Room 9112C
San Diego, CA 92103

Noriko Yokosawa
Dept. of Microbiology
Sapporo Medical College
South 1, West 17
Sapporo, 060
JAPAN

INDEX

686

Vitamin A
  antibody response to tetanus toxoid, 483
  effect of deficiency, 483, 485, 487
  relation with immune response, 483
  (retinol) as adjuvant, 484, 488

Wallerian degeneration, 53
World Health Organization (WHO), 1
World War II, 515

Xenopus oocytes
  injection of synaptic vesicles, 145
  release of ACh from, 148

Zn^{++}
  in active site, *xi*
  binding domain, *xi*
  in binding of L-chain to substrate, *xi*
  chelators, *xi*
  protease, *xi*
  protease inhibitor, *xi*
  as stabilizer of L-chain, *xi*
Zinc-dependent metalloproteases, 366
Zinc protease neurotoxins, 3